Textbook of
Quantum
Mechanics

Second Edition

Textbook of Quantum Mechanics

Second Edition

AK Saxena MSc, PhD

Senior Lecturer
Department of Physics
APS University
Rewa, MP

CBS

CBS Publishers & Distributors Pvt. Ltd.

New Delhi • Bengaluru • Chennai • Kochi • Kolkata • Mumbai
Hyderabad • Nagpur • Patna • Pune • Vijayawada

Textbook of
Quantum
Mechanics

ISBN: 978-81-239-1899-0

First Edition: 2007
Second Edition: 2010
Reprint: 2012, 2017

Published by **Satish Kumar Jain** and produced by **Varun Jain** for
CBS Publishers & Distributors Pvt. Ltd.,
4819/XI Prahlad Street, 24 Ansari Road, Daryaganj, New Delhi - 110002
delhi@cbspd.com, cbspubs@airtelmail.in • www.cbspd.com
Ph.: 23289259, 23266861, 23266867 • Fax: 011-23243014

Corporate Office: 204 FIE, Industrial Area, Patparganj, Delhi - 110 092
Ph: 49344934 • Fax: 011-49344935
E-mail: publishing@cbspd.com • publicity@cbspd.com

Branches:
• **Bengaluru:** 2975, 17th Cross, K.R. Road, Bansankari 2nd Stage,
  Bengaluru - 70 • Ph: +91-80-26771678/79 • Fax: +91-80-26771680
  E-mail: cbsbng@gmail.com, bangalore@cbspd.com
• **Chennai:** No. 7, Subbaraya Street, Shenoy Nagar, Chennai - 600030
  Ph: +91-44-26681266, 26680620 • Fax: +91-44-42032115
  E-mail: chennai@cbspd.com
• **Kochi:** Ashana House, 39/1904, A.M. Thomas Road, Valanjambalam,
  Ernakulum, Kochi • Ph: +91-484-4059061-65
  Fax: +91-484-4059065 • E-mail: cochin@cbspd.com
• **Kolkata:** 6-B, Ground Floor, Rameshwar Shaw Road, Kolkata - 700014
  Ph: +91-33-22891126/7/8 • E-mail: kolkata@cbspd.com
• **Mumbai:** 83-C, Dr. E. Moses Road, Worli, Mumbai - 400018
  Ph: +91-9833017933, 022-24902340/41 • E-mail: mumbai@cbspd.com

Representatives:

• Hyderabad: 0-9885175004   • Nagpur: 0-9021734563
• Patna: 0-9334159340   • Pune: 0-9623451994
• Vijayawada: 0-9000660880

Printed at:
India Binding House, Noida, UP (India)

to
Prof A. N. Mitra

Preface to the Second Edition

I am thankful to the readers for their interest and applause offered for this book. The second edition has been revised and the following topics have been included: Physical significance of spherical harmonics, field and coordinates of the field, Lagrangian formulation for the fields, commutation relations for field operators, quantisation of non-relativistic Schrödinger field, the neutral Klein Gordon field, the Dirac field, the neutral Klein Gordon field in terms of plane waves, the charged KG field, the Einstein-Podolsky-Rosen problem and the Aharonov-Bohm effect.

I express my sincere thanks to Mr YN Arjuna, Publishing Director, CBS Publishers and Distributors Pvt Ltd, for publishing promptly the second edition.

Further suggestions from the teachers as well as the students for further improvement are highly welcome.

Ajay Kumar Saxena

Preface to the First Edition

This book aims at presenting basic principles of *Quantum Mechanics* to graduate and postgraduate students of physics at various Universities. The text has been divided into 23 chapters for the convenience of the students. Solved examples and questions have been incorporated to enable the students in gaining a better understanding of the concepts involved.

The first chapter entails the basic historic theoretical developments involved in the emergence of this branch of physics. The Chapter 2 is about Schrödinger's wave mechanics covering barrier penetration, alpha decay, free electron theory of metals and concept of Fermi energy. The Chapter 3 is on operator-formalism, observations and expectation values and basic postulates of quantum mechanics. Chapter 4 involves Fourier methods, Dirac delta function and wave packets. Chapter 5 discusses the problem of one dimensional harmonic oscillator based on Hermite polynomials. Chapter 6 focusses on the solutions of Schrödinger equation as applied to spherically symmetric problems such as the hydrogen atom and free particle. Chapter 7 contains matrix formulation of quantum mechanics and Chapter 8 is on Dirac representation theory. Chapter 9 discusses the linear harmonic oscillator problem based on ladder operators. The three quantum mechanical pictures, viz. Schrödinger, Heisenberg and Interaction, pictures are dealt in Chapter 10 with an introduction of time evolution operator and scattering matrix. The orbital angular momentum has been given in Chapter 11 and Chapter 12 is related to spin angular momentum. Chapter 13 lays an emphasis on LS and jj coupling, Hund's rules, spectroscopic notation, total angular momentum and their addition, Clebsch-Gordan coefficients, recurrence relations and Wigner Eckart theorem. Chapter 14 is related to symmetries, conservation laws and physical situations related with symmetries such as Bose and Fermi particles, exchange degeneracy, two-electron spin wave functions, and Heitler and London theory of H_2 molecule. Chapter 15 deals with time independent perturbation theory, normal and anomalous Zeeman effect and Stark effect in hydrogen atom. Chapter 16 contains the variation method as applied to the hydrogen and helium atoms, the hydrogen molecular ion H_2^+ and Van der Waal's interaction between two hydrogen-like atoms. WKB approximation has been discussed in the Chapter 17. Chapter 18 covers time dependent perturbation theory, Fermi's golden rule, adiabatic and sudden approximations, interaction of atoms with electromagnetic waves, electric dipole approximation, Einstein's coefficients and scattering matrix. Chapter 19 introduces many electron atoms emphasizing, Hartree's, Hartree Fock and Thomas Fermi methods. Chapter 20 deals with scattering theory, Born approximation, partial wave analysis, S-matrix and Breit Wigner formula. Relativistic theory has been described in Chapter 21, comprising Klein Gordon wave equation, Dirac's theory and hydrogen atom in relativistic case, charge conjugation (for spin-zero and spin-half particles) and Lamb-shift. Chapter 22 discusses quantization and second quantization. Chapter 23 presents a brief description of the hidden variables, Einstein's locality principle and the Einstein-Podolsky-Rosen paradox with an introduction of Bell's inequality.

There are a few appendices and bibliography for further consultation.

I am grateful to my teachers, Prof AN Mitra and Prof SN Biswas (Delhi University), Prof ON Srivastava (BHU, Varanasi) and Prof AK Ghatak (IIT Delhi) who have been constantly a source of inspiration to me.

I am grateful to our Honorable Vice Chancellor, Prof ADN Bajpai and Dr AP Mishra, Head, Department of Physics for boosting my morale.

I am thankful to writers and publishers of various books which have been consulted by me.

I wish to thank Prof DP Tiwari, Dr SL Agrawal and Mr NK Dube, for providing moral encouragement.

Further, I wish to thank Mr Rajeev Prithyani who took all the pains for preparing the typescript and to Mr HS Poplai, CBS Publishers for efficient handling of various details encountered in bringing out the book.

In the last, but not the least I also thank my wife Alka and sons Ankur and Akshat for maitaining endurance during completion of this task.

Although care has been taken to keep off errors, suggestions of readers in pointing out errors and omissions would be highly welcome.

Ajay Kumar Saxena

Contents

4. The Fourier Transform, Dirac Delta Function and Wave Packets 161

5. Linear Harmonic Oscillator 208

1

Foundation of 'Quantum' – Concept

1.1 Introduction

Newton's law of gravitation indicates that law of terrestrial gravity is same as the laws governing gravity of celestial objects. Similarly, scientists in the nineteenth century like Mayer, Clausius, Boltzmann and Gibbs unified thermodynamics and mechanics (since heat was found to be vibrational motion of constituent molecules or atoms of solids, liquids and gases).

For a long time, light was thought of as exhibiting wave-like (by Fresnel and Huygen) or corpuscular behaviour (by Newton). But Faraday and Maxwell combined the phenomena of electricity, magnetism and light indicating that light is nothing but electromagnetic radiation.

We know that classical physics, as represented by newtonian mechanics and Maxwell's laws of electromagnetism, work very well for the analysis of macroscopic objects in terms of empirically determined laws of force. However, as soon as we enter the world of the atom, we find that newtonian mechanics and wave aspect of light begin to fail badly, seriously requiring new concepts for their analysis and description.

1.2 Inadequacy of Classical Mechanics

There were no reasons to disregard the Newtonian laws and Maxwell's theory as both explained very well the motion of material bodies and behaviour of light respectively. However, in the realm of the world of the atom there were experimental observations such as photoelectric effect, Compton effect, etc. which could not be explained by the laws of classical mechanics. Apart from these observed phenomena, the foremost set back of applying the laws of classical physics to the submicroscopic world of atom was that it set a question mark about the stability of the atoms. Experiments of Rutherford (in 1910) had established that an atom consisted of a heavy positively charged nucleus having dimensions very small compared to the atom itself, and is surrounded by negatively charged electrons of very little mass. If the classical picture is adopted, according to Earnshaw, the system would be stable only if the negatively charged electrons

1

keep on revolving around the positively charged heavy nucleus in a manner similar to planets revolving round the sun. But unlike planets, according to electromagnetic theory, the moving electron being an accelerated charged particle, must radiate energy given by

$$\frac{dE}{dt} = \frac{2e^2 \dot{v}^2}{3c^3} \tag{1}$$

where $-\dfrac{dE}{dt}$ is the rate at which the energy E of the particle is converted into a radiant energy). Consequently electron's velocity would decrease and moving spirally round the nucleus of gradually decreasing radius, it might fall into the nucleus. This shows that an atom has to be unstable, which contradicts the observed fact of the stability of the atom. Thus classical mechanics fails to explain the stability of the atoms.

The classical mechanics is also unable to explain the observed spectrum of the hydrogen atom. It has been observed experimentally that the spectrum of hydrogen atom consists of a discrete set of lines but according to classical mechanics atoms of hydrogen should emit electromagnetic radiations of all wavelengths continuously (continuous energy changes). Thus, it could not explain the origin of discrete spectra of atoms.

Further, classical mechanics was unable to offer an explanation for the nature of distribution of energy in the spectrum of radiation from a black body and variation of specific heats of metals and gases.

It was however, Planck's concept of energy 'quanta' (and later established experimentally by Einstein's observation and explanation of photoelectric effect) which necessitated the need for supplementing the classical concepts—for instance, the need for supposition of harmonic oscillators (radiation oscillators in black body spectrum and material oscillators in Einstein's theory of specific heats) and that these oscillators can take only discrete energy values called quanta of energy.

1.3 Line-spectra, Rydberg's Constant and Ritz Combination Principle

It has been observed that

1. Most atoms when isolated from radiation or other atoms, they remain stable indefinitely (orbiting electrons do not collapse into the positively charged nucleus).
2. All atoms are typically of a few angstroms in diameter ($1\text{Å} = 10^{-10}$ m) with remarkably little difference in size between the lightest and the heaviest (i.e. there is not a wide range of sizes).
3. When atoms are excited electrically or by collisions or otherwise, they emit radiation of discrete wavelengths (line spectra) characteristic of the kind of atoms excited.

From the stand point of classical physics, the only reasonable explanation is that the observed wavelengths are an expression of characteristic vibrations within the atom. Historically, the famous experimental evidence was observation of Balmer series of hydrogen (Table 1.1 and Fig. 1.1).

Table 1.1: Balmer series of hydrogen

Line	*Colour*	*Wavelength (λ in Å)*
H_α	Red	6563
H_β	Turquoise	4861
H_γ	Blue	4341
H_δ	Violet	4102
H_ϵ	Extreme violet	3970

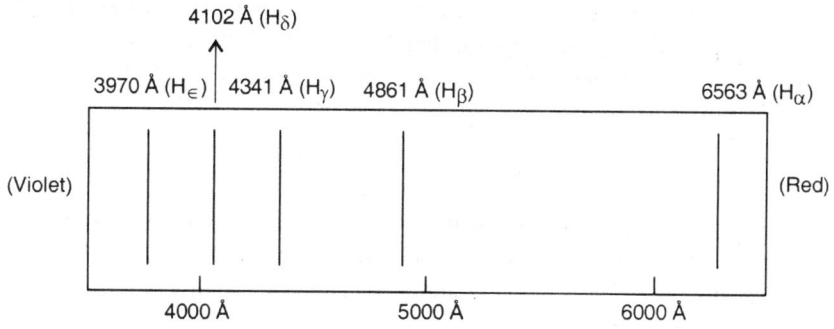

Fig. 1.1: The Balmer series of hydrogen spectrum

Attempts were made to explain such line-spectra in analogy of overtones in acoustics to find harmonic relations in lines found in the spectrum of a given element. In this regard, these attempts were in vain but some relations of another sort were discovered. In 1880, Liveing and Dewar emphasized the physical similarities in the spectra of similar elements such as alkali metals. They attempted to focus the attention on successive pairs of lines in the arc spectrum of sodium atom and pointed out that these pairs were alternately sharp and diffuse and were more closely spaced towards the short wavelength end of the spectrum. This hinted towards a certain sort of series relation which they were however unable to discover. In 1883, Hartley discovered that if frequencies (instead of wavelengths) are used, it was observed that *the difference in frequency between the components of a multiplet (viz. doublet or a triplet) in a particular spectrum is the same for all similar multiplets of lines in that spectrum.* This law enabled to isolate from the large number of lines in any given spectrum those groups of lines which were related to one another.

In 1885, the wavelengths of the nine then known lines in the spectrum of hydrogen atom could be expressed by the following formula given by Balmer:

$$\lambda = b\,\frac{n^2}{n^2 - 4} = 3646\left(\frac{n^2}{n^2 - 4}\right)\text{Å} \tag{2}$$

where integer n can take values $n = 3, 4, 5, \ldots$ for respectively the first (beginning at the red), second, third... line in the spectrum. But this did not provide an explanation for the mechanism of atomic radiation. In this respect, Balmer proposed that his formulat might be a special case of a more general formula applicable to other series of lines in other elements. Rydberg made an attempt to discover such formula using Hartley's law of constant wave number separation as applied to comparatively large mass wavelength data of alkalis. In all cases, the series showed a tendency to converge to some limit in the ultraviolet. There were two types of such series, viz. that having comparatively sharp lines which he called *sharp series* and another having comparatively broad lines, called *diffuse series*. Both of these occurred in the arc spectrum of the same element. It was found that there existed also a third type of series in which doublet spacing decreased as the frequency of the line increased, as if tending to vanish at the convergence limit of the series, he called this the *principal series* because they contained the brightest and the most persistent lines in the spectrum.

It was discovered by Rydberg that many observed series could be fitted closely by the relation:

$$\overline{V}_m = \overline{V}_\infty - \frac{R}{(n+\mu)^2} \tag{3}$$

where μ and $\overline{V}_\infty$ are constants which vary from one series to another. The constant $\overline{V}_\infty$ represents the high frequency limit to which the lines in the series ultimately converge. The Balmer formula is a special case of the above Rydberg formula as we can write:

$$\overline{V} = \frac{1}{\lambda} = \frac{1}{b} - \frac{4}{bn^2} = \frac{1}{b} - \frac{R}{n^2} = \overline{V}_\infty - \frac{R}{n^2} \tag{4}$$

with $\qquad\qquad R = 4/b$ and $\overline{V}_\infty = 1/b = \frac{1}{4} R$

or $\qquad\qquad \overline{V} = \frac{R}{4} - \frac{R}{n^2} = R\left[\frac{1}{2^2} - \frac{1}{n^2}\right]$

$$= 5R/36, \ 3R/16, \ 21R/100 \ \ldots \tag{5}$$

and $\qquad\qquad n = 3, 4, 5, \ldots$

This is of the type of Eqn. (3) with $\mu = 0$ and $(\frac{1}{4})$ R represents the convergence limit, corresponding to $n = \infty$. This corresponds to state of ionization (i.e. complete removal of electron from the nucleus. Thus spectral lines obtained were correlated empirically by Rydberg. The constant R in Eqn. (3), now called the *Rydberg constant* was found to have the same value for a large group of series for each substance and very nearly the same for all substances. Its slight variation from one atom to another is now known to be due to difference in the atomic weights. The value of R for the hydrogen atom as derived from spectroscopic observations is

$$R = 109677.6 \text{ cm}^{-1}$$
$$= 1.09678 \times 10^7 \text{ m}^{-1} \tag{6}$$

It is to be noted here that these studies were made long before the atomic theory was given by Bohr, and appearance of Rydberg's constant in the expression for line-spectra was an outcome of experimental observations.

Each line series (such as Balmer series or Lyman series) shows a continuous emission (or absorption) to high wave numbers of the convergence limits.

The convergence limit corresponds to atomic electron absorbing sufficient energy to escape from the nucleus with zero velocity. However, more energy could be absorbed by the electron and hence escape with higher velocities. Now since the electron has escaped its energy is not quantized and any energy higher than convergence limit can be absorbed, so spectrum in this region is continuous.

Rydberg noticed several remarkable relationships between different spectral series belonging to the same element. On the basis of these observations, the following formulas for the principal, sharp and diffuse series were obtained

$$\bar{v}_p = \frac{R}{(1+S)^2} - \frac{R}{(n+P)^2}; n = 1, 2, 3, \ldots$$

$$\bar{v}_s = \frac{R}{(1+P)^2} - \frac{R}{(n+S)^2}; n = 2, 3, 4, \ldots \tag{7}$$

$$\bar{v}_d = \frac{R}{(1+P)^2} - \frac{R}{(n+D)^2}; n = 2, 3, 4, \ldots$$

for the alkali metals, S, P, D are constants that are characteristic of the particular series. From these formulas, Rydberg conjectured that the first term on the right hand side might also vary similar to second term giving rise to additional series of lines as for instance, we might get a series represented by

$$\bar{v} = \frac{R}{(2+S)^2} - \frac{R}{(n+P)^2}; n = 3, 4, 5, \ldots \tag{8}$$

Lines or series of this sort were discovered later by Ritz. The values of $\bar{v}$ for such lines in a given spectrum were equal to the sums or differences of the reciprocal wavelengths of other lines taken in pairs. Such lines are called *intercombination lines* and possibility of their occurrence is known as the *Ritz combination principle*. Many examples of these are known. For instance, series other than the Balmer series in hydrogen spectrum were predicted well before their actual discovery by Paschen and Brackett. Taking the first two lines H_α and H_β of the Balmer series, we may represent them by

$$\bar{v}_\alpha = R\left[\frac{1}{2^2} - \frac{1}{3^2}\right] \text{ and } \bar{v}_\beta = R\left[\frac{1}{2^2} - \frac{1}{4^2}\right]$$

Combining these two as

$$\bar{v}_\beta - \bar{v}_\alpha = R\left\{\left(\frac{1}{2^2} - \frac{1}{4^2}\right) - \left(\frac{1}{2^2} - \frac{1}{3^2}\right)\right\}$$

$$= R\left[\frac{1}{3^2} - \frac{1}{4^2}\right]$$

This represents a new line, the first line of a new series in the infra-red discovered by Paschen. Similarly the secondline of the same series can be obtained by forming the difference between H_γ and H_α. Likewise, another series in the infra-red discovered by Brackett, can be obtained.

The Ritz principle gave the clue to interpret atomic spectra in terms of the quantum theory.

1.4 Bohr's Theory of the Atom and Spectrum of Hydrogen

Consider an electron of mass m and charge e revolving in a circular orbit of radius r about the stationary nucleus, so that electrostatic force of attraction (between electron and nucleus) equals the inward centripetal force

$$\frac{1}{4\pi \epsilon_0} \frac{Ze^2}{r^2} = \frac{mv^2}{r} \tag{9}$$

where Ze is the nuclear charge.

Kinetic energy of the electron $W_K = \frac{1}{2} mv^2$

$$= \frac{1}{8\pi \epsilon_0} \frac{Ze^2}{r} \tag{10}$$

Potential energy

$$W_P = -\frac{1}{4\pi \epsilon_0} \frac{Ze^2}{r} \tag{11}$$

Total energy

$$W = W_K + W_P$$

or

$$W = \frac{1}{8\pi \epsilon_0} \frac{Ze^2}{r} - \frac{1}{4\pi \epsilon_0} \frac{Ze^2}{r}$$

or

$$W = -\frac{1}{8\pi \epsilon_0} \frac{Ze^2}{r} \tag{12}$$

According to classical electrodynamics, the electron moving under acceleration, should radiate energy; consequently W should decrease. This means that r must decrease continuously, making the electron approach nucleus and finally fall into it. To overcome this difficulty, Bohr introduced following postulates:

I Postulate: Out of the innumerable circular orbits possible, the electron is permitted to revolve only in those which satisfy the quantum condition:

$$mvr = n\hbar \tag{13}$$

where mvr is the angular momentum, $\hbar$ is $(h/2\pi)$ and n is any positive integer known as principal quantum number. Eliminating v from Eqns. 9 and 13:

$$\frac{1}{4\pi \epsilon_0} \frac{Ze^2}{r} = mv^2 = m \left(\frac{n\hbar}{mr} \right)^2$$

or

$$\frac{1}{4\pi \epsilon_0} \frac{Ze^2}{r} = \frac{mn^2\hbar^2}{m^2 r^2}$$

or

$$r_n = \frac{4\pi \epsilon_0 . n^2 \hbar^2}{mZe^2} \tag{14}$$

Substituting the value of m in expression (12), we get total energy of the electron in n^{th} orbit.

$$W_n = -\frac{1}{8\pi \epsilon_0} \frac{Ze^2}{r_n} = -\frac{1}{8\pi \epsilon_0} Ze^2 \frac{mZe^2}{4\pi \epsilon_0 n^2 \hbar^2}$$

$$= \frac{-mZ^2 e^4}{32\pi^2 \epsilon_0^2 n^2 \dfrac{h^2}{4\pi^2}}$$

or $$W_n = \frac{-mZ^2e^4}{8\,\epsilon_0^2\,n^2h^2} \quad (15)$$

II Postulate: (It is concerned with the origin of spectral lines). It states that when an electron jumps from an outer (initial) orbit of higher energy to an inner (final) orbit of lower energy, the energy difference of the two orbits is radiated in the form of quantum of radiation ($h\nu$) whose frequency is given by

$$\nu = \frac{W_i - W_f}{h} = -\frac{me^4}{8\,\epsilon_0^2\,h^3}\left[\frac{1}{n_i^2} - \frac{1}{n_f^2}\right]$$

(because for hydrogen, $Z = 1$)

or

$$\nu = \frac{W_i - W_f}{h} = -\frac{me^4}{8\,\epsilon_0^2\,h^3}\left[\frac{1}{n_f^2} - \frac{1}{n_i^2}\right] \quad (16)$$

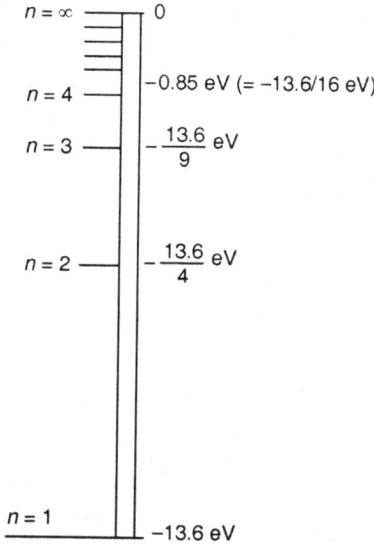

Fig. 1.2: Orbits corresponding to energies specified by Eqn. (15)

In terms of photon wavelength $\lambda = c/\nu$, we have

$$\bar{\nu} = \frac{1}{\lambda} = \frac{\nu}{c} = \frac{me^4}{8\,\epsilon_0^2\,ch^3}\left(\frac{1}{n_f^2} - \frac{1}{n_i^2}\right) = \frac{2\pi^2me^4}{(4\pi\epsilon_0)^2\,ch^3}\left[\frac{1}{n_f^2} - \frac{1}{n_i^2}\right] \quad (17)$$

This equation states that the radiation emitted by excited H-atom should contain certain wavelengths only. These wavelengths fall into definite sequences that depend on the quantum number n_f of the final energy level of the electron. Since n_i must always be greater than n_f (in order that there be an excess of energy to be given off as photon), the calculated formulae for the first five series are:

$$n_f = 1 \; : \quad \frac{1}{\lambda} = \frac{me^4}{8\,\epsilon_0^2\,ch^3}\left(\frac{1}{1^1} - \frac{1}{n^2}\right) \quad n = 2,3,4\ldots \text{ Lyman}$$

$$n_f = 2 \; : \quad \frac{1}{\lambda} = \frac{me^4}{8\,\epsilon_0^2\,ch^3}\left(\frac{1}{2^2} - \frac{1}{n^2}\right) \quad n = 3,4,5\ldots \text{ Balmer}$$

$$n_f = 3 \; : \quad \frac{1}{\lambda} = \frac{me^4}{8\,\epsilon_0^2\,ch^3}\left(\frac{1}{3^2} - \frac{1}{n^2}\right) \quad n = 4,5,6\ldots \text{ Paschen}$$

$$n_f = 4 \; : \quad \frac{1}{\lambda} = \frac{me^4}{8\,\epsilon_0^2\,ch^3}\left(\frac{1}{4^2} - \frac{1}{n^2}\right) \quad n = 5,6,7\ldots \text{ Brackett}$$

$$n_f = 5 \; : \quad \frac{1}{\lambda} = \frac{me^4}{8\,\epsilon_0^2\,ch^3}\left(\frac{1}{5^2} - \frac{1}{n^2}\right) \quad n = 6,7,8\ldots \text{ P. fund} \quad (18)$$

(the Lyman transitions involve the largest changes of energy, produce the highest frequencies and provide the (bluest) ultraviolet light. Paschen and Brackett series lie in the infra red and Pfund series in the far infrared region. Balmer series contain wavelength in the visible portion of H spectrum. Now, when we compare Eqn. (17) with Eqn. (5), we find

$$R = \frac{2\pi^2 me^4}{(4\pi\varepsilon_0)^2 ch^3}$$

Substituting the values for all the constant quantities, we get $R = 1.09738 \times 10^7$ m^{-1}. This result is in very good agreement wih the experimental value of 1.09678×10^7 m^{-1}.

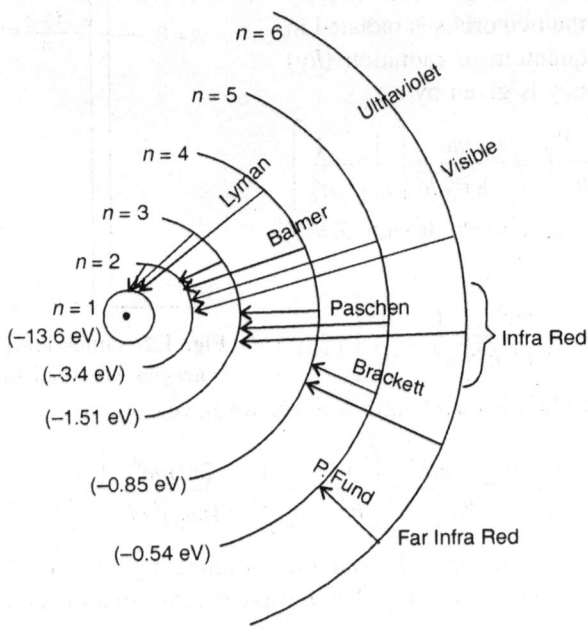

Fig. 1.3: Relation between spectral lines and energy levels for hydrogen spectrum

1.5 Reduced Mass of Electron and Nuclear Motion

In Bohr's theory it is assumed that the nucleus remains stationary, i.e. it is assumed to be infinitely massive compared with the mass of the electron. But actually the mass of the nucleus is finite. Both nucleus and the electron revolve around their common centre of mass.

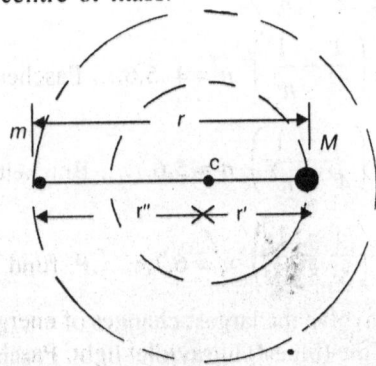

Fig. 1.4: Pertaining to effect of nuclear motion

Let $\quad c$ = centre of mass of the system

m = mass of the electron

M = mass of the nucleus

r = distance between electron and nucleus

r' = distance from c to the nucleus

r'' = distance from c to the electron

Then $\quad r = r' + r''$ and $mr'' = Mr'$

$\therefore \qquad r'' = \dfrac{M}{m}r \quad$ or $\quad r - r' = \dfrac{M}{m}r'$

or $\qquad r' = \dfrac{r}{1 + \dfrac{M}{m}} = \dfrac{mr}{m + M}$

Similarly $\quad r'' = \dfrac{Mr}{m + M}$

The total angular momentum about the centre of mass c is

$$= mr''^2\omega + Mr'^2\omega$$

$$= m \cdot \dfrac{M^2 r^2}{(m + M)^2}\omega + M\dfrac{m^2 r^2}{(m + M)^2}\omega = \dfrac{mM^2 r^2 + Mm^2 r^2}{(m + M)^2}\omega$$

$$= \dfrac{mM}{m + M}r^2\omega$$

Thus the nucleus-electron system is equivalent to a single particle of mass $m' = \dfrac{mM}{m + M}$ which appears to move about the nucleus, with its distance from the nucleus = distance of the orbital electron from the nucleus. Also $m' < m$, $\therefore$ m' is called the reduced mass of the electron.

Thus equation 17 becomes

$$\dfrac{1}{\lambda} = \dfrac{\nu}{c} = \dfrac{\mu e^4 Z^2}{8 \,\epsilon_0^2\, ch^3}\left(\dfrac{1}{n_f^2} - \dfrac{1}{n_i^2}\right) \qquad (19)$$

where $\qquad \mu = \dfrac{mM}{m + M}$.

1.6 Rydberg Constant for Hydrogen and Helium Atoms

$$R_H = \dfrac{\mu e^4}{8 \,\epsilon_0^2\, ch^3} = \dfrac{M_H m e^4}{(m + M_H)8\, \epsilon_0^2\, ch^3}$$

Here, $\quad m$ = mass of electron

M_H = mass of hydrogen nucleus

R_H = Rydberg's constant for H-atom.

$$R_{He} = \dfrac{M_{He} m e^4}{8\, \epsilon_0^2\, ch^3 (m + M_{He})} \qquad \text{(for singly ionized helium atom)}$$

$$\therefore \qquad \frac{R_{He}}{R_H} = \left[\frac{M_{He}m}{m + M_{He}}\right]\frac{(m + M_H)}{mM_H} = \frac{M_{He}}{M_H}\left(\frac{m + M_H}{m + M_{He}}\right) \qquad (20)$$

We know that $M_{He} = 4M_H$, hence

$$\frac{R_{He}}{R_H} = \frac{4(M_H)}{(M_H)}\left(\frac{m + M_H}{m + 4M_H}\right)$$

Since $\qquad 4(m) + (4M_H) > m + (4M_H)$

Therefore, $R_{He} > R_H$ $\qquad\qquad\qquad\qquad\qquad\qquad\qquad (21)$

i.e. Rydberg's constant for helium atom is greater than that for the hydrogen atom.

1.7 Sommerfeld's Theory

The simple Bohr's theory of circular orbit of an electron around the central nucleus predicts that lines in H spectrum should each have a well defined λ given by

$$\overline{v} = \frac{1}{\lambda} = \frac{me^4}{8\,\epsilon_0^2\,ch^3}\left(\frac{1}{n}\frac{1}{n_1^2} - \frac{1}{n_2^2}\right)$$

corresponding to a transition from n_1^{th} orbit to n_2^{th} orbit.

A close inspection of lines of H spectrum (under high resolution spectroscope) reveals that the individual lines are not single but consist of very fine lines close together. This fine structure could be explained only if corresponding to any principal quantum number, there may be more than one allowed orbits having slightly different energies. Sommerfeld modified Bohr's theory to explain experimental results. He postulated that an electron going around the nucleus under central (attractive) inverse square field force, must be describing an ellipse with nucleus at one of the foci. The polar equation of the ellipse is given by

$$\frac{1}{r} = \frac{1 + \epsilon\cos\theta}{a(1 - \epsilon^2)}$$

where a is the semimajor axis and ϵ is eccentricity.

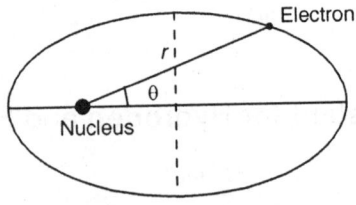

Fig. 1.5: An elliptical orbit with nucleus at one of the two foci of the ellipse

There are obviously two degrees of freedom associated with such a motion—the position of the electrons depending on two coordinates (r, θ) accordingly each degree of freedom should seperately be quantised i.e.

$$\oint p_r dr = n_r h \qquad (22)$$

$$\oint p_\theta d\theta = n_\theta h \qquad (23)$$

where n_r = radial quantum number and $p_r = \left(m\dfrac{dr}{dt} \right)$ radial momentum.

n_θ = azimuthal quantum nuumber and $p_\theta = \left(mr^2\dfrac{d\theta}{dt} \right)$ angular momentum.

also

$$n = (n_r + n_\theta) \tag{24}$$

Total energy of the system

$$W = -\frac{Ze^2}{4\pi \in_0 r} + \frac{1}{2}m\left(\frac{dr}{dt} \right)^2 + \frac{1}{2}mr^2\left(\frac{d\theta}{dt} \right)^2 \tag{25}$$

From the fundamental Eqns. (22), (23) and (24) and the Eqn. of ellipse, it can be shown that

$$1 - \varepsilon^2 = \frac{b^2}{a^2} = \frac{n_\theta^2}{n^2} \tag{26}$$

and

$$W = -\frac{me^4 Z^2}{8 \in_0^2 h^2}\left(\frac{1}{n_r + n_\theta} \right)^2 = -\frac{me^4}{8 \in_0^2 n^2 h^2} \tag{27}$$

where

ε = eccentricity of the ellipse
a = semimajor axis
b = semiminor axis

For a given principal quantum number (n) and corresponding major axis ($2a$), there can only be a limited number of elliptical orbits (e.g. $n = 4$ corresponds to n_θ = 4, 3, 2, 1, four ellipses having same major axis but with semiminor axes $b = a$, $(\frac{3}{4})a$, $(\frac{2}{4})a$, $(\frac{1}{4})a$.

However, according to the energy formula all these orbits have same energy, thus apparently no new energy levels are introduced inspite of the introduction of multiplicity of orbits.

The solution lies in the fact that for an electron moving in an elliptical orbit, the velocity and hence the mass of electron does not remain constant but changes with the position of the electron v is highest when the electron is nearest to the nucleus and mass varies as $m = m_0 / \sqrt{1 - v^2/c^2}$. The relativistic effects are particularly pronounced in the elliptical orbits. The energy of the electron in a more eccentric orbit would be greater than that in a less eccentric orbit even if the major axis is same.

Fig. 1.6: A precessing ellipse-rosette path of the electron

When relativistic variation is taken into consideration, the equation of the path changes to:

$$\frac{1}{r} = \frac{1 + \varepsilon \cos \phi\theta}{a(1 - \varepsilon^2)} \tag{28}$$

where

$$\phi^2 = 1 - \left(\frac{Ze^2}{4\pi \in_0 cp_\theta} \right)^2$$

∴ the path of the electron becomes rosette–a precessing ellipse, major axis turning slowly in the plane of the orbit with constant angular velocity.

The total energy W of the system (corrected for relativistic variation) is given by:

$$W = -\frac{me^4}{8 \in_0^2 n^2 h^2} \left[1 + \frac{Z^2 \alpha^2}{n^2} \left(\frac{n}{n_\theta} - \frac{3}{4} \right) \right] \tag{29}$$

where

$$\alpha = \left(\frac{2\pi e^2}{4\pi \in_0 ch} \right) \tag{30}$$

is a dimensionless constant called *fine structure constant*. It's value is 1/137. The first term on the r.h.s. is the energy of the electron in the orbit with the principal quantum number n according to Bohr's theory. The second term is Sommerfeld's relativity correction arising from rosette motion of the electron orbit. The fine structure is thus explained by the fact that corresponding to any principal quantum number n there are n subenergy levels (corresponding to n permitted values of n_θ) very close to each other (because α^2 is a small number). Transition may take place from any of the higher energy sublevel to lower energy sublevel giving rise to many spectral lines having slightly different wavelengths. However not all transitions are allowed. Only such transitions are allowed for which $\Delta n_\theta = \pm 1$ (selection rule).

1.8 Ionization and Resonance Potentials

Normally an atom exists in the state where its potential energy is least i.e. the electrons occupy those shells and subshells which are most closely bound to the nucleus. When energy is given to the atom, it raises an electron in the atom from its normal orbit to an orbit of higher energy, the atom is then excited. The least energy in electron volts, which is necessary to excite a free neutral atom from its ground state to a higher state is called a *critical potential* of the atom. A critical potential is called *ionization potential* when absorption of energy by the unexcited atom just raises it to a higher state but does not ionize it. Excitation potentials are also called *resonance potentials* because the atoms get excited to any higher state from the ground state by absoption of spectral radiations having the same energy as the corresponding critical potential.

1.9 Franck and Hertz Experiment

The gas of the element under study is contained in a closed glass tube at a pressure of about 1 mm of mercury. A three-electrode structure consisting of filament F, grid G and plate P is mounted in the tube. F is heated by a steady current from a battery and emitted electrons are accelerated towards G by a potential difference V maintained between F and G (V can be varied). A small opposing potential difference V_0 is maintained between G and P. A galvanometer measures the plate current. The pressure of the gas is so adjusted that the electrons undergo several collisions in the space between the filament and the grid. The current registered by ammeter A is plotted against the accelerating potential V. The curve obtained exhibits a series of regularly spaced peak values (Figs 1.7 and 1.8).

In the initial stage, since there is potential V_0 between grid and plate, only electrons having energies greater than a certain minimum contribute to the current. As V is increased, more and more electrons reach the plate and the current increases. After a certain critical electron energy (4.9 *eV*) is reached the plate current drops abruptly—an electron colliding with one of the atoms gives up some or all of its kinetic energy in exciting the atom to the level above ground state. The critical electron energy 4.9 *eV* corresponds to the excitation energy of the atom. (This increase of electron energy agrees with the quantum energy for one of the most prominent lines in the mercury spectrum—an ultraviolet line of wavelength 2537 Å).

$$E_{photon} = \frac{hc}{\lambda} = \frac{1.24 \times 10^4 \, eV}{2537} \approx 4.9 eV \qquad (31)$$

Then as V is raised further, the plate current again increases, since the electrons now have a sufficient energy left after experiencing an inelastic collision to reach

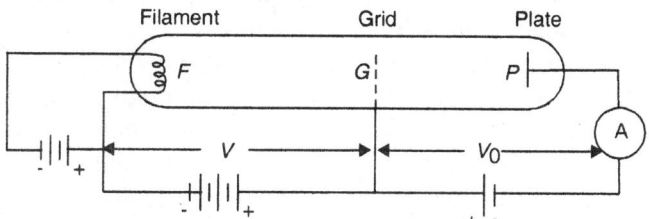

Fig. 1.7: Apparatus for Franck and Hertz experiment

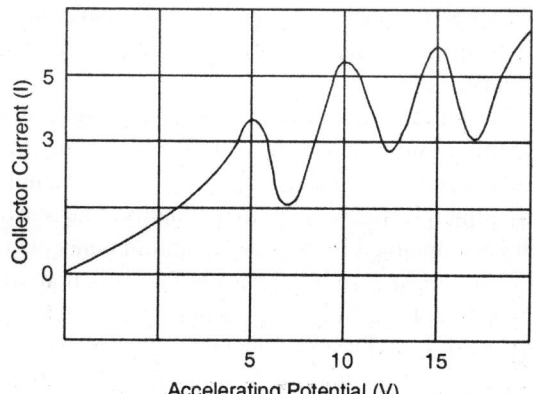

Fig. 1.8: Collector current versus accelerating potential

the plate. Another sharp drop in plate current occurs due to excitation of energy level. Thus a series of critical potentials is obtained in this way for a particular atomic species. The experiment confirmed the existence of discrete energy levels in the mercury atom. Franck and Hertz were awarded the Nobel Prize in 1925 for this work.

1.10 Achievements and Limitations of Old Quantum Theory

(I) Successes of old quantum theory

1. It successfully provided the explanation of the spectrum of the hydrogen atom including its fine structure.
2. It could also explain energy states of the alkali atoms.
3. It could explain X-ray spectra and normal Zeeman effect.
4. Recognition of the quantum character of the magnitude and direction of angular momentum was one of its finest achievements.

(II) Shortcomings of the old quantum theory

The old quantum theory was based on the Bohr's postulates and Wilson Sommerfeld's quantization for periodic systems. Although it was able to resolve some problems associated with the classical theory of matter and radiation, it had only limited applicability. It was lacking in following respects:

1. Although Franck-Hertz experiment gave direct support to Bohr's concept, the theory could not explain the mechanism in Franck-Hertz experiment as the behaviour of the collision phenomenon (scattering of electrons by atoms) are outside the purview of the theory.
2. It led to an ambiguous prescription that the components of angular momentum along any axis passing through the origin of the system having spherical symmetry must be an integral multiple of h.
3. The theory could be applied only to a periodic or multiply-periodic systems and there exists no method in this theory to quantize a periodic motion. Further, the theory could not explain the processes connected with the spin of electrons and Pauli's exclusion principle. This theory could not explain Anomalous Zeeman effect and Stark effect.
4. It could not explain spectra obtained from H_2 molecule or He-atom. The intensities of different lines could not be understood at all on the basis of this old theory.
5. The theory implied that a particle was possessing at each instant a well defined position and momentum and that these quantities were varying continuously in course of time, so it was not in agreement with Heisenberg's uncertainty principle and wave particle duality. The reason was that it was a hybrid theory obtained by grafting quantum concepts onto the classical mechanics. The most incomprehensible fact was that while retaining the classical picture of well defined particle orbits, why certain orbits are completely stable and others are not allowed at all. This was later resolved by realizing that particle states at the microscopic level are not describable in terms of well defined orbits but correspond to wave like nature.

The complex spectra of atoms and their relation to atomic structure was explained by vector model of the atom. Two distinct features of the vector atom model are (i) the concept of spatial quantization and (ii) the spinning electron hypothesis.

1.11 Atomic Space Quantization

The Bohr Sommerfeld orbits are all quantised as regards their magnitude i.e. their size and form. But quantum theory demands more than this viz. quantisation also of direction or orientation of the orbit in space as for instance, in presence of some applied field. The introduction of such a spatial quantisation makes the orbit vector quantities.

The orbital motion of an electron is equivalent to a current in a loop of wire, so each orbit has a magnetic moment. The magnetic moment vector is perpendicular to plane of orbit but antiparallel to $\vec{l}$ (because of negative charge on electron).

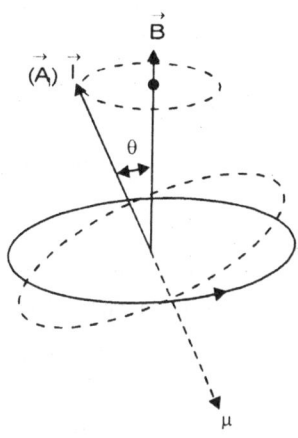

When the atom is placed in an external field, each electron orbit will be subject to a torque which tends to make $\vec{l}$ -vector parallel to the field. Because of the torque of the field on the system, the $\vec{l}$ vector precesses about $\vec{B}$. The motion of the orbit called *Larmor precession* introduces additional energy states into the atomic system. The amount of energy depends on θ (the angle which $\vec{l}$ makes with $\vec{B}$). If all

Fig. 1.9: Larmor precession

values of θ could occur, there would be an infinite number of new energy states. Zeeman effect shows however only a few additional lines therefore only a few values of θ are allowed and these are those for which $\vec{l} \cos \theta$ (i.e. projection of l on the direction of $\vec{B}$) is an integer. This introduces another quantum number called orbital magnetic quantum number m_l. It can have any integral value from $-l$ through 0 to $+l$. Since m_l is limited to whole numbers, the component of orbital angular momentum along the magnetic field is restricted to integral multiples of $h/2\pi$.

Because of the restriction on the orientations of electron orbits, they are said to be space quantised. Choosing the z-axis along the external (applied) magnetic field provides a meaningul reference direction to quantify L_z. Since the appearance of spatial orientations occurs when a magnetic field is present, m_l is called *magnetic quantum number*.

1.12 Bohr Magneton

A loop bearing a current I has a magnetic moment m equal to I times the area of the loop. Similarly, an electron moving in a Bohr orbit constitutes an atomic current which has associated with it a magnetic moment. For the nth Bohr orbit, the area is πr_n^2 while the current

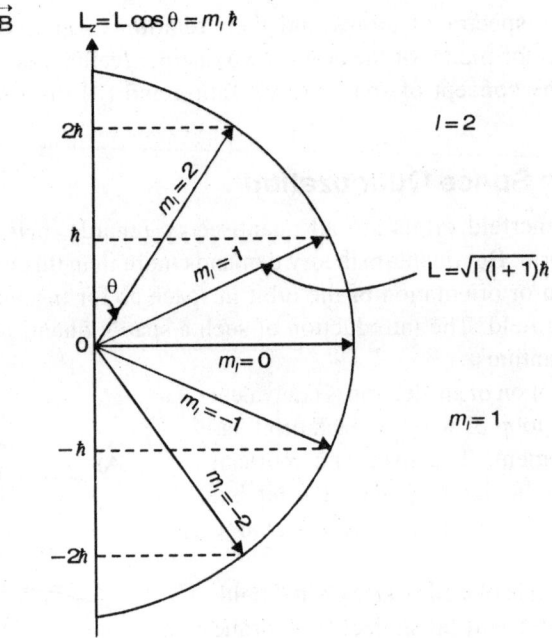

Fig. 1.10: Space quantization of orbital angular momentum $\vec{L}$.
The component of angular momentum L_z is quantized in units of $\hbar$

$$I = -e \times \frac{\text{number of revolutions made by electron}}{\text{time}}$$

$$= -e \frac{\omega}{2\pi} = -\frac{er_n \omega}{2\pi r_n} = -\frac{ev}{2\pi r_n}$$

$$\therefore \quad \mu = I\left(\pi r_n^2\right) = -\frac{ev}{2\pi r_n} \pi r_n^2 = -\frac{e}{2} v r_n \qquad (32)$$

Multiplying the numerator and denominator by m, we get

$$\mu = -\frac{e}{2m} m v r_n$$

or
$$\vec{\mu}_l = -\frac{e}{2m} \vec{A}_l \qquad (33)$$

where
$\mu_l \rightarrow$ vector magnetic moment
$A_l \rightarrow$ vector angular momentum.

The subscript l is used to designate "orbital" angular momentum and magnetic moment.

The ratio
$$\frac{\mu_l}{A_l} = -\frac{e}{2m} \qquad (34)$$

is called *gyromagnetic ratio*.

The negative sign shows these vectors to be in opposite directions for a negative revolving charge (see Fig. 1.9).

For a Bohr hydrogen atom in its ground state, $n = 1$, and since angular momentum $A = n\hbar$

$$\therefore \qquad A_l\big|_{n=1} = \hbar \quad (= h/2\pi)$$

The magnetic moment predicted by eqn. (33) for this orbit ($n = 1$) is called the *Bohr magneton* and is given by

$$\mu_B = \frac{e\hbar}{2m_e} = 9.27 \times 10^{-24} \; Am^2 \tag{35}$$

The electronic magnetic moments are expressed in units of Bohr magneton.

1.13 The Sodium Spectrum, the Notations for Orbitals and Concept of Electron Spin

The eleven electrons of the normal Na-atom arrange themselves around the nucleus so that the total energy is a minimum, the K-shell ($n = 1$) filled with 2 electrons, L shell ($n = 2$) filled with 8 electrons and the remaining one electron lies in the M ($n = 3$) shell. The nucleus together with filled K- and L-shells acts as an Na^+ ion and the outermost (valence) electron in the M shell ($n = 3$) only plays an important role in the production of optical spectrum. It is called an optical electron. It has been shown that if a single electron alone is active in production of spectra, then the orbital azimuthal momentum is $lh/(2\pi)$ and not $kh/(2\pi)$, (where l is called orbital quantum number and k, the azimuthal quantum number)

$$l = (k - 1)$$

therefore l is used instead of k.

Four series are observed in the arc spectra of alkali metals–sharp, principal diffuse and fundamental.

Sharp series arises from an electron falling to an orbit with $k = 1$ ($l = 0$). Therefore, an atom with an electron in a state where $k = 1$ is described as possessing an electron in the s-(orbital) state.

Similarly, for $k = 2$ (i.e. $l = 1$), a p-state of electron exists, for $k = 3$ a d-state and for $k = 4$, a f-state.

For example, a $2s$ electron is one for which $n = 2$ and $l = 0$. For a given value of n, the possible values of l are $(n–1)$, $(n–2)$, $(n–3)$ ...0.

Selection rule : $\Delta l = \pm 1$.

It s found in sodium spectrum that a series of doublets (pairs of lines close together in wave number) occur in the principal series. This is explained in terms of *Goudsmit and Uhlenbeck's hypothesis of electron spin*.

The electron itself rotates about its own axis with an angular momentum of $\left(\frac{h}{2}\right)\big/2\pi$. In doing so it is associated with a magnetic moment $e\hbar/2m$. This magnetic $\vec{j}$ moment reacts with the magnetic moment due to orbital motion of the electron in a quantized way so that spin can be either parallel or antiparallel. It is designated by quantum number s and magnetic interaction is given by

$$\vec{j} = \vec{l} \pm \vec{s} \tag{36}$$

+ and – signs correspond respectively to the cases when the spin is parallel to l and antiparallel to l. $\vec{j}$ is termed the *total angular momentum quantum number.*

Consequently, for each value of l other than zero, there are two values of j given by $(l + \frac{1}{2})$ and $(l - \frac{1}{2})$ and corresponding two energy level terms arise (parallel addition and antiparallel addition respectively).

$2p$ state corresponds to $n = 2$ and $l = 1$. An electron formerly lying in level $2p$ now can exist in two states close together in energy and responsible for doublets in spectrum on transitions. $2p$ state is a case where

$$n = 2, \qquad l = 1, \qquad j = \tfrac{1}{2} \qquad (j = l - \tfrac{1}{2})$$

and $\qquad n = 2, \qquad l = 1, \qquad j = \tfrac{3}{2} \qquad (j = l + \tfrac{1}{2})$

The notation now gets modified, – the designation of the electron being either $2p_{1/2}$ or $2p_{3/2}$. For the lower (s) state $l = 0, j = \tfrac{1}{2}$, only one term $2s_{1/2}$ is possible.

The transitions that can take place between the two terms of the p state and the single term of the s-state are

$$2p_{1/2} \rightarrow 2s_{1/2}$$

and $\qquad\qquad\qquad\qquad 2p_{3/2} \rightarrow 2s_{1/2}$ \hfill (37)

Applying the selection rules

$$\Delta l = \pm 1 \quad \text{and} \quad \Delta j = 0 \quad \text{or} \quad \pm 1. \tag{38}$$

Both the transitions are allowed which explains the doublet fine structure of the sodium D line. These are due to transitions of the atomic state, written in the form

$$2p_{1/2} \rightarrow 2s_{1/2} \leftrightarrow D_1; \lambda = 5896 \text{ Å}$$

$$2p_{3/2} \rightarrow 2s_{1/2} \leftrightarrow D_2; \lambda = 5890 \text{ Å} \tag{39}$$

D_2 is more intense than D_1 because

for $\qquad\qquad\qquad D_2, \Delta l = -1; \Delta j = -1$

while for $\qquad\qquad\qquad D_1, \Delta l = -1; \Delta j = 0$ \hfill (40)

1.14 Stern and Gerlach Experiment (Verification of the concept of space quantization)

The electron orbits of the atom can only lie along certain predetermined directions, incapable of being explained in classical physics, but given in quantum theory by

$$\cos\theta = \frac{m_j}{j}$$

$(m_j$ = magnetic total angular momentum quantum number)

$\qquad j$ = total angular quantum number

m_j has $2j + 1$ vlaues from $+j$ to $-j$ (except 0)

where θ is the angle between the vector $\bar{j} = \bar{l} + \bar{s}$ and the magnetic field direction. This result was confirmed by *Stern and Gerlach experiment*.

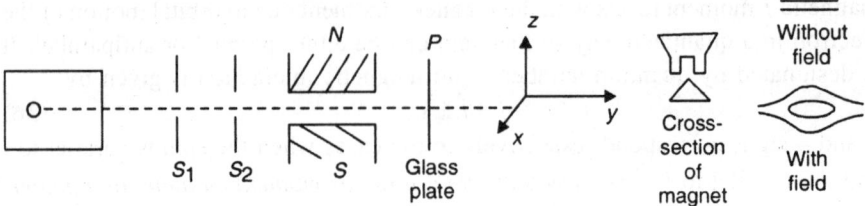

Fig. 1.11: Stern and Gerlach experiment

A beam of Ag atoms was passed through an inhomogeneous magnetic field.

To produce the beam Ag was vaporised by heating it in a small electric oven which was furnished with an exit aperture. Two further slits s_1 and s_2 then collimated the Ag atoms into a narrow beam, which travelled along the x-axis into the space between the magnetic pole pieces N and S.

The arrangement was erected in a highly evacuated glass vessel. The magnetic field between the specially shaped pole pieces had a large space rate of variation provided by having one pole in the form of a knife edge and the other in the form of a channel. The magnetic field was of much greater intensity near the knife edge than anywhere else in the gap. The beam impinged on a cold glass plate immediately after emerging from the field.

With no magnetic field a thin line trace was obtained on P. With the field on, the trace was divided into two lines, except at the ends where the Ag passed through the field outside the region of strongest field near the knife edge shaped pole piece.

This result confirms the existence of electron spin and the postulate of space quantisation

Explanation

The force that deflects the Ag atoms is not the Lorentz force $q\vec{v} \times \vec{B}$ since the (silver) atoms are electrically neutral.

For silver $Z = 47$, the outermost electron will be in 5s state and all the inner shells and subshells are closed. Ag is therefore monovalent with a single optical electron this 5s electron alone will be responsible for the magnetic dipole moment of the atom. This causes individual atomic magnetic dipoles to interact with the spatially varying magnetic field having a gradient along the vertical (z-) direction. If the orbit of the 5s electron of the Ag atom would lie in any plane in space then for the millions of atoms in the beam, a continuous distribution of directions in space would be obtained and not distinct directions as shown by the experiment.

The electron in the Ag is normally in $2\,s_{1/2}$ state and for s state $l = 0$, $j = l + s = \frac{1}{2}$ and magnetic moment is Bohr magneton,

m_j will have $(2j + 1) = 2$ values, one of which will be $\frac{1}{2}$ and other $-\frac{1}{2}$.

The angles of setting of $\vec{j}$ to the magnetic field will be then

$$\cos\theta = \frac{m_j}{j} = 1 \text{ or } -1$$

$$\therefore \qquad \theta = 0 \text{ or } 180^0$$

i.e. the magnetic moment will be directed either along field direction or opposite to it.

Expression for splitting of beam

Let an atom of magnetic moment $\vec{\mu}$ enter a nonuniform magnetic field $\vec{B}$ gradually varying in space along z-direction. Then its potential energy $U = \vec{\mu} \cdot \vec{B}$, and

force
$$\vec{F} = -\nabla U = \frac{\partial}{\partial z}(\vec{\mu} \cdot \vec{B}) \tag{41}$$

Suppose the angle between $\vec{\mu}$ and $\vec{B}$ is θ, then

$$F = \mu \cos\theta \frac{\partial B}{\partial z} \qquad (42)$$

Classically θ is not defined, but quantum mechanically, it may have two possible values. When atom enters the non-uniform field it experiences an acceleration

$$a_z = \frac{F}{M_0} \qquad (43)$$

where M_0 is the mass of the atom. If l is the length of the magnetic field and v is the velocity of atom along the direction of beam then time taken in magnetic field

$$t = l/v \qquad (44)$$

If we assume the acceleration along z-direction to be constant, then displacement

$$Z = \frac{1}{2}a_z t^2 \qquad (45)$$

$$= \frac{F}{M_0}\left(\frac{l}{v}\right)^2$$

$$= \frac{1}{2}\frac{\mu\cos\theta\left(\dfrac{\partial B}{\partial z}\right)}{M_0}\left(\frac{l}{v}\right)^2$$

$$= \frac{1}{2}\mu\cos\theta\left(\ell^2\Big/M_0 v^2\right)\frac{\partial B}{\partial z}$$

Due to space quantization we have $\cos\theta = \pm 1$. Therefore,

$$Z = \pm\frac{1}{2}\frac{\mu\ell^2}{M_0 v^2}\frac{\partial B}{\partial z} \qquad (46)$$

1.15 Pauli's Exclusion Principle

Four quantum number

$\quad\quad\quad n$ (total quantum number)

$\quad\quad\quad l$ (orbital quantum number)

$\quad\quad\quad m_l$ (orbital magnetic quantum number)

and $\quad\quad\quad m_s$ (spin magnetic quantum number)

decide completely the state of the electron in the atom. According to Pauli's principle, no two electrons in an atom can exist in the same quantum state i.e. no two electrons can have the same set of four quantum numbers n, l, m_l and m_s.

Further, one atomic state corresponds to the assignment of a set of quantum states decided by four quantum numbers.

1.16 Photo-Electric Effect

Figures 1.12 shows schematic setup for observing the photoelectric effect. The figure shows an evacuated quartz tube containing two electrodes. The thin wire A is held at a positive potential with respect to the metal plate C. When light of

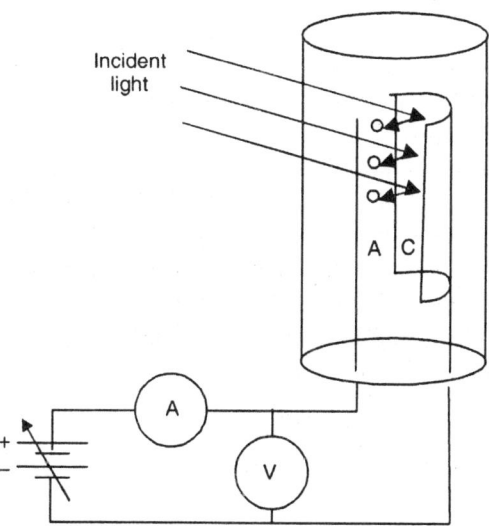

Incident
light

A C

A

V

Fig. 1.12: Schematic apparatus for observing the photoelectric effect. When light strikes a metal plate C, electrons are ejected that are collected at A, causing a current

a given frequency strikes place *C* it ejects electrons from its surface. These electrons are called *photoelectrons*. These are attracted by the positive electric potential and produce a measurable current.

This liberation of electrons from certain metals under the influence of a high frequency electromagnetic radiation is known as *photo electric effect*.

If we gradually reduce the potential applied to *A* until it becomes negative with respect to the plate *C*, some of the ejected electrons will not have enough kinetic energy to reach the wire. They will be repelled back to the plate reducing the current. As the applied potential is made further negative, the current decreases until it becomes zero at a potential called the *stopping potential*. This potential difference V_0 multiplied by the electronic charge *e* gives a measure of the maximum kinetic energy which the emitted photoelectrons can acquire ($K_{max} = eV_0$). It is observed that stopping potential is independent of the intensity of the incident light and is different for different materials. The maximum kinetic energy of the liberated photoelectrons is also independent of intensity of incident light.

Experimental study further reveals following facts:

(*i*) For a given photosensitive material, there is a certain minimum frequency called the *threshold frequency*, below which there takes place no photo-electron emission, no matter how great is the intensity of light.

(*ii*) For light of a given frequency, the photoelectric current is directly proportional to the intensity of light, provided the frequency is above the threshold frequency (More intense light means more photoelectrons).

(*iii*) The maximum kinetic energy of the photoelectrons is found to increase with the increase in the frequency of the incident light provided the frequency exceeds the threshold limit. The maximum kinetic energy is, however, found to be independent of intensity of incident light.

(*iv*) The photoelectron emission is an instantaneous process. The emission starts immediately without any time-lag (Classically one may expect photoelectrons will absorb energy over a period of time and then gain enough energy to escape).

In 1905, A. Einstein proposed a theory of the photoelectric effect based upon Planck's idea of 'quanta' of radiation. According to him, monochromatic light of frequency v consists of photons of energy hv where h is Plank's constant.

When light of sufficient energy strikes a photo sensitive metal, a part of its energy (known as the *work function* W_0 of the material) is used up in liberating the electron from the metal surface and the remaining is spent in imparting kinetic energy to the liberated electron.

$$hv = W_0 + \frac{1}{2}mv^2 \qquad (47)$$

This is known as *Einstein's photoelectric equation*. The maximum kinetic energy of the emitted electrons

$$\frac{1}{2}mv^2 = \left(hv - W_0\right)$$

If v_c is the lowest (threshold) frequency which just causes the emission of electrons, then,

$$hv_c = W_0 \qquad (48)$$

and
$$\frac{1}{2}mv^2 = h(v - v_c) \qquad (49)$$

Evidently, no photoelectric effect could be observed if $v < v_c$. v_c is also known as the cut off frequency. It is a characteristic of the metal. The product h time v_c equals the work function. Table 1.2 lists the photoelectric work function for some metals.

Table 1.2

Metal	Work function W_0 (eV)
Na	2.75
Ag	4.73
Cu	4.94
Au	5.30
Pt	5.65

If v is reduced below v_c, hv will be smaller than W_0 and the individual photons, no matter, how many they are (i.e. no matter what the intensity of radiation is) will not be able to eject photoelectrons.

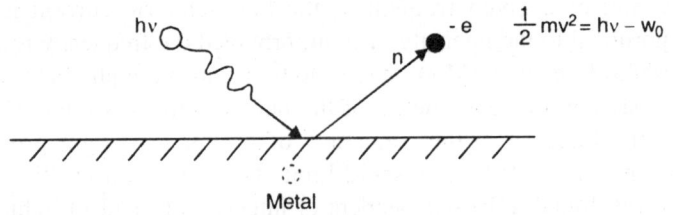

Fig. 1.13: Schematic photoelectric effect

Thus according to Einstein, the emission of a photoelectron is the result of the interaction of a single quantum of radiation (i.e. a photon) with an electron. For this discovery, he received the Noble prize of Physics in 1921.

We can rewrite

$$h\nu = K_{max} + W_0 = eV_0 + W_0$$

or

$$V_0 = \left(\frac{h}{e}\right)\nu - \left(\frac{W_0}{e}\right).$$

This predicts a linear relationship between V_0 and ν as can be verified experimentally. The slope of the experimental curve should give h/e. We can find the value of h experimentally by multiplying the slope of the experimental curve by e. From a careful analysis of experiments on Lithium surfaces, Millikan found $h = 6.57 \times 10^{-34}$ j sec. +6This agrees very well with the value of h derived from Planck's radiation formula. This together with existence of ν_c provides a striking verification of Einstein's photon-concept.

1.17 Short Wavelength Limit in X-Rays

When a metal target is bombarded with electrons that have been accelerated through some considerable potential (say 5–50 kV), one obtains what is called *Bremstrahlung* (German, for "deceleration radiation"). The spectrum of this radiation is continuous and covers a wide range of wavelengths. The observed spectra cannot be explained by classical mechanics—a sharp cut off is obtained at a certain minimum wavelength λ_m, which is the same for all metals, varying systematically with the accelerating voltage V_0 according to the law:

$$\frac{1}{\lambda_m} = \frac{e}{hc}V_0 \tag{50}$$

Then, defining $\nu_m = c/\lambda_m$, we have (Maximum photon energy) $= h\nu_m = V_0e$ = (kinetic energy of incident electron). Experimental data on this relationship are shown in Fig. 1.14.

This phenomenon resembles the photoelectric effect in reverse.

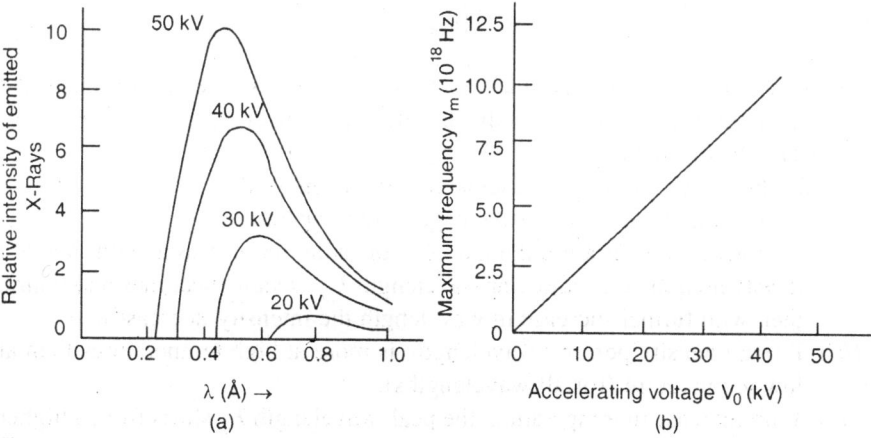

Fig. 1.14: (*a*) Bremstrahlung spectra produced by electrons of various energies striking a metal target. (*b*) The maximum frequency of emission is proportional to the accelerating voltage V_0

1.18 Black Body Radiation

1.18.1 Early Radiation Laws and Difficulties with Classical Theory

A *perfectly black body* is defined as one which absorbs all the heat radiations (corresponding to all wavelengths) falling on it. Experimentally, such a body is represented by a hollow spherical cavity having a narrow hole and it is painted dull black from inside. When such a body is heated, it emits radiation of all wavelengths.

In 1859, *Kirchoff* formulated two *laws* concerning the properties of a black body:

1. A black body, not only absorbs all the radiation falling on it but also acts as a perfect radiator when heated.

2. The radiation given out by a black body is dependent on the temperature of the cavity and is independent of the nature of the interior material.

In 1879, Stefan found empirically that the power emitted per unit area by a black body is proportional to the fourth power of the absolute temperature or

$$R_B = \sigma T^4 \tag{51}$$

where R_B = power radiated per unit area

σ = Stefan's constant = 0.5669×10^{-7} W/m^2 – K^4

T = absolute temperature

Five years later, Boltzmann deduced the relation (51) from thermodynamic considerations and this relation is known as *Stefan-Boltzmann law.*

The radiant energy may be analysed by passing it through a prism and breaking it up into various wavelengths. The energy associated with different wavelengths can be measured. In the study of black body radiation, it is a general practice to study the spectral distribution (i.e. distribution of energy among different wavelengths). We define the *monochromatic emissive power* e_λ as the radiant power emitter per unit area in the spectral range λ to $\lambda + d\lambda$ and is expressed as $e_\lambda \, d\lambda$. Obviously power radiated per unit area

$$R = \int_0^\infty e_\lambda d\lambda \tag{52}$$

The spectral distribution of black body radiation was studied by Lummer and Pringsheim (in 1899) experimentally at different temperatures. Some of these results are shown in Fig. 1.15.

From the Fig. 1.15 following conclusion can be inferred:

(*i*) e_λ is a function of both wavelength and temperature.

(*ii*) At a given temperature, the intensity of radiation increases with increase in wavelength, at a particular wavelength λ_m its value becomes maximum, then with further increase in wavelength the intensity decreases.

(*iii*) Energy density per unit wavelength is more at high temperatures than at low temperature (for all wavelengths).

(*iv*) With increase in temperature, the peak wavelength λ_m shifts from a higher to a lower value as indicated by the dotted curve.

(*v*) The area under each curve represents the total power emitted for the complete spectrum at a particular temperature.

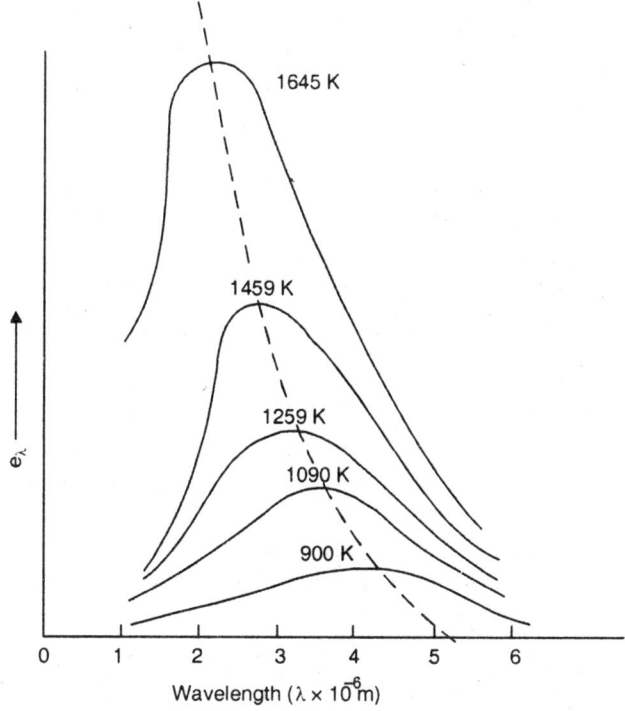

Fig. 1.15: The spectral distribution of black body radiation

On the basis of classical theory, attempts were made to explain qualitatively and quantitatively the nature of the curves of Fig. 1.15. Two fundamental laws regarding the nature of the curves were established. These laws are:

      1. Stefan Boltzmann Law

and   2. Wien's Displacement Law

We have already stated above the Stefan Boltzmann law. According to *Wien's displacement law*, the wavelength at the maximum of the spectral distribution λ_m is inversely proportional to the temperature of emission T. i.e.

$$\lambda_m \times T = \text{constant} = 2.898 \times 10^{-3} \text{ m} -^0\text{K} \tag{53}$$

for all temperatures.

This indicates that λ_m decreases with increase in temperature.

It is evident from Fig. 1.15 that e_λ is a function of both λ and temperature. In 1893, Wien also showed that for any black body, e_λ is given by

$$e_\lambda = T^5 f(\lambda T) = \lambda^{-5} F(\lambda T) \tag{54}$$

where $F(\lambda T) = (\lambda T)^5 f(\lambda T)$. Thus e_λ/T^5 is the same for all black bodies. This is shown schematically in Fig. 1.16 which corresponds to three temperatures of emission.

Thus a single curve served to represent black body radiation at all temperatures. Equivalently, if λ_m is the wavelength at which e_λ is maximum then

$$\lambda_m T = \text{constant} \tag{55}$$

for all temperatures.

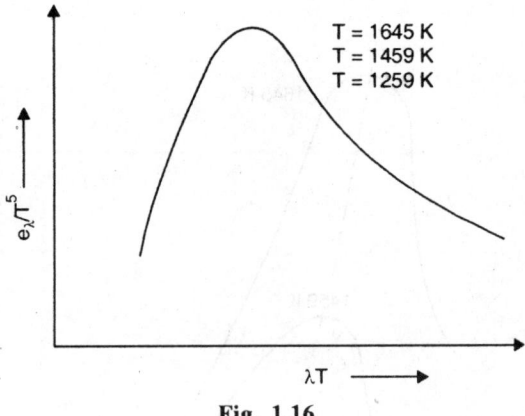

Fig. 1.16

Thus, by classical reasoning, the problem of black body radiation reduced to the determination of a single unknown function $F(\lambda T)$. All attempts to obtain the correct form of the function $F(\lambda T)$ from classical theory failed.

However, two attempts on a classical basis deserve to be mentioned here. Wien derived the following formula based on special assumptions regarding emission and absorption of radiation.

Energy density for wavelengths between λ and $\lambda + d\lambda$

$$\equiv U_\lambda \, d\lambda = \tfrac{1}{4} \, c \, e_\lambda \, d\lambda = A_1 \, \lambda^{-5} \, e^{-A_2/\lambda T} \, d\lambda \tag{56}$$

where c is the speed of light in free space and A_1 and A_2 are two undetermined constants. By proper choice of constants A_1 and A_2, values from Eqn. (55) can be made to fit the data of Fig. 1.16. But at higher values of λT, it fails to fit and predicts very low value.

1.18.2 Rayleigh Jean's Formula and Ultraviolent Catastrophe

In 1900 Rayleigh applied the principle of equipartition of energy to black body radiation. This accompanied by a contribution from Jeans, led to a formula for e_λ which fitted very well with the high $-\lambda T$ part of the e_λ curve but rose to infinity with decreasing λT (Fig. 1.17). The reasoning employed by Rayleigh and Jeans, has applications in modern physics. It is discussed as given below.

Allowed Vibrations in an Enclosure

Let us consider a cubical box with perfectly reflecting walls each of length 'a', enclosing radiation which produces stationary waves. α, β and γ are the angles that a standing wave makes with three walls of the cubical box. If this vibration is to be a standing wave, it is necessary that the total number of modes of vibrations measured along each side be an integral value. This requires that the path lengths be integral multiple of half the wavelength or

$$a \cos \alpha = n_1 \frac{\lambda}{2}$$

$$a \cos \beta = n_2 \frac{\lambda}{2} \tag{57}$$

$$a \cos \gamma = n_3 \frac{\lambda}{2}$$

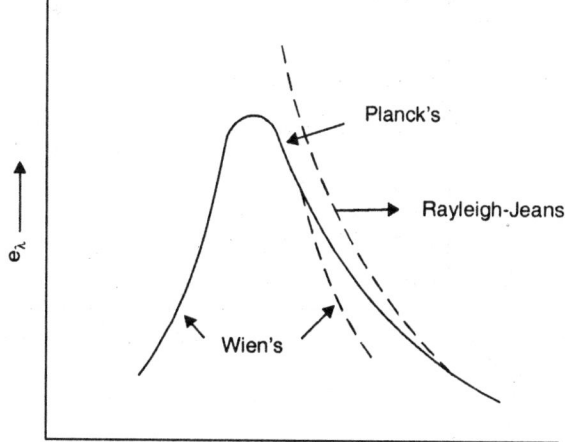

Fig. 1.17: Curve indicating the fitting of the experimental curve by the three laws viz., Wien's law, Rayleigh Jeans law and Planck's law

where n_1, n_2 and n_3 are integers.

Squaring and adding Eqns. (56), we get

$$a^2(\cos^2\alpha + \cos^2\beta + \cos^2\gamma) = \frac{\lambda^2}{4}(n_1^2 + n_2^2 + n_3^2)$$

Since the direction cosines satisfy the relation

$$\cos^2\alpha + \cos^2\beta + \cos^2\gamma = 1$$

we find,

$$(n_1^2 + n_2^2 + n_3^2)\frac{1}{a^2} = \frac{4}{\lambda^2} = \frac{4v^2}{c^2} \tag{58}$$

or

$$v = (n_1^2 + n_2^2 + n_3^2)^{1/2}\frac{c}{2a}$$

where v is the frequency and c the velocity of radiation (= speed of light). Each set of positive values of n_1, n_2, n_3 gives one allowed value of v. The number of possible stationary vibrations between the frequency limits v and $v + dv$ can now be found if we imagine an n-space in which each point ($n_1/2a$, $n_2/2a$, $n_3/2a$) represents a possible standing wave. Thus we have a cubic lattice of points in which the distance from the origin $n \equiv \sqrt{n_1^2 + n_2^2 + n_3^2}$ when multiplied by a constant ($c/2a$) is equal to v (Fig. 1.18). Now we try to calculate the total number of combinations of $(n_1^2 + n_2^2 + n_3^2)^{\frac{1}{2}}$ $c/2a$ between v and $v + dv$. The volume of the complete spherical shell is $4\pi v^2 dv$. Therefore, the volume occupied by one lattice point is $(c/2a)^3$ and we are interested only in the octant of this shell that includes positive values of n_1, n_2 and n_3. The volume of this octant is (1/8) $4\pi v^2 dv$. Hence the total number of possible independent vibration's (in the cube of volume a^3) is

$$\frac{(1/8)\,4\pi v^2 dv}{(c/2a)^3} = \frac{4\pi v^2 dv}{c^3}a^3 \tag{59}$$

and the number of vibrations (dN) per unit volume is

$$dn = \frac{4\pi v^2 dv}{c^3} \tag{60}$$

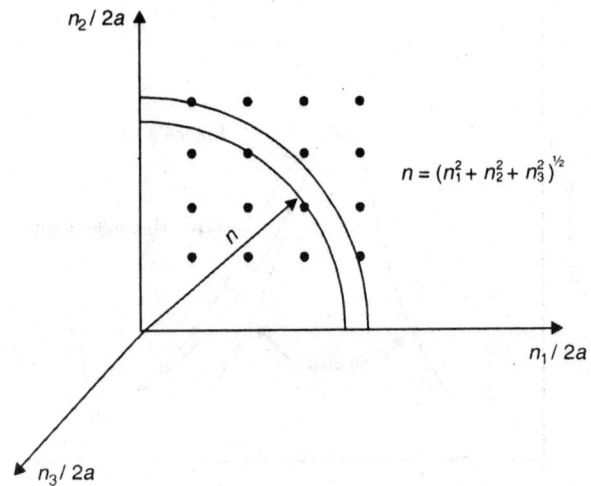

Fig. 1.18: Each point in the n-space corresponds to a possible standing wave

Now, according to law of equipartition of energy, the average energy per degree of freedom of an oscillator in a system of oscillators in thermal equilibrium at absolute temperature T is $(½)\, kT$. Each standing wave in a radiation filled cavity corresponds to two degrees of freedom, because each wave originates in an atomic oscillator having two degrees of freedom – one kinetic and the other potential energy. Or, in other words, an electromagnetic wave has two modes of energy possession – electric and magnetic energy, hence two degrees of freedom. Hence, the number of modes of vibrations per unit volume for standing waves within the range v and $v + dv$ is

$$dN = \frac{8\pi v^2 dv}{c^3} \tag{61}$$

and the mean energy of an oscillator $= kT$. Consequently energy density

$$e_v dv = \frac{8\pi v^2}{c^3} kT \, dv \tag{62}$$

This is *Rayleigh Jeans formula*. This agrees, fairly well with experimental results in the long wavelength region. But as $v \to \infty$ (short wavelength), energy density becomes infinity without going through a maximum. Thus corresponding to ultraviolet regions of spectrum, $e_v\, dv \to \infty$ in the limit of very high frequency. In reality, however, $e_v\, dv \to \infty$ as $v \to 0$ as well as for $v \to \infty$. This discrepancy between theory and experimental results is known as *ultraviolet catastrophe*. Thus black body radiation cannot be explained using classical mechanics.

1.18.3 Planck's Quantum Hypothesis

In 1900, Max Planck gave the first successful theory for the spectral distribution of black body radiation. According to Planck:

1. An oscillator absorbs or emits radiation discontinuously, in the form of energy packets called *quanta*. These behave like a stream of particles having mass, energy and momentum. A *quantum of radiation is a photon*. Its energy is

$$E = h\nu \qquad (63a)$$

where ν is the frequency of radiation and h is a constant known as Planck's constant ($h = 6.627 \times 10^{-34}$ J-s). The magnitude of a quantum is not fixed but depends on the frequency of the oscillator.

2. The oscillator has definite amounts of energies in discrete levels (called energy levels). The energy of the nth level is an integral multiple of a quantum i.e.

$$E_n = nh\nu \qquad (63b)$$

where $n = 0, 1, 2, \ldots.$

1.18.4 The Average Energy of a Radiation Oscillator

Suppose we have a black body radiator consisting of N simple harmonic oscillators having fundamental frequency ν (i.e. these oscillators can take up energies in multiples of $h\nu$). Let, out of N, N_0 oscillators take energy 0, N_1 take energy $h\nu$, N_2 take energy $2h\nu$, ... N_i take energy $ih\nu$ ... respectively. The relative probability that at a given temperature T, an oscillator will have the energy E_n is given by the Boltzmann factor e^{-E_n/k_BT}. Therefore,

$$N = N_0 + N_1 + N_2 + \ldots \qquad (64)$$

and

$$N_1 = N_0\, e^{-h\nu/k_BT}$$

$$N_2 = N_0\, e^{-2h\nu/k_BT}$$

$$\ldots \qquad \ldots$$
$$\ldots \qquad \ldots$$

$$N_i = N_0\, e^{-ih\nu/k_BT}$$

$$\therefore \quad N = N_0 + N_0\, e^{-h\nu/k_BT} + N_0\, e^{-2h\nu/k_BT} + \ldots$$

$$= N_0\, [1 + e^{-h\nu/k_BT} + e^{-2h\nu/k_BT} + \ldots]$$

$$= N_0 \sum_{i=0}^{x} e^{-ih\nu/k_BT} \qquad (65)$$

Total energy of the oscillators:

$$E = \sum_{i=0}^{\infty} E_i N_i$$

$$= 0N_0 + h\nu\, N_0\, e^{-h\nu/k_BT} + 2h\nu\, N_0\, e^{-2h\nu/k_BT} + \ldots$$

$$= N_0\, h\nu\, [0 + e^{-h\nu/k_BT} + 2\, e^{-2h\nu/k_BT} + \ldots]$$

$$= N_0 h\nu \sum_{i=0}^{\infty} i\, e^{-ih\nu/k_BT} \qquad (66)$$

The average energy of an oscillator,

$$\bar{E} = \frac{E}{N} = \frac{N_0 h\nu \sum_{i=0}^{\infty} i\, e^{-ih\nu/k_BT}}{N_0 \sum_{i=0}^{\infty} e^{-ih\nu/k_BT}} \qquad (67)$$

Putting $e^{-h\nu/k_BT} = x$, we have

$$\overline{E} = h\nu \sum_{i=0}^{\infty} i x^i \Big/ \left(\sum_{i=0}^{\infty} x^i \right)$$

Now
$$\sum_{i=0}^{\infty} x^i = (1 + x + x^2 + ...) = \frac{1}{1-x}$$

and
$$\sum_{i=0}^{\infty} i x^i = \left(0 + x + 2x^2 + 3x^3 + ...\right) = x\left(1 + 2x + 3x^2 + ...\right) = \frac{x}{(1-x)^2}$$

Consequently,
$$\overline{E} = \frac{h\nu x(1-x)}{(1-x)^2} = \frac{h\nu x}{(1-x)} \tag{68}$$

or
$$E = \frac{h\nu e^{-h\nu/K_B T}}{\left(1 - e^{-h\nu/K_B T}\right)}$$

or
$$\overline{E} = \frac{h\nu}{(e^{h\nu/k_B T} - 1)} \tag{69}$$

1.18.5 Planck's Radiation Law

The atoms exchange radiant energy in discrete multiples (called quantas). Inside a perfectly black body when e.m. radiation enters through a pinhole and interacts at the inner surface, the particle nature of radiation is evinced and photons come into the picture. Photons are Bose-particles. Let us now derive an expression for frequency distribution of radiation emitted by a black body.

Consider a box containing electromagnetic radiation (i.e. photons) at a certain temperature T. These photons would be absorbed and re emitted by the container walls. In thermal equilibrium we can apply the B-E statistics. There is an important difference in the case of photons, they are created and reabsorbed; so, there number is not conserved. This is apparent because walls can absorb a photon of energy $E = h\nu$ and can emit photons of energy $h\nu_1$ and $h\nu_2$ such that $E_1 + E_2 = E$. There are elastic collisions with the wall so total energy is conserved but there is no restriction on photon number. There is a correspondence between the wave-nature and the photons inside the black-body chamber. The incoming radiation can be considered to be a super position of some harmonic waves which are like standing waves in a vibrating string (normal modes). These normal modes maintain an equilibrium situation at a definite temperature.

Since photons are bosons, the distribution $N(E)\, dE$ for them is described by

$$N(E)\, dE = g(E)\, dE \frac{1}{\exp\left(\dfrac{E}{k_B T}\right) - 1} \tag{70a}$$

where $g(E)\, dE$ is the number of quantum-states available to a photon between an energy E and $E + dE$. On the r.h.s. of Eqn. (70) the term $\exp(\alpha)$ is missing because there is no restriction on photon number. We rewrite

$$N(E)\, dE = g(E)\, dE \frac{1}{\exp\left(\dfrac{h\nu}{k_B T}\right) - 1} \tag{70b}$$

For photons
$$E \neq \frac{p^2}{2m} \tag{71}$$

but
$$p = E/c.$$

We know that

$$g(E)\, dE = \frac{2V}{h^3} \int dp_x dp_y dp_z \tag{72}$$

the integral is over the entire volume in a spherical momentum space.

The reason that factor 2 appears here is that a photon of definite momentum p can be in two permissible modes of polarization which are mutually perpendicular. The integral can be easily evaluated if we write in spherical coordinates

$$g(E)\, dE = \frac{2V}{h^3} \int\limits_{p}^{p+dp} p^2 dp \int\limits_{0}^{\pi} \sin\theta\, d\theta\, (2\pi) \tag{73}$$

$$= \frac{2V}{h^3} \frac{4\pi}{3} \left[(p+dp)^3 - p^3 \right]$$

(using $(E + dE)^3 = E^3 + (dE)^3 + 3E\, dE\, (E + dE)$) and neglecting $(dE)^3$ and $3E$ $(dE)^2$).

$$= \frac{2V}{h^3} \frac{4\pi}{3} \left[\left(\frac{E+dE}{c} \right)^3 - \left(\frac{E}{c} \right)^3 \right]$$

$$\cong \frac{8\pi V}{h^3 c^3} E^2\, dE = \frac{8\pi V}{c^3} v^2\, dv \qquad (\because\ E = h\nu) \tag{74}$$

Thus,
$$N(E) = \frac{8\pi V}{h^3 c^3} \frac{E^2}{\exp\left(\dfrac{h\nu}{k_B T} \right) - 1} \tag{75}$$

$N(E)\, dE$ gives the photon distribution. We write the energy of photons per unit volume in a unit frequency range

$$e_\nu\, d\nu = \frac{EN(E)}{V} dE \tag{76}$$

$$= \frac{8\pi}{h^3 c^3} \frac{E^3 dE}{\exp\left(\dfrac{h\nu}{k_B T} \right) - 1}$$

$$= \frac{8\pi h\nu^3}{c^3} \frac{1}{e^{h\nu/k_B T} - 1} \tag{77a}$$

This energy distribution function is known as *Planck's law* and it agrees well with experiments.

Now, since
$$\nu = \frac{c}{\lambda} \text{ and}$$

$$|d\nu| = \frac{c}{\lambda^2} d\lambda, \text{ hence}$$

$$e_\lambda d\lambda = \frac{8\pi hc}{\lambda^5} \left(\frac{1}{e^{hc/\lambda k_B T} - 1} \right) \tag{77b}$$

This equation is found to be in agreement with the experimental results of Lummer and Pringsheim.

The Planck's equation (77b) may be used to account for the Wien and Rayleigh Jean's laws as follows:

When $\qquad hc \gg \lambda \, k_B T$ (ie. λ or T small)

In that case, $e^{hc/\lambda k_B T} \gg 1$, then -1 in the denominator of Eqn. (77b) may be neglected. Then

$$e_\lambda d_\lambda = \frac{8\pi hc}{\lambda^5} e^{-hc/\lambda k_B T} d\lambda$$

$$= A_1 \lambda^{-5} e^{-A_2/\lambda T} d\lambda \qquad (78)$$

$$\left(\therefore A_1 = 8\pi hc \quad \text{and} \quad A_2 = \frac{hc}{k_B} \right)$$

which is the Wien's equation.

When $hc \ll \lambda k_B T$ (i e. T or λ large), then

$$e^{hc/\lambda k_B T} = 1 + \frac{hc}{\lambda k_B T} + \frac{1}{2!} \left(\frac{hc}{\lambda k_B T} \right)^2 + \dots \cong 1 + \frac{hc}{\lambda k_B T}$$

$$\therefore e_\lambda d\lambda = \frac{8\pi hc}{\lambda^5} \frac{d\lambda}{(1 + \frac{hc}{\lambda k_B T} - 1)}$$

$$= \frac{8\pi hc}{\lambda^5} \cdot \frac{\lambda k_B T \, d\lambda}{hc} = \frac{8\pi}{\lambda^4} k_B T d\lambda \qquad \left(\because \lambda = \frac{c}{v} \quad \text{and} \quad |d\lambda| = \frac{c}{v^2} d\lambda \right)$$

$$\left(\lambda^4 = \frac{c^4}{v^4} \right)$$

or $$e_\lambda d\lambda = \frac{8\pi v^2}{c^3} k_B T dv \qquad (79)$$

which is Rayleigh Jean's equation. We can also derive Wien's displacement law and Stefan Boltzmann law from Planck's radiation formula as follows:

We have

$$e_\lambda = \frac{8\pi hc}{\lambda^5 (e^{hc/\lambda kT} - 1)}$$

The maximum value of e_λ is determined by setting

$$\left. \frac{de_\lambda}{d\lambda} \right|_{\lambda = \lambda_m} = 0$$

and then solve for λ_m.

Let us put $x = \dfrac{hc}{\lambda kT}$ such that λ_m corresponds to $x_0 = \dfrac{hc}{\lambda_m kT}$. Then

$$e_\lambda = \frac{8\pi (kT)^5}{h^4 c^4} \frac{x^5}{(e^x - 1)} \qquad (80)$$

$$dx = \frac{hc}{kT}(-\lambda^{-2}) \, d\lambda = -\frac{x}{\lambda} d\lambda$$

or
$$d\lambda = -\frac{\lambda}{x}dx \qquad \therefore \quad \frac{d}{d\lambda} = -\frac{x}{\lambda}\frac{d}{dx}$$

so that
$$\frac{de_\lambda}{d\lambda} = -\frac{x}{\lambda}\frac{de_\lambda}{dx} = 0$$

or
$$\frac{8\pi(kT)^4}{h^3c^3}\frac{1}{\lambda^2}\frac{d}{dx}\left(\frac{x^5}{e^x-1}\right) = 0$$

or
$$\frac{1}{\lambda^2}\frac{d}{dx}\left(\frac{x^5}{e^x-1}\right) = 0$$

or
$$\frac{1}{\lambda^2}[5x^4(e^x-1)^{-1} + x^5(-1)(e^x-1)^{-2}.e^x] = 0$$

$$\frac{5x^4}{\lambda^2}\left(\frac{1}{e^x-1}\right) = \frac{x^5 \cdot e^x}{\lambda^2(e^x-1)^2}$$

or
$$\frac{5}{x_0} = \frac{e^{x_0}}{e^{x_0}-1} \quad \text{or} \quad \frac{x_0}{5} = 1-e^{-x_0} \tag{81}$$

This transcedental equation can only be solved by successive approximations for x_0 and yields:

$$x_0 = 4.965 \quad \text{or} \quad \frac{hc}{\lambda_m kT} = 4.965$$

Thus,
$$\lambda_m T = \left(\frac{hc}{4.965\,k}\right)$$

$$\equiv (2.898 \times 10^{-3})\ \text{m} - \text{K} \ (\textit{Wien's constant}) \tag{82}$$

Now, the total energy density R within the cavity is obtained by integrating the energy density over all frequencies:

$$R = \int_0^\infty e_v dv = \int_0^\infty \frac{8\pi h v^3 dv}{c^3(e^{hv/kT}-1)} = \int_0^\infty \frac{8\pi(kT)^4}{(hc)^3}\frac{x^3\,dx}{(e^x-1)}$$

Using
$$\int_0^\infty \frac{x^3\,dx}{(e^x-1)} = \frac{\pi^4}{15},$$

we get
$$R = \left(\frac{8\pi^5 k^4}{15c^3 h^3}\right)T^4$$

But total emissivity (power radiated per unit area) is

$$R_B = \frac{c}{4} \times R$$

$$\left(\frac{2\pi^5 k^4}{15c^2 h^3}\right)T^4 \tag{83}$$

$$\equiv \sigma\,T^4$$

where $\sigma = (5.625 \times 10^{-8})$ W.m^{-2}.K^{-4} (*Stefan's constant*) $\tag{84}$

Thus, we see that Planck's theory successfully explains all the facts of black body radiation.

1.19 Davisson and Germer Experiment

While studying the reflection of electrons from a polycrystalline Ni target mounted in a high vacuum, it was found that vacuum system broke accidentally and air oxidised the metal surface. To reduce the oxide to pure *Ni*, target was baked in a high temperature oven. When the apparatus was re-assembled, the results obtained were greatly modified, instead of a continuous variation of scattered electron intensity with angle, distinct maxima and minima were observed whose positions depended on the electron energy.

Principle: The effect of heating a *Ni* block at high temperature is to cause the many small individual crystals (of which it is normally composed) to form into a single large crystal all of whose atoms are arranged in a regular lattice. The maxima and minima of electron intensity are due to diffraction of electron waves from parallel planes of atoms in this single crystal.

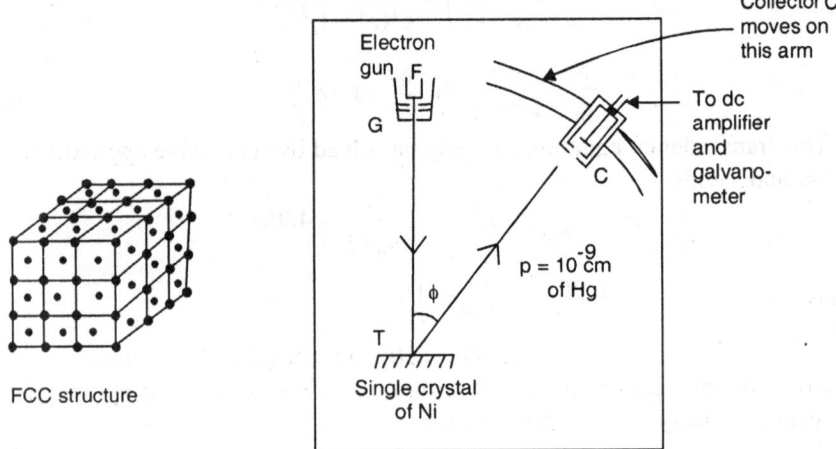

Fig. 1.19: Schematic diagram of apparatus for Davisson and Germer experiment

By using an electron gun *G*, electrons from a heated tungsten filament *F* are collimated and since the potential difference between *F* and *G* is *V* volts, the energy of the electron is *V* electron volts. These mono energetic electrons impinge upon the Ni target *T* and some of the scattered electrons enter the Faraday cylinder *C* which has a retarding field established between two walls, magnitude of this field being such that only electrons with energy just less than *V* eV can enter the cylinder. The collector current is amplified and measured with a sensitive galvanometer. The current is proportional to the number of electrons arriving per second at the aperture of the collector which can be shifted to receive electrons scattered in a direction defined by the angle ϕ called *colatitude angle* (complement of the angle between the collector axis and crystal face). The distribution with angle of the scattered electrons intensity is investigated for various values of *V*.

The method of plotting the polar graphs shown in Fig. 1.20 is such that the intensity at any angle is proportional to the distance of the curve at that angle from the point of scattering.

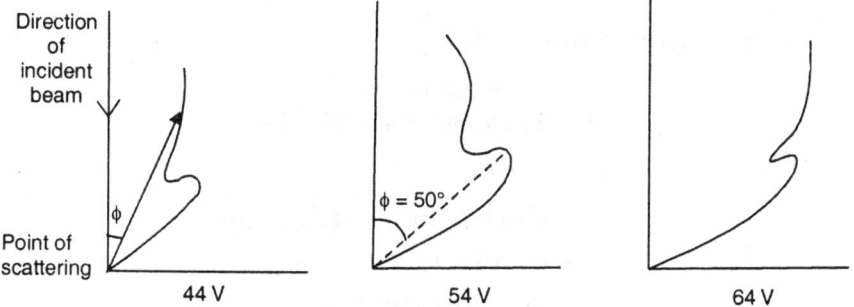

Fig. 1.20: Results of Davisson and Germer experiment

All these curves show a spur which is a maximum when the applied potential difference is 54V and colatitude angle is 50°.

Figure 1.21 is drawn to show the reinforcement of electron waves from certain planes in the Ni crystal.

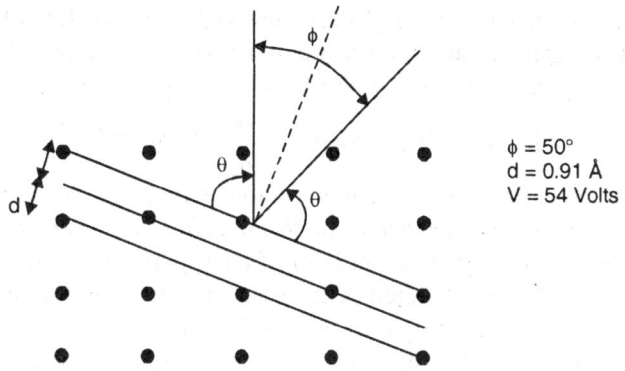

$\phi = 50°$
$d = 0.91$ Å
$V = 54$ Volts

Fig. 1.21: Diffraction of electron waves resulting in reinforcement from planes of atoms in Ni crystal

For a Bragg reflection to occur, the planes responsible for the reflection must be set so that they are perpendicular to the line bisecting ϕ. The Bragg angle $\theta = \left(\dfrac{\pi}{2} - \dfrac{\phi}{2}\right)$. If spacing is d, reinforcement will occur if

$$n\lambda = 2d\sin\theta = 2d\sin\left(\frac{\pi}{2} - \frac{\phi}{2}\right) \qquad (85i)$$

$$= 2d\sin\left(\frac{180}{2} - \frac{50}{2}\right)$$

$$= 2d\sin 65°$$

d measured by x-ray diffraction is 0.91Å. Therefore, the wavelength of the diffracted electron waves is

$$\lambda = 2 \times 0.91 \times \sin 65°$$

so $$\lambda = 1.65\text{Å} \qquad (85ii)$$

Using de Broglie's formula; $\lambda = \dfrac{h}{mv}$

$$K = \tfrac{1}{2}\, mv^2,\ 2mK = m^2 v^2$$
$$K = 54 \text{ eV}, m = 9.1 \times 10^{-31} \text{ kg}$$

$\therefore \qquad mv = \sqrt{2mK}$

$$\sqrt{2 \times 9.1 \times 10^{-31} \times 54 \times 1.6 \times 10^{-19}}$$
$$= 4 \times 10^{-24} \text{ kg-m/s}$$

$\therefore \qquad \lambda = \dfrac{h}{mv} = \dfrac{6.63 \times 10^{-34}\,\text{J-s}}{4 \times 10^{-24}\,\text{kgm/s}}$

$$= 1.66 \text{ Å} \qquad\qquad (85iii)$$

which agrees well with the observed value.

1.20 Compton Effect

Compton found that when monochromatic X-rays are incident on carbon atoms, for each scattering angle, two peaks are observed in the scattered beam–one at the incident wavelength, and the second at a wavelength longer by an amount $\Delta\lambda$ depending upon the scattering angle θ according to relation

$$\Delta\lambda = 0.02426(1 - \cos\theta)\mathring{A} = \frac{h}{m_0 c}(1 - \cos\theta) \qquad (86)$$

This effect is explained by assuming particle properties of e.m. radiation–*scattering process can be treated as an elastic collision between photon and an electron* (governed by conservation laws of energy and momentum)

At each angle θ, the scattered X-ray also include a substantial proportion having the initial wavelength which is explained as follows:

In derivation of $\Delta\lambda = 0.02426\ (1 - \cos\theta)$ Å, it was assumed that scattering particle is able to move freely ($\because$ many of the electrons in matter are only loosely bound to their parent atoms) other electrons, however are very tightly bound, and when struck by a photon, the entire atom recoils instead of the single electron.

In this effect, the value of m_0 to use in Eqn. is that of the entire atom which is tens of thousands of times greater than that of an electron and the resulting compton shift is so minute as to be undetectable.

Derivation of Relation for wavelength shift

Let a photon (of energy $h\nu$) collide with an electron initially at rest. The photon is scattered through an angle θ, while the electron recoils in a direction ϕ. Kinetic energy given to electron

$$K = (m - m_0)\, c^2 \qquad\qquad (87)$$

where m_0 = rest mass of electron and m is its relativistic mass.

Let ν' = frequency of scattered photon

Energy conservation requires that

energy of incident photon = K + energy of scattered photon

i.e. $\qquad\qquad h\nu = h\nu' + (m - m_0)\, c^2 \qquad\qquad (88)$

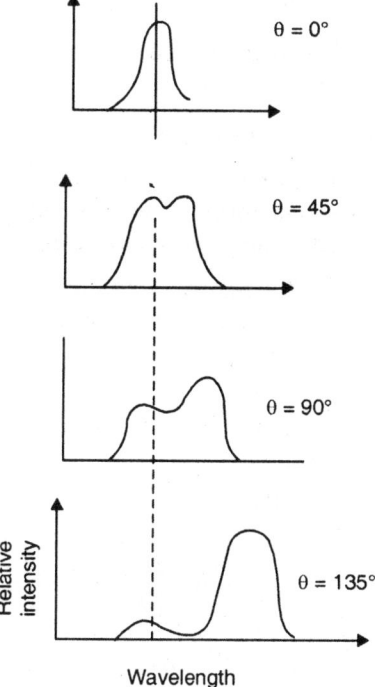

Fig. 1.22

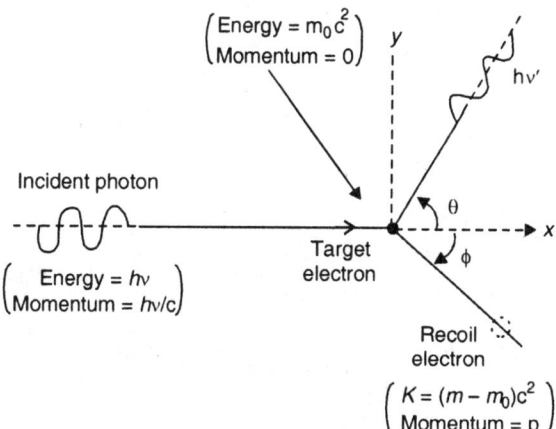

Fig. 1.23: Compton effect

or
$$h\frac{v}{c} = h\frac{v'}{c} + (m - m_0)c$$

or
$$\frac{h}{\lambda} = \frac{h}{\lambda'} + (m - m_0)c \qquad (89)$$

From momentum conservation

(i) in x direction $\dfrac{h\nu}{c} = \dfrac{h\nu'}{c}\cos\theta + p\cos\phi$ $\qquad\qquad$ (90a)

(ii) in y direction $\quad 0 = \dfrac{h\nu'}{c}\sin\theta + p\sin\phi$ $\qquad\qquad$ (90b)

(p = momentum of recoil electron)

It is to noted here that + sign appears with ϕ because it is assumed all angles are positive when measured counter clockwise from x axis.

$\therefore$ $\qquad\qquad \dfrac{h}{\lambda} = \dfrac{h}{\lambda'}\cos\theta + p\cos\phi$ $\qquad\qquad$ (91)

$$0 = \dfrac{h}{\lambda'}\sin\theta + p\sin\phi$$ (92)

$$p\cos\phi = \dfrac{h}{\lambda} - \dfrac{h}{\lambda'}\cos\theta$$ (93a)

$$p\sin\phi = -\dfrac{h}{\lambda'}\sin\theta$$ (93b)

$$p^2 = \dfrac{h^2}{\lambda^2} + \dfrac{h^2}{\lambda'^2}\cos^2\theta - \dfrac{2h^2}{\lambda\lambda'}\cos\theta + \dfrac{h^2}{\lambda'^2}\sin^2\theta$$

or $\qquad\qquad p^2 = \dfrac{h^2}{\lambda^2} + \dfrac{h^2}{\lambda'^2} - \dfrac{2h^2}{\lambda\lambda'}\cos\theta$ $\qquad\qquad$ (94)

since momentum $\quad p = \dfrac{m_0 v}{\sqrt{1 - {v^2}/{c^2}}}$, we have

$$p^2 = \dfrac{m_0^2 v^2}{1 - v^2/c^2} = m_0^2 c^2 \left[\dfrac{1 - 1 + v^2/c^2}{1 - v^2/c^2} \right]$$

$$= \dfrac{m_0^2 c^2}{1 - v^2/c^2} - m_0^2 c^2$$ (95)

hence, $\qquad\qquad \dfrac{h^2}{\lambda^2} + \dfrac{h^2}{\lambda'^2} - \dfrac{2h^2}{\lambda\lambda'}\cos\theta = \dfrac{m_0^2 c^2}{1 - v^2/c^2} - m_0^2 c^2$

$$\dfrac{h^2}{\lambda^2} + \dfrac{h^2}{\lambda'^2} - \dfrac{2h^2}{\lambda\lambda'}\cos\theta + m_0^2 c^2 = \dfrac{m_0^2 c^2}{1 - v^2/c^2}$$ (96)

From Eqn. (89)

$$\left(\dfrac{h}{\lambda} - \dfrac{h}{\lambda} + m_0 c \right)^2 = m^2 c^2$$

or $\qquad \dfrac{h^2}{\lambda^2} + \dfrac{h^2}{\lambda'^2} + m_0^2 c^2 + \dfrac{2h}{\lambda} m_0 c - \dfrac{2h}{\lambda'} m_0 c - \dfrac{2h^2}{\lambda\lambda'} = \dfrac{m_0^2 c^2}{1 - v^2/c^2}$

$$\dfrac{h^2}{\lambda^2} + \dfrac{h^2}{\lambda'^2} + m_0^2 c^2 - \dfrac{2h^2}{\lambda\lambda'} = \dfrac{m_0^2 c^2}{1 - v^2/c^2} - 2hm_0 c \left(\dfrac{1}{\lambda} - \dfrac{1}{\lambda'} \right)$$ (97)

Subtracting eqn. (97) from Eqn. (96), we get:

$$\frac{2h^2}{\lambda\lambda'}(1-\cos\theta) = 2hm_0c\left(\frac{1}{\lambda} - \frac{1}{\lambda'}\right)$$

or
$$\Delta\lambda = (\lambda' - \lambda) = \frac{h}{m_0c}(1-\cos\theta) \qquad (98a)$$
$$h = 6.63 \times 10^{-34} \text{J-s}$$
$$m_0 = 9.1 \times 10^{-31} \text{ kg}$$
$$c = 3 \times 10^8 \text{ m/s}$$

∴
$$\Delta\lambda = (\lambda' - \lambda) = 0.02426 \,(1 - \cos\theta) \text{ Å} \qquad (98b)$$

The quantity $\left(\dfrac{h}{m_0c}\right)$ is known as the *compton wavelength of electron*. It correspond to the wavelength of a photon whose energy is equal to the rest energy m_0c^2 of the electron i.e. if the electron with rest mass m_0 were considered as a de Broglie wave, then its wavelength will be $\left(\dfrac{h}{m_0c}\right)$.

1.21 Photons and Wave-Particle Duality

The wave nature of electromagnetic radiations was quite evident from the phenomena of interference and diffraction and relation $\nu = c/\lambda$ incorporates the wave concept in itself. M. Planck (Ann. Phys. 1,69(1900)) propounded that radiant energy is quantised in units called photons which have energy $E = hc \times \dfrac{1}{\lambda}$

$\equiv h\nu$ (i.e. inversely proportional to wavelength), where ν is frequency and h is Planck's constant.

$$h = 6.6262 \times 10^{-27} \text{ erg-sec} = 6.6262 \times 10^{-34} \text{ j-sec} \qquad (99)$$

It is the Planck's constant which can be said to characterise quantum physics. Although Max Planck first introduced the constant in 1900 to account for the shape of the continuous spectrum of radiation from incandescent objects, it was Einstein (Ann. Phys. 17,132 (1905)) who proposed the photon concept and postulated Eqn. $E = h\nu$, and verified it by photoelectric effect. The Compton effect also indicated the particle property of radiation.

When dealing with radiations of visible wavelength, we are not normally able to measure the frequency it is rather too high,

$$\nu = \frac{c}{\lambda} = \frac{3 \times 10^8 \text{ m/s}}{5000 \times 10^{-10} \text{ m}} = 10^{15} \text{Hz} \qquad (100)$$

But we can measure wavelength through the use of diffraction gratings. Thus it is often convenient to rewrite

$$E_{photon} = h\nu = h\frac{c}{\lambda} \qquad (101)$$

We most often measure energies in electron volts:

$$1 \text{ eV} = 1.602 \times 10^{-19} \text{ joule} = 1.602 \times 10^{-12} \text{ ergs} \qquad (102)$$

It is also standard practice to express the wavelength λ of atomic spectra in angstroms. It is therefore convenient to express the quantities (h) and (hc) in corresponding units. We find

$$h = 4.136 \times 10^{-15} \text{ eV-sec} \tag{103}$$
$$hc = 12,400 \text{ eV-Å} \tag{104}$$

We can rewrite Eqn. (101) numerically as

$$E_{photon} = \frac{12,400}{\lambda(\text{Å})} \text{ (in electron volts)} \tag{105}$$

The visible spectrum (4000–7000 Å) thus represents a range of photon energies from about 1.8 to 3.0 eV.

The Planck's constant denotes an elementary quantum or quantity of action. The dimensions of h are

$$h = \frac{[\text{Energy}]}{[\text{Frequency}]} = \frac{ML^2 T^{-2}}{T^{-1}} = [ML^2 T^{-1}]$$

$$= [L]\,[M]\,[LT^{-1}] = [\text{Angular Momentum}] \tag{106}$$

i.e. h is defined as the smallest quantum of angular momentum of a particle. h endows the radiation with discreteness at quantum scales. Also, for a relativistic particle

$$E^2 = p^2 c^2 + m^2 c^4$$

Since photons have to move at the speed of light, their rest mass is zero and

$$E = pc \text{ or } h\nu = pc \text{ or } p = \frac{h\nu}{c} \tag{107}$$

or

$$\frac{hc}{\lambda} = pc$$

or

$$p = \frac{h}{\lambda} \tag{108}$$

That is from Eqn. (107), $m_{ef} c = \dfrac{h\nu}{c}$ or effective mass of a photon $m_{ef} = \dfrac{h\nu}{c^2}$ and light of a given frequency has a fixed momentum (particle like property) and Eqn. (108) implies it is also a wave having a definite wavelength. This is expressed by stating that a photon is a wave or a particle.

The graininess of light quantum is apparent only at very high frequencies. The reason is that the energy of a single photon at high frequencies is comparable to that required for various atomic phenomena and a single photon can then show itself in an event. On the other hand, at very low frequencies, the energy of a single photon is so low that it requires a large number of them to produce some noticeable change. Thus photoelectric effect's discovery by Einstein was the first experimental evidence for quantum nature of radiation as it is associated with existence of a lower cut off frequency below which no photoelectron emission is possible. A single photon of a frequency equal to the threshold frequency is able to eject a photoelectron from a metal.

On the other had, while dealing with atomic systems, when incident wavelength (say e.g. from an X-ray source) are comparable to atomic spacings wave mechanics is applicable. If wave properties are ascribed to an electron[*] (evinced from

(*) De Broglie's Hypothesis: Not only light but moving particle, irrespective of its physical nature has wave properties associated with it.

Davisson and Germer experiment), then a sort of resonance might occur if the circumference of electron's orbit is equal to an integral multiple of the wavelength of the wave associated with an electron then $\lambda = \dfrac{h}{p}$ and $2\pi r = n\lambda$.

$$\Rightarrow \qquad 2\pi r = \frac{nh}{p} \text{ or } J = rp = \frac{nh}{2\pi} = n\hbar$$

which is the same as Bohr's postulate for quantization of angular momentum in atomic orbits.

1.22 Heisenberg's Uncertainty Relation

According to de Broglie a moving particle has a wavelike nature. According to Newtonian mechanics a particle has both position and momentum, each of which can be determined with a precision limited only by that of the measuring instrument. Since, however a wave extends across a region of space, the determination of the location, e.g. that of an electron considered as a wave is not too easy. There is, in fact, not merely an instrumental limit, but a fundamental limit to the accuracy with which the position of the electron can be determined. Similarly there is a limit to the accuracy with which the position of the electron can be determined. Similarly there is a limit to the accuracy with which the wavelength of the wave and therefore the momentum ($\lambda = h/p$) can be determined. This problem was tackled by Heisenberg and the solution lies in the uncertainty principle enunciated by Hesienberg. It may be stated as follows:

The product of the uncertainties Δx and Δp in the position coordinate and its conjugate momentum coordinate respectively are such that

$$\Delta x \cdot \Delta p \geq a\hbar \tag{109}$$

where $\hbar = h/2\pi$ and a is a numerical factor depending upon the statistical definition of an uncertainty in a measurement. When it is defined as the root mean square deviation of a set of observations from their mean value = 0.5, it is often taken to be unity.

1.23 Derivation of Uncertainty Relation from de Broglie's Concept

The wave nature of a moving particle leads one to think what he means by the position of the particle, i.e. where it could be located ($\because$ We know that the classical particles are localized). There should be a continuity in classical idea and the microscopic (atomic) realm. For this we require a super position of waves of various wavelengths so that we have a localized packet of waves which moves with the particle, i.e. the particle lies somewhere in the wave packet and is represented by wave packet formed by super position of waves of definite wavelengths, which are large in small region of space and nearly zero elsewhere (In fact a single wave is unable to assign a position to the particle. We may consider formation of a wave group from combination of two waves having same amplitude A but differing in angular frequency by $d\omega$ and in propagation constant by dk. The original waves may be represented by

$$y_1 = A \cos (\omega t - kx)$$
$$y_2 = A \cos [(\omega + d\omega)t - (k + dk) x]$$

The resultant displacement y at any time t and any position x is given by

$$y = y_1 + y_2$$

with the help of the identity $\cos C + \cos D = \cos\dfrac{C+D}{2}\cos\dfrac{C-D}{2}$ and the relation $\cos (-\theta) = \cos \theta$
we find that $y = 2A \cos \{(\frac{1}{2}) [(2\omega + d\omega)t - (2k+dk)x]\} \cos [\frac{1}{2} (d\omega t - dkx)]$

But $\qquad\qquad\qquad 2\omega + d\omega \cong 2w \text{ and } 2k + dk \cong 2k$

So, $\qquad\qquad y = 2A \cos (\omega t - kx) \cos \left(\dfrac{d\omega}{2}t - \dfrac{dk}{2}x\right)$

This represents a wave of angular frequency ω and propagation constant k that has superimposed upon it a modulation of angular frequency $d\omega/2$ and propagation constant $dk/2$. The effect of the modulation is to produce successive wave groups. The wave velocity is ω/k and velocity of the wave group is $(d\omega/2)/(dk/2) = d\omega/dk$.

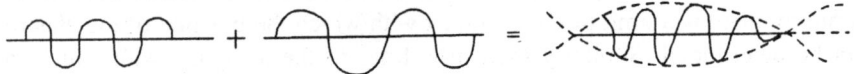

Fig. 1.24: Superposition of two waves
(of slightly different frequencies) to form a wave packet

In general, a wave packet is formed by a number of waves of slightly different frequencies. When a wave packet is formed, the size of the wave packet and the range of the wave numbers required to form this are related according to Fourier analysis by,

$$\Delta x \cdot \Delta k \cong 1 \qquad\qquad\qquad (110)$$

Now, according to de Broglie, wavelength is related to momentum associated with the wave by

$$\lambda = \dfrac{h}{p} \qquad \therefore \dfrac{p}{h} = \dfrac{1}{\lambda} \qquad\qquad (111)$$

Thus, a range of wave numbers Δk can be related to a range of momenta Δp by

$$\Delta k = 2\pi\Delta\left(\dfrac{1}{\lambda}\right) \qquad \left(\therefore k = \dfrac{2\pi}{\lambda}\right)$$

$$= 2\pi\dfrac{\Delta p}{h} \qquad\qquad\qquad (112)$$

so, $\qquad\qquad \Delta x = \dfrac{1}{\Delta k} = \dfrac{h}{2\pi\Delta p}$

or $\qquad\qquad \Delta x.\Delta p = \dfrac{h}{2\pi} = \hbar \qquad\qquad (113)$

The more accurately the position of the particle is specified, the greater will be the spread of possible particle momenta, and, in the limit where the position is specified with infinite accuracy, the momentum of the particle can have any value, i.e. it would be indeterminate and vice versa.

The uncertainty principle can also be expressed in the time domain. We know that for any wave train

$$\Delta\omega \,.\, \Delta t \approx 1 \qquad (114)$$
$$(\omega = 2\pi\nu)$$

Since $E = h\nu$, we have $\Delta E = h\nabla\nu$

or
$$\Delta\nu = \frac{\Delta E}{h} \qquad (115)$$

or
$$\frac{\Delta(2\pi\nu)}{2\pi} = \frac{\Delta E}{h} \text{ or } \frac{\Delta\omega}{2\pi} = \frac{\Delta E}{h}$$

or
$$\frac{1}{2\pi\Delta t} = \frac{\Delta E}{h} \qquad \text{(using Eqn. 114)}$$

or
$$\Delta E.\Delta t = \frac{h}{2\pi} = \hbar \qquad (116)$$

It should be noted here that the distinction between classical (large) systems and quantum (small) systems is not made on the basis of spatial extension only, but in units of $\hbar$. This has the dimensions of "action" (ML^2T^{-1}) which is (length × momentum) or (time × energy) and it is in terms of the typical action that the "size" of 'a' (in the uncertainty relation) should be judged e.g. in a one electron atom (H-atom), typical action is the product of the Bohr radius ~ 10^{-10} m and the momentum in the Bohr orbit $\cong 10^{-24}$ mks units. This is just of the order of $\hbar$.

1.24 Heisenberg's Gamma Ray Microscope

Classically, momentum and coordinates of a particle can be measured sufficiently accurately. However, in case of microscopic particle eg. an electron, there is a limit to the accuracy with which we can measure simultaneously position and momentum. Heisenberg proposed a hypothetical experiment known as Heisenberg's gamma ray microscope to evaluate the order of limitation in the simultaneous measurement of position and momentum of an electron considering the Compton scattering of a gamma ray photon incident on an electron by the help of an ultramicroscope shown schematically in the Fig. 1.25a.

The electron is initially stationary. It is bombarded by photons from the left. In order to locate the electron, photons strike it and get scattered into the microscope. The best resolving power of the lens L (ideal microscope) is known to provide an accuracy $\Delta x = \lambda/(2 \sin \theta)$ for the measurement of electron's position, where λ is the wavelength of light rays used and θ is the half angle subtended at P by the lens.

One of the photons of momentum $h\nu/c$ strikes the electron at rest. It transfers a momentum mv to the electron. The component of momentum transfer (in x-direction) to the electron is given by

$$\frac{h\nu}{c} = \frac{h\nu'}{c} \cos \alpha + mv \cos \beta$$

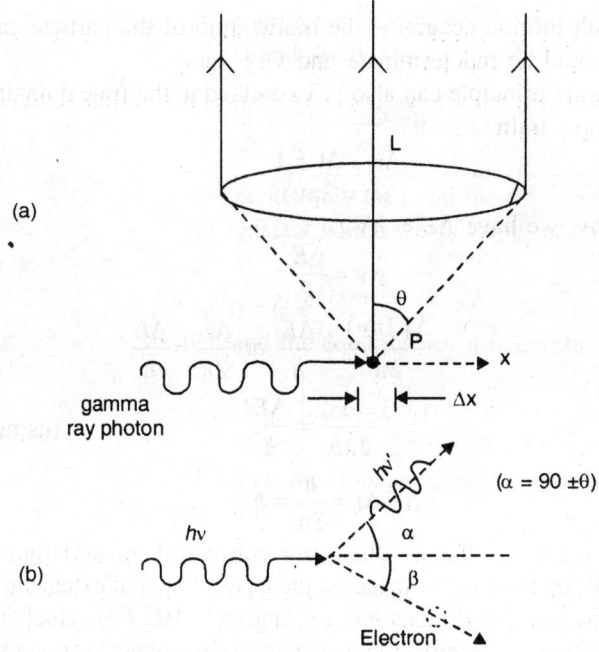

Fig. 1.25. (a) Heisenberg microscope, (b) Photon interacting by Compton scattering with an electron

where α is the angle which scattered photon makes with the x-direction.

$$p_x = mv \cos \beta = \frac{h}{c}(v - v' \cos \alpha)$$

In a limiting situation when $v' \to v$, we get a head on collision and

$$p_x = \frac{h}{\lambda}(1 - \cos \alpha).$$

The angle α will lie within the microscope in the limits $(90 - \theta)$ to $(90 + \theta)$, so, the spread in the x component of momentum will be

$$\frac{h}{\lambda}(1 - \sin \theta) \le p_x \le \frac{h}{\lambda}(1 + \sin \theta)$$

Consequently

$$\Delta p_x \Delta x \approx \frac{\lambda}{2 \sin \theta} \times \frac{2h}{\lambda} \sin \theta$$

or
$$\Delta p_x \cdot \Delta x \approx h \qquad \qquad (h \text{ is Planck's constant})$$

1.25 Bohr's Correspondence Principle

Newtonian mechanics and classical electrodynamics are based on thoroughly established experimental evidence. Therefore, one may expect that the quantum theory should yield in every instance results that become identical with those of classical physics if the masses and dimensions of the system under consideration are made to approach those of classical system.

It is evident from the energy level diagram for the H-atom (Fig. 1.2) that energy difference between one energy level and the next is very small for large values of the quantum number n. Similarly, the difference between the radius of one Bohr orbit and the next is also very small. Thus, if an electron jumps down from an orbit (say $n = 3000$) step by step to the next lower state ($n = 2999$) and then to next ($n = 2998$) etc. the steps in energy and radius are very small and the changes in energy and radius will appear to proceed almost continuously. Likewise the changes in angular momentum (e.g. from $L = 3000\ \hbar$ to $2999\ \hbar$ etc.) will also appear to be continuous. Thus we see that there is a concordance between classical and quantum mechanics in the limit of large quantum numbers. Bohr generalised this concordance in mechanical behaviour of electron to also the electromagnetic behaviour (emission of radiation). He enunciated in 1923 that:

"In the limiting case of large quantum numbers, the frequencies and the intensities of radiation calculated from classical theory must agree with those of quantum theory."

This is known as *Bohr's correspondence principle*. It can be easily verified in case of frequencies of light emitted by an electron in a hydrogen atom;

According to Eqn. (16), the frequency for a transition from (n) to $(n-1)$ is

$$\nu = \frac{e^4}{4\pi\left(4\pi\,\epsilon_0\right)^2}\frac{m}{\hbar^3}\left[\frac{1}{\left(n-1\right)^2}-\frac{1}{n^2}\right]$$

$$= \frac{e^4}{4\pi\left(4\pi\,\epsilon_0\right)^2}\frac{m}{\hbar^3}\left(\frac{2n-1}{n^2\left(n-1\right)^2}\right)$$

If n is very large,

$$\frac{\left(2n-1\right)}{n^2\left(n-1\right)^2}\cong\frac{2n}{n^4}=\frac{2}{n^3},\ \text{then}$$

$$\nu \cong \frac{e^4}{2\pi\left(4\pi\,\epsilon_0\right)^2}\frac{m}{\hbar^3}\frac{1}{n^3} \tag{117}$$

This is the frequency according to quantum theory. To find the frequency according to classical theory, we note that for an accelerated charge, classical electrodynamics predicts that the frequency of the emitted light coincides with the frequency of the motion. Using Eqn. (13) and (14)

$$\nu_{classical} = \frac{v}{2\pi r} = \frac{n\hbar/mr}{2\pi r} = \frac{n\hbar}{2\pi m}\frac{1}{r^2} = \frac{n\hbar}{2\pi m}\left(\frac{e^2 m}{4\pi\,\epsilon_0\,n^2\hbar^2}\right)^2 \tag{118}$$

simplifying this, we find that this expression is same as Eqn. (117). Hence, for large n, the result of the quantum mechanical calculation agrees with that provided by classical calculation.

Thus quantum physics gives the same result as classical physics in the limit of large quantum number and the greater the quantum number, the closer quantum physics approaches classical physics.

Example 1.1: (Diffraction through a slit)
Consider the Example of diffraction through a slit. If we propose to regard the passage of an electron through a slit and the observation of diffraction pattern as

simultaneous measurement of position and momentum (from the standpoint of corpuscular concept), then breadth of the slit gives uncertainty (Dx) in specification of position perpendicular to the direction of flight (because it is indefinite at which point in the slit the emergence of electron takes place). Again, from the standpoint of corpuscular concept, the diffraction pattern on the screen is due to deflection suffered by the electron at the slit (upwards or downwards).

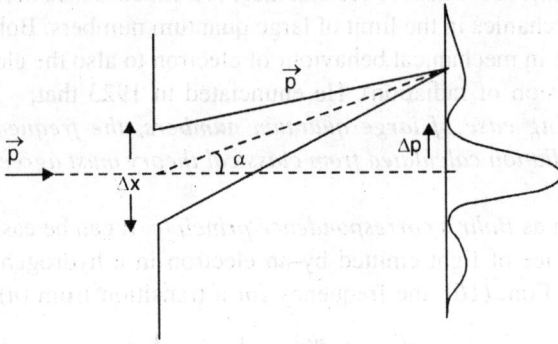

Fig. 1.26: Diffraction of an electron at a slit

Mean value of $\Delta p \cong p \sin \alpha$, where α is mean angle of deflection. We know that experimental results can be obtained satisfactorily on the basis of the wave representation according to which α is connected with slit width Δx and the wavelength λ by

$$\Delta x \sin \alpha \cong \lambda \text{ or } \Delta x . \sin \alpha = \frac{h}{p}$$

Thus, the mean added momentum in direction parallel to slit is

$$\Delta p = p \sin \alpha \cong \frac{h}{\Delta x} \quad \therefore \quad \Delta p . \Delta x \cong h.$$

Thus h represents an absolute limit to the simultaneous measurement of position and momentum.

Example 1.2: The radius of the Bohr's First orbit)
If Δq and Δp are the uncertainties in the position and momentum of the electron in first orbit, then

$$\Delta q . \Delta p \approx \hbar \text{ or } \Delta p \approx \frac{\hbar}{\Delta q} \qquad (i)$$

The uncertainty in the kinetic energy

$$\Delta T = \frac{1}{2} m (\Delta v)^2 = \frac{1}{2} \frac{(m \Delta v)^2}{m}$$

$$= \frac{1}{2} \frac{(\Delta p)^2}{m} = \frac{1}{2m} \left(\frac{\hbar}{\Delta q} \right)^2 = \frac{\hbar^2}{2m(\Delta q)^2} \qquad (ii)$$

The uncertainty in the potential energy

$$\Delta V = \frac{-Ze^2}{\Delta q}$$

so, the uncertainty in the total energy $\Delta E = \Delta T + \Delta V$

$$= \frac{\hbar^2}{2m(\Delta q)^2} - \frac{Ze^2}{\Delta q} \qquad (iii)$$

ΔE will be minimum if $\dfrac{d(\Delta E)}{d(\Delta q)} = 0$ and $\dfrac{d^2(\Delta E)}{d(\Delta q)^2} = +\text{ive}$

Equation (iii) yields

$$\frac{d(\Delta E)}{d(\Delta q)} = -\frac{\hbar^2}{m(\Delta q)^3} + \frac{Ze^2}{(\Delta q)^2} \qquad (iv)$$

For E to be minimum, we must have

$$-\frac{\hbar^2}{m(\Delta q)^3} + \frac{Ze^2}{(\Delta q)^2} = 0 \quad \Rightarrow \quad \Delta q = \frac{\hbar^2}{mZe^2} \qquad (v)$$

Differentiating Eqn. (iv), we find,

$$\frac{d^2(\Delta E)}{d(\Delta q)^2} = +\frac{3\hbar^2}{m(\Delta q)^4} - \frac{2Ze^2}{(\Delta q)^3}$$

$$= \frac{3\hbar^2}{m(\Delta q)^3 \left(\dfrac{\hbar^2}{mZe^2}\right)} - \frac{2Ze^2}{(\Delta q)^3} \qquad \text{(using Eqn. v)}$$

$$\approx \frac{3Ze^2}{(\Delta q)^3} - \frac{2Ze^2}{(\Delta q)^3} \approx \frac{Ze^2}{(\Delta q)^3}$$

$$= a +\text{ive quantity}$$

So, Eqn. (v) represents the condition of minimum uncertainty in the first orbit. Hence,

$$r = \Delta q = \frac{\hbar^2}{mZe^2}$$

the radius of Bohr's first orbit.

Example 1.3: The ground state energy of a Linear Hormonic oscillator
For a linear harmonic oscillator, the total energy is given by

$$E = \frac{p_x^2}{2m} + \frac{1}{2}m\omega^2 x^2$$

Assume that the particle is confined in a region of dimension $\sim a$, i.e.

$$x \sim \Delta x \cong a$$

then, according to uncertainty principle

$$p_x \sim \Delta p_x \cong \frac{\hbar}{2a} \qquad (\because \text{ displacement in a period} = 2a)$$

Thus
$$E = \frac{\hbar^2}{8ma^2} + \frac{1}{2}m\omega^2 a^2$$

The ground state will be the state of lowest energy for which $dE/da = 0$. So,

$$\frac{dE}{da} = -\frac{\hbar^2}{4ma^3} + m\omega^2 a = 0$$

$$\therefore \qquad a^4 \cong \left[\frac{\hbar^2}{4m^2\omega^2}\right] \quad \text{or } a \cong \left(\frac{\hbar}{2m\omega}\right)^{\frac{1}{2}}$$

The minimum energy is given by

$$E = \frac{\hbar^2}{8m} \times \frac{2m\omega}{\hbar} + \frac{1}{2}m\omega^2 a^2$$

$$= \frac{\hbar\omega}{4} + \frac{1}{2}m\omega^2 \times \frac{\hbar}{2m\omega} = \frac{1}{2}\hbar\omega$$

This is really the ground state energy of the linear harmonic oscillator (known as the zero-point energy). It is a consequence of the uncertainty principle.

Example 1.4: Using uncertainty principle, calculate the size and energy of the ground state of hydrogen atom.

(See Example 1.2)

$$\Delta x = \frac{\hbar^2}{me^2} = \frac{\left(1.05 \times 10^{-27}\right)^2}{9.1 \times 10^{-28} \times \left(4.8 \times 10^{-10}\right)^2} \approx 0.5 \times 10^{-8} \text{cm} \quad \text{(using CGS units)}$$

$$\Delta E = \frac{\hbar^2}{2m(\Delta x)^2} - \frac{e^2}{\Delta x}$$

$$= \frac{\hbar^2}{2m} \frac{m^2 e^4}{\hbar^4} - e^2 \frac{me^2}{\hbar^2}$$

$$= -\frac{me^4}{2\hbar^2} = -13.6 \text{ eV}$$

i.e. radius of the first Bohr's orbit, which is of the order of the uncertainty in the electron's position, is $a_0 \cong \Delta x = 0.5 \times 10^{-8}$ cm and the corresponding energy is −13.6 eV.

Example 1.5: An electron of energy 200 eV is passed through a circular hole of radius 10^{-3} cm. What is the uncertainty introduced in the angle of emergence?

Solution: $\qquad 1 \text{ eV} = 1.6 \times 10^{-19} \text{ J} = 1.6 \times 10^{-12} \text{ ergs}$

$$p = \sqrt{2mE} = \left(2 \times 0.9 \times 10^{-27} \times 3.2 \times 10^{-10}\right)^{1/2}$$

$$\cong 8 \times 10^{-19} \text{ g.cm.s}^{-1}$$

$$\Delta p \cong \frac{\hbar}{\Delta x} \cong \frac{10^{-27} \text{ erg.s}}{2 \times 10^{-3} \text{cm}} = 5 \times 10^{-25} \text{ g.cm.s}^{-1}$$

$$\theta = \frac{\Delta p_x}{p} = \frac{5 \times 10^{-25}}{8 \times 10^{-19}} \cong 6 \times 10^{-7} \text{radians.}$$

Example 1.6: Calculate the uncertainty in the momentum of a proton which is confined to a nucleus of radius 10^{-13} cm. From this result estimate the kinetic energy of the proton inside the nucleus.

Solution: The proton is confined within a sphere of radius $r_0 \cong 10^{-13}$ cm. Thus the uncertainty in the momentum must be at least of the order of $\hbar/r_0$ or

$$p \sim \frac{\hbar}{r_0}$$

Therefore, the kinetic energy of the proton will be given by

$$E = \frac{p^2}{2m_p} \cong \frac{\hbar^2}{2m_p r_0^2}$$

where m_p is the mass of the proton. On substitution of numerical values, we get

$$E \cong \frac{\left(1.05 \times 10^{-27} \text{ erg} - \text{sec}\right)^2}{\left(2 \times 1.67 \times 10^{-24} \text{ g}\right) \times (10^{-13} \text{ cm})^2}$$

$$\cong 3 \times 10^{-5} \text{ ergs} \cong 20 \text{ MeV}$$

Example 1.7: An electron has a speed 1.05×10^4 m/s within the accuracy of 0.02%. Calculate the uncertainty in the position of the electron.

$$(\hbar = 1.05 \times 10^{-34} \text{ joule-}s, \, m = 9 \times 10^{-31} \text{ kg})$$

Solution: The uncertainty in velocity

$$= \frac{0.02}{100} \times 1.05 \times 10^4 \text{ m/s}$$

$$\Delta p = m \Delta v = 9 \times 10^{-31} \times \frac{0.02}{100} \times 1.05 \times 10^4$$

$$\Delta p . \Delta x = \hbar$$

$$\therefore \quad \Delta x = \frac{\hbar}{\Delta p} = \frac{1.05 \times 10^{-34}}{9 \times 10^{-31} \times \frac{0.02}{100} \times 1.05 \times 10^4} \text{ metres}$$

$$= \frac{1.05 \times 100 \times 10^{-34}}{9 \times 10^{-31} \times 0.02 \times 1.05 \times 10^4} \text{ m}$$

$$= 0.55 \times 10^{-4} \text{ metres.}$$

Example 1.8: What is the energy of a photon having a wavelength 1.5Å (given $h = 6.627 \times 10^{-34}$ Js)

Solution:

$$E = h\nu = \frac{hc}{\lambda}$$

$$= \frac{6.627 \times 10^{-34} \times 3 \times 10^8}{1.5 \times 10^{-10} \text{ m}} \text{ Js×ms}^{-1}$$

$$= 13.254 \times 10^{-16} \text{ J}$$

Example 1.9: What potential difference must be applied to stop the fastest photoelectrons emitted by a nickel surface under the action of ultraviolet light of wavelength 2000 Å? The work function of Nickel is 5.01 eV.

Solution: Energy of photon $= \dfrac{hc}{\lambda}$

$$= \frac{(6.63 \times 10^{-34} \text{ J.s.})(3 \times 10^8 \text{ m/s})}{2000 \times 10^{-10} \text{ m}}$$

$$= 9.95 \times 10^{-19} \text{J} = \frac{9.95}{1.6} \text{eV} = 6.21 \text{ eV}$$

From the photoelectric equation, the energy of the fastest emitted electron is

$$6.21 \text{ eV} - 5.01 \text{ eV} = 1.20 \text{ eV}.$$

Hence a negative (i.e. retarding) potential of 1.2s0 V is needed.

Example 1.10: Will photoelectrons be emitted by a copper surface of work function 4.4 eV when illuminated by visible light?

Solution: $$h\nu_c = W_0 \quad \text{or} \quad \frac{hc}{\lambda} = W_0$$

∴ Threshold wavelength $\lambda = \dfrac{hc}{W_0}$

$$= \frac{(6.63 \times 10^{-34} \text{ J.s.})(3 \times 10^8 \text{ m/s})}{4.4 \times 1.6 \times 10^{-19} \text{ J}} = 282 \text{ nm.}$$

Hence, visible light (400 nm < λ < 700 nm) cannot eject photoelectrons from copper. (since $h\nu_c > (h\nu)_{\text{visible}}$)

Example 1.11: If θ and ϕ are the angles of scattering of photon and recoil electron then for compton scattering, show that

$$\cot\phi = \left(1 + \frac{h\nu}{m_0 c^2}\right) \tan\frac{\theta}{2}$$

Solution: We have $$\lambda' = \lambda + \frac{h}{m_0 c}(1 - \cos\theta)$$

$$= \lambda + \frac{2h\sin^2\dfrac{\theta}{2}}{m_0 c}$$

or $$\frac{c}{\lambda'} = \frac{c}{\lambda + \dfrac{2h}{m_0 c}\sin^2\dfrac{\theta}{2}} = \frac{c}{\lambda\left[1 + \dfrac{2h}{m_0 c\lambda}\sin^2\dfrac{\theta}{2}\right]}$$

or $$\nu' = \frac{\nu}{1 + \dfrac{2h\nu}{m_0 c^2}\sin^2\dfrac{\theta}{2}}$$

Now $$p\sin\phi = \frac{h}{\lambda'}\sin\theta$$

and $$p\cos\phi = \frac{h}{\lambda} - \frac{h}{\lambda'}\cos\theta$$

∴ $$\cot\phi = \frac{\dfrac{h}{\lambda} - \dfrac{h}{\lambda'}\cos\theta}{\dfrac{h}{\lambda'}\sin\theta} = \frac{\dfrac{\nu}{c} - \dfrac{\nu'}{c}\cos\theta}{\dfrac{\nu'}{c}\sin\theta}$$

$$= \frac{v - v'\cos\theta}{v'\sin\theta} = \frac{v - \dfrac{v\cos\theta}{1 + 2\alpha\sin^2\theta/2}}{\dfrac{v\sin\theta}{1 + 2\alpha\sin^2\theta/2}} \qquad \left(\text{where } \alpha = \frac{hv}{m_0 c^2} \right)$$

or

$$\cot\phi = \frac{v\left(1 + 2\alpha\sin^2\theta/2\right) - v\cos\theta}{v\sin\theta}$$

$$= \frac{1 + 2\alpha\sin^2(\theta/2) - 1 + 2\sin^2(\theta/2)}{2\sin(\theta/2)\ \cos(\theta/2)}$$

$$= \frac{2\sin^2(\theta/2)\ (1 + \alpha)}{2\sin(\theta/2)\ \cos(\theta/2)} = \left(1 + \frac{hv}{m_0 c^2}\right)\tan(\theta/2) \cdot$$

Example 1.12: Derive expressions for the kinetic energy of recoil electron in Compton effect.

Solution: Kinetic energy of the scattered electron

$$= (m - m_0)\, c^2 = h(v - v')$$

$$= hv\left(1 - \frac{v'}{v}\right) = hv\left(1 - \frac{\lambda}{\lambda'}\right)$$

$$= hv\frac{\Delta\lambda}{\lambda + \Delta\lambda} = \frac{hv\dfrac{h}{m_0 c\lambda}(1 - \cos\theta)}{1 + \dfrac{h}{m_0 c\lambda}(1 - \cos\theta)}$$

$$= \frac{hv\alpha\left(1 - \cos\theta\right)}{1 + \alpha\left(1 - \cos\theta\right)} \qquad\qquad (i)$$

where

$$\alpha = \frac{hv}{m_0 c^2} \quad \text{and} \quad \frac{1}{\lambda} = \frac{v}{c}$$

θ is the scattering angle of photon. We can also express the kinetic energy of the recoil electron in terms of its scattering angle (viz. ϕ) (see Fig. 1.23).

$$\text{K.E. of the electron} = hv\frac{\alpha\left(1 - \cos\theta\right)}{1 + \alpha\left(1 - \cos\theta\right)}$$

$$= hv\frac{\alpha.2\sin^2\theta/2}{1 + \alpha.2\sin^2\theta/2} = \frac{hv.2\alpha}{(\operatorname{cosec}^2\theta/2) + 2\alpha}$$

$$= \frac{hv.2\alpha}{(\cot^2\theta/2) + 1 + 2\alpha} \qquad\qquad (ii)$$

From Example (1.11), $(1 + \alpha)\tan\theta/2 = \cot\phi$

or

$$(1 + \alpha)\tan\phi = \cot\theta/2 \qquad\qquad (iii)$$

Using (*iii*), (*ii*) becomes

$$K.E. = \frac{hv.2\alpha}{(1+\alpha)^2 \tan^2\phi + 1 + 2\alpha}$$

$$= hv \frac{2\alpha \cos^2\phi}{(1+\alpha)^2 \sin^2\phi + (1+2\alpha)\cos^2\phi}$$

$$= hv \frac{2\alpha.\cos^2\phi}{(1+\alpha)^2 - \alpha^2 \cos^2\phi}$$

Example 1.13: Show that the velocity of the electron in the first orbit of hydrogen atom is $(1/137)$ c, where c is the velocity of light.

Solution: We have $\qquad \dfrac{mv^2}{r} = \dfrac{1}{4\pi\,\epsilon_0}\dfrac{Ze^2}{r^2}$ $\qquad\qquad$ (*i*)

and $\qquad\qquad\qquad mvr = n\hbar = \dfrac{nh}{2\pi}$ $\qquad\qquad$ (*ii*)

Dividing Eqn. (*i*) by (*ii*),

$$v_n = \frac{Ze^2}{4\pi\,\epsilon_0}\frac{2\pi}{nh}$$

$$\therefore \qquad v_1 = \frac{Ze^2}{4\pi\,\epsilon_0}\frac{2\pi}{h} = \left(\frac{2\pi e^2}{4\pi\,\epsilon_0\, ch}\right)c = \alpha c = (\frac{1}{137})c \quad (\because Z = 1)$$

Example 1.14: In the Stern Gerlach experiment using Ag atoms the oven temperature is 1000 K and $l \approx 25$ cm, $\dfrac{\partial B_z}{\partial z} \approx 10^3$ Tesla/m. Calculate the separation of the two components.

Solution: Separation $\qquad z = \pm\dfrac{1}{2M_0}\mu\dfrac{\partial B}{\partial Z}\left(\dfrac{\ell}{v}\right)^2$

The atoms (because of collimation) entering the inhomogenous magnetic field have their velocities in *y*-direction and

$$\frac{1}{2}M_0 v_y^2 \cong \frac{3}{2}kT \qquad\qquad \left(\because \frac{1}{2}Mv^2 = \frac{3}{2}kT\right)$$

Thus $\qquad\qquad\qquad v_y \cong \sqrt{\dfrac{3kT}{M_0}} \qquad\qquad (M_0 \text{ is mass of Ag atom})$

$$\therefore \qquad z \cong \pm\frac{1}{2M_0}\mu\frac{\partial B}{\partial Z}\times\ell^2\times\frac{M_0}{3kT} = \frac{\ell^2}{6kT}\mu\frac{\partial B}{\partial Z}$$

We know that $\qquad\qquad \mu = \dfrac{e\hbar}{2m} = 9.27\times10^{-24}$ J/T

$$\therefore \qquad z = \pm\frac{(0.25)^2}{6\times1.38\times10^{-23}\times1000}\times9.27\times10^{-24}\times10^3\,\text{m} \cong \pm\,0.7 \text{ cm.}$$

Thus separation is about 1.4 cm.

Example 1.15: Apply Wilson-Sommerfeld quantization condition to a harmonic oscillator and show that its energy levels are discrete.

Solution: The one dimensional harmonic oscillator consists of a particle of mass m, bound to the equilibrium position $x = 0$ by a restoring force proportional to the displacement x, i.e. $F = -kx$ and the particle is constrained to move along the x-axis. k is the force constant.

The potential energy of the oscillator is given by

$$V(x) = -\int F dx = \int kx dx = \frac{1}{2}kx^2$$

The kinetic energy

$$= \frac{1}{2}mv^2 = \frac{p_x^2}{2m} \qquad (\because p_x = mv)$$

The total energy

$$E = \frac{p_x^2}{2m} + \frac{1}{2}kx^2 \qquad (i)$$

According to Wilson Sommerfeld quantization rule,

$$\oint p_x dx = nh \qquad (ii)$$

where

$$n = 0, 1, 2, 3, \ldots$$

Now we try to evaluate $\oint p_x dx$, the phase integral. The phase integral represents an area in the phase-space. The particle will execute a simple harmonic vibration in one dimension and this will be represented by a closed curve in phase space described by two coordinates x and p_x such that

$$\oint p_x dx = \text{area of the ellipse} = \pi ab \qquad (iii)$$

Here 'a' represents semimajor axis and 'b', semiminor axis in phase space (x, p_x). The axes are given by

if

$$p_x = 0, x = a; x = 0, p_x = b \qquad (iv)$$

Substituting these in Eqn. (i), we obtain,

$$E = \frac{1}{2}ka^2 \quad \text{or} \quad a = \sqrt{\frac{2E}{k}}$$

and

$$E = \frac{b^2}{2m} \quad \text{or} \quad b = \sqrt{2mE}$$

Substituting these values in Eqn. (iii), we get $\oint p_x dx = \pi \sqrt{\frac{2E}{k}} \sqrt{2mE}$

But $\sqrt{\frac{k}{m}} = \omega$ for a hamonic oscillator, therefore

$$\oint p_x dx = \pi \sqrt{\frac{2E}{m\omega^2}} \sqrt{2mE}$$

$$= \frac{2\pi E}{\omega} \qquad (v)$$

Comparing equation (ii) and (v), we have

$$\frac{2\pi E}{\omega} = nh$$

or
$$E = \frac{nh\omega}{2\pi} = nh\omega = nh\nu \qquad (vi)$$

where ω is the angular frequency and ν is the (linear) frequency.

Thus it is evident from equation (vi) that the energy levels of a harmonic oscillator are not continuous but integral multiples of $h\nu$. It is illustrated in the Fig. 1.27. However, we will see in Chapter 6 that the new quantum theory gives the more exact result

$$E_n = \left(n + \frac{1}{2}\right)\hbar\omega = \left(n + \frac{1}{2}\right)h\nu \qquad (vii)$$

where,
$$n = 0, 1, 2, \ldots$$

The selection rule $\Delta n = \pm 1$ permits equally separated energy levels.

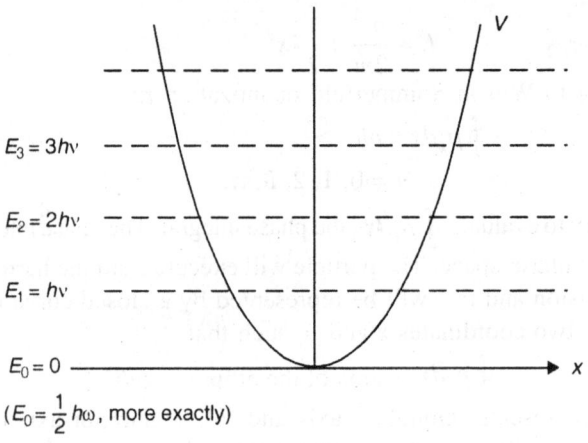

$E_3 = 3h\nu$

$E_2 = 2h\nu$

$E_1 = h\nu$

$E_0 = 0$

$\left(E_0 = \frac{1}{2}h\omega,\ \text{more exactly}\right)$

Fig. 1.27

Example 1.16: If K is kinetic energy of a particle which is not small compared with m_0c^2, then prove that its de Broglie wavelength

$$\lambda = ch\left[K\left(K + 2m_0c^2\right)\right]^{-1/2}$$

Solution: Total energy $E = \left(p^2c^2 + m_0^2c^4\right)^{1/2}$ \qquad (relativistic expression)

$$= K + m_0c^2$$

Squaring both sides,

$$p^2c^2 + m_0^2c^4 = K^2 + 2Km_0c^2 + m_0^2c^4$$

or
$$p^2c^2 = 2m_0Kc^2\left(1 + \frac{K}{2m_0c^2}\right)$$

or
$$p^2 = 2m_0K\left(1 + \frac{K}{2m_0c^2}\right)$$

$$= \frac{2m_0K(2m_0c^2 + K)}{2m_0c^2} = \frac{K}{c^2}(K + 2m_0c^2)$$

or
$$p = \frac{1}{c}\left[K(K + 2m_0c^2)\right]^{1/2}$$

de Broglie wavelength $\quad \lambda = \frac{h}{p} = hc\left[K(K + 2m_0c^2)\right]^{-1/2}$.

Example 1.17: The average time that an atom retains excess excitation energy before re emitting it in the form of electromagnetic radiation is 10^{-8} sec. Calculate the limit of accuracy with which the excitation energy of the emitted radiation can be determined

Solution: According to uncertainty principle
$$\Delta E \cdot \Delta t \cong \hbar$$

Given $\quad\quad \Delta t = 10^{-8}\,\text{s} \quad \therefore \Delta E = \frac{\hbar}{\Delta t} = \frac{1.05 \times 10^{-34}\,\text{J s}}{10^{-8}\,\text{s}}$

or
$$\Delta E = 1.05 \times 10^{-26}\,\text{J}$$

$$= \frac{1.05 \times 10^{-26}}{1.6 \times 10^{-19}}\,\text{eV} = 6.56 \times 10^{-8}\,\text{eV}$$

$\therefore \quad$ Uncertainty in energy $= 6.56 \times 10^{-8}$ eV.

Example 1.18: The earth receives solar radiation at a rate of 8.2 J/(cm²-min). Assuming that the sun radiates like a black body, calculate the surface temperature of the sun. The angle subtended by the sun on the earth is 0.53^0 and the Stefan constant $s = 5.625 \times 10^{-8}$ W/(m² – K⁴).

Solution:

Let D denote sun's diameter and R its distance from the earth.

Now, $\quad\quad \dfrac{D}{R} = 0.53 \times \dfrac{\pi^c}{180} = 9.25 \times 10^{-3}$ Radians.

The radiation emitted by the surface of the sun per unit time is
$$4\pi\left(\frac{D}{2}\right)^2 \sigma T^4 = \pi D^2 \sigma T^4$$

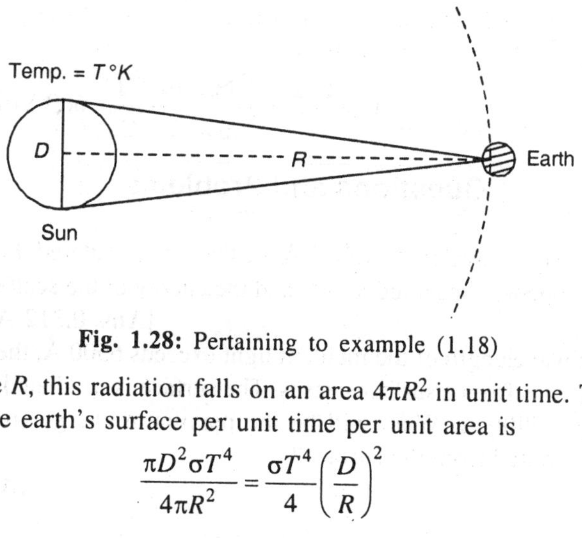

Fig. 1.28: Pertaining to example (1.18)

At distance R, this radiation falls on an area $4\pi R^2$ in unit time. The radiation received at the earth's surface per unit time per unit area is
$$\frac{\pi D^2 \sigma T^4}{4\pi R^2} = \frac{\sigma T^4}{4}\left(\frac{D}{R}\right)^2$$

that is

$$\frac{\sigma T^4}{4}\left(\frac{D}{R}\right)^2 = 8.2 \text{ J}/(\text{cm}^2 - \text{min})$$

or

$$\frac{1}{4} \times \left[5{,}625 \times 10^{-8} \frac{W}{m^2 - K^4}\right] \times T^4 \times \left(9.25 \times 10^{-3}\right)^2 = \frac{8.2}{10^{-4} \times 60} \frac{W}{m^2}$$

$$\Rightarrow \qquad \cdot T^4 = \frac{8.2 \times 4 \times 10^{18}}{60 \times 5.625 \times (9.25)^2} \frac{K^4}{} = 0.011358 \times 10^{17} \ K^4$$

$$\therefore \qquad\qquad T = 5805 \text{ K}.$$

Example 1.19: The average lifetime of an excited atomic state is 10^{-9} s. If the spectral line associated with the decay of this state is 6500 Å, estimate the width of the line.

Solution:
$$E = h\nu = \frac{hc}{\lambda}, \ \lambda = 6500 \text{ Å}, \ \Delta t = 10^{-9} \text{ s}$$

$$\Delta E = \frac{hc \, \Delta\lambda}{\lambda^2} \Rightarrow \Delta E \cdot \Delta t = \frac{hc}{\lambda^2} \Delta\lambda \cdot \Delta t \cong h$$

$$\Delta\lambda \cdot \Delta t = \frac{\lambda^2}{c}$$

or

$$\Delta\lambda = \frac{\lambda^2}{c\Delta t} = \frac{\left(6500 \times 10^{-10} \text{ m}\right)^2}{\left(3 \times 10^8 \text{ m/s}\right)\left(10^{-9}\text{ s}\right)} = \frac{\left(6.5 \times 10^{-7}\right)^2}{\left(3 \times 10^{-1}\right)} = 14.08 \times 10^{-13} \text{ m}$$

Example 1.20: What potential difference is required in an electron microscope to give electrons a wavelength of 0.5 Å.

Solution: Kinetic energy of electron = ½ mv^2

$$\frac{1}{2} m \left(\frac{h}{m\lambda}\right)^2 = \frac{h^2}{2m\lambda^2} \qquad\qquad \text{(using } \lambda = \frac{h}{mv}\text{)}$$

or K.E. = 9.66×10^{-17} J (Substituting the values of constants and λ)

But
$$\text{K.E.} = eV.$$

$$\therefore \qquad\qquad V = \frac{\text{K.E.}}{e} = \frac{9.66 \times 10^{-17}\text{J}}{1.6 \times 10^{-19}\text{C}} = 600 \text{ Volts}$$

Questions and Problems

1. X-rays having a wavelength of 0.3 Å are Compton scattered. Find the wavelength of a photon scattered at 60° and the energy of the scattered electron.
   [Ans. 0.312 Å, 1.56 KeV**]**

2. When the wavelength of the incident light exceeds 6500 Å, the emission of photoelectrons from a surface ceases. The surface is irradiated with light of wavelength 3900 Å. What will be the maximum energy (in eV), of the electrons emitted from the surface?
   [Ans. 1.27 eV**]**

3. Describe Bohr's theory of hydrogen spectra and show how Rydberg's constant was obtained in terms of fundamental constants. Why Rydberg's constant has different values for isotopes of hydrogen?

4. Explain photoelectric effect. Define the terms work function, threshold frequency and critical wavelength. Give the relations between them.

5. Describe Davisson and Germer's experiment. How does it prove the wave nature of particle?

6. What is Heisenberg's uncertainty principle? Use it to find (*i*) Bohr's first orbit radius and (*ii*) binding energy of an electron in the hydrogen atom.

7. An electron is moving with velocity 10^6 m/s. Calculate the wavelength associated with it. **[Ans. 7.3 Å]**

8. Show that the relativistic expression for the de Broglie wavelength associated with a particle of rest mass m_0, charge e, when accelerated through V volt is given by

$$\lambda = \frac{h}{\sqrt{2m_0 eV[1 + (eV/2m_0 c^2)]}}$$

9. What potential difference is required to accelerate an electron so that it acquires a de Broglie wavelength of 0.5 Å. **[Ans. 600 V]**

10. Find the energy of neutron in eV whose de Broglie wavelength is 1 Å. **[Ans. 0.0815 eV]**

11. An electron is accelerated through a potential difference of 40 volts. The error in measurement of its energy is 1 eV. What is the error in measurement of its position? **[Ans. 2.46 × 10^9 m)**

[**Hint:** momentum $p = \sqrt{2mE}$]

12. Assuming the nuclear force between two protons to have a range $\cong 10^{-13}$ cm due to π-meson exchange, use the uncertainty relation $\Delta E . \Delta t \cong h$ to calculate the approximate mass of the π-meson in MeV.
[**Hint:** $\Delta t = \Delta x/c$] **[Ans. ~1240 MeV]**

13. Excited nuclear states may have life time as short as 10^{-15} sec. Calculate the sharpness (the natural line width) of such a state in ergs. Given $h = 6.62 \times 10^{-27}$ erg-s. **[Ans. ~6.62 × 10^{-12} erg-s]**

14. How early radiation laws failed to account for spectral distribution of energy density in the black body radiation? What is Rayleigh Jean's law? What do you understand by ultraviolet catastrophe?

15. Enumerate the basic experimental results obtained about photoelectric effect?

16. How particle nature of radiation was confirmed by the photoelectric effect and the Compton effect?

17. The uncertainity in the velocity of a particle is equal to its velocity. Show that the uncertainity in its location is equal to its de Broglie's wavelength.

18. What is the work function of a metal if the threshold wavelength for it is 570 nm? If light of 465 nm wavelength falls on the metal, what is its stopping potential?

2

Wave Mechanics

2.1 Introduction

There is a close parallelism between geometrical optics and Newtonian mechanics developed by Hamilton. The laws of geometrical optics can be deduced from the laws of wave optics i.e. from Maxwell's equations; they represent an approximation which hold provided that the wavelength of the light is very short in comparison to the size of the refracting or reflecting objects in the field. By analogy, Newtonian mechanics can be considered to be a similar approximation of Quantum Mechanics.

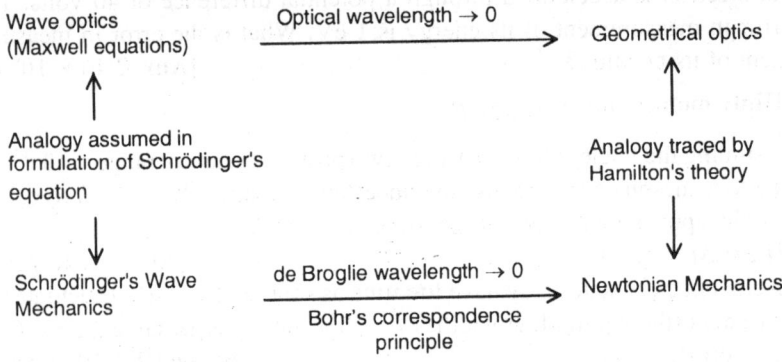

Fig. 2.1

In fact quantum mechanics is the description of the behaviour of matter and light and their interaction on atomic scales, subject to the restriction that de Broglie wavelength be small compared to the dimensions of the physical objects involved. Informations about atomic and nuclear scale behaviour were gathered during the 20th century particularly in the years 1926 and 1927 by Schrödinger, Heisenberg, Born and Dirac. The old quantum theory was generalized and given a rigorous mathematical foundation by these scientists during 1925 to 1929.

The development of quantum mechanics took place along two streams of thought viz. the Wave Mechanics due to *E.* Schrödinger and Matrix Mechanics developed by Heisenberg.

2.2 The Double Slit Experiment in Wave Mechanics

The wave vs particle nature can be clearly understood if we consider the two slit experiment of light interference with a slight modification, the screen replaced by a photoelectric emitter, so that when a photon arrives, a speck of light produced is detected.

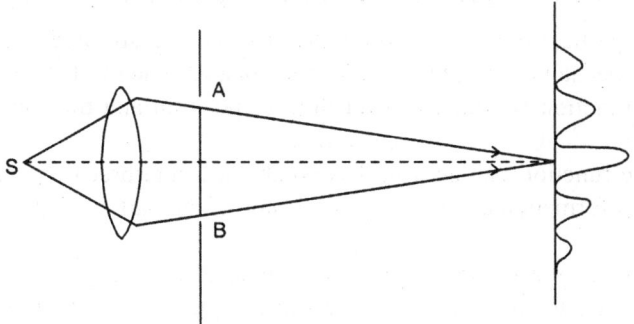

Fig. 2.2: Two slit experiment

It is observed experimentally that photoelectrons tend to be emitted from the photoemissive screen only at the locations of the bright fringe (and not at the black fringe locations). Although an incoming photon appears at a bright fringe, it is difficult to tell in advance which fringe, so that the intensity distribution over the given fringe serves only as a spatial probability distribution for the arrival of any given photon. The probability of finding a photon at a certain place and time depends on the value of $\left|\vec{E}(x,t)\right|^2$ where $\vec{E}(x,t)$ is the electric field vector component. We have to regard each photon as having a wave associated with it so that the intensity of this wave ($I = \left|\vec{E}(x,t)\right|^2$) at a given place on the screen determines the likelihood that a photon will arrive their.

Inferences from this experiment

1. Considering light as a stream of photons reveals an associated wave, the amplitude of which plays the role of a probability amplitude. The square of the amplitude gives a measure of probability of finding a photon at a particular point.

$$I = \left|\vec{E}(x,t)\right|^2$$

$$\vec{E} = \vec{E}_1(x, t) + \vec{E}_2(x, t)$$

so $\quad I = |\vec{E}_1(x, t)|^2 + |\vec{E}_2(x, t)|^2 + 2\,\mathrm{Re}(\vec{E}_1^* \cdot \vec{E}_2) \neq I_1 + I_2$ $\qquad$ (1)

(The third term is the interference term)

Thus probability enters quantum mechanics in a fundamental non classical way.

2. For all particles, the probability amplitude propagates as a typical wave ($\vec{E}(x, t) \propto E_0 \exp(-i\omega t)$). The wave intensity is interpreted as a photon probability density.

3. For photons, the wave amplitude (in general, represented by ψ) contains all the information about the photon probability distribution i.e. it completely describes the state of the photon. Hence the wave plays the role of a probability amplitude in the probabilistic description of the particle.

2.3 The Wave Function ψ and its Physical Significance

From wave particle duality, we know that a wave is associated with a particle. In quantum mechanics, the physical condition of a particle i.e. its state is described by a wave function ψ which is a function of position and time so that

$$\psi = \psi(\vec{r}, t) \tag{2}$$

The wave function though itself has no physical interpretation yet it contains all relevant information about the physical state of the particle and thus describes it completely.

ψ in general is a complex quantity so that ψ* ψ (i.e. ψ multiplied by its complex conjugate) is a physically meaningful quantity and $\psi^*\psi = |\psi(\vec{r}, t)|^2$ defines the probability density. It is a real positive quantity which satisfies

$$\int_{-\infty}^{\infty} |\psi|^2 \, dV = 1 \tag{3}$$

meaning, that particle exists somewhere at all the times. This equation is known as normalization condition. The wave function $\psi(\vec{r}, t)$ satisfies a differential equation which determines the character of the motion of the particle, called the Schrödinger equation. It plays the same role in Wave Mechanics as Newton's equation in classical mechanics.

A wave function would be admissible as a mathematical representation of state of a particle if and only if it satisfies the following conditions:

1. ψ is continuous and single valued every where
2. $\partial\psi / \partial x$, $\partial\psi / \partial y$ and $\partial\psi/\partial z$ are continuous and single valued everywhere.
3. It must be normalizable. For this ψ must go to 0 as $x \to \pm\infty$, $y \to \pm\infty$, $z \to \pm\infty$ in order that $\int |\psi(\vec{r}, t)|^2 \, d\vec{r}$ over all splace be a finite constant.

It is to be noted here that the quantity $|\psi(\vec{r}, t)|^2 \, d\vec{r}$ is proportional to the probability that as the result of our measurement, the values of the coordinates of the particles lie within the interval $\vec{r}$, $\vec{r} + d\vec{r}$ at instant t. If

$$\int |\psi|^2 \, d\vec{r} = N \tag{4}$$

then a new function ψ' differing by an arbitrary non vanishing complex factor such that

$$\psi' = \frac{1}{\sqrt{N}} \psi \tag{5}$$

can be chosen corresponding to the same state in such a way that it satisfies

$$\int |\psi'(\vec{r})|^2 \, d\vec{r} = 1 \tag{6}$$

2.4 The Schrödinger Wave Equation

Erwin Schrödinger put forward his wave mechanics in 1926. Starting with the general wave equation and keeping in mind de Broglie's concept of matter waves, he derived the time dependent and time independent wave equations for a particle.

He used these equations to solve the problems of hydrogen atom, harmonic oscillator, rigid rotator, Stark effect etc. In March 1926, he proved that his wave mechanics and Heisenberg's matrix mechanics are equivalent. He was awarded Nobel prize for his work in 1933.

The general wave equation of a wave propagated in x direction is

$$\frac{\partial^2 y}{\partial x^2} = \frac{1}{v^2}\frac{\partial^2 y}{\partial t^2} \tag{7}$$

where v is the velocity of propagation. Its solution is of the form $y = f\left(t \pm \frac{x}{v}\right)$

For an unrestricted wave (i.e. for a potential free region), the solution will be

$$y = Ae^{-i\omega\left(t - \frac{x}{v}\right)} \tag{8}$$

and, for the wave associated with a particle

$$\frac{\partial^2 \psi}{\partial x^2} = \frac{1}{v^2}\frac{\partial^2 \psi}{\partial t^2} \tag{9}$$

and,

$$\psi(x, t) = Ae^{-i\omega(t - \frac{x}{v})} \tag{10}$$

for one dimensional case. Here,

$\omega = 2\pi v$

and $v = v\lambda$

$$\tag{11}$$

$$\therefore \quad \psi(x, t) = Ae^{-2\pi i\left(vt - \frac{x}{\lambda}\right)} \tag{12}$$

Now we know that v and λ are related to the total energy E and momentum p of the particle by

$$v = \frac{E}{h} = \frac{E}{2\pi\hbar} \tag{13}$$

and

$$\lambda = \frac{h}{p} = \frac{2\pi\hbar}{p} \tag{14}$$

so that

$$\psi = Ae^{-2\pi i\left(\frac{Et}{2\pi\hbar} - \frac{xp}{2\pi\hbar}\right)}$$

or

$$\psi = Ae^{-i(Et - px)/\hbar} \tag{15}$$

or

$$\psi(x,t) = A\,\exp\left[\frac{i}{\hbar}\left(px - \frac{p^2}{2m}t\right)\right] \tag{16}$$

This is the wave function for a particle of momentum p and energy $p^2/2m$ moving freely in x-direction. Thus

$$\psi(\vec{r},t) = A\,\exp\left[\frac{i}{\hbar}(\vec{p}\cdot\vec{r} - Et)\right] \tag{17}$$

represents a plane wave.

But, however, we are interested to certain motion of the particle subjected to certain restrictions, e.g. an electron bound to an atom by the electric field of the nucleus. So, let us try to obtain a fundamental differential equation for ψ which can be solved in a specific situation.

Differentiating Eqn. (15) twice with respect to x gives

$$\frac{\partial^2 \psi}{\partial x^2} = -\frac{p^2}{\hbar^2} \psi \tag{18}$$

Differentiating Eqn. (15) once with respect to t gives

$$\frac{\partial \psi}{\partial t} = -\frac{iE}{\hbar} \psi \tag{19}$$

In general, v is small compared to c, the velocity of light, therefore the total energy of the particle

$$E = \frac{p^2}{2m} + V \tag{20}$$

in presence of a potential V. Post-multiplying both sides of this equation by ψ

$$E\psi = \frac{p^2}{2m}\psi + V\psi \tag{21}$$

From Eqns (18) and (19) respectively, we get

$$\frac{p^2 \psi}{2m} = -\frac{\hbar^2}{2m}\frac{\partial^2 \psi}{\partial x^2} \tag{22}$$

$$E\psi = -\frac{\hbar}{i}\frac{\partial \psi}{\partial t} = i\hbar\frac{\partial \psi}{\partial t} \tag{23}$$

Substituting these values in Eqn. (21), we get

$$i\hbar\frac{\partial \psi}{\partial t} = -\frac{\hbar^2}{2m}\frac{\partial^2 \psi}{\partial x^2} + V\psi \tag{24}$$

or

$$\frac{\partial^2 \psi}{\partial x^2} = \left(-i\frac{2m}{\hbar}\right)\frac{\partial \psi}{\partial t} + \left(\frac{2m}{\hbar^2}V\right)\psi$$

or in general, in case of three dimensions,

$$i\hbar\frac{\partial \psi}{\partial t} = \left\{-\frac{\hbar^2}{2m}\nabla^2 + V(\vec{r},t)\right\}\psi \tag{25}$$

This is the *time-dependent Schrödinger equation.* To obtain the time independent equation, we note that the one dimensional wave function ψ of an unrestricted particle may be written as

$$\psi = Ae^{-i(Et-px)/\hbar}$$
$$= Ae^{-iEt/\hbar}\, e^{ipx/\hbar}$$
$$\equiv e^{-(iE/\hbar)t}\, \psi(x) \tag{26}$$

i.e. ψ is the product of a time dependent function and a position dependent function.

$$E\,\psi(x)\,e^{-(iE/\hbar)t} = -\frac{\hbar^2}{2m}e^{-(iE/\hbar)t}\frac{\partial^2 \psi(x)}{\partial x^2} + Ve^{-(iE/\hbar)t}\,\psi(x)$$

$$\therefore \qquad \frac{\hbar^2}{2m}\frac{\partial^2 \psi(x)}{\partial x^2} + (E-V)\psi(x) = 0$$

or in general

$$\nabla^2\psi(\vec{r}) + \frac{2m}{\hbar^2}(E-V)\psi(\vec{r}) = 0 \tag{27}$$

This is the *time independent or steady state form of the Schrödinger's equation.* ∇^2 is called the Laplacian operator.

2.5 Free Particle in a Box Shaped Volume

An Example of a wave function which cannot be normalized by condition Eqn. (6) is the wave function

$$\psi(\vec{r},t) = Ae^{i\left[(\vec{k}.\vec{r})-\omega t\right]} \tag{28}$$

corresponding to the state of free motion of a particle with a well defined momentum $\vec{p} = \hbar\vec{k}$. However, if we consider a definite volume, say of box shape (L^3), we can normalize this wave function. The wave functions must satisfy some boundary conditions on the surface of this volume. If L be sufficiently large (quantum mechanically, $L >> 10^{-6}$ cm) the influence of the boundary conditions on the character of the motion of the particle in the volume $V = L^3$ will be very small. We can choose the boundary conditions in an arbitrary, sufficiently simple form. We may take periodic boundary conditions with period L, and require that the wave functions satisfy the conditions

$$\begin{aligned} \psi \ (x, \ y, \ z) &= \psi \ (x+L, \ y \ ,z) \\ &= \psi \ (x, \ y + L, \ z) \\ &= \psi \ (x, \ y, \ z + L) \end{aligned} \tag{29}$$

We are interested in studying the state at time t = 0. We can then verify by substituting Eqn. (29) in Eqn. (28) that the periodicity conditions are satisfied. If the function (28) normalized in the volume V, is of the form

$$\psi_i(\vec{r}) = \frac{1}{\sqrt{V}} e^{i(\vec{k}.\vec{r})} \tag{30}$$

where
$$k_x = \frac{2\pi}{L} n_x, \ k_y = \frac{2\pi}{L} n_y, \ k_z = \frac{2\pi}{L} n_z \tag{31}$$

where n_x, n_y and n_z are positive or negative integers.

The boundary conditions Eqn. (29) reduce now to the requirement that the vector $\vec{k}$ run through a discrete set of values, determined by condition Eqn. (31). While passing to the limit $L \to \infty$, the distance between two k values tends to zero ($\because \vec{k}$ space being reciprocal space), and we return to the free motion of a particle in unbounded space.

The totality of functions (30) corresponding to all possible values of $\vec{k}$ satisfying Eqn. (31) form a set of functions satisfying the condition

$$\int_V \psi_{k'}^* \psi_k d^3\vec{r} = \delta_{k'k} \tag{32}$$

where $\delta_{k'k} = \delta_{k'_x k_x} \delta_{k'_y k_y} \delta_{k'_z k_z}$ and $d^3r = dx \, dy \, dz$

The functions (Eqn. 30) form a complete set of functions, that is, any wave function can be expanded as

$$\psi(\vec{r}) = \sum_k a_k \psi_k(\vec{r}) \tag{33}$$

The coefficients a_k are evaluated as

$$a_k = \int_V \psi(\vec{r}) \psi_k^*(\vec{r}) d^3r \tag{34}$$

If the functions $\psi(\vec{r})$ are normalized in the volume V, we find by substituting (Eqn. 33) into the normalization condition and using Eqn. (32)

$$\int \psi * (\vec{r})\psi(\vec{r})d^3\vec{r} = \sum_k |a_k|^2 = 1 \tag{35}$$

It follows from Eqn. (33) that the coefficients a_k determine in how far the state with well defined momentum $\vec{p} = \hbar\vec{k}$ takes part in the general state $\psi(\vec{r})$; the absolute square of a_k determines the probability of observing the value of the momentum $\vec{p} = \hbar\vec{k}$ for a system which is in the state ψ. Then Eqn. (35) expresses the condition that the sum of the probabilities for all possible values of the momentum must equal unity.

2.6 Probability Current Density

We have seen that time dependent Schrödinger equation is

$$i\hbar \frac{\partial \psi}{\partial t} = \left\{ -\frac{\hbar^2}{2m}\nabla^2 + V(\vec{r},t)\right\}\psi$$

This is a complex equation which is satisfied separately by its real and imaginary parts. The complex conjugate equation is obtained by changing the sign of the imaginary part everywhere

$$-i\hbar \frac{\partial \psi *}{\partial t} = \left\{ -\frac{\hbar^2}{2m}\nabla^2 + V(\vec{r},t)\right\}\psi *$$

The above two equations may be used to evaluate the time rate of change of the probability density at any point

$$\frac{\partial}{\partial t}(\psi * \psi) = \left(\frac{\partial \psi *}{\partial t}\right)\psi + \psi * \left(\frac{\partial \psi}{\partial t}\right)$$

$$= -\frac{i\hbar}{2m}\left\{(\nabla^2\psi *)\psi - \psi * (\nabla^2\psi)\right\} \tag{36}$$

The expression on the right-hand side is closely connected with a vector $\vec{j}$ defined by

$$j(\vec{r},t) = \frac{i\hbar}{2m}\left\{(grad\ \psi *)\psi - \psi * (grad\ \psi)\right\} \tag{37i}$$

and in fact, by direct differentiation

$$div\ \vec{j} = \frac{i\hbar}{2m}\left\{(\nabla^2\psi *)\psi - \psi * (\nabla^2\psi)\right\}$$

$$= -\frac{\partial}{\partial t}(\psi * \psi) \tag{37ii}$$

This is called the continuity equation, its physical significance may be seen by integrating both sides through an arbitrary volume V enclosed by the surface S. With the help of Gauss's divergence theorem of vector analysis,

$$\frac{\partial}{\partial t}\int_V \psi * \psi\ dv = -\int_V div\ \vec{j}\ dv$$

$$= -\int_s \vec{j} \cdot d\vec{s} \tag{38}$$

where the element $\vec{ds}$ is a vector of magnitude ds directed along the outward normal to the surface. The equation shows that the rate of change of the chance of finding the particle in the volume V is equal to the inward flux of the vector $\vec{j}$ through the surface S. In fact $\vec{j} \cdot \vec{ds}$ may be interpreted as the chance per unit time that a particle passes through $\vec{ds}$; $\vec{j}$ is therefore called the *probability current density*. The decrease of probability arises due to the change of ψ with time (since the particle is in motion). The situation here is quite analoguous to continuity equation of hydrodynamics.

Further we can write

$$\psi(\vec{r},t) = \sqrt{\rho(r,t)} \exp\left[\frac{iS(\vec{r},t)}{\hbar}\right] \qquad (39)$$

so that $\left[\sqrt{\rho(r,t)}\right]^2$ denotes the probability density $|\psi|^2$ ($\rho > 0$ and S is real). $S(\vec{r},t)$ is called the *phase function*. We note that

$$\nabla\psi = (\nabla\sqrt{\rho})\exp\frac{iS}{\hbar} + \sqrt{\rho}\frac{i}{\hbar}\nabla S e^{iS/\hbar}$$

$$\psi*\nabla\psi = \sqrt{\rho}\nabla(\sqrt{\rho}) + \frac{i}{\hbar}\rho(\nabla S)$$

Since
$$\vec{j} = \frac{\hbar}{m}\mathrm{Im}(\psi*\nabla\psi) \qquad \text{(from Eqn. (37}i\text{))}$$

$$= \frac{\hbar}{m}\times\frac{1}{\hbar}\rho(\nabla S) = \frac{\rho(\nabla S)}{m} \qquad (40)$$

Thus $\rho(\nabla S)$ is akin to momentum density flux with $\left(\dfrac{\nabla S}{m}\right)$ as some kind of velocity, which when multiplied by the probability density equals probability current density. The larger and steeper the phase variation (∇S), the more intense the flux. The direction of $\vec{j}$ at some point is normal to the surface of a constant phase at that point.

Example 2.1: Using $\psi = u + iv$ *and* $\psi* = u - iv$ in the Schrödinger equation

$$i\hbar\frac{\partial y(\vec{r},t)}{\partial t} = \left[-\frac{\hbar^2}{2m}\nabla^2 + V(\vec{r})\right]\psi(\vec{r},t)$$

and its complex conjugate form, find the coupled equations satisfied by the real functions u and v.

Solution: $\psi = u + iv$, $\psi* = u - iv$ (u and v are real functions)

$$-i\hbar\frac{\partial}{\partial t}\psi*(\vec{r},t) = \left(-\frac{\hbar^2}{2m}\nabla^2 + V(\vec{r})\right)\psi*(\vec{r},t)$$

$$i\hbar\frac{\partial}{\partial t}(u + iv) = \frac{-\hbar^2}{2m}(\nabla^2 u + i\nabla^2 v) + V(\vec{r})(u + iv)$$

or
$$i\hbar\frac{\partial u}{\partial t} - \hbar\frac{\partial v}{\partial t} = -\frac{\hbar^2}{2m}(\nabla^2 u + i\nabla^2 v) + V(\vec{r})u + iV(\vec{r})v \qquad (i)$$

Similarly,

$$-i\hbar\frac{\partial u}{\partial t} - \hbar\frac{\partial v}{\partial t} = -\frac{\hbar^2}{2m}(\nabla^2 u - i\nabla^2 v) + V(\vec{r})u - iV(\vec{r})v \qquad (ii)$$

Adding Eqns *(i)* and *(ii)*

$$-2\hbar\frac{\partial v}{\partial t} = -\frac{\hbar^2}{m}\nabla^2 u + 2V(\vec{r})u$$

or

$$\frac{\partial v}{\partial t} = \left[\frac{\hbar}{2m}\nabla^2 - \frac{1}{\hbar}V(\vec{r})\right]u \qquad (iii)$$

Subtracting Eqns *(ii)* from *(i)*, we get

$$\frac{\partial u}{\partial t} = \left[-\frac{\hbar}{2m}\nabla^2 + \frac{1}{\hbar}V(\vec{r})\right]v \qquad (iv)$$

Eqns *(iii)* and *(iv)* are the required coupled equations satisfied by the real functions *u* and *v*.

Example 2.2: Normalize the wave function

$$\psi(x) = e^{-|x|}\sin\alpha x$$

Solution: Let the normalized wave function be

$$\psi'(x) = Ne^{-|x|}\sin\alpha x$$

then

$$\int_{-\infty}^{+\infty}\psi'^*\psi'dx = |N|^2\int_{-\infty}^{+\infty}\psi^*\psi dx = 1$$

Now

$$\psi(x) = e^{-x}\sin\alpha x \text{ for } x > 0$$

$$= e^{x}\sin\alpha x \text{ for } x < 0$$

$$\therefore \quad \int_{-\infty}^{+\infty}\psi^*\psi\,dx = \int_{-\infty}^{0}e^{x}\sin\alpha x\, e^{x}\sin\alpha x\,dx + \int_{0}^{\infty}e^{-x}\sin\alpha x\, e^{-x}\sin\alpha x\,dx$$

$$\equiv I_1 + I_2$$

Now,

$$I_1 \equiv \int_{-\infty}^{0}e^{2x}\sin^2\alpha x\,dx = \int_{-\infty}^{0}e^{2x}\frac{(1-\cos 2\alpha x)}{2}dx$$

$$= \frac{1}{2}\int_{-\infty}^{0}e^{2x} - \frac{1}{2}\int_{-\infty}^{0}e^{2x}\cos 2\alpha x\,dx$$

$$= \frac{1}{2}\times\frac{1}{2} - \frac{1}{2}I_1'$$

where

$$I_1' = \int_{-\infty}^{0}\cos 2\alpha x \cdot e^{2x}\,dx$$

$$\left|\cos 2\alpha x \cdot \frac{e^{2x}}{2}\right|_{-\infty}^{0} + \int_{-\infty}^{0}2\alpha\sin 2\alpha x \cdot \frac{e^{2x}}{2}dx$$

$$= \frac{1}{2} + \alpha\left\{\sin 2\alpha x \cdot \frac{e^{2x}}{2}\Big|_{-\infty}^{0} - 2\alpha\int_{-\infty}^{0}\cos 2\alpha x \cdot \frac{e^{2x}}{2}dx\right\}$$

$$= \frac{1}{2} + 0 - \alpha^2 \int\limits_{-\infty}^{0} \cos 2\alpha x \ e^{2x} \ dx$$

or $\qquad I_1' = \frac{1}{2} - \alpha^2 I_1' \qquad$ so that $\qquad I_1' = \frac{1}{2(1+\alpha^2)}$

$\therefore \qquad I_1 = \frac{1}{4} - \frac{1}{4(1+\alpha^2)} = \left(\frac{1}{4}\right)\frac{\alpha^2}{(1+\alpha^2)}$

Similarly, by symmetry,

$$I_2 = \left(\frac{1}{4}\right)\frac{\alpha^2}{(1+\alpha^2)}$$

Consequently, $\quad |N|^2 = \int\limits_{-\infty}^{+\infty} \psi * \psi \ dx = 1 \Rightarrow |N| = \frac{\sqrt{2(1+\alpha^2)}}{\alpha}$

Example 2.3: Give the mathematical expressions for a spherically outgoing wave (travelling away from a point) and evaluate its pobability current density.

Solution: $\quad \psi(r) = \frac{A}{r} e^{ikr}$ $\qquad\qquad\qquad$ *(k is the wave vector)*

Probability current density

$$j = \frac{i\hbar}{2m}\left[\psi \ grad \ \psi * - \psi * \ grad \ \psi\right]$$

$$= \frac{i\hbar}{2m}|A|^2\left[\frac{e^{ikr}}{r}\nabla\left(\frac{e^{-ikr}}{r}\right) - \frac{e^{-ikr}}{r}\nabla\left(\frac{e^{ikr}}{r}\right)\right]$$

$$= \frac{i\hbar}{2m}|A|^2\left[\frac{e^{ikr}}{r}\left(-\frac{ik}{r}e^{-ikr} - \frac{e^{-ikr}}{r^2}\right) - \frac{e^{-ikr}}{r}\left(\frac{ik}{r}e^{ikr} - \frac{e^{ikr}}{r^2}\right)\right]$$

$$= \frac{i\hbar}{2m}|A|^2\left(-\frac{2ik}{r^2} - \frac{1}{r^3} + \frac{1}{r^3}\right)$$

$$= \frac{\hbar k}{mr^2}|A|^2$$

2.7 Solution of the Schrödinger's Equation for a Free Particle

For a free particle $V = 0$, and Schrödinger's equation becomes,

$$\nabla^2\psi + \frac{2m}{\hbar^2}E\psi = 0 \qquad\qquad (41i)$$

or $\qquad \dfrac{\partial^2 \psi}{\partial x^2} + \dfrac{\partial^2 \psi}{\partial y^2} + \dfrac{\partial^2 \psi}{\partial z^2} + \dfrac{2m}{\hbar^2}E\psi = 0 \qquad\qquad (41ii)$

This equation is a partial differential equation in three variables x, y and z and can be solved by the method of separation of variables. Let the solution be

$$\psi \ (x, \ y, \ z) = X \ (x). \ Y \ (y). \ Z \ (z) \qquad\qquad (42)$$

where $X(x)$ is a function of x only, $Y(y)$, a function of y alone and $Z(z)$, a function of z alone. Substituting this in Eqn. (41ii) and dividing by $X(x). \ Y(y). \ Z(z)$, we get

$$\frac{1}{X}\frac{\partial^2 X}{\partial x^2} + \frac{1}{Y}\frac{\partial^2 Y}{\partial y^2} + \frac{1}{Z}\frac{\partial^2 Z}{\partial z^2} + \frac{2m}{\hbar^2}E = 0 \tag{43}$$

or
$$\frac{1}{X}\frac{\partial^2 X}{\partial x^2} = -\frac{1}{Y}\frac{\partial^2 Y}{\partial y^2} - \frac{1}{Z}\frac{\partial^2 Z}{\partial z^2} - \frac{2m}{\hbar^2}E \tag{44}$$

In this Eqn. l.h.s is a function of x only while r.h.s. is a function of y and z and is independent of x. This is possible only when the quantity on either side is equal to a constant which is independent of x, y and z, i.e.

$$\frac{1}{X}\frac{\partial^2 X}{\partial x^2} = k_1 \tag{45}$$

and
$$-\frac{1}{Y}\frac{\partial^2 Y}{\partial y^2} - \frac{1}{Z}\frac{\partial^2 Z}{\partial z^2} - \frac{1}{Y}\frac{\partial^2 Y}{\partial y^2} - \frac{1}{Z}\frac{\partial^2 Z}{\partial z^2} - \frac{2m}{\hbar^2}E = k_1 \tag{46}$$

Eqn. (33) may be written as

$$\frac{1}{Y}\frac{\partial^2 Y}{\partial y^2} = -k_1 - \frac{1}{Z}\frac{\partial^2 Z}{\partial z^2} - \frac{2mE}{\hbar^2} \tag{47}$$

In the above expression, the quantity on the r.h.s. is function of y alone and the r.h.s. is independent of y. therefore, it is possible only when the quantity on each side is equal to a constant (say k_2) i.e.

$$\frac{1}{Y}\frac{\partial^2 Y}{\partial y^2} = k_2 \tag{48}$$

$$-k_1 - \frac{1}{Z}\frac{\partial^2 Z}{\partial z^2} - \frac{2m}{\hbar^2}E = k_2 \tag{49}$$

Eqn. (36) may be written as

$$\frac{1}{Z}\frac{\partial^2 Z}{\partial z^2} = -k_1 - \frac{2m}{\hbar^2}E - k_2$$

In above expression, r.h.s. is constant. Let this constant bé k_3, so

$$\frac{1}{Z}\frac{\partial^2 Z}{\partial z^2} = k_3 \tag{50}$$

and
$$-k_1 - \frac{2m}{\hbar^2}E - k_2 = k_3$$

or
$$k_1 + k_2 + k_3 = -\frac{2m}{\hbar^2}E \tag{51}$$

For sake of convenience, let us put

$$k_1 = \frac{2m}{\hbar^2}E_x \tag{52}$$

Then,
$$\frac{\partial^2 X}{\partial x^2} + \frac{2m}{\hbar^2}E_x X = 0 \tag{53}$$

(from Eqn. (45))

The general solution of this equation can be written as

$$X(x) = N_x \sin\left\{\frac{\sqrt{2mE_x}}{\hbar}(x - x_0)\right\}$$

where N_x and x_0 are arbitraty constants. Similarly, by substituting

$$k_2 = -\frac{2m}{\hbar^2}E_y \text{ and } k_3 = -\frac{2m}{\hbar^2}E_z$$

in equations (48) and (50), we obtain

$$\frac{\partial^2 Y}{\partial y^2} + \frac{2m}{\hbar^2}E_y Y = 0 \text{ and } \frac{\partial^2 Z}{\partial z^2} + \frac{2m}{\hbar^2}E_z Z = 0$$

The general solutions of the above equations are

$$Y(y) = N_y \sin\left\{\frac{\sqrt{2mE_y}}{\hbar}(y - y_0)\right\} \tag{54}$$

and

$$Z(z) = N_z \sin\left\{\frac{\sqrt{2mE_z}}{\hbar}(z - z_0)\right\} \tag{55}$$

Substituting values of k_1, k_2 and k_3 in Eqn. (51)

$$E_x + E_y + E_z = E \tag{56}$$

Since, any sine function is finite, single valued and continuous for real values of its argument, hence for the real particles, i.e. for finite values of ψ (or X, Y, Z), the values of E_x, E_y, E_z (i.e. E) will be positive.

Thus the eigen functions of the free particle are

$$\psi = X \, Y \, Z$$

or

$$\psi = N \sin\left\{\frac{\sqrt{2mE_x}}{\hbar}(x - x_0)\right\} \sin\left\{\frac{\sqrt{2mE_y}}{\hbar}(y - y_0)\right\}$$

$$\times \sin\left\{\frac{\sqrt{2mE_z}}{\hbar}(z - z_0)\right\} \tag{57}$$

where $N = N_x N_y N_z$ is a normalization constant. The energy eigen values are

$$E = E_x + E_y + E_z = -\frac{\hbar^2}{2m}(k_1 + k_2 + k_3) \tag{58}$$

clearly, there are continuous energy levels. The complete time dependent wave functions are given as

$$\psi = N \sin\left\{\frac{\sqrt{2mE_x}}{\hbar}(x - x_0)\right\} e^{-iE_x t / \hbar} \sin\left\{\frac{\sqrt{2mE_y}}{\hbar}(y - y_0)\right\} e^{-iE_y t / \hbar}$$

$$\times \sin\left\{\frac{\sqrt{2mE_z}}{\hbar}(z - z_0)\right\} e^{-iE_z t / \hbar}$$

or
$$\psi = N \sin\left\{\frac{\sqrt{2mE_x}}{\hbar}(x - x_0)\right\} \sin\left\{\frac{\sqrt{2mE_y}}{\hbar}(y - y_0)\right\}$$

$$\times \sin\left\{\frac{\sqrt{2mE_z}}{\hbar}(z - z_0)\right\} e^{-iEt/\hbar} \tag{59}$$

where $\quad E = E_x + E_y + E_z$

Though the free particle has continuous enrergy levels, we will see that the energy values are quantized if the particle is constrained to remain in a box.

Example 2.4: The normalized wave function for 1s electron may be put as

$$\psi_{1s} = \left(\frac{1}{\pi a_0^3}\right)^{1/2} e^{-r/a_0}$$

Calculate the probability of the electron in a sphere of radius $r = a_0$.

Solution: The probability $P = \int \psi_{1s}^* \psi_{1s} dV$

$$dV = (r\, dr)\, (r\sin\theta\, d\theta)\, (d\phi) = r^2\, dr\, \sin\theta\, d\theta\, d\phi$$

$$P = \frac{1}{\pi a_0^3} \int_0^{a_0} e^{-2r/a_0} r^2 dr \int_0^{\pi} \sin\theta\, d\theta \int_0^{2\pi} d\phi$$

The $d\phi$ integral contributes 2π and θ integral contributes 2, so

$$P = \frac{4}{a_0^3} \int_0^{a_0} e^{-2r/a_0} r^2\, dr$$

$$= \frac{4}{a_0^3}\left\{\left[r^2 \int e^{-2r/a_0} dr\right]_0^{a_0} - \int_0^{a_0} 2r\left(\int_0^{a_0} e^{-2r/a_0} dr\right) dr\right\} \quad \text{(Integrating by parts)}$$

$$= \frac{4}{a_0^3}\left\{\left[-\frac{a_0}{2} r^2 e^{-2r/a_0}\right]_0^{a_0} - 2\int_0^{a_0} r\left(-\frac{a_0}{2}\right) e^{-2r/a_0} dr\right\}$$

$$= \frac{4}{a_0^3}\left\{-\frac{a_0^3}{2} e^{-2} + \left[a_0 r\left(-\frac{a_0}{2}\right) e^{-2r/a_0}\right]_0^{a_0} - a_0\int_0^{a_0} \frac{dr}{dr}\left(-\frac{a_0}{2}\right) e^{-2r/a_0} dr\right\}$$

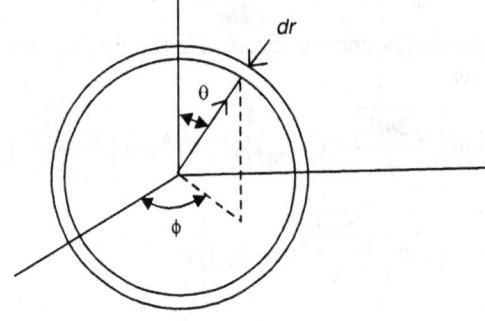

Fig. 2.3

$$= -2e^{-2} - 2e^{-2} + \frac{4a_0}{a_0^3}\left(\frac{a_0}{2}\right)\left(-\frac{a_0}{2}\right)\left[e^{-2r/a_0}\right]_0^{a_0}$$

$$= (-5e^{-2} + 1) = 0.323 \textbf{ Ans.}$$

Example 2.5: Consider the one dimensional wave function

$$\psi(x) = A\left(\frac{x}{x_0}\right)^n e^{-x/x_0},$$

where A, n and x_0 are constants. Using Schrödinger equation, find the potential $V(x)$ and energy E for which this wave function is an eigen function. Assume that as $x \to \infty$, $V(x) \to 0$.

Solution:

$$\psi(x) = A\left(\frac{x}{x_0}\right)^n e^{-x/x_0}$$

$$\therefore \qquad \frac{d\psi(x)}{dx} = A\frac{n}{x_0}\left(\frac{x}{x_0}\right)^{n-1} e^{-x/x_0} + A\left(\frac{x}{x_0}\right)^n \left(-\frac{1}{x_0}\right)e^{-x/x_0}$$

and $\qquad \frac{d^2\psi(x)}{dx^2} = A\frac{n(n-1)}{x_0^2}\left(\frac{x}{x_0}\right)^{n-2} e^{-x/x_0} - \frac{An}{x_0^2}\left(\frac{x}{x_0}\right)^{n-1} e^{-x/x_0}$

$$-\frac{A}{x_0}\frac{n}{x_0}\left(\frac{x}{x^0}\right)^{n-1} e^{-x/x_0} + \frac{A}{x_0^2}\left(\frac{x}{x_0}\right)^n e^{-x/x_0}$$

$$= \left[\frac{n(n-1)}{x^2} - \frac{2n}{x_0 x} + \frac{1}{x_0^2}\right]\psi(x)$$

Substituting this value of $\dfrac{d^2\psi}{dx^2}$ in the time independent Schrödinger equation

$$\left[-\frac{\hbar^2}{2m}\frac{d^2}{dx^2} + V(x)\right]\psi(x) = E\,\psi(x)$$

we have

$$E - V(x) = -\frac{\hbar^2}{2m}\left[\frac{n(n-1)}{x^2} - \frac{2n}{xx_0} + \frac{1}{x_0^2}\right]$$

Since $V(x) \to 0$ when $x \to \infty$, we have

$$E = \frac{-\hbar^2}{2m}\left(\frac{1}{x_0^2}\right)$$

Hence $\qquad V(x) = \frac{\hbar^2}{2m}\left[\frac{n(n-1)}{x^2} - \frac{2n}{x_0 x}\right]$

2.8 A Free Particle in a Box

Consider a single particle of mass m enclosed in a rectangular three dimensional box. The sides of the box along X, Y and Z axes are a, b, c respectively.

The particle can freely move from 0 to a along the X-axis, 0 to b along the Y-axis and 0 to c along the Z-axis. There is no force on the particle inside the box so that $V = 0$ inside. The walls of the box are perfectly rigid, hence the value of V abruptly increases at the walls ($V = V_0$). Since infinite work is done by the particle to come out of the box, so $V = \infty$ outside. Let E denote the total energy of the particle and the particle remains inside the box (i.e. $V = 0$). The time independent Schrödinger equation is

$$\frac{\partial^2 \psi}{\partial x^2} + \frac{\partial^2 \psi}{\partial y^2} + \frac{\partial^2 \psi}{\partial z^2} + \frac{2m}{\hbar^2} E\psi = 0 \tag{60}$$

This equation can be solved by the method of separation of variables. Let the solution be

$$\psi(x, y, z) = X\, Y\, Z \tag{61}$$

where X is the function of x only, Y is the function of Y only and Z is the function of Z only.

Substituting the value of ψ from Eqn. (61) in Eqn. (60) and dividing both sides by $X\, Y\, Z$, we get

$$\frac{1}{X}\frac{\partial^2 X}{\partial x^2} + \frac{1}{Y}\frac{\partial^2 Y}{\partial y^2} + \frac{1}{Z}\frac{\partial^2 Z}{\partial z^2} + \frac{2m}{\hbar^2} E = 0$$

$$\frac{1}{X}\frac{\partial^2 X}{\partial x^2} = -\frac{1}{Y}\frac{\partial^2 Y}{\partial y^2} - \frac{1}{Z}\frac{\partial^2 Z}{\partial z^2} - \frac{2m}{\hbar^2} E$$

In this expression, left hand side is a function of only x while the right hand side is the function of y and z. This is possible only if the quantity on either side is equal to a constant, say k_1 which is independent of x, y and z, i.e.

$$\frac{1}{X}\frac{\partial^2 X}{\partial x^2} = k_1 \tag{62}$$

and

$$-\frac{1}{Y}\frac{\partial^2 Y}{\partial y^2} - \frac{1}{Z}\frac{\partial^2 Z}{\partial z^2} - \frac{2m}{\hbar^2} E = k_1$$

or

$$-\frac{1}{Y}\frac{\partial^2 Y}{\partial y^2} = -k_1 - \frac{1}{Z}\frac{\partial^2 Z}{\partial z^2} - \frac{2mE}{\hbar^2} \tag{63}$$

In Eqn. (63), the quantity on the l.h.s. is a function of only y while the quantity on the r.h.s. is the function of only z, hence each side of Eqn. (63) must be equal to a constant, say k_2, i.e.

$$\frac{1}{Y}\frac{\partial^2 Y}{\partial y^2} = k_2 \tag{64}$$

and

$$-k_1 - \frac{1}{Z}\frac{\partial^2 Z}{\partial z^2} - \frac{2m}{\hbar^2} E = k_2$$

or

$$\frac{1}{Z}\frac{\partial^2 Z}{\partial z^2} = -k_1 - k_2 - \frac{2mE}{\hbar^2} = k_3 \text{ (say)} \tag{65}$$

so,

$$k_1 + k_2 + k_3 = -\frac{2mE}{\hbar^2} \tag{66}$$

For the sake of convenience, let

$$k_1 = -\frac{2mE_x}{\hbar^2}; \quad k_2 = -\frac{2mE_y}{\hbar^2}; \quad k_3 = -\frac{2mE_z}{\hbar^2} \tag{67}$$

then

$$E = E_x + E_x + E_z$$

Thus Eqns (62), (64) and (65) become

$$\frac{\partial^2 X}{\partial x^2} + \frac{2mE_x}{\hbar^2} X = 0 \tag{68i}$$

$$\frac{\partial^2 Y}{\partial y^2} + \frac{2mE_y}{\hbar^2} Y = 0 \tag{68ii}$$

$$\frac{\partial^2 Z}{\partial z^2} + \frac{2mE_z}{\hbar^2} Z = 0 \tag{68iii}$$

Now, let us consider the differential equation along X-direction.

$$\frac{\partial^2 X}{\partial x^2} + \frac{2m}{\hbar^2} E_x X = 0$$

Its general solution is

$$X(x) = A \sin (Bx + C) \tag{69}$$

A, B and C are constants which can be determined by the boundary conditions as follows:

The probability of finding the particle inside the box is proportional to $|\psi|^2$, hence the quantity $|X|^2$ gives the probability of the particle to be at a point along X-direction, inside the box. Since potential at the walls is very large, the probability of finding the particle at the walls of the box is $\cong 0$. Thus,

$$|X|^2 = 0 \text{ at } x = 0 \text{ and } x = a$$

or

$$X = 0 \text{ at } x = 0 \text{ and } x = a$$

On applying this boundary condition, we get $\sin C = 0$ and $\sin (Ba + C) = 0$

$$\therefore \qquad C = 0 \text{ and } \sin Ba = 0$$

or

$$Ba = n_1 \pi \text{ or } B = \frac{n_1 \pi}{a} \tag{70}$$

where n_1 is a positive integer.

Substituting the values of B and C in Eqn. (69), the general solution of differential equation in X-direction is

$$X(x) = A \sin \left(\frac{n_1 \pi x}{a} \right) \tag{71}$$

Applying the normalization condition

$$\int_0^a |X(x)|^2 dx = 1 \text{ or } \int_0^a \left| A \sin \left(\frac{n_1 \pi x}{a} \right) \right|^2 dx = 1$$

or $A^2 \displaystyle\int_0^a \sin^2 \left(\frac{n_1 \pi x}{a} \right) dx = 1$ or $A^2 \times \dfrac{a}{2} = 1$ or $A = \sqrt{\dfrac{2}{a}}$

Hence,

$$X(x) = \sqrt{\frac{2}{a}} \sin \left(\frac{n_1 \pi x}{a} \right) \tag{72}$$

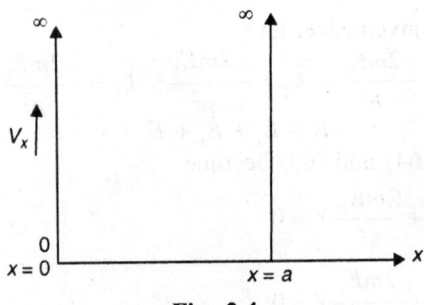

Fig. 2.4

Probability of finding the particle inside the box between x and $x + dx$ is

$$P(x)dx = \left|X(x)\right|^2 dx = \frac{2}{a}\sin^2\left(\frac{n_1\pi x}{a}\right)dx \qquad (73)$$

So, the probability density $P(x)\, dx = \frac{2}{a}\sin^2\left(\frac{n_1\pi x}{a}\right)$

Figure 2.5 a Shows the wave function $X(x)$ and probability density $|X(x)|^2$ for $n_1 = 1, 2, 3$. For the maxima of probability density

$$\frac{n_1\pi x}{a} = \frac{\pi}{2}, \frac{3\pi}{2}\cdots, \quad \text{or} \quad x = \frac{a}{2n_1}, \frac{3a}{2n_1}, \cdots$$

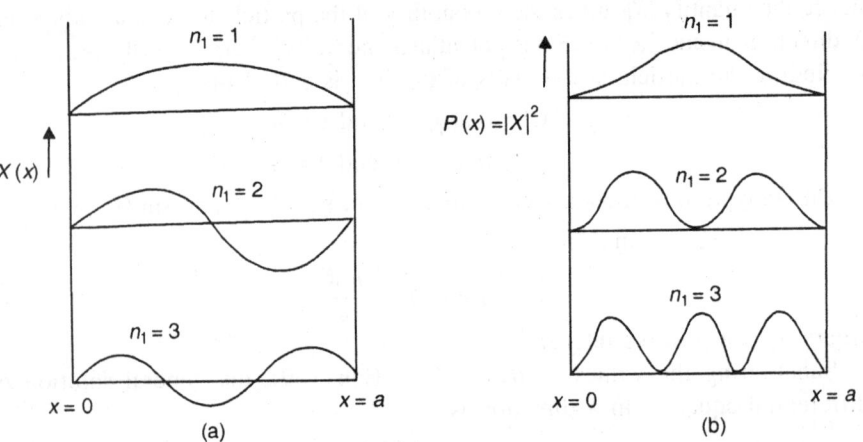

Fig. 2.5: (a) X component of wavefunction and (b) the probability density

Similarly, the normalized wave functions in Y and Z directions will be

$$Y(y) = \sqrt{\frac{2}{b}}\sin\left(\frac{n_2\pi y}{b}\right) \text{ and } Z(z) = \sqrt{\frac{2}{c}}\sin\left(\frac{n_3\pi z}{c}\right)$$

Hence, the eigen functions for the free paricle insider the box, is

$$\psi(x,y,z) = 2\sqrt{\frac{2}{abc}}\sin\left(\frac{n_1\pi x}{a}\right)\sin\left(\frac{n_2\pi y}{b}\right)\sin\left(\frac{n_3\pi z}{c}\right) \qquad (74)$$

Energy Eigenvalues: Differentiating Eqn. (72) twice w.r.t. x, we get

$$\frac{\partial^2 X(x)}{\partial x^2} = -\left(\frac{n_1 \pi}{a}\right)^2 X(x)$$

From Eqn. (68i)

$$\frac{\partial^2 X(x)}{\partial x^2} + \frac{2m}{\hbar^2} E_x X = 0$$

or

$$\frac{2m}{\hbar^2} E_x X = \left(\frac{n_1 \pi}{a}\right)^2 X$$

or

$$E_x = \left(\frac{n_1 \pi \hbar}{a}\right)^2 \frac{1}{2m} = \frac{n_1^2 h^2}{8ma^2} \quad \left(\because h = \frac{h}{2\pi}\right) \tag{75i}$$

Similarly

$$E_y = \frac{n_2^2 h^2}{8mb^2} \; ; \; E_z = \frac{n_3^2 h^2}{8mc^2} \tag{75ii}$$

Total energy

$$E = E_x + E_y + E_z = \frac{h^2}{8m}\left(\frac{n_1^2}{a^2} + \frac{n_2^2}{b^2} + \frac{n_3^2}{c^2}\right) \tag{76}$$

From Eqn. (75i); For

$$n_1 = 0, \; E_x = 0$$

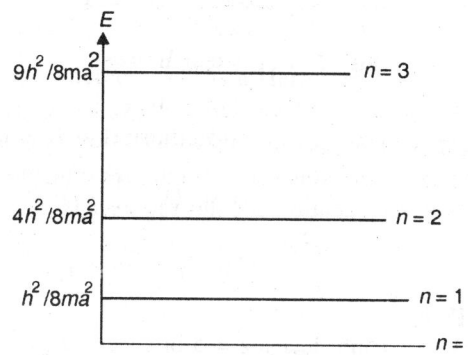

Fig. 2.6: Energy levels for particle enclosed in a one dimension box

$$n_1 = 1, \; E_x = \frac{h^2}{8ma^2}$$

$$n_1 = 2, \; E_x = 4\frac{h^2}{8ma^2}$$

$$n_1 = 3, \; E_x = 9\frac{h^2}{8ma^2}$$

Accidental Degeneracy

For a cube

$$E_{n_x,n_y,n_z} = \frac{h^2}{8mL^2}\left(n_z^2 + n_y^2 + n_z^2\right)$$

$$E_{122} = \frac{9h^2}{8mL^2} = E_{003}$$

Thus, two apparently quite different quantum states have the same energy. This is referred to as *accidental degeneracy*.

2.9 A Potential Step Barrier

It is represented by the function

$$V(x) = \begin{cases} 0 & \text{for } x < 0 \\ V_0 & \text{for } x > 0 \end{cases} \tag{77}$$

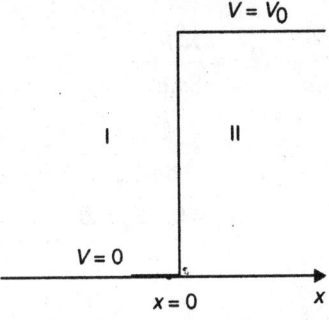

Fig. 2.7: A potential step

Let particles of energy E move from left to right along the positive direction of x-axis. The situation is analogous to a light wave moving from air to glass plate where it suffers a partial reflection and a fractional transmission.

The Schrödinger's wave equation for the region I is

$$\frac{d^2\psi}{dx^2} + \frac{2m}{\hbar^2} E\psi = 0 \tag{78}$$

and for the region II; it is

$$\frac{d^2\psi}{dx^2} + \frac{2m}{\hbar^2}(E - V_0)\psi = 0 \tag{79}$$

We can rewrite (i) and (ii) as

$$\frac{d^2\psi}{dx^2} + k_1^2\psi = 0 \; ; \; k_1 = \sqrt{\frac{2m}{\hbar^2}} \;\; (x < 0) \tag{80}$$

and

$$\frac{d^2\psi}{dx^2} + k_2^2\psi = 0 \; ; \; k_2 = \sqrt{\frac{2m(E - V_0)}{\hbar^2}} \;\; (x > 0) \tag{81}$$

General solutions of these equations are

$$\psi_1 = Ae^{ik_1x} + Be^{-ik_1x} \; ; \; (x < 0) \tag{82}$$

$$\psi_2 = Ce^{ik_2x} + De^{-ik_2x} \; ; \; (x > 0) \tag{83}$$

ψ_1 and ψ_2 are wave functions for regions I and II respectively, A, B, C and D are constants to be determined by the boudary conditions. In Eqn. (82), the term e^{ik_1x} represents a wave moving along positive x-direction (i.e. incident wave) and e^{-ik_1x} represents a wave moving in the negative x-direction (i.e.

reflected wave). In Eqn. (83), the term e^{ik_2x} represents a wave moving in the positive x-direction i.e. transmitted wave and e^{-ik_2x} represents a wave moving in the negative x-direction i.e. reflected wave. Since there is no barrier to the right of plane $x = 0$, there can be no reflection towards left, so the term e^{-ik_2x} should be zero and

$$y_2 = C e^{ik_2x} \tag{84}$$

Applying the boundary conditions to Eqns (82) and (84) (that ψ an $d\psi/dx$ be continuous at $x = 0$), we get

$$A + B = C \tag{85}$$

and $$ik_2C = ik_1 (A - B) \tag{86}$$

Solving these two equations, we get

$$C = \left(\frac{2k_1}{k_1 + k_2}\right) A \tag{87}$$

and $$B = \left(\frac{k_1 - k_2}{k_1 + k_2}\right) A \tag{88}$$

Here it is to be noted that B and C represent the amplitudes of reflected and transmitted beams respectively in terms of A (the amplitude of incident wave).

Now, there arise two cases:

Case I: $E > V_0$ i.e., $k_2 = \sqrt{\dfrac{2m(E - V_0)}{\hbar^2}}$ is real. we would now derive the probability current densities in the regions I and II.

In region I, $$\psi_1 = Ae^{ik_1x} + Be^{-ik_1x} \tag{89}$$

and $$\psi_1^* = A^* e^{-ik_1x} + B^* e^{ik_1x} \tag{90}$$

So, $$\frac{\partial\psi_1}{\partial x} = ik_1 \left[Ae^{ik_1x} - Be^{-ik_1x} \right] \tag{91}$$

and $$\frac{\partial\psi_1^*}{\partial x} = -ik_1 \left[A^* e^{-ik_1x} - B^* e^{ik_1x} \right] \tag{92}$$

The probability current

$$\bar{J} = \frac{\hbar}{2im}[\psi * \nabla\psi - \psi\nabla\psi *]$$

In the present case (particles moving along x-axis), for region I:

$$(J_x)_I = \frac{\hbar}{2im}\left[\psi_1 * \frac{\partial\psi_1}{\partial x} - \psi_1 \frac{\partial\psi_1 *}{\partial x} \right]$$

Using Eqns (89), (90), (91) and (92), we get

$$(J_x)_I = \frac{\hbar}{2im}\left[\left\{ \left(A^* e^{-ik_1x} + B^* e^{ik_1x}\right) \times ik_1 \left(Ae^{ik_1x} - Be^{-ik_1x}\right)\right\} \right.$$
$$\left. - \left\{\left(Ae^{ik_1x} + Be^{-ik_1x}\right)\times(-ik_1)\times\left(A^* e^{-ik_1x} - B^* e^{ik_1x}\right)\right\}\right]$$
$$= \frac{\hbar k_1}{m}\left(AA^* - B^*B\right) = \frac{\hbar k_1}{m}\left[|A|^2 - |B|^2\right] \tag{93}$$

In this expression, the first term represents the incident wave while the second term represents the reflected wave.

In region II:
$$\psi_2 = Ce^{ik_2x}$$

and
$$\psi_2^* = C^* e^{-ik_2x}$$

So,
$$\frac{\partial \psi_2}{\partial x} = ik_2 Ce^{ik_2x}$$

and
$$\frac{\partial \psi_2^*}{\partial x} = -ik_2 C^* e^{-ik_2x} \tag{94}$$

The probability current for region II is

$$(J_x)_{II} = \frac{\hbar}{2im}\left[\psi_2^* \frac{\partial \psi_2}{\partial x} - \psi_2 \frac{\partial \psi_2^*}{\partial x}\right]$$

$$\frac{\hbar k_2}{2m}\left[C^*C + CC^*\right] = \frac{\hbar k_2}{m}|C|^2 \tag{95}$$

In this region there is only transmitted wave.

Now, we can obtain the coefficient of reflection (reflectance, R) and coefficient of transmission (transmittance, T) as follows:

$$R = \frac{\text{magnitude of reflected current}}{\text{magnitude of incident current}}$$

$$= \frac{|B|^2 \, \hbar k_1/m}{|A|^2 \, \hbar k_1/m} = \frac{(k_1 - k_2)^2}{(k_1 + k_2)^2} \quad \text{(using Eqn. (88))} \tag{96}$$

$$T = \frac{\text{magnitude of transmitted current}}{\text{magnitude of incident current}}$$

$$= \frac{|C|^2 \, \hbar k_2/m}{|A|^2 \, \hbar k_1/m} = \frac{4k_1 k_2}{(k_1 + k_2)^2} \quad \text{(using Eqn. (87))} \tag{97}$$

From Eqns (96) and (97)

$$R + T = \frac{(k_1 - k_2)^2}{(k_1 + k_2)^2} + \frac{4k_1 k_2}{(k_1 + k_2)^2} = 1$$

It is to be noted here that $R \to 0$ as $k_2 \to k_1$ and $R \to 1$ as $k_2 \to 0$. Now k_2 will approach k_1 if $V_0 = 0$. So, there must be some reflectance even if $E \gg V_0$. This property of sudden reflection at the potential step arises from the wave nature and does not arise in classical mechanics if $E > V_0$. (It is purely a quantum mechanical effect.)

Case II: $E < V_0$ i.e. $k_2 = \sqrt{\dfrac{2m(E - V_0)}{\hbar^2}} = i\sqrt{\dfrac{2m(V_0 - E)}{\hbar^2}}$

(i.e. k_2 is imaginary)

Now, we have $\quad k_2^* = -i\sqrt{\dfrac{2m(V_0 - E)}{\hbar^2}} = -k_2$

The probability current associated with wave function ψ_2 is evaluated as follows:

$$\psi_2 = Ce^{ik_2x}, \quad \psi_2^* = C^* e^{-ik_2x}$$

$$\frac{\partial \psi_2}{\partial x} = ik_2 C e^{ik_2 x}$$

$$\frac{\partial \psi_2^*}{\partial x} = -ik_2^* C^* e^{-ik_2 x}$$

$$\therefore \quad (J_x)_{II} = \frac{\hbar}{2im}\left[\psi_2 {}^* \frac{\partial \psi_2}{\partial x} - \psi_2 \frac{\partial \psi_2^*}{\partial x} \right]$$

$$= \frac{\hbar}{2im}\left[C^* e^{-ik_2^* x} (ik_2) C e^{ik_2 x} + C e^{ik_2 x} (-ik_2) C^* e^{-ik_2^* x} \right]$$

Putting $k_2^* = -k_2$, we get

$$(J_x)_{II} = \frac{\hbar}{2im}\left[C^* e^{ik_2 x} (ik_2) C e^{ik_2 x} - CC^* (ik_2) e^{ik_2 x} e^{ik_2 x} \right]$$

i.e. the transmitted current is zero, or $T = 0$

$$\therefore \qquad\qquad R = 1 - T = 1$$

Example 2.6: A beam of electrons impinges on an energy step barrier of height 0.035eV. Calculate the fraction of electrons reflected and transmitted at the barrier when the energy of the electrons is
(i) 0.045 eV (ii) 0.020 eV.

Solution: *(i)* $E = 0.045$ eV $= 0.045 \times 1.6 \times 10^{-12}$ ergs $= 7.2 \times 10^{-14}$ ergs

$V_0 = 0.035$ eV $= 0.035 \times 1.6 \times 10^{-12}$ ergs $= 5.6 \times 10^{-14}$ ergs

$$p_1 = \sqrt{2mE} = \sqrt{2 \times 9 \times 10^{-28} \times 7.2 \times 10^{-14}} \text{ c.g.s. units}$$

$$(\because m = 9 \times 10^{-28} \text{ gm})$$

$$= 1.1384 \times 10^{-20} \text{ g cm/s}$$

$$p_2 = \sqrt{2m(E - V_0)} = \sqrt{2 \times 9 \times 10^{-28}(7.2 \times 10^{-14} - 5.6 \times 10^{-14})}$$

$$= \sqrt{2 \times 9 \times 10^{-28} \times 1.6 \times 10^{-14}}$$

$$= 5.3665 \times 10^{-21} \text{ g cm /s}$$

Reflectance $R = \dfrac{\left(k_1 - k_2\right)^2}{\left(k_1 + k_2\right)^2} = \dfrac{\left(\hbar k_1 - \hbar k_2\right)^2}{\left(\hbar k_1 + \hbar k_2\right)^2} = \dfrac{\left(p_1 - p_2\right)^2}{\left(p_1 + p_2\right)^2}$

or $\qquad R = \dfrac{\left[1.1384 \times 10^{-20} - 5.3665 \times 10^{-21}\right]^2}{\left[1.1384 \times 10^{-20} + 5.3665 \times 10^{-21}\right]^2}$

$$= \frac{3.6210 \times 10^{-41}}{2.8057 \times 10^{-40}} = 0.129$$

Transmittance $T = \dfrac{4k_1 k_2}{\left(k_1 + k_2\right)^2} = \dfrac{4(\hbar k_1)(\hbar k_2)}{\left(\hbar k_1 + \hbar k_2\right)^2} = \dfrac{4 p_1 p_2}{\left(p_1 + p_2\right)^2}$

or
$$T = \frac{4 \times 1.1384 \times 10^{-20} \times 5.3665 \times 10^{-21}}{\left(1.1384 \times 10^{-20} + 5.3665 \times 10^{-21}\right)^2}$$

$$= \frac{2.444 \times 10^{-40}}{2.806 \times 10^{-40}} = 0.871$$

(ii) In this case $\quad E < V_0. \therefore T = 0$

and $\qquad\qquad\qquad\qquad R = 1 - T = 1 - 0 = 1$

2.10 Barrier Penetration

Consider that a beam of particles (each of energy E) comes across a barrier of potential P_0 and width w. Classical physics predicts that if $E > P_0$, each and every particle will be able to cross over the barrier and transmitted to region III, but if $E < P_0$, none of the particles would be able to do so. However, according to Quantum Mechanics, even if $E < P_0$, there is a small but finite probability that the particle is able to penetrate the barrier. This is known as the *tunnel effect*. In one dimension, the Schrödinger equation is

$$\frac{d^2\psi}{dx^2} + \frac{2m}{\hbar^2}(E - V)\psi = 0 \tag{98}$$

In regions I and III, $V = 0$, so for regions I & III

$$\frac{d^2\psi}{dx^2} + \frac{2m}{\hbar^2}\, E\psi = 0 \tag{99}$$

For region II, $V = P_0$, so

$$\frac{d^2\psi}{dx^2} + \frac{2m}{\hbar^2}(E - P_0)\psi = 0 \; .$$

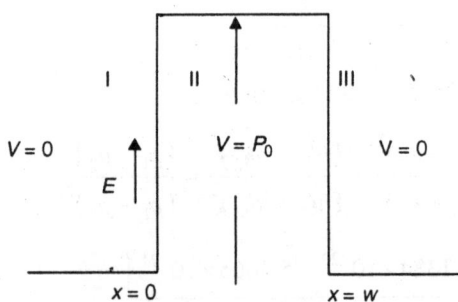

Fig. 2.8: A potential barrier

Since $P_0 > E$, therefore

$$\frac{d^2\psi}{dx^2} - \frac{2m}{\hbar^2}(P_0 - E)\psi = 0 \tag{100}$$

The general nature of solutions of Schrödinger equation in various regions would be

Region I: $\qquad \psi_1 = Ae^{ik_1x} + Be^{-ik_1x}$

Region II: $\qquad \psi_2 = Ce^{\alpha x} + De^{-\alpha x}$

Region III: $\qquad \psi_3 = Ge^{ik_1x} + He^{-ik_1x}$

where $\qquad k_1^2 = 2mE/\hbar^2 \text{ and } \alpha = \sqrt{\dfrac{2m}{\hbar^2}(P_0 - E)}.$

A and B are the amplitudes of the incident and reflected waves respectively in region I. Similarly C and D are the amplitudes in region II and G and H in region III. Since there is no reflection $\therefore H = 0$. The transmission coefficient T is given by $|G^2/A^2|$. Applying the condition that ψ and its derivative are continuous at $x = 0$ and $x = w$, we get

At $\qquad \psi_1|_{x=0} = \psi_2|_{x=0} \Rightarrow A + B = C + D$ $\hfill (101)$

$$\frac{d\psi_1}{dx}\Big|_{x=0} = \frac{d\psi_2}{dx}\Big|_{x=0} \Rightarrow ik_1(A - B) = \alpha(C - D) \hfill (102)$$

At $x = w$: $\qquad\qquad\qquad \psi_2|_{x=w} \quad \psi_3|_{x=w}$

$$\Rightarrow Ce^{\alpha w} + De^{-\alpha w} = Ge^{ik_1w} \hfill (103)$$

$$\frac{d\psi_2}{dx}\Big|_{x=w} = \frac{d\psi_3}{dx}\Big|_{x=w} \Rightarrow \alpha(Ce^{\alpha w} - De^{-\alpha w}) = ik_1Ge^{ik_1w} \hfill (104)$$

Adding Eqns (101) and (102), we get

$$2A = \frac{\alpha}{ik_1}(C - D) + (C + D) \hfill (105)$$

Adding Eqns (103) and (104), we get

$$2C\,e^{\alpha w} = \frac{ik_1}{\alpha}\,Ge^{ik_1w} + Ge^{ik_1w}$$

or $\qquad C = \dfrac{G}{2}e^{ik_1w}\left(\dfrac{ik_1}{\alpha} + 1\right)e^{-\alpha w}$ $\hfill (106)$

Subtracting Eqn (104) from Eqn (103), we get

$$D = \frac{G}{2}e^{ik_1w}\left(1 - \frac{ik_1}{\alpha}\right)e^{\alpha w} \hfill (107)$$

So, $\qquad C - D = \dfrac{G}{2}e^{ik_1w}\left[\dfrac{ik_1}{\alpha}e^{-\alpha w} + e^{-\alpha w} - e^{\alpha w} + \dfrac{ik_1}{\alpha}e^{\alpha w}\right]$ $\hfill (108)$

or $\qquad \dfrac{\alpha}{ik_1}(C - D) = \dfrac{G}{2}e^{ik_1w}\left[e^{-\alpha w} + \left(\dfrac{\alpha}{ik_1}e^{-\alpha w} - \dfrac{\alpha}{ik_1}e^{\alpha w}\right) + e^{\alpha w}\right]$

and $\qquad C + D = \dfrac{G}{2}e^{ik_1w}\left[\dfrac{ik_1}{\alpha}e^{-\alpha w} + e^{-\alpha w} + e^{\alpha w} - \dfrac{ik_1}{\alpha}e^{\alpha w}\right]$ $\hfill (109)$

Substituting Eqn. (108) and (109) in Eqn. (105), we get

$$A = \frac{G}{4}e^{ik_1w}\left[\left(\frac{\alpha}{ik_1} + \frac{ik_1}{\alpha}\right)(e^{-\alpha w} - e^{\alpha w}) + 2(e^{\alpha w} + e^{-\alpha w})\right]$$

$$= Ge^{ik_1 w}\left[\frac{1}{2}\left(\frac{\alpha}{ik_1}+\frac{ik_1}{\alpha}\right)\frac{(e^{-\alpha w}-e^{\alpha w})}{2}+\frac{(e^{\alpha w}+e^{-\alpha w})}{2}\right]$$

or

$$A = Ge^{ik_1 w}\left[-\frac{i}{2}\left(\frac{\alpha}{k_1}-\frac{k_1}{\alpha}\right)\sin h(\alpha w)+\cos h(\alpha w)\right]$$

or

$$A = Ge^{ik_1 w}\left[\cos h(\alpha w)+\frac{i}{2}\left(\frac{k_1}{\alpha}-\frac{\alpha}{k_1}\right)\sin h(\alpha w)\right] \quad (110)$$

$$A^* = G^* e^{-ik_1 w}\left[\cos h(\alpha w)-\frac{i}{2}\left(\frac{k_1}{\alpha}-\frac{\alpha}{k_1}\right)\sin h(\alpha w)\right] \quad (111)$$

$$\therefore \quad |AA^*| = |GG^*|\left[\cos h^2(\alpha w)-\frac{i^2}{2^2}\left(\frac{k_1}{\alpha}-\frac{\alpha}{k_1}\right)^2\sin h^2(\alpha w)\right]$$

or

$$|A^2| = |G^2|\left[1+\sin h^2(\alpha w)+\frac{1}{4}\left(\frac{k_1}{\alpha}-\frac{\alpha}{k_1}\right)^2\sin h^2(\alpha w)\right]$$

$$= |G^2|\left\{1+\left[1+\frac{1}{4}\left(\frac{k_1}{\alpha}-\frac{\alpha}{k_1}\right)^2\right]\sin h^2(\alpha w)\right\}$$

$$\left(\because \cos h^2\theta = 1+\sin h^2\theta\right).$$

$$= |G^2|\left\{1+\frac{1}{4}\left(\frac{k_1}{\alpha}+\frac{\alpha}{k_1}\right)^2\sin h^2(\alpha w)\right\} \quad (112)$$

Substituting

$$k_1 = \sqrt{\frac{2mE}{\hbar^2}}, \quad \alpha = \sqrt{\frac{2m(P_0-E)}{\hbar^2}}$$

$$\therefore \quad \frac{k_1}{\alpha} = \sqrt{\frac{2mE}{\hbar^2}\frac{\hbar^2}{2m(P_0-E)}} = \sqrt{\frac{E}{P_0-E}}$$

$$\frac{k_1}{\alpha}+\frac{\alpha}{k_1} = \sqrt{\frac{E}{P_0-E}}+\sqrt{\frac{P_0-E}{E}} = \frac{E+P_0-E}{\sqrt{E(P_0-E)}}$$

$$\therefore \quad \frac{1}{T} = \frac{|A^2|}{|G^2|} = 1+\frac{P_0^2}{4E(P_0-E)}\sin h^2\left[\sqrt{\frac{2m}{\hbar^2}(P_0-E)}w\right] \quad (113)$$

We can draw following conclusions from this expression:

(*i*) Since the hyperbolic function increases very rapidly with respect to its argument, therefore the transmission coefficient decreases very rapidly with the increase in size P_0 of the barrier.

(ii) If $E > P_0$, the transmission coefficient oscillates between a steadily increasing lower envelop and unity whenever $\alpha w = \pi, 2\pi, 3\pi, ...$

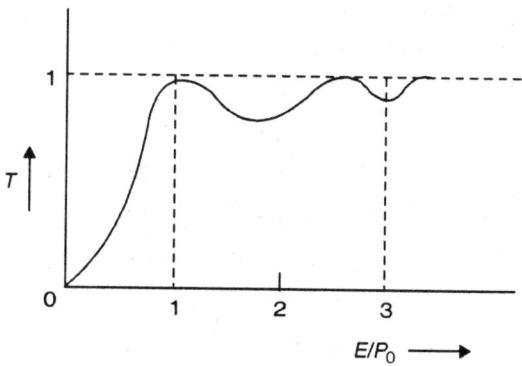

Fig. 2.9

(iii) If the energy $E \to P_0$, then $\alpha \to 0$

and
$$\sinh \alpha w \to \alpha w$$

so
$$T \cong \left[1 + \frac{P_0^2}{4E(P_0 - E)} \times \frac{2m(P_0 - E)w^2}{\hbar^2}\right]^{-1}$$

$$= \frac{1}{1 + \frac{2mP_0^2 w^2}{4E\hbar^2}}$$

For $E = P_0$, it becomes
$$T = \left[1 + \frac{mP_0 w^2}{2\hbar^2}\right]^{-1}$$

(iv) If the width of the potential barrier w is very large, then
$$\alpha w \gg 1 \therefore e^{\alpha w} \gg e^{-\alpha w}$$

$$\sin h^2(\alpha w) = \left(\frac{e^{\alpha w} - e^{-\alpha w}}{2}\right)^2 \cong \frac{1}{4}e^{2\alpha w}$$

From Eqn (117), Transmittance
$$T = \frac{1}{1 + \frac{P_0^2}{4E(P_0 - E)} \times \frac{1}{4}e^{2\alpha w}}$$

In the denominator of this expression, the second term is very large as compared to 1, so 1 can be neglected. Therefore,
$$T = \frac{16E(P_0 - E)}{P_0^2} e^{-2\sqrt{2m(P_0-E)}\ w/\hbar} \tag{114}$$

Obviously, even if the width w of the potential barrier is very large, the transmittance is not zero, but decreases exponentially with increase in width w of the barrier.

Example 2.7: Calculate the probability of transmission of 1.0 eV alpha particle from a rectangle potential barrier of height 2.0 eV and width 1.0 Å. (Mass of α particle = 6.4×10^{-27} Kg).

Solution: Transmission probability

$$T = \frac{16E(P_0 - E)}{P_0^2} e^{-2\sqrt{2m(P_0 - E)} \, w/\hbar}$$

Here,
$$P_0 = 2\text{eV} = 2 \times 1.6 \times 10^{-19} = 3.2 \times 10^{-19} \text{ J}$$
$$E = 1.0 \text{ eV} = 1.6 \times 10^{-19} \text{ J}$$
$$w = 1.0 \text{ Å} = 10^{-10} \text{ metre}$$
$$m = 6.4 \times 10^{-27} \text{ kg}$$
$$\hbar = 1.05 \times 10^{-34} \text{ J-s}$$

Now
$$2\sqrt{2m(P_0 - E)} \, w/\hbar$$

$$= \frac{2\sqrt{2 \times 6.4 \times 10^{-27} \times (3.2 - 1.6) \times 10^{-19}} \times 10^{-10}}{1.05 \times 10^{-34}} = 86.2$$

$$\therefore \quad T = \frac{16 \times 1.6 \times 10^{-19} \times (3.2 - 1.6) \times 10^{-19}}{\left(3.2 \times 10^{-19}\right)^2} e^{-86.2}$$

$$= \frac{16 \times 1.6 \times 1.6}{(1.6)^2 \times 4} e^{-86.2}$$

$$= 4e^{-86.2} \approx 1.46 \times 10^{-37}$$

Example 2.8: An alpha particle of energy 10 MeV is incident towards a potential barrier of height 30 MeV. If the transmission coefficient is 2×10^{-3}, find the width of the potential barrier (Mass of the α-particle = 6.68×10^{-27} Kg).

Solution:
$$T = \frac{16E}{P_0}\left(1 - \frac{E}{P_0}\right) e^{-2\alpha w}$$

where
$$\alpha = \sqrt{2m(P_0 - E)}/\hbar$$

$$e^{2\alpha\omega} = \frac{16E}{P_0 T}\left(1 - \frac{E}{P_0}\right)$$

or
$$e^{2\alpha w} = \log_e\left[\frac{16E}{P_0 T}\left(1 - \frac{E}{P_0}\right)\right]$$

$$= 2.3026 \log_{10}\left[\frac{16E}{P_0 T}\left(1 - \frac{E}{P_0}\right)\right]$$

$$w = \frac{1}{2\alpha} \times 2.3026 \, \log_{10}\left[\frac{16E}{P_0 T}\left(1 - \frac{E}{P_0}\right)\right]$$

Now
$$E = 10 \text{ MeV} = 10 \times 1.6 \times 10^{-13} \text{ J} = 1.6 \times 10^{-12} \text{ J}$$
$$P_0 = 30 \text{ MeV} = 30 \times 1.6 \times 10^{-13} \text{ J} = 4.8 \times 10^{-12} \text{ J}$$
$$P_0 - E = (4.8 - 1.6) \times 10^{-12} = 3.2 \times 10^{-12} \text{ J}$$

$$\frac{E}{P_0} = \frac{1}{3}; \text{ Transmission coefficient } T = 2 \times 10^{-3}$$

$$\alpha = \frac{\sqrt{2m(P_0 - E)}}{\hbar} = \frac{\sqrt{2 \times 6.68 \times 10^{-27} \times 3.2 \times 10^{-12}}}{1.05 \times 10^{-34}} = \frac{20.67 \times 10^{-20}}{1.05 \times 10^{-34}}$$

$\therefore$
$$w = \frac{1.05 \times 10^{-34}}{2 \times 20.67 \times 10^{-20}} \times 2.3026 \log_{10} \left[\frac{16\left(1 - \dfrac{1}{3}\right)}{2 \times 10^{-3} \times 3} \right]$$

$$= 2.54 \times 10^{-16} \times 2.3026 \ \log_{10} \left(\frac{16 \times 10^3}{2 \times 3} \times \frac{2}{3} \right)$$

$$= 1.9 \times 10^{-15} \text{ m}$$

2.11 A Particle in a One Dimensional Square Well Potential

Consider a particle having potential energy function in the shape of a "well" defined by

$V(x) = 0$	for $x < -a$	(Region I)
$V(x) = -V_0$	for $-a < x < a$	(Region II) (115)
$V(x) = 0$	for $x > a$	(Region III)

It is further assumed that $E < 0$ so that particle's kinetic energy is

$$E - V(x) \equiv V_0 - |E|$$

so that the particle classically cannot enter regions I and III and is bound by the potential between $-a < x < a$. Quantum mechanically, the stationary states are described by solutions of Schrödinger's equation

$$\left[-\frac{\hbar^2}{2m} \nabla^2 + V(x) \right] \psi(x) = E\psi(x) \tag{116}$$

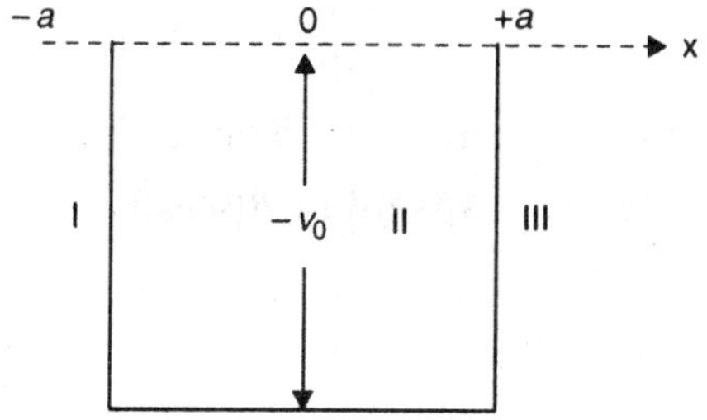

Fig. 2.10: The square well potential

This equation takes following two forms in the different regions

$$-\frac{\hbar^2}{2m}\frac{d^2\psi}{dx^2} = E\psi \quad \text{for } |x| > a \tag{117}$$

$$(\because V = 0)$$

and $$-\frac{\hbar^2}{2m}\frac{d^2\psi}{dx^2} - V_0\psi = E\psi \quad \text{for } |x| < a \tag{118}$$

Equations (117) and (118) can be trivially solved. Since $E < 0$, we will have bound state solutions. Let us put

$$\frac{2mE}{\hbar^2} = -\alpha^2 \tag{119}$$

and $$\frac{2m}{\hbar^2}(E + V_0) = \beta^2 \tag{120}$$

with $\alpha > 0$ and $\beta > 0$.

We can rewrite Eqns (117) and (118) as

$$\frac{\partial^2\psi}{\partial x^2} - \alpha^2\psi = 0; \quad |x| > a \tag{121}$$

$$\frac{\partial^2\psi}{\partial x^2} + \beta^2\psi = 0; \quad |x| < a \tag{122}$$

The general solution of the Eqn. (122) is

$$\psi^{II}(x) = A\cos\beta x + B\sin\beta x \tag{123}$$

where A and B are arbitrary constants, and ψ^{II} is the solution valid in region II (i.e. $|x| < a$). The equation which holds in region I (i.e. $-\infty < x < a$) is Eqn. (121). Its general solution is $Ce^{\alpha x} + De^{-\alpha x}$. However, as $x \to -\infty$, $De^{-\alpha x}$ diverges in region I so that the admissible solution in region I is

$$\psi^{I}(x) = Ce^{\alpha x} \tag{124}$$

Region III ($a < x < \infty$) is also governed by Eqn. (121), and admissible solution in this region is

$$\psi^{III}(x) = De^{-\alpha x} \tag{125}$$

Here C and D are arbitrary constants. Hence, the solution $\psi(x)$ has three different forms in the three regions. We now apply the boundary conditions that $\psi(x)$ and $d\psi/dx$ are continuous at the boundaries $x = -a$ and $x = a$. Therefore,

$$\psi^{I}\Big|_{x=-a} = \psi^{II}\Big|_{x=-a}$$

and $$\frac{d\psi^{I}}{dx}\Big|_{x=-a} = \frac{d\psi^{II}}{dx}\Big|_{x=-a}$$

or $$Ce^{-\alpha a} = A\cos\beta a - B\sin\beta a \tag{126}$$

and $$C\alpha e^{-\alpha a} = A\beta\sin\beta a + B\beta\cos\beta a \tag{127}$$

Similarly $$\psi^{II}\Big|_{x=a} = \psi^{III}\Big|_{x=a}$$

and $$\frac{d\psi^{II}}{dx}\Big|_{x=a} = \frac{d\psi^{III}}{dx}\Big|_{x=a}$$

$$\Rightarrow \qquad D\,e^{-\alpha a} = A\cos\beta a + B\sin\beta a \qquad (128)$$

and
$$-D\,\alpha\,e^{-\alpha a} = A\beta\sin\beta a + B\beta\cos\beta a \qquad (129)$$

From Eqns (126) and (128), we get

$$2A\,\cos\beta a = (C + D)e^{-\alpha a} \qquad (130)$$

and
$$2B\sin\beta a = -(C - D)e^{-\alpha a} \qquad (131)$$

Similarly from Eqns (127) and (129)

$$2A\beta\sin\beta a = (C + D)\alpha e^{-\alpha a} \qquad (132)$$

$$2B\beta\cos\beta a = (C - D)\alpha e^{-\alpha a} \qquad (133)$$

Eqns (130) and (132) show that if $C + D \neq 0$, then $A \neq 0$ and also

$$\frac{2A\beta\sin\beta a}{2A\cos\beta a} = \frac{(C + D)\alpha e^{-\alpha a}}{(C + D)e^{-\alpha a}} \text{ or } \frac{\alpha}{\beta} = \tan(\beta a) \qquad (134)$$

Putting the value of α from Eqn. (134) in Eqn. (133), we get

$$2B\beta\cos\beta a = (C - D)\beta\,\tan\beta a \cdot e^{-\alpha a}$$

Multiplying both sides by $(\sin\beta a)$, we have

$$2B\beta\,\sin\beta a\cos\beta a = (C - D)\beta\frac{\sin^2\beta a}{\cos\beta a}e^{-\alpha a}$$

Using Eqn. (131) on l.h.s. of this, we get

$$-(C - D)e^{-\alpha a}\beta\,\cos^2\beta a = (C - D)\beta\,\sin^2\beta a\,e^{-\alpha a}$$

or
$$(C - D)(-\cos^2\beta a) = (C - D)\sin^2\beta a$$

Now, since $\sin^2\beta a$ cannot be equal to the negative quantity $-\cos^2\beta a$, hence this can only be true if either side is equal to zero or $C = D$. This implies $B = 0$ (from Eqn. (131)). Consequently, from Eqn. (130)

$$2A\,\cos\beta a = 2De^{-\alpha a} \text{ or } D = Ae^{\alpha a}\cos\beta a \qquad (135)$$

This is one type of solution of our problem. However, if $C \neq D$ (and $B \neq 0$), then from Eqns (131) and (133)

$$\frac{\alpha}{\beta} = -\cot(\beta a) \qquad (136)$$

Using the value of α from this Eqn. in Eqn. (132) we get

$$2A\beta\,\sin\beta a = -(C + D)\beta\cot\beta a \cdot e^{-\alpha a}$$

Multiplying both sides of this by $(\cos\beta a)$, we get

$$2A\beta\,\sin\beta a\cos\beta a = -(C + D)\beta\frac{\cos^2\beta a}{\sin\beta a}e^{-\alpha a}$$

Using Eqn. (130), this becomes

$$(C + D)e^{-\alpha a}\sin^2\beta a = -(C + D)\cos^2\beta a \cdot e^{-\alpha a}$$

$$\Rightarrow \qquad C + D = 0 \text{ or } C = -D$$

$\Rightarrow$ (using Eqn. 130) $A = 0$ and so from Eqn. (131), we get

$$2B\sin\beta a = 2D\,e^{-\alpha a}$$

or
$$D = Be^{\alpha a}\sin\beta a \qquad (137)$$

From Eqns (125), (135) and (137), we have

$$\psi^{III}(x) = (Ae^{\alpha a}\cos\beta a)e^{-\alpha x} \qquad (138i)$$

or
$$\psi^{III}(x) = (Be^{\alpha a}\sin\beta a)e^{-\alpha x} \qquad (138ii)$$

$$\psi''(x) = A\cos\beta x$$

or

$$\psi''(x) = B\sin\beta x \tag{139}$$

From Eqns (124), (135) and (137)

$$\psi'(x) = (Ae^{\alpha a}\cos\beta a)\, e^{\alpha x} \tag{140}$$

or

$$\psi'(x) = -(Be^{\alpha a}\sin\beta a)\, e^{\alpha x} \tag{141}$$

The Energy Eigenvalues

According to Eqns. (134) and (136), we see that there are two types of solutions. From Eqns. (119) and (120), we get

$$\left(\alpha^2 + \beta^2\right)\, a^2 = \frac{2mV_0}{\hbar^2}a^2 \equiv \frac{V_0}{\Delta} \tag{142}$$

where

$$\Delta = \frac{\hbar^2}{2ma^2} \tag{143}$$

The parameter Δ is interpreted as follows:

The half width a of the potential well indicates the uncertainity in the position of a particle confined to the well. Associated with this, there is an uncertainity of the order of $(\hbar/a)$ in the momentum.

The corresponding energy $\Delta \equiv (\hbar/a)^2/2m$ is a natural unit to measure the depth of the potential. Thus V_0/Δ is a dimensionless parameter and gives the strength of the potential.

Case I $\qquad\qquad\qquad \dfrac{\alpha}{\beta} = \tan\beta a$

Since α and β are defined as positive, therefore, $\tan\beta a$ must be positive, therefore only values of βa lying in the intervals

$$2m\frac{\pi}{2} \le \beta a \le (2m+1)\frac{\pi}{2} \tag{144}$$

are allowed. Further, substituting $\alpha = \beta\tan\beta a$ in Eqn. (142) leads to the condition

$$\frac{V_0}{\Delta} = \beta^2 a^2 \sec^2\beta a$$

or

$$\left(\frac{\Delta}{V_0}\right)^{1/2}\beta a = |\cos\beta a| \tag{145}$$

The modulus sign is used because l.h.s. is positive.

Case II $\qquad\qquad\qquad \dfrac{\alpha}{\beta} = -\cot\beta a$

This leads to $\qquad (2m-1)\dfrac{\pi}{2} \le \beta a \le 2m\dfrac{\pi}{2} \qquad (m = 1, 2, 3, \ldots) \tag{146}$

and

$$\frac{V_0}{\Delta} = \beta^2 a^2 \cos ec^2\beta a \text{ or } \left(\frac{\Delta}{V_0}\right)^{1/2}\beta a = |\sin\beta a| \tag{147}$$

Eqns (144) and (146) can be satisfied only by certain specific discrete values of β which can be found graphically. These specific values β_n are determined by

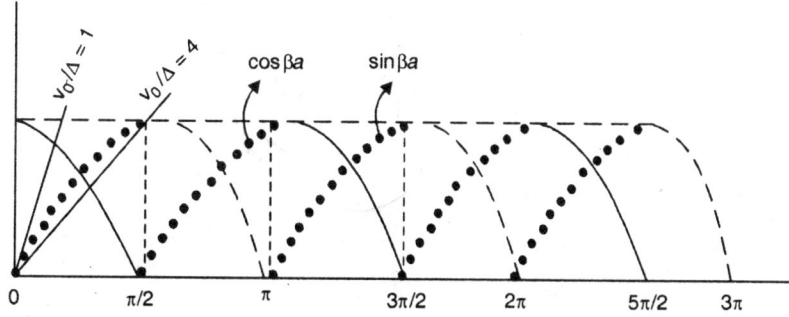

Fig. 2.11: Graphical solution for allowed values of β

the intersection of the straight line $(\Delta/V_0)^{\frac{1}{2}} \beta a$ with the curves for $|\cos \beta a|$ and $|\sin \beta a|$. The parts of $|\cos \beta a|$ and $|\sin \beta a|$ which lie within the respective allowed intervals conditions (144) and (146) are shown as solid lines and dashed lines respectively in Fig. 2.11.

The parts to be ignored are indicated by dotted lines. If the intersections occur at $\beta = \beta_n$ (n = 0, 1, 2, 3, ...), the corresponding allowed values of the energy are obtained from Eqn. (120) as

$$E_n = \left(\frac{\hbar^2}{2m}\right)\beta_n^2 - V_0 = \left[(\beta_n a)^2 \frac{\Delta}{V_0} - 1\right]V_0 \qquad (148)$$

The figure indicates that the number of these energy levels is finite. Inspection of Fig. (2.11) shows that if $(\Delta/V_0)^{\frac{1}{2}} \beta a$ reaches the value unity, for a value of βa in the interval $\frac{\pi}{2}N \leq \beta a < \frac{\pi}{2}(N+1)$, then there are $(N + 1)$ intersections or, in other words, there are $(N + 1)$ discrete energy levels if

$$\frac{\pi}{2}N(\Delta/V_0)^{1/2} \leq 1 < \frac{\pi}{2}(N+1)(\Delta/V_0)^{1/2} \qquad (149)$$

The Energy Eigen functions: The energy levels E_n with n = 0, 2, 4, ... correspond to solutions having the following forms in the three regions:

$$\psi_n^I(x) = (Ae^{\alpha_n a} \cos\beta_n a)e^{\alpha_n x} \qquad (x < -a)$$

$$\psi_n^{II}(x) = A\cos\beta_n x \qquad (-a < x < a) \qquad (150)$$

$$\psi_n^{III}(x) = (Ae^{\alpha_n a} \cos\beta_n a)e^{-\alpha_n x} \qquad (x > a)$$

These are illustrated in Fig. 2.12a.

The energy levels E_n (with n = 1, 3, 5,...) are characterised by eigenfunctions which are given by

$$\psi_n^I(x) = -(Be^{\alpha_n a} \sin\beta_n a)e^{\alpha_n x} \qquad (x < -a)$$

$$\psi_n^{II}(x) = B\sin\beta_n x \qquad (-a < x < a) \qquad (151)$$

$$\psi_n^{III}(x) = (Be^{\alpha_n a} \sin\beta_n a)e^{-\alpha_n x} \qquad (x > a)$$

These are illustrated in Fig. 2.12b.

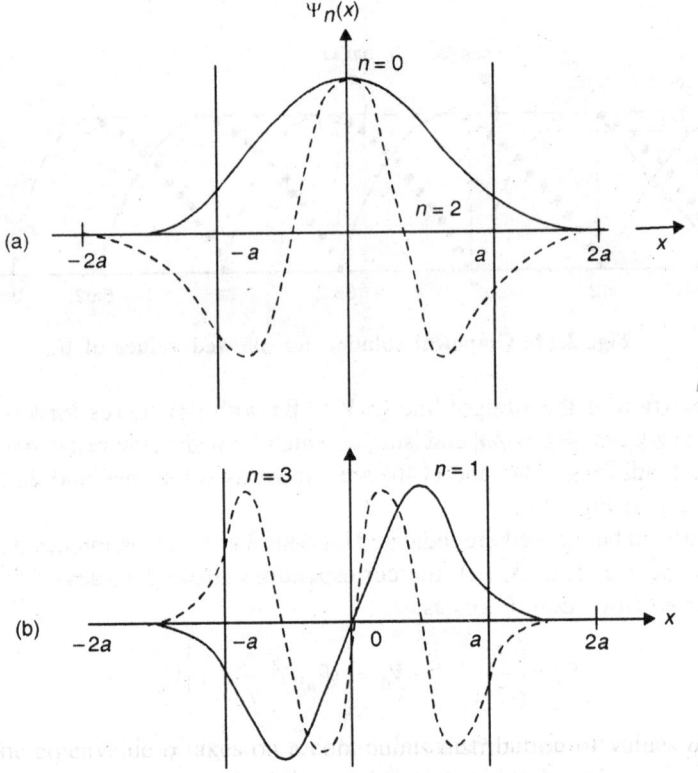

Fig. 2.12(a and b): The energy eigenfunctions of a particle
confined to potential well of Fig. 2.10.

Example 2.9: A particle of mass m moves in a one dimensional potential well of depth V and width $2a$. Prove that its energy levels are given by

$$E = \frac{\hbar^2}{2ma^2}\xi^2$$

where the values of ξ are given by the points of intersection of the circle $\xi^2 + \eta^2 = \frac{2mV_0 a^2}{\hbar^2}$ with the curves $\eta = \xi \tan \xi$ *and* $\eta = -\xi \cot \xi$. Give the parities of the different states.

Solution:

$$V(x) = 0, \text{ for } -a < x < a$$
$$= V_0, \text{ for } |x| > a$$

Increase in potential energy at the walls is abrupt but finite.

$$\frac{-\hbar^2}{2m}\frac{d^2 u}{dx^2} = Eu \qquad (\text{for } -a < x < a) \qquad\qquad (i)$$

$$\frac{-\hbar^2}{2m}\frac{d^2 u}{dx^2} + V_0 u = Eu \qquad (\text{for } |x| > a \text{ and } |x| < a) \qquad (ii)$$

Any particle with $E < V_0$ cannot come out of the potential well and is bound between $-a$ and $+a$.

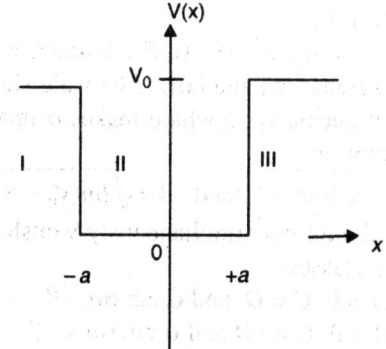

Fig. 2.13

From Eqns (*i*) and (*ii*)

$$\frac{d^2u}{dx^2} + \frac{2mE}{\hbar^2}u = 0 \qquad \text{(for } -a < x < a) \qquad (iii)$$

$$\frac{d^2u}{dx^2} - \frac{2m(V_0 - E)}{\hbar^2}u = 0 \qquad \text{(for } |x| > a \text{ and } |x| < a) \qquad (iv)$$

For bound states ($E < V_0$):

The general solutions of Eqns (*iii*) and (*iv*) are

$$u(x) = A\sin\alpha x + B\cos\alpha x \text{ for } -a < x < a \quad (\alpha = \sqrt{2mE/\hbar^2}) \qquad (v)$$

$$u(x) = Ce^{-\beta x} + De^{\beta x} \text{ for } |x| > a \text{ \& } |x| < a, \quad \left(\beta = \sqrt{2m(V_0 - E)/\hbar^2}\right) \qquad (vi)$$

The boundary conditions at $x = \pm a$ require that $D = 0$ if Eqn. (*vi*) is to represent the solution for $x > a$ and $C = 0$ if Eqn. (*vi*) is to represent the solution for $x < -a$. (because $u(x)$ must be finite)

Applying the condition that u must be continuous at $x = \pm a$, we get

for $\qquad x = a$: $\qquad A\sin\alpha a + B\cos\alpha a = Ce^{-\beta a}$

for $\qquad x = -a$: $\qquad -A\sin\alpha a + B\cos\alpha a = De^{-\beta a} \qquad (vii)$

Applying the condition that $\dfrac{du}{dx}$ be continuous at $x = \pm a$, we get

for $x = a$: $\quad \alpha A\cos\alpha a - \alpha B\sin\alpha a = -\beta Ce^{-\beta a}$

for $x = -a$: $\quad \alpha A\cos\alpha a + \alpha B\sin\alpha a = \beta De^{-\beta a} \qquad (viii)$

From Eqns (*vii*),

$$2A\sin\alpha a = (C - D)e^{-\beta a} \qquad (ix\ a)$$

$$2B\cos\alpha a = (C + D)e^{-\beta a} \qquad (x\ a)$$

and from Eqn (*viii*),

$$2\alpha A\cos\alpha a = -\beta(C - D)e^{-\beta a} \qquad (ix\ b)$$

$$2\alpha B\sin\alpha a = -\beta(C + D)e^{-\beta a} \qquad (x\ b)$$

From Eqns (*ix a*) and (*ix b*)

$$\alpha\cot\alpha a = -\beta \text{ (if } A \neq 0 \text{ and } C \neq D) \qquad (xi)$$

and from Eqns (*x a*) and (*xb*)

$$\alpha \tan \alpha a = \beta \quad (\text{if } B \neq 0 \text{ and } C \neq -D) \tag{xii}$$

Now it is possible for Eqns. (*xi*) and (*xii*) to be valid simultaneously. Because, if we eliminate β we get $\tan^2 \alpha a = -1$ which makes α imaginary and β negative which is contrary to equations

$$\alpha = \sqrt{2mE/\hbar^2} \quad \text{and} \quad \beta = \sqrt{2m(V_0 - E)/\hbar^2}$$

Also, A, B, C and D should not simultaneously vanish, therefore the solution can be divided into two classes:

$$A = 0, C = D \text{ and } \alpha \tan \alpha a = \beta \tag{xiii a}$$
$$B = 0, C = -D \text{ and } \alpha \cot \alpha a = -\beta \tag{xiii b}$$

Energy levels

Let

$$\xi = \alpha a \text{ and } \eta = \beta a \tag{xiv}$$

$$\xi^2 = \alpha^2 a^2 \quad \text{or} \quad \xi^2 = \frac{2mE}{\hbar^2} a^2$$

$$\therefore \qquad E = \frac{\hbar^2}{2ma^2} \xi^2 \tag{xv}$$

From (*xii*) and (*xiv*)

$$\xi \tan \xi = \eta \tag{xvi}$$

and

$$\xi^2 + \eta^2 = (\alpha^2 + \beta^2) a^2$$

$$= \left(\frac{2mE}{\hbar^2} + \frac{2m(V_0 - E)}{\hbar^2} \right) a^2$$

or

$$\xi^2 + \eta^2 = \frac{2mV_0 a^2}{\hbar^2} \tag{xvii}$$

Since ξ and η are restricted to positive values, the values of ξ for Eqn. (*xv*) are found from intersections in the first quadrant of the curve $\xi \tan \xi$ (plotted against ξ) with the circles of known radius $\sqrt{2m(V_0 a^2)/\hbar^2}$. From Eqn. (xiiib)

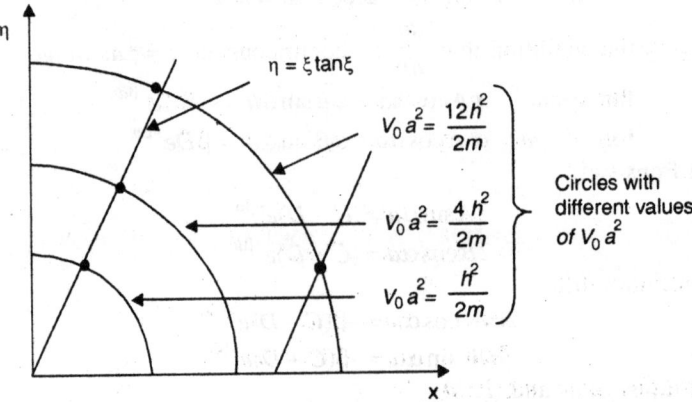

Fig. 2.14

$$\xi \cot \xi = -\eta \quad \text{or} \quad \eta = -\xi \cot \xi$$

$$\therefore \qquad \xi^2 + \eta^2 = \frac{2mV_0 a^2}{\hbar^2} \qquad \qquad (xviii)$$

In this case, the values of ξ are found from the intersections in the first quadrant of the $-\xi \cot \xi$ (plotted against ξ) with the same circles.

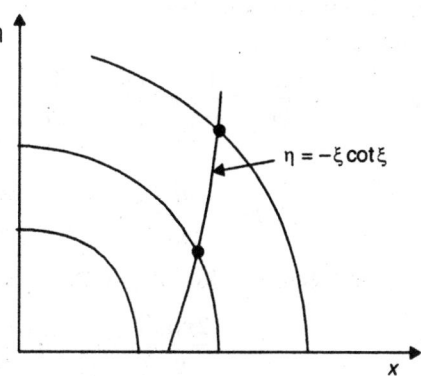

Fig. 2.15

Parity: The eigenfunctions of the first class are even with respect to change in sign of x [$u(-x) = u(x)$] while the eigenfunctions of the second class are odd [$u(-x) = -u(x)$]. $V(x)$ is symmetric about $x = 0$.

$$-\frac{\hbar^2}{2m} \frac{d^2 u(x)}{dx^2} + V(x) u(x) = E u(x) \qquad \qquad (xix)$$

If we change $x \to -x$, (if $V(-x) = V(x)$) then

$$-\frac{\hbar^2}{2m} \frac{d^2 u(-x)}{dx^2} + V(x) u(-x) = E u(-x) \qquad \qquad (xx)$$

Therefore $u(x)$ and $u(-x)$ are solutions of the same wave equation with the same eigenvlaue E. If there is no other linearly independent eigenfunction corresponding to this energy level, these two solutions differ only by a multiplicative constant, i.e.

$$u(-x) = \epsilon\, u(x)$$
$$\Rightarrow \qquad u(x) = \epsilon\, u(-x)$$
$$\therefore \qquad \epsilon^2 = 1 \text{ or } \epsilon = \pm 1.$$

Thus all such eigenfunctions of a symmetric potential are either even or odd with respect to change of sign of x. Such wavefunctions are said to have even or odd parity.

2.12 Rectangular Barrier and Spherically Symmetric Probability Currents

Consider a three dimensional system having spherical symmetry. In such a case, there is an equivalent one dimensional Schrödinger equation

$$-\frac{\hbar^2}{2m} \frac{d^2 u}{dr^2} + V(r) u = E u \qquad \qquad (152)$$

where $\qquad\qquad u\,(r) = r\,\psi\,(r).$

Now, the radial probability current density is given by

$$j(r) = -\frac{i\hbar}{2m}\left(\psi*\frac{\partial\psi}{\partial r} - \psi\frac{\partial\psi^*}{\partial r}\right)$$

In this three dimensional case, the total probability current J through the surface of a sphere of radius r is equal to $4\pi r^2\,j(r)$.

Since $\psi = u/r$, we have

$$\psi*\frac{\partial\psi}{\partial r} - \psi\frac{\partial\psi^*}{\partial r} = \frac{1}{r^2}\left(u*\frac{\partial u}{\partial r} - u\frac{\partial u^*}{\partial r}\right)$$

Hence

$$j(r) = -\frac{i\hbar}{2mr^2}\left(u*\frac{\partial u}{\partial r} - u\frac{\partial u^*}{\partial r}\right)$$

The net current flow $J(r)$, equal to $4\pi r^2\,j(r)$ is thus given by

$$J = -\frac{2\pi i\hbar}{m}\left(u*\frac{\partial u}{\partial r} - u\frac{\partial u^*}{\partial r}\right) \qquad (153)$$

Consider a spherical volume bounded by a surface of a given radius r_0. The integral of the probability density $\psi*\psi$ within the volume is given by

$$P = \int_0^{r_0}(\psi*\psi)4\pi r^2\,dr = 4\pi\int_0^{r_0}(u*u)\,dr$$

$$\frac{\partial P}{\partial t} = -J(r_0) + J(0) \qquad (154)$$

That is

$$4\pi\frac{\partial}{\partial t}\int_0^{r_0}(u*u)\,dr = 4\pi\frac{i\hbar}{2m}\left(u*\frac{\partial u}{\partial r} - u\frac{\partial u^*}{\partial r}\right)\Bigg|_{0}^{r_0}$$

$$= -J\,(r_0) + J\,(0) \qquad (155)$$

In this case there is a very important point that does not arise in one dimensional discussions. The current $J(0)$ at $r = 0$ must be zero unless there is a source or a sink of particles at the origin. In the absence of any such source or sink we have two possibilities:

1. A time-independent situation, in which the probability current $J\,(r_0)$ vanishes for all r_0.
2. $J(r_0) \neq 0$, in which case then the integrated probability between $r = 0$ and $r = r_0$ must change with time.

The second possibility can be used to describe alpha-particle emission from nuclei. The next article is devoted to this problem. But, for a first rough analysis, we shall use the simplified potential shown in Fig. 2.16. At any given instant, the amplitudes on the two sides of the barrier are related by the usual time independent methods. Thus:

$$u_I(r) = r\psi_I(r) = A_0 e^{ik_0 r} + A e^{-ik_0 r}$$
$$u_{III}(r) = r\psi_{III}(r) = D e^{ikr}$$

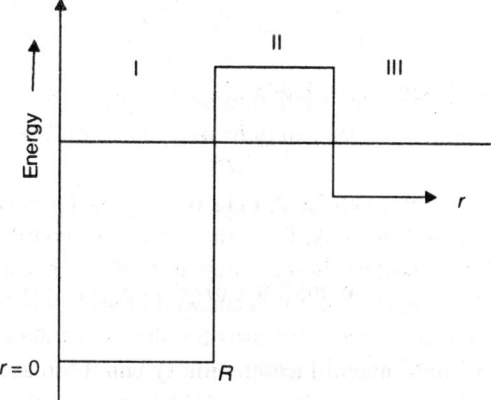

Fig. 2.16: Simple rectangular barrier model for analysis of
nuclear alpha particle emission

Since the transmission coefficient is extremely small for alpha-emitting nuclei,
we have $|A| \approx |A_0|$. The integrated probability P within the sphere of nuclear radius
R is then given by

$$P = 4\pi \int_0^R (u*u)\, dr = 4\pi\, [2|A_0|^2 R]$$

The total probability current outside the nuclear barrier is given by

$$J_{III} = -4\pi \frac{ih}{2m} |D|^2\, (2ik) = 4\pi v_{III}\, |D|^2$$

where $v_{III}\ (= \hbar k/m)$ is the velocity in the exterior region. By the conservation
condition (Eqn. 154) we then have

$$2R \frac{d}{dt} |A_0|^2 = -v_{III}\, |D|^2$$

Dividing both sides of this Eqn. by $|A_0|^2$, and rearranging, we have

$$\frac{d|A_0|^2}{|A_0|^2} = -\frac{v_{III}}{2R} \frac{|D|^2}{|A_0|^2}\, dt$$

Since the probability P of finding the alpha particle inside the nucleus is
directly proportional to $|A_0|^2$, the last equation tells us how P varies with time:

$$\frac{dP}{P} = -\frac{v_{III}}{2R} \frac{|D|^2}{|A_0|^2}\, dt \equiv -\gamma\, dt$$

i.e. $$P(t) = P(0)\, e^{-\gamma t} \qquad \text{(radioactive decay)}$$

where the decay constant is given by

$$\gamma = \frac{v_{III}}{2R} \frac{|D|^2}{|A_0|^2} \qquad\qquad (156)$$

This expression for γ can be put into an interesting form, if we introduce the
barrier penetration coefficient. For this rectangular barrier, with different potential
levels on the two sides, the transmission coefficient (the ratio of emergent to
incident probability currents) is given by

$$T = \frac{v_{III}}{v_I} \frac{|D|^2}{|A_0|^2}$$

(157)

Substituting this in Eqn. (156) leads to

$$\gamma = \frac{v_I}{2R} T$$

(158)

If, we could picture now an alpha particle as a classical point particle, bouncing back and forth across the nucleus, the value of $v_I/2R$ would be the number of times per second that it strikes the potential wall at $r = R$ and T represents the chance, for each such impact, that it will succeed in tunneling through the barrier to the outside. Thus, Eqn. (158) expresses the decay constant γ as a product of Newtonian and wave mechanical factors. But we cannot treat alpha particle as a point object because its de Broglie wavelength inside the nucleus would be comparable to the nuclear radius.

Further, the major defect of the above analysis is that the assumed shape of the barrier is completely different from the true potential presented by a nucleus to a charged particle. So, in the next article, we calculate how the decay rate might be expected to vary with alpha particle energy when the correct barrier shape is used.

2.13 Quantitative Treatment of Alpha Decay and Geiger Nuttal Law

The emission of alpha particles in natural radio activity was the subject of one of the great triumphs of wave mechanics in the early days. A staggering range of mean life times was observed for alpha-emitting nuclei from billions of years down to microseconds. It was also observed that there is a strong correlation between life times and the corresponding alpha particle energies–the higher the energy, the shorter the life time. But, whereas the life times vary by a factor as huge as 10^{23}, the energies all lie within 4 to 9 MeV (approx.). In 1928, Gamow, Gurney and Condon showed that this remarkable variation can be understood in terms of the quantum mechanical theory of barrier penetration (G. Gamow, Z. Phys., 51, 204 (1928)).

The nuclear potential as seen by an alpha particle can be represented by Fig. 2.17a. An origin is placed at the center of the nucleus. Within a radial distance R from this center, the nucleus is regarded as providing a constant negative potential. The R = nuclear radius = the limit of the nuclear forces and the radius within which the nuclear particles are confined. Outside the range of the nuclear forces, an alpha particle experiences only the repulsive coulomb potential of the form $q_1 q_2/r$. Putting $q_1 = (Z - 2)e$, $q_2 = 2e$, we have

$$V(r) = \frac{2(Z-2)e^2}{r} \text{ (cgs units) } (r > R)$$

(159 i)

The maximum height of this barrier V_0 is attained at $r = R$

$$V_0 = \frac{2(Z-2)e^2}{R}$$

(159 ii)

As rough values we can put $Z - 2 \cong 90$, $R \cong 10^{-12}$ cm, we then have (with $e = 4.8 \times 10^{-10}$ esu)

$$V_0 = \frac{180 \times 2.3 \times 10^{-19}}{10^{-12}} \text{ erg} \cong 4 \times 10^{-5} \text{erg}$$

Since 1 MeV $= 1.6 \times 10^{-6}$ erg, we have

$$V_0 \approx 25 \text{ MeV}$$

There is thus no question that the Wbarrier height is much greater than the energies of the emitted alpha particles, which range from 4 to 9 MeV.

The extent to the barrier is from the inner radius $r = R$ to the radius r_1 at which an alpha particle of any given energy E breaks through into the region of positive kinetic energy (Fig. 2.17b). This radius is clearly a function of E; it is defined by the condition

$$\frac{2(Z-2)e^2}{r_1} = E$$

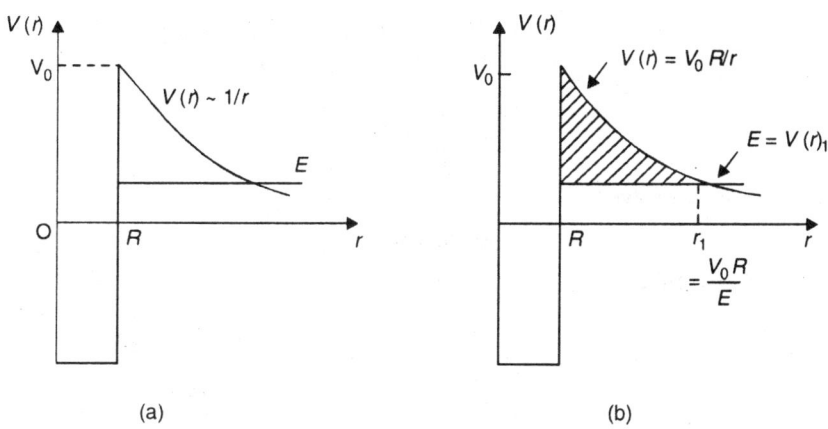

(a)   (b)

Fig. 2.17: Alpha particle emission, (a) Potential energy of the α-particle in the vicinity of a nucleus, (b) The thickness of the classically forbidden region can be expressed by $r_1 - R = \left(\dfrac{V_0 - E}{E} \right) R$. The shaded region in the diagram is the barrier through which an escaping α-particle must tunnel

Since the maximum height of the barrier V_0 is given by $V_0 = 2(Z - 2)e^2/R$, we can express r_1 simply through the equation

$$r_1 = \frac{V_0}{E} R \tag{160}$$

Using Eqn. (159), we can also write the potential $V(r)$ at any point across the barrier in the form

$$V(r) = \frac{R}{r} V_0 \tag{161}$$

We now proceed to calculate the penetrability T of the barrier as given approximately by Eqn.

$$T \approx \exp \left\{ -2 \int_R^{r_1} \frac{\sqrt{2mV(r) - E}}{\hbar} \, dr \right\}$$

$$= \exp\left\{\frac{-2\sqrt{2m}}{\hbar}\int_R^{r_1}\left(\frac{R}{r}V_0 - E\right)^{\frac{1}{2}}dr\right\}$$

which can conveniently be rewritten as follows:

$$T \approx \exp\left\{\frac{-2\sqrt{2mE}}{\hbar}\int_R^{r_1}\left(\frac{r_1}{r} - 1\right)^{\frac{1}{2}}dr\right\} \tag{162}$$

The integal in Eqn. (162) is easily evaluated. Since $r \le r_1$, we can put $r = r_1 \sin^2\theta$ and we have

$$\int_R^{r_1}\left(\frac{r_1}{r} - 1\right)^{\frac{1}{2}}dr = \int_{r=R}^{r_1}\left(\operatorname{cosec}^2\theta - 1\right)^{\frac{1}{2}}d(r_1\sin^2\theta)$$

$$= r_1\int_{r=R}^{r_1}\cot\theta \cdot 2\sin\theta\,\cos\theta\,d\theta$$

$$= r_1\int_{r=R}^{r_1}2\,\cos^2\theta\,d\theta = r_1\int_{r=R}^{r_1}(1 + \cos 2\theta)\,d\theta$$

$$= r_1\left[\theta + \frac{1}{2}\sin 2\theta\right]_{r=R}^{r_1}$$

Now, at $r = r_1$, we have

$$\theta = \frac{\pi}{2}, \quad \sin 2\theta = 0$$

and at $r = R$, we have

$$\theta = \sin^{-1}\left(\frac{R}{r_1}\right)^{\frac{1}{2}}; \quad \sin 2\theta = 2\left(\frac{R}{r_1}\right)^{\frac{1}{2}}\left(1 - \frac{R}{r_1}\right)^{\frac{1}{2}}$$

Therefore,

$$\int_R^{r_1}\left(\frac{r_1}{r} - 1\right)^{\frac{1}{2}}dr = r_1\left[\frac{\pi}{2} - \sin^{-1}\left(\frac{R}{r_1}\right)^{\frac{1}{2}} - \left(\frac{R}{r_1}\right)^{\frac{1}{2}}\left(1 - \frac{R}{r_1}\right)^{\frac{1}{2}}\right] \tag{163}$$

For $R \ll r_1$ (which corresponds to $E \ll V_0$), the right hand side of Eqn. (163) can be simplified by the approximations:

$$\sin^{-1}\left(\frac{R}{r_1}\right)^{\frac{1}{2}} \approx \left(\frac{R}{r_1}\right)^{\frac{1}{2}} \quad \text{and} \quad \left(1 - \frac{R}{r_1}\right)^{\frac{1}{2}} \approx 1$$

We then find

$$\int_R^{r_1}\left(\frac{r_1}{r} - 1\right)^{\frac{1}{2}}dr = r_1\left[\frac{\pi}{2} - 2\left(\frac{R}{r_1}\right)^{\frac{1}{2}}\right] \tag{164}$$

Using Eqn. (160), this gives

$$\int_R^{r_1}\left(\frac{r_1}{r} - 1\right)^{\frac{1}{2}}dr \approx \frac{\pi}{2}\frac{V_0}{E}R - 2\left(\frac{V_0}{E}\right)^{\frac{1}{2}}R$$

Substituting this in Eqn. (162) gives

$$T \approx \exp\left\{-\frac{2\sqrt{2mE}}{\hbar}R\left[\frac{\pi}{2}\frac{V_0}{E} - 2\left(\frac{V_0}{E}\right)^{1/2}\right]\right\}$$

Multiplying this, we have finally

$$T \approx \exp\left\{-\frac{\pi\sqrt{2m}}{\hbar}\cdot\frac{V_0 R}{E^{1/2}} + \frac{4\sqrt{2mV_0}}{\hbar}R\right\} \qquad (165)$$

From Eqn. (165) our approximation to the penetrability has the form

$$T(E) \approx A\, e^{-C/E^{1/2}} \qquad (166)$$

where

$$C = \frac{\pi\sqrt{2m}\,V_0 R}{\hbar} = \frac{\pi\sqrt{2m}}{\hbar}\cdot 2(Z-2)e^2$$

The discussion in the previous article led to the result that the decay constant

$$\gamma = \frac{v_I}{2R}T \qquad (167)$$

where v_I is the alpha particle velocity inside the nucleus and thus increases with E. However, the variation of γ with alpha particle energy is almost completely dominated by the exponential factor in $T(E)$ because the exponent $-C/E^{1/2}$ is numerically very large.

For α-particle emission from the heavy radioactive nuclei, such as radium and uranium, we have

$$Z - 2 \approx 90$$
$$m \approx 6.6 \times 10^{-24}\text{ g}$$

Substituting these and the values of e and $\hbar$ in the expression for C in Eqn. (166) gives

$$C \approx 0.45\text{ erg}^{1/2} \approx 360\text{ MeV}^{1/2}$$

Our test of the success of the theory is to plot the logarithm of the decay constant against $E^{-1/2}$. If we assume $\gamma \sim T$, then from Eqn. (167), we have

Table 2.1. (*) Some data for alpha-particle emitters

Parent	Nucleus	E (MeV)	$T_{1/2}$	γ (sec^{-1})	$\log_{10}\gamma$	$1/\sqrt{E}$
Th^{232}	4.05		1.39×10^{10} y	1.5×10^{-18}	-17.80	0.497
Ra^{226}	4.88		1.62×10^{3} y	1.36×10^{-11}	-10.9	0.452
Th^{228}	5.52		1.9 y	1.16×10^{-8}	-7.9	0.425
Em^{222} (Rn)	5.59		3.83 d	2.1×10^{-6}	-5.7	0.423
Po^{218}	6.12		3.05 m	3.78×10^{-3}	-2.4	0.404
Po^{216}	6.89		0.16 s	4.33	0.6	0.381
Po^{214}	7.83		1.5×10^{-4} s	4.23×10^{3}	3.6	0.358
Po^{212}	8.95		3.0×10^{-7} s	2.31×10^{6}	6.4	0.335

(*) *Adapted from:* I. Kaplan, Nuclear Physics, Addison Wesley (1955)

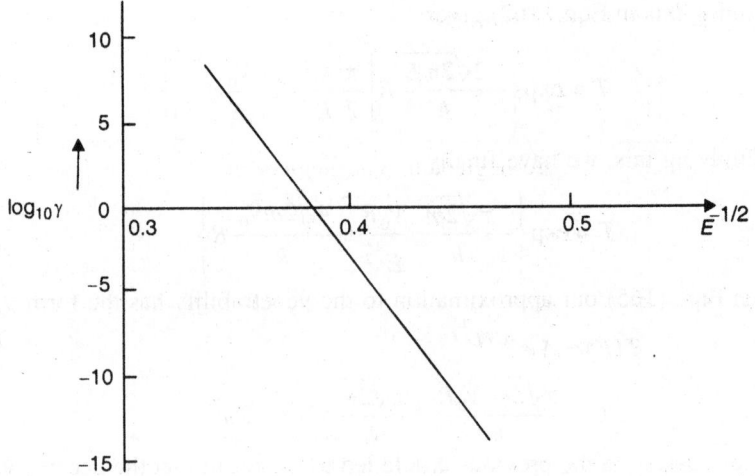

Fig. 2.18: Test of the theoretical relationship between decay constant and energy for α-particle emitters (γ in seconds $^{-1}$ and E – in Mev)

$$\log_{10} \gamma = \text{const} - \frac{C \log_{10} e}{E_{MeV}^{1/2}}$$

Substituting $C = 360 \ (\text{MeV})^{1/2}$, $\log_{10} e = 0.4343$, we have

$$\log_{10} \gamma = \text{const} - \frac{156}{E_{MeV}^{1/2}} \qquad (168)$$

If $T_{1/2}$ is measured, γ can be inferred from $e^{-\gamma T_{1/2}} = \dfrac{1}{2}$

or
$$\gamma = \frac{\log_e 2}{T_{1/2}} = \frac{0.693}{T_{1/2}}$$

If we plot the graph of $\log_{10}\gamma$ against $E^{-1/2}$ from experimentally measured values, it is observed that the empirical slope of the straight line graph obtained is only within a few percent of that given by Eqn. (168) giving a striking demonstration of the ability of wave-mechanics to describe the alpha particle emission by radioactive nuclei.

$$\log_{10} 0.693 - \log_{10} T_{1/2} = \text{const} - 156 \ E^{-1/2}$$

or
$$\log_{10} T_{1/2} = \text{const} + BE^{-1/2} \qquad (169)$$

where B is a constant.

This is known as *Geiger Nuttal Law*.

2.14 Free Electron Theory of Metals

Metals are an important class of solids and more than a two-third of elements are metals, therefore various attempts to model a metal have been made right form the beginning of the twentieth century. The formulation of the theory began well before the Pauli's enunciation of exclusion principle (1926) and the advent of quantum mechanics. The first model was proposed by Drude in 1900, three years after the discovery of electron by J.J. Thomson.

2.14.1 The Drude Model

Drude constructed his theory on the basis of classical kinetic theory of gases to metals, imagining the metals to be containing a gas of free electrons. Although there is only one kind of particle present in the gases, in a metal there are two-electrons which are negatively charged and to compensate their negative charge are relatively heavier particles which are immobile. At Drude's time there was no precise motion of the origin of light, mobile free electrons (which are now called valence electrons) and the concept of nucleus (surrounded by electron-orbitals). However, we assume in this context that when atoms of a metallic element are bought together to form a metal, the outer-most valence electrons behave like free electrons (they can wander free by through the metal, e.g. for Mg, the configuration is $1s^2\,2s^2\,2p^6\,3s^2$ so the two $3s^2$ electrons are free electrons), and the metallic ions (nucleus plus core electrons) remain intact and play the role of immobile positive particles of Drude's theory. The valence electrons are called conduction electrons since they are responsible for conduction of electricity in metals. When the isolated atoms condense to form a metal, the core electrons remain bound to the nucleus (to form the metallic ion) but the valence electrons are free to wander away from their parent atoms but within the metal. An estimate of the density of the free electrons can be made as follows:

A metallic element contains 6.023×10^{23} atoms/mole (Avogadro number) and ρ_m/A moles/cm^3 (where ρ_m is the mass density in gms per cubic centimeter and A is the atomic mass of the element). Let each atom contribute Z free electrons, then the number of free electrons/cm^3 ($n = N/V$) is

$$n = 6.023 \times 10^{23} \times \frac{Z\rho_m}{A} \qquad (170)$$

The free electron densities for some metals is given in Table 2.2

Table 2.2. Free Electron Densities of Some Metallic Elements

Element	Z	n (10^{22} per cm^3)
Li	1	4.70
Na	1	2.65
K	1	1.40
Rb	1	1.15
Cs	1	0.91
Cu	1	8.47
Ag	1	5.86
Au	1	5.90
Mg	2	8.61
Ca	2	4.61

Drude applied kinetic theory to this "gas" of conduction electrons which (in contrast to the molecules of an ordinary gas) move against a background of heavy immobile ions. The basic assumptions are as follows:

1. In the presence of an externally applied field, each (free) electron moves as determined by Newtons's law of motion, but additional complications produced by the other electrons and ions are neglected. The neglect of electron-electron interactions between collisions is known as the *independent electron approximation*. The neglect of electron-ion interactions is known as the *free-electron approximation*.

2. Collisions in the Drude's model (as in kinetic theory) are instantaneous events that abruptly alter the velocity of an electron (The electron's during an impact with ion cores, bounces off and moves in a straight line between two successive collisions). Electron-electron scatterings are unimportant.

3. All (free) electrons move with the root mean squared speed of a Maxwell-Boltzmann distribution, representing the random thermal velocity of electrons. The average electron velocity immediately after the collision is zero.

2.14.2 Lorentz Modification of Drude's Model

Lorentz modified the oversimplified model of Drude with following considerations:

1. The assumption that all electrons move with the same thermal velocity is abandoned.

2. The classical Maxwell-Boltzmann velocity distribution is perturbed by the presence of an electric field or a thermal gradient. (These tend to displace the equilibrium velocity distribution and distort the symmetry).

3. Boltzmann transport equation is applied to describe charge transport and kinetic energy of electrons.

The Drude Lorentz theory had several impressive successes and several remarkable failures. It was successful in derivation of Ohm's law which connects electric current with the electric field. It also led to the relation between the electrical and thermal conductivities of metal (Wiedemann Franz law).

2.14.3 Shortcomings of Drude-Lorentz Theory

The important physical problem which could not be explained by the Drude-Lorentz theory are mainly the following:

(*i*) Why is the Debye theory of lattice specific heats which ignores the electronic contribution completely, so accurately valid for metals? Experiments at liquid helium temperatures reveal that the specific heat of silver could be represented in the form

$$C_v = \alpha T^3 + \gamma T \tag{171}$$

where αT^3 is the Debye factor and γT is a linear term which could only be attributed to the electrons. This electronic contribution is very small.

(*ii*) The paramagnetic susceptibility of metals does not obey Curie's law and is practically independent of temperature.

(*iii*) The Hall effect of some divalent metals is positive (showing that charge carriers are positively charged).

(*iv*) Metals with relatively poor electrical conductivity like Sn, Hg and Pb become superconducting (zero resistance) when cooled below a critical temperature T_c which is a characteristic of a particular metal.

2.14.4 Sommerfeld's Theory

The first successful attempt in explaining the above properties of metals was made by Sommerfeld in 1928. Sommerfeld, with the advent of quantum mechanics realized that the above problems were due to the use of classical Boltzmann statistics and could be resolved by the use of Fermi-Dirac statistics. He investigated the behaviour of a free electron gas taking into the account quantum (Fermi-Dirac) statistics and the Pauli-exclusion principle.

An Electron in a Periodic Metallic Crystal

The electron-ion interaction cannot be totally ignored because conduction (valence) electrons remain confined within the volume of the metal. Since no emission of elecrtons is observed at room temperature, the potential energy of an electron at rest inside the metal must be lower than that of an electron at rest outside. An electron needs to be excited to a certain level so as to leave the metal and escape to infinity. It is called the vacuum level. In *one electron* approximation, an electron in a metal finds itself in the field of all nuclei and all other electrons. The potential energy for such an electron may therefore be expected to be periodic, the periodicity being that of the lattice as in Fig. 2.19.

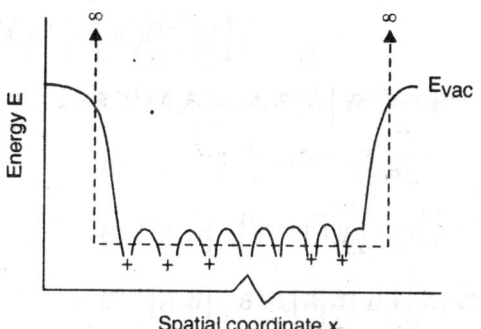

Fig. 2.19: Qualitative form of the potential energy of an electron in a one dimensional periodic lattice of positive ion cores (+).

The Sommerfeld Model

In the model employed by Sommerfeld and Bethe (1933) however, it is assumed that the free electrons, find themselves in a potential which is constant everywhere inside the metal. In this model, a metal crystal (a cube of edge L) is described by a three dimensional potential box of depth E_s with an infinite barrier at the surfaces. Thus, electrons remaining confined to the crystal, it is a gross oversimplification given that work function ϕ values are lying in the range of ~5eV (see Fig. 2.20) E_s is the order of 10eV (for nickel E_s=14.8eV). Despite its simplicity, this model yielded a much improved understanding of many of the electronic properties of metals, in good agreement with experiments.

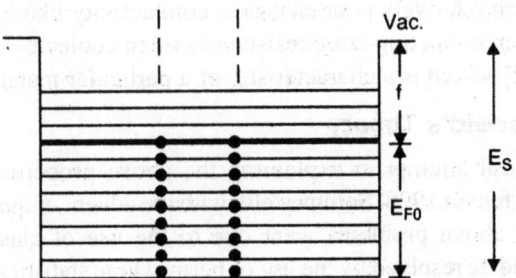

Fig. 2.20. The Sommerfeld free electron model at $T = 0^0$K. All energy levels upto E_f are filled and all higher ones are empty. The work function $\phi = E_s - E_f$. The solid circles (•) denote electrons

The Allowed Energies

Let $V(\vec{r})$ denote the potential of an electron at point $\vec{r}$ and E' be its total energy. Then time independent Schröndinger equation in the one electron approximation is

$$\left[-\frac{\hbar^2}{2m} + V(\vec{r})\right] \psi(\vec{r}) = E' \, \psi(\vec{r}) \qquad (172)$$

Sommerfeld defined $V(\vec{r})$ as

$$V(\vec{r}) = V_0 \qquad \text{(a constant) for } 0 \le x, y, z \le L$$

$$= \infty \text{ elsewhere} \qquad (173)$$

$$\left(\vec{r} = x\hat{i} + y\hat{j} + z\hat{k}\right)$$

Let the kinetic energy of the electron be denoted by E (i.e. $E = E' - V_0$) then

$$-\frac{\hbar^2}{2m} \nabla^2 \, \psi(\vec{r}) = E \, \psi(\vec{r}) \qquad (174)$$

or

$$\nabla^2\psi + \frac{2m}{\hbar^2} E \, \psi = 0$$

or

$$\frac{\partial^2\psi}{\partial x^2} + \frac{\partial^2\psi}{\partial y^2} + \frac{\partial^2\psi}{\partial z^2} + \frac{2mE}{\hbar^2}\psi = 0 \qquad (175)$$

Using the method of separation of variables, we write

$$\psi\,(x,y,z) = X(x)\,Y(y)\,Z(z) \qquad (176)$$

Substituting this, and dividing throughout by XYZ, we get

$$\frac{1}{X}\frac{\partial^2\psi}{\partial x^2} + \frac{1}{Y}\frac{\partial^2\psi}{\partial y^2} + \frac{1}{Z}\frac{\partial^2\psi}{\partial z^2} + \frac{2mE}{\hbar^2}(E_z + E_y + E_z) = 0 \qquad (177)$$

Here we have put $E = E_x + E_y + E_z$ as the scalar sum of components of E along x, y, z directions. Since x, y, z are independent variables, each coordinate depedent term must separately be equal to a constant:

$$\frac{\partial^2\psi}{\partial x^2} = -\frac{2mE}{\hbar^2} \, X \, E_x \qquad (178a)$$

$$\frac{\partial^2\psi}{\partial y^2} = -\frac{2mE}{\hbar^2} \, YE_y \qquad (178b)$$

and
$$\frac{\partial^2 \psi}{\partial z^2} = -\frac{2mE}{\hbar^2} ZE_z \tag{178c}$$

The solutions are of the form
$$X(X) = Ae^{ixk_x} + Be^{-ixk_x} \tag{179}$$

On substituting Eqn. (179) in Eqn. (178a), we get
$$k_x^2 = \frac{2m}{\hbar^2} E_x \tag{180}$$

The infinite barrier at the boundary surfaces (x, y, $z = 0$ and 1) keeps the electron confined within the box and gives the following boundary conditions for ψ:

$$\psi = 0 \text{ for } x = 0 \text{ and } L; \text{ and } 0 < y, z < L$$
$$y = 0 \text{ and } L; \text{ and } 0 < z, x < L \tag{181}$$
$$z = 0 \text{ and } L; \text{ and } 0 < x, y < L$$

Hence $\psi = 0$ or $X = 0$ at $x = 0$ and L gives
$$A + B = 0 \tag{182}$$

and
$$Ae^{ixk_x^2} + Be^{-ixk_x^2} = 0 \tag{183}$$

These are satisfied if
$$k_x = \frac{\pi n_x}{L}, \quad n_s = 1, 2, \dots \tag{184}$$

Substituting this in Eqn. (180) gives
$$E_x = \frac{\hbar^2 k_x^2}{2m} = \frac{\hbar^2 \pi^2 n_x^2}{2mL^2} \tag{185}$$

and
$$X(x) = C \sin x k_x \tag{186}$$

Normalization condition yields
$$C^2 \int_0^L \sin^2 \frac{\pi x n_x}{L} dx = 1$$

or
$$C = \sqrt{\frac{2}{L}} \tag{187}$$

Hence
$$X(x) = \sqrt{\frac{2}{L}} \sin \frac{\pi x n_x}{L} \tag{188}$$

corresponding to the energy
$$E_x = \frac{\hbar^2 \pi^2 n_x^2}{2mL^2}, \quad n_x = 1, 2, \dots$$

with n_x as the quantum number describing state X.

Hence, the electron wave function is
$$\psi(x,y,z) = \left(\frac{2}{L}\right)^{3/2} \sin \frac{\pi x n_x}{L} \sin \frac{\pi y n_y}{L} \sin \frac{\pi z n_z}{L} \tag{189}$$

with corresponding (allowed) energies
$$E = \frac{\hbar^2 \pi^2}{2mL^2}\left(n_x^2 + n_y^2 + n_z^2\right) = \frac{\hbar^2}{2m}|\vec{k}|^2 \tag{190}$$
$$(n_x = 1, 2, \dots; \ n_y = 1, 2, \dots; \ n_z = 1, 2, \dots).$$

where wave factor

$$\vec{k} = \hat{i}\,\frac{\pi n_x}{L} + \hat{j}\,\frac{\pi n_y}{L} + \hat{k}\,\frac{\pi n_z}{L} \tag{191}$$

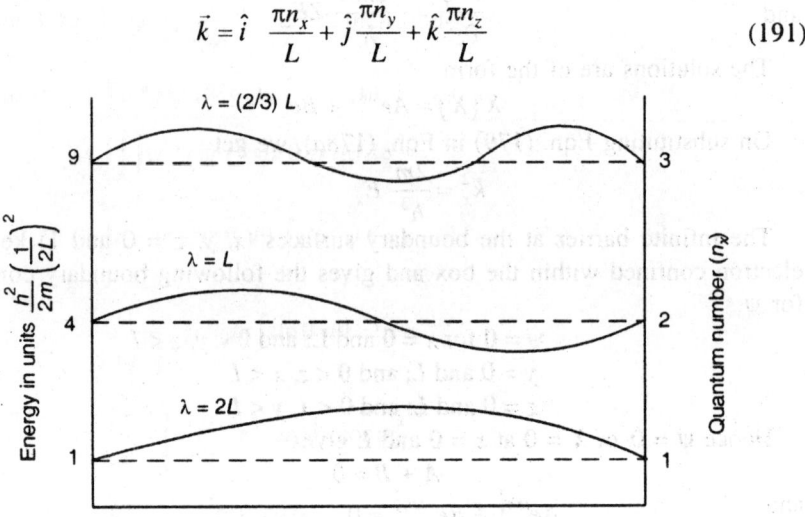

Fig. 2.21. First few allowed wave functions (—) and allowed energy levels (– – –) of a free electron confined to a potential box of length L (one dimensional case)

2.14.5 Density of states

A funciton that gives the number of available electron states per unit volume in an energy interval between E and $E + dE$ is designated by $N(E)\,dE$ where $N(E)$ is known as the *density of states function* and represents the number of states per unit volume per electron volt. Thus $N(E)\,dE$ is the density of states. To derive an expression for $N(E)$, we need to consider phase space (Fig. 2.22). We assume that $E(k)$ curve is parabolic (as evident from Eqn. (190) also) and for a given value of k, the constant energy surfaces are spherical, with radii

$$\sqrt{k_x^2 + k_y^2 + k_z^2} = \left(2mE\big/\hbar^2\right)^{1/2}.$$

According to the quantum prescription, the phase space is divided into cells each of volume h^3 and there are 2 electrons per cell.

The volume of k-space between constant energy surfaces E and $E + dE$ (i.e. having radii k and $k + dk$) is $4\pi k^2 dk$ and so the number of states contained in it is

$$N(k)dk = V \times \frac{4\pi k^2 dk}{h^3} = \frac{4\pi k^2 dk}{(2\pi\hbar)^3} V \tag{192}$$

But

$$E = \frac{\hbar^2 k^2}{2m}$$

∴

$$dE = \frac{\hbar^2}{m} k\, dk \tag{193}$$

and

$$k = \left(\frac{2mE}{\hbar^2}\right)^{1/2}$$

so that

$$k^2 \, dk = k.k \, dk = \left(\frac{2mE}{\hbar^2}\right)^{1/2} \frac{m}{\hbar^2} dE$$

$$= \frac{\sqrt{2} m^{3/2} E^{1/2}}{\hbar^3} dE \qquad (194)$$

Substituting the value of $k^2 \, dk$ in Eqn. (192) we have writing

$$N(k) \, dk = N(E) \, dE \qquad (195)$$

$$\Rightarrow \qquad N(E) dE = V \times \frac{4\pi}{8\pi^3 \hbar^3} \frac{\sqrt{2}.m^{3/2} E^{1/2}}{\hbar^3} dE$$

$$= \frac{V}{4\pi^2} \frac{1}{\hbar^3} (2m)^{3/2} E^{1/2} dE \qquad (196)$$

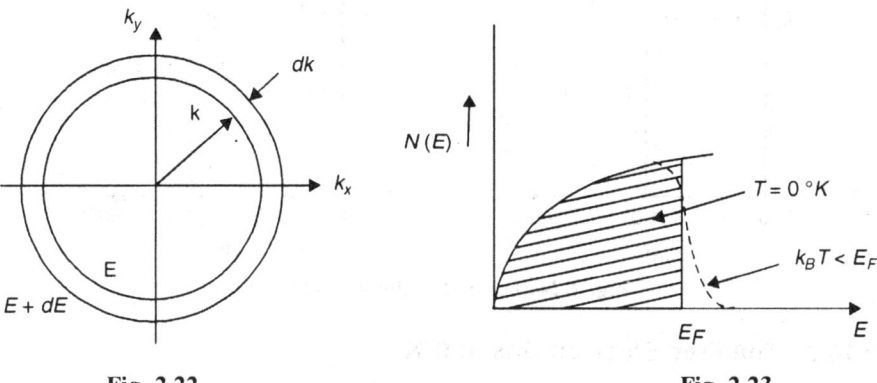

Fig. 2.22   Fig. 2.23

The density of available electron states is twice of this because each cell can contain two electron with opposite spins, hence

$$N(E) dE = \frac{V}{2\pi^2 \hbar^3} (2m)^{3/2} \sqrt{E} \, dE \qquad (197)$$

Figure 2.23 shows variation of the density of single particle states as a function of temperature for a free electron gas.

2.14.6 Concept of Fermi Energy and Fermi Level

The Fermi level is defined as the highest energy level which the electrons assume in a metal at $0\,°K$. In Sommerfeld's theory, the conduction electrons in a metal behave like a free electron-gas (fermions), and they obey Fermi-Dirac statistics. The probability of occupancy of an energy level at any temperature $T\,°K$ is given by the Fermi-Dirac distribution function

$$F(E) = \frac{1}{e^{(E-E_F)/k_B T} + 1} \qquad (198)$$

where E is the energy, k_B is the Boltzmann constant. At $T = 0\,°K$,

For $E < E_F$ $\qquad F(E) = \dfrac{1}{e^{-\infty} + 1} = \dfrac{1}{0+1} = 1$

For $E > E_F$
$$F(E) = \frac{1}{e^\infty + 1} = \frac{1}{\infty} = 0$$

i.e. all energy levels upto $E = E_F$ are occupied and none will be occupied above E_F.

Figure 2.24 (a) shows a plot of Fermi distribution function, (b) shows that the distribution function changes only for the levels of energies close to the Fermi energy E_F.

For $E = E_F$
$$F(E) = F(E_F) = \frac{1}{e^0 + 1} = \frac{1}{2} = 0.5$$

i.e. the *Fermi-energy* of given material is the energy of that quantum state which has the probability of ½ of being occupied by the conduction electrons.

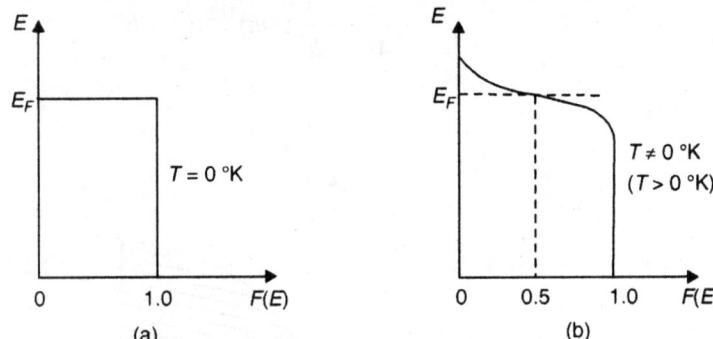

Fig. 2.24: Fermi distribution function

2.14.7   The Free Electron Gas at 0°K

The state of the conduction electrons at the 0 °K temperature forms the basis for studying the properties of metals. At 0 °K temperature all the levels upto E_F are full and all above E_F are empty. Using this fact we can calculate the total number of electrons (N).

$$N = \int_0^{E_F} N(E)\, F(E)\, dE \tag{199}$$

$$\frac{V}{2\pi^2}\frac{(2m)^{3/2}}{\hbar^3}\int_0^{E_F} E^{1/2}\, dE \qquad (\because \text{ for } E < E_F,\ F(E) = 1)$$

or
$$N = \frac{V}{3\pi^2}\frac{(2mE_F)^{3/2}}{\hbar^3} \tag{200}$$

$\therefore$
$$\ln N = \tfrac{3}{2}\ln E_F + a \text{ constant} \tag{201}$$

Differentiating, we get
$$\frac{dN}{dE_F} = \frac{3N}{2E_F} = N(E_F) \tag{202}$$

Denoting $N/V = n$, we have
$$n = \frac{N}{V} = \frac{1}{3\pi^2}\left(\frac{2mE_F}{\hbar^2}\right)^{3/2} \tag{203}$$

$$\Rightarrow \qquad E_F = \frac{\hbar^2}{2m}\left(3\pi^2 n\right)^{2/3} \tag{204}$$

Thus, knowing the electron density n, we can calculate the Fermi energy. E_F is also a measure of the Fermi wave vector k_F which is the radius of the spherical Fermi energy surface in the $\vec{k}$ -space. ($E_F = \hbar k_F$)

$$\therefore \qquad k_F = \left(3\pi^2 n\right)^{1/3} \tag{205}$$

The electron moves on the Fermi surface with a constant speed given by

$$v_F = \frac{\hbar k_F}{m} = \frac{\hbar}{m}\left(3\pi^2 n\right)^{1/3} \tag{206}$$

We can estimate the energy of the free electron at 0^0K per unit volume by

$$E_0 = \frac{1}{V}\int_0^{E_F} E\, N(E)\, dE \tag{207a}$$

Using the expression for $N(E)$, we get

$$E_0 = \frac{E_F^{5/2}}{5\pi^2}\left(\frac{2m}{\hbar^2}\right)^{3/2} \tag{207b}$$

The average kinetic energy per free electron is

$$\frac{E_0}{n} = \frac{3}{5}E_F \tag{208}$$

Questions and Problems

1. Discuss the double slit experiment with reference to probabilistic description of photons. What inferences are drawn from this experiment.
2. Discuss the physical signficance of the wave function ψ. Describe the general properties of the wave function.
3. Derive both time dependent and time independent Schrödinger wave equation for a non relativistic particle in a potential $V(\vec{r}\,)$.
4. What is meant by normalization of a wave function? Obtain the condition of normalization of a wave function.
5. Obtain eigenfunctions and energy eigenvalues with the help of Schrödinger's equation for a particle enclosed in a box and prove that the eigen values are discrete.
6. What do you mean by tunneling through a barrier? A particle travelling with energy E along X-axis meets a potential barrier defined by

$$V(x) = \begin{cases} 0 & \text{for } x < 0 \\ V_0 & \text{for } 0 < x < w \\ 0 & \text{for } x > w \end{cases}$$

Derive the expression for the reflection and transmission coefficients of the particle. (**Hint:** Reflection coefficient $R = 1$ - Transmission coefficient)

7. A particle is moving in the one dimensional potential well:

$$V(x) = \begin{cases} 0 & \text{for} \quad x < -a \\ -V_0 & \text{for} \quad -a < x < a \\ 0 & \text{for} \quad x > a \end{cases}$$

Find out the energy eigenvalues and the eigenfunctions. Also plot the first four eigenfunctions as a function of x.

8. Discuss quantitatively, the barrier penetration in alpha decay.

9. An electron is moving in a one dimensional box of infinite height and width 1 Å. Find the difference in its two lowest energy states.

[**Ans.** 112.125 eV]

10. The width of a rectangular potential barrier is 1 Å and its height is 2 eV. A proton of energy 1 eV is incident on it. Calculate the probability of its transmission through the barrier. [**Ans.** 7.47×10^{-19}]

11. Write down the Schrödinger wave equation for a free particle ($V = 0$) and solve it. Show that its energy values are continuous.

12. Solve the Schrödinger wave equation for a particle having energy E in a square well potential defined by

$$V(x) = \begin{cases} V_0 & \text{for } x < -a \\ 0 & \text{for } -a < x < a \\ V_0 & \text{for } x > a \end{cases}$$

where $E < V_0$.

13. Explain the probabilistic interpretation of wave function and show that the probability density P and probability current density J obey the following equation of continuity

$$\frac{dP}{dt} + \nabla \cdot \vec{J} = 0$$

Explain the physical significance of this equation.

14. Normalize the wavefunction $\psi = Ax\, e^{-\alpha x^2}$ between $-\infty \leq x \leq \infty$.

$$\left[\textbf{ Ans. } A = \left(\frac{32\alpha^3}{\pi} \right)^{1/4} \right]$$

15. Consider a potential barrier of the form
$$V(x) = V_0 \text{ for } |x| < a/2$$
$$= 0 \text{ for } |x| > a/2$$
so that in the region $|x| > a/2$, the solution of the Schrödinger equation can be written as

$$\psi(x) = Ae^{ikx} + Be^{-ikx} \;; x < -a/2$$

$$= Ce^{ikx} + De^{-ikx} \;; x > + a/2$$

Using continuity conditions at $x = \pm\, a/2$ show that

$$\begin{pmatrix} A \\ B \end{pmatrix} = \begin{pmatrix} R_{11} & R_{12} \\ R_{21} & R_{22} \end{pmatrix} \begin{pmatrix} C \\ D \end{pmatrix}$$

where
$$R_{11} = \left(\cosh k'a + \frac{i\sigma}{2} \sinh k'a \right) e^{ika}$$

$$R_{22} = \left(\cosh k'a - \frac{i\sigma}{2} \sinh k'a \right) e^{-ika}$$

$$R_{12} = \left(\frac{i\tau}{2} \right) \sinh k'a = -R_{21}$$

$$\sigma = \left(\frac{k'}{k} - \frac{k}{k'} \right); \quad \tau = \left(\frac{k'}{k} + \frac{k}{k'} \right)$$

and k and k' are defined through equations
$$k = \sqrt{\frac{2mE}{\hbar^2}} \; ; \; k' = \sqrt{\frac{2m(V_0 - E)}{\hbar^2}} \, .$$

For $E < V_0$, k' is real and for $E > V_0$, k' is purely imaginary.

16. Show that for a charged particle in the presence of electromagnetic potentials, the current density is
$$\vec{S} = -\frac{i\hbar}{2m} \left(\psi * \nabla \psi - \psi \nabla \psi * -2i \frac{e}{\hbar c} \vec{A} \psi * \psi \right)$$

17. For a particle in the state $n = 2$ of a one dimensional box of length L, show that the probability of finding the particle in the region $0 \le x \le L/4$ is 0.25.

[**Hint:** $P = \frac{2}{L} \int_0^{L/4} \sin^2 \frac{n\pi x}{L} dx$ and $\sin^2 \theta = \frac{1 - \cos 2\theta}{2}$]

18. For molecules of a gas of molecular weight 60 confined to a cube of edge 5 cm, what is the quantum number of the stationary state whose energy is equal to the average molecular kinetic energy at 300^0K.

[**Hint:** $E_{n_x, n_y, n_z} = \frac{h^2}{8mL^2} \left(n_z^2 + n_y^2 + n_z^2 \right) = kT$ and $m = \frac{\text{mol.wt.}}{6.02 \times 10^{23}}$ grams].

3

Operator—Formalism

Mathematically non-relativistic quantum mechanics is a linear theory. According to the superposition principle, the wave functions representing various states of a physical system can be combined additively, and the resulting functions represent new states. In this way, a mathematically complex wave function can be written as a linear combination of simpler functions; e.g. a wave packet can be represented as a linear combination of monochromatic components.

3.1 Linear Operators

The result obtained after performing a given mathematical operation upon a wave function ψ is symbolized as

$$\phi = A_{op}\psi \tag{1}$$

This means that the function ϕ is the result of applying ψ the operator denoted by A_{op}. e.g. if $A_{op} = d/dx$, ϕ is the derivative of ψ with respect to x. the subscript 'op' with A is used to specify that it is not a multiplicative factor with A. Let

$$\frac{d}{dx}(fg) = f\frac{dg}{dx} + \frac{df}{dx}g \tag{2}$$

then

$$A_{op}(fg) = f(A_{op}g) + (A_{op}f)g \tag{3}$$

in which $A_{op} = d/dx$. It may be noted here that in contrast to the ordinary multiplication, the above identity is quite different since, in ordinary multiplication

$$A(fg) = (Af).g = f(Ag) \tag{4}$$

where A is a number.

A_{op} is called a *linear operator* if it satisfies

$$A_{op}(\psi_1 + \psi_2) = A_{op}\psi_1 + A_{op}\psi_2 \tag{5}$$

$$A_{op}(c\psi_1) = c(A_{op}\psi_1) \tag{6}$$

where ψ_1 and ψ_2 are two given functions and c is a constant (may be a complex number as well).

Examples of linear operators: x_i, $\partial/\partial x_i$, ∇^2, $\partial/\partial t$.

An algebra of linear operators may be constructed by defining the terms "product", "sum", "power" etc. for example, we write two linear operators of fundamental importance the *null* or *zero operator* defined by $Z_{op}\psi = 0$ which annihilates the function ψ and the unit operator 1_{op} defined by $1_{op}\psi = \psi$, which produces no change in the wave function. Similarly

$$C_{op} = A_{op} + B_{op} \tag{7}$$

and

$$A_{op}(B_{op}\psi) = C_{op}\psi \tag{8a}$$

where

$$C_{op} = A_{op}B_{op} \tag{8b}$$

An important point to be noted here is that

$$A_{op}B_{op} \neq B_{op}A_{op} \tag{9}$$

By combining the process of addition and multiplication, a function of an operator can be formed e.g., $\dfrac{d^2y}{dx^2} + k^2y = 0$ is the same as $(A_{op}^2 + k^2 . 1_{op})y = 0$ where $A_{op} = d/dx$. An algebraic function of a linear operator is itself a linear operator.

If two operators A_{op} and B_{op} are related as

$$A_{op}B_{op} = B_{op}A_{op} = 1_{op}$$

then they are *reciprocal* to each other, i.e.

$$A_{op}^{-1} = B_{op} \; ; \; B_{op}^{-1} = A_{op}$$

An operator for which a reciprocal exists is *non singular* (i.e. it does not exist singularly, it appears in conjunction with another operator). A non singular operator can be inverted, e.g. if $\phi = A_{op}\psi$ and A_{op} has a reciprocal, then

$$\psi = A_{op}^{-1}\phi$$

However, if there is some non-zero function ψ such as

$$A_{op}\psi = 0$$

then A_{op} has no reciprocal and then A_{op} is *singular* (for ψ). *The question whether a given operator appearing in conjunction with another operator is non singular depends on whether the pair of operators form a canonically conjugate pair, as for example position and momentum, energy and time etc. Such a pair does not commute* i.e.

$$A_{op}B_{op} \neq B_{op}A_{op} \text{ and } A_{op}(B_{op}\psi) \neq B_{op}(A_{op}\psi) \tag{10}$$

3.2 The Commutator [A, B]

For convenience we drop the suffixes op and let A, B be operators, then the commutator is defined by

$$[A,B] = AB - BA \tag{11}$$

Example 3.1:
$$\left[x_i, \frac{\partial}{\partial x_j} \right]\psi = \left(x_i \frac{\partial}{\partial x_j} - \frac{\partial}{\partial x_j}x_i \right)\psi$$

$$= x_i \frac{\partial}{\partial x_j}\psi - \left(\frac{\partial}{\partial x_j}x_i \right)\psi$$

$$= x_i \frac{\partial}{\partial x_j} \psi - \frac{\partial x_i}{\partial x_j} \psi - x_i \left(\frac{\partial}{\partial x_j} \right) \psi$$

$$= - \delta_{ij} \psi$$

$$\Rightarrow \qquad \left[x_i, \frac{\partial}{\partial x_j} \right] = -\delta_{ij} \qquad \therefore \quad \left[\frac{\partial}{\partial x_i}, x_j \right] = \delta_{ij}$$

Here δ_{ij} is the kronecker delta.

Example 3.2:
$$\left[f(x), \frac{\partial}{\partial x_j} \right] \psi = \left(f(x) \frac{\partial}{\partial x_j} - \frac{\partial}{\partial x_j} f \right) \psi$$

$$= f \frac{\partial}{\partial x_j} \psi - \left(\frac{\partial}{\partial x_j} f \right) \psi - f \left(\frac{\partial}{\partial x_j} \psi \right)$$

(Note: $\dfrac{\partial}{\partial x_j} f(x)$ is like a non commuting operator.)

$$= - \left(\frac{\partial}{\partial x_j} f \right) \psi$$

Hence,
$$\left[f(x), \frac{\partial}{\partial x_j} \right] = - \frac{\partial}{\partial x_j} f(x)$$

Example 3.3: $\left[x_i, x_j \right] = 0$ (Real numbers commute)

Example 3.4: $\left[\dfrac{\partial}{\partial x_i}, \dfrac{\partial}{\partial x_j} \right] = 0$ (order of differentiation also commutes)

The operator $\dfrac{d}{dx}$ is singular with respect to the class of differentiable functions of x; which includes the constant c because $\dfrac{d}{dx} c = 0$ and the number c cannot be reconstructed. If however, the class is restricted to functions which are integrable in a proper sense the function c is not a number of this class and $\dfrac{d}{dx}$ is non singular.

Example 3.5: An operator O_1 is defined by $O_1 \psi(x) = x \dfrac{d}{dx} \psi(x)$.
Check for its linearity.
Solution: An operator O is said to be linear if

$$O_1 \{ c_1 f_1(x) + c_2 f_2(x) \} = c_1 O_1 f_1(x) + c_2 O_1 f_2(x)$$

For the given operator O,

$$O_1 \left[c_1 f_1(x) + c_2 f_2(x) \right] = x \frac{d}{dx} \{ c_1 f_1(x) + c_2 f_2(x) \}$$

$$= x \left\{ c_1 \frac{df_1}{dx} + c_2 \frac{df_2}{dx} \right\} = c_1 x \frac{df_1}{dx} + c_2 x \frac{df_2}{dx} \qquad (i)$$

Again $c_1 O_1 f_1(x) + c_2 O_1 f_2(x)$

$$= c_1 x \frac{df_1}{dx} + c_2 x \frac{df_2}{dx} \qquad\qquad (ii)$$

Hence operator O is linear.

Example 3.6: Check for the linearity of an operator O defined by

$$O_2 \psi(x) = x^3 \psi(x)$$

Solution: $\quad O_2\left[c_1 f_1(x) + c_2 f_2(x)\right] = x^3\left[c_1 f_1(x) + c_2 f_2(x)\right]$

$$= c_1 x^3 f_1(x) + c_2 x^3 f_2(x)$$

$$= c_1 O_2 f_1(x) + c_2 O_2 f_2(x)$$

Hence operator O_2 is linear.

Example 3.7: An operator O_3 is defined by

$$O_3 \psi(x) = e^{\psi(x)}$$

Check whether it is a linear operator or not.

Solution: Let $\qquad\qquad y(x) = c_1 f_1(x) + c_2 f_2(x)$

$$O_3 \psi(x) = e^{\psi(x)}$$

$\Rightarrow \qquad\qquad O_3\left[c_1 f_1(x) + c_2 f_2(x)\right] = \exp\left[c_1 f_1(x) + c_2 f_2(x)\right]$

$$= e^{c_1 f_1(x)} \cdot e^{c_2 f_2(x)} \qquad\qquad (i)$$

But L.H.S. gives

$$O_3\left[c_1 f_1(x) + c_2 f_2(x)\right] = c_1 O_3 f_1(x) + c_2 O_3 f_2(x)$$

$$= c_1 e^{f_1(x)} + c_2 e^{f_2(x)} \qquad\qquad (ii)$$

From (i) and (ii) we see that r.h.s. of (i) and (ii) are not equal hence O_3 is not a linear operator.

3.3 Eigen Functions and Eigen Values

If A_{op} is a given operator, the function

$$\phi = A_{op} \psi \qquad\qquad (12)$$

which results from applying the operator A_{op} to ψ will in general, be linearly independent of ψ. It may happen, however, that for some function ψ

$$A_{op} \psi = \alpha \psi \qquad\qquad (13)$$

where α is a complex number. In this case, if ψ is a member of the class of physically meaningful functions, it is an eigenfunction of the operator A_{op}. Except for multiplication by the number α, ψ is not changed by the operator A_{op} (it is invariant). The number α is called the *eigenvalue* of A_{op} associated with the eigenfunction or state function ψ. For example, $H\psi = E\psi$, where $E = p^2 / 2m + V(\vec{r})$.

Example 3.8: Show that the wave function $\psi = Nx\exp\left(-\dfrac{x^2}{2}\right)$ is an eigen

function of $H = -d^2/dx^2 + x^2 3$. Also find out the normalization constant N and the eigen value.

$$H\psi = \left(-\frac{d^2}{dx^2} + x^2\right) N x \exp\left(-\frac{x^2}{2}\right)$$

$$= N\, x^3 \exp(-x^2/2) - N\frac{d}{dx}\left[\exp\left(-\frac{x^2}{2}\right) - x^2\exp\left(-\frac{x^2}{2}\right)\right]$$

$$= 3N\, x\, \exp\left(-\frac{x^2}{2}\right)$$

$$= 3\psi.$$

Hence the eigen value is 3.

From normalization condition,

$$N^2 \int_{-\infty}^{+\infty} x^2 \exp(-x^2)\,dx = 1$$

$$N^2 \frac{\sqrt{\pi}}{2} = 1 \text{ or } N = \left(\frac{2}{\sqrt{\pi}}\right)^{1/2}$$

3.4 Simultaneous Eigenfunctions

If A and B are linear operators, and ψ is a function satisfying both the equations:

$$A\psi = \alpha\psi \tag{14}$$

$$B\psi = \beta\psi \tag{15}$$

then ψ is a simultaneous eigenfunction of A and B belonging to the eigenvalues α and β. The above Eqn. implies that

$$BA\psi = B\alpha\psi = \alpha B\psi = \alpha\beta\psi$$

and

$$AB\psi = A(\beta\psi) = \beta A\psi = \beta\alpha\psi$$

then by subtraction

$$(AB - BA)\,\psi = 0 \tag{16}$$

This Eqn. shows that ψ is also an eigenfunction of the operator $[A, B]$. This is represented by $[A, B]\psi = 0$. $[A, B]$ is called the *commutator*. Thus, if the commutator of two operators is zero, they are said to commute and they are said to have simultaneous eigenfunctions.

We can see that eigenfunctions of commuting operators can always be constructed in such a way that they are simultaeous eigenfunctions. Suppose ψ is an eigenfunction of A that is

$$A\psi = \alpha\psi \tag{17}$$

and that A and B commute. By multiplying this equation on the left with B, we obtain

$$BA\psi = AB\psi = B\alpha\psi = \alpha B\psi \tag{18}$$

or

$$A(B\psi) = \alpha(B\psi) \tag{19}$$

This shows (if we assume $B\psi \neq 0$) that the function $B\psi$ is an eigenfunction of A belonging to the same eigenvalue α. Now two basically different cases arise:

(1) If the eigenvalue α is non degenerate, i.e. if there is only one state ψ corresponding to α, then the function $B\psi$ can differ from ψ only by a constant multiplier, that is,

$$B\psi = \beta\psi \tag{20}$$

β is a constant. Hence if ψ is a non degenerate eigenfunction of A, then Eqn. (16) is both a necessary and a sufficient condition for ψ to be a simultaneous eigenfunction of A and B.

(2) The eigenvalue α may be degenerate. In this the argument leading to equation (20) cannot be made.

Example 3.9: Let a complete set of eigenfunctions $\psi_n(x)$; $n = 1, 2, \dots$ of operators A and B be given, with eigenvalues a_n and b_n, respectively. Then it can be shown that A and B commute.

Since

$$[A, B]\psi_n = (AB - BA)\psi_n$$
$$= (a_n b_n - b_n a_n)\psi_n = 0,$$

it follows for arbitrary

$$\psi = \sum_n c_n \psi_n \text{ that}$$
$$[A, B]\psi = 0$$

and thus

$$[A, B] = 0$$

Examples of commuting operators are x_1, x_2, x_3 or p_1, p_2, p_3 or x_1, p_2, p_3 but not x_1, p_1.

Example 3.10: Consider the commutator $[H, p]$ where

$$H = \frac{p^2}{2m} \tag{i}$$

and p is linear momentum. This corresponds to the case of a free particle. If $\psi(x)$ is an eigenfuncton of H belonging to the energy eigenvalue, E i.e.

$$H\psi(x) = E\psi(x) \tag{ii}$$

then, by symmetry the function $\psi(x + \varepsilon)$ is also an eigenfunction belonging to the same energy E:

$$H\psi(x+\varepsilon) = E\psi(x+\varepsilon) \tag{iii}$$

Here, ε is a real number. Now if ε is small, we have

$$\psi(x+\varepsilon) = \psi(x) + \varepsilon\frac{d\psi}{dx} = \psi(x) + \varepsilon\frac{i}{\hbar}p\psi \tag{iv}$$

and Eqn. (*iii*) becomes

$$H\psi(x) + \varepsilon\frac{i}{\hbar}Hp\psi(x) = E\psi(x) + \varepsilon\frac{i}{\hbar}pE\psi(x)$$

Hence from Eqn.

$$\varepsilon\frac{i}{\hbar}Hp\psi(x) = \varepsilon\frac{i}{\hbar}pH\psi(x)$$

or, if we transpose and cancel a common factor, then

$$(Hp - pH)\, \psi(x) = 0 \qquad (v)$$

or

$$[H, p] = 0 \qquad (vi)$$

that is eigenfunctions of H must also be the eigenfunctions of p.

This is a consequence of the invariance of the system to the linear transformation:

$$x \rightarrow x + \varepsilon.$$

Example 3.11: According to uncertainty principle there is a direct connection between position and momentum and we may write a commutator

$$[x, p] = \alpha \hbar \qquad (i)$$

where α is a number to be determined. Since

$$\left[x_i, \frac{\partial}{\partial x_i} \right] = -1$$

therefore

$$\left[x, -\alpha \hbar \frac{\partial}{\partial x} \right] = \alpha \hbar \qquad (ii)$$

Comparing (i) and (ii) we have operator

$$p \equiv -\alpha \hbar \frac{\partial}{\partial x}$$

The eigenvlaue equation for momentum operator

$$p \psi_p(x) = \left(-\alpha \hbar \frac{\partial}{\partial x} \right) \psi_p(x)$$

The eigenfunctions are thus

$$\psi_p = \exp\left(-\frac{px}{\alpha \hbar} \right)$$

If we take $\alpha = i$, we then have the space part of a de Broglie wave $\psi = A e^{-i(Et - px)/\hbar}$ appearing automatically as the state function of a particle of momentum p.

Hence we may write

$$[x, p] = i\hbar \qquad (21)$$

where position and momentum operators are x and $-i\hbar\, \partial/\partial x$ respectively. Any two operators $\hat{A}$ and $\hat{B}$ which satisfy the commutation relation

$$\left[\hat{A}, \hat{B} \right] = i\hbar \qquad (22)$$

the observables A and B are said to form a *canonically conjugate pair* of observables.

3.5 Observations and Expectation Values

To each physical observable in a particular state of a dynamical system, we can assign (in classical mechanics) a particular mathematical variable, such as position energy, momentum, etc. due to the fact that classically, all physical observables have definite values in classical states. However, in quantum mechanics, the statistical interpretation of wave function describing dynamical state of a system implies uncertainty in the measurement of any two canonically conjugate observables. Thus we talk of *expectation value* of an operator assigned to any physical observable, in quantum mechanics.

Let $\psi_n(x)$ denote the state function of a dynamical system and $\hat{A}$ be an operator such that the possible result of operation of $\hat{A}$ on ψ_n gives the eigenvalue a_n.

Now let the dynamical system (described by $\psi(x)$) be characterized by a set of systems each one in an arbitrary state $\psi_n(x)$ then expectation value of an operator $\hat{A}$ is obtained as

$$\bar{a}_{\psi_n} = \frac{\displaystyle\int_{-\infty}^{+\infty} \psi_n^*(x)\hat{A}\psi_n(x)\,dx}{\displaystyle\int_{-\infty}^{+\infty} \psi_n^*(x)\psi_n(x)\,dx} \tag{23}$$

This implies that repeated observations on a set of system (each in the state ψ_n) will produce a statistical distribution of the different eigenvalues a_{ψ_n} for

$$a_{\psi_n} = \int_{-\infty}^{+\infty} \psi_n^*\hat{A}\psi_n\,dx ,$$

such that the average taken over the entire distribution of functions $\psi_n(x)$ gives the expectation value of the observable $\bar{a}_{\psi_n}$ corresponding to the operator $\hat{A}$. This average will be some number constructed from operator $\hat{A}$ on the state ψ_n such that in special case

$$\psi(x) = \psi_n(x)$$

with ψ_n's satisfying

$$\hat{A}\Psi_n(x) = a_n\psi_n(x)$$

Physically, $\bar{a}_{\psi_n}$ signifies the most expected outcome of a consequence of operation by the operator $\hat{A}$ on all the possible states $\psi_n(x)$ of the wave function $\psi(x)$.

3.6 Postulates of Quantum Mechanics

We have the following postulates of quantum mechanics:

Postulate 1: For a system consisting of a particle moving in a conservative field of force (produced by an external potential) there is a associated *wave function* ψ. This wave function determines everything that can be known about the system and it is a single valued function of the coordinates of the particle and of time.

In general it is a complex function, and may be multiplied by an arbitrary complex number without changing its physical significance.

Postulate 2: With every physical observable (e.g. the energy of the system, the x position coordinate of the particle) there is associated an operator. Let us denote by Q the operator associated with the observable q. Then a measurement of q gives a result which is one of the eigenvaleus of the Eqn:

$$Q\psi_n = q_n\psi_n \tag{24}$$

If initially, the wave function is not a eigen solution of Eqn. (24), it is impossible to predict with certainty which of several possible results will be obtained.

The measurement constitutes an interaction between the system and the measuring apparatus. If the result q_n is obtained, the interaction changes the state of the system to the state described by the wave function ψ_n. Thus immediately after a measurement yielding the value q_n, the state function is ψ_n. Observables represented by operators multiply, differentiate or otherwise act on the wave function to produce a new function.

Postulate 3: Any operator associated with a physically measurable quantity is Hermitian.

Postulate 4: The set of functions ψ_j which are eigenfunctions of the eigenvalue equation

$$Q \, \psi_j = q_j \, \psi_j \qquad (25)$$

form (in general) an infinite set of linearly independent functions. A linear combination of these functions of the form

$$\psi = \sum_j c_j \psi_j \qquad (26)$$

can be used to express an infinite number of possible functions. This infinite set of linearly independent functions can be used to expand any arbitrary function ψ.

Postulate 5: If a system is described by a wave function ψ, the expectation value of any observable q with corresponding operator Q is given by

$$\langle q \rangle = \int \overline{\psi} Q \psi \, d^3 r \qquad (27)$$

if ψ is a normalized function.

Postulate 6: The development (in time) of the wave function ψ, given its form at an initial time (and assuming the system is left undisturbed) is determined by the Schrödinger equation

$$H\psi = i\hbar \frac{\partial}{\partial t} \psi \qquad \cdot (28)$$

where the Hamiltonian operator $H = -\hbar^2 \nabla^2 + V$.

Postulate 7: The operators of quantum theory are such that their commutators are proportional to the corresponding classical Poisson brackets according to

$$[Q, R] \equiv (QR - RQ) \Leftrightarrow i\hbar \, \{q, r\} \qquad (29)$$

where $\{q, r\}$ is the classical Poisson bracket for the observables q and r.

(The operators Q and R must be Hermitian and the coordinates and momenta must be expressed in cartesian coordinates.)

Example 3.12: The wave function in $1s$ state of hydrogen atom is given as

$$\psi_{1s} = \frac{1}{\sqrt{\pi}} \left(\frac{1}{a_0} \right)^{3/2} e^{-r/a_0}$$

where $a_0 = \dfrac{(4\pi\varepsilon_0)h^2}{4\pi^2 me^2}$ is the Bohr radius. Find the expectation value of potential energy of electron in the $1s$ – state.

Solution: Potential energy $V(R) = -\dfrac{1}{4\pi\varepsilon_0}\dfrac{e^2}{r}$

The expectation value, $\quad \langle V \rangle = \dfrac{\iiint \psi_{1s}^* \, V \, \psi_{1s} \, dv}{\iiint \psi_{1s}^* \, \psi_{1s} \, dv}$

Volume element $\qquad dv = r^2 \, dr \, \sin\theta \, d\theta \, d\phi.$

$\therefore \qquad \langle V \rangle = \dfrac{\iiint \left(\dfrac{-e^2}{4\pi\varepsilon_0 r}\right)\dfrac{1}{\pi a_0^3} e^{-2r/a_0} r^2 \, dr \, \sin\theta \, d\theta \, d\phi}{\iiint \dfrac{1}{\pi a_0^3} e^{-2r/a_0} r^2 \, dr \, \sin\theta \, d\theta \, d\phi}$

$= \dfrac{-e^2}{4\pi\varepsilon_0} \dfrac{\displaystyle\int_{-\infty}^{+\infty} e^{-2r/a_0} r \, dr \int_0^\pi \sin\theta \, d\theta \int_0^{2\pi} d\phi}{\displaystyle\int_{-\infty}^{+\infty} e^{-2r/a_0} r^2 \, dr \int_0^\pi \sin\theta \, d\theta \int_0^{2\pi} d\phi} = \dfrac{-e^2}{4\pi\varepsilon_0} \dfrac{\displaystyle\int_{-\infty}^{+\infty} e^{-2r/a_0} r \, dr}{\displaystyle\int_{-\infty}^{+\infty} e^{-2r/a_0} r^2 \, dr}$

Now, we have the standard integral $\displaystyle\int_0^\infty x^n e^{-\alpha x} \, dx = \dfrac{n!}{\alpha^{n+1}}$

So, $\qquad \langle V \rangle = \dfrac{-e^2 \times 2 \times [1!/(2/a_0)^2]}{4\pi\varepsilon_0 \times 2 \times [2!/(2/a_0)^3]} = \dfrac{-e^2}{(4\pi\varepsilon_0)a_0}$

Substituting the value of $a_0 = \dfrac{(4\pi\varepsilon_0)h^2}{4\pi^2 me^2}$, we get

$$\langle V \rangle = \dfrac{-e^2 \times 4\pi^2 me^2}{(4\pi\varepsilon_0)^2 h^2} = \dfrac{-4\pi^2 me^4}{(4\pi\varepsilon_0)^2 h^2}$$

Substituting the values of the constants,

$$\langle V \rangle = \dfrac{4 \times (3.14)^2 \times (9.1 \times 10^{-31}) \times (1.6 \times 10^{-19})^4}{(6.6 \times 10^{-34})^2} \times (9 \times 10^9)^2$$

$$\left(\because \quad \dfrac{1}{4\pi\varepsilon_0} = 9 \times 10^9\right)$$

or $\qquad \langle V \rangle = -43.5 \times 10^{-19} \text{ J} = \dfrac{-43.5 \times 10^{-19}}{1.6 \times 10^{-19}} \text{ eV} = -27.2 \text{ eV}$

Table 3.1 summarizes various operators corresponding to different physical quantities.

Table 3.1: The simple operators in quantum mechanics

Physical quantity	Operator
Coordinate	$\vec{r}$
x, y, z	x, y, z
Momentum $\vec{p}$	$-i\hbar\nabla$
p_x, p_y, p_z	$-i\hbar\dfrac{\partial}{\partial x}, -i\hbar\dfrac{\partial}{\partial y}, -i\hbar\dfrac{\partial}{\partial z}$
Angular Momentum $\vec{L} = \vec{r} \times \vec{p}$	$\hat{L} = -i\hbar\vec{r} \times \nabla$
$L_x = yp_z - zp_y$	$\hat{L}_x = -i\hbar\left(y\dfrac{\partial}{\partial z} - z\dfrac{\partial}{\partial y}\right)$
$L_y = zp_x - xp_z$	$\hat{L}_y = -i\hbar\left(z\dfrac{\partial}{\partial x} - x\dfrac{\partial}{\partial z}\right)$
$L_z = xp_y - yp_x$	$\hat{L}_z = -i\hbar\left(x\dfrac{\partial}{\partial y} - y\dfrac{\partial}{\partial x}\right)$
Energy in the non relativistic approximation	$E = \dfrac{p^2}{2\mu} + V(\vec{r})$	$\hat{H} = -\dfrac{\hbar^2}{2\mu}\nabla^2 + V(\vec{r})$
Hamiltonian	H	$\hat{H} = i\hbar\dfrac{\partial}{\partial t}$

Example 3.13:   Given that ψ is the Gaussian function

$$\psi = \left(\frac{1}{\sigma\sqrt{\pi}}\right)^{1/2} \exp\left(-\frac{x^2}{2\sigma^2}\right)$$

Find $\left\langle x^n \right\rangle$.

Solution:

$$\left\langle x^n \right\rangle = \int_{-\infty}^{+\infty} \psi * x^n \psi \, dx$$

$$\sigma\sqrt{\pi}\left\langle x^n \right\rangle = \int_{-\infty}^{+\infty} x^n e^{-x^2/\sigma^2} \, dx$$

$$= \int_{0}^{\infty} x^n e^{-x^2/\sigma^2} \, dx + \int_{-\infty}^{0} x^n e^{-x^2/\sigma^2} \, dx$$

Substituting $x = -x$ in the second integral, we get

$$\sigma\sqrt{\pi}\left\langle x^n \right\rangle = \left\{1 + (-1)^n\right\} \int_{0}^{\infty} x^n e^{-x^2/\sigma^2} \, dx \qquad (i)$$

Thus $<x^n>$ is zero if n is odd. This is due to the fact that the integrand $x^n e^{-x^2/\sigma^2}$ is of odd parity for odd n and the integral from $-\infty$ to $+\infty$ of any odd parity function is zero. If n is an even integer say $n = 2n$, we substitute $x^2/\sigma^2 = s$. The integral becomes

$$\sigma\sqrt{\pi}\left\langle x^n \right\rangle = \int_{-\infty}^{+\infty} x^n e^{-x^2/\sigma^2} \, dx$$

or
$$\sigma\sqrt{\pi}\langle x^n \rangle = \int_{-\infty}^{+\infty} x^n e^{-x^2/\sigma^2}\, dx$$

$$\left(\because x\, dx = \frac{1}{2}\sigma^2\, ds \quad \text{and} \quad x = \sigma\, s^{1/2} \right)$$

or
$$\sqrt{\pi}\langle x^{2n'} \rangle = \frac{(\sigma)^{2n'}}{2} \times 2\int_0^\infty e^{-s} \cdot s^{\left(n'+\frac{1}{2}-1\right)}\, ds$$

$$= (\sigma)^{2n'}\, \Gamma\!\left(n' + \tfrac{1}{2}\right) \qquad\qquad (ii)$$

$$\Gamma\!\left(n' + \tfrac{1}{2}\right) = \left(n' + \tfrac{1}{2} - 1\right)!$$

$$= \left(n' - \tfrac{1}{2}\right)! \qquad\qquad (iii)$$

Now we make use of the formula [*]

$$n!\left(n + \tfrac{1}{2}\right)! = \left(\sqrt{\pi}\right)\frac{(2n+1)!}{2^{2n+1}} \qquad\qquad (iv)$$

by substituting $n = (n' - 1)$; so that

$$\left(n' - \tfrac{1}{2}\right)! = \frac{\sqrt{\pi}}{2} \times \frac{(2n'-1)!}{2^{2n'-2}(n'-1)!}$$

$$= \left(\sqrt{\pi}\right)\frac{(2n'-1)!}{(n'-1)!} \times \frac{1}{2^{2n'-1}}$$

$$\sqrt{\pi}\, \frac{2n' \cdot (2n'-1)!}{2n' \cdot (n'-1)!} \cdot \frac{1}{2^{2n'-1}}$$

$$= \sqrt{\pi}\, \frac{(2n')!}{n'! \times 2^{2n'}} \qquad\qquad (v)$$

Hence, from (ii), (iii) and (v)

$$\langle x^{2n'} \rangle = \left(\frac{\sigma}{2}\right)^{2n'} \frac{(2n')!}{n'!}$$

This implies $\langle x^n \rangle = 1$ for $n = 2n' = 0$.
This verifies that the given wave function is properly normalized.

Example 3.14: Consider a particle whose normalized wave function is
$$\psi(x) = 2\alpha\sqrt{\alpha}\, x\, e^{-\alpha x}, \quad x > 0 = 0,\ x < 0$$
(a) For what value of x does $|\psi(x)|^2$ peak?
(b) Calculate $\langle x \rangle$ and $\langle x^2 \rangle$.

(c) What is the probability that the particle is found between $x = 0$ and $x = \dfrac{1}{\alpha}$?

Solution: (a) The peak in $P(x) \equiv |\psi(x)|^2$ occurs when $dP(x)/dx = 0$ i.e. when
$$\frac{d}{dx}(x^2 e^{-2\alpha x}) = 0$$

---

[*] See Mathematical Methods for Physicists by Arfken and Weber, Academic Press (2001), p.no. 656.

or
$$2xe^{-2\alpha x} + x^2 (-2\alpha) e^{-2\alpha x} = 0$$

or
$$x = 1/\alpha.$$

(b)
$$\langle x \rangle = \int_0^\infty dx \cdot x \left(4\alpha^3 x^2 e^{-2\alpha x} \right)$$

Let
$$2\alpha x = y \qquad \therefore dx = \frac{dy}{2\alpha}$$

$$x = \frac{y}{2\alpha}$$

$$\therefore \qquad \langle x \rangle = 4\alpha^3 \int_0^\infty \frac{dy}{2\alpha} \frac{y^3}{8\alpha^3} e^{-y}$$

or
$$\langle x \rangle = \frac{1}{4\alpha} \int_0^\infty dy. \, y^3 \, e^{-y} = \frac{3!}{4\alpha} = \frac{3}{2\alpha}$$

(c) Probability

$$P = \int_0^{1/\alpha} dx \cdot 4\alpha^3 x^2 e^{-2\alpha x}$$

Let
$$2\alpha x = y \quad \therefore dx = \frac{dy}{2\alpha}$$

$$\therefore \qquad P = 4\alpha^3 \int_0^{1/\alpha} \frac{dy}{2\alpha} \frac{y^2}{4\alpha^2} e^{-y}$$

$$= \frac{1}{2} \int_0^2 dy. \, y^2 \, e^{-y}$$

or
$$P = \frac{1}{2} \left\{ \left[y^2 (-e^{-y}) \right]_0^2 - \int_0^2 2y(-e^{-y}) dy \right\}$$

$$= \frac{1}{2} \left\{ \left[-y^2 e^{-y} \right]_0^2 + 2 \int_0^2 y e^{-y} dy \right\}$$

$$= -\frac{1}{2} \left[4e^{-2} \right] + \frac{2}{2} \left\{ \left[y(-e^{-y}) \right]_0^2 - \int_0^2 1(-e^{-y}) dy \right\}$$

$$= -2e^{-2} - 2e^{-2} + \int_0^2 e^{-y} dy$$

$$= -4e^{-2} + \left[-e^{-y} \right]_0^2 = -4e^{-2} + e^{-2} + 1 = 1 - 5e^{-2}$$

$$1 - \frac{5}{e^2} = 1 - \frac{5}{(2.7183)^2} = 1 - 0.6766 = 0.3234$$

3.7 The Ehrenfest Theorem

The expectation value of momentum operator is

$$\langle p \rangle = \int_V \psi * (-i\hbar \nabla) \psi \, d^3 r \qquad (30)$$

$$\Rightarrow \qquad \frac{d}{dt}\langle p_x \rangle = -i\hbar \frac{d}{dt}\left\{ \int_V \psi^* \frac{\partial}{\partial x} \psi \, d^3r \right\}$$

$$= -i\hbar \left\{ \int_V \left(\frac{\partial \psi^*}{\partial t} \right)\left(\frac{\partial \psi}{\partial x} \right) d^3r + \int_V \psi^* \frac{\partial}{\partial t} \frac{\partial \psi}{\partial x} \, d^3r \right\}$$

$$= \left\{ \int_V \left(-\frac{\hbar^2}{2m} \nabla^2 \psi^* + V\psi^* \right) \frac{\partial \psi}{\partial x} d^3r - \int_V \psi^* \frac{\partial}{\partial x}\left(-\frac{\hbar^2}{2m} \nabla^2\psi + V\psi \right) d^3r \right\}$$

(using time dependent Schrödinger equation)

$$= \left[-\frac{\hbar^2}{2m} \int_V \left(\nabla^2 \psi^* \frac{\partial \psi}{\partial x} - \psi^* \frac{\partial}{\partial x}(\nabla^2\psi) \right) d^3r \right.$$

$$\left. + \int_V \left\{ V\psi^* \frac{\partial \psi}{\partial x} - \psi^* \frac{\partial}{\partial x}(V\psi) \right\} d^3r \right]$$

We now put $\dfrac{\partial \psi}{\partial x} = \phi_1$ and $\psi^* = \phi_2$, then the integral is

$$\int_V \left(\phi_1 \nabla^2 \phi_2 - \phi_2 \nabla^2 \phi_1 \right) d^3r = \int_V \left(\nabla \cdot \phi_1 \nabla \phi_2 - \phi_2 \nabla \phi_1 \right) d^3r$$

$$= \int_s \left(\phi_1 \nabla \phi_2 - \phi_2 \nabla \phi_1 \right) \cdot ds = 0$$

so that

$$\frac{d}{dt}\langle p_x \rangle = \int \left[V\psi^* \frac{\partial \psi}{\partial x} - \psi^* \frac{\partial}{\partial x}(V\psi) \right] d^3r$$

$$= \int -\psi^* \frac{\partial V}{\partial x} \psi \, d^3r \qquad (31)$$

or

$$\frac{d}{dt}\langle p \rangle = -\langle \nabla V \rangle \qquad (32)$$

This is identical with the classical result $\dfrac{d\vec{p}}{dt} = -\text{grad } V$.

Hence average motion of a wave packet is identical with the classical motion of the corresponding particle.

3.8 Energy Eigenfunctions

The time dependent Schrödinger equation is

$$i\hbar \frac{\partial \psi}{\partial t} = -\frac{\hbar^2}{2m} \nabla^2 \psi + V(\vec{r},t)\psi \qquad (33)$$

This describes the motion of a particle of mass m in a force field given by

$$\vec{F}(\vec{r},t) = -\nabla V(\vec{r},t) \qquad (34)$$

Consider the case when the potential energy V does not depend on time.

Separation of the Wave Equation

We consider a particular solution of Eqn. (33) ($V(\vec{r},t) \equiv V(r)$) that can be written as

$$\psi(\vec{r},t) = u(\vec{r}) f(t) \qquad (35)$$

Then a general solution can be written as a sum of such separated solutions. It we substitute this into Eqn. (33) and divide through by Eqn. (35), we get

$$\frac{i\hbar}{f}\frac{df}{dt} = \frac{1}{u}\left[-\frac{\hbar^2}{2m}\nabla^2 u + V(\vec{r})u\right] \tag{36}$$

Since the l.h.s. of this Eqn. depends only on t and the r.h.s. only on $\vec{r}$, both sides must be equal to the same constant which we call E. Then the equation for f can be readily integrated to give

$$f(t) = C\,e^{-iEt/\hbar} \tag{37}$$

where C is an arbitrary constant and the equation for u becomes

$$\left[-\frac{\hbar^2}{2m}\nabla^2 + V(\vec{r})\right]u(\vec{r}) = E\,u(\vec{r}) \tag{38}$$

Since Eqn. (38) is homogenous in u, the constant C may be chosen to normalize u. Then a particular solution is

$$\psi(\vec{r},t) = u(\vec{r})\,e^{-iEt/\hbar} \tag{39}$$

The Schrödinger equation may be rewritten as

$$H\psi = E\psi \tag{40}$$

where

$$H \rightarrow i\hbar\frac{\partial}{\partial t}$$

and

$$E \rightarrow p^2/2m + V$$

or

$$i\hbar\frac{\partial}{\partial t}\psi = E\psi \tag{41}$$

This is called an *eigenvalue equation*; ψ is an *eigenfunction* of the operator that appears on the left and the multiplying constant (on the r.h.s.) i.e. E is the corresponding *eigenvalue*. Thus Eqn. (41) means that when the state of the particle is described by an energy eigenfunction, the energy of the particle has a definite value, given by the eigenvalue E.

Expansion in Energy Eigenfunctions

All the eigenfunctions $u_E(\vec{r})$ of the total energy operator constitute a complete set of functions. Consequently an arbitrary continuous function can be expanded in terms of them:

$$\psi(\vec{r}) = \sum_E A_E u_E(\vec{r}) \tag{42}$$

where it is assumed that $\psi(\vec{r})$ at a particular instant of time is normalized in a box (L^3) and obeys periodic boundary conditions at the walls of the box.

3.9 General Solution of the Schrödinger Equation

If the potential energy $V(\vec{r})$ is independent of t and we know the solution of the Schrödinger equation (33) at a particular time, we can write a formal solution at any time. We expand $\psi(\vec{r},t)$ in energy eigenfunctions at the time t, in which case the expansion coefficients depend on the time:

$$\psi(\vec{r},t) = \sum_E A_E(t) \, u_E(\vec{r}) \tag{43}$$

$$A_E(t) = \int u_E^*(\vec{r}) \, \psi(\vec{r},t) \, d^3r \tag{44}$$

Substitution of Eqn. (43) into the wave equation (41) gives

$$i\hbar \sum_E u_E(\vec{r}) \frac{d}{dt} A_E(t) = \sum_E A_E(t) \, E \, u_E(\vec{r}) \tag{45}$$

Then using the orthonormal property[*] of the u_E's, Eqn. (45) implies

$$i\hbar \frac{\partial}{\partial t} A_E(t) = E \, A_E(t)$$

Integrating, we get

$$A_E(t) = A_E(t_0) \, e^{-iE(t-t_0)/\hbar} \tag{46}$$

We note that energy probability $P(E) \equiv |A_E|^2$ is constant in time. Thus if $\psi(\vec{r}, t)$ is known at the time $t = t_0$, the solution at any time t is given in terms of Eqns (43), (44) and (46) by

$$\psi(\vec{r},t) = \sum_E A_E(t_0) \, e^{-iE(t-t_0)/\hbar} \, u_E(\vec{r}) \tag{47}$$

where

$$A_E(t_0) = \int u_E^*(\vec{r}') \, \psi(\vec{r}',t_0) \, d^3r \tag{48}$$

Example 3.15: A particle of mass m moves in a one dimensional box of length L with the potential

$$V = \infty \quad x < 0$$
$$V = 0 \quad 0 < x < L$$
$$V = \infty \quad x > L$$

At a certain instant $t = 0$, the wave function of this particle is known to have the form

$$\psi = \sqrt{\frac{30}{L^5}} \, x(L-x) \text{ for } 0 < x < L = 0 \text{ otherwise.}$$

Find an expression for $\psi(x, t > 0)$ as a series and expressions for the coefficients in the series.

Solution: The eigenfunctions and the corresponding energy eigenvaleus are

$$\psi_n(x) = \sqrt{\frac{2}{L}} \sin\left(\frac{n\pi x}{L}\right) \quad E_n = \frac{\hbar^2}{2m} \left(\frac{\pi n}{L}\right)^2 \quad n = 1, 2, 3...$$

We also know that

$$\psi(x, t > 0) = \sum_n A_n(t = 0) \, e^{-iE_n t/\hbar} \, \psi_n(x)$$

(*) (i) If $E \neq E'$, u_E and $u_{E'}$ are orthogonal.

(ii) $\sum_E u_E^*(\vec{r}') \, u_E(\vec{r}) = 0$ for $\vec{r}' \neq \vec{r}$

otherwise $\int \sum u_E^*(\vec{r}') \, u_E(\vec{r}) \, d^3r' = 1$.

See also article (3.12) for definition of orthonormality

where
$$A_n(t=0) = \int_0^L \psi_n^*(x)\,\psi(x, t=0)\,dx$$

or
$$A_n(t=0) = \int_0^L \sqrt{\frac{2}{L}}\,\sin\left(\frac{n\pi x}{L}\right)\sqrt{\frac{30}{L^5}}\,x(L-x)\,dx$$

$$= \sqrt{\frac{2}{L}}\sqrt{\frac{30}{L^5}}\int_0^L \sin\left(\frac{n\pi x}{L}\right)(xL - x^2)\,dx$$

$$= \sqrt{\frac{60}{L^6}}\left[\left|\frac{-\cos(n\pi x/L)}{n\pi/L}(xL - x^2)\right|_0^L + \int_0^L \frac{L}{n\pi}\cos\left(\frac{n\pi x}{L}\right)(L - 2x)\,dx\right]$$

$$= \sqrt{\frac{60}{L^6}}\left[\left\{-\frac{L}{n\pi}(xL - x^2)\cos\left(\frac{n\pi x}{L}\right)\right\}_0^L + \left\{\left(\frac{L}{n\pi}\right)^2\sin\left(\frac{n\pi x}{L}\right)(L - 2x)\right\}_0^L\right.$$

$$\left. -\left(\frac{L}{n\pi}\right)^2\int_0^L \sin\left(\frac{n\pi x}{L}\right)(-2)\,dx\right]$$

$$= \frac{2\sqrt{15}}{L^3}\left\{\text{(The first and the second term contribute zero)}\right.$$

$$\left. +2\left(\frac{L}{n\pi}\right)^3\left[-\cos\left(\frac{n\pi x}{L}\right)\right]_0^L\right\}$$

$$= \frac{4\sqrt{15}}{(n\pi)^3}\{1 - \cos n\pi\} = \frac{4\sqrt{15}}{(n\pi)^3}\left\{1 - (-1)^n\right\}$$

Hence
$$\psi(x, t>0) = \sum_{n=1}^{\infty} A_n(t=0)\,\psi_n(x)\exp\left(-\frac{iE_n t}{\hbar}\right)$$

$$= \sum_{n'=1}^{\infty} 4\sqrt{15}\left[1 - (-1)^{n'}\right]\frac{1}{(n\pi)^3}\sqrt{\frac{2}{L}}\sin\left(\frac{n'\pi x}{L}\right)\times\exp\left[-\frac{i\hbar}{2m}\frac{n'^2\pi^2}{L^2}t\right]$$

Let
$$n' = 2n + 1$$

$$\therefore \quad \left[1 - (-1)^{2n+1}\right] = \left[1 - (-1)^{2n}(-1)\right] = \left[1 - (-1)\right] = 2$$

So,
$$\psi(x, t>0) = \sum_{n=0}^{\infty} 8\sqrt{\frac{30}{L}}\frac{1}{(2n+1)^3\pi^3}\sin\frac{(2n+1)\pi x}{L}$$

$$\times\exp\left[-\frac{i\hbar}{2m}\left(\frac{(2n+1)\pi}{L}\right)^2 t\right]$$

3.10 Hermitian Operators and their Properties

An operator A_{op} associated with a dynamical varibale is said to be Hermitian if its average or expectation value in any eigenstate ψ is real.

The average value of an operator A for the normalized eigenfunction ψ is expressed as

$$\langle A \rangle = \int \psi * A\psi \, dx \qquad (49)$$

If <A> is to be real, it means that the imaginary part of $\int \psi * A\psi \, dx$ must be zero i.e.

Imaginary part of $\int \psi * A\psi \, dx = 0$ $\qquad (50)$

The complex conjugate of the function $\int \psi * A\psi \, dx$ is

$$\int (\psi * A\psi) * dx = \int \psi (A\psi) * dx = \int A * \psi * \psi \, dx$$

The condition expressed by Eqn. (50) is satisfied only if the function $\int \psi * A\psi \, dx$ is equal to the function $\int A * \psi * \psi \, dx$. Thus the condition for an operator A to be Hermitian is

$$\int \psi * A\psi \, dx = \int A * \psi * \psi \, dx \qquad (51)$$

Example 3.16: Show that the momentum operator $p = -i\hbar\nabla$ is a Hermitian operator.

Solution: If the momentum operator is Hermitian, it should satisfy the condition

$$\psi * (-i\hbar\nabla)\psi \, d^3r = \int (-i\hbar\nabla\psi) * \psi \, d^3r$$

or $\qquad -i\hbar \int \psi * (\nabla\psi) \, d^3r = i\hbar \int (\nabla\psi*)\psi \, d^3r$ $\qquad (i)$

Now, $\qquad\qquad \nabla(\psi*\psi) = \psi*(\nabla\psi) + (\nabla\psi*)\psi$

or $\qquad\qquad \psi*(\nabla\psi) = \nabla(\psi*\psi) - (\nabla\psi*)\psi$

Integrating both sides, we get

$$\int \psi * (\nabla\psi) \, d^3r = \int \nabla(\psi * \psi) \, d^3r - \int (\nabla(\psi*)\psi \, d^3r$$

$$= \nabla \int \psi * \psi \, d^3r - \int (\nabla\psi*)\psi \, d^3r$$

$$= 0 - \int (\nabla\psi*)\psi \, d^3r$$

$$\left(\because \int \psi * \psi \, d^3r = N, \text{a constant} \right)$$

Multiplying both sides by $-i\hbar$, we get

$$\int \psi * (-i\hbar\nabla\psi) \, d^3r = \int (i\hbar\nabla\psi*)\psi \, d^3r$$

or $\qquad \int \psi * (-i\hbar\nabla\psi) \, d^3r = \int (-i\hbar\nabla\psi) * \psi \, d^3r$

Hence, the momentum operator is a Hermitian operator.

Properties of Hermitian Operators

[1] Any set of mutually orthogonal eigenfunctions is linearly independent.

[2] They have real eigenvalues.

[3] Two eigenfunctions of a Hermitian operator belonging to different eigenvalues are orthogonal.

[4] If two Hermitian operators commute, their product is also a Hermitian operator.

Proof of [1]: Let the orthogonal set of wave functions be (ψ_i) and suppose that

$$c_1\psi_1 + c_2\psi_2 + \ldots + c_n\psi_n = 0$$

is satisfied. Multiply this equation by ψ_j^* and integrate over whole space.

$$\sum_{i \neq j} c_i \int \psi_j^* \psi_i \, d^3r + c_j \int \psi_j^* \psi_j \, d^3r = 0$$

As the functions are orthogonal, the first term vanishes, therefore

$$c_j \int \psi_j^* \psi_j \, d^3r = 0$$

Obviously the integral is non zero so $c_j = 0$. This is true for all j hence the orthogonal functions are linearly independent.

Proof of [2]: Let λ be an eigenvalue of the Hermitian operator A in the state described by the normalized wave function ψ. The eigenvalue equation is

$$A\psi = \lambda\psi \tag{52}$$

Its complex conjugate is

$$A^*\psi^* = \lambda^*\psi^* \tag{53}$$

Now, the condition for Hermitian operator is

$$\int \psi^* A\psi \, dx = \int A^*\psi^*\psi \, dx \tag{54}$$

using Eqns. (52) and (53), we get from Eqn. (54)

$$\int \psi^* \lambda\psi \, dx = \int \lambda^*\psi^*\psi \, dx$$

or

$$\lambda \int \psi^*\psi \, dx = \lambda^* \int \psi^*\psi \, dx$$

or

$$(\lambda - \lambda^*) \int \psi^*\psi \, dx = 0$$

But $\int \psi^*\psi \, dx = 1$ for a normalized wave function ψ

$\therefore$
$$\lambda - \lambda^* = 0$$

or
$$\lambda = \lambda^*$$

which is possible only if λ is a real number.

Therefore, eigenvalue of a Hermitian operator is real.

Proof of [3]: Let A be any Hermitian operator and ψ_1 and ψ_2 be two normalized eigenfunctions of operator A. If λ_1 and λ_2 are the distinct eigenvalues of operator A corresponding to the eigenfunctions ψ_1 and ψ_2, then the eigenvalue equations are

$$A\psi_1 = \lambda_1\psi_1 \text{ and } A\psi_2 = \lambda_2\psi_2 \tag{55}$$

Taking complex conjugate:

$$A^*\psi_1^* = \lambda_1^*\psi_1^* \text{ or } A^*\psi_1^* = \lambda_1\psi_1^* \tag{56}$$

and
$$A^*\psi_2^* = \lambda_2^*\psi_2^* \text{ or } A^*\psi_2^* = \lambda_2\psi_2^*$$

Since λ_1 and λ_2 being the eigenvalues of Hermitian operator, are real ($\lambda_1{}^*=\lambda_1$ and $\lambda_2{}^* = \lambda_2$).

The general condition for Hermitian operator is

$$\int \psi_2^* A \psi_1 \ dx = \int A * \psi_2^* \psi_1 \ dx$$

Therefore from Eqns (55) and (56), we get

$$\int \psi_2^* \lambda_1 \psi_1 \ dx = \int \lambda_2 \psi_2^* \psi_1 \ dx$$

or

$$\lambda_1 \int \psi_2^* \psi_1 \ dx = \lambda_2 \int \psi_2^* \psi_1 \ dx$$

or

$$(\lambda_1 - \lambda_2) \int \psi_2^* \psi_1 \ dx = 0$$

Since $\lambda_1 \neq \lambda_2$ therefore $\int \psi_2^* \psi_1 \ dx = 0$. This implies that ψ_1 and ψ_2 are orthogonal eigenfunctions.

Proof of [4]: Let A and B be two Hermitian operators and ψ_1 and ψ_2 be two normalized eigenfunctions. The operators A and B commute, i.e. $AB = BA$ and $A*B* = B*A*$. We have to prove now that AB is also a Hermitian operator. Now let us consider $\int \psi_1^* AB \ \psi_2 \ dx$.

If A is Hermitian, then

$$\int \psi_1^* AB \ \psi_2 \ dx = \int \psi_1^* A(B\psi_2) \ dx = \int A * \psi_1^* (B \ \psi_2) \ dx \qquad (57)$$

Again if B is Hermitian, then

$$\int A * \psi_1^* B \ \psi_2 \ dx = \int B * A * \psi_1 \psi_2 \ dx \qquad (58)$$

Thus from Eqns (57) and (58)

$$\int \psi_1^* AB \ \psi_2 \ dx = \int B * A * \psi_1 \psi_2 \ dx = \int A * B * \psi_1 \psi_2 \ dx$$

$$\text{(since } B*A* = A*B*)$$

This is the condition for the product operator AB to be Hermitian. Thus if A and B are commuting Hermitian operators, then their product operator AB is also Hermitian.

Definition: An eigenvalue q *of an eigenvalue equation is* m^{th} *order degenerate if there are* m *linearly independent eigenfunctions corresponding to this eigenvalue.*

3.11 Completeness of Eigenfunction Sets

A set of functions constitutes a complete set of linearly independent eigenfunctions corresponding to the eigenvalue q if with any other function of eigenvalue q the set becomes linearly dependent. In other words, the set of functions (ψ_j) is *complete* if there is no other function, which lies in this set of linearly independent functions. It is to be noted here that if ψ is a state function which does not fall in the complete set, then ψ can be expressed as a linear combination of ψ_j's.

Theorem: If ψ_j ($j = 1, 2, ..., m$) constitute a complete set of eigenfunctions with the eigenvalue q of mth order degeneracy for some operator, then any other eigenfunction of this eigenvalue may be expanded in terms of this complete set.

Proof: Let

$$a\psi - \sum_{j=1}^{m} c_j \psi_j = 0$$

If $a = 0$, then all the c's in this equation must be zero, since ψ_j are linearly independent. If this were the only possibility, ψ would constitute a member of the linearly independent set of functions. Since, however, it was assumed that the set of functions ψ_1 through ψ_m was linearly independent and complete, there must be a solution to the equation with $a \neq 0$. If $a \neq 0$, then

$$\psi = \frac{1}{a} \sum_j c_j \psi_j$$

Theorem: If ψ_j constitute a complete set of eigenfunctions with eigenvalues q of mth order degeneracy then linear combinations of ψ_j may be taken to form a set of m mutually orthogonal functions. These m mutually orthogonal functions are also linearly independent, and can be used to expand any other eigenfunctions corresponding to a particular eigenvalue. This can be verified by Schmidt orthogonalization procedure.

3.12 Schmidt Orthogonalization Procedure

Designate the set of independent functions corresponding to the eigenvalue q as ψ_j $(j = 1, 2, ..., m)$. Choose any one of these functions, say ψ_1 as the first member of a new set u_j $(j = 1, 2, ..., m)$

$$u_1 \equiv \psi_1 \tag{59}$$

Designate $\qquad \int |u_1|^2 \, d^3\vec{r} \equiv c_{11} \qquad \int \bar{u}_1 \psi_2 \, d^3\vec{r} \equiv c_{12}$

Take $\qquad\qquad u_2 \equiv \dfrac{c_{12}}{c_{11}} u_1 - \psi_2 \tag{60}$

Then $\qquad\qquad \int \bar{u}_1 u_2 \, d^3 r \equiv \dfrac{c_{12}}{c_{11}} \int |u_1|^2 \, d^3 r - \int \bar{u}_1 \psi_2 \, d^3 r$

$$= c_{12} - c_{12} = 0$$

Designate $\int |u_2|^2 \, d^3 r \equiv c_{22}; \quad \int \bar{u}_1 \psi_3 \, d^3\vec{r} \equiv c_{13} \quad \int \bar{u}_2 \psi_3 \, d^3 r = c_{23}$

Take $\qquad\qquad u_3 \equiv \dfrac{c_{13}}{c_{11}} u_1 + \dfrac{c_{23}}{c_{22}} u_2 - \psi_3 \tag{61}$

Then $\qquad \int \bar{u}_1 u_3 \, d^3 r \equiv \dfrac{c_{13}}{c_{11}} \int |u_1|^2 \, d^3 r + \dfrac{c_{23}}{c_{22}} \int \bar{u}_1 u_2 \, d^3 r - \int \bar{u}_1 \psi_3 \, d^3 r$

$$= c_{13} + \dfrac{c_{23}}{c_{22}} 0 - c_{13} = 0$$

Similarly $\qquad \int \bar{u}_2 u_3 \, d^3\vec{r} = 0$

Thus, we have obtained u_1, u_2 and u_3 and the procedure can be extended to obtain u_4, u_5, ... u_m. Since eigenfunctions corresponding to different eigenvalues are already orthogonal, the above procedure can be used to obtain a complete orthogonal set of eigenfunctions u_j $(j = 1, 2, ..., m)$ for any Hermitian operator.

If
$$\int \bar{\psi}_j \psi_k \, d^3r = \delta_{jk} \tag{62}$$

(ψ_i forming a complete set for an operator)

then the set of eigenfunction (ψ_j) is said to form a *complete orthonormal set*.

The above process of obtaining orthogonal functions u_j's from a given set of independent eigenfunctions ψ_j's is known as *Schmidt orthogonalization procedure*.

3.13 Compatibility

If there exists a complete set of linearly independent state functions ψ_j such that ψ_j is an eigenfunction of both the operators R and S corresponding to physical observables, the corresponding observables are said to be *compatible*.

By "compatible observables" is meant that both R and S are completely predictable for the complete set of states ψ_j. (Position and momentum measurements are not compatible). If two observables r and s are compatible, they are simul-taeously measurable in a particular state and then if either is measured a unique result ψ_j is obtained. This implies that the state function ψ_j is an eigenfunction of the representative operators R and S i.e.

$$R\psi_j = r_j \psi_j$$
$$S\psi_j = s_j \psi_j$$

Theorem: If two observables are compatible, their operators commute.

Proof:
$$S\psi_j = s_j \psi_j \quad R\psi_j = r_j \psi_j$$
$$RS\psi_j = R(s_j \psi_j) = s_j R\psi_j = s_j r_j \psi_j$$
$$SR\psi_j = S(r_j \psi_j) = r_j S\psi_j = r_j s_j \psi_j$$

$\therefore$
$$(RS - SR)\psi_j = 0$$
$$\sum_j c_j (RS - SR)\psi_j = 0$$

or
$$(RS - SR)\sum_j c_j \psi_j = 0$$

or
$$\left(RS - SR\right)\psi = 0$$

ψ is any arbitrary wave function.

$\therefore$
$$[R, S] \equiv RS - SR = 0.$$

Theorem: If two operators R and Q commute and if one of these operators has non degenerate eigenvalues, then its eigenfunctions are also the eigenfunctions of the other operator.

Proof:
$$Q\psi_j = q_j \psi_j \tag{63}$$

(q is assumed to be non degenerate)

then
$$Q(R\psi_j) = R(Q\psi_j)$$
$$= R(q_j \psi_j) = q_j(R\psi_j) \tag{64}$$

This equation states that the function $R\psi_j$ is an eigenfunction of the operator Q. But the operator Q has been assumed to have only non degenerate eigenvalues.

Consequently, function $R\psi_j$ can differ from the original eigenfunction ψ_j by at most a multiplicative constant, (because Eqn. (64)

$$\Rightarrow \qquad Q\psi_j = q_j\psi'_j \qquad (65)$$

$$(\psi'_j = R\psi_j)$$

Eqns (63) and (65) imply that eigenvalues are the same but eigenfunction are different, which is not possible).

$$\therefore \qquad R\psi_j = r_j\psi_j$$

$\therefore$ The wavefunction ψ_j is simultaeously an eigenfunction of both Q and R.

Theorem: If Q and R are operators which commute with each other, there exists a complete set of eigenstates which are simultaneously eigenstates of both Q and R.

Proof:

    Case 1 : Non-degenerate eigenvaleus (as treated in the preceding theorem)

    Case 2 : Degenerate eigenvalues

Let

$$Q\psi_j = q\psi_j \qquad (66)$$

    where q is the mth order degenerate eigenvalue of Q. Operating on Eqn. (66) with R.

$$R(Q\psi_j) = q(R\psi_j) \quad (Q \text{ and } R \text{ commute } \because RQ = QR)$$

or

$$Q(R\psi_j) = q(R\psi_j) \qquad (67)$$

$\therefore$ $R\psi_j$ is an eigenfunction of Q, and can be expanded in the set of functions ψ_j

$$R\psi_j = \sum_{k=1}^{m} q_{jk}\psi_k \qquad (68)$$

Multiplying by a constant c_j and summing over j, this becomes

$$R\sum_{j=1}^{m} c_j\psi_j = \sum_{j,k} c_j q_{jk}\psi_k \qquad (69)$$

Now, assume that

$$\sum_j c_j q_{jk} \text{ or } \sum_j q_{jk}c_j = rc_k \qquad (70)$$

then

$$R\sum_j c_j\psi_j = \sum_k r\,c_k\psi_k$$

or

$$\sum_{j=1}^{m} c_j q_{jk} = r\sum_{j=1}^{m} c_j\delta_{jk} \qquad (71)$$

Eqn. (70) constitutes a set of m linear equations ($k = 1, 2,..., m$) in m unknowns. c_k are m unknowns c_1 ...... c_m.

Eqn. (71) has a nonzero solution for the c_k provided the constant r satisfies the characteristic equation

$$\left| q_{jk} - r\delta_{jk} \right| = 0 \qquad \left(\because \sum_j c_j \left(q_{jk} - r\delta_{jk} \right) = 0 \right)$$

$$
\text{or} \quad
\begin{vmatrix}
q_{11}-r & q_{12} & .. & .. & q_{1m} \\
q_{21} & q_{22}-r & .. & .. & q_{2m} \\
.. & & & & \\
.. & & & & \\
q_{m1} & & .. & .. & q_{mm}-r
\end{vmatrix}
= 0
\tag{72}
$$

Expansion of this determinant leads to an mth order equation equation for r, which has m roots. With each root r_ℓ there is associated a solution $c_k^{(\ell)}$ for the c's.

$$
\sum_j q_{jk} c_j^{(\ell)} = r_\ell c_k^{(\ell)}
\tag{73}
$$

or

$$
R\left(\sum_j c_j^{(\ell)} \psi_j \right) = r_\ell \left(\sum_k c_k^{(\ell)} \psi_k \right)
\tag{74}
$$

or

$$
R u_\ell = r_\ell u_\ell
$$

where

$$
u_\ell = \sum_j c_j^{(\ell)} \psi_j
\tag{75}
$$

From Eqn. (66)

$$
Q \sum_j c_j^{(\ell)} \psi_j = q \sum_j c_j^{(\ell)} \psi_j
$$

or

$$
Q u_\ell = q u_\ell
$$

The m functions given by Eqn. (75) are linearly independent.

$\therefore$ u_m constitute a complete set of simultaneous eigenfunctions of R and Q.

Example 3.17: Can we measure simultaneously the kinetic and potential energies of a particle with an arbitrary precision.

Solution

Operator for kinetic energy $\quad \hat{K} = -\dfrac{\hbar^2}{2m} \nabla^2$

Operator for potential energy $\hat{V} = V(r)$

$$
\left[\hat{K}, \hat{V} \right] \psi = \left[-\frac{\hbar^2}{2m} \nabla^2, V \right] \psi
$$

$$
= -\frac{\hbar^2}{2m} \nabla^2 (V\psi) - V\left(-\frac{\hbar^2}{2m} \nabla^2 \right)\psi
$$

$$
= -\frac{\hbar^2}{2m} V\nabla^2\psi - \frac{\hbar^2}{2m}(\nabla^2 V)\psi + \frac{\hbar^2}{2m} V\nabla^2\psi
$$

$$
= -\frac{\hbar^2}{2m}(\nabla^2 V)\psi
$$

Since this is not zero, the operators $\hat{K}$ and $\hat{V}$ do not commute, so their simultaneous measurement is not possible. This is possible if $\nabla^2 V = 0$ or $V =$ constant.

Example 3.18: Find the value of x for which the following set of vectors is linearly dependent:
$(1, 2, 3)$, $(4, 5, 6)$ and $(x, 8, 9)$.

Solution: Let $u_1 = (1, 2, 3)$, $u_2 = (4, 5, 6)$ and $u_3 = (x, 8, 9)$

$$\alpha_1 u_1 + \alpha_2 u_2 + \alpha_3 u_3 = 0$$

$\Rightarrow \qquad \alpha_1(1,2,3) + \alpha_2(4,5,6) + \alpha_3(x,8,9) = (0,0,0)$

or
$$\alpha_1 + 4\alpha_2 + x\alpha_3 = 0 \qquad\qquad (i)$$
$$2\alpha_1 + 5\alpha_2 + 8\alpha_3 = 0 \qquad\qquad (ii)$$
$$3\alpha_1 + 6\alpha_2 + 9\alpha_3 = 0 \qquad\qquad (iii)$$

For the given set to be linearly dependent the above three equations should be satisfied simultaneously without all the α's being zero. For this there should be a non zero solution so that the determinant of the coefficients of α's should vanish, $3\alpha_1 + 6\alpha_2 + 9\alpha_3 = 0\,h$, i.e.

$$\begin{vmatrix} 1 & 4 & x \\ 2 & 5 & 8 \\ 3 & 6 & 9 \end{vmatrix} = 0 \qquad\qquad (iv)$$

or for
$$\begin{vmatrix} a_1 & b_1 & c_1 \\ a_2 & b_2 & c_2 \\ a_3 & b_3 & c_3 \end{vmatrix} = 0$$

$$a_1(b_2 c_3 - b_3 c_2) - a_2(b_1 c_3 - b_3 c_1) + a_3(b_1 c_2 - b_2 c_1) = 0$$

or from (iv)
$$1(45 - 48) - 2(36 - 6x) + 3(32 - 5x) = 0$$
$$-3 - 72 + 12x + 96 - 15x = 0$$

or
$$-3x = -24 + 3$$

or
$$x = 7$$

Example 3.19: Obtain a set of four orthonormal vectors from the following:
$$u_1 = (1, 1, 0, 1)$$
$$u_2 = (2, 0, 0, 1)$$
$$u_3 = (0, 2, 3, -2)$$
$$u_4 = (1, 1, 1, -5)$$

Solution: We use Schmidt's method for this.

Let
$$v_1 = u_1 = (1, 1, 0, 1)$$

then
$$v_2 = u_2 + a_{21}v_1$$

where
$$a_{21} = -\frac{(v_1, u_2)}{(v_1, v_1)}$$

$$-\frac{(1.2 + 1.0 + 0.0 + 1.1)}{(1.1 + 1.1 + 0.0 + 1.1)} = -\frac{3}{3} = -1$$

$$\therefore \qquad v_2 = u_2 - v_1 = (2, 0, 0, 1) - (1, 1, 0, 1)$$
$$= (1, -1, 0, 0)$$

Now,
$$v_3 = u_3 + a_{31}v_1 + a_{32}v_2$$

where
$$a_{31} = -\frac{(v_1, u_3)}{(v_1, v_1)} = 0 \text{ and } a_{32} = -\frac{(v_2, u_3)}{(v_2, v_2)}$$

or
$$a_{32} = 1 \text{ on evaluation.}$$

$$\therefore \qquad v_3 = u_3 + 0 + v_2$$
$$= (1, 1, 3, -2)$$

Taking
$$v_4 = u_4 + a_{41}v_1 + a_{42}v_2 + a_{43}v_3$$

where
$$a_{41} = -\frac{(v_1, u_4)}{(v_1, v_1)} = 1, \ a_{42} = -\frac{(v_2, u_4)}{(v_2, v_2)} = 0, \ a_{43} = -\frac{(v_3, u_4)}{(v_3, v_3)} = -1$$

$$\therefore \qquad v_4 = u_4 + v_1 + 0 + (-v_3) = (1, 1, -2, -2).$$

The vectors v_1, v_2, v_3, v_4 are orthogonal to each other. In order to find the orthonormal set, we divide each of these by its norm, we get

$$v_1' = \frac{1}{\sqrt{3}}(1,1,0,1); \quad v_2' = \frac{1}{\sqrt{2}}(1,-1,0,0);$$

$$v_3' = \frac{1}{\sqrt{15}}(1,1,3,2); \quad v_4' = \frac{1}{\sqrt{10}}(1,1,-2,-2).$$

3.14 The Commutation Rule and Poisson Brackets

The commutation rule
$$[x, p] = i\hbar$$
is a fundamental rule and forms the basis for the formal structure of quantum theory. This point was emphasized by Dirac. The commutation rule for x and p is, in a sense, a precise formulation of the correspondence principle. This is clearly apparent in the Hamiltonian formulation of classical mechanics.

The Poisson bracket
$$\{f, g\} = \frac{\partial f}{\partial x}\frac{\partial g}{\partial p} - \frac{\partial f}{\partial p}\frac{\partial g}{\partial x} \qquad (76)$$

in which $f(x, p)$ and $g(x, p)$ are functions of the coordinate x and momentum p (for a particle) plays a fundamental role in this theory. For example, the equations of motion for a particle whose Hamiltonian is
$$H = \frac{p^2}{2m} + V(x)$$

are
$$\frac{\partial x}{\partial t} = \{x, H\} \qquad (77)$$

and
$$\frac{\partial p}{\partial t} = \{p, H\} \qquad (78)$$

Further, it is evident that
$$\{x, p\} = 1 \qquad (79)$$

The parallelism between Poisson brackets and quantum mechanical commutator allows an elegant expression of the correspondence principle first formulated by Dirac:

The quantum mechanical operators f_{op} and g_{op} which in quantum theory, replace the classically defined functions f and g, must always be such that the commutator of f_{op} and g_{op} corresponds to the Poisson bracket of f and g according to

$$i\hbar \{f, g\} \to [f_{op}, g_{op}] \tag{80}$$

By this assumption, the relation

$$[x, p] = i\hbar$$

is a consequence of Eqn. (79).

Every classical equation in Poisson brackets has its quantum counterpart in the operator formalism. Thus, the operator equation $[H_{op}, P_{op}] = 0$ which is true for a free particle, implies that

$$\{H, p\} = 0 \tag{81}$$

where $H = p^2/2m$ is the Hamiltonian function for a free particle. As a consequence $\dot{p} = 0$, that is, the momentum of a free particle is a constant of the motion in the classical theory. This can be generalised to three dimensional case. If the coordinates of the particle are denoted by x_1, x_2, x_3 and the momenta by p_1, p_2, p_3, then the Poisson bracket is

$$\{f, g\} = \sum_{i=1}^{3} \left(\frac{\partial f}{\partial x_i} \frac{\partial g}{\partial p_i} - \frac{\partial f}{\partial p_i} \frac{\partial g}{\partial x_i} \right) \tag{82}$$

The functions f and g depend upon the six independent variables x_i, p_i and the partial derivative signs refer to this set of variables. In this notation the Hamiltonian function for a particle of potential energy $V(x,y,z) = V(x_1, x_2, x_3)$ is

$$H = \sum_i \frac{p^2}{2m} + V(x_1 + x_2 + x_3)$$

and the equations of motion in Poisson brackets are

$$\frac{\partial x_i}{\partial t} = \dot{x}_i = \{x_i, H\} \text{ and } \dot{p}_i = \{p_i, H\} \tag{83}$$

By means of the identity (82), the fundamental Poisson bracket relations

$$\{x_i, p_j\} = \delta_{ij} \tag{84}$$

are easily obtained with

$$\{x_i, p_j\} = i\hbar \delta_{ij} \tag{85}$$

Commutator Algebra

In analogy with Poisson brackets we have

$$[A, B] = -[B, A] \tag{i}$$

$$[A, B + C] = [A, B] + [A, C] \tag{ii}$$

$$[A+B, C] = [A, C] + [B, C] \tag{iii}$$

$$[A, BC] = [A, B]C + B[A, C] \tag{iv}$$

$$[AB, C] = [A, C]B + A[B, C] \tag{v}$$

$$[A, [B, C]] + [B, [C, A]] + [C, [A, B]] = 0 \tag{vi}$$

$$\tag{86}$$

3.14.1 Commutation Relation between Position and Momentum

Example 3.20:
$$x \rightarrow \hat{x}$$

and
$$p_x \rightarrow \hat{p} = \frac{\hbar}{i} \frac{\partial}{\partial x}$$

Consider the operation of commutator $\left[\hat{x}, \hat{p}_x\right]$ on a function $\psi(x)$

i.e.
$$\left[\hat{x}, \hat{p}_x\right] \psi(x) = \left(\hat{x} \, \hat{p}_x - \hat{p}_x \, \hat{x}\right) \psi(x) \qquad (i)$$

We have
$$\hat{x} \hat{p}_x \psi = x \frac{\hbar}{i} \frac{\partial \psi}{\partial x}$$

and
$$\hat{p}_x \hat{x} \, \psi = \frac{\hbar}{i} \frac{\partial}{\partial x}(x\psi) = \frac{\hbar}{i}\left(\psi + x \frac{\partial \psi}{\partial x}\right)$$

$\therefore$ Eqn. (*i*) gives

$$\left[\hat{x}, \hat{p}_x\right] = \frac{\hbar}{i} \frac{\partial \psi}{\partial x} - \frac{\hbar}{i}\left(\psi + x \frac{\partial \psi}{\partial x}\right)$$

$$= \frac{\hbar}{i}\left(x \frac{\partial \psi}{\partial x} - \psi - x \frac{\partial \psi}{\partial x}\right) = -\frac{\hbar}{i} \psi$$

$$= i\hbar \psi$$

so, $\left[\hat{x}, \hat{p}_x\right] = i\hbar$

Example 3.21: $\left[\hat{x}^2, \hat{p}_x\right] = \left[\hat{x} \, \hat{x}, \hat{p}_x\right]$

$$= \left[\hat{x}, \hat{p}_x\right]\hat{x} + \hat{x}\left[\hat{x}, \hat{p}_x\right]$$

$$= i\hbar x + x i\hbar \qquad\qquad (\text{as } \hat{x} \rightarrow x)$$

$$= 2i\hbar x$$

By induction, we may show that
$$\left[\hat{x}, \hat{p}_x\right] = ni\hbar x^{n-1}$$

Example 3.22: $\left[\hat{x}, \hat{p}_x^2\right] = \left[\hat{x}, \hat{p}_x \, \hat{p}_x\right] = \hat{x} \, \hat{p}_x \, \hat{p}_x - \hat{p}_x \, \hat{p}_x \, \hat{x}$

$$= \hat{x} \, \hat{p}_x \, \hat{p}_x - \hat{p}_x \hat{x} \, \hat{p}_x + p_x x \, p_x - \hat{p}_x \, \hat{p}_x \, \hat{x} = \left[\hat{x} \, \hat{p}_x\right]\hat{p}_x + \hat{p}_x\left[\hat{x} \, \hat{p}_x\right]$$

$$= i\hbar \hat{p}_x + \hat{p}_x \, i\hbar = 2i\hbar \hat{p}_x$$

Similarly
$$\left[\hat{x}, \hat{p}_x^3\right] = 3i\hbar \hat{p}_x^2$$

and by induction
$$\left[\hat{x}, \hat{p}_x^n\right] = ni\hbar \hat{p}_x^{n-1}$$

Example 3.23: Let us consider $\left[\hat{x}, \hat{p}_y\right]$

$$\left[\hat{x}, \hat{p}_y\right]\psi = \left(\hat{x}, \hat{p}_y - \hat{p}_y \, \hat{x}\right)\psi$$

$$= x\frac{\hbar}{i}\frac{\partial \psi}{\partial y} - \frac{\hbar}{i}\frac{\partial}{\partial y}(x\psi)$$

$$= x\frac{\hbar}{i}\frac{\partial \psi}{\partial y} - \frac{\hbar}{i}\frac{\partial \psi}{\partial y}x - \frac{\hbar}{i}\Psi\frac{\partial x}{\partial y} \qquad \left(\because \frac{\partial x}{\partial y} = 0\right)$$

$$= 0$$

$$\Rightarrow \qquad \left[\hat{x}, \hat{p}_y\right] = 0$$

Similarly $\qquad \left[\hat{p}_x, \hat{p}_y\right] = 0$

3.14.2 Commutation Relation between Momentum and Hamiltonian

$$\hat{H} = \frac{\hat{p}_x^2}{2m} + \hat{V}(x)$$

where $\hat{V}(x)$ is potential energy operator

$$\hat{V}(x) \to V(x)$$

$$\left[\hat{H}, \hat{p}_x\right] = \left[\frac{\hat{p}_x^2}{2m} + \hat{V}(x), \hat{p}_x\right]$$

$$= \left[\frac{\hat{p}_x^2}{2m}, \hat{p}_x\right] + \left[\hat{V}(x), \hat{p}_x\right]$$

$$= \frac{1}{2m}\left[\hat{p}_x^2, \hat{p}_x\right] + \left[\hat{V}(x), \hat{p}_x\right]$$

$$= 0 + \left[\hat{V}(x), \hat{p}_x\right]$$

Now consider the operation of $\left[\hat{H}, \hat{p}_x\right]\psi(x)$

$$= \left[\hat{V}(x), \hat{p}_x\right]\psi(x)$$

$$= \left\{V(x)\hat{p}_x - \hat{p}_x V(x)\right\}\psi(x)$$

$$= \left\{V(x)\frac{\hbar}{i}\frac{\partial}{\partial x} - \frac{\hbar}{i}\frac{\partial}{\partial x}V(x)\right\}\psi(x)$$

$$= V(x)\frac{\hbar}{i}\frac{\partial \psi(x)}{\partial x} - \frac{\hbar}{i}\frac{\partial}{\partial x}(V(x)\,\psi(x))$$

$$= V(x)\frac{\hbar}{i}\frac{\partial \psi(x)}{\partial x} - \frac{\hbar}{i}V(x)\frac{\partial \psi(x)}{\partial x} - \frac{\hbar}{i}\psi(x)\frac{\partial V}{\partial x}$$

$$= -\frac{\hbar}{i}\frac{\partial V(x)}{\partial x}\psi(x)$$

So, $\qquad \left[\hat{H}, \hat{p}_x\right] = i\hbar\frac{\partial V(x)}{\partial x}$

Case (i): For a free particle $V(x) = 0$ and then

$$\left[\hat{H}, \hat{p}_x\right] = 0$$

Case (ii): If momentum operator is three dimensional and potential is $V(\vec{r})$, then $\left[\hat{H}, \hat{p}\right] = i\hbar \nabla V$.

3.15 Operators for Conserved Quantities

Let O be an operator, then

$$\frac{d}{dt}\langle O \rangle = \frac{d}{dt} \int \psi * O\psi \, dx \tag{87}$$

$$= \int \psi * \frac{\partial O}{\partial t} \psi \, dx + \int \frac{\partial \psi *}{\partial t} O\psi \, dx + \int \psi * O \frac{\partial \psi}{\partial t} \, dx$$

$$= \int \psi * \frac{\partial O}{\partial t} \psi \, dx - \frac{1}{i\hbar} \int (H * \psi*)O\psi \, dx + \frac{1}{i\hbar} \int \psi * O(H\psi) \, dx$$

$$= \int \psi * \frac{\partial O}{\partial t} \psi \, dx - \frac{1}{i\hbar} \int \psi * HO\psi \, dx + \frac{1}{i\hbar} \int \psi * OH\psi \, dx$$

$$= \int \psi * \left\{ \frac{\partial O}{\partial t} - \frac{1}{i\hbar}(HO - OH) \right\} \psi \, dx \tag{88}$$

Here we have used the fact that H is Hermitian. In the Schrödinger representation, the time depedence is contained in the wave function itself, so the time derivative of a variable does not have an obvious quantum mechanical interpretation. However, such representation is feasible if we define the differentiation of an operator with respect to time by the relation

$$\frac{d}{dt}\langle O \rangle = \langle \dot{O} \rangle = \int \psi * \dot{O}\psi \, dx \tag{89}$$

Comparing Eqns (88) and (89), we get

$$\dot{O} = \frac{\partial O}{\partial t} + \frac{i}{\hbar}(HO - OH) = \frac{\partial O}{\partial t} + \frac{1}{i\hbar}[O, H] \tag{90}$$

If the operator O for a physical quantity does not explicitly depend on time and also commutes with H, then by Eqn. (90), the quantity will be conserved. Thus for a system which conserves energy

$$\dot{H} = \frac{i}{\hbar}(HH - HH) = 0 \tag{91}$$

the Hamiltonian being independent of time. The operator of any quantity that is conserved commutes with the Hamiltonian and so this quantity can be measured simultaneously with the energy.

3.16 The Virial Theorem

From equation (89)

$$\frac{d}{dt}\langle O \rangle = \int \psi * \dot{O}\psi \, d\tau \tag{92}$$

and from Eqn. (67)

$$\dot{O} = \frac{\partial O}{\partial t} + \frac{1}{i\hbar}[O, H]$$

Hence

$$ih\frac{d<O>}{dt} = \int\psi*[O,H]\psi \, d\tau + ih\int\psi*\frac{\partial O}{\partial t}\psi \, d\tau \tag{93}$$

Now let us consider the equation of motion for the operator $\vec{r}\cdot\vec{p}$. According to the above relation, we have

$$ih\frac{d}{dt}\langle\vec{r}\cdot\vec{p}\rangle = \int\psi*[\vec{r}\cdot\vec{p},H]\psi \, d\tau \tag{94}$$

But

$$[xp_x, H] = \left[xp_x, \frac{p_x^2}{2m}+V\right] \tag{95}$$

$$= \left[xp_x, \frac{p_x^2}{2m}\right] + [xp_x,V]$$

$$= \frac{1}{2m}(xp_x p_x^2 - p_x^2 xp_x) + x[p_x,V]$$

$$= \frac{1}{2m}(xp_x^2 p_x - p_x^2 xp_x) + x[p_x,V]$$

$$= \frac{1}{2m}\left[x,p_x^2\right]p_x + x[p_x,V]$$

$$= \frac{1}{2m}\left\{p_x xp_x^2 - p_x^2 xp_x + xp_x^3 - p_x xp_x^2\right\} + x[p_x,V]$$

$$= \frac{1}{2m}\left\{p_x[x,p_x]p_x + [x,p_x]p_x^2\right\} + x[p_x,V]$$

$$= ih\frac{p_x^2}{m} + x[p_x,V] \qquad (\because [x,p_x]=ih)$$

$$= ih\frac{p_x^2}{m} - ih\left[\frac{\partial}{\partial x},V\right] \tag{96}$$

Similar relations hold for y- and z-components also. Combining them, we get

$$[\vec{r}\cdot\vec{p},H] = ih\frac{\vec{p}^2}{m} - ih\vec{r}\cdot[\nabla,V] \tag{97}$$

Therefore,

$$ih\frac{d}{dt}\langle\vec{r}\cdot\vec{p}\rangle = \int\psi*[\vec{r}\cdot\vec{p},H]\psi \, d\tau \tag{98}$$

$$= ih\left\langle\frac{\vec{p}^2}{m}\right\rangle - ih\int\psi*\vec{r}\cdot[\nabla V - V\nabla]\psi \, d\tau$$

or

$$\frac{d}{dt}\langle\vec{r}\cdot\vec{p}\rangle = \left\langle\frac{\vec{p}^2}{m}\right\rangle - \int\psi*\vec{r}\cdot[\nabla(V\psi) - V\nabla\psi] \, d\tau$$

$$= \left\langle\frac{\vec{p}^2}{m}\right\rangle - \int\psi*\vec{r}\cdot[(\nabla V)\psi + V\nabla\psi - V\nabla\psi] \, d\tau$$

$$= \left\langle \frac{\vec{p}^2}{m} \right\rangle - \int \psi * \vec{r} \cdot (\nabla V) \psi \, d\tau \qquad (99)$$

For a stationary state,

$$\frac{d}{dt} \langle \vec{r} \cdot \vec{p} \rangle = 0$$

then

$$\left\langle \frac{\vec{p}^2}{m} \right\rangle = \int \psi * \vec{r} \cdot (\nabla V) \psi \, d\tau$$

or

$$2 \langle K \rangle = \langle \vec{r} . \nabla V \rangle \qquad (100)$$

where K is the kinetic energy. This relation is known as the *Virial theorem*.

Example 3.24: The parity operator π is defined by $\pi \psi (x) = \psi (-x)$ for any function $\psi (x)$. Show that

(*i*) It is a linear operator.

(*ii*) It is Hermitian.

(*iii*) Find out its eigenvalues.

(*iv*) If the potential energy is an even function, then show that p commutes with the Hamiltonian

$$H = \frac{p^2}{2m} + V(x)$$

Solution: (*i*) For functions $\psi_1(x)$ and $\psi_2(x)$,

$$\pi \left[\psi_1(x) + \psi_2(x) \right] = \psi_1(-x) + \psi_2(-x)$$

$$= \pi \psi_1(x) + \pi \psi_2(x) \qquad (a)$$

and

$$\pi \left[c\psi_1(x) \right] = c\psi_1(-x) = c\pi \psi_1(x) \qquad (b)$$

(*a*) and (*b*) imply that π is a linear operator.

(*ii*) To show that π is Hermitian, we have to show that

$$(\psi, \pi \phi) = (\pi \psi, \phi).$$

Now

$$(\psi, \pi \phi) = \int \psi * (x) \pi \phi(x) \, dx$$

$$= \int \psi * (x) \phi(-x) \, dx$$

Replacing $x' = -x$, we get

$$(\psi, \pi \phi) = \int \psi * (-x') \phi(x') \, dx'$$

$$= \int \psi * (-x) \phi(x) \, dx$$

$$= (\pi \psi, \phi)$$

Thus, π is a Hermitian operator and it will have real eigenvalues.

(*iii*) To find the eigenvalues, we write the eigenvalue equation

$$\pi \psi = \lambda \psi$$

where λ is the eigenvalue of π in the state ψ. Operating again by π, we get

$$\pi^2 \psi = \pi \lambda \psi = \lambda \pi \psi = \lambda^2 \psi \qquad (c)$$

(because λ is a number).

Again from the definition of π

$$\pi\psi(x) = \psi(-x)$$

and

$$\pi^2\psi(x) = \pi\,(\pi\psi(x)) = \pi\psi(-x) = \psi(x)$$

Hence from (c)

$$\psi(x) = \lambda^2\psi(x)$$

i.e., eigenvalues satisfy the equation $\lambda^2 = 1$, so $\lambda = \pm 1$. The eigenfunctions belonging to the eigenvalue $\lambda = +1$ of the parity operator are the even functions ψ_e and satisfy

$$\psi_e(x) = \psi_e(-x)$$

The eigenfunctions belonging to the eigenvalue $\lambda = -1$ are the odd functions ψ_0 and satisfy

$$\psi_0(x) = -\psi_0(-x)$$

Any function $\psi(x)$ can be expressed as

$$\psi(x) = \tfrac{1}{2}\,[\psi(x) + \psi(-x)] + \tfrac{1}{2}\,[\psi(x) - \psi(-x)]$$

$$\equiv \tfrac{1}{2}\,\psi_e(-x) + \tfrac{1}{2}\,\psi_0(x)$$

(iv) To show that π commutes with H,

$$H\psi(x) = -\frac{\hbar^2}{2m}\frac{d^2\psi(x)}{dx^2} + V(x)\psi(x)$$

Then

$$\pi H\psi(x) = -\frac{\hbar^2}{2m}\pi\frac{d^2\psi(x)}{dx^2} + \pi V(x)\psi(x)$$

$$= -\frac{\hbar^2}{2m}\frac{d^2\psi(-x)}{dx^2} + V(x)\psi(-x)$$

(Given that V is an even function)

$$= -\frac{\hbar^2}{2m}\frac{d^2\psi(-x)}{dx^2} + V(x)\psi(-x)$$

$$= H\psi(-x) = H\pi\,\psi(x) \;.$$

Hence $\pi H\psi(x) - H\pi\psi(x) = 0$ \hfill (d)

or $[\pi H - H\pi]\,\psi(x) = 0$ or $[\pi, H] = 0$

because Eqn. (d) is true for any arbitrary $\psi(x)$.

3.17 The Adjoint of an Operator

Consider the integral

$$\int\phi^* A\psi\, dv \equiv (\phi,\, A\psi) \tag{101}$$

which in a special case reduces to

$$<A> = (\psi,\, A\psi) \text{ for } \phi = \psi.$$

It can be shown that in relation to Eqn. (101) one can always find another operator called the adjoint of A (denoted by $A^\dagger$, and read as "A dagger") such that

$$\int\phi^* A\psi\, dv = \int(A^\dagger\phi)^*\,\psi\, dv \tag{102i}$$

or $(\phi,\, A\psi) = (A^\dagger\phi,\, \psi)$ \hfill (102ii) i.e., as far as the value of the integral (in Eqn. 102*i*) is concerned, it is immaterial whether *A* acts on ψ or $A^\dagger$ acts on ϕ. The Eqn. 102*i* or *ii* serves to define the adjoint of an operator *A*.

An operator is said to be self adjoint if its adjoint is equal to itself i.e. if $A^\dagger = A$. A self adjoint operator is Hermitian (see also Chapter 7). Hence for a Hermitian operator

$$\int \phi^* A\psi \, dv = \int (A\phi)^* \psi \, dv \qquad (102\ iii)$$

Example 3.25: Prove that $a = \dfrac{1}{\sqrt{2}}(q+ip)$ and $a^\dagger = \dfrac{1}{\sqrt{2}}(q-ip)$ are adjoint of each other.

Solution: The operators *P* and *Q* are adjoint of each other if they satisfy

$$\int \overline{\psi}_a P\psi_b \, dx = \int (\overline{Q\psi}_a)\psi_b \, dx$$

and

$$\int \overline{\psi}_a Q\psi_b \, dx = \int (\overline{P\psi}_a)\psi_b \, dx$$

Now let

$$\frac{1}{\sqrt{2}}(q+ip) = P \text{ and } \frac{1}{\sqrt{2}}(q-ip) = Q$$

$$p = -i\hbar \frac{\partial}{\partial x} \qquad \therefore \quad P = \frac{1}{\sqrt{2}}\left(q + \hbar \frac{\partial}{\partial x}\right)$$

and

$$Q = \frac{1}{\sqrt{2}}\left(q - \hbar \frac{\partial}{\partial x}\right)$$

$$\int (\overline{Q\psi}_a)\psi_b \, dx = \int \frac{1}{\sqrt{2}}\left(q - \hbar \frac{\partial}{\partial x}\right)\overline{\psi}_a \psi_b \, dx$$

$$= \frac{1}{\sqrt{2}}\int q\overline{\psi}_a\psi_b \, dx - \frac{\hbar}{\sqrt{2}}\int \left(\frac{\partial}{\partial x}\overline{\psi}_a\right)\psi_b \, dx$$

$$= \frac{1}{\sqrt{2}}\int q\overline{\psi}_a\psi_b \, dx - \frac{\hbar}{\sqrt{2}}\left[\psi_b\overline{\psi}_a\Big|_{-\infty}^{+\infty} - \int \frac{\partial \psi_b}{\partial x}\overline{\psi}_a \, dx\right]$$

(integrating by parts)

$$= \frac{1}{\sqrt{2}}\int q\overline{\psi}_a\psi_b \, dx + \frac{\hbar}{\sqrt{2}}\int \overline{\psi}_a \frac{\partial \psi_b}{\partial x} \, dx \qquad (i)$$

and

$$\int \overline{\psi}_a P\psi_b \, dx = \int \overline{\psi}_a \frac{1}{\sqrt{2}}\left(q + \hbar \frac{\partial}{\partial x}\right)\psi_b \, dx$$

$$= \frac{1}{\sqrt{2}}\int q\overline{\psi}_a\psi_b \, dx + \frac{\hbar}{\sqrt{2}}\int \overline{\psi}_a \frac{\partial \psi_b}{\partial x} \, dx \qquad (ii)$$

From (*i*) and (*ii*)

$$\int \overline{\psi}_a P\psi_b \, dx = \int (\overline{Q\psi}_a)\psi_b \, dx$$

i.e. $P = \dfrac{1}{\sqrt{2}}(q+ip)$ and $Q = \dfrac{1}{\sqrt{2}}(q-ip)$ are adjoint of each other.

3.18 Unitary Operator

An operator U is said to be unitary if it satisfies

$$U^{-1} = U^\dagger \qquad (103)$$

or
$$UU^\dagger = U^\dagger U = I \qquad (104)$$

A unitary operator may be constructed from an ordinary Hermitian operator A,

$$U = \frac{1+iA}{1-iA}, \quad U^\dagger = \frac{1-iA}{1+iA}; \quad UU^\dagger = I$$

We can construct a unitary operator U also in the following manner.

$$U = \frac{U+U^\dagger}{2} + i\frac{U-U^\dagger}{2i} \equiv C + iD \qquad \text{(say)}$$

where
$$C = \frac{U+U^\dagger}{2} \quad \text{and} \quad D = \frac{U-U^\dagger}{2i} \qquad (105)$$

$$\therefore \qquad C^\dagger = \left(\frac{U+U^\dagger}{2}\right)^\dagger = \frac{U^\dagger+U}{2} = \frac{U+U^\dagger}{2} = C$$

and
$$D^\dagger = \left(\frac{U-U^\dagger}{2i}\right)^\dagger = \frac{U^\dagger - U}{-2i} = \frac{U-U^\dagger}{2i} = D$$

i.e. C and D are Hermitian. Now we will see that C and D commute with each other. For this, we note,

$$CD = \frac{U+U^\dagger}{2} \cdot \frac{U-U^\dagger}{2i} = \frac{UU - UU^\dagger + U^\dagger U - U^\dagger U^\dagger}{4i}$$

$$= \frac{UU - U^\dagger U^\dagger}{4i} \qquad \left(\because UU^\dagger = U^\dagger U = I\right)$$

and
$$DC = \frac{U-U^\dagger}{2i} \cdot \frac{U+U^\dagger}{2} = \frac{UU + UU^\dagger - U^\dagger U - U^\dagger U^\dagger}{4i}$$

$$= \frac{UU - U^\dagger U^\dagger}{4i} .$$

Hence $CD = DC$.

i.e. C and D commute with each other. Thus we can find a simultaneous eigenfunction ψ for C and D:

$$C\psi = C'\psi$$
$$D\psi = D'\psi$$

Where C' and D' are real numbers because C and D are Hermitian.
Now,

$$U\psi = (C + iD)\,\psi$$
$$= (C' + iD')\psi = U'\psi \qquad \text{(say)}$$

i.e. the eigenvalue of U is $U' = C' + iD'$.

Further,

$$C^2 + D^2 = \frac{1}{4}(U + U^\dagger)^2 - \frac{1}{4}(U - U^\dagger)^2$$

$$= U^\dagger U = I$$

$\therefore$ $$\left(C^2 + D^2\right)\psi = \left(C'^2 + D'^2\right)\psi = 1\psi$$

$\Rightarrow$ $$C'^2 + D'^2 = 1$$

i.e. $$|U'| = |C' + iD'| = \sqrt{C'^2 + D'^2} = 1$$

Hence, the eigenvalue of any arbitrary unitary operator U has the absolute value unity. Therefore, we may express

$$U' = \exp(iA') \tag{106}$$

where A' is a real number.

3.19 The Closure Property

In article (3.9), we had expansion in terms of energy eigenfunctions

$$\psi(\vec{r}) = \sum_E A_E u_E(\vec{r}) \tag{107i}$$

Premultiplying both sides by u_E^*, and integrating

$$\int u_{E'}^*(\vec{r}) \, \psi(\vec{r}) \, d\tau = \sum_E A_E \int u_{E'}^*(\vec{r}) \, u_E(\vec{r}) \, d\tau$$

$$= \sum_E A_E \delta_{EE'} = A_{E'} \tag{107ii}$$

(since energy eigenfunctions form an orthonormal set).

Substituting Eqn. (ii) for A_E in Eqn. (i) gives

$$\psi(\vec{r}) = \sum_E \left[\int u_{E'}^*(\vec{r})\psi'(\vec{r}) \right] u_E(\vec{r}) \, d\tau$$

$$= \sum_E \int u_E^*(\vec{r}')\psi(\vec{r}') \, u_E(\vec{r}) \, d\tau'$$

$$= \int \psi(\vec{r}') \left[\sum_E u_E^*(\vec{r}') \, u_E(\vec{r}) \right] d\tau'$$

$\psi(\vec{r})$ being an arbitrary function of $\vec{r}$, the two sides are equal under the condition that term within square brackets is zero unless $\vec{r}' = \vec{r}$, otherwise the value of ψ (at $\vec{r}' \neq \vec{r}$) would get changed. Further the value of the integral must be unity when the volume of integration includes the point $\vec{r}' = \vec{r}$. So we must have

and $$\sum_E u_E^*(\vec{r}') \, u_E(\vec{r}) = 0, \text{ for } \vec{r}' \neq \vec{r} \tag{107 iii}$$

$$\int \psi(\vec{r}') \, \delta(\vec{r}' - \vec{r}) \, d\tau' = \psi(r)$$

i.e. $$\sum_E u_E^*(\vec{r}') \, u_E(\vec{r}) = \delta_{rr'}$$

The Eqns. (107iii) embody the closure property of the orthonormal set of functions $u_E(\vec{r})$ and is a consequence of completeness of these functions.

Example 3.26: Show directly that $i(p^2x - xp^2)$ is Hermitian.

Solution: Let

$$i(p^2x - xp^2) = A$$

or

$$A = i(ppx - xpp)$$

then

$$A^\dagger = -i\left[(ppx)^\dagger - (xpp)^\dagger\right]$$

$$= -i\left[x^\dagger p^\dagger p^\dagger - p^\dagger p^\dagger x^\dagger\right]$$

$$= -i[xpp - ppx] \quad \text{(because } x \text{ and } p \text{ are Hermitian}$$
$$x^\dagger = x, p^\dagger = p, \text{ etc)}$$

$$= -i\left[xp^2 - p^2x\right] = i(p^2x - xp^2) = A$$

Thus the operator $i(p^2x - xp^2)$ is Hermitian.

Example 3.27: Show that every operator can be written as the combination of two operators, each of which is Hermitian.

Solution: Let A be any operator and $A^\dagger$, its adjoint. We can write A as

$$A = \frac{A + A^\dagger}{2} + i\frac{A - A^\dagger}{2i} \equiv C + iD \quad \text{(say)}$$

where

$$C = \frac{A + A^\dagger}{2} \quad \text{and} \quad D = \frac{A - A^\dagger}{2i}$$

Now,

$$C^\dagger = \frac{A^\dagger + \left(A^\dagger\right)^\dagger}{2} = \frac{A^\dagger + A}{2} = \frac{A + A^\dagger}{2} = C$$

and

$$D^\dagger = \frac{A^\dagger - \left(A^\dagger\right)^\dagger}{-2i} = \frac{A^\dagger - A}{-2i} = \frac{A - A^\dagger}{2i} = D$$

Hence, C and D are Hermitian and A can be written in terms of them as $A = C + iD$.

Example 3.28: Show that xp_x is not self adjoint while $(xp_x + p_xx)$ is a self adjoint operator.

Solution: We know that both x and p_x are self adjoint operators. Their product xp_x can be self adjoint if they commute with each other. But

$$[x, p_x] = i\hbar \neq 0$$

i.e. they do not commute, hence xp_x cannot be self adjoint.

Now,

$$\left(xp_x + p_xx\right)^\dagger = \left(xp_x\right)^\dagger + \left(p_xx\right)^\dagger$$

$$= p_x^\dagger x^\dagger + x^\dagger p_x^\dagger$$

$$= p_xx + xp_x$$

$$= \left(xp_x + p_xx\right)$$

i.e. the adjoint of $\left(xp_x + p_xx\right)$ is equal to itself, hence it is self adjoint.

Example 3.29: Consider the motion of a particle of mass m, charge e and velocity v in an electromagnetic field, described by the scalar potential V and the vector potential $\vec{A}$. Find the Hamiltonian operator associate with the classical expression of the total energy

$$H = \frac{1}{2m}(\vec{p} + e\vec{A})^2 - eV$$

Solution: We have

$$(p + e\vec{A})^2 = (p + e\vec{A}) \cdot (\vec{p} + e\vec{A})$$
$$= p^2 + e(\vec{p} \cdot \vec{A} + \vec{A} \cdot \vec{p}) + e^2 A^2$$

$$\therefore \qquad H = \frac{1}{2m} p^2 + \frac{e}{2m}(\vec{p} \cdot \vec{A} + \vec{A} \cdot \vec{p}) + \frac{e^2}{2m} A^2 - eV$$

Using $\qquad p = -i\hbar\nabla$, we have

$$H\psi = -\frac{\hbar^2}{2m}\nabla^2 - \frac{ie\hbar}{2m}\nabla.(A\psi) - \frac{ie\hbar}{2m}\vec{A}.\nabla\psi + \frac{e^2}{2m}A^2$$

Using $\nabla.(A\psi) = \vec{A}.\nabla\psi + (\nabla.\vec{A})\psi$, we get

$$H = -\frac{\hbar^2}{2m}\nabla^2 - \frac{ie\hbar}{m}\vec{A}.\nabla - \frac{ie\hbar}{2m}\nabla.\vec{A} + \frac{e^2}{2m}A^2 - eV$$

Example 3.30: Show that if the operators L and M are Hermitian, then the operators $F = \frac{1}{2}(ML + LM)$ and $f = \frac{1}{2}(LM - ML)$ are also Hermitian.

Solution: An operator A is Hermitian if

$$\int \bar{\psi}_a A\psi_b \, dr = \int \overline{A\psi_a}\psi_b \, dr$$

For any operator L,

$$\int \bar{\psi}_a LM\psi_b \, dr = \int \bar{\psi}_a L(M\psi_b) \, dr = \int \overline{L^\dagger \psi_a}(M\psi_b) \, dr$$

$$= \int \overline{L\psi_a}(M\psi_b) \, dr \qquad (\because L \text{ being Hermitian } L^\dagger = L)$$

$$= \int \overline{M^\dagger L\psi_a}\psi_b \, dr$$

$$= \int \overline{ML\psi_a}\psi_b \, dr \qquad\qquad (\because \ M^\dagger = M)$$

This shows LM is Hermitian. Similarly ML is Hermitian. Therefore $\frac{1}{2}(ML + LM)$ is Hermitian, because, for two Hermitian operators P and Q

$$\int \bar{\psi}_a(P + Q)\psi_b \, dr = \int \bar{\psi}_a P\psi_b \, dr + \int \bar{\psi}_a Q\psi_b \, dr$$

$$= \int \overline{P\psi_a}\psi_b \, dr + \int \overline{Q\psi_a}\psi_b \, dr$$

$$= \int \overline{(P + Q)\psi_a}\psi_b \, dr$$

$(P + Q)$ is Hermitian.

Now $\qquad\qquad f = \frac{i}{2}(LM - ML)$

$$f^\dagger = -\frac{i}{2}(LM - ML)^\dagger$$

$$= -\frac{i}{2}\left[(LM)^\dagger - (ML)^\dagger\right] \qquad \left(\because (LM)^\dagger = M^\dagger L^\dagger\right)$$

$$= -\frac{i}{2}\left[M^\dagger L^\dagger - L^\dagger M^\dagger\right]$$

$$= -\frac{i}{2}\left[ML - LM\right]$$

$$= \frac{i}{2}(LM - ML) = f$$

Hence operator f is Hermitian.

Example 3.31: Show that $\left(\dfrac{\partial^n}{\partial x^n}\right)^\dagger = (-1)^n \dfrac{\partial^n}{\partial x^n}$

Hence find the Hermitian conjugate of the operator $e^{i\alpha \,\partial/\partial\phi}$.

Solution: Let

$$\frac{\partial}{\partial x} = Q \quad \therefore \left(\frac{\partial}{\partial x}\right)^\dagger = Q^\dagger$$

$$\int \bar{\psi}_a Q \psi_b \, dx = \int \overline{Q^\dagger \psi_a} \psi_b \, dx$$

(Note: bar represents complex conjugate)

$$\int_{-\infty}^{+\infty} \bar{\psi}_a \frac{\partial}{\partial x} \psi_b \, dx = \bar{\psi}_a \psi_b \Big|_{-\infty}^{+\infty} - \int_{-\infty}^{+\infty} \frac{\partial}{\partial x} \bar{\psi}_a \psi_b \, dx \quad \text{(integrating by parts)}$$

the 1st term gives zero.

$$= -\int_{-\infty}^{+\infty} \frac{\partial}{\partial x} \bar{\psi}_a \psi_b \, dx = \int_{-\infty}^{+\infty} \left(-\frac{\partial}{\partial x} \psi_a\right) \psi_b \, dx \quad \left(\text{because } \frac{\partial}{\partial x} \text{ is real}\right)$$

Now,

$$\int_{-\infty}^{+\infty} \bar{\psi}_a \frac{\partial^2}{\partial x^2} \psi_b \, dx = \bar{\psi}_a \frac{\partial}{\partial x} \psi_b \Big|_{-\infty}^{+\infty} - \int_{-\infty}^{+\infty} \frac{\partial}{\partial x} \bar{\psi}_a \frac{\partial}{\partial x} \psi_b \, dx$$

(integrating by parts, taking $\bar{\psi}_a$ as first function)

$$= \underbrace{\bar{\psi}_a \frac{\partial}{\partial x} \psi_b \Big|_{-\infty}^{+\infty}}_{0} - \left\{ \underbrace{\frac{\partial}{\partial x} \bar{\psi}_a \psi_b \Big|_{-\infty}^{+\infty}}_{0} + \int \frac{\partial^2}{\partial x^2} \bar{\psi}_a \psi_b \, dx \right\}$$

$$= \int (-1)^2 \frac{\partial^2}{\partial x^2} \psi_a \, \psi_b \, dx$$

Therefore, if $Q = \dfrac{\partial^2}{\partial x^2}$ then $Q^\dagger = \left(\dfrac{\partial^2}{\partial x^2}\right)^\dagger = (-1)^2 \dfrac{\partial^2}{\partial x^2}$ or $Q^\dagger = (-1)^n Q$

Hence by mathematical induction

$$\left(\frac{\partial^n}{\partial x^n}\right)^\dagger = (-1)^n \frac{\partial^n}{\partial x^n}$$

Now,

$$Q = e^{ia\,\partial/\partial\phi} = 1 + ia\frac{\partial}{\partial\phi} + \frac{(ia)^2}{2!}\frac{\partial^2}{\partial\phi^2} + \dots$$

$$Q^\dagger = 1 - ia\left(\frac{\partial}{\partial\phi}\right)^\dagger + \frac{(ia)^2}{2!}\left(\frac{\partial^2}{\partial\phi^2}\right)^\dagger + \dots$$

$$= 1 - ia\frac{\partial}{\partial\phi}(-1) + \frac{(ia)^2}{2!}\left(\frac{\partial^2}{\partial\phi^2}\right)^\dagger(-1)^2 + \dots$$

$$= \sum_n \frac{(-ia)^n}{n!}(-1)^n\frac{\partial^n}{\partial\phi^n}$$

$$= \sum_n \frac{(ia)^n}{n!}\frac{\partial^n}{\partial\phi^n} = Q \qquad \left(\because\ (-1)^{2n} = 1\right)$$

Example 3.32: Suppose that an energy level E_0 is three fold degenerate. If the orthormal degenerate eigenfunctions corresponding to this energy level are ψ_1, ψ_2 and ψ_3. Then show that

$$\psi = \frac{1}{\sqrt{5}}(\psi_1 + 2\psi_2 + \psi_3)$$

is also an eigenfunction of H corresponding to the eigenvalue E_0.

Solution: Eigenvalue E_0 is degenerate

$$H\psi_1 = E_0\psi_1$$
$$H\psi_2 = E_0\psi_2$$
$$H\psi_3 = E_0\psi_3$$

$$H\psi = H\left[\frac{1}{\sqrt{5}}\psi_1 + \frac{2}{\sqrt{5}}\psi_2 + \frac{1}{\sqrt{5}}\psi_3\right]$$

$$= \left[\frac{1}{\sqrt{5}}\psi_1 + \frac{2}{\sqrt{5}}\psi_2 + \frac{1}{\sqrt{5}}\psi_3\right]$$

$$= \frac{1}{\sqrt{5}}E_0\psi_1 + \frac{2}{\sqrt{5}}E_0\psi_2 + \frac{1}{\sqrt{5}}E_0\psi_3$$

$$= E_0\left[\frac{1}{\sqrt{5}}\psi_1 + \frac{2}{\sqrt{5}}\psi_2 + \frac{1}{\sqrt{5}}\psi_3\right]$$

$$= E_0\,\psi$$

Therefore, $\psi = \frac{1}{\sqrt{5}}(\psi_1 + 2\psi_2 + \psi_3)$ is also an eigenfunction of H corresponding to eigenvalue E_0.

Example 3.33: If $\psi = \left(\dfrac{1}{\pi a^3}\right)^{1/2} e^{-r/a}$ then find $<r>$.

Solution: $\langle r \rangle = \int \psi * r \psi \, dv$

$$= \iiint \psi * r \psi \; r^2 \sin\theta \, dr \, d\theta \, d\phi$$

$$= \int_0^\infty r^3 \, dr \int_0^\pi \sin\theta \, d\theta \int_0^{2\pi} d\phi \; \psi * r\psi$$

$$= \frac{1}{(\pi a^3)} \int_0^\infty r^3 \, dr \int_0^\pi \sin\theta \, d\theta \; 2\pi \, e^{-2r/a}$$

$$= \frac{2\pi}{\pi a^3} \int_0^\infty e^{-2r/a} r^3 \, dr \; |-\cos\theta|_0^\pi$$

$$= \frac{2\pi}{\pi a^3} \times (2) \int_0^\infty r^3 e^{-2r/a} \, dr$$

Let $\quad (2r/a) = x \qquad\qquad\qquad\qquad\qquad \therefore \; dr = (a/2) \, dx$

$$r^3 = \frac{a^3}{8} x^3$$

$$\therefore \quad \langle r \rangle = \frac{4}{a^3} \left(\frac{a}{2}\right) \int_0^\infty x^3 e^{-x} \, dx \; \frac{a^3}{8}$$

or $\quad \langle r \rangle = \frac{a}{4} \int_0^\infty x^{4-1} e^{-x} \, dx = \frac{a}{4} \Gamma(4) = \frac{a}{4} (3 \,!) = \frac{3a}{2}$

Example 3.34: Show that for a 3-dimensional wave packet

$$\frac{d}{dt}\langle x^2 \rangle = \frac{1}{m}\left(\langle xp_x + p_x x \rangle\right)$$

Solution: In general

$$\frac{d}{dt}\langle x^2 \rangle = \left\langle \frac{i}{\hbar}\left[H, x^2\right] \right\rangle \qquad\qquad \left(\text{because } \frac{\partial}{\partial t}(x^2) = 0 \right)$$

Now $\qquad\qquad H = \dfrac{p_x^2 + p_y^2 + p_z^2}{2m} + V(x, y, z)$

$$\therefore \qquad \left[H, x^2\right] = \frac{1}{2m}\left[p_x^2 + p_y^2 + p_z^2, x^2\right] + \left[V, x^2\right]$$

Now $\quad \dfrac{1}{2m}\left[p_x^2, x^2\right] = \dfrac{1}{2m}\left[p_x p_x, x^2\right]$

$$= \frac{1}{2m}\left(p_x\left[p_x, x^2\right] + \left[p_x, x^2\right]p_x\right)$$

$$= \frac{1}{2m} \left\{ p_x \left([p_x, x] x + x [p_x, x] \right) + \left([p_x, x] x + x [p_x, x] \right) p_x \right\}$$

(using Eqn. 86(*iv*))

$$= -\frac{i\hbar}{2m} (p_x x + p_x x + x p_x + x p_x) \qquad (\because [p_x, x] = -i\hbar)$$

$$= -\frac{i\hbar}{2m} (2) \cdot (p_x x + x p_x)$$

Similarly $\left[p_y^2, x^2 \right] = 0 = \left[p_z^2, x^2 \right]$ and $\left[V, x^2 \right] = 0$

Hence
$$\frac{d}{dt} \left\langle x^2 \right\rangle = \left\langle \frac{i}{\hbar} \left(-\frac{i\hbar}{m} \right) (p_x x + x p_x) \right\rangle$$

$$= \frac{1}{m} (< x p_x + p_x x >)$$

Example 3.35: Show that

$$[A, B^n] = n B^{n-1} [A, B]$$

under the assumption $\left[[A, B], B \right] = 0$.

Solution: $\left[[A, B], B \right] = 0 \Rightarrow [A, B] B = B [A, B]$ (*i*)

Now
$$[A, B^n] = \left[A, B^{n-1} \cdot B \right]$$

$$= B^{n-1} [A, B] + \left[A, B^{n-1} \right] B$$

$$= B^{n-1} [A, B] + \left[A, B^{n-2} \cdot B \right] B$$

$$= B^{n-1} [A, B] + \left\{ B^{n-2} [A, B] + \left[A, B^{n-2} \right] B \right\} B$$

$$= B^{n-1} [A, B] + B^{n-2} [A, B] B + \left[A, B^{n-2} \right] B^2$$

(using (*i*) in 2nd term)

$$= 2 B^{n-1} [A, B] + \left[A, B^{n-2} \right] B^2$$

$$= 2 B^{n-1} [A, B] + \left[A, B^{n-3} \cdot B \right] B^2$$

$$= 2 B^{n-1} [A, B] + \left\{ B^{n-3} [A, B] + \left[A, B^{n-3} \right] B \right\} B^2$$

$$= 2 B^{n-1} [A, B] + B^{n-3} [A, B] B^2 + \left[A, B^{n-3} \right] B^3$$

Using (*i*) again, we get

$$= 2 B^{n-1} [A, B] + B^{n-2} [A, B] B + \left[A, B^{n-3} \right] B^3$$

$$= 3 B^{n-1} [A, B] + \left[A, B^{n-3} \right] B^3$$

Proceeding further similarly, we get

$$= pB^{n-1}[A,B] + \left[A, B^{n-p}\right]B^p$$

(where p is an integer $<n$)

Letting $p \to (n-1)$, we get

$$\left[A, B^n\right] = (n-1)\,B^{n-1}[A,B] + [A,B]\,B^{n-1} \qquad (ii)$$

Now, $\quad [A,B]B^{n-1} = [A,B]\,B \cdot B^{n-2}$

$$= B[A,B]\,B^{n-2} \qquad \text{(using } (i)\text{)}$$

$$= B[A,B]\,B \cdot B^{n-3}$$

$$= B\{B[A,B]\,\}B^{n-3}$$

$$= B^2[A,B]\,B^{n-3}$$

$$= - - - - - - - - - - - - -$$

$$= B^{q-1}[A,B]\,B^{n-q}$$

Letting $q \to n$, we get

$$[A,B]B^{n-1} = B^{n-1}[A,B] \qquad (iii)$$

From (ii) and (iii)

$$\left[A, B^n\right] = (n-1)B^{n-1}[A,B] + B^{n-1}[A,B]$$

$$= nB^{n-1}[A,B]$$

Example 3.36: Prove that for two operators A and B and a real number λ:

$$e^{\lambda A}Be^{-\lambda A} = B + \frac{\lambda}{1!}[A,B] + \frac{\lambda^2}{2!}[A,[A,B]] + \frac{\lambda^3}{3!}\left[A,[A,[A,B]]\right] + ...$$

Solution: Let us represent $e^{\lambda A}Be^{-\lambda A}$ by $f(\lambda)$ and write Taylor expansion

$$e^{\lambda A}Be^{-\lambda A} \equiv f(\lambda) = f(0) + \frac{\lambda}{1!}\left(\frac{df}{d\lambda}\right)_{\lambda=0} + \frac{\lambda^2}{2!}\left(\frac{d^2f}{d\lambda^2}\right)_{\lambda=0} + ...$$

$$\equiv f(0) + \frac{\lambda}{1!}f'(0) + \frac{\lambda^2}{2!}f''(0) + ... \qquad (i)$$

For $\lambda = 0$, we have

$$f(0) = e^{\lambda A}Be^{-\lambda A}\Big|_{\lambda=0} = B \qquad (ii)$$

$$f'(\lambda) = \frac{df}{d\lambda} = Ae^{\lambda A}Be^{-\lambda A} + e^{\lambda A}Be^{-\lambda A}(-A)$$

$$= Ae^{\lambda A}Be^{-\lambda A} - e^{\lambda A}Be^{-\lambda A}A$$

$$= Af(\lambda) - f(\lambda)A = [A, f(\lambda)]$$

and $\qquad\qquad f'(0) = [A, f(0)] = [A, B] \qquad (iii)$

For $f''(0)$, we note that

$$f'(\lambda) = e^{\lambda A} AB e^{-\lambda A} - e^{\lambda A} BA e^{-\lambda A}$$

$$= e^{\lambda A}(AB - BA)e^{-\lambda A}$$

$$= e^{\lambda A}[A, B]e^{-\lambda A}$$

Hence
$$f''(\lambda) = e^{\lambda A}[A, B]e^{-\lambda A} - e^{\lambda A}[A, B]A e^{-\lambda A}$$

$$= e^{\lambda A}[A,[A, B]]e^{-\lambda A}$$

Thus,
$$f''(0) = [A,[A, B]] \qquad (iv)$$

Similarly,
$$f'''(0) = [A, [A, [A, B]]] \qquad (v)$$

Substituting values of $f(0)$, $f'(0)$, $f''(0)$, $f'''(0)$ etc, in Eqn. (i) we get

$$e^{\lambda A} B e^{-\lambda A} = B + \frac{\lambda}{1!}[A, B] + \frac{\lambda^2}{2!}[A,[A, B]] + \frac{\lambda^3}{3!}[A,[A,[A, B]]] + ...$$

Example 3.37: If A and B are two operators which satisty

$$[A, [A, B]] = 0 = [B, [A, B]]$$

then prove

$$e^A e^B = \exp\left(A + B + \frac{1}{2}[A, B]\right).$$

Solution: Consider the function

$$f(\lambda) = e^{\lambda A} e^{\lambda B} \qquad (i)$$

$$\frac{df}{d\lambda} = A e^{\lambda A} e^{\lambda B} + e^{\lambda A} B e^{\lambda B}$$

$$= A f(\lambda) + e^{\lambda A} B e^{-\lambda A} e^{\lambda A} e^{\lambda B}$$

$$= A f(\lambda) + e^{\lambda A} B e^{-\lambda A} f(\lambda)$$

$$= \left(A + e^{\lambda A} B e^{-\lambda A}\right) f(\lambda)$$

But we know that

$$e^{\lambda A} B e^{-\lambda A} = B + \frac{\lambda}{1!}[A, B] + \frac{\lambda^2}{2!}[A,[A, B]] + ... \qquad (iii)$$

Also it is given that

$$[A,[A, B]] = 0 \text{ and } [B,[A, B]] = 0$$

Therefore, Eqn. (iii) becomes

$$e^{\lambda A} B e^{-\lambda A} = B + \lambda(A, B)$$

Consequently, from Eqn. (ii)

$$\frac{df}{d\lambda} = (A + B + \lambda[A, B]) f(\lambda)$$

or
$$\frac{df}{f(\lambda)} = (A + B + \lambda [A, B]) d\lambda$$

Integrating this expression, we get

$$\log_e f(\lambda) = (A+B)\lambda + \frac{\lambda^2}{2}[A, B]$$

or

$$f(\lambda) = \exp\left\{(A+B)\lambda + \frac{\lambda^2}{2}[A, B]\right\}$$

or

$$e^{\lambda A} e^{\lambda B} = \exp\left\{(A+B)\lambda + \frac{\lambda^2}{2}[A, B]\right\}$$

Putting $\lambda = 1$, we get

$$e^A e^B = \exp\left(A + B + \tfrac{1}{2}[A, B]\right)$$

Example 3.38: A and B are two operators such that $[A, B] \neq 0$. If $[A, B]$ commutes with A and B separately then prove that

$$e^A e^B e^{-A} = e^B e^{[A,B]}$$

Solution:

$$e^{\lambda A} B e^{-\lambda B} = B + \frac{\lambda}{1!}[A, B] + \frac{\lambda^2}{2!}\left[A, [A, B]\right] + \ldots$$

So we can write

$$e^A e^B e^{-A} = e^B + \left[A, e^B\right] + \frac{1}{2}\left[A, \left[A, e^B\right]\right] + \ldots \qquad (i)$$

Now

$$[A, B^2] = [A, B]B + B[A, B]$$

$$= 2B[A, B] \ (\because [[A, B], B] = 0)$$

or

$$[A, f(B)] = f'(B)[A, B]$$

Therefore

$$\left[A, e^B\right] = e^B[A, B] \qquad (ii)$$

$$\left[A, \left[A, e^B\right]\right] = \left[A, e^B[A, B]\right] = A e^B[A, B] - e^B[A, B]A$$

$$= A e^B\left[A, e^B\right] - e^B\left[A, e^B\right]A \qquad (\because [[A, B], A] = 0$$

$$\therefore [A, B]A = A[A, B])$$

$$= \left(A e^B - e^B A\right)[A, B]$$

$$= \left[A, e^B\right][A, B]$$

$$= e^B[A, B][A, B] \qquad \text{(using Eqn. } (ii))$$

$$= e^B[A, B]^2 \qquad (iii)$$

Substituting Eqns. (ii), (iii) etc. in Eqn. (i)

$$e^A e^B e^{-A} = e^B + e^B[A, B] + \frac{e^B}{2}[A, B]^2 + \ldots$$

$$= e^B e^{[A,B]}$$

Example 3.39: If A and B are two operators such that $[A, B] = \alpha B$, where α is a number then show that

$$e^A B e^{-A} = e^\alpha B$$

Solution: We have

$$e^A B e^{-A} = B + [A, B] + \frac{1}{2!}[A,[A, B]] + ... \qquad (i)$$

Now,

$$[A, B] = \alpha B$$

$$\therefore \qquad [A,[A, B]] = [A, \alpha B] = \alpha[A, B] = \alpha^2 B$$

and $\qquad [A,[A,[A, B]]] = [A, \alpha^2 B] = \alpha^2[A, B] = \alpha^3 B$

Thus, from Eqn. (*i*), we get

$$e^A e^B e^{-A} = B + \alpha B + \frac{1}{2!}\alpha^2 B + \frac{1}{3!}\alpha^3 B + ...$$

$$= \left[1 + \alpha + \frac{\alpha^2}{2!} + \frac{\alpha^3}{3!} + ... \right] B$$

$$= e^\alpha B$$

Example 3.40: Check whether $\dfrac{d^2}{dx^2}$ is a Hermitian operator.

Solution: An operator A is Hermitian if

$$\int \psi^*(A\phi)\ dx = \int (A\psi)^* \phi\ dx$$

Here $\qquad\qquad A = \dfrac{d^2}{dx^2}$

Let $\qquad\qquad \psi = e^{ix}$ and $\phi = \sin x$. Then,

$$\int e^{-ix} \frac{d^2}{dx^2}(\sin x)dx = \int e^{-ix}(-\sin x)dx \qquad (i)$$

and $\quad \int \left(\dfrac{d^2}{dx^2}e^{ix} \right)^* \sin x\ dx = \int \sin x \left(\dfrac{d^2}{dx^2}e^{ix} \right)^* dx$

$$= \int \sin x \left(i^2 \right)e^{-ix}\ dx$$

$$= -\int \sin x\ e^{-ix}\ dx \qquad (ii)$$

Since (*i*) and (*ii*) are equal, hence the operator $A = \dfrac{d^2}{dx^2}$ is Hermitian.

Questions and Problems

1. (a) What do you mean by linear operators?
   (b) Define expectation value of an operator associated with an observable.
2. (a) What is Schmidt orthogonalization procedure?
   (b) What is an orthonormal set?
3. What is Hermitian operator? What are their properties?
4. (a) Show that Hermitian operators have real eigenvalues.
   (b) Prove that two eigenfunctions of a Hermitian operator belonging to different eigenvalues are orthogonal.
5. Prove that if Q and R are operators which commute with each other, then there exists a complete set of eigenstates which are simultaneously eigenstates of both Q and R.
6. Derive the quantum law if a physical quantity is conserved.
7. (a) Prove that $[A^n, B] = n\, A^{n-1}\, [A, B]$ under the condition $[A, [A, B]] = 0$.
   (b) Prove that $[A, [B, C]] + [B, [C, A]] + [C, [A, B]] = 0$.
8. Given the following operators

   (a) $O_1\, \psi(x) = x^3\psi(x)$

   (b) $O_2\, \psi(x) = x\dfrac{d}{dx}\,\psi(x)$]

   (c) $O_3\, \psi(x) = \lambda\, \psi^*(x)$

   (d) $O_4\, \psi(x) = e^{\psi(x)}$

   (e) $O_5\, \psi(x) = \left(x + \dfrac{d}{dx}\right)^2 \psi(x)$

   Which of these are linear operators.

   Check whether $-i\hbar x\dfrac{d}{dx}$ is a Hermitian operator.

   The operators a and $a^\dagger$ are defined as

   $$a = \sqrt{\frac{m\omega}{2\hbar}}\left(x - i\,\frac{p}{m\omega}\right) \text{ and } a^\dagger = \sqrt{\frac{m\omega}{2\hbar}}\left(x + i\,\frac{p}{m\omega}\right)$$

   Find the value of $[a, a^\dagger]$. [**Ans.** -1]

9. (a) Show that for any observable A,

   $$i\hbar\frac{d\langle A\rangle}{dt} = \langle[A, H]\rangle + i\hbar\left\langle\frac{\partial A}{\partial t}\right\rangle$$

   Give physical significance of this equation.
   (b) Using the above equation, show that

   $$\frac{d}{dt}\langle p_x\rangle = \left\langle-\frac{\partial V}{\partial x}\right\rangle$$

10. Show that $\left[H, p_x^n\right] = [H, p_x]p_x^{n-1} + p_x[H, p_x^{n-1}]$

11. If $u_1(\vec{r})$ and $u_2(\vec{r})$ are degenerate eigenfunctions of the Hamiltonian

    $H = \left(\dfrac{p^2}{2m}\right) + V(\vec{r})$, show that $\int u_1^*(\vec{r})(xp_x + p_x x)\, u_2(\vec{r})\, d^3r = 0$.

12. Show that if the operators L, M are Hermitian, the operators

$$F = \frac{1}{2}(ML + LM) \text{ and } f = \frac{i}{2}(LM - ML)$$

are also Hermitian.

13. Prove that $a = \frac{1}{\sqrt{2}}(q + ip)$ and $a^\dagger = \frac{1}{\sqrt{2}}(q - ip)$ are adjoint of each other.

14. Two Hermitian operators A and B anticommute i.e. $\{A, B\} = AB + BA = 0$. Is it possible to have a simultaneous eigenfunction of A and B? Prove or illustrate your assertion.

15. Show that $[x, p_r] = i\hbar$, where $p_r = -i\hbar \frac{1}{r}\frac{\partial}{\partial r}r$.

16. Consider a particle in three dimensions whose Hamiltonian is given by

$$H = \frac{\vec{p}^2}{2m} + V(\vec{x})$$

By calculating $[\vec{x}.\vec{p}, H]$ obtain

$$\frac{d}{dt}\langle \vec{x}.\vec{p}\rangle = \left\langle \frac{\vec{p}^2}{m}\right\rangle - \langle \vec{x}.\nabla V\rangle$$

17. An electron in an oscillating electric field is described by the Hamiltonian operator

$$H = \frac{p^2}{2m} - (eE_0 \cos \omega t)x$$

Calculate $\frac{dx}{dt}$, $\frac{dp}{dt}$ and $\frac{dH}{dt}$.

18. If $\psi_1(x)$ and $\psi_2(x)$ are degenerate eigenfunctions of the hamiltonian

$$H = \frac{\vec{p}^2}{2m} + V(\vec{x}),$$

Show that $\int(\psi_1^*(\vec{x})xp_x + p_x x \ \psi_2(x))\,d\tau = 0$.

19. If A and B are any two operators and $C = e^{\alpha A} B e^{-\alpha A}$, show that

$$\frac{dC}{d\alpha} = [A, C]$$

Hence show that the Taylor expansion of C in powers of a is

$$e^{\alpha A} B e^{-\alpha A} = B + \alpha[A, B] + \frac{\alpha^2}{2!}[A,[A,B]] + ...$$

20. Prove that for a particle in a potential V,

$$\frac{d}{dt}\langle \vec{L}\rangle = \langle \vec{T}\rangle \text{ where } \vec{T} \equiv -\vec{r} \times \nabla V$$

is the torque. (Note that $\vec{T}$ vanishes if V is a central potential)

21. Show that the wavefunction $\psi = e^{i\vec{k}.\vec{r}}$ is simultaneously an eigenfunction of the operators $-i\hbar\nabla$ and $-\hbar^2\nabla^2$.

22. Enumerate the different postulates of quantum mechanics.

23. Show that $\left(\dfrac{\partial}{\partial x}+x\right)\left(\dfrac{\partial}{\partial x}-x\right) \neq \left(\dfrac{\partial}{\partial x}-x\right)\left(\dfrac{\partial}{\partial x}+x\right)$

24. Show that $\left[\hat{p},\hat{V}(x)\right]=-i\hbar\,\dfrac{\partial V(x)}{\partial x}$

25. Prove that the Hamiltonian operator and time form canonically conjugate pair.

    [**Hint:** Prove $[H, t] = i\hbar$, where $\hat{H}=i\hbar\dfrac{\partial}{\partial t}$.]

4

The Fourier Transform, Dirac Delta Function and Wave Packets

4.1 The Fourier Series and Fourier Transform

According to Fourier's theorem, a periodic function with period L i.e.

$$f(x + nL) = f(x), \quad n = 0, \pm 1, \pm 2, \ldots \tag{1}$$

can be expanded in the form

$$f(x) = \frac{a_0}{2} + \sum_{n=1}^{\infty} \left[a_n \cos\left(\frac{2\pi nx}{L}\right) + b_n \sin\left(\frac{2\pi nx}{L}\right) \right] \tag{2}$$

called the *Fourier series*.

The coefficients a_n and b_n can be easily evaluated by using the following properties of trigonometric functions

$$\frac{2}{L} \int_{-L/2}^{L/2} \cos\left(\frac{2\pi mx}{L}\right) \cos\left(\frac{2\pi nx}{L}\right) dx = \delta_{mn} \tag{3}$$

$$\frac{2}{L} \int_{-L/2}^{L/2} \sin\left(\frac{2\pi mx}{L}\right) \sin\left(\frac{2\pi nx}{L}\right) dx = \delta_{mn} \tag{4}$$

and

$$\frac{2}{L} \int_{-L/2}^{L/2} \cos\left(\frac{2\pi mx}{L}\right) \sin\left(\frac{2\pi nx}{L}\right) dx = 0 \tag{5}$$

where δ_{mn} is the Kronecker delta defined by

$$\delta_{mn} = \begin{cases} 0 & \text{if } m \neq n \\ 1 & \text{if } m = n \end{cases} \tag{6}$$

If the expression for $f(x)$ is multiplied by $(2/L) \cos(2\pi mx/L)$ and integrated over the interval $(-L/2, L/2)$, the result is

$$\frac{2}{L} \int_{-L/2}^{L/2} f(x) \cos\left(\frac{2\pi mx}{L}\right) dx = \frac{a_0}{2} \frac{2}{L} \int_{-L/2}^{L/2} \cos\left(\frac{2\pi mx}{L}\right) dx$$

$$+\sum_{n=1}^{\infty} a_n \frac{2}{L} \int_{-L/2}^{L/2} \cos\left(\frac{2\pi mx}{L}\right) \cos\left(\frac{2\pi nx}{L}\right) dx$$

which yields, for $m = 0$

$$a_0 = \frac{2}{L} \int_{-L/2}^{L/2} f(x)\ dx \tag{7}$$

and for $m \neq 0$

$$a_m = \frac{2}{L} \int_{-L/2}^{L/2} f(x) \cos\left(\frac{2\pi mx}{L}\right) dx \tag{8}$$

Similarly multiplying by $\dfrac{2}{L}\left[\sin\left(\dfrac{2\pi mx}{L}\right)\right]$, the coefficients b_n are obtained:

$$a_n = \frac{2}{L} \int_{-L/2}^{L/2} f(x)\ \cos\left(\frac{2\pi nx}{L}\right) dx\ ;\ \ n = 1, 2, ... \tag{9}$$

and

$$b_n = \frac{2}{L} \int_{-L/2}^{L/2} f(x)\ \sin\left(\frac{2\pi nx}{L}\right) dx\ ;\ \ n = 1, 2, ... \tag{10}$$

If we now substitute these values in the expression (2), we obtain

$$f(x) = \frac{1}{L} \int_{-L/2}^{+L/2} f(x')dx' + \sum_{n=1}^{\infty} \left[\frac{2}{L}\cos\left(\frac{2\pi nx}{L}\right) \int_{-L/2}^{+L/2} f(x')\cos\left(\frac{2\pi nx'}{L}\right)dx' \right.$$

$$\left. + \frac{2}{L}\sin\left(\frac{2\pi nx}{L}\right) \int_{-L/2}^{+L/2} f(x')\sin\left(\frac{2\pi nx'}{L}\right)dx' \right] \tag{11}$$

or $\quad f(x) = \dfrac{\Delta k}{2\pi} \displaystyle\int_{-\pi/\Delta k}^{+\pi/\Delta k} f(x')dx' + \sum_{n=1}^{\infty} \left\{ \dfrac{\Delta k}{\pi} \int_{-\pi/\Delta k}^{+\pi/\Delta k} f(x')\cos\left[n\Delta k(x' - x)\right]dx' \right\}$ (12)

where $\Delta k = \dfrac{2\pi}{L}$ (and using $\cos A \cos B + \sin A \sin B = \cos (A - B)$.

Now let $L \to \infty$ so that $\Delta k \to 0$. Thus, now if the integral $\displaystyle\int_{-\infty}^{+\infty} f(x')\ dx'$ exists

(i.e. it has a finite value) then the first term on the r.h.s. of Eqn. (12) would go to zero. Further also

$$\int_0^{\infty} F(k)dk = \underset{\Delta k \to 0}{\mathrm{Lim}} \sum_{n=1}^{\infty} F(n\Delta k)\Delta k \tag{13}$$

so, in the limit $\Delta k \to 0$, we have from Eqn. (12)

$$f(x) = \underset{\Delta k \to 0}{\mathrm{Lim}} \sum_{n=1}^{\infty} \left\{ \frac{\Delta k}{\pi} \int_{-\infty}^{+\infty} f(x')\ \cos\left[n\Delta k(x' - x)\right]dx' \right\}$$

or
$$f(x) = \frac{1}{\pi} \int_0^\infty \left\{ \int_{-\infty}^{+\infty} f(x') \cos\left[k(x'-x)\right] dx' \right\} dk \tag{14}$$

This equation is known as the *Fourier Integral*. Since the cosine function is an even function of k,

$$\therefore \quad f(x) = \frac{1}{2\pi} \int_{-\infty}^{+\infty} \left\{ \int_{-\infty}^{+\infty} f(x') \cos\left[k(x'-x)\right] dx' \right\} dk \tag{15}$$

Further, $\sin[k(x'-x)]$ being an odd function of k, hence

$$0 = \frac{1}{2\pi} \int_{-\infty}^{+\infty} \left\{ \int_{-\infty}^{+\infty} f(x') \sin[k(x'-x)] dx' \right\} dk \tag{16}$$

From Eqns (15) and (16), we get

$$f(x) = \frac{1}{2\pi} \int_{-\infty}^{+\infty} \int_{-\infty}^{+\infty} f(x') \exp\left[\pm ik(x'-x)\right] dx' dk \tag{17}$$

using
$$\cos\left[k(x'-x)\right] \pm i \sin\left[k(x'-x)\right] = \exp\left[\pm ik(x'-x)\right]$$

Thus, if

$$F(k) = \frac{1}{\sqrt{2\pi}} \int_{-\infty}^{+\infty} f(x') \exp(-ikx') dx' \tag{18}$$

then from Eqn. (17) we have

$$f(x) = \frac{1}{\sqrt{2\pi}} \int_{-\infty}^{\infty} F(k) e^{+ikx} dk \tag{19}$$

The function $F(k)$ is known as the *Fourier transform* of $f(x)$ and vice versa. In terms of wave functions, the fourier transform $\phi(k)$ of a wave function $\psi(x)$ is defined as

$$\phi(k) = \frac{1}{\sqrt{2\pi}} \int_{-\infty}^{+\infty} \psi(x) e^{-ikx} dx \tag{20}$$

with
$$\psi(x) = \frac{1}{\sqrt{2\pi}} \int_{-\infty}^{+\infty} \phi(k) e^{ikx} dk \tag{21}$$

The Fourier transform as defined by Eqn. (20) exists only when

$$\int_{-\infty}^{+\infty} |\psi(x)|^2 dx < \infty . \tag{22}$$

If we put the value of $\phi(k)$ from Eqn. (20) into Eqn. (21), we get

$$\psi(x) = \left(\frac{1}{\sqrt{2\pi}}\right)^2 \int_{-\infty}^{+\infty} \left(\int_{-\infty}^{+\infty} \psi(x') e^{-ikx'} dx' \right) e^{ikx} dk$$

or
$$\psi(x) = \left(\frac{1}{2\pi}\right) \int_{-\infty}^{+\infty} \int_{-\infty}^{+\infty} \psi(x') e^{ik(x-x')} dx' dk \tag{23}$$

This relation is known as *Fourier integral theorem.*

For a three dimensional space $\psi = \psi(x, y, z)$ then,

$$\psi(x,y,z) = \frac{1}{(2\pi)^{3/2}} \int\limits_{-\infty}^{\infty} \int\limits_{-\infty}^{\infty} \int\limits_{-\infty}^{\infty} \phi(k_x, k_y, k_z) e^{ik_x x} e^{ik_y y} e^{ik_z z} dk_x dk_y dk_z \quad (24)$$

If the integral is considered as a volume integral over a three dimensional k-space above expression simplifies to

$$F^{-1}[f(k)] = \psi(\vec{r}) = \frac{1}{(2\pi)^{3/2}} \int\limits_{-\infty}^{+\infty} \phi(\vec{k}) e^{i\vec{k}\cdot\vec{r}} d^3k \quad (25)$$

$d^3\vec{k}$ refers to an element of volume in k-space. Similarly

$$F[f(x)] = \phi(\vec{k}) = \frac{1}{(2\pi)^{3/2}} \int\limits_{-\infty}^{+\infty} \psi(\vec{r}) e^{-i\vec{k}\cdot\vec{r}} d^3\vec{r} \quad (26)$$

Here $F[\ ]$ denotes the Fourier transform and $F^{-1}[\ ]$, the inverse Fourier transform.

4.2 Dirac Delta Function

Consider Eqn. (26). This can be thought of as an expansion of an arbitrary function $f(\vec{r})$ in terms of exponential periodic functions (plane waves) of $\vec{k}\cdot\vec{r}$. Since for such functions the condition of Eqn. (22) is not satisfied, it is not possible to obtain a Fourier transform of this exponential function.

Although a plane wave $\left(e^{\pm i\vec{k}\cdot\vec{r}}\right)$ does not have a true Fourier transform, we can define the improper Dirac delta function so as to perform the role of such a transform. To do this we write the exponential periodic function as

$$f(x) = e^{ik_0 x} = \lim_{\alpha \to 0} e^{-\alpha x^2} e^{ik_0 x} \quad (27)$$

where α is a finite, real positive-number. For a finite, real, positive value of α, the Fourier integral of Eqn. (27) exists and this allows the calculation of Fourier transform of Eqn. (27) as

$$F(k) = \lim_{\alpha \to 0} \frac{1}{\sqrt{2\pi}} \int\limits_{-\infty}^{+\infty} e^{-\alpha x^2} e^{ik_0 x} e^{-ikx} dx \quad (28)$$

The resulting limit vanishes for $k \neq k_0$ and diverges for $k = k_0$, giving the form of the improper function as

$$\left. \begin{array}{l} F(k) = 0 \ \text{ for } k \neq k_0 \\ \quad\quad = \infty \ \text{ for } k = k_0 \end{array} \right| \quad (29)$$

Although this function is singular, yet it is possible to define $\int\limits_{-\infty}^{+\infty} F(k)\, dk$ by performing one integration before taking the limit. From Eqns (20), (21) and (23)

$$f(x) = \int\limits_{-\infty}^{+\infty} dx' \ f(x') \frac{1}{2\pi} \int\limits_{-\infty}^{\infty} e^{ik(x-x')} dk$$

$f(x)$ is single valued and satisfies Eqn. (22). This equation holds for any function $f(x)$, this suggests the definition of the Dirac delta function:

(i) $\qquad \delta(x - x') = \dfrac{1}{2\pi} \displaystyle\int\limits_{-\infty}^{+\infty} e^{ik(x-x')} dk \Bigg|$

(ii) $\qquad \delta(x - x') = 0 \quad \text{if } x \neq x'$ $\qquad\qquad$ (30)

(iii) $\qquad \displaystyle\int \delta(x - x') dx = 1$

(i) $\qquad \delta(k) = \dfrac{1}{2\pi} \displaystyle\int\limits_{-\infty}^{+\infty} e^{ikx} dx$

(ii) $\qquad \delta(k) = 0 \qquad \text{if } k \neq 0$ $\qquad\qquad$ (31)
$\qquad\qquad\qquad = \infty \qquad \text{if } k = 0$

(iii) $\qquad \displaystyle\int\limits_{-\infty}^{+\infty} \delta(k) dk = 1$

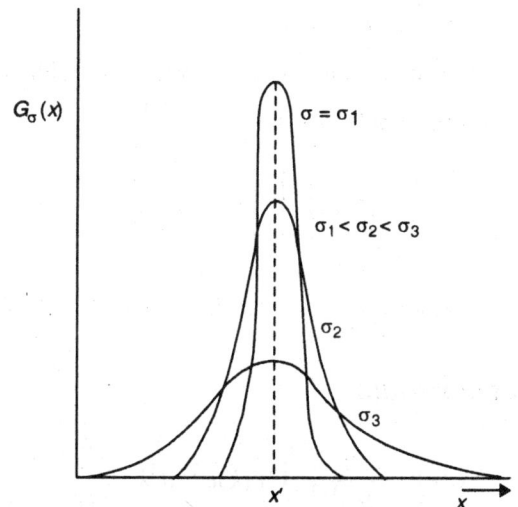

Fig. 4.1: Delta function as a limit of the Gaussian function with half width $\sigma \to 0$

The Gaussian $G_\sigma = \dfrac{1}{\sigma\sqrt{2\pi}} e^{\dfrac{-(x-x')^2}{2\sigma^2}}$ represents the δ-function and has unit area

$\displaystyle\int\limits_{-\infty}^{+\infty} G_\sigma dx = 1$ (independent of σ) and is peaked about $x = x'$ with a width $\sim\sigma$.

For $\sigma \to 0$, the function would have an infinite value at $x = x'$ and zero elsewhere.

Properties of the Delta Function

(i) $\delta(x) = \delta(-x)$

(ii) $\displaystyle\int_{-\infty}^{\infty} f(x) \, \delta(x-a) dx = f(a)$

(iii) $\delta(ax) = \dfrac{1}{a}\delta(x); \quad a > 0$ (32)

(iv) $\displaystyle\int \delta(x - x_1) \, \delta(x_1 - x_2) \, dx_1 = \delta(x - x_2)$

(v) $f(x) \, \delta(x - a) = f(a) \, \delta(x - a)$

Definition: $\delta'(k) = \dfrac{1}{2\pi}\displaystyle\int_{-\infty}^{\infty} ik \, e^{ikx} dx$ (33)

(vi) $\delta'(x) = -\delta'(-x)$ (34)

(vii) $\displaystyle\int_{-\infty}^{\infty} f(x) \, \delta'(x-a) \, dx = f(x) \, \delta(x-a)\Big|_{-\infty}^{+\infty} - \int_{-\infty}^{+\infty} f'(x) \, \delta(x-a) \, dx$

$$= -f'(a) \qquad (35)$$

In 3-dimensions, the delta funciton of the vector variable $\vec{k}$ is

$$\delta(\vec{k}) = \delta(k_x)\delta(k_y)\delta(k_z)$$

$$= \frac{1}{(2\pi)^3} \int_{-\infty}^{\infty}\int_{-\infty}^{\infty}\int_{-\infty}^{\infty} e^{i(k_x x + k_y y + k_z z)} \, dx \, dy \, dz$$

or $\delta(\vec{k}) = \dfrac{1}{(2\pi)^3}\displaystyle\int_{-\infty}^{\infty} e^{i\vec{k}\cdot\vec{r}} d^3\vec{r}; \quad \left(\left|d^3\vec{r}\right| = dx \, dy \, dz\right)$ (36)

4.3 Parseval's Formula

We have

$$\psi(\vec{r}) = \frac{1}{(2\pi)^{3/2}} \int_{-\infty}^{\infty} \Psi(\vec{k}) e^{i\vec{k}\cdot\vec{r}} d^3\vec{k} \qquad (37)$$

If ψ is normalised, $\displaystyle\int_{-\infty}^{+\infty} |\psi(r)|^2 \, d^3r = 1$

$$\psi^*(r) = \frac{1}{(2\pi)^{3/2}} \int_{-\infty}^{\infty} \Psi^*(\vec{k}') e^{-i\vec{k}'\cdot\vec{r}} d^3\vec{k}' \qquad (38)$$

$$\therefore \qquad \psi(r) \, \psi^*(r) = \frac{1}{(2\pi)^3} \int_{-\infty}^{+\infty}\int_{-\infty}^{\infty} \Psi(\vec{k}) \, \Psi^*(\vec{k}') e^{i(\vec{k}-\vec{k}')\cdot\vec{r}} d^3\vec{k} \, d^3\vec{k}'$$

$$\int \psi(r) \, \psi^*(r) \, d^3r = \frac{1}{(2\pi)^3} \int_{-\infty}^{+\infty}\int_{-\infty}^{\infty} \Psi(\vec{k}) \, \Psi^*(\vec{k}') \, d^3\vec{k} \, d^3\vec{k}' \int_{-\infty}^{+\infty} e^{i(\vec{k}-\vec{k}')\cdot\vec{r}} \, d^3r$$

or
$$1 = \frac{1}{(2\pi)^3} \int\limits_{-\infty}^{+\infty} \int\limits_{-\infty}^{\infty} \Psi(\vec{k})\, \Psi^*(\vec{k}')\, \delta(\vec{k}-k')d^3k\; d^3\vec{k}' \qquad (39)$$

Therefore,

$$\int\limits_{-\infty}^{+\infty} \psi(k)\; \psi*(k)\, d^3k = 1 \qquad (40)$$

because
$$\delta(k-k') = \delta(k'-k) \text{ and}$$

$$\int\limits_{-\infty}^{+\infty} \Psi*(\vec{k}')\; \delta(k'-k)\; d^3k' = \Psi^*(\vec{k}) \qquad (41)$$

Hence,
$$\int\limits_{-\infty}^{+\infty} |\psi(\vec{r})|^2\; d^3r = \int\limits_{-\infty}^{+\infty} |\Psi(\vec{k})|^2\; d^3\vec{k} = 1 \qquad (42)$$

It is known as the *Parseval's formula*.

Example 4.1: Find the Fourier transform of the Dirac $\delta(x)$ function.

Solution: According to Eqn. (18), the Fourier transform of $\delta(x)$ is

$$F_\delta(k) = \frac{1}{\sqrt{2\pi}} \int\limits_{-\infty}^{+\infty} \delta(x)\, \exp(-ikx)\, dx \qquad (i)$$

We know that the delta function is defined by
$$\delta(x) = 0 \text{ if } x \neq 0$$

and
$$\int\limits_{-\infty}^{+\infty} f(x)\, \delta(x-a)\, dx = f(a)$$

$$\therefore \qquad \int\limits_{-\infty}^{+\infty} f(x)\, \delta(x)\, dx = f(0) \qquad (ii)$$

From (i)
$$F_\delta(k) = \frac{1}{\sqrt{2\pi}} \int\limits_{-\infty}^{+\infty} \exp(-ikx)\, \delta(x)\; dx = \frac{1}{\sqrt{2\pi}} \left[\exp(-ikx)\right]_{x=0}$$

$$= \frac{1}{\sqrt{2\pi}} \qquad \text{(a constant)}$$

The form of $\delta(x)$ can be obtained as (using Eqn. 19)

$$\delta(x) = \frac{1}{\sqrt{2\pi}} \int\limits_{-\infty}^{\infty} F_\delta(k)\, e^{ikx}\, dk$$

or
$$\delta(x) = \frac{1}{2\pi} \int\limits_{-\infty}^{\infty} e^{ikx}\, dk$$

which is same as Eqn. (30(i)).

Example 4.2: Show that the Dirac δ-function can be represented by

$$\delta(x) = \lim_{\alpha \to 0} \frac{1}{\alpha\sqrt{\pi}} e^{-x^2/\alpha^2}$$

Solution: We have to show that the function so defined satisfies

$$\int_{-\infty}^{+\infty} f(x)\, \delta(x)\, dx = f(0)$$

Thus, for an arbitrary function $f(x)$

$$\int_{-\infty}^{+\infty} f(x)\, \delta(x)\, dx = \lim_{\alpha \to 0} \int_{-\infty}^{+\infty} f(x) \frac{1}{\alpha\sqrt{\pi}} e^{-x^2/\alpha^2}\, dx$$

$$= \lim_{\alpha \to 0} \frac{1}{\sqrt{\pi}} \int_{-\infty}^{+\infty} f(\alpha u)\, e^{-u^2}\, du$$

$$= \frac{1}{\sqrt{\pi}} \int_{-\infty}^{+\infty} \lim_{\alpha \to 0} f(\alpha u)\, e^{-u^2}\, du$$

$$= \frac{1}{\sqrt{\pi}} \int_{-\infty}^{+\infty} f(0)\, e^{-u^2}\, du = f(0)\, \frac{1}{\sqrt{\pi}} \int_{-\infty}^{+\infty} e^{-u^2}\, du$$

$$= f(0)$$

Hence the given representation is that of Dirac δ-function.

4.4 Continuous Distribution of Eigenvalues

The eigenvalue equation can be written as

$$Q\psi_q = q\psi_q \tag{43}$$

Here the eigenvalue q takes on a continuous distribution of values q is also used as a subscript to designate the eigenfunction with which it is associated.

Orthogonality condition, corresponding to different eigenvalues q and q' takes the form

$$\int \overline{\psi}_{q'} \psi_q\, d^3r = 0, \quad q \neq q' \tag{44}$$

For $q' = q$, the integral is divergent, because now the eigenfunctions are unnormalizable, therefore, we can define the orthogonality integral as

$$\int \overline{\psi}_{q'} \psi_q\, d^3r = \delta(q - q') \tag{45}$$

Any wave function ψ can be expanded as

$$\psi = \int u(q)\, \psi_q\, dq \tag{46}$$

In general, for an operator Q, the eigenvalues may have a range of values over which they are continuous and a separate range over which the values form a discrete set. Then

$$\psi = \sum_q u_q \psi_q + \int u(q)\, \psi(q)\, dq \tag{47}$$

The summation is over the discrete range of eigenvalues and the integration is over the continuous range. If the wave function ψ is normalized to unity

$$\int |\psi|^2\, d^3r = \sum_q |u_q|^2 + \int |u(q)|^2\, dq = 1 \tag{48}$$

Illustration for the case in which there is only a continuous range of eigenvalues:

$$\psi = Ae^{i(\vec{k}\cdot\vec{x})} \quad \left(\vec{k} = \vec{p}/\hbar\right)$$

$$\psi_p(x) = Ae^{i\left(\frac{px}{\hbar}\right)}$$

$$\psi_{p'} = Ae^{i\left(\frac{p'x}{\hbar}\right)} \tag{49}$$

$$\int \bar{\psi}_{p'}\psi_p \, dx = |A|^2 \int e^{i(p-p')\frac{x}{\hbar}} dx \tag{50}$$

Let $x' = x/\hbar$

$$\therefore \qquad dx' = \frac{dx}{\hbar} \text{ or } dx = \hbar dx'$$

Eqn. (50) becomes

$$\int \bar{\psi}_{p'}\psi_p \, dx = |A|^2 \hbar \int e^{i(p-p')x'} dx'$$

$$= 2\pi\hbar|A|^2 \frac{1}{2\pi} \int e^{i(p-p')x'} dx'$$

$$= 2\pi\hbar \, |A|^2 \, \delta(p-p') \tag{51}$$

$$\therefore \quad \int \bar{\psi}_{p'}\psi_p \, dx = \delta(p-p') \text{ if } |A| = \sqrt{\frac{1}{2\pi\hbar}} \tag{52}$$

4.5 Closure Relation

Consider the expansion of an arbitrary wave function in terms of eigenfunctions of some particular operator as

$$\psi(r) = \sum_q u_q \psi_q + \int u(q) \, \psi_q dq \tag{53}$$

Multiplying both sides by $\bar{\psi}_{q'}(r)$ and integrating

$$\int \bar{\psi}_{q'}\psi \, d^3r = \int \sum_q u_q \bar{\psi}_{q'}\psi_q \, d^3r \; + 0$$

$$= \sum_q u_q \int \bar{\psi}_{q'}\psi_q \, d^3r$$

$$= \sum_q u_q \, \delta_{qq'} = u_{q'}$$

$$\int \bar{\psi}_{q'}\psi \, d^3r = u_q \tag{54}$$

$\therefore$ Eqn. (53) becomes

$$\psi(r) = \sum_q \left(\int \bar{\psi}_q\psi \, d^3r'\right) \psi_q(r) + \int \left(\int \bar{\psi}_q\psi \, d^3r'\right) \psi_q \, dq$$

$$= \int d^3r' \psi(r') \left(\sum_q \bar{\psi}_q(r')\psi_q(r) + \int \bar{\psi}_q(r')\psi_q(r) \, dq\right) \tag{55}$$

From the form of this equation, it follows that the term in brackets under the integral sign is a delta function, i.e.

$$\sum_q \bar{\psi}_q(r')\psi_q(r) + \int \bar{\psi}_q(r')\psi_q(r)\, dq = \delta(r - r') \tag{56}$$

because then Eqn. (55) satisfies

$$\psi(r) = \int \psi(r')\, \delta(r' - r)\, d^3 r'$$

The relation given by equation (56) is known as the *closure relation*.

4.6 Physical Significance of the Expectation Value

For a normalized wave function, the expectation value

$$\langle q \rangle = \int \bar{\psi} Q \psi\, d^3 r \tag{57}$$

To see its physical significance, we expand the wave function in the eigenfunctions of Q in accordance with

$$\psi = \sum_j c_j \psi_j \qquad Q\psi_j = q_j \psi_j \tag{58}$$

Assuming that eigenfunctions given in Eqn. (58) are orthogonal and normalized to unity.

$$\int \bar{\psi}_j \psi_k\, d^3 r = \delta_{jk}$$

If ψ is also normalised, then $\int \bar{\psi}\psi\, d^3 r = 1$

or
$$\sum_{jk} \bar{c}_j\, c_k \int \bar{\psi}_j \psi_k\, d^3 r = 1$$

or
$$\sum_{jk} \bar{c}_j\, c_k \delta_{jk} = 1$$

or
$$\sum_j \left| c_j \right|^2 = 1 \tag{59}$$

$$(Q\psi) = Q\left(\sum_k c_k\, \psi_k \right)$$

From Eqn. (58),
$$\bar{\psi} Q \psi = \sum_j \bar{c}_j\, \bar{\psi}_j \sum_k c_k q_k\, \psi_k$$

$$\langle q \rangle = \int \bar{\psi} Q \psi\, d^3 r = \int \sum_{jk} \bar{c}_j c_k q_k\, \bar{\psi}_j\, \psi_k\, d^3 r$$

$$= \sum_{jk} \bar{c}_j c_k \dot{q}_k \int \bar{\psi}_j\, \psi_k\, d^3 r = \sum_{jk} \bar{c}_j\, c_k q_k \delta_{jk}$$

or
$$\langle q \rangle = \sum_j \left| c_j \right|^2 q_j \tag{60}$$

From relations (59) and (60) it is evident that $|c_j|^2$ is the probability of finding the system in the state (ψ_j) designated by subscript j, $\therefore$ in a measurement in which q is determined, the probability that the result q_j will be obtained is given by

$$p_j = \left| c_j \right|^2 \tag{61}$$

4.7 Completeness Condition in Terms of the Dirac Delta Function

Let $\psi_n(x)$ form a complete set of orthonormal functions in the range $-\infty < x < \infty$. The orthonormality condition would require

$$\int_{-\infty}^{+\infty} \psi_m^*(x) \ \psi_n(x) \ dx = \delta_{mn} \tag{62}$$

Since $\psi_n(x)$ form a complete set of functions, we can expand an arbitrary "well behaved" function $\phi(x)$ in terms of $\psi_n(x)$.

$$\phi(x) = \sum_{n=0,1,2,...}^{\infty} c_n \psi_n(x) \tag{63}$$

In order to determine c_n, multiply the above equation by $\psi_m^*(x)$ and integrate:

$$\int_{-\infty}^{+\infty} \psi_m^*(x) \ \phi(x) \ dx = \sum_{n=0,1,2,...}^{\infty} c_n \int_{-\infty}^{+\infty} \psi_m^*(x) \ \psi_n(x) \ dx$$

$$= \sum_{n=0,1,2,...}^{\infty} c_n \delta_{mn} = c_m$$

Thus

$$c_n = \int_{-\infty}^{+\infty} \psi_n^*(x) \ \phi(x) \ dx \tag{64}$$

Substituting in Eqn. (63), we get

$$\phi(x) = \sum_{n=0,1,2,...}^{\infty} \left[\int_{-\infty}^{+\infty} \psi_n^*(x') \ \phi(x') \ dx' \right] \psi_n(x) \tag{65}$$

Here we have put a prime on the integration variable so that it does not create confusion with the variable of $\psi_n(x)$. On taking the summation first, we obtain

$$\phi(x) = \int_{-\infty}^{+\infty} dx' \ \phi(x') \sum_{n=0,1,2,...}^{\infty} \psi_n^*(x') \ \psi_n(x) \tag{66}$$

Comparing this with the following equation

$$\phi(x') = \int_{-\infty}^{+\infty} \phi(x) \ \delta(x - x') \ dx$$

we get

$$\sum_{n=0,1,2,...}^{\infty} \psi_n^*(x') \ \psi_n(x) = \delta(x - x') \tag{67}$$

This is known as *completeness condition*.

4.8 A Localized Wave-Packet

One cannot think of a configuration of particles giving rise to a wave-behaviour. However, on the other hand, we can imagine a group of waves, when superimposed on one another, may form a wave-packet that is localized.

As an example, consider the function defined by

$$f(x) = \int_{-\infty}^{+\infty} dk \; g(k) \; e^{ikx} \tag{68}$$

The real part of $f(x)$ is given by $\int_{-\infty}^{+\infty} dk \; g(k) \cos kx$. This is a linear super-position of waves of wavelength $\lambda \; (= 2\pi/k)$ where there is a range of k values characterized by the function $g(k)$. Let us choose the Gaussian form:

$$g(k) = e^{-\alpha(k-k_0)^2} \tag{69i}$$

and put $k - k_0 = k'$. Then

$$f(x) = \int_{-\infty}^{+\infty} dk \; g(k) \; e^{i(k-k_0)x} e^{ik_0 x}$$

$$= e^{ik_0 x} \int_{-\infty}^{+\infty} dk' \; e^{ik' x} e^{-\alpha k'^2} \tag{69ii}$$

$$= e^{ik_0 x} \int_{-\infty}^{+\infty} dk' \; e^{-\alpha\left[k' - (ix/2\alpha)\right]^2} e^{-x^2/4\alpha}$$

Because

$$ik'x - \alpha k'^2 = -\alpha\left[k'^2 - \frac{2ik'x}{2\alpha} + \left(\frac{ix}{2\alpha}\right)^2\right] - \frac{x^2}{4\alpha}$$

or

$$f(x) = e^{ik_0 x} e^{-x^2/4\alpha} \int_{-\infty}^{+\infty} ds \; e^{-\alpha s^2} \quad \left(s = k' - \frac{ix}{2\alpha}\right) \tag{70a}$$

or

$$f(x) = \sqrt{\frac{\pi}{\alpha}} e^{ik_0 x} e^{-x^2/4\alpha} \quad \left[\because \int_{-\infty}^{+\infty} ds \; e^{-\alpha s^2} = \sqrt{\frac{\pi}{\alpha}}\right] \tag{70b}$$

The factor $e^{ik_0 x}$ is the phase factor (Note that $\left|e^{ik_0 x}\right|^2 = 1$). Thus

$$\left|f(x)\right|^2 = \frac{\pi}{\alpha} e^{-x^2/2\alpha} \tag{71}$$

This represents a wave packet. It is a function which peaks at $x = 0$ and the wave packet is broad if α is large, and narrow if α is small. The width of the packet may be taken to be $2\sqrt{2\alpha}$. Thus, one may consider $\left|f(x)\right|^2$ as representing a particle.

Example 4.3: Find the Fourier transform of the function

$$F(x) = A e^{-\alpha|x|} \quad (\alpha > 0)$$

Hence show that

$$\int_{-\infty}^{+\infty} \frac{e^{ikx}}{\alpha^2 + k^2} dk = \frac{\pi}{\alpha} e^{-\alpha|x|}.$$

Solution: The Fourier transform of $f(x)$ is given by

$$F(k) = \frac{1}{\sqrt{2\pi}} \int_{-\infty}^{+\infty} f(x) \, e^{-ikx} dx$$

or

$$F(k) = \frac{A}{\sqrt{2\pi}} \int_{-\infty}^{+\infty} e^{-\alpha|x|} e^{-ikx} dx$$

$$= \frac{A}{\sqrt{2\pi}} \left[\int_{-\infty}^{0} e^{-\alpha|x|} e^{-ikx} dx + \int_{0}^{\infty} e^{-\alpha|x|} e^{-ikx} dx \right]$$

$$= \frac{A}{\sqrt{2\pi}} \left[\int_{\infty}^{0} e^{-\alpha|t|} e^{ikt} d(-t) + \int_{0}^{\infty} e^{-\alpha|x|} e^{-ikx} dx \right]$$

$$= \frac{A}{\sqrt{2\pi}} \left[-\int_{\infty}^{0} e^{-(\alpha-ik)t} dt + \int_{0}^{\infty} e^{-(\alpha+ik)x} dx \right]$$

$$= \frac{A}{\sqrt{2\pi}} \left\{ -\frac{e^{-(\alpha-ik)t}}{-(\alpha-ik)} \Big|_{\infty}^{0} + \frac{e^{-(\alpha+ik)x}}{-(\alpha+ik)} \Big|_{0}^{\infty} \right\}$$

$$\frac{A}{\sqrt{2\pi}} \left[\frac{1}{\alpha-ik} - 0 + 0 + \frac{1}{\alpha+ik} \right] = \frac{A}{\sqrt{2\pi}} \frac{2\alpha}{\left(\alpha^2 + k^2\right)}$$

Now,

$$f(x) = \frac{1}{\sqrt{2\pi}} \int_{-\infty}^{+\infty} F(k) \, e^{ikx} dk \equiv A e^{-\alpha|x|}$$

or

$$\frac{1}{\sqrt{2\pi}} \int_{-\infty}^{+\infty} \frac{A}{\sqrt{2\pi}} \left(\frac{2\alpha}{\alpha^2 + k^2} \right) e^{ikx} dk = A e^{-\alpha|x|}$$

or

$$\frac{\alpha}{\pi} \int_{-\infty}^{+\infty} \frac{e^{ikx}}{\alpha^2 + k^2} \, dk = e^{-\alpha|x|}$$

or

$$\int_{-\infty}^{+\infty} \frac{e^{ikx}}{\alpha^2 + k^2} \, dk = \frac{\pi}{\alpha} e^{-\alpha|x|}.$$

4.9 Group Velocity of a Wave Packet

A one-dimensional wave packet is formed by the superposition of plane waves

$$\psi(r, t) = \frac{1}{\sqrt{2\pi}} \int_{-\infty}^{+\infty} A(k) \, e^{i(kx-\omega t)} dk \tag{72}$$

where the frequency $\qquad \omega = \omega(k) \tag{73}$

Eqn. (73) is known as the *dispersion relation*. Form example, for a free (non relativistic) particle

$$E = p^2/(2m) \tag{74}$$

$$\Rightarrow \qquad \hbar\omega = \frac{\hbar^2 k^2}{2m}$$

or
$$\omega(k) = \frac{\hbar}{2m}k^2 \tag{75}$$

Now, for a purely monochromatic wave

$$A(k) = A_0 \, \delta \, (k - k_0) \tag{76}$$

If we substitute the above expression for $A(k)$ in Eqn. (72), we would get

$$\psi(x, t) = \frac{A_0}{\sqrt{2\pi}} \, e^{i(k_0 x - \omega_0 t)} \tag{77}$$

where
$$\omega_0 = \omega \, (k_0) \tag{78}$$

Eqn. (77) is never achieved in practice. Usually $A(k)$ is a very sharply peaked function around $k = k_0$. Therefore, we assume that there exists only a small range Δk (around $k = k_0$) where the amplitudes of the component waves $A(k)$ have significant values; beyond $k = k_0 \pm \Delta k$, $A(k)$ is assumed to be negligible. Under such an assumption, one can assign an average velocity of the wave packet provided $\omega(k)$ is a smoothly varying function in the domain $k_0 - \Delta k \leq k \leq k_0 + \Delta k$. This average velocity is known as the *group velocity*. We write

$$A(k) = |A(k)| \, e^{i\phi(k)} \tag{79}$$

and assume that $|A(k)|$ is very sharply peaked when k lies in a small interval Δk aroung $k = k_0$ and negligible everywhere else so that we have

$$\psi(x, t) \approx \frac{1}{\sqrt{2\pi}} \int_{\Delta k} dk \, |A(k)| \, e^{i[kx - \omega(k)t + \phi(k)]} \tag{80}$$

where the integration extends over the interval Δk where $|A(k)|$ has appreciable values. Now, we make Taylor expansions of $\omega(k)$ and $\phi(k)$ around $k = k_0$:

$$\omega(k) = \omega(k_0) + \frac{d\omega}{dk}\bigg|_{k=k_0} (k - k_0) + \frac{1}{2}\frac{d^2\omega}{dk^2}\bigg|_{k=k_0} (k - k_0)^2 + \dots \tag{81}$$

$$\phi(k) = \phi(k_0) + \frac{d\phi}{dk}\bigg|_{k=k_0} (k - k_0) + \frac{1}{2}\frac{d^2\phi}{dk^2}\bigg|_{k=k_0} (k - k_0)^2 + \dots \tag{82}$$

We substitute these expressions in Eqn. (80) and assume that in the domain of integration Δk, $\omega(k)$ and $\phi(k)$ do not vary significantly so that the terms involving quadratic and higher powers of $(k - k_0)$ can be neglected. Under such an approximation, we obtain

$$\psi(x, t) = f(x, t) \, e^{i[k_0 x - \omega_0 t + \phi_0]} \tag{83}$$

where $\omega_0 = \omega(k_0)$, $\phi_0 = \phi(k_0)$ and $f(x, t)$ which represents the envelope of the wave packet is given by

$$f(x, t) = \frac{1}{\sqrt{2\pi}} \int_{\Delta k} dk \, |A(k)| \, e^{i(k - k_0)(x - x_0 - v_g t)} \tag{84}$$

where
$$x_0 = -\left.\frac{d\phi}{dk}\right|_{k=k_0}$$

and group velocity
$$v_g = -\left.\frac{d\omega}{dk}\right|_{k=k_0} \tag{85}$$

Since the function $f(x, t)$ depends on x and t only, the wave packet propagates (without any distortion) with velocity v_g.

Now, from Eqn. (84), we have

$$f(x, 0) = \frac{1}{\sqrt{2\pi}} \int_{\Delta k} |A(k)| \, e^{i(k-k_0)(x-x_0)} dk \tag{86}$$

Thus at $x = x_0$, the integrand is everywhere positive and the value of the integral is maximum. For $|x - x_0| \geq 1/\Delta k$, the exponential function oscillates rapidly in the domain of integraion and the value of the integral is very small. This implies

$$\Delta k. \Delta x \geq 1 \tag{87}$$

which when multiplied by $\hbar$ leads to the uncertainty relation

$$\Delta p . \Delta x \geq \hbar \tag{88}$$

Thus at $t = 0$, the wave packet is sharply peaked at $x = x_0$ and as t increases, the centre of the packet $x_c(t)$ moves accordingly to the equation

$$x_c(t) = x_0 + v_g t$$

Now since
$$E = \hbar\omega \text{ and } p = \hbar k$$

so we get
$$v_g = \frac{d\omega}{dk} = \frac{dE}{dp}$$

Thus for a non relativistic particle for which $E = p^2/2m$, we have

$$v_g = \frac{dE}{dp} = \frac{p}{m} = v \tag{89}$$

Further, for a relativistic particle,

$$E^2 = p^2 c^2 + m^2 c^4$$

(where m is rest mass), we have

$$\frac{dE}{dp} = \frac{pc^2}{E} \text{ but } E = \frac{mc^2}{\sqrt{1 - v^2/c^2}}$$

and
$$p = \frac{mv}{\sqrt{1 - v^2/c^2}} \text{ gives } \frac{pc^2}{E} = \frac{mv}{\sqrt{1 - v^2 c^2}} \cdot \frac{\sqrt{1 - v^2 c^2}}{m} = v$$

Thus
$$v = \frac{dE}{dp} = v_g \tag{90}$$

Eqns (89) and (90) tell us that the group velocity of the packet is to be associated with the velocity of the particle. The *phase velocity* is defined by $v_p = \omega/k$ and represents the velocity of propagation of an infinitely long monochromatic plane wave. Only when ω is proportional to k are v_p and v_g equal. In general only v_g is of physical significance.

Example 4.4: Find the relation between the phase and group velocities of a wave packet formed by superposition of three waves:

$$\psi = A\, e^{i(kx-\omega t)}$$

$$\psi_1 = \frac{A}{2}\, e^{i\left[(k+dk)x-(\omega-d\omega)t\right]}$$

and

$$\psi_2 = \frac{A}{2}\, e^{i\left[(k-dk)x-(\omega-d\omega)t\right]}$$

all travelling in the x-direction.

Solution: Superposition of the three waves gives

$$\psi + \psi_1 + \psi_2 = A\, e^{i(kx-\omega t)}\left[1 + \frac{1}{2}e^{i(xdk-td\omega)} + \frac{1}{2}e^{-i(xdk-td\omega)}\right]$$

$$= A\, e^{i(kx-\omega t)}\left[1 + \cos\left(xdk - td\omega\right)\right]$$

$$= 2A\,\cos^2\left(\frac{dk}{2}x - \frac{d\omega}{2}t\right)e^{i(kx-\omega t)}$$

The wavefront of the wave defined by $kx - \omega t = $ constant or $x = \dfrac{\omega t}{k} + $ constant travels with phase velocity $u = \omega/k$. The envelope of the wave packet is modulated by the cosine term. Each point of the envelope (e.g. its maximum) which is defined by $x\, dk - t\, d\omega = 0$ propagates with the group velocity given by

$$v = \frac{x}{t} = \frac{d\omega}{dk}$$

The phase velocity is given by

$$u = \frac{\omega}{k}$$

$\therefore$ Substituting $\omega = ku$ gives

$$v = u + k\frac{du}{dk} \qquad (i)$$

We can also derive the relation in terms of wavelength λ $(=2\pi/k)$ as

$$\frac{du}{dk} = \frac{du}{d\lambda}\cdot\frac{d\lambda}{dk} = -\frac{\lambda^2}{2\pi}\cdot\frac{du}{d\lambda}$$

But from (i)

$$\frac{du}{dk} = \frac{1}{k}(v-u) = \frac{\lambda}{2\pi}(v-u)$$

So

$$\frac{\lambda}{2\pi}v = \frac{\lambda}{2\pi}u - \frac{\lambda^2}{2\pi}\frac{du}{d\lambda}$$

or

$$v = u - \lambda\frac{du}{d\lambda} \qquad (ii)$$

4.10 The Wave-Packet Spreads as it Travels

Propagation of a wave packet in time depends on the propagation of individual constituent wave. A simple plane wave has only a spatial variation in x, and is given by $e^{ikx-i\omega t}$. We know that $\omega = 2\pi\nu$ and $k = 2\pi/\lambda$, so we have

$$e^{2\pi i\left[(x/\lambda)-\nu t\right]} \tag{91}$$

For light waves, $\omega = kc$, but for waves describing particles, ω is, in general, a function of k, so that

$$f(x,t) = \int dk \, g(k) \, e^{ikx-i\omega(k)t} \tag{92}$$

Now, let us consider a wave packet that is strongly localized in k-space about a value k_0. Then we may expand $\omega(k)$ about k_0 assuming that $\omega(k)$ is not a very rapidly varying function of k:

$$\omega(k) \cong \omega(k_0) + (k - k_0)\left(\frac{d\omega}{dk}\right)_{k_0} + \frac{1}{2}(k - k_0)^2\left(\frac{d^2\omega}{dk^2}\right)_{k_0}$$

Here we have $\left(\dfrac{d\omega}{dk}\right)_{k_0} = v_g$ and we write

$$\left(\frac{d^2\omega}{dk^2}\right)_{k_0} = 2\beta \tag{93}$$

With $g(k) = e^{-\alpha k'^2}$, we have (for $(k - k_0) \equiv k'$

$$f(x,t) = \int_{-\infty}^{+\infty} dk' e^{-\alpha k'^2} e^{ik_0 x} e^{ik'x} e^{-i\omega(k_0)t} e^{-ik'v_g t} e^{-ik'^2 \beta t}$$

$$= e^{ik_0 x - i\omega(k_0)t} \int_{-\infty}^{+\infty} dk' e^{-\alpha k'^2} e^{ik'(x-v_g t)} e^{-ik'^2 \beta t}$$

$$= e^{ik_0 x - i\omega(k_0)t} \int_{-\infty}^{+\infty} dk' e^{ik'(x-v_g t)} e^{-(\alpha+i\beta t)k'^2} \tag{94}$$

This is similar to integral in Eqn. 69 (*ii*) with $x - v_g t$ in place of x and $\alpha + i\beta t$ in place of α. So, we get

$$f(x,t) = e^{i[k_0 x - \omega(k_0)t]}\left(\frac{\pi}{\alpha + i\beta t}\right)^{1/2} \exp\left[-\frac{(x - v_g t)^2}{4(\alpha + i\beta t)}\right] \tag{95}$$

Consequently

$$|f(x,t)|^2 = \left[\frac{\pi^2}{\alpha^2 + \beta^2 t^2}\right]^{1/2} \exp\left[-\frac{(x - v_g t)^2}{2(\alpha^2 + \beta^2 t^2)}\right] \tag{96i}$$

This represents a wave packet which does not have a definite width. The quantity that was α at $t = 0$ now becomes $\alpha + \left(\dfrac{\beta^2 t^2}{\alpha}\right)$ at time t:

$$|f(x,t)|^2 = \left[\frac{\pi^2}{\alpha^2 + \beta^2 t^2}\right]^{1/2} \exp\left[-\frac{(x - v_g t)^2}{2\alpha\left(\alpha + \frac{\beta^2 t^2}{\alpha}\right)}\right] \tag{96ii}$$

The width of the packet is

$$2 \times \sqrt{2\alpha}\left[\alpha + \frac{\beta^2 t^2}{\alpha}\right]^{1/2} = 2 \times \sqrt{2\alpha}\sqrt{\alpha}\left[1 + \frac{\beta^2 t^2}{\alpha^2}\right]^{1/2}$$

$$= \left(2\sqrt{2}\right)\alpha\left[1 + \frac{\beta^2 t^2}{\alpha^2}\right]^{1/2} \tag{97}$$

Thus rate of spreading will be small if α is large i.e. if the packet is spatially large in the beginning.

Example 4.5: Find the Fourier transform of the guassian function

$$\psi(x) = \frac{1}{\left(\sigma\sqrt{\pi}\right)^{1/2}} e^{-\left(x^2/2\sigma^2\right)}$$

Solution: The Fourier transform of $f(x)$ is given by

$$F(k) = \frac{1}{\sqrt{2\pi}} \int_{-\infty}^{+\infty} f(x) e^{-ikx} dx$$

So, the Fourier transform of $\psi(x)$ is

$$\phi(k) = \frac{1}{\sqrt{2\pi}} \frac{1}{\left(\sigma\sqrt{\pi}\right)^{1/2}} \int_{-\infty}^{+\infty} e^{-\left(x^2/2\sigma^2\right)} e^{-ikx} dx$$

$$= \frac{1}{\sqrt{2\pi}} \frac{1}{\left(\sigma\sqrt{\pi}\right)^{1/2}} \int_{-\infty}^{+\infty} \exp\left[-\frac{x^2}{2\sigma^2} - ikx - \frac{\sigma^2 k^2}{2} + \frac{\sigma^2 k^2}{2}\right] dx$$

$$= \frac{1}{\sqrt{2\pi}} \frac{e^{-\sigma^2 k^2/2}}{\left(\sigma\sqrt{\pi}\right)^{1/2}} \int_{-\infty}^{+\infty} \exp\left[-\frac{1}{2\sigma^2}\left(x^2 + 2ixk\sigma^2 + i^2 k^2 \sigma^4\right)\right] dx$$

$$= \frac{e^{-\sigma^2 k^2/2}}{\left(2\pi\sigma\sqrt{\pi}\right)^{1/2}} \int_{-\infty}^{+\infty} \exp\left[-\frac{1}{2\sigma^2}\left(x + ik\sigma^2\right)^2\right] dx \tag{i}$$

The integral in this equation is evaluated as follows: Consider the integral

$$\int_C \exp\left[-\frac{1}{2\sigma^2}\left(z + ik\sigma^2\right)^2\right] dz \tag{ii}$$

in which C is the closed rectangular contour shown in Fig. 4.2.

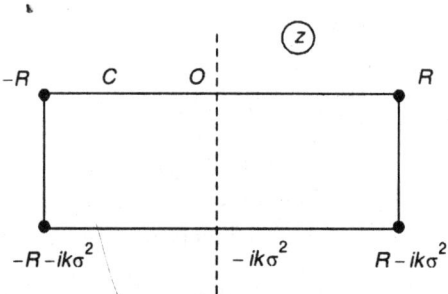

Fig. 4.2: Pertaining to example 4.5

By Cauchy's theorem, the integral (*ii*) is zero since the integrand is regular at every point within and on *C*. Breaking up the path of integration into its straight line parts, we have

$$\int_{-R}^{R} \exp\left[-\frac{1}{2\sigma^2}\left(z + ik\sigma^2\right)^2\right] dz$$

$$= -\left[\int_{R}^{R-ik\sigma^2} + \int_{R-ik\sigma^2}^{-R-ik\sigma^2} + \int_{-R-ik\sigma^2}^{-R}\right] \exp\left[-\frac{1}{2\sigma^2}\left(z + ik\sigma^2\right)^2\right] dz$$

Now if we allow *R* to become very large ($R \to \infty$), the integrals along the vertical parts of the path approach zero, and we have

$$\int_{-\infty}^{+\infty} \exp\left[-\frac{1}{2\sigma^2}\left(z + ik\sigma^2\right)^2\right] dz = \int_{-\infty-ik\sigma^2}^{\infty-ik\sigma^2} \exp\left[-\frac{1}{2\sigma^2}\left(z + ik\sigma^2\right)^2\right] dz \qquad (iii)$$

Substituting $\quad u = \frac{1}{\sigma\sqrt{2}}\left(z + ik\sigma^2\right)$ and $du = \frac{dz}{\sigma\sqrt{2}}$

And, the integral (*iii*) becomes

$$\sigma\sqrt{2} \int_{-\infty}^{+\infty} e^{-u^2} du = \sigma\sqrt{2} \times \sqrt{\pi} = \sigma\sqrt{2\pi}$$

Hence $\qquad\qquad \phi(k) = \left(\frac{\sigma}{\sqrt{\pi}}\right)^{1/2} e^{-\sigma^2 k^2 / 2}$

This is also a gaussian function.

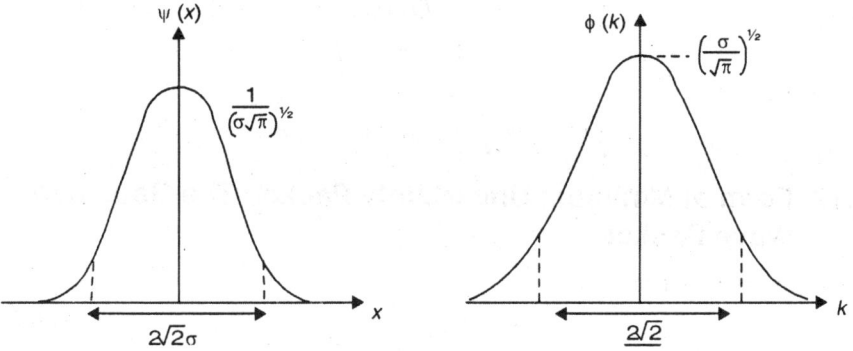

Fig. 4.3: The gaussian function $\psi(x)$ and its Fourier transform $\phi(k)$ (example 4.5)

4.11 Schwarz Inequality

Statement: It states that if f and g are two eigen vectors, then

$$\int f * f dv \int g * g dv \ge \left| \int f * g dv \right|^2 \qquad (98i)$$

i.e.
$$(f, f)\,(g,\,g) \ge |(f,\,g)|^2 \qquad (98ii)$$

where bracket notation $(f,\,f)$ means the scalar product $\int f^* f\, dv$.

The equality sign holds only if the functions f and g are multiples of each other.

Proof: Consider a vector $\psi = f + \lambda g$ where λ is an undetermined parameter (may be real or complex).

Since the norm of a vector is always greater than or equal to zero, we must have

$$N = (\psi, \psi) = (f + \lambda g) * (f + \lambda g) \ge 0$$

i.e.
$$N = (f, f) + \lambda * (g, f) + \lambda (f, g) + \lambda \lambda * (g, g) \ge 0 \qquad (99)$$

The best inequality is obtained if λ is chosen so as to minimize the l.h.s i.e.

$$\frac{\partial N}{\partial \lambda} = 0 \text{ or } \frac{\partial N}{\partial \lambda *} = 0$$

i.e.
$$\frac{\partial}{\partial \lambda *} \left[(f, f) + \lambda * (g, f) + \lambda (f, g) + \lambda \lambda * (g, g) \right] = 0$$

i.e.
$$(g, f) + \lambda\, (g,\, g) = 0$$

This gives
$$\lambda = -\frac{(g, f)}{(g, g)}$$

Substituting this value of λ in Eqn. (99), we get

$$(f, f) + \lambda * (g, f) + \left[-\frac{(g, f)}{(g, g)} \right](f, g) + \lambda * \left[-\frac{(g, f)}{(g, g)} \right](g, g) \ge 0$$

or
$$(f, f) + \lambda * (g, f) - \frac{(g, f)(f, g)}{(g, g)} - \lambda * (g, f) \ge 0$$

or
$$(f, f) - \frac{(g, f)(f, g)}{(g, g)} \ge 0$$

or
$$(f,f)\,(g,g) - (g,f)\,(f,g) \ge 0$$

or
$$(f,f)\,(g,g) \ge (g,f)\,(f,g) \ge (f,g)* \,(f,g)$$

or
$$(f,f)\,(g,g) \ge |(f,g)|^2$$

Rewriting

$$\int f * f\, dv \int g * g\, dv \ge \left| \int f * g\, dv \right|^2$$

4.12 Form of Minimum Uncertainty Packet: The Gaussian Wave Packet

From uncertainty principle

$$\Delta x \,.\, \Delta p \ge \frac{\hbar}{2} \qquad (100i)$$

So, the minimum uncertainty product is

$$\Delta xzx.\Delta p = \frac{\hbar}{2} \tag{100ii}$$

We know that the minimum uncertainty product will be obtained when the functions $f(p)$ and $\phi(x)$ are so chosen that there is sign of equality in Schwarz inequality. However, the sign of equality will only be obtained when the functions f and ϕ are proportional to each other i.e.

$$f \propto \phi \text{ or } f = \beta\phi \tag{101}$$

where β is a constant of proportionality which is so chosen to get the sign of equality in Schwarz inequality.

Further,

$$f = -i\hbar\frac{d\psi}{dx} \quad \left(f^* = i\hbar\frac{d\psi^*}{dx}\right)$$

and

$$\phi = x\psi$$

so,

$$-i\hbar\frac{d\psi}{dx} = \beta x\psi \tag{102}$$

In order to get the minimum uncertainty product, we must have $\int f^*\phi\,dx$ or

$$\int i\hbar\frac{d\psi^*}{dx}x\psi\,dx = \text{purely imaginary}$$

i.e. $\beta^* \int x\psi^* x\psi\,dx = \text{purely imaginary}$

[by substituting complex conjugate of Eqn. (102)]

or $\int \beta^* x\psi^* x\psi\,dx = \text{purely imaginary}$

This shows that β^* must be purely an imaginary number, so, for convenience, we can put

$$\beta = \frac{i\hbar}{\alpha} \tag{103}$$

where α is purely a real number. If this is substituted in Eqn. (102), we obtain,

$$-i\hbar\frac{d\psi}{dx} = \frac{i\hbar}{\alpha}x\psi$$

$$\therefore \qquad \frac{d\psi}{dx} = -\frac{x}{\alpha}\psi \tag{104}$$

If this differential equation is solved, we get

$$\psi = Ne^{-x^2/2\alpha} \tag{105}$$

N is to be determined from normalization condition. Since the integral $\int \psi^*\psi\,dx$ must converge, therefore α must be positive.

$$\int \psi^*\psi\,dx = 1 \Rightarrow \int_{-\infty}^{+\infty} N^2 e^{-x^2/\alpha}dx = 1$$

or

$$N^2\sqrt{\alpha\pi} = 1 \quad \therefore \quad N = \frac{1}{(\alpha\pi)^{1/4}} \tag{106}$$

Thus,

$$\psi = \frac{1}{(\alpha\pi)^{1/4}} e^{-x^2/2\alpha} \tag{107}$$

Now, in order to obtain the value of α, we note that (in analogy with standard deviation of statistics), the uncertainty

$$\Delta x = \sqrt{<(x-<x>)^2>}$$

or
$$(\Delta x)^2 = <x^2> \qquad\qquad \text{(since <x> = 0)}$$

or
$$\int_{-\infty}^{+\infty} \psi^* x^2 \psi \ dx = (\Delta x)^2$$

or
$$\int_{-\infty}^{+\infty} x^2 \frac{1}{(\alpha\pi)^{1/2}} e^{-x^2/\alpha} dx = (\Delta x)^2$$

or
$$\frac{1}{(\alpha\pi)^{1/2}} \int_{-\infty}^{+\infty} x^2 e^{-x^2/\alpha} dx = (\Delta x)^2$$

or
$$\frac{1}{(\alpha\pi)^{1/2}} \frac{1}{2}\sqrt{\pi\alpha^3} = (\Delta x)^2 ; \ \therefore \ \alpha = 2(\Delta x)^2 \tag{108}$$

Substituting the value of α in Eqn. (107), we get

$$\psi(x,0) = \frac{1}{\left[2\pi(\Delta x)^2\right]^{1/4}} e^{-x^2/4(\Delta x)^2} \tag{109}$$

This equation shows that the normalized minimum wave packet is having the shape of a Gaussian curve. If the assumption <x> = <p> = 0 is not made, then we will get

$$\psi(x,0) = \frac{1}{\left[2\pi(\Delta x)^2\right]^{1/4}} e^{-(x-<x>)^2/4(\Delta x)^2} e^{+i<p>x/\hbar} \tag{110i}$$

This represents a wave packet having Gaussian distribution about <x> which is moving with average momentum <p>. The quantity Δx gives the width of the packet.

If it is assumed that the wave packet is centered at the origin, and moves with average momentum <p>, then

$$\psi(x,0) = \frac{1}{\left[2\pi(\Delta x)^2\right]^{1/4}} e^{-x^2/4(\Delta x)^2} e^{+i<p>x/\hbar} \tag{110ii}$$

and position probability density is

$$|\psi(x,0)|^2 = \frac{1}{\left[2\pi(\Delta x)^2\right]^{1/2}} e^{-x^2/2(\Delta x)^2} \tag{110iii}$$

Let us put
$$C = \left[2\pi(\Delta x)^2\right]^{-1/4} \quad \text{so that}$$

$$\psi(x) = C \exp\left[-\frac{(x-<x>)^2}{4(\Delta x)^2} + \frac{i}{\hbar}<p>x \right] \tag{111}$$

4.13 The Momentum Space Wave Function Corresponding to the Minimum Uncertainty Wave-Packet

This can be computed as follows:

$$\phi(p) = \frac{1}{(2\pi\hbar)^{1/2}} \int_{-\infty}^{+\infty} \psi(x) \, e^{-ipx/\hbar} dx \tag{112}$$

$$= \frac{C}{(2\pi\hbar)^{1/2}} \int_{-\infty}^{+\infty} \exp\left[-\frac{(x-<x>)^2}{4(\Delta x)^2} \right] \exp\left[-\frac{i}{\hbar}(p-<p>)x \right] dx$$

$$= \frac{C}{(2\pi\hbar)^{1/2}} \exp\left[-\frac{i}{\hbar}(p-<p>)<x> \right] \int_{-\infty}^{+\infty} \exp\left[-\frac{(x-<x>)^2}{4(\Delta x)^2} \right]$$

$$\times \exp\left[-\frac{i}{\hbar}(p-<p>)(x-<x>) \right] d(x-<x>)$$

The integral can be evaluated by using

$$\int_{-\infty}^{+\infty} e^{-\alpha x^2 + \beta x} dx = \sqrt{\frac{\pi}{\alpha}} \, \exp\left(\frac{\beta^2}{4\alpha} \right) \tag{113}$$

where $\qquad \alpha = \frac{1}{4(\Delta x)^2}$ and $\beta = -\frac{i}{\hbar}(p-<p>)$

Consequently,

$$\phi(p) = \frac{1}{\left[2\pi(\Delta x)^2 \right]^{1/4}} \times \frac{1}{(2\pi\hbar)^{1/2}} \exp\left[-\frac{i}{\hbar}(p-<p>)<x> \right] \times \sqrt{4\pi(\Delta x)^2}$$

$$\times \exp\left[-(\Delta x)^2 \left(\frac{p-<p>}{\hbar} \right)^2 \right]$$

$$= \frac{1}{(2\pi)^{1/4}(2\pi)^{1/2}} \times \frac{\sqrt{2}}{(\hbar \Delta x)^{1/2}} \times (2\pi)^{1/2}$$

$$\times (\Delta x) \exp\left[-(\Delta x)^2 \left(\frac{p-<p>}{\hbar} \right)^2 \right] \times \exp\left[-\frac{i}{\hbar}(p-<p>)\langle x \rangle \right]$$

(Here we have used $\Delta x = \dfrac{\hbar}{2(\Delta p)}$ from uncertainty relation)

$$\Rightarrow (\Delta x)^{1/2} = \frac{(\hbar)^{1/2}}{\sqrt{2}(\Delta p)^{1/2}}$$

Therefore

$$\phi(p) = \frac{1}{\left[2\pi(\Delta p)^2\right]^{1/4}} \exp\left[-\frac{(p-<p>^2}{4(\Delta p)^2}\right] \exp\left[-\frac{i}{\hbar}(p-<p>)<x>\right] \quad (114)$$

and probability density in momentum space

$$\phi*(p)\phi(p) = \frac{1}{\left[2\pi(\Delta p)^2\right]^{1/2}} \exp\left[-\frac{(p-<p>)^2}{2(\Delta p)^2}\right] \quad (115)$$

4.14 Time Variation of the Minimum Uncertainty Free Particle (Gaussian) Wave Packet

The minimum uncertainty wave function at $t = 0$ is

$$\psi(x,0) = \frac{2}{\left[2\pi(\Delta x)^2\right]^{1/4}} e^{-x^2/(2\Delta x)^2} \quad (116)$$

The time dependent wave function is the Fourier transform

$$\psi(x,t) = \frac{1}{\sqrt{2\pi\hbar}} \int_{-\infty}^{+\infty} \phi(p,0) e^{i(px-Et)/\hbar} dp \quad (117)$$

where

$$\phi(p,0) = \frac{1}{\sqrt{2\pi\hbar}} \int_{-\infty}^{+\infty} \psi(x,0) e^{-i(p-<p>)x/\hbar} dx \quad (118)$$

(for free particle $E = p^2/2m$)

Now,

$$\phi(p,0) = \frac{1}{\sqrt{2\pi\hbar}} \int_{-\infty}^{+\infty} \frac{1}{\left[2\pi(\Delta x)^2\right]^{1/4}} e^{-x^2/4(\Delta x)^2} \times \exp\left[-\frac{i(p-<p>)x}{\hbar}\right] dx$$

The integral can be evaluated by using

$$\int_{-\infty}^{+\infty} e^{-\alpha x^2 + \beta x} dx = \sqrt{\frac{\pi}{\alpha}} \exp\left(\frac{\beta^2}{4\alpha}\right)$$

where $\alpha = \frac{1}{4(\Delta x)^2}$ and $\beta = -\frac{i}{\hbar}(p-<p>)$.

Consequently,

$$\phi(p,0) = \frac{\sqrt{4\pi(\Delta x)^2}}{\sqrt{2\pi\hbar\Delta x\sqrt{2\pi}}} \exp\left[-(\Delta x)^2(p-<p>)^2/\hbar^2\right]$$

$$= \sqrt{\frac{2\Delta x}{\hbar\sqrt{2\pi}}} e^{-(\Delta x)^2(p-<p>)^2/\hbar^2} \quad (119)$$

Hence we get

$$\psi(x,t) = \frac{1}{\sqrt{2\pi\hbar}} \sqrt{\frac{2\Delta x}{\hbar\sqrt{2\pi}}} \int_{-\infty}^{+\infty} e^{-(\Delta x)^2 (p-<p>)^2 / \hbar^2} \times e^{i\left[px - (p^2/2m)t)/\hbar\right]} dp$$

$$= \frac{1}{\sqrt{2\pi\hbar}} \sqrt{\frac{2\Delta x}{\hbar\sqrt{2\pi}}} \int_{-\infty}^{+\infty} e^{-(\Delta x)^2 (p^2 + <p>^2 - 2p<p>)/\hbar^2} e^{ipx} \times e^{-ip^2 t/2m\hbar} dp \qquad (120)$$

$$= \frac{1}{\sqrt{2\pi\hbar}} \sqrt{\frac{2\Delta x}{\hbar\sqrt{2\pi}}} \int_{-\infty}^{+\infty} \exp\left[-\left(\frac{(\Delta x)^2}{\hbar^2} + \frac{it}{2m\hbar}\right)p^2\right]$$

$$\times \exp\left[\left(2<p>\frac{(\Delta x)^2}{\hbar^2} + \frac{ix}{\hbar}\right)p\right] \times \exp\left[-<p>^2 \frac{(\Delta x)^2}{\hbar^2}\right] dp \qquad (121)$$

For minimum uncertainty

$$\Delta p \cdot \Delta x = \frac{\hbar}{2} \quad \text{or} \quad \frac{\Delta x}{\hbar} = \frac{1}{2(\Delta p)} \equiv \frac{1}{2a}$$

Then

$$\psi(x,t) = \frac{1}{\sqrt{2\pi\hbar}} \sqrt{\frac{2}{2a\sqrt{2\pi}}} \int_{-\infty}^{+\infty} \exp\left[-\left(\frac{1}{4a^2} + \frac{it}{2m\hbar}\right)p^2\right]$$

$$\exp\left[\left(\frac{<p>}{2a^2} + \frac{i}{\hbar}x\right)p\right] \times \exp\left[-\frac{<p>^2}{4a^2}\right] dp$$

The last exponential in the integral is independent of $\bar{p}$ and may be taken outside the integral. We can complete the square in p in the integrand by multiplying the integrand by the $\bar{p}$ independent term $e^{-\beta^2}$ and the whole integral by $e^{+\beta^2}$, so then

$$\psi(x,t) = \frac{1}{\sqrt{2\pi\hbar a}} \sqrt{\frac{1}{\sqrt{2\pi}}} \exp\left(-\frac{<p>^2}{4a^2}\right) e^{\beta^2} \int_{-\infty}^{+\infty} \exp\left[-(\alpha p + \beta)^2\right] dp \qquad (122)$$

where

$$\alpha = \left(\frac{1}{4a^2} + \frac{it}{2m\hbar}\right)^{1/2} \qquad (123)$$

and

$$\beta = -\frac{\left(\frac{<p>}{2a^2} + \frac{ix}{\hbar}\right)}{2\alpha} \qquad (124)$$

Let us denote $\lambda = \alpha p + \beta$, therefore

$$\psi(x,t) = \frac{1}{\sqrt{2\pi\hbar a}} \frac{1}{(2\pi)^{1/4}} \exp\left(-\frac{<p>^2}{4a^2} + \beta^2\right) \frac{1}{\alpha} \int_{-\infty}^{+\infty} e^{-\lambda^2} d\lambda \qquad (125)$$

The value of the integral is $\sqrt{\pi}$

$$\therefore \qquad \psi(x,t) = \frac{1}{\sqrt{2\pi\hbar}\sqrt{a}} \frac{\sqrt{\pi}}{(2\pi)^{1/4}} \frac{1}{\alpha} \exp\left(-\frac{<p>^2}{4a^2} + \beta^2\right) \qquad (126)$$

The β^2 will contain the terms in $<p>^2$, x^2 and $i<p>x/\hbar$. If we substitute the values of α and β from Eqns. (123) and (124) in Eqn. (126) and simplify, we obtain

$$\psi(x,t) = \frac{1}{\sqrt{2\pi\hbar}} \frac{1}{\sqrt{\Delta p}} \frac{\sqrt{\pi}}{(2\pi)^{1/4}} \frac{1}{\left(\dfrac{1}{4a^2} + \dfrac{it}{2m\hbar}\right)^{1/2}}$$

$$\times \exp\left[-\frac{\left(x - \dfrac{<p>t}{m}\right)^2}{4\hbar^2\left(\dfrac{1}{4a^2} + \dfrac{it}{2m\hbar}\right)}\right] \times \exp\left[\left(\frac{i<p>}{\hbar}\right)\left(x - \frac{<p>}{m}t\right)\right]$$

But $\qquad \Delta p = \dfrac{\hbar}{2\Delta x}$ and $4a^2 = (2\Delta p)^2 = \dfrac{\hbar^2}{(\Delta x)^2}$

$$\Rightarrow \qquad \psi^*(x,t) = \frac{1}{\sqrt{2\pi\hbar}} \frac{1}{\sqrt{\Delta p}} \frac{\sqrt{\pi}}{(2\pi)^{1/4}} \frac{1}{\left(\dfrac{1}{4a^2} - \dfrac{it}{2m\hbar}\right)^{1/2}}$$

$$\times \exp\left[-\frac{\left(x - \dfrac{<p>t}{m}\right)^2}{4\hbar^2\left(\dfrac{1}{4a^2} - \dfrac{it}{2m\hbar}\right)}\right] \times \exp\left[-\frac{i<p>}{\hbar}\left(x - \frac{<p>t}{m}\right)\right]$$

$$|\psi(x,t)|^2 = \psi^*\psi = \frac{1}{2\pi\hbar} \frac{1}{\Delta p} \frac{\pi}{(2\pi)^{1/2}} \frac{1}{\left[\left(\dfrac{\Delta x}{\hbar}\right)^4 + \left(\dfrac{it}{2m\hbar}\right)^2\right]^{1/2}}$$

$$\times \exp\left[-\frac{\left(x - \dfrac{<p>t}{m}\right)^2}{4\hbar^2\left(\dfrac{(\Delta x)^2}{\hbar^2} + \dfrac{it}{2m\hbar}\right)} - \frac{\left(x - \dfrac{<p>t}{m}\right)^2}{4\hbar^2\left(\dfrac{(\Delta x)^2}{\hbar^2} - \dfrac{it}{2m\hbar}\right)}\right]$$

$$= \frac{1}{2\pi\hbar} \frac{2\Delta x}{\hbar} \frac{\pi}{(2\pi)^{1/2}} \frac{1}{\left[\dfrac{1}{\hbar^4}\left\{(\Delta x)^4 + \left(\dfrac{t\hbar}{2m}\right)^2\right\}\right]^{1/2}}$$

$$\times \exp\left[-\frac{\left(x-\frac{<p>t}{m}\right)^2}{4}\left\{\frac{1}{(\Delta x)^2+\frac{it\hbar}{2m}}+\frac{1}{(\Delta x)^2-\frac{it\hbar}{2m}}\right\}\right]$$

$$=\frac{\Delta x}{\hbar^2}\frac{1}{(2\pi)^{1/2}}\frac{1}{\frac{1}{\hbar^2}}\left\{\frac{1}{(\Delta x)^4+\left(\frac{t\hbar}{2m}\right)^2}\right\}^{1/2}$$

$$\times \exp\left[-\frac{\left(x-\frac{<p>t}{m}\right)^2}{4}\left\{\frac{(\Delta x)^2+(\Delta x)^2}{(\Delta x)^4+\left(\frac{t\hbar}{2m}\right)^2}\right\}\right]$$

$$=\frac{\Delta x}{(2\pi)^{1/2}}\left\{\frac{1}{(\Delta x)^4+\left(\frac{t\hbar}{2m}\right)^2}\right\}^{1/2}\times \exp\left[-\frac{\left(x-\frac{<p>t}{m}\right)^2(\Delta x)^2}{2\left\{(\Delta x)^4+\left(\frac{t^2\hbar^2}{4m^2}\right)\right\}}\right]$$

$$=\frac{\Delta x}{(2\pi)^{1/2}}\left\{\frac{1}{(\Delta x)^4+\left(\frac{t\hbar}{2m}\right)^2}\right\}^{1/2}\times \exp\left[-\frac{\left(x-\frac{<p>t}{m}\right)^2}{2(\Delta x)^2\left\{1+\left(\frac{t^2\hbar^2}{4m^2(\Delta x)^4}\right)\right\}}\right]$$

Thus the packet remains Gaussian in shape but spreads with time, since the effective width is given by

$$(\Delta x)_{eff}=(\Delta x)_{t=0}\left[1+\frac{t^2\hbar^2}{\left\{2m(\Delta x)^2\right\}^2}\right]^{1/2}$$

Example 4.6: Give a packet whose spectrum of k-values is the Gaussian function

$$\phi(k)=Ae^{-k^2/2\sigma^2}$$

where σ is the standard deviation of the distribution (σ is a measure of the spread of the packet in k-space). Find the shape of the packet in x-space.

Solution: $\psi(x)$ is the Fourier transform of $\phi(k)$

i.e.
$$\psi(x)=\frac{A}{\sqrt{2\pi}}\int_{-\infty}^{+\infty}e^{-k^2/2\sigma^2}e^{ikx}dk$$

Completing the square in the potential (by adding and subtracting $\dfrac{x^2\sigma^2}{2}$), we get

$$\psi(x) = \frac{A}{\sqrt{2\pi}} \int_{-\infty}^{+\infty} dk\ \exp\left[-\frac{k^2}{2\sigma^2} + ikx + \frac{x^2\sigma^2}{2} - \frac{x^2\sigma^2}{2}\right]$$

$$= \frac{A}{\sqrt{2\pi}} \exp\left(-\frac{x^2\sigma^2}{2}\right) \int_{-\infty}^{+\infty} dk\ \exp\left[-\frac{1}{2\sigma^2}(k^2 - 2ikx\sigma^2 - x^2\sigma^4)\right]$$

$$= \frac{A}{\sqrt{2\pi}} \exp\left(-\frac{x^2\sigma^2}{2}\right) \int_{-\infty}^{+\infty} dk\ \exp\left[-\frac{1}{2\sigma^2}(k - ix\sigma^2)^2\right] \tag{i}$$

Now let
$$\frac{k - ix\sigma^2}{\sqrt{2}\sigma} = u \ \ \text{so}\ dk = \sqrt{2}\sigma du$$

and therefore expression (*i*) becomes

$$\psi(x) = \frac{A\sqrt{2}\sigma}{\sqrt{2\pi}} e^{-x^2\sigma^2/2} \int_{-\infty}^{+\infty} e^{-u^2}\, du$$

The value of the integral is $\sqrt{\pi}$, hence

$$\psi(x) = A\sigma\, e^{-x^2\sigma^2/2} \tag{ii}$$

Thus expression (*ii*) also represents a Gaussian curve. We can rewrite

$$\psi(x) = A\sigma \exp\left[-\frac{x^2}{(\sqrt{2}/\sigma)^2}\right]$$

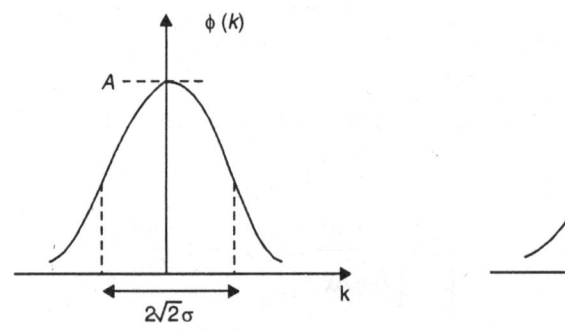
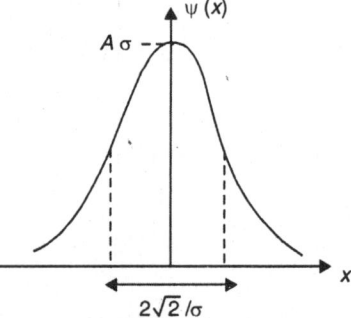

Fig. 4.4: Pertaining to example (4.2). Each packet is the Fourier transform of the other

Therefore the spread of the packet in *x*-space is $2\sqrt{2}/\sigma$. The spread of the packet in *k*-space is $2\sqrt{2}\sigma$. This illustrates the reciprocal nature of Δk and Δx as required by the uncertainty principle (Fig. 4.4).

Example 4.7: Find the group velocity of the Gaussian wave packet:

$$A(k) = a_0 \sigma^{1/2} \exp\left[-\frac{1}{2}\sigma^2(k - k_0)^2\right]$$

show that the packet expands as it travels.

Solution: $E = \dfrac{p^2}{2m} = \hbar\omega; \quad p = \hbar k$

$\therefore \quad \dfrac{\hbar^2 k^2}{2m} = \hbar\omega \quad \text{or} \quad \omega = \dfrac{\hbar k^2}{2m}$

group velocity $\dfrac{d\omega}{dk} = \dfrac{\hbar k}{m}$

$$\psi(x,t) = \int\limits_{-\infty}^{+\infty} A(k)e^{i(kx-\omega t)}\,dk$$

$$= a_0\sigma^{1/2}\int\limits_{-\infty}^{+\infty}\exp\left[-\frac{1}{2}\sigma^2(k-k_0)^2\right]\exp i\left[kx - \frac{\hbar k^2 t}{2m}\right]dk$$

$$= a_0\sigma^{1/2}\int\limits_{-\infty}^{+\infty}\exp\left[-\frac{1}{2}\sigma^2 k^2 - \frac{1}{2}\sigma^2 k_0^2 + \sigma^2 k k_0\right]\exp\left[ikx - \frac{i\hbar k^2 t}{2m}\right]dk$$

$$= a_0\sigma^{1/2}\int\limits_{-\infty}^{+\infty}\exp\left(-\frac{1}{2}\sigma^2 k_0^2\right)\exp\left[-\frac{1}{2}\left(\sigma^2 k^2 + \frac{i\hbar k^2 t}{m}\right)\right]e^{ik(x-i\sigma^2 k_0)}\,dk$$

$$= a_0\sigma^{1/2}\exp\left(-\frac{1}{2}\sigma^2 k_0^2\right)\int\limits_{-\infty}^{+\infty} e^{-\frac{1}{2}\left[\sigma^2 + \frac{i\hbar t}{m}\right]k^2}\, e^{i\left[k(x-i\sigma^2 k_0)\right]}\,dk$$

$$\equiv a_0\sigma^{1/2}\exp\left(-\frac{1}{2}\sigma^2 k_0^2\right).I \tag{i}$$

Now, let

$$\frac{1}{2}\left(\sigma^2 + \frac{i\hbar t}{m}\right) \equiv \alpha^2$$

and

$$x - i\sigma^2 k_0 = \beta$$

Then, the integral I in Eqn. (i) is

$$I = \int\limits_{-\infty}^{+\infty} e^{-\alpha^2 k^2}e^{i\beta k}\,dk$$

$$= \sum_{n=0}^{\infty}\int\limits_{-\infty}^{+\infty} k^n e^{-\alpha^2 k^2}\frac{(i\beta)^n}{n!}\,dk$$

$$= \sum_{n=0}^{\infty}\frac{(i\beta)^n}{n!}\int\limits_{-\infty}^{+\infty} k^n e^{-\alpha^2 k^2}\,dk \tag{ii}$$

Now,

$$I' \equiv \int\limits_{-\infty}^{+\infty} k^n e^{-\alpha^2 k^2}\,dk = \int\limits_{0}^{\infty} k^n e^{-\alpha^2 k^2}\,dk + \int\limits_{-\infty}^{0} k^n e^{-\alpha^2 k^2}\,dk$$

(substituting $k = k'$ in 2nd integral)

$$= \left[1 + (-1)^n\right] \int_0^\infty k^n e^{-\alpha^2 k^2} dk$$

$$= 0 \qquad\qquad\qquad\qquad \text{for odd values of } n$$

$$= 2 \int_0^\infty k^{2s} e^{-\alpha^2 k^2} dk \qquad \text{(for even } n = 2s) \ (\equiv I') \qquad\qquad (iii)$$

To solve the integral in Eqn. (iii), let us put

$$\alpha^2 k^2 = \omega \ \therefore \ 2\alpha^2 k\, dk = d\omega \text{ or } dk = \frac{d\omega}{2\alpha^2} k^{-1}$$

$$k^{2s} = \frac{\omega^s}{\alpha^{2s}} \ \therefore \ k^{2s-1} = \frac{\omega^{(s-\frac{1}{2})}}{\alpha^{2s-1}}$$

$$I' = \frac{2}{2\alpha^2} \int_0^\infty k^{2s} e^{-\omega} d\omega k^{-1} = \frac{1}{\alpha^2 . \alpha^{2s-1}} \int_0^\infty \omega^{s-\frac{1}{2}} e^{-\omega} d\omega$$

$$= \frac{1}{\alpha^{2s+1}} \Gamma(s + \tfrac{1}{2}) = \frac{1}{\alpha^{2s+1}} (s - \tfrac{1}{2})!$$

$$= \frac{1}{\alpha^{2s+1}} \left[(s - \tfrac{1}{2})(s - \tfrac{3}{2})....\tfrac{1}{2}\right] \Gamma\!\left(\frac{1}{2}\right)$$

$$= \frac{1}{\alpha^{2s+1}} \left[\frac{(2s - 1)}{2} \frac{(2s - 3)}{2}....\frac{1}{2}\right] \Gamma\!\left(\frac{1}{2}\right)$$

$$= \frac{1}{\alpha^{2s+1}} \left[\frac{2s(2s - 1)(2s - 2)(2s - 3)...\Gamma\!\left(\frac{1}{2}\right)}{2^s . 2s(2s - 2)...2}\right]$$

$$= \frac{1}{\alpha^{2s+1}} \frac{(2s)!}{2^{2s} . s!} \sqrt{\pi} \qquad \left(\because \Gamma\!\left(\tfrac{1}{2}\right) = \sqrt{\pi}\right)$$

$\therefore$ Integral of Eqn. (ii) is

$$I = \sqrt{\pi} \sum_{s=0}^\infty \frac{(i\beta)^{2s}}{(2s)!} \frac{1}{\alpha^{2s+1}} \frac{(2s)!}{2^{2s} . s!}$$

$$= \sqrt{\frac{\pi}{\alpha}} \sum_{s=0}^\infty \frac{1}{s!} \left(\frac{-\beta^2}{4\alpha^2}\right)^s$$

$$= \sqrt{\frac{\pi}{\alpha}} \, e^{-\beta^2/4\alpha^2} \qquad\qquad\qquad\qquad (iv)$$

Hence, from Eqns. (i) and (iv), we get

$$\psi(x, t) = a_0 \frac{\sqrt{\sigma} \exp\left(-\dfrac{1}{2}\sigma^2 k_0^2\right) \sqrt{\pi}}{\left[\dfrac{1}{2}\left(\sigma^2 + \dfrac{i\hbar t}{m}\right)\right]^{1/2}} \exp\left[-\frac{(x - i\sigma^2 k_0)^2}{2\left(\sigma^2 + \dfrac{i\hbar t}{m}\right)}\right]$$

$$= a_0 \left[\frac{2\pi\sigma}{\sigma^2 + \dfrac{i\hbar t}{m}} \right]^{1/2} \exp\left[-\frac{(x - i\sigma^2 k_0)^2}{2\left(\sigma^2 + \dfrac{i\hbar t}{m} \right)} \right] \exp\left(-\frac{\sigma^2 k_0^2}{2} \right)$$

As t increases, the first exponential term in the above expression decreases and so the packet expands as it travels.

Example 4.8: For an arbitrary normalized wave function $\psi(x)$, we define

$$\phi(p) = \frac{1}{\sqrt{2\pi\hbar}} \int_{-\infty}^{+\infty} \psi(x) \exp\left[-\frac{i}{\hbar} px \right] dx$$

Show that

$$\langle p \rangle = \int_{-\infty}^{+\infty} \phi^*(p) \, p \, \phi(p) \, dp$$

Solution: We know that

$$\psi(x) = \frac{1}{\sqrt{2\pi\hbar}} \int_{-\infty}^{+\infty} \phi(p) \exp\left[\frac{i}{\hbar} px \right] dp$$

and

$$\psi^*(x) = \frac{1}{\sqrt{2\pi\hbar}} \int_{-\infty}^{+\infty} \phi^*(p) \exp\left[-\frac{i}{\hbar} px \right] dp$$

$$\langle p \rangle = \int_{-\infty}^{+\infty} \psi^*(-i\hbar) \frac{\partial \psi}{\partial x} \, dx$$

$$= (-i\hbar) \int_{-\infty}^{+\infty} \psi^*(x) \left[\frac{1}{\sqrt{2\pi\hbar}} \int_{-\infty}^{+\infty} \frac{i}{\hbar} p' \, \phi(p') \exp\left(\frac{i}{\hbar} p'x \right) dp' \right] dx$$

$$= \frac{1}{\sqrt{2\pi\hbar}} \int_{-\infty}^{+\infty} \psi^*(x) \left\{ \int_{-\infty}^{+\infty} p' \, \phi(p') \exp\left(\frac{i}{\hbar} p'x \right) dp' \right\} dx$$

Putting the value of $\psi^*(x)$

$$= \frac{1}{(2\pi\hbar)} \int_{-\infty}^{+\infty} \int_{-\infty}^{+\infty} \phi^*(p) \exp\left[-\frac{i}{\hbar} px \right] dp \left\{ \int_{-\infty}^{+\infty} p' \, \phi(p') \exp\left(\frac{i}{\hbar} p'x \right) dp' \right\} dx$$

$$= \frac{1}{(2\pi\hbar)} \int_{-\infty}^{+\infty} dp \, \phi^*(p) \int_{-\infty}^{+\infty} \int_{-\infty}^{+\infty} p' \, \phi(p') \exp\left[\frac{i}{\hbar}(p' - p)x \right] dp' \, dx$$

Using

$$\delta(p' - p) = \frac{1}{2\pi\hbar} \int_{-\infty}^{+\infty} e^{\left[i(p'-p)\frac{x}{\hbar} \right]} dx$$

we have

$$\langle p \rangle = \int_{-\infty}^{+\infty} dp \, \phi^*(p) \int_{-\infty}^{+\infty} p' \, \phi(p') \, \delta(p' - p) \, dp'$$

Again using

$$\int_{-\infty}^{+\infty} f(x)\, \delta(x-a)\, dx = f(a)$$

We get

$$\langle p \rangle = \int_{-\infty}^{+\infty} \phi^*(p)\, p\, \phi(p)\, dp$$

4.15 Momentum in Coordinate Space

We can write wave functions in momentum and coordinate representations which are related to each other by Fourier transform as follows:

$$\phi(\vec{p},t) = \frac{1}{(2\pi\hbar)^{3/2}} \int_{-\infty}^{+\infty} \psi(\vec{r},t) e^{-i\vec{p}\cdot\vec{r}/\hbar} d^3\vec{r} \tag{127}$$

and

$$\psi(\vec{r},t) = \frac{1}{(2\pi\hbar)^{3/2}} \int_{-\infty}^{+\infty} \phi(\vec{p},t) e^{i\vec{r}\cdot\vec{p}/\hbar} d^3\vec{p} \tag{128}$$

This is for one particular time t. The momentum measurement is made at that particular time at which the wave function has the particular value given in Eqn. (112) $(\vec{p}/\hbar = \vec{k}) \cdot |\phi(p)|^2$ is the probability per unit volume that particle has a given momentum p.

The expectation value of momentum is defined by

$$\langle \vec{p} \rangle = \int_{-\infty}^{+\infty} \phi^*(\vec{p},t)\vec{p}\ \phi(\vec{p},t) d^3p \tag{129}$$

Substituting for $\phi(\vec{p},t)$ and $\phi^*(\vec{p},t)$

$$\langle \vec{p} \rangle = \frac{1}{(2\pi\hbar)^3} \int d^3p \int d^3r' \psi^*(\vec{r}',t) e^{i\vec{p}\cdot\vec{r}'/\hbar}\vec{p} \int d^3r\ \psi(\vec{r},t) e^{-i\vec{p}\cdot\vec{r}/\hbar}$$

$$= \frac{\int d^3p}{(2\pi\hbar)^3} \int d^3r'\ e^{i\vec{p}\cdot\vec{r}'/\hbar}\psi^*(\vec{r}',t) \int d^3r \left[-i\hbar\nabla e^{-i\vec{p}\cdot\vec{r}/\hbar} \right] \psi(\vec{r},t)$$

$$= \int d^3r \int d^3r' \frac{1}{(2\pi\hbar)^3} \psi^*(\vec{r}',t)\left(\frac{\hbar}{i}\nabla\psi(\vec{r},t) \right)$$

$$\times \int d^3p\ \exp\left\{ \frac{i}{\hbar}(\vec{r}'-\vec{r})\cdot\vec{p} \right\} \tag{130}$$

Now, we know that the last integral is

$$\int d^3p\ \exp\left\{ \frac{i}{\hbar}(\vec{r}'-\vec{r})\cdot\vec{p} \right\} = (2\pi\hbar)^3 \delta^3(\vec{r}'-\vec{r}) \tag{131}$$

So, finally

$$\langle p \rangle = \int d^3r \int d^3r' \psi^*(\vec{r}',t)\frac{\hbar}{i}\nabla\psi(\vec{r},t)\delta^3(\vec{r}'-\vec{r})$$

$$= \int d^3r\ \psi^*(\vec{r},t)\frac{\hbar}{i}\nabla\psi(\vec{r},t) \tag{132}$$

Because of this relation, $\dfrac{\hbar}{i}\nabla$ is called the momentum operator in coordinate space.

4.16 Schrödinger Equation in Momentum Representation

The time independent Schrödinger equation may be written as

$$-\frac{\hbar^2}{2m}\nabla^2\psi + V\psi = i\hbar\frac{\partial\psi}{\partial t} \tag{133}$$

The wave function of any plane wave is

$$e^{i(\vec{p}\,\cdot\,\vec{r}\text{-Et})/\hbar}$$

The superposition of such waves forms a wave packet whose wave function using Fourier theorem is written as

$$\psi(\vec{r},t) = \frac{1}{(2\pi\hbar)^{3/2}}\int\phi(\vec{p})e^{i(\vec{p}\cdot\vec{r}-Et)/\hbar}d^3p \tag{134}$$

Differentiating Eqn. (134), w.r.t. t, we get

$$\frac{\partial\psi}{\partial t} = \frac{1}{(2\pi\hbar)^{3/2}}\int\frac{-iE}{\hbar}\phi(\vec{p})e^{i(\vec{p}\cdot\vec{r}-Et)/\hbar}d^3p \tag{135}$$

Eqn. (134) may be rewritten as

$$\psi = \frac{1}{(2\pi\hbar)^{3/2}}\int\phi(\vec{p})e^{i\left[(p_xx+p_yy+p_zz)-Et\right]/\hbar}d^3p$$

Differentiating this Eqn. w.r.t. x

$$\frac{\partial\psi}{\partial x} = \frac{1}{(2\pi\hbar)^{3/2}}\int\frac{ip_x}{\hbar}\phi(\vec{p})\exp\left\{i\left[(p_xx+p_yy+p_zz)-Et\right]/\hbar\right\}d^3p$$

Differentiating again

$$\frac{\partial^2\psi}{\partial x^2} = \frac{1}{(2\pi\hbar)^{3/2}}\int\left(\frac{ip_x}{\hbar}\right)^2\phi(\vec{p})\exp\left\{i\left[(p_xx+p_yy+p_zz)-Et\right]/\hbar\right\}d^3p$$

$$= \frac{1}{(2\pi\hbar)^{3/2}}\int\left(-\frac{p_x^2}{\hbar^2}\right)\phi(\vec{p})\exp\left\{i(\vec{p}.\vec{r}-Et)/\hbar\right\}d^3p \tag{136}$$

Similarly,

$$\frac{\partial^2\psi}{\partial y^2} = \frac{1}{(2\pi\hbar)^{3/2}}\int\left(-\frac{p_y^2}{\hbar^2}\right)\phi(\vec{p})\exp\left\{i(\vec{p}.\vec{r}-Et)/\hbar\right\}d^3p \tag{137}$$

and

$$\frac{\partial^2\psi}{\partial z^2} = \frac{1}{(2\pi\hbar)^{3/2}}\int\left(-\frac{p_z^2}{\hbar^2}\right)\phi(\vec{p})\exp\left\{i(\vec{p}.\vec{r}-Et)/\hbar\right\}d^3p \tag{138}$$

Substituting values of $\dfrac{\partial\psi}{\partial t}$ and $\nabla^2\psi$ from Eqn. (135) and Eqns (136), (137) and (138) in Eqn. (133), we get

$$= \frac{\hbar^2}{2m} \frac{1}{(2\pi\hbar)^{3/2}} \int \left(-\frac{p^2}{\hbar^2} \right) \phi(\vec{p}) \exp\{i(\vec{p}.\vec{r} - Et)/\hbar\} \, d^3p + V\psi$$

$$= \frac{i\hbar}{(2\pi\hbar)^{3/2}} \int -\frac{iE}{\hbar} \phi(\vec{p}) \exp\{i(\vec{p}.\vec{r} - Et)/\hbar\} \, d^3p$$

or
$$\frac{1}{(2\pi\hbar)^{3/2}} \int_{-\infty}^{+\infty} \left(\frac{p^2}{2m} - E \right) \phi(\vec{p}) e^{i(\vec{p}.\vec{r} - Et)/\hbar} \, d^3p + V\psi = 0 \qquad (139)$$

This is Schrödinger's equation in momentum representation $\phi(\vec{p})$ is related to $\psi(\vec{r}, t)$ via the relation:

$$\phi(\vec{p}, t) = \frac{1}{(2\pi\hbar)^{3/2}} \int_{-\infty}^{+\infty} \psi(\vec{r}, t) \, e^{-i(\vec{p}.\vec{r} - Et)/\hbar} d^3r \qquad (140)$$

For a free particle, $V = 0$, then, Schrödinger's equation in momentum representation, for a free particle becomes

$$\frac{1}{(2\pi\hbar)^{3/2}} \int_{-\infty}^{+\infty} \left(\frac{p^2}{2m} - E \right) \phi(\vec{p}) e^{i(\vec{p}.\vec{r} - Et)/\hbar} \, d^3p = 0 \qquad (141)$$

4.17 Position Vector Operator in Momentum Representation

Just as the operator $\hat{p} \ (\equiv -i\hbar\nabla)$ corresponds to the momentum, determining its eigenfunctions $\left(\psi_p \equiv e^{(i/\hbar)\vec{p}.\vec{r}} \right)$ in the coordinate representation satisfying

$$\psi(\vec{r}) = \frac{1}{(2\pi\hbar)^{3/2}} \int \phi(\vec{p}) \, \psi_p(\vec{r}) \, d^3p \qquad (142)$$

and
$$\phi(\vec{p}) = \frac{1}{(2\pi\hbar)^{3/2}} \int \psi(\vec{r}) \, \psi_p^*(\vec{r}) \, d^3r \qquad (143)$$

we can introduce the position vector operator $\hat{r}$ of the coordinates of the particle in the momentum representation. It must be so defined that its expectation value can be represented in the form

$$\langle r \rangle = \frac{1}{(2\pi\hbar)^{3/2}} \int \phi^*(\vec{p}) \, \hat{r} \, \phi(p) \, d^3p \qquad (144)$$

On the other hand, this mean value is determined from the wave function $\psi(\hat{r})$ by

$$\langle r \rangle = \int \psi^* r \, \psi \, d^3r \qquad (145)$$

We have, using Eqn. (142),

$$\vec{r} \, \psi(\vec{r}) = (2\pi\hbar)^{-3/2} \int \vec{r} \, \phi(\vec{p}) \, e^{(i/\hbar)\vec{p}.\vec{r}} \, d^3p$$

$$= (2\pi\hbar)^{-3/2} \int e^{(i/\hbar)\vec{p}.\vec{r}} \left(\vec{r} \, \phi(\vec{p}) \right) \, d^3p$$

Using $\displaystyle\int_{-\infty}^{+\infty} uv \, dx = u \int_{-\infty}^{+\infty} v \, dx - \int_{-\infty}^{+\infty} \frac{du}{dx} \int v \, dx$

with $\qquad$ $u = r\phi(\vec{p})$ and $v = e^{(i/\hbar)\vec{p}.\vec{r}}$

we get

$$\vec{r}\ \psi(\vec{r}) = (2\pi\hbar)^{-3/2} \left\{ \vec{r}\ \phi(\vec{p}) \left[\frac{e^{(i/\hbar)\vec{p}.\vec{r}}}{ir/\hbar} \right]_{-\infty}^{+\infty} - r \int \frac{\partial\phi(\vec{p})}{\partial p} \left(\frac{\hbar}{ir} \right) e^{(i/\hbar)\vec{p}.\vec{r}}\ d^3p \right\}$$

$$= (2\pi\hbar)^{-3/2} \int i\hbar e^{(i/\hbar)\vec{p}.\vec{r}} \left\{ \frac{\partial\phi(\vec{p})}{\partial p} \right\} d^3p$$

Using Eqn. (145), we get

$$\langle r \rangle = (2\pi\hbar)^{-3/2} \int\int \psi*(\vec{r})\ i\hbar \left[\frac{\partial\phi(\vec{p})}{\partial p} \right] e^{(i/\hbar)\vec{p}.\vec{r}} d^3p\ d^3r$$

$$= (2\pi\hbar)^{-3/2}(2\pi\hbar)^{-3/2} \int\int \phi*(\vec{p})\ i\hbar \left[\frac{\partial\phi(\vec{p})}{\partial p} \right] \psi_p^*(\vec{r})\ \psi_p(\vec{r}) \times d^3r\ d^3p$$

(using Eqn. 142)

$$\langle r \rangle = \frac{1}{(2\pi\hbar)^3} \int i\hbar\phi*(\vec{p}) \left[\frac{\partial\phi(\vec{p})}{\partial p} \right] d^3p \qquad (146)$$

Hence, the position vector operator in the momentum representation is

$$\hat{r} = i\hbar \frac{\partial}{\partial p} \qquad (147)$$

4.18 Principle of Causality

General meaning of causality is that the effect cannot precede the cause. Physical systems obey causality. Thus we can determine the wave function $\psi\ (x,\ t)$ at a later instant if its value is known at an earlier instant.

We know that the expression for $\psi(\vec{r},t)$ in terms of its Fourier component $\phi(\vec{p})$ is

$$\psi(\vec{r},t) = \frac{1}{(2\pi\hbar)^{3/2}} \int \phi(\vec{p})\exp\left[\frac{i}{\hbar}\left(\vec{p}\cdot\vec{r} - \frac{p^2}{2m}t \right) \right] d\vec{p} \qquad (148)$$

where the momentum amplitude function is independent of time and is given by

$$\phi(\vec{p}) = \frac{1}{(2\pi\hbar)^{3/2}} \int \phi(\vec{p})\exp\left[\frac{i}{\hbar}\left(\vec{p}\cdot\vec{r} - \frac{p^2}{2m}t \right) \right] d\vec{p}$$

where $\qquad\qquad dr'=dx'\ dy'\ dz'$

Substituting this value of $\phi(p)$ into Eqn. (148), we get

$$\psi(\vec{r},t) = \int K(\vec{r},t,\vec{r}',t')\psi(\vec{r}'t')d\vec{r}' \qquad (149)$$

where $K(\vec{r},t,\vec{r}',t') = \frac{1}{(2\pi\hbar)^3} \int \exp\left\{ \frac{i}{\hbar}(\vec{p}\cdot(\vec{r}-\vec{r}') - \frac{p^2}{2m}(t-t') \right\} d\vec{p}.$

In Eqn. (149), wave function at any time t is expressed linearly in terms of its values at the earlier time t'. This is an explicit expression of the principle of causality. It is a quantum analogue of the classical principle that the future behaviour of a particle can be predicted if its position and velocity are known at an early instant.

Thus complete description of a system (or a particle) is embodied in its wave function and the Schrödinger's equation, which contains the time only in the first derivative, yields the rate of change of ψ in terms of its value at a given instant. This calculation leads to a formulation of wave mechanical properties of a free particle. Relativistic generalization of this formulation forms the starting point for modern theoretical work on the physics of fundamental particles.

Example 4.9: $\psi(x, t)$ is a solution of the Schrödinger equation for a free particle of mass m in one dimension and

$$\psi(x, 0) = A\exp\left(-x^2/a^2\right)$$

(a) At time $t = 0$ find the probability amplitude in momentum space.

(b) Find $\psi(x, t)$.

Solution: (a) At time $t = 0$, the probability amplitude in momentum space is

$$\psi(p, 0) = \frac{1}{\sqrt{2\pi\hbar}} \int\limits_{-\infty}^{+\infty} e^{-ipx/\hbar}\, \psi(x,0)\, dx$$

$$= \frac{A}{\sqrt{2\pi\hbar}} \int\limits_{-\infty}^{+\infty} \exp\left(-\frac{x^2}{a^2} - \frac{ipx}{\hbar}\right) dx . \text{ Now,} \qquad (i)$$

$$\frac{x^2}{a^2} + \frac{ipx}{\hbar} = \left(\frac{x}{a}\right)^2 + 2\left(\frac{x}{a}\right)\left(\frac{ipa}{2\hbar}\right) + \left(\frac{ipa}{2\hbar}\right)^2 - \left(\frac{ipa}{2\hbar}\right)^2$$

$$= \left(\frac{x}{a} + \frac{ipa}{2\hbar}\right)^2 - \left(\frac{ipa}{2\hbar}\right)^2$$

or $\qquad -\dfrac{x^2}{a^2} - \dfrac{ipx}{\hbar} = -\left(\dfrac{x}{a} + \dfrac{ipa}{2\hbar}\right)^2 + \left(\dfrac{ipa}{2\hbar}\right)^2 = -\left(\dfrac{x}{a} + \dfrac{ipa}{2\hbar}\right) - \left(\dfrac{p^2a^2}{4\hbar^2}\right)$

or $\ \exp\left(-\dfrac{x^2}{a^2} - \dfrac{ipx}{\hbar}\right) = \exp\left[-\left(\dfrac{x}{a} + \dfrac{ipa}{2\hbar}\right)^2\right]\exp\left(-\dfrac{p^2a^2}{4\hbar^2}\right) \qquad (ii)$

Substituting this in Eqn. (i)

$$\psi(p, 0) = \frac{A}{\sqrt{2\pi\hbar}} \exp\left(-\frac{p^2a^2}{4\hbar^2}\right) \int\limits_{-\infty}^{+\infty} \exp\left[-\left(\frac{x}{a} + \frac{ipa}{2\hbar}\right)^2\right] dx$$

Let $\qquad \dfrac{x}{a} + \dfrac{ipa}{2\hbar} = t \quad \therefore \quad dx = a\, dt$

$\therefore \qquad \psi(p, 0) = \dfrac{A}{\sqrt{2\pi\hbar}} \exp\left(-\dfrac{p^2a^2}{4\hbar^2}\right) \int\limits_{-\infty}^{+\infty} a\, e^{-t^2}\, dt$

$$= \frac{Aa}{\sqrt{2\pi\hbar}} \exp\left(-\frac{p^2 a^2}{4\hbar^2}\right) 2 \int_0^{+\infty} e^{-t^2} dt \quad \left(\because \int_0^{+\infty} e^{-t^2} dt = \frac{\sqrt{\pi}}{2}\right)$$

$$\therefore \qquad \psi(p,0) = \frac{Aa}{\sqrt{2\hbar}} \exp\left(-\frac{p^2 a^2}{4\hbar^2}\right)$$

(b) The Schrödinger equation in momentum space for a free particle

$$i\hbar \frac{\partial \psi(p,t)}{\partial t} = \hat{H}\, \psi(p,t) = \frac{p^2}{2m} \psi(p,t)$$

gives

$$\psi(p,t) = B\, \exp\left(-\frac{iEt}{\hbar}\right)$$

$$= B\, \exp\left(-\frac{ip^2 t}{2m\hbar}\right)$$

At time $t = 0$, we have $B = \psi(p,0)$.
Hence,

$$\psi(p,t) = \frac{Aa}{\sqrt{2\hbar}} \exp\left(-\frac{p^2 a^2}{4\hbar^2} - \frac{ip^2 t}{2m\hbar}\right)$$

and

$$\psi(x,t) = \frac{1}{\sqrt{2\pi\hbar}} \int_{-\infty}^{+\infty} \exp\left(\frac{ipx}{\hbar}\right) \psi(p,t) dp$$

or

$$\psi(x,t) = \frac{Aa}{\sqrt{2\pi\hbar}} \frac{1}{\sqrt{2\hbar}} \int_{-\infty}^{+\infty} \exp\left(\frac{ipx}{\hbar}\right) \exp\left(-\frac{p^2 a^2}{4\hbar^2}\right) \exp\left(-\frac{ip^2 t}{2m\hbar}\right) dp$$

$$= \frac{Aa}{\sqrt{2\pi\hbar}} \frac{1}{\sqrt{2\hbar}} \int_{-\infty}^{+\infty} \exp\left[-p^2\left(\frac{a^2}{4\hbar^2} + \frac{it}{2m\hbar}\right) + \left(\frac{ix}{\hbar}\right)p\right] dp$$

$$= \frac{Aa}{\sqrt{2\pi\hbar}} \frac{1}{\sqrt{2\hbar}} \int_{-\infty}^{+\infty} \exp\left[-\left\{\frac{p^2}{4\hbar^2}\left(a^2 + \frac{2it\hbar}{m}\right) - \left(\frac{ix}{\hbar}\right)p\right\}\right] dp$$

$$= \frac{Aa}{\sqrt{2\pi\hbar}} \frac{1}{\sqrt{2\hbar}} \int_{-\infty}^{+\infty} \exp A' \, dp \qquad (iii)$$

where

$$A' = -\left\{\frac{p^2}{4\hbar^2}\left(a^2 + \frac{2it\hbar}{m}\right) - \left(\frac{ix}{\hbar}\right)p\right\}$$

$$= -\frac{1}{4\hbar^2}\left(a^2 + \frac{2it\hbar}{m}\right)\left[p^2 - \frac{4ix\hbar}{\left(a^2 + \frac{2it\hbar}{m}\right)}p\right] \qquad (iv)$$

Now,

$$p^2 - \left(\frac{4ix\hbar}{a^2 + \frac{2it\hbar}{m}}\right)p = \left(p - \frac{2ix\hbar}{a^2 + \frac{2it\hbar}{m}}\right)^2 - \left(\frac{2ix\hbar}{a^2 + \frac{2it\hbar}{m}}\right)^2 \qquad (v)$$

Let
$$\frac{2ixh}{a^2 + \dfrac{2ith}{m}} = s$$

∴ Eqn. (*v*) is $\quad p^2 - 2sp = (p-s)^2 - s^2$

Hence
$$A' = -\frac{1}{4\hbar^2}\left(a^2 + \frac{2ith}{m}\right)\left[(p-s)^2 - s^2\right]$$

and integral of Eqn. (*iii*) is

$$\text{Integral} = \int_{-\infty}^{+\infty} \exp\left[-\frac{1}{4\hbar^2}\left(a^2 + \frac{2ith}{m}\right)\left\{(p-s)^2 - s^2\right\}\right] dp$$

$$= \int_{-\infty}^{+\infty} \exp\left[\frac{s^2}{4\hbar^2}\left(a^2 + \frac{2ith}{m}\right) - \frac{(p-s)^2}{4\hbar^2}\left(a^2 + \frac{2ith}{m}\right)\right] dp$$

$$= \exp\left[\frac{s^2}{4\hbar^2}\left(a^2 + \frac{2ith}{m}\right)\right]\int_{-\infty}^{+\infty} \exp\left[-\frac{1}{4\hbar^2}\left(a^2 + \frac{2ith}{m}\right)(p-s)^2\right] dp$$

Let
$$\sqrt{\frac{1}{4\hbar^2}\left(a^2 + \frac{2ith}{m}\right)}\,(p-s) = u$$

∴
$$\sqrt{\frac{1}{4\hbar^2}\left(a^2 + \frac{2ith}{m}\right)}\, dp = du$$

Hence

$$\text{Integral} = \exp\left[\frac{s^2}{4\hbar^2}\left(a^2 + \frac{2ith}{m}\right)\right]\int_{-\infty}^{+\infty} \exp\left[-u^2\right] du \times \frac{1}{\sqrt{\dfrac{1}{4\hbar^2}\left(a^2 + \dfrac{2ith}{\tilde{m}}\right)}}$$

(on putting the value of *dp*)

or
$$\text{Integral} = \frac{2\hbar}{\sqrt{a^2 + \dfrac{2ith}{m}}}\exp\left[\frac{s^2}{4\hbar^2}\left(a^2 + \frac{2ith}{m}\right)\right]\times 2 \times \frac{\sqrt{\pi}}{2}$$

Substituting this value in Eqn. (*iii*), we get

$$\psi(x, t) = \frac{Aa}{2\hbar\sqrt{\pi}}\frac{2\hbar\sqrt{\pi}}{\sqrt{a^2 + \dfrac{2ith}{m}}}\exp\left\{\left(\frac{2ixh}{a^2 + \dfrac{2ith}{m}}\right)^2\frac{\left(a^2 + \dfrac{2ith}{m}\right)}{4\hbar^2}\right\}$$

$$= \frac{Aa}{\sqrt{a^2 + \dfrac{2ith}{m}}}\exp\left[-\frac{x^2}{a^2 + \dfrac{2ith}{m}}\right] \qquad (\because i^2 = -1)$$

Example 4.10: Calculate an explicit expression representing the envelop of a free particle wave packet given by

$$\psi(x,t) = \int_{-\infty}^{+\infty} \exp\left[-\frac{(k-k_0)^2}{a^2}\right] e^{-i(kx-\omega t)} dk$$

Solution: For a small range of values in the vicinity of k_0,

$$\omega = \omega_0 + (k-k_0)\left(\frac{d\omega}{dk}\right)_{k_9}$$

or

$$\omega = \omega_0 + (k-k_0)v_g \qquad (i)$$

$$k = k_0 + (k-k_0) \equiv k_0 + \beta$$

$$\psi(x,t) = \int_{-\infty}^{+\infty} \exp\left[-\frac{\beta^2}{a^2} - i(kx-\omega t)\right] dk$$

$$= \int_{-\infty}^{+\infty} \exp\left[-\frac{\beta^2}{a^2}\right] e^{-ikx} e^{i\omega t} dk$$

$$= \int_{-\infty}^{+\infty} \exp\left[-\frac{\beta^2}{a^2}\right] e^{-ik_0 x} e^{-i(k-k_0)x} e^{i\left[\omega_0 + (k-k_0)v_g\right]t} dk$$

$$= e^{-i(k_0 x - \omega_0 t)} \int_{-\infty}^{+\infty} \exp\left[-\frac{\beta^2}{a^2} - i\beta(x-v_g t)\right] dk$$

$$= e^{-i(k_0 x - \omega_0 t)} \int_{-\infty}^{+\infty} \exp\left\{-\frac{1}{a^2}\left[\beta^2 + i\beta a^2(x-v_g t)\right]\right\} d\beta$$

$$= e^{-i(k_0 x - \omega_0 t)} \int_{-\infty}^{+\infty} \exp\left\{-\frac{1}{a^2}\left[\beta^2 + 2\beta i \frac{a^2}{2}(x-v_g t) - \frac{a^4}{4}(x-v_g t)^2\right]\right\}$$

$$\times \exp\left[-\frac{a^2}{4}(x-v_g t)^2\right] d\beta$$

(Completing the square in the bracket)

$$= e^{-i(k_0 x - \omega_0 t)} \exp\left[-\frac{a^2(x-v_g t)^2}{4}\right] \int_{-\infty}^{+\infty} \exp\left[-\frac{1}{a^2}\left\{\beta + \frac{ia^2}{2}(x-v_g t)\right\}^2\right] d\beta$$

Substituting $\dfrac{1}{a^2} = \alpha$ and $\left[\beta + \dfrac{i(x-v_g t)}{2\alpha}\right] = \lambda$ we get,

$$\psi(x,t) = e^{-i(k_0 x - \omega_0 t)} \exp\left[-\frac{(x-v_g t)^2}{4\alpha}\right] \int_{-\infty}^{+\infty} e^{-\alpha\lambda^2} d\lambda$$

Since $\int\limits_{-\infty}^{+\infty} \exp\left[-\alpha\lambda^2\right] d\lambda = \sqrt{\dfrac{\pi}{\alpha}}$, therefore

$$\psi(x,t) = \sqrt{\dfrac{\pi}{\alpha}} \, \exp\left[-\dfrac{\left(x - v_g t\right)^2}{4\alpha}\right] e^{-i\left(k_0 x - \omega_0 t\right)}$$

The envelop function is given by

$$\sqrt{\dfrac{\pi}{\alpha}} \, \exp\left[-\dfrac{\left(x - v_g t\right)^2}{\left(2\sqrt{\alpha}\right)^2}\right] \text{ where } \alpha = \dfrac{1}{a^2}$$

Example 4.11: The ground state wave function of hydrogen atom in the coordinate representation is given by

$$\psi(\vec{r}) = \dfrac{1}{\left(\pi a_0^3\right)^{1/2}} e^{-r/a_0}$$

where a_0 is a constant. Find the corresponding in momentum representation.

Solution: $\phi(\vec{p}) = \dfrac{1}{(2\pi\hbar)^{3/2}} \int e^{i\,\vec{p}\cdot\vec{r}/\hbar} \psi(\vec{r}) d^3r$

$$= \dfrac{1}{(2\pi\hbar)^{3/2}} \dfrac{1}{(\pi a_0^3)^{1/2}} \int\int e^{-i\,\vec{p}\cdot\vec{r}/\hbar} \int e^{-r/a_0} \times r^2 dr \sin\theta \, d\theta \, d\phi \qquad (i)$$

$$\left(\because d^3r = r^2 dr \sin\theta \, d\theta \, d\phi\right)$$

In order to evaluate the integral in Eqn. (i), we consider the integral

$$I = \int e^{i\,\vec{k}\cdot\vec{r}} f(r) r^2 dr \sin\theta \, d\theta \, d\phi$$

$$= 2\pi \int\limits_0^\infty \int\limits_{-1}^{+1} e^{i\,k\,r\cos\theta} \cdot f(r) r^2 dr \, d(\cos\,\theta) \qquad \left(\because \int d\phi = 2\pi\right)$$

$$= 2\pi \int\limits_0^\infty 2 f(r) \, r^2 \, dr \left(\dfrac{e^{ikr} - e^{-ikr}}{2ikr}\right)$$

$$= \dfrac{4\pi}{\cdot k} \int\limits_0^\infty \sin kr \, f(r) \, r \, dr \qquad (ii)$$

Thus, we can write

$$\phi(\vec{p}) = \dfrac{1}{(2\pi\hbar)^{3/2}} \dfrac{1}{(\pi a_0^3)^{1/2}} \dfrac{4\pi}{(-p/\hbar)} \int\limits_0^\infty \sin\left(-\dfrac{pr}{\hbar}\right) r \, e^{-r/a_0} dr$$

$$\left(\because p/\hbar = k \text{ and } f(r) \equiv e^{-r/a_0}\right)$$

$$= \dfrac{1}{(2\pi\hbar)^{3/2}} \dfrac{1}{(\pi a_0^3)^{1/2}} \dfrac{4\pi\hbar}{p} \int\limits_0^\infty \sin\left(\dfrac{pr}{\hbar}\right) e^{-r/a_0} r \, dr \qquad (iii)$$

Now

$$\int_0^\infty \sin\left(\frac{pr}{\hbar}\right) e^{-r/a_0} r\, dr = \text{Im} \int_0^\infty e^{-r(1/a_0 - ip/\hbar)} r\, dr$$

$$= \text{Im}\left[\frac{1}{\left(\dfrac{1}{a_0} - \dfrac{ip}{\hbar}\right)^2}\right] \quad \left(\text{using } \int_0^\infty r^n e^{-\alpha r}\, dr = \frac{n!}{\alpha^{n+1}}\right)$$

$$= \text{Im}\left[\frac{1}{\left(\dfrac{1}{a_0} - \dfrac{ip}{\hbar}\right)\left(\dfrac{1}{a_0} - \dfrac{ip}{\hbar}\right)}\right] = \text{Im}\left[\frac{\left(\dfrac{1}{a_0} + \dfrac{ip}{\hbar}\right)^2}{\left(\dfrac{1}{a_0^2} - \dfrac{i^2 p^2}{\hbar^2}\right)\left(\dfrac{1}{a_0^2} - \dfrac{i^2 p^2}{\hbar^2}\right)}\right]$$

$$= \text{Im}\left[\frac{(1/a_0)^2 + (ip/\hbar)^2 + 2ip/(\hbar a_0)}{\left(1/a_0^2 + p^2/\hbar^2\right)}\right] = \frac{2p/(\hbar a_0)}{\left(1/a_0^2 + p^2/\hbar^2\right)^2}$$

Hence

$$\phi(\vec{p}) = \frac{1}{(2\pi\hbar)^{3/2}} \frac{1}{(\pi a_0^3)^{1/2}} \frac{4\pi\hbar}{P} \frac{2p}{(\hbar a_0)\left(1/a_0^2 + p^2/\hbar^2\right)^2}$$

$$= \frac{1}{(2\pi\hbar)^{3/2}} \frac{1}{(\pi a_0^3)^{1/2}} \frac{8\pi}{a_0} \frac{a_0^4 \hbar^4}{\left(\hbar^2 + p^2 a_0^2\right)^2}$$

$$= \frac{1}{2^{3/2}} \frac{1}{(\hbar)^{3/2}} \frac{1}{\pi} \frac{2^3}{a_0^{3/2}} \frac{a_0^3 \hbar^4}{\left(\hbar^2 + p^2 a_0^2\right)^2}$$

or

$$\phi(p) = \frac{1}{\pi}\left(\frac{2a_0}{\hbar}\right)^{3/2} \frac{\hbar^4}{\left(\hbar^2 + p^2 a_0^2\right)^2}$$

or

$$\phi(p) = \frac{2^{3/2}}{\pi} \frac{a_0^{3/2} \hbar^{5/2}}{\left(\hbar^2 + p^2 a_0^2\right)^2}$$

Example 4.12: For the Gaussian wave packet given by

$$\psi(x, 0) = \frac{1}{\left(\pi\sigma_0^2\right)^{1/4}} e^{-x^2/2\sigma_0^2} \exp\left(\frac{ip_0 x}{\hbar}\right),$$

evaluate $<x>$, $<x^2>$, $<p>$, $<p^2>$ and show that

$$\Delta x.\Delta p = \frac{\hbar}{2}$$

Solution:
$$\langle x \rangle = \int\limits_{-\infty}^{+\infty} \psi^*(x,0)\, x\, \psi(x,0)\, dx$$

$$= \frac{1}{\sqrt{\pi\sigma_0^2}} \int\limits_{-\infty}^{+\infty} x\, e^{-x^2/\sigma_0^2}\, dx = 0$$

(since the integrand is an odd function of x)

$$\langle x^2 \rangle = \frac{1}{\sqrt{\pi\sigma_0^2}} \int\limits_{-\infty}^{+\infty} x^2\, e^{-x^2/\sigma_0^2}\, dx \qquad (i)$$

Let
$$\frac{x^2}{\sigma_0^2} = t \; ; \quad \therefore \;\; 2x\,dx = \sigma_0^2\, dt$$

or
$$dx = \frac{\sigma_0^2}{2}\, dt \times (\sigma_0)^{-1}\, t^{-1/2}$$

The integral in Eqn. (i) is

$$I = \int\limits_{-\infty}^{+\infty} \sigma_0^2 t . e^{-t} . \frac{\sigma_0}{2}\, t^{-1/2}\, dt$$

$$= \frac{\sigma_0^3}{2} \int\limits_{-\infty}^{+\infty} t^{1/2} . e^{-t}\, dt$$

$$= \frac{\sigma_0^3}{2} \times 2 \int\limits_{0}^{\infty} t^{1/2} . e^{-t}\, dt$$

$$\left(\text{Using } \Gamma(z) = \int\limits_{0}^{\infty} x^{z-1} e^{-x}\, dx, \;\; \Gamma\!\left(\frac{3}{2}\right) = \int\limits_{0}^{\infty} t^{1/2} e^{-t}\, dt \right)$$

$$= \frac{\sigma_0^3}{2} \times 2 \times \Gamma\!\left(\frac{3}{2}\right)$$

$$= \sigma_0^3 \times \frac{1}{2}\sqrt{\pi} \qquad (ii)$$

So,
$$\langle x^2 \rangle = \frac{1}{\sqrt{\pi}\sigma_0} \times \frac{\sigma_0^3 \sqrt{\pi}}{2} = \frac{1}{2}\sigma_0^2 \qquad (iii)$$

Again $<p> = 0$ (because the integrand would be an odd function)

$$\langle p^2 \rangle = \int\limits_{-\infty}^{+\infty} \psi^*(x,0)\, p^2\, \psi(x,0)\, dx$$

$$\langle p^2 \rangle = \frac{-\hbar^2}{\sqrt{\pi\sigma_0^2}} \int\limits_{-\infty}^{+\infty} e^{-x^2/2\sigma_0^2} \frac{d^2}{dx^2}\left(e^{-x^2/2\sigma_0^2} \right) dx$$

$$= \frac{-\hbar^2}{\sqrt{\pi\sigma_0^2}} \times I \qquad (iv)$$

$$\left(\because \ p = -i\hbar \frac{\partial}{\partial x} \ \text{and} \ p^2 = -\hbar^2 \frac{\partial^2}{\partial x^2} \right)$$

Now $I = \left[e^{-x^2/2\sigma_0^2} \int \frac{d^2}{dx^2} \left(e^{-x^2/2\sigma_0^2} \right) dx \right]_{-\infty}^{+\infty}$

$$+ \int_{-\infty}^{+\infty} \left\{ \frac{2x}{2\sigma_0^2} \ e^{-x^2/2\sigma_0^2} \int \frac{d^2}{dx^2} \left(e^{-x^2/2\sigma_0^2} \right) dx \right\} dx$$

$$\left(\because \ \text{Using} \int uv \ dx = \left[u \int v \ dx \right] - \int \frac{du}{dx} \left(\int v \ dx \right) dx \right)$$

$$= \left[e^{-x^2/2\sigma_0^2} \frac{d}{dx} \left(e^{-x^2/2\sigma_0^2} \right) \right]_{-\infty}^{+\infty} + \frac{1}{\sigma_0^2} \int_{-\infty}^{+\infty} x \ e^{-x^2/2\sigma_0^2} \frac{d}{dx} \left(e^{-x^2/2\sigma_0^2} \right) dx$$

$$= e^{-x^2/2\sigma_0^2} \left(-\frac{2x}{2\sigma_0^2} \right) e^{-x^2/2\sigma_0^2} \Big|_{-\infty}^{+\infty} + \frac{1}{\sigma_0^2} \left[x \ e^{-x^2/2\sigma_0^2} \int \frac{d}{dx} \left(e^{-x^2/2\sigma_0^2} \right) dx \right.$$

$$\left. - \int \left\{ \frac{d}{dx} \left(x \ e^{-x^2/2\sigma_0^2} \right) \int \frac{d}{dx} \left(e^{-x^2/2\sigma_0^2} \right) dx \right\} dx \right]$$

$$= \left(-\frac{x}{\sigma_0^2} e^{-x^2/\sigma_0^2} \right) + \left(\frac{x}{\sigma_0^2} e^{-x^2/\sigma_0^2} \right)$$

$$- \frac{1}{\sigma_0^2} \int \left[e^{-x^2/2\sigma_0^2} - \frac{2x^2}{2\sigma_0^2} e^{-x^2/2\sigma_0^2} \right] e^{-x^2/2\sigma_0^2} dx$$

$$= \frac{1}{\sigma_0^2} \left[-\int_{-\infty}^{+\infty} e^{-x^2/2\sigma_0^2} e^{-x^2/2\sigma_0^2} \ dx + \int_{-\infty}^{+\infty} \frac{x^2}{\sigma_0^2} e^{-x^2/\sigma_0^2} \ dx \right]$$

$$= \frac{1}{\sigma_0^2} (I_1 + I_2) \qquad\qquad\qquad (v)$$

But we know that $\qquad \int_{-\infty}^{+\infty} e^{-\alpha x^2} dx = \sqrt{\frac{\pi}{\alpha}}$

Therefore $\qquad\qquad I_1 = -\sigma_0 \sqrt{\pi}$

and $\qquad\qquad I_2 = \frac{1}{\sigma_0^2} \times \sigma_0^3 \frac{\sqrt{\pi}}{2} = \sigma_0 \frac{\sqrt{\pi}}{2}$

So, $\qquad\qquad I = -\frac{\sqrt{\pi}}{2\sigma_0}$

Hence, from Eqn. (*iv*) $<p^2> = \frac{\hbar^2}{2\sigma_0^2}$

We know that $\quad \Delta x = \sqrt{\langle x^2 \rangle - \langle x \rangle^2} = \dfrac{\sigma_0}{\sqrt{2}}$

and $\qquad\qquad \Delta p = \sqrt{\langle p^2 \rangle - \langle p \rangle^2} = \dfrac{\hbar}{\sqrt{2}\sigma_0}$

$\therefore \qquad\qquad \Delta x.\Delta p = \dfrac{\hbar}{2}$

4.19 Heisenberg's Uncertainty Principle from Schwarz Inequality

Schwarz's inequality can also be written as

$$\int f^* f dv \int g^* g dv \geq \frac{1}{4}\left[\int (f^* g + g^* f) dv\right]^2 \tag{150}$$

Now assume a system to be in a state ϕ. We are interested in the result of measurements on the observations belonging to two operators $\hat{P}$ and $\hat{Q}$ (at present unspecified). Let us introduce the functions

$$f = (\hat{P} - \bar{p})\phi \quad \text{and} \quad g = i(\hat{Q} - \bar{q})\phi \tag{151}$$

where $\qquad \bar{p} = \lim_{v \to \infty} \dfrac{\displaystyle\int_v \phi^* \hat{P} \phi dv}{\displaystyle\int_v \phi^* \phi dv} \tag{152}$

and $\qquad \bar{q} = \lim_{v \to \infty} \dfrac{\displaystyle\int_v \phi^* Q \phi dv}{\displaystyle\int_v \phi^* \phi dv} \tag{153}$

Substituting the values of f and g from Eqn. (151) in (150) we get

$$\int (\hat{P} - \bar{p})^* \phi^* (\hat{P} - \bar{p})\phi dv \int (\hat{Q} - \bar{q})^* \phi^* (\hat{Q} - \bar{q})\phi dv$$

$$\geq \frac{1}{4}\left[i\int (\hat{P} - \bar{p})^* \phi^* (\hat{Q} - \bar{q})\phi dv - i\int (\hat{Q} - \bar{q})^* \phi^* (\hat{P} - \bar{p})\phi dv\right]^2$$

or $\quad \displaystyle\int \phi^* (\hat{P} - \bar{p})^2 \phi dv \int \phi^* (\hat{Q} - \bar{q})^2 \phi dv$

$$\geq \frac{1}{4}\left[i\int \phi^* (\hat{P} - \bar{p})(\hat{Q} - \bar{q})\phi dv - i\int \phi^* (\hat{Q} - \bar{q})(\hat{P} - \bar{p})\phi dv\right]^2$$

(using the hermitian property of $\hat{P}$ and $\hat{Q}$). Since $\bar{p}$ and $\bar{q}$ are constants therefore

$$\int \phi^* (\hat{P} - \bar{p})^2 \phi dv \int \phi^* (\hat{Q} - \bar{q})^2 \phi dv$$

$$\geq -\frac{1}{4}\left[\int \phi^* (\hat{P}\hat{Q} - \hat{Q}\hat{P})\phi dv\right]^2 \tag{154}$$

Let us consider the meaning of the quantity

$$\int \phi^* \left(\hat{P} - \bar{p} \right)^2 \phi \, dv$$

Now

$$\int \phi^* \left(\hat{P} - \bar{p} \right)^2 \phi \, dv = \left\langle \left(\hat{P} - \bar{p} \right)^2 \right\rangle$$

$$= \left\langle P^2 \right\rangle - \left\langle 2 \hat{P} \bar{p} \right\rangle + \left\langle \bar{p}^2 \right\rangle$$

$$= \left\langle P^2 \right\rangle - 2 \bar{p} \left\langle \hat{P} \right\rangle + \left\langle \bar{p}^2 \right\rangle$$

$$= \left\langle P^2 \right\rangle - 2 \bar{p} \cdot \bar{p} + \bar{p}^2$$

$$= \left\langle P^2 \right\rangle - \left\langle p \right\rangle^2$$

$$= (\Delta p)^2 \qquad \text{(by standard deviation)}$$

Similary,

$$\int \phi^* \left(\hat{Q} - \bar{q} \right)^2 \phi \, dv = (\Delta q)^2$$

Hence from Eqn. (154)

$$(\Delta p)^2 \, (\Delta q)^2 \geq -\frac{1}{4} \left\{ \int \phi^* \left(\hat{P} \hat{Q} - \hat{Q} \hat{P} \right) \phi \, dv \right\}^2 \qquad (155)$$

If operators $\hat{P}$ and $\hat{Q}$ commute, the right hand side is zero. When $\hat{P}$ and $\hat{Q}$ do not commute, the relation (155) sets a lower limit for the product of the dispersions called uncertainties.

Suppose $\quad \hat{P} = -i\hbar \dfrac{\partial}{\partial x}$ or $\dfrac{\hbar}{i} \dfrac{\partial}{\partial x}$ and $\hat{Q} = x$

then $\qquad (\hat{P} \hat{Q} - \hat{Q} \hat{P}) \psi = \left(\dfrac{\hbar}{i} \dfrac{\partial}{\partial x} x - x \dfrac{\hbar}{i} \dfrac{\partial}{\partial x} \right) \psi$

$$= \frac{\hbar}{i} \left[\frac{\partial}{\partial x} (x \psi) - x \frac{\partial \psi}{\partial x} \right]$$

$$= \frac{\hbar}{i} \left[x \frac{\partial \psi}{\partial x} + \psi - x \frac{\partial \psi}{\partial x} \right]$$

$\Rightarrow \qquad \hat{P} \hat{Q} - \hat{Q} \hat{P} = \dfrac{\hbar}{i}$

$\Rightarrow \qquad (\Delta p)^2 \, (\Delta q)^2 \geq -\dfrac{1}{4} \left(\dfrac{\hbar}{i} \right)^2$ or $\dfrac{\hbar^2}{4}$

Thus $\qquad \Delta p \cdot \Delta q \geq \dfrac{\hbar}{2}$

This is uncertainty relation for position and momentum.

Questions and Problems

1. (a) Show that the Fourier transform of the function $f(x) = Ae^{-\alpha|x|}$, $\alpha > 0$ is given by

$$F(k) = \frac{A}{\sqrt{2\pi}} \frac{2\alpha}{\alpha^2 + k^2}$$

   (b) Hence show that

$$\int_{-\infty}^{+\infty} \frac{e^{ikx}}{\alpha^2 + k^2} dk = \frac{\pi}{\alpha} e^{-\alpha|x|}$$

2. (a) What do you mean by a wave packet? Discuss the spreading of a wave packet.

   (b) For matter waves prove that group velocity is equal to the particle velocity.

3. (a) Describe the shape of the wave packet for which the uncertainty product attains the theoretical minimum value.

   (b) Discuss the spreading with time of the above wave packet.

4. The following wave packet describes the one dimensional motion of a free particle of mass m:

$$\psi(x, t) = \frac{1}{(2\pi)^{1/4}} \left(a + \frac{i\hbar t}{2ma} \right) \exp \left[\frac{x^2}{4a^2 + \dfrac{2i\hbar t}{m}} \right]$$

   Show that the above wave function is properly normalized.

5. At time $t = 0$, show that the wave function of problem (4) yields the uncertainty relation

$$(\Delta \bar{x})^2 (\Delta \bar{p})^2 = \frac{\hbar^2}{4}$$

   where $\Delta x = x - <x>$ and $(\Delta \bar{x})^2 = \langle (x - \langle x \rangle)^2 \rangle$, $(\Delta \bar{p})^2$ is similarly defined.

6. Describe briefly how the wave packet of problem (4) spreads in time, given that

$$\int_{-\infty}^{+\infty} e^{-\alpha x^2} dx = \sqrt{\frac{\pi}{\alpha}}$$

   and

$$\int_{-\infty}^{+\infty} e^{-\alpha x^2} x^2 dx = \frac{1}{2} \sqrt{\frac{\pi}{\alpha^3}}$$

7. A particle in free space is initially in a wave packet described by

$$\psi(x) = \left(\frac{\alpha}{\pi} \right)^{1/4} e^{-\alpha x^2/2}$$

   (a) What is the probability that its momentum is in the range between p and $p + dp$.

   (b) What is the expectation value of the energy?

8. Determine the probability density of the momentum values for the ground state of the hydrogen atom.

9. The 2s-state of hydrogen has the wave function

$$\psi(r) = \frac{1}{\sqrt{32\pi}}(r-2)e^{-r/2}$$

in which r is measured in units of $\hbar^2/me^2$. Find the momentum representation for this state.

10. Find the spectral density $\phi(k)$ for the wave packet described by $\psi(x) = Ae^{ik_0x}$, $|x| < a$ and $\psi(x) = 0; |x| > a$.

$$\textbf{Ans.}\quad \frac{2Aa}{\sqrt{2\pi}}\ \frac{\sin(k_0-k)a}{(k_0-k)a}$$

11. A linear quantum oscillator in its ground state has a wave function

$$\psi(x) = \frac{1}{\sqrt{a}}\pi^{-\frac{1}{4}}e^{-x^2/(2a^2)}$$

Show that the corresponding momentum function is

$$\psi(p) = \frac{\sqrt{a}}{\sqrt{\hbar}}\pi^{-\frac{1}{4}}e^{-a^2p^2/(2\hbar^2)}$$

5

Linear Harmonic Oscillator

5.1 Introduction

If a particle is slightly displaced from its stable equilibrium position, it executes small amplitude oscillations. The potential energy can be expanded in a Taylor series as:

$$V(x) = V(x_0 + x - x_0)$$

$$= V(x_0) + (x - x_0)\left(\frac{\partial V}{\partial x}\right)_{x_0} + \frac{1}{2}(x - x_0)^2\left(\frac{\partial^2 V}{\partial x^2}\right)_{x_0} + \dots \tag{1}$$

Since $\left(\dfrac{\partial V}{\partial x}\right)_{x_0} = 0$ at a point x_0 of stable equilibrium

$$\therefore \quad V(x) = V(x_0) + \frac{1}{2}k(x - x_0)^2, \quad \left(k = \frac{\partial^2 V}{\partial x^2}\bigg|_{x=x_0}\right) \tag{2}$$

since zero level of potential energy is arbitrary

$$\therefore \quad V(x) = \frac{1}{2}kx^2 \tag{3}$$

Eqn. of motion:

$$\vec{F} = -\nabla V; \quad F_x = -\frac{\partial V}{\partial x} = -kx; \quad m\frac{d^2x}{dt^2} = -kx \tag{4i}$$

and displacement $x = A\cos(\omega t + \phi)$

$$\left(\omega^2 = \frac{k}{m}\right) \tag{4ii}$$

Total energy $\quad E = \dfrac{1}{2}m\dot{x}^2 + \dfrac{1}{2}m\omega^2 x^2 \tag{5}$

$$= \frac{1}{2}mA^2\omega^2 \sin^2(\omega t + \phi) + \frac{1}{2}mA^2\omega^2 \cos^2(\omega t + \phi)$$

or
$$E = \frac{1}{2}mA^2\omega^2 = m\omega^2 \frac{A^2}{2}$$ (6)

$$\langle x^2 \rangle = \langle A^2 \cos^2(\omega t + \phi) \rangle = \frac{A^2}{2}$$

$$\therefore \qquad E = m\omega^2 \langle x^2 \rangle$$

in n^{th} state $\quad E_n = m\omega^2 \langle x^2 \rangle_n$ (7)

5.2 The Eigenvalue Equation and Energy Eigenvalues

The stationary state energies E_n are solutions of the eigenvalue equation

$$Hu = Eu$$ (8)

or
$$\left[-\frac{\hbar^2}{2m}\frac{\partial^2}{\partial x^2} + \frac{1}{2}kx^2 \right] u = Eu \quad \left(\because V(x) = \tfrac{1}{2}kx^2 \right)$$ (9)

Dividing this equation by $\dfrac{\hbar\omega}{2}$, we get

$$-\frac{\hbar}{m\omega}\frac{\partial^2 u}{\partial x^2} + \left(\frac{m\omega}{\hbar}x^2 \right) u = \frac{2E}{\hbar\omega}u \quad (k = \omega^2 m)$$ (10)

Let
$$\left(\frac{2E}{\hbar\omega} \right) = \lambda \text{ and } \left(\frac{m\omega}{\hbar} \right)^{1/2} x = y$$

$$\therefore \qquad \frac{\partial u}{\partial y} = \frac{\partial u}{\partial x}\cdot\frac{\partial x}{\partial y} = \frac{\partial u}{\partial x} \Big/ \left(\frac{m\omega}{\hbar} \right)^{1/2}$$

$$\therefore \qquad \frac{\partial^2 u}{\partial y^2} = \frac{\partial^2 u}{\partial x^2}\cdot\frac{\partial x}{\partial y}\left(\frac{m\omega}{\hbar} \right)^{-1/2} = \frac{\partial^2 u}{\partial x^2}\cdot\left(\frac{m\omega}{\hbar} \right)^{-1}$$

Eqn. (3) reduces to

$$-\frac{d^2 u}{dy^2} + y^2 u = \lambda u$$

or
$$\frac{d^2 u}{dy^2} + \left(\lambda - y^2 \right) u = 0$$ (11)

We have to find the solutions of this equation, which do not diverge at infinity. The Eqn. shows that $e^{-y^2/2}$ is an exact solution when $\lambda = 1$. For other values of λ also, the dominant asymptotic behaviour of $u(y)$ would be of the same type, because as $y \to \pm\infty$ the constant λ becomes insignificant compared to y^2 in Eqn. (11) and hence its value does not matter too much. Therefore we can write

$$u(y) = e^{-y^2/2}H(y)$$ (12)

$$u' = \frac{\partial u}{\partial y} = e^{-y^2/2}H' + \left(-2\frac{y}{2} \right)e^{-y^2/2}H$$

$$u'' = \frac{\partial^2 u}{\partial y^2} = e^{-y^2/2} H'' - 2ye^{-y^2/2} H' - H\left[e^{-y^2/2} + y(-y)e^{-y^2/2}\right]$$

$$= e^{-y^2/2} H'' - 2ye^{-y^2/2} H' - He^{-y^2/2}\left[1 - y^2\right]$$

Substituting the values of u' and u'' in Eqn. (11)

$$= e^{-y^2/2}\left\{\left[H'' - 2yH' + \left(y^2 - 1\right) H\right] + \left(\lambda - y^2\right) H\right\} = 0$$

or $\qquad H'' - 2yH'(y) + (\lambda - 1) H(y) = 0 \qquad\qquad$ (13)

This equation is solved by the power series method. We seek a solution of the form

$$H = y^s (a_0 + a_1 y + a_2 y^2 + ...) = y^s \sum_{v=0}^{\infty} a_v y^v \qquad (14)$$

with $a_0 \neq 0$ and $s \geq 0$

$$= \sum_{v=0}^{\infty} a_v y^{s+v}$$

$$\frac{dH}{dy} = \sum_v a_v (s + v) y^{s+v-1}; \quad \frac{d^2 H}{dy^2} = \sum_v a_v (s + v)(s + v - 1) y^{s+v-2}$$

Substituting these values in Eqn. (13) we have

$$\sum_v a_v (s+v)(s+v-1) y^{s+v-2} - 2y \sum_v a_v (s+v) y^{s+v-1} + (\lambda - 1) \sum_v a_v y^{s+v} = 0$$

or $\quad \sum_v a_v (s+v)(s+v-1) y^{s+v-2} - 2 \sum_v a_v (s+v) y^{s+v} + (\lambda - 1) \sum_v a_v y^{s+v} = 0$

Equating to zero the coefficients of $y^{(s-2)}$, $y^{(s-1)}$, y^s and $y^{(s+1)}$ we get

$$s(s-1)a_0 = 0 \qquad\qquad\qquad (i)$$

$$(s+1)s\ a_1 = 0 \qquad\qquad\qquad (ii)$$

$$(s+2)(s+1)a_2 - (2s+1-\lambda)a_0 = 0$$

$$(s+3)(s+2)a_3 - (2s+3-\lambda)a_1 = 0 \qquad\qquad (15)$$

———— ———— ———— ———— ————

———— ———— ———— ———— ————

$$(s+v+2)(s+v+1)a_{v+2} - (2s+2v+1-\lambda)a_v = 0$$

where v is an integer. We have recurrence formula

$$a_{v+2} = \frac{2s+2v+1-\lambda}{(s+v+1)(s+v+2)} a_v \qquad (16)$$

Since a_0 cannot be zero.

∴ $\qquad\qquad\qquad s = 0 \text{ or } s = 1 \qquad\qquad$ (from 15 i)

From Eqn. (15) (*ii*) $s = 0$ or $a_1 = 0$, or both s and a_1 zero.

The function $u(y)$ and hence also $H(y)$ can be chosen to be either even or odd in y. Then from Eqn. (14) a_1 and all the other odd-subscript coefficients are zero and the wave function is even or odd according as $s = 0$ or $s = 1$.

Eqn. (15) shows that the presence in the series (14) of a finite or an infinite number of terms depends on s and λ.

For arbitrary values of the energy parameter $\lambda \left(= \dfrac{2E}{\hbar\omega} \right)$, the series contains an infinite number of terms.

Also then

$$\underset{v\to\infty}{\text{Lim}}\frac{a_{v+2}}{a_v} = \underset{v\to\infty}{\text{Lim}}\frac{2s+2v+1-\lambda}{(s+v+2)(s+v+1)} = \underset{v\to\infty}{\text{Lim}}\frac{\dfrac{2s}{v}+2+\dfrac{1}{v}-\dfrac{\lambda}{v}}{\left(\dfrac{s}{v}+1+\dfrac{2}{v}\right)(s+v+1)}$$

$$\to \frac{2}{v} \quad (\because v \gg s) \tag{17}$$

Consider the series expansion of e^{y^2}

$$e^{y^2} = 1+y^2+\frac{y^4}{2!}+\frac{y^6}{3!}+...+\frac{y^v}{(v/2)!}+\frac{y^{v+2}}{(\frac{v}{2}+1)!}+...$$

$$= b_0 + b_2 y^2 + b_4 y^4 + ... + b_v y^v + b_{v+2} y^{v+2} + ...$$

$$\therefore \quad \frac{b_{v+2}}{b_v} = \frac{\dfrac{1}{\left(\frac{v}{2}+1\right)!}}{\dfrac{1}{\left(\frac{v}{2}\right)!}} = \frac{\left(\frac{v}{2}\right)!}{\left(\frac{v}{2}+1\right)!} = \frac{1}{\frac{v}{2}+1}$$

or

$$\lim_{v\to\infty}\frac{b_{v+2}}{b_v} = \frac{2}{v} \tag{18}$$

Eqns (17) and (18) show that $H(y)$ diverges (approx.) as e^{y^2} and the product $H(y)\ e^{-y^2/2}$ will behave like $e^{+y^2/2}$, thus making it unacceptable wave function.

The only way in which this situation can be avoided is to choose λ in such a way that the series for $H(y)$ cuts off after some terms, making $H(y)$ a polynomial $H_n(y)$.

The value of λ which causes the series to break off after the vth term is

$$\lambda = 2s + 2v + 1$$

Since s is either 0 or 1, correspondingly λ is $2v + 1$ or $2v + 3$ (v is an even integer). Both cases can be expressed in terms of a quantum number n

For
$$s = 0, n = 0, 2, 4, ... \qquad\qquad (n = s + v)$$
$$s = 1, n = 1, 3, 5, ...$$
$$\lambda = 2n + 1$$

then
$$\frac{2E_n}{\hbar\omega} = 2n + 1 \tag{19}$$

or
$$E_n = (n+\frac{1}{2})\hbar\omega \qquad\qquad n = 0, 1, 2, ...$$

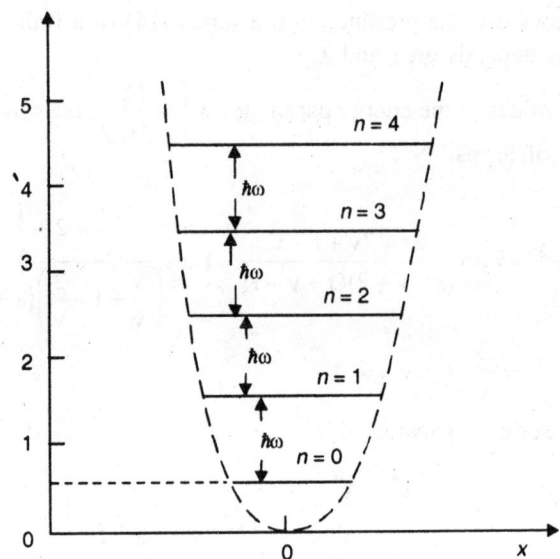

Fig. 5.1: Energy levels of the harmonic oscillator.
The dashed parabola gives the potential energy of the particle

The energy levels given by this Eqn. are discrete. They have equal spacing = ½ $\hbar\omega$.

When $n = 0$, the finite value of ground state energy is (½ $\hbar\omega$). This is known as *zero point energy*. Associated with this minimum energy state is the "*zero point motion*" of the particle, the particle does not settle down to zero excursion about the origin but remains spread out as though in a state of small but finite agitation. This zero point motion predicted by quantum mechanics is not a mathematical artifact but a real physical phenomenon. This explains a number of experimental facts which are otherwise inexplicable on a classical model.

Example 5.1: Using the function

$$\psi(x) = A\,e^{-bx^2}$$

and the linear harmonic oscillator wave equation show that $\psi(x)$ satisfies the oscillator equation only if $E = \dfrac{1}{2}\hbar\omega$ and $b = \dfrac{1}{2}\dfrac{m\omega}{\hbar}$.

Solution:
$$\frac{d\psi}{dx} = \frac{d}{dx}(Ae^{-bx^2})$$

$$= -2bxAe^{-bx^2} = -2bx\psi(x)$$

$$\frac{d^2\psi}{dx^2} = -2b\psi(x) - 2bx\frac{d\psi(x)}{dx}$$

$$= -2b\psi(x) - 2bx(-2bx)Ae^{-bx^2}$$

$$= -2b\psi(x) + 4b^2x^2\psi(x)$$

Now the linear harmonic oscillator wave equation is

$$\frac{d^2\psi}{dx^2} + \frac{2m}{\hbar^2}\left[E - \frac{1}{2}m\omega^2 x^2\right]\psi(x) = 0$$

$$\Rightarrow \quad 4b^2 x^2 \psi(x) - 2b\psi(x) + \frac{2m}{\hbar^2}\left[E - \frac{1}{2}m\omega^2 x^2\right]\psi(x) = 0$$

Since $\psi(x) \neq 0$, therefore

$$4b^2 x^2 - 2b + \frac{2m}{\hbar^2}\left[E - \frac{1}{2}m\omega^2 x^2\right] = 0$$

or $\qquad 4b^2 x^2 - 2b + \frac{2mE}{\hbar^2} - \frac{m^2}{\hbar^2}\omega^2 x^2 = 0 \qquad\qquad (i)$

$\therefore$ The coefficient of x^2 must vanish (since there is no x^2 term on the r.h.s.)

or

$$4b^2 = \frac{m^2\omega^2}{\hbar^2} \text{ or } b = \frac{1}{2}\frac{m\omega}{\hbar} \qquad\qquad (ii)$$

Putting this value of b in Eqn. (i)

$$4 \times \frac{1}{4}\frac{m^2\omega^2 x^2}{\hbar^2} - \frac{m\omega}{\hbar} + \frac{2mE}{\hbar^2} - \frac{m^2\omega^2 x^2}{\hbar^2} = 0$$

or

$$\frac{2mE}{\hbar^2} = \frac{m\omega}{\hbar}$$

or

$$E = \frac{1}{2}\hbar\omega \qquad\qquad (iii)$$

Hence the wave equation is satisfied only if b and E are given by equations (ii) and (iii) respectively.

5.3 Hermite Polynomials

When $\lambda = 2n + 1$, Eqn. (13) becomes

$$H_n''(y) - 2yH_n'(y) + 2nH_n(y) = 0 \qquad\qquad (20)$$

This is Hermite's equation.

The polynomials $H_n(y)$ are known as the Hermite polynomials. They can be conveniently expressed in terms of a generating function $G(y, s)$:

$$G(y, s) = e^{y^2-(s-y)^2} = e^{(-s^2+2sy)}$$

$$= 1 - s^2 + 2sy + \frac{(-s^2 + 2sy)^2}{2!} + \dots$$

$$= 1 + 2sy + (4y^2 - 2)\frac{s^2}{2!} + \dots$$

$$= H_0 + H_1 s + H_2 \frac{s^2}{2!} + \dots$$

or $\qquad e^{(-s^2+2sy)} = \sum_{n=0}^{\infty} \frac{H_n(y)s^n}{n!} \qquad\qquad (21)$

In order to prove that $H_n(y)$ will satisfy the differential equation (20), we differentiate both sides of equation (21) first with respect to y and then with respect to s.

$$\frac{\partial G}{\partial y} = 2s\, e^{(-s^2+2sy)} = \sum_n \frac{s^n}{n!} H'_n(y) \qquad \text{(using Eqn. 21 on l.h.s.)}$$

$$= \sum_n \frac{2s^{n+1}}{n!} H_n(y) = \sum_n \frac{s^n}{n!} H'_n(y)$$

$$\Rightarrow \quad \frac{2s^2}{1!} H_1(y) + \frac{2s^3}{2!} H_2(y) + \dots + \frac{2s^n H_{n-1}(y)}{(n-1)!} + \dots$$

$$= \frac{s}{1!} H'_1(y) + \frac{s^2}{2!} H'_2(y) + \dots + \frac{s^n H'_n(y)}{n!} + \dots$$

Equating the terms having equal powers of s (say s^n) on the two sides, we get

$$\frac{2H_{n-1}(y)}{(n-1)!} = \frac{H'_n(y)}{n!}$$

or $\qquad\qquad H'_n(y) = 2nH_{n-1}(y)$ (Recurrence relation i) $\qquad\qquad$ (22)

$$\frac{\partial G}{\partial s} = (-2s+2y)\, e^{(-s^2+2sy)} = \sum_n \frac{s^{n-1}}{(n-1)!} H_n(y)$$

or $\qquad\qquad \sum_n (-2s+2y) \frac{s^n}{n!} H_n(y) = \sum_n \frac{s^{n-1}}{(n-1)!} H_n(y)$

or $\qquad \sum_n \left[-\frac{2s^{n+1}}{n!} H_n(y) + \frac{2s^n y}{n!} H_n(y) \right] = \sum_n \frac{s^{n-1}}{(n-1)!} H_n(y)$

$$\Rightarrow \quad -\frac{2s^2}{1!} H_1(y) - \frac{2s^3}{2!} H_2(y) - \dots - \frac{2s^{(n-1)+1}}{(n-1)!} H_{n-1}(y) + \dots + \frac{2s^n y}{n!} H_n(y) + \dots$$

$$= \frac{s}{1!} H_2(y) + \frac{s^2}{2!} H_3(y) + \dots + \frac{s^{(n+1)-1}}{n!} H_{n+1}(y) + \dots$$

Equating the terms of equal powers on both the sides (say s^n) we get,

$$-\frac{2}{(n-1)!} H_{n-1}(y) + \frac{2y}{n!} H_n(y) = \frac{H_{n+1}(y)}{n!}$$

or $\qquad\qquad -2H_{n-1}(y) + \frac{2y}{n!} H_n(y) = \frac{H_{n+1}(y)}{n!}$

or $\qquad\qquad H_{n+1} = 2yH_n - 2nH_{n-1}$ (Recurrence relation ii) $\qquad\qquad$ (23)

Differentiating Eqns. (22) and (23) w.r.t.y, we get

$$H''_n = 2nH'_{n-1}$$

and $\qquad\qquad H'_{n+1} = 2yH'_n - 2nH'_{n-1} + 2H_n$

Adding these two equations,

$$H''_n + H'_{n+1} = 2yH'_n + 2H_n \qquad\qquad (24)$$

But $$H'_{n+1} = 2(n+1)H_n \text{ from Eqn. (22), so}$$

$$H''_n + 2nH_n + 2H_n = 2yH'_n + 2H_n$$

or $\quad H''_n - 2yH'_n + 2nH_n = 0$

Thus $H_n(y)$ as defined by Eqn. (21) satisfy the Eqn. (20), hence these are Hermite polynomials.

Now, $$H_n(y) = \frac{\partial^n G(y,s)}{\partial s^n}$$

$$= \frac{\partial^n}{\partial s^n}(e^{-s^2+2sy}) = \frac{\partial^n}{\partial s^n}\left[e^{[y^2-(s-y)^2]}\right] = e^{y^2}\frac{\partial^n}{\partial s^n}e^{-(s-y)^2}$$

$$= (-1)^n e^{y^2}\frac{\partial^n}{\partial y^n}e^{-(s-y)^2}$$

This equation represents n^{th} Hermite polynomial. Setting $s = 0$, we get

$$H_n(y) = (-1)^n e^{y^2}\frac{\partial^n}{\partial y^n}e^{-y^2} \tag{25}$$

The first few Hermite polynomials are

$$H_0 y = 1$$
$$H_1(y) = 2y$$
$$H_2(y) = 4y^2 - 2$$
$$H_3(y) = 8y^3 - 12y$$
$$H_4(y) = 16y^4 - 48y - 12$$

The peculiarity of the Hermite polynomials is that the coefficient of highest power of y are 2^n.

5.4 Harmonic Oscillator Wave Functions

The solution of the equation

$$H''_n(y) - 2yH''_n(y) + 2nH_n(y) = 0$$

will be

$$u_n(y) = H_n(y)e^{-y^2/2} \tag{26i}$$

The general solutions will be

$$\psi_n(y) = N_n H_n(y)e^{-y^2/2} \tag{26ii}$$

where N_n is the normalization constant, N_n may be evaluated as follows.
We know that

$$e^{-s^2+2sy} = \sum_{n=0}^{\infty}\frac{H_n(y)}{n!}s^n$$

Similarly

$$e^{-t^2+2ty} = \sum_{m=0}^{\infty}\frac{H_m(y)}{m!}t^m$$

Hence,

$$\exp(-s^2 + 2sy)\exp(-t^2 + 2ty)dy = \sum_{n=0}^{\infty} \sum_{m=0}^{\infty} \frac{s^n t^m}{n! \, m!} H_n(y) H_m(y) \, dy$$

If this equation is first multiplied by e^{-y^2} and then integrated w.r.t. y from $-\infty$ to $+\infty$, we get

$$\int_{-\infty}^{+\infty} \exp(-s^2 + 2sy)\exp(-t^2 + 2ty)\exp(-y^2) \, dy$$

$$= \sum_{n=0}^{\infty} \sum_{m=0}^{\infty} \frac{s^n t^m}{n! \, m!} \int_{-\infty}^{+\infty} H_n(y) H_m(y) e^{-y^2} dy \tag{27}$$

Let us consider the l.h.s. of this equation.

$$\int_{-\infty}^{+\infty} \exp(-s^2 + 2sy)\exp(-t^2 + 2ty)\exp(-y^2) \, dy$$

$$= \int_{-\infty}^{+\infty} \exp(-s^2 - t^2 + 2sy + 2ty - y^2) \, dy$$

$$= e^{2st} \int_{-\infty}^{+\infty} \exp\left\{-(y - s - t)^2\right\} d(y - s - t)$$

$$= e^{2st} \cdot \sqrt{\pi}$$

$$= (\pi)^{1/2} \left\{ 1 + \frac{2st}{1!} + \frac{(2st)^2}{2!} + \dots + \frac{(2st)^n}{n!} \right\}$$

$$= (\pi)^{1/2} \sum_{n=0}^{\infty} \frac{(2st)^n}{n!} \tag{28}$$

So, we have

$$(\pi)^{1/2} \sum_{n=0}^{\infty} \frac{(2st)^n}{n!} = \sum_{n=0}^{\infty} \sum_{m=0}^{\infty} \frac{s^n t^m}{n! \, m!} \int_{-\infty}^{+\infty} H_n(y) H_m(y) e^{-y^2} dy \tag{29i}$$

On equating equal powers of s and t in this series, we obtain

$$\int_{-\infty}^{+\infty} H_n^2(y) e^{-y^2} dy = \sqrt{\pi} \cdot 2^n \, n! \text{ for } n = m$$

and

$$\int_{-\infty}^{+\infty} H_n(y) H_m(y) e^{-y^2} dy = 0 \tag{29ii}$$

By normalization condition

$$\int_{-\infty}^{+\infty} |\psi_n(\alpha, x)|^2 \frac{d(\alpha x)}{\alpha} = 1$$

or

$$\frac{1}{\alpha} \int_{-\infty}^{+\infty} |\psi_n(y)|^2 \, dy = 1 \quad \left(\because \ y = \alpha x \ \text{and} \ \alpha = \left(\frac{m\omega}{\hbar} \right)^{1/2} \right)$$

or

$$\frac{|N_n|^2}{\alpha} \int_{-\infty}^{+\infty} H_n^2(y)e^{-y^2}\, dy = 1$$

or

$$\frac{|N_n|^2}{\alpha} \pi^{1/2} 2^n (n!) = 1$$

Thus

$$N_n = \left\{ \frac{\alpha}{\sqrt{\pi}.2^n\, n!} \right\}^{1/2}$$

Hence from Eqn. (26*ii*), we get

$$\psi_n(y) = \left\{ \frac{\alpha}{2^n\, n!\sqrt{\pi}} \right\}^{1/2} H_n(y)\, e^{-y^2/2} \tag{30}$$

5.5 Physical Interpretation of Harmonic Oscillator Wave Functions

The H_n's in expression (30) represents a polynomial of degree n (Fig. 5.2). The various wave functions for $n = 0, 1, 2,...$ are also shown. The ψ_n's have n zeros i.e. n points where cuts the $\psi = 0$ line.

The probability $|\psi_n|^2$ of finding the particle at these points will be zero. For the ground state $n = 0$, the wave function is given by

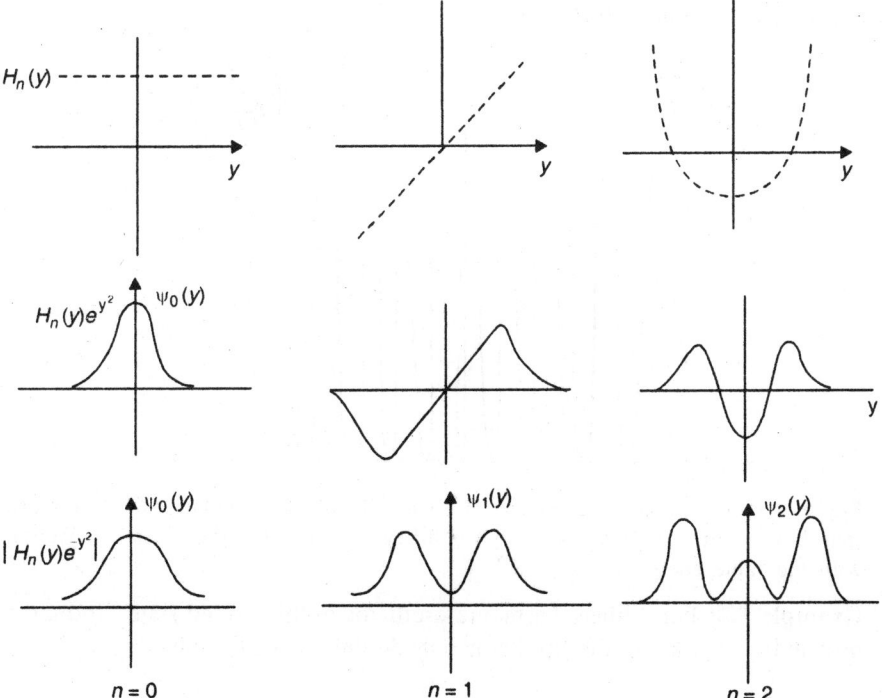

Fig. 5.2: The Hermite polynomials and harmonic oscillator wave functions (for $n = 0, 1, 2$) as a function of y

$$\psi_0(y) = \left(\frac{\alpha}{\sqrt{\pi}}\right)^{1/2} \text{ and } \alpha = \left(\frac{m\omega}{\hbar}\right)^{1/2} \tag{31}$$

The Fig. 5.3 gives the probability distribution for the ground state.

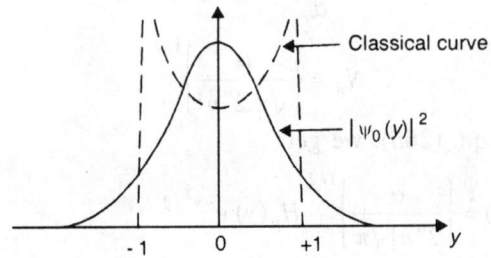

Fig. 5.3: Probability distribution for the ground state

The figure shows that it is not having any agreement with the classical curve for an oscillator with the same energy. It is seen that though there is a small but finite probability outside the classically permitted region. This is a consequence of energy-time uncertainty. The maximum time is spent near the classical limits. At these points Dt is minimum and DE is maximum. Outside the classically permitted regions, the potential energy would become higher than the total energy and the particle would be having a negative kinetic energy–which is a classically forbidden result.

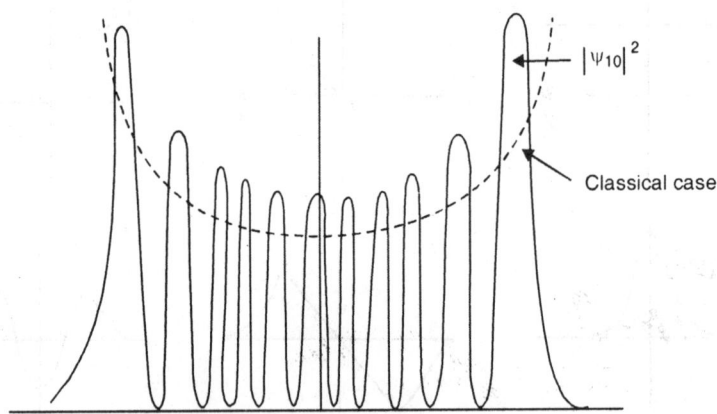

Fig. 5.4: Plot of $|\psi_{10}(y)|^2$ (solid curve) for $n = 10$ of the harmonic oscillator. Dashed (parabolic) curve represents the probability function for the classical oscillator with the same energy

Example 5.2: For a linear harmonic oscillator in the ground state, find out the probability of finding the oscillator outside the classical limits.

$$\psi_0(y) = \left(\frac{\alpha}{\sqrt{\pi}}\right)^{1/2} e^{-y^2/2}$$

Solution: Let a denote the amplitude of oscillations. Then

$$P = \int_{-a}^{+a} |\psi_0|^2 \, dx = 2\int_0^a \frac{\alpha}{\sqrt{\pi}} e^{-y^2} \, dx$$

$$= 2\int_0^a \frac{\alpha}{\sqrt{\pi}} e^{-\alpha^2 x^2} \, dx$$

According to quantum mechanics,

$$E = \frac{1}{2}\hbar\omega_0 = \frac{1}{2}\hbar\sqrt{\frac{k}{m}} = \frac{1}{2}kx^2$$

or

$$x^2 = \frac{\hbar}{\sqrt{mk}} = \frac{\hbar}{m\omega} = \frac{1}{\alpha^2}$$

i.e. x tends to vary from $-\dfrac{1}{\alpha}$ to $+\dfrac{1}{\alpha}$.

$$\therefore \qquad P = 2 \int_0^{1/\alpha} \frac{\alpha}{\sqrt{\pi}} e^{-\alpha^2 x^2} \, dx$$

Put $ax = y$

$$P = \frac{2}{\sqrt{\pi}} \int_0^1 e^{-y^2} \, dy$$

$$= \frac{2}{\sqrt{\pi}} \int_0^1 \left(1 - \frac{y^2}{1!} + \frac{y^4}{2!} - \frac{y^6}{3!} + \frac{y^8}{4!} + \ldots\right) dy$$

$$= \frac{2}{\sqrt{\pi}} \left[y - \frac{y^3}{3\times 1!} + \frac{y^5}{5\times 2!} - \frac{y^7}{7\times 3!} + \ldots\right]_0^1$$

$$= \frac{2}{\sqrt{\pi}} \left[1 - \frac{1}{3\times 1!} + \frac{1}{5\times 2!} - \frac{1}{7\times 3!} + \ldots\right]$$

$$= 0.83 = 83\%$$

Therefore, the probability of finding the oscillator outside the classical limits will be $(1 - 0.83) \times 100 = 17\%$.

5.6 Selection Rules for the Harmonic Oscillator

We have the recurrence relation

$$H_{n+1} = 2yH_n - 2n\,H_{n-1}$$

We know that

$$\psi_n(y) = \left\{\frac{\alpha}{\sqrt{\pi}.2^n n!}\right\}^{1/2} H_n(y)\, e^{-y^2/2}$$

Putting $n \to n+1$ and $n \to n-1$, we get

$$\psi_{n+1}(y) = \left\{\frac{1}{\sqrt{\pi}.2^{n+1}(n+1)!}\right\}^{1/2} H_{n+1}(y)\, e^{-y^2/2} \qquad (32)$$

and
$$\psi_{n-1}(y) = \left\{ \frac{1}{\sqrt{\pi}.2^{n-1}(n-1)!} \right\}^{1/2} H_{n-1}(y)\, e^{-y^2/2} \qquad (33)$$

If, from these expressions, we substitute the values of H_{n-1}, H_n and H_{n+1} in the recurrence relation, we get

$$\sqrt{(n+1)}\ \psi_{n+1} = \sqrt{2}y\ \psi_n - \sqrt{n}\ \psi_{n-1} \qquad (34)$$

Multiplying both sides by ψ_n and integrating, we get

$$\sqrt{(n+1)}\ \int_{-\infty}^{+\infty} \psi_{n+1}\psi_n dy = \sqrt{2}\int_{-\infty}^{+\infty}\psi_n y\psi_n dy - \sqrt{n}\int_{-\infty}^{+\infty}\psi_{n-1}\psi_n dy$$

Using Eqns. (29ii), we have

$$0 = \sqrt{2}\langle y\rangle - 0 \qquad \left(\because\ \int_{-\infty}^{+\infty}\psi_m\psi_n dy = \delta_{mn}\right)$$

or
$$\langle y\rangle = 0$$

Now, pre-multiplying Eqn. (34) by ψ_{n+1}, we get on integration

$$\sqrt{(n+1)}\ \int_{-\infty}^{+\infty}\psi_{n+1}\psi_{n+1}dy = \sqrt{2}\int_{-\infty}^{+\infty}\psi_{n+1}^* y\psi_n dy - \sqrt{n}\int_{-\infty}^{+\infty}\psi_{n+1}\psi_{n-1}dy$$

The integral on the l.h.s. equals unity, whereas 2nd integral on the r.h.s. is zero. So,

$$\sqrt{(n+1)}\ = \sqrt{2}\int_{-\infty}^{+\infty}\psi_{n+1}y\psi_n dy$$

or
$$\int_{-\infty}^{+\infty}\psi_{n+1}y\psi_n dy = \sqrt{\left(\frac{n+1}{2}\right)} \qquad (35)$$

Replacing n by $n-1$, we get

$$\int_{-\infty}^{+\infty}\psi_n y\psi_{n-1}dy = \sqrt{\frac{n}{2}} \qquad (36)$$

The two integrals in Eqns (35) and (36) may be expressed in the combined form as

$$\int_{-\infty}^{+\infty}\psi_m y\psi_n dy \neq 0 \ \text{ or finite}$$

for
$$m = n \pm 1$$
or
$$m - n = \pm 1 \text{ or } \Delta n = \pm 1 \qquad (37)$$

In these transitions, the oscillator changes in its adjacent energy levels thereby giving the emission or absorption of radiation between the energy states E_m and E_n.

5.7 Uncertainty Relation for Harmonic Oscillator

We have from Eqn. (34)

$$\sqrt{(n+1)}\ \psi_{n+1} = \sqrt{2}y\ \psi_n - \sqrt{n}\ \psi_{n-1}$$

Post multiplying this Eqn. by $y\psi_n$ and integrating, we get

$$\sqrt{(n+1)}\int_{-\infty}^{+\infty}\psi_{n+1}y\psi_n dy = \sqrt{2}\int_{-\infty}^{+\infty}\psi_n y^2\psi_n dy - \sqrt{n}\int_{-\infty}^{+\infty}\psi_{n-1}y\psi_n dy$$

Using Eqns (35) and (36), this changes to

$$\sqrt{(n+1)}.\sqrt{\left(\frac{n+1}{2}\right)} = \sqrt{2}<y^2> -\sqrt{n}\frac{\sqrt{n}}{\sqrt{2}}$$

or

$$\langle y^2\rangle = \frac{n+1}{2}+\frac{n}{2} = \left(n+\tfrac{1}{2}\right) \tag{38}$$

But

$$y^2 = \alpha^2 x^2$$

∴

$$\langle x^2\rangle = \frac{\langle y^2\rangle}{\alpha^2} = \left(n+\tfrac{1}{2}\right)\frac{\hbar}{m\omega} \quad \left(\because \ \alpha^2 = \frac{m\omega}{\hbar}\right) \tag{39}$$

Now we know (from Eqn. (11)) that

$$\frac{d^2\psi_n}{dy^2}+(\lambda-y^2)\psi_n = 0$$

or

$$\frac{d^2\psi_n}{dy^2}+(2n+1-y^2)\psi_n = 0$$

or

$$\psi''_n + \psi_n(2n+1-y^2) = 0 \tag{40}$$

∴

$$\int_{-\infty}^{+\infty}\psi''_n\psi_n dy + \int_{-\infty}^{+\infty}\psi_n(2n+1-y^2)\psi_n dy = 0$$

or

$$-\int_{-\infty}^{+\infty}\psi''_n\,\psi_n dy = (2n+1)\int_{-\infty}^{+\infty}\psi_n\psi_n dy - \int_{-\infty}^{+\infty}\psi_n y^2\psi_n dy$$

or

$$-\int_{-\infty}^{+\infty}\psi''_n\psi_n dy = (2n+1)-\left(n+\frac{1}{2}\right) \qquad \text{using Eqn. (38)}$$

$$=\left(n+\frac{1}{2}\right) \tag{41}$$

Also

$$\langle p^2\rangle = \int_{-\infty}^{+\infty}\psi_n(-\hbar^2)\frac{\partial^2\psi_n}{\partial x^2}dy = \int_{-\infty}^{+\infty}(-\hbar^2)\psi_n\frac{\partial^2\psi_n}{\partial(\alpha^2 x^2)}\alpha^2 dy$$

$$= \hbar^2\alpha^2\left[-\int_{-\infty}^{+\infty}\psi_n\psi''_n\,dy\right] = \left(n+\tfrac{1}{2}\right)\hbar^2\cdot\frac{m\omega}{\hbar}$$

$$=\left(n+\tfrac{1}{2}\right)\hbar m\omega \tag{42}$$

Now

$$\Delta x = \left(\langle x^2\rangle\right)^{1/2} = \left[\left(n+\tfrac{1}{2}\right)\frac{\hbar}{m\omega}\right]^{1/2}$$

$$\Delta p = \left(\langle p^2 \rangle\right)^{1/2} = \left[\left(n+\tfrac{1}{2}\right)\hbar m\omega\right]^{1/2}$$

Consequently $\Delta x.\Delta p = \left(n+\tfrac{1}{2}\right)\hbar$. For the ground state ($n = 0$), we obtain $\Delta x.\Delta p = \dfrac{\hbar}{2}$.

Example 5.3: A body of mass 1 gm tied to a spring is stretched by 1 cm by a force of 0.1 newton. If the body can vibrate only along the x-axis, calculate the zero point energy of the system.

Solution: Force constant $\quad k = \dfrac{F}{x} = \dfrac{0.1}{(1\times 10^{-2})} = 10$ N/m

$$m = 1 \text{ gm} = 10^{-3} \text{ kg}$$

$\therefore \quad$ Angular frequency $\quad \omega = \sqrt{\dfrac{k}{m}} = 100$ rad/s.

Zero point energy $E_0 = \dfrac{1}{2}\hbar\omega = \dfrac{1}{2} \times 1.05 \times 10^{-34} \times 100 = 5.25 \times 10^{-33}$ J

Example 5.4: The angular frequency of a one dimensional harmonic oscillator is 5×10^{14} rad/s. Calculate the zero point energy. If the simple harmonic oscillator is an electron, what will be the classical limits of its motion in the state $n = 0$?

Solution: Zero point energy $E_0 = \dfrac{1}{2}\hbar\omega = \dfrac{1}{2} \times 1.05 \times 10^{-34} \times 5 \times 10^{14}$

$$= 2.625 \times 10^{-20} \text{ J}$$

$$= \dfrac{2.625 \times 10^{-20}}{1.6 \times 10^{-19}} \text{ eV} = 0.164 \text{ eV}$$

Energy of classical oscillator $= \dfrac{1}{2}mA^2\omega^2$

Energy of oscillator in the state $n = 0$ is $E_0 = \dfrac{1}{2}\hbar\omega$

$\therefore \quad \dfrac{1}{2}mA^2\omega^2 = \dfrac{1}{2}\hbar\omega$ or $A = \sqrt{\dfrac{\hbar}{m\omega}} = \sqrt{\dfrac{1.05 \times 10^{-34}}{9.1 \times 10^{-31} \times 5 \times 10^{14}}} = 4.6 \times 10^{-10}$ m

Example 5.5: The vibrational energy gap for hydrogen molecule is 0.5 eV. Find the vibrational energy gap for deuterium.

Solution: $\Delta E = E_{n+1} - E_n$

$$\left\{(n+1)+\dfrac{1}{2}\right\}\hbar\omega - \left(n+\dfrac{1}{2}\right)\hbar\omega = \hbar\omega$$

$\omega = \sqrt{k/\mu}$ where reduced mass $\mu = \dfrac{m_1 m_2}{m_1 + m_2}$

For the hydrogen molecule, $\quad \mu_{H_2} = \dfrac{m_H \times m_H}{m_H + m_H} = \dfrac{m_H}{2}$

For the deuterium molecule, $\mu_{D_2} = \dfrac{m_D \times m_D}{m_D + m_D} = \dfrac{2m_H \times 2m_H}{2m_H + 2m_H} = m_H$

$\therefore$ $\dfrac{(\Delta E)_{D_2}}{(\Delta E)_{H_2}} = \dfrac{(\hbar\omega)_{D_2}}{(\hbar\omega)_{H_2}} = \sqrt{\dfrac{\mu_{H_2}}{\mu_{D_2}}} = \sqrt{\dfrac{m_H/2}{m_H}} = \dfrac{1}{\sqrt{2}}$

or $\qquad (\Delta E)_{D_2} = \dfrac{(\Delta E)_{H_2}}{\sqrt{2}} = \dfrac{0.5 \, eV}{\sqrt{2}}$

Example 5.6: The energy gap of two consecutive vibrational energy levels for HCl is 0.36 eV. (i) Calculate the wavelength of radiation emitted in the vibration spectra. (ii) If atomic weights of H and Cl are respectively 1 and 35, calculate the force constant of the molecule. $\hbar = 1.05 \times 10^{-34} J$ - s

Solution: $\Delta E = \hbar\omega = 0.36 eV = 0.36 \times 1.6 \times 10^{-19} J = 0.576 \times 10^{-19} J$

(*i*) Wavelength $\qquad \lambda = \dfrac{c}{\nu} = \dfrac{2\pi c \hbar}{\Delta E}$

$$= \dfrac{2 \times 3.14 \times 3 \times 10^8 \times 1.05 \times 10^{-34}}{0.576 \times 10^{-19}} = 3.43 \times 10^{-6} m$$

(*ii*) Since $\qquad \omega = \sqrt{k/\mu}$

$\therefore$ Force constant $k = \mu\omega^2 = \mu\left(\dfrac{\Delta E}{\hbar}\right)^2$

Here $\mu = \dfrac{m_H m_{Cl}}{m_H + m_{Cl}} = \dfrac{1 \times 35}{1 + 35} = \dfrac{35}{36} a.m.u. = \dfrac{35}{36} \times 1.6 \times 10^{-27} kg$

$= 1.556 \times 10^{-27} kg$

$\therefore$ $\qquad k = 1.556 \times 10^{-27} \times \left(\dfrac{0.576 \times 10^{-19}}{1.05 \times 10^{-34}}\right)^2 = 468.2$ N/m.

Example 5.7: Find the expectation value of energy when the state of the harmonic oscillator is described by the following wave function

$$\psi(x, t) = \dfrac{1}{\sqrt{2}}\left[\psi_0(x, t) + \psi_1(x, t)\right]$$

where $\psi_0(x, t)$ and $\psi_1(x, t)$ are the wavefunctions for the ground state and the first excited state respectively.

Solution: We know that $\langle E \rangle = \int \psi^*(x,t) \, E \, \psi(x,t) dx$

Here $\qquad \psi(x,t) = \dfrac{1}{\sqrt{2}}\left[\psi_0(x,t) + \psi_1(x,t)\right]$

$$\psi^*(x,t) = \dfrac{1}{\sqrt{2}}\left[\psi_0^*(x,t) + \psi_1^*(x,t)\right]$$

Now,
$$\psi_0(x,t) = \left(\frac{\alpha}{\sqrt{\pi}}\right)^{1/2} e^{-\alpha^2 x^2/2}$$

$$\psi_1(x,t) = \left(\frac{\alpha}{2\sqrt{\pi}}\right)^{1/2} e^{-\alpha^2 x^2/2} 2\alpha x$$

and
$$\psi_0^*(x,t) = \psi_0(x,t)$$

$$\psi_1^*(x,t) = \psi_1(x,t)$$

Therefore

$$\psi^*(x,t)\psi(x,t) = \frac{1}{\sqrt{2}}\left[\psi_0^*(x,t) + \psi_1^*(x,t)\right] \times \frac{1}{\sqrt{2}}\left[\psi_0(x,t) + \psi_1(x,t)\right]$$

$$= \frac{1}{2}\left[\psi_0(x,t) + \psi_1(x,t)\right]^2$$

$$= \frac{1}{2}\left[\psi_0^2(x,t) + \psi_1^2(x,t) + 2\psi_0(x,t)\psi_1(x,t)\right]$$

Consequently, expectation value of energy is

$$\langle E \rangle = \frac{1}{2}\left[\int_{-\infty}^{+\infty}\psi_0^2(x,t)E_0 dx + \int_{-\infty}^{+\infty}\psi_1^2(x,t)E_1 dx + 2\int_{-\infty}^{+\infty}\psi_0(x,t)\psi_1(x,t)E\,dx\right]$$

Now
$$E_0 = \tfrac{1}{2}\hbar\omega,\ E_1 = \tfrac{3}{2}\hbar\omega \text{ and } \int_{-\infty}^{+\infty}\psi_0(x,t)\psi_1(x,t)\,dx = 0$$

$$\therefore \quad \langle E \rangle = \frac{1}{4}\hbar\omega\int_{-\infty}^{+\infty}\frac{\alpha}{\sqrt{\pi}}e^{-\alpha^2 x^2}dx + \frac{3}{4}\hbar\omega\int_{-\infty}^{+\infty}\frac{\alpha}{2\sqrt{\pi}}e^{-\alpha^2 x^2}4\alpha^2 x^2 dx$$

$$= \frac{1}{4}\hbar\omega\frac{\alpha}{\sqrt{\pi}}\cdot\frac{\sqrt{\pi}}{\alpha} + \frac{3}{4}\hbar\omega\frac{\alpha}{2\sqrt{\pi}}4\alpha^2\frac{1}{2}\frac{\sqrt{\pi}}{\alpha^3} = \frac{1}{4}\hbar\omega + \frac{3}{4}\hbar\omega = \hbar\omega$$

5.8 Parity of the Oscillator Wave Functions

The harmonic oscillator wave functions are given by
$$\psi_n(x) = C\,H_n(\alpha x)\,e^{-\alpha^2 x^2/2}$$

Since $e^{-\alpha^2 x^2/2}$ is an even function of x (it does not change sign for the transformation $x \to -x$), therefore function $\psi_n(x)$ has a parity dependent on the Hermite polynomial H_n, which is of degree n in $y\,(=\alpha x)$. Since $H_n(y)$ is wholly even or odd according as n is even or odd, therefore $\psi_n(x)$ will have the parity of n i.e. the oscillator wave function will be an even (odd) function if n is an even (odd) function.

Hence,

$$\psi_n(x)\big|_{x\to(-x)} = \psi_n(-x) = \psi_n(x) \text{ for even } n \tag{43}$$

and
$$\psi_n(x)\big|_{x\to(-x)} = \psi_n(-x) = -\psi_n(x) \text{ for odd } n \tag{44}$$

The operation $x \rightarrow -x$ is materialized by defining a parity operator P accordingly as

$$P\ \psi(x) = \psi(-x)$$
$$\Rightarrow \quad\quad P^2\ \psi(x) = P\ \psi(-x) = \psi(x)$$
$$\therefore \quad\quad\quad\quad P = \pm 1$$
or $\quad\quad\quad\quad\quad \psi(x) = \pm\psi(-x)$ \hfill (45)

The change of sign under a parity transformation does not affect the Hamiltonian and the states $\psi(x)$ and $P\psi\ (x)$ are degenerate. It is also to be noted that plot of the wavefunctions ψ_n for n even are symmetric for even values of n and antisymmetric for odd values of n. We will discuss further about parity in chapter 14.

Example 5.8: Determine the probability distribution of the various values of the momentum for an oscillator.

Solution: The Schrödinger equation for a harmonic oscillator is

$$H\psi = E\psi$$

where $\quad\quad\quad\quad H = \dfrac{p^2}{2m} + \dfrac{1}{2}m\omega^2 x^2$ \hfill (i)

or $\quad\quad\quad\quad -\dfrac{\hbar^2}{2m}\dfrac{\partial^2\psi}{\partial x^2} + \dfrac{1}{2}m\omega^2 x^2\psi = E\psi$

or $\quad\quad\quad\quad \dfrac{\partial^2\psi}{\partial x^2} + \dfrac{2m}{\hbar^2}\left(E - \dfrac{1}{2}m\omega^2 x^2\right)\psi = 0$ \hfill (ii)

Its solutions are

$$\psi_n(x) = \left(\frac{\alpha}{2^n n!\sqrt{\pi}}\right)^{1/2} e^{-\alpha^2 x^2/2} H_n(\alpha x)$$

Since $\quad\quad\quad\quad \alpha = \sqrt{\dfrac{m\omega}{\hbar}}$

Therefore

$$\psi_n(x) = \left(\frac{m\omega}{\pi\hbar}\right)^{1/4} \frac{1}{2^{n/2}\sqrt{n!}} e^{-m\omega x^2/2\hbar} \times H_n(x\sqrt{m\omega/\hbar})$$

and $\quad\quad\quad |\psi_n(x)|^2 = \sqrt{\dfrac{m\omega}{\pi\hbar}}\dfrac{1}{2^n n!}e^{-m\omega x^2/\hbar}H_n^2(x\sqrt{m\omega/\hbar})$

In an analogous way, we may write the Schrödinger's equation in the momentum representation. Using the coordinate operator $\hat{x} = i\hbar\ \partial/\partial p$, we obtain the Hamiltonian in the momentum representation

$$\hat{H} = \frac{p^2}{2m} - \frac{1}{2}m\omega^2\hbar^2\frac{\partial^2}{\partial p^2}$$

The corresponding Schrödinger's equation $\hat{H}\phi(p) = E\phi(p)$ for the wave function $\phi(p)$ in the momentum representation is

$$\frac{p^2}{2m}\phi(p) - \frac{1}{2}m\omega^2\hbar^2\frac{\partial^2\phi(p)}{\partial p^2} = E\phi(p)$$

or

$$\frac{\partial^2\phi(p)}{\partial p^2} + \frac{2}{m\omega^2\hbar^2}(E - \frac{p^2}{2m})\phi(p) = 0 \qquad (iii)$$

This equation is of the same form as Eqn. (ii). In analogy with coordinate space, the solutions of Eqn. (iii) are

$$\phi_n(p) = \left(\frac{\alpha'}{2^n n!\sqrt{\pi}}\right)^{1/2} e^{-\alpha'^2 p^2/2} H_n(\alpha'p)$$

Since

$$\alpha' = \frac{1}{\sqrt{m\omega\hbar}}$$

Hence the required probability distribution is

$$|\phi_n(p)|^2 dp = \frac{\alpha'}{2^n n!\sqrt{\pi}} e^{-\alpha'^2 p^2} H_n^2(\alpha'p) \, dp$$

$$\frac{1}{2^n n!\sqrt{\pi m\omega\hbar}} e^{-p^2/m\omega\hbar} H_n^2\left(\frac{p}{\sqrt{m\omega\hbar}}\right) dp$$

Example 5.9: Consider a particle of mass m subject to a one dimensional potential of the form

$$V = \frac{1}{2}m\omega^2 x^2 \quad \text{for } x > 0$$

$$= \infty \text{ for } x < 0$$

What is the zero point energy. Find the energy levels.

Solution: For $x > 0$ (region I), we have the harmonic oscillator potential. The solutions are

$$\psi_n = N_n e^{-\frac{1}{2}\alpha x^2} H_n\left(\sqrt{\alpha}\, x\right), \quad n = 0, 1, 2, \ldots$$

For region II, $(V = \infty)$, $\psi \equiv 0$ for $x < 0$. The two solutions must match at $x = 0$. Only $\psi_n(x)$ for n odd gives $\psi_n(x) = 0$; since $e^{-\frac{1}{2}\alpha x^2}$ is an even function of x, so ψ_n is zero only if $H_n\left(\sqrt{\alpha}\, x\right) = 0$ (at $x = 0$) which is possible only if $n = $ odd $\equiv 2m + 1$, $(m = 0, 1, 2, \ldots)$.

Hence the solution is

$$\psi_n(x) = N_n e^{-\frac{1}{2}\alpha x^2} H_n(\sqrt{\alpha}x)$$

where $n = 2m + 1$, $m = 0, 1, 2\ldots$

The energy levels are

$$E_n = \hbar\omega\left(n + \frac{1}{2}\right)$$

$$= \hbar\omega\left(2m+1+\frac{1}{2}\right), \; m = 1, 2, 3, \ldots$$

So the zero point energy is $\hbar\omega\left(1+\frac{1}{2}\right) = \frac{3}{2}\hbar\omega$.

Example 5.10: Show that for a simple harmonic oscillator

$$\int \psi_n^*(x) x \psi_m(x) dx = \sqrt{\frac{\hbar}{2m\omega}}\left\{\sqrt{m+1}\, \delta_{n,m+1} + \sqrt{m}\, \delta_{n,m-1}\right\}$$

Solution: We know that

$$\psi_n(x) = \left(\frac{\alpha}{2^n n! \sqrt{\pi}}\right)^{1/2} e^{-\alpha^2 x^2/2} H_n(\alpha x)$$

where $\alpha = \sqrt{m\omega/\hbar}$ and $H_n(\alpha x)$ is the Hermite polynomial of degree n.

We have the following recurrence relation for H_n's:

$$H_{n+1}(\xi) = 2\xi H_n(\xi) - 2n\, H_{n-1}(\xi)$$

We can write $\int \psi_n^*(x) x \psi_m(x) dx$

$$= \frac{\alpha}{\sqrt{2^{n+m} n! m! \pi}} \int e^{-\alpha^2 x^2} H_n(\alpha x) \frac{\alpha x}{\alpha} H_m(\alpha x) dx$$

$$= \frac{1}{\sqrt{2^{n+m} n! m! \pi}} \int e^{-\alpha^2 x^2} H_n(\alpha x) \times \left[\frac{1}{2}H_{m+1}(\alpha x) + m H_{m-1}(\alpha x)\right] dx$$

(using the recurrence relation)

$$= \frac{1}{2} \times \frac{1}{\sqrt{2^{n+m} n! m! \pi}} \int e^{-\alpha^2 x^2/2} H_n(\alpha x) e^{-\alpha^2 x^2/2} H_{m+1}(\alpha x) dx$$

$$+ \frac{m}{\sqrt{2^{n+m} n! m! \pi}} \int e^{-\alpha^2 x^2/2} H_n(\alpha x) H_{m-1}(\alpha x) e^{-\alpha^2 x^2/2} dx$$

$$= \frac{\sqrt{m+1}}{\sqrt{2} \times \alpha} \int \left\{\sqrt{\frac{\alpha}{2^n n! \sqrt{\pi}}} e^{-\alpha^2 x^2/2} H_n(\alpha x)\right\}$$

$$\times \left\{\sqrt{\frac{\alpha}{2^{m+1}(m+1)! \sqrt{\pi}}} e^{-\alpha^2 x^2/2} H_{m+1}(\alpha x)\right\} dx$$

$$+ \sqrt{\frac{m}{2}} \frac{1}{\alpha} \int \left\{\sqrt{\frac{\alpha}{2^n n! \sqrt{\pi}}} e^{-\alpha^2 x^2/2} H_n(\alpha x)\right\}$$

$$\times \left\{\sqrt{\frac{\alpha}{2^{m-1}(m-1)! \sqrt{\pi}}} e^{-\alpha^2 x^2/2} H_{m-1}(\alpha x)\right\} dx$$

($\because m! = m(m-1)!$ in the 2nd term)

$$= \sqrt{\frac{(m+1)}{2}} \frac{1}{\alpha} \int \psi_n(x)\psi_{m+1}(x)dx + \sqrt{\frac{m}{2}} \frac{1}{\alpha} \int \psi_n(x)\psi_{m-1}(x)dx$$

$$= \frac{1}{\sqrt{2}\alpha} \left[\sqrt{(m+1)}\ \delta_{n,\,m+1} + \sqrt{m}\ \delta_{n,\,m-1} \right]$$

$$= \sqrt{\frac{\hbar}{2m\omega}} \left[\sqrt{(m+1)}\ \delta_{n,\,m+1} + \sqrt{m}\ \delta_{n,\,m-1} \right]$$

Example 5.11: Determine the energy levels and eigen functions for a three dimensional harmonic oscillator in a potential field $V = \frac{1}{2}\mu\omega^2 r^2$.

Solution:
$$V = \frac{1}{2}\mu\omega^2 r^2 \equiv \frac{1}{2}(k_1 x^2 + k_2 y^2 + k_3 z^2)$$

$$\equiv \frac{1}{2}k r^2$$

where, by symmetry (for isotropic oscillator) $k_1 = k_2 = k_3 \equiv k$ and $x^2 + y^2 + z^2 \equiv r^2$

Further
$$\omega_i = \sqrt{\frac{k_i}{\mu}} \quad (i = x, y, z)$$

In analogy with the one-dimensional harmonic oscillator, the Shrödinger equation for the system is

$$-\frac{\hbar^2}{2\mu}\nabla^2\psi(\vec{r}) + V\ \psi(\vec{r}) = E\psi(\vec{r})$$

By the method of separation of variables

$$\psi(\vec{r}) = X(x)Y(y)Z(z)$$

This implies

$$\frac{d^2 X}{dx^2} + \frac{2\mu}{\hbar^2}\left(E_x - \frac{1}{2}\mu\omega_x^2 x^2 \right)X = 0$$

$$\frac{d^2 Y}{dy^2} + \frac{2\mu}{\hbar^2}\left(E_y - \frac{1}{2}\mu\omega_y^2 y^2 \right)Y = 0$$

$$\frac{d^2 Z}{dz^2} + \frac{2\mu}{\hbar^2}\left(E_z - \frac{1}{2}\mu\omega_z^2 z^2 \right)Z = 0$$

(and $E_x + E_y + E_z = E$).

Each of these three equations resembles the Schrödinger equation for a one dimensional harmonic oscillator, hence

$$E_x = (n_x + 1/2)\hbar\omega_x \quad n_x = 0,1,2...$$
$$E_y = (n_y + 1/2)\hbar\omega_y \quad n_y = 0,1,2...$$
$$E_z = (n_z + 1/2)\hbar\omega_z \quad n_z = 0,1,2...$$

$$E = \left[(n_x + n_y + n_z) + \frac{3}{2} \right]\hbar\omega \equiv \left(n + \frac{3}{2} \right)\hbar\omega$$

(if the oscillator is isotropic, $\omega_x = \omega_y = \omega_z = \omega$ and $n = n_x + n_y + n_z$)

The eigen functions are given by

$$\psi_{n_x n_y n_z} = N \, H_{nx}(\alpha_1 x) H_{ny}(\alpha_2 y) H_{nz}(\alpha_3 z)$$

$$\times \exp\left[-\frac{1}{2}\left(\alpha_1^2 x^2 + \alpha_2^2 y^2 + \alpha_3^2 z^2 \right) \right]$$

where

$$\alpha_1 = \left(\frac{m\omega_x}{\hbar} \right)^{1/2} ; \quad \alpha_2 = \left(\frac{m\omega_y}{\hbar} \right)^{1/2} ; \quad \alpha_3 = \left(\frac{m\omega_z}{\hbar} \right)^{1/2}$$

and N is the normalization constant given by

$$N = \frac{\alpha_1^{1/2} \alpha_2^{1/2} \alpha_3^{1/2}}{\pi^{3/4} \left(2^{n_x + n_y + n_z} \, n_x! \, n_y! \, n_z! \right)^{1/2}}$$

Example 5.12: In case of a three dimensional harmonic oscillator, if the oscillator is isotropic, determine the degree of degeneracy.

Solution: For an isotropic oscillator, the eigenvalues are given by

$$E_n = \left(n + \frac{3}{2} \right) \hbar\omega, \quad n = n_x + n_y + n_z$$

and the levels are degenerate.

Now, assuming n_z as fixed, the sum $(n_x + n_y)$ takes the definite value $n - n_z$. When n_x is given values from 0 to $n - n_z$, we obtain $n - n_z + 1$ values of the sum, which completely determine n_x and n_y. Hence, for each n_z value, there will be $n - n_z + 1$ distinct values for n, and thus for E_n. Summing over all the allowed n_z values, from 0 to n, we get

$$\sum_{n_z=0}^{m} \left(n - n_z + 1 \right) = \sum_{n_z=0}^{n} (n+1) - \sum_{n_z=0}^{n} n_z$$

$$= (n+1)^2 - \frac{n(n+1)}{2} = \frac{1}{2}(n+1)(n+2)$$

which is the degree of degeneracy of E_n.

Example 5.13: Find the wave functions of the states of a linear oscillator that minimize the uncertainty relation i.e. in which the standard deviations of the coordinate and momentum in the wave packet are related by $\delta p \cdot \delta x = \hbar/2$.

Solution: The required wave functions must have the form

$$\psi(x,t) = \frac{1}{(2\pi)^4 (\delta x)^{1/2}} \exp\left\{ \frac{i\bar{p}x}{\hbar} - \frac{(x - \bar{x})^2}{4(\delta x)^2} - i\phi(t) \right\} \tag{i}$$

For linear oscillator

$$\dot{\bar{p}} = -m\omega^2 \bar{x} \tag{ii}$$

and

$$\bar{p} = \bar{p}(t) = m\dot{\bar{x}}(t) \tag{iii}$$

The unknowns (δx) and $\phi(t)$ in Eqn. (i) are found by substituting (i) in the Schrödinger wave equation:

$$-\frac{\hbar^2}{2m}\frac{\partial^2 \psi}{\partial x^2} + \frac{1}{2}m\omega^2 x^2 \psi = i\hbar\frac{\partial \psi}{\partial t}$$

or

$$-\frac{1}{2}\frac{\partial^2 \psi}{\partial x^2} + \frac{m^2\omega^2 x^2}{2\hbar^2}\psi - \frac{im}{\hbar}\frac{\partial \psi}{\partial t} = 0 \qquad (iv)$$

We rewrite (i) as

$$\psi(x,t) = A\exp\left\{\frac{i\overline{p}x}{\hbar} - \frac{(x-\overline{x})^2}{4(\delta x)^2}\right\}\exp(-i\phi(t)) \qquad (v)$$

So,

$$\frac{\partial \psi}{\partial x} = \left(\frac{i\overline{p}}{\hbar} - \frac{2(x-\overline{x})}{4(\delta x)^2}\right)\psi$$

and

$$\frac{\partial^2 \psi}{\partial x^2} = \left(0 - \frac{2}{4(\delta x)^2}\right)\psi + \left(\frac{i\overline{p}}{\hbar} - \frac{2(x-\overline{x})}{4(\delta x)^2}\right)\frac{\partial \psi}{\partial x}$$

$$= -\frac{2\psi}{4(\delta x)^2} + \left(\frac{i\overline{p}}{\hbar} - \frac{(x-\overline{x})}{2(\delta x)^2}\right)^2\psi$$

$$= \left[-\frac{2}{4(\delta x)^2} + \frac{i^2\overline{p}^2}{\hbar^2} + \frac{1}{4}\frac{(x-\overline{x})^2}{(\delta x)^4}\right]\psi$$

$$\therefore \quad -\frac{1}{2}\frac{\partial^2 \psi}{\partial x^2} = \left[\frac{1}{4(\delta x)^2} + \frac{1}{2}\frac{m^2\dot{\overline{x}}^2}{\hbar^2} - \frac{1}{8(\delta x)^4}\left(x^2 + \overline{x}^2 - 2x\overline{x}\right)\right]\psi$$

$$= \left[\frac{1}{4(\delta x)^2} + \frac{1}{2}\frac{m^2\dot{\overline{x}}^2}{\hbar^2} - \frac{x^2}{8(\delta x)^4} - \frac{\overline{x}^2}{8(\delta x)^4} + \frac{x\overline{x}}{4(\delta x)^4}\right]\psi \qquad (vi)$$

$$\frac{\partial \psi}{\partial t} = \left(\frac{i\overline{p}}{\hbar}\left(\frac{\partial \overline{x}}{\partial t}\right) + \frac{i\dot{\overline{p}}x}{\hbar} - i\dot{\phi}\right)\psi$$

$$\therefore \quad -\frac{im}{\hbar}\frac{\partial \psi}{\partial t} = \left[\frac{m(-m\omega^2\overline{x})x}{\hbar^2} - \frac{m}{\hbar}\dot{\phi}(t)\right]\psi \qquad (vii)$$

from (iv), (vi) and (vii)

$$\frac{1}{2}\frac{m^2\omega^2 x^2}{\hbar^2} + \frac{1}{4(\delta x)^2} + \frac{1}{2}\frac{m^2\dot{\overline{x}}^2}{\hbar^2} - \frac{1}{2}\times\frac{x^2}{4(\delta x)^4} - \frac{\overline{x}^2}{8(\delta x)^4} + \frac{x\overline{x}}{4(\delta x)^4}$$

$$-x\overline{x}\frac{m^2\omega^2}{\hbar^2} - \frac{m}{\hbar}\dot{\phi}(t) = 0$$

or $\quad \left(\dfrac{1}{2}x^2 - x\bar{x}\right)\left(\dfrac{m^2\omega^2}{\hbar^2} - \dfrac{1}{4(\delta x)^4}\right) + \left[\dfrac{m^2\dot{\bar{x}}^2}{2\hbar^2} - \dfrac{\bar{x}^2}{8(\delta x)^4} + \dfrac{1}{4(\delta x)^2} - \dfrac{m}{\hbar}\dot{\phi}\right] = 0$

Both of these two terms must be separately equal to zero. Therefore

$$\dfrac{m^2\omega^2}{\hbar^2} = \dfrac{1}{4(\delta x)^4}$$

or $\quad 2(\delta x)^2 = \dfrac{\hbar}{m\omega} \text{ or } (\delta x)^2 = \dfrac{\hbar}{2m\omega}$

and $\quad \dfrac{m^2\dot{\bar{x}}^2}{2\hbar^2} - \dfrac{\bar{x}^2}{8(\delta x)^4} + \dfrac{1}{4(\delta x)^2} - \dfrac{m}{\hbar}\dot{\phi} = 0$

or $\quad \dfrac{m^2\dot{\bar{x}}^2}{2\hbar^2} - \dfrac{\bar{x}^2 m^2\omega^2}{2\hbar^2} + \dfrac{m\omega}{2\hbar} = \dfrac{m}{\hbar}\dot{\phi}(t)$

or $\quad \dfrac{\partial\phi}{\partial t} = \dfrac{m}{2\hbar}\left(\dot{\bar{x}}^2 - \bar{x}^2\omega^2\right) + \dfrac{\omega}{2}$

$\qquad = \dfrac{1}{2\hbar}\left(m\dot{\bar{x}}\,\dot{\bar{x}} - m\omega^2\bar{x}^2\right) + \dfrac{\omega}{2}$

$\qquad = \dfrac{1}{2\hbar}\left(\bar{p}\,\dot{\bar{x}} - m\omega^2\bar{x}^2\right) + \dfrac{\omega}{2}$

$\qquad = \dfrac{1}{2\hbar}\dfrac{\partial}{\partial t}\left(\bar{p}\,\bar{x}\right) + \dfrac{\omega}{2}$

Hence $\phi(t) = \dfrac{1}{2\hbar}\bar{p}\,\bar{x} + \dfrac{1}{2}\omega t$; Consequently

$$\psi(x,t) = \left(\dfrac{m\omega}{\pi\hbar}\right)^{1/4} \exp\left\{\dfrac{i\bar{p}x}{\hbar} - \dfrac{m\omega(x-\bar{x})^2}{2\hbar}\right\}\exp\left\{-\dfrac{1}{2}i\omega t - \dfrac{i\bar{p}\,\bar{x}}{2\hbar}\right\}$$

Questions and Problems

1. Establish Schrödinger equation for a linear harmonic oscillator and solve it to obtain its eigen values and eigen functions. Discuss the significance of zero point energy.
2. A particle of mass m is confined to a harmonic oscillator potential given by $V = \frac{1}{2}\omega^2 mx^2$ where $\omega^2 = k/m$ (k is force constant). The particle is in a state described by the wave function

$$\psi = N \exp\left(-\dfrac{\omega mx^2}{2\hbar}\right) e^{-i\omega t/2}$$

Verify that ψ satisfies Schrödinger's equation.

3. Normalize the ground state wave function for the one dimensional oscillator so that the total probability of finding the particle is unity

$$\left[\textbf{Ans. }\psi_0 = \left(\frac{\omega m}{\pi\hbar}\right)^{1/4} e^{-\omega mx^2/2\hbar}\right]$$

4. Find the expectation value of x^2 for the ground state of the harmonic oscillator.

$$\left[\textbf{Ans.} \langle x^2 \rangle = \frac{1}{2}\frac{\hbar}{m\omega}\right]$$

5. The state of an oscillator of angular frequency ω is represented by

$$\psi(x) = \exp\left(-\frac{m\omega x^2}{2\hbar}\right)$$

   Find the probability that the magnitude of the momentum is larger than $(m\hbar\omega)^{1/2}$.

   [**Hint:** $\phi(p) = \frac{1}{(2\pi\hbar)^{1/2}} \int\limits_{-\infty}^{+\infty} \psi(x)e^{-ipx/\hbar}dx$ and desired probability

$$P = \int\limits_{(m\hbar\omega)^{1/2}}^{\infty} |\phi(p)|^2 dp \ ]$$

6. Find the expectation value of p_x and p_x^2 for the ground state of the one dimensional harmonic oscillator.

$$\left[\textbf{Ans.} \langle p_x \rangle = 0; \ \langle p_x^2 \rangle = \frac{\omega m\hbar}{2}\right]$$

7. Show that the average value of the kinetic energy of the linear harmonic oscillator is equal to one half of the total energy of the oscillator when it is in the state of definite energy.

8. Use the generating function for the Hermite polynomials to evaluate

$$\int\limits_{-\infty}^{+\infty} u_n^*(x)x^2 u_m(x) \ dx$$

   where the u's are normalized harmonic oscillator wave functions.

9. Show that corresponding to oscillator wave function

$$\psi_n(x) = \left(\frac{1}{2^n n!\pi^{1/2}}\right)^{1/2} \cdot e^{-x^2/2} \cdot H_n(x),$$

   the wave function in the momentum space is

$$\phi_n(p) = \frac{1}{i^n (2^n n!\pi^{1/2})^{1/2}} e^{-p^2/2} \cdot H_n(p)$$

   H_n being Hermite polynomials.
   [**Hint:** Use the generating function to calculate the integral.]

10. Evaluate $\langle xp + px \rangle$ explicitly for any stationary state of the harmonic oscillator.

11. A harmonic oscillator is in the ground state
     (*i*) where is the probability density maximum?
     (*ii*) What is the value of maximum probability density?

$$\left[\textbf{Ans. } (i) \; x = 0, \; (ii) \; \left(\frac{m\omega}{\pi\hbar}\right)^{1/2}\right]$$

12. Write the Schrödinger equation for a harmonic oscillator in momentum representation and obtain its solution.

13. Show that in cylindrical coordinates

$$\nabla^2 = \frac{1}{r}\frac{\partial}{\partial r}\left(r\frac{\partial}{\partial r}\right) + \frac{1}{r^2}\frac{\partial^2}{\partial\phi^2} + \frac{\partial^2}{\partial z^2}$$

6

Solution of Schrödinger Equation for Spherically Symmetric Problems

6.1 The Rigid Rotator

If a system consists of two mass particles situated at a fixed distance and can rotate about an axis passing through their centre of mass and perpendicular to the line joining them, then it is called a *rigid rotator*.

If the rotation of two constituent particles of a rotator takes place only in a plane then the axis of rotation is fixed and it is then called a *rotator with a fixed axis*. However, if its axis of rotation can change, then it is called *rotator with a free axis*. For instance, in a rigid diatomic molecule, both atoms of the molecule can rotate about an axis passing through their centre of mass and perpendicular to line joining them; it is a rotator with a fixed axis.

Let m_1 and m_2 be two mass particles separated by a distance r and C be their centre of mass distant r_1 from m_1 and r_2 from m_2 so that

$$r_1 + r_2 = r$$

and $$m_1 r_1 = m_2 r_2 \qquad (1)$$

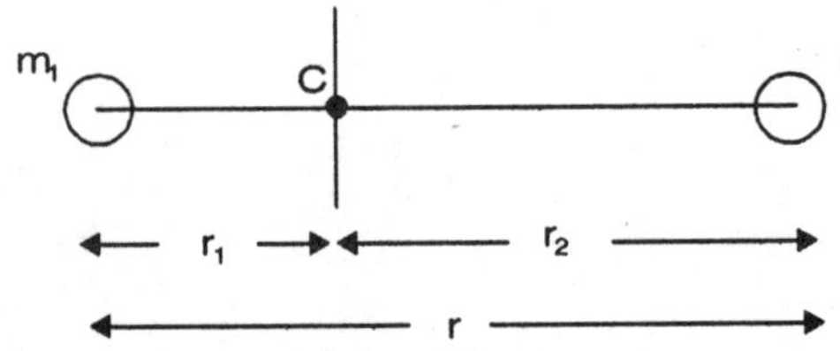

Fig. 6.1: A rigid rotator

$$\Rightarrow \qquad r_1 = \frac{m_2 r}{m_1 + m_2}, \; r_2 = \frac{m_1 r}{m_1 + m_2} \qquad (2)$$

The moment of inertia of the system about an axis passing through C and perpendicular to the line joining m_1 and m_2 is:

$$I = m_1 r_1^2 + m_2 r_2^2 = m_1 \left(\frac{m_2 r}{m_1 + m_2} \right)^2 + m_2 \left(\frac{m_1 r}{m_1 + m_2} \right)^2$$

or
$$I = \frac{m_1 m_2}{m_1 + m_2} r^2 = \mu r^2 \tag{3}$$

where $\mu \equiv \dfrac{m_1 m_2}{m_1 + m_2}$ represents the reduced mass of the system. If this rotator is rotating with an angular velocity ω, the angular momentum is $J = I\omega$ and the rotational kinetic energy is $E = \frac{1}{2} I \omega^2$. Since the distance r is fixed, the potential energy can be assumed to be zero and

$$\text{total energy} = E = \frac{1}{2} I \omega^2$$

$$= \frac{1}{2} I \left(\frac{J}{I} \right)^2 = \frac{J^2}{2I}$$

or
$$J = \sqrt{2EI} \tag{4}$$

Case I: Eigen functions for the Free Axis rotator

The time independent Schrödinger equation is

$$\nabla^2 \psi + \frac{2\mu}{\hbar^2} (E - V)\psi = 0 \tag{5}$$

But, we have $V = 0$, so,

$$\nabla^2 \psi + \frac{2\mu E}{\hbar^2} \psi = 0 \tag{6}$$

In terms of spherical polar coordinates, this equation becomes

$$\frac{1}{r^2} \frac{\partial}{\partial r}\left(r^2 \frac{\partial \psi}{\partial r} \right) + \frac{1}{r^2 \sin\theta} \frac{\partial}{\partial \theta}\left(\sin\theta \frac{\partial \psi}{\partial \theta} \right) + \frac{1}{r^2 \sin^2\theta} \frac{\partial^2 \psi}{\partial \phi^2} + \frac{2\mu E}{\hbar^2} \psi = 0 \tag{7}$$

But since r is fixed, ψ will not vary with r and will depend only on θ and ϕ:

$$\frac{1}{\sin\theta} \frac{\partial}{\partial \theta}\left(\sin\theta \frac{\partial \psi}{\partial \theta} \right) + \frac{1}{\sin^2\theta} \frac{\partial^2 \psi}{\partial \phi^2} + \frac{2\mu r^2}{\hbar^2} E\psi = 0$$

or $\quad \dfrac{1}{\sin\theta} \dfrac{\partial}{\partial \theta}\left(\sin\theta \dfrac{\partial \psi}{\partial \theta} \right) + \dfrac{1}{\sin^2\theta} \dfrac{\partial^2 \psi}{\partial \phi^2} + \dfrac{2I}{\hbar^2} E\psi = 0$ (since $\mu r^2 = I$) (8)

This wave equation can be solved by the method of separation of variables, as follows.

Let

$$\psi = \Theta(\theta)\, \Phi(\phi)$$

where Θ is a function of θ only and Φ is a function of ϕ only. Then,

$$\frac{\partial \psi}{\partial \theta} = \Phi \frac{\partial \Theta}{\partial \theta} \quad \text{and} \quad \frac{\partial \psi}{\partial \phi} = \Theta \frac{\partial \Phi}{\partial \phi}$$

Now,

$$\frac{\partial}{\partial \theta}\left(\sin\theta \frac{\partial \psi}{\partial \theta} \right) = \frac{\partial}{\partial \theta}\left(\sin\theta . \Phi \frac{\partial \Theta}{\partial \theta} \right) = \Phi\left[\sin\theta \frac{\partial^2 \Theta}{\partial \theta^2} + \cos\theta \frac{\partial \Theta}{\partial \theta} \right] \tag{i}$$

and
$$\frac{\partial^2 \psi}{\partial \phi^2} = \Theta \frac{\partial^2 \Phi}{\partial \phi^2} \tag{ii}$$

Substituting the values from (i) and (ii) in Eqn. (8), we get

$$\frac{1}{\sin\theta}\Phi\left[\sin\theta\frac{\partial^2\Theta}{\partial\theta^2}+\cos\theta\frac{\partial\Theta}{\partial\theta}\right]+\frac{1}{\sin^2\theta}\frac{\partial^2\Phi}{\partial\phi^2}+\frac{2IE}{\hbar^2}\Theta\Phi=0 \qquad (9)$$

or

$$\Phi\left[\frac{\partial^2\Theta}{\partial\theta^2}+\frac{\cos\theta}{\sin\theta}\frac{\partial\Theta}{\partial\theta}\right]+\frac{\Theta}{\sin^2\theta}\frac{\partial^2\Phi}{\partial\phi^2}+\frac{2IE}{\hbar^2}\Theta\Phi=0$$

or

$$\Phi\left[\frac{\partial^2\Theta}{\partial\theta^2}+\cot\theta\frac{\partial\Theta}{\partial\theta}\right]+\frac{2IE}{\hbar^2}\Theta\Phi=-\frac{\Theta}{\sin^2\theta}\frac{\partial^2\Phi}{\partial\phi^2}$$

or

$$\left(\sin^2\theta\right)\Phi\left[\frac{\partial^2\Theta}{\partial\theta^2}+\cot\theta\frac{\partial\Theta}{\partial\theta}\right]+\left(\sin^2\theta\right)\frac{2IE}{\hbar^2}=-\Theta\frac{\partial^2\Phi}{\partial\phi^2}$$

Dividing both sides by $\Theta\Phi$, we get,

$$\frac{1}{\Theta}\sin^2\theta\left[\frac{\partial^2\Theta}{\partial\theta^2}+\cot\theta\frac{\partial\Theta}{\partial\theta}\right]+\left(\sin^2\theta\right)\frac{2IE}{\hbar^2}=-\frac{1}{\Phi}\frac{\partial^2\Phi}{\partial\phi^2} \qquad (10)$$

In this expression, the l.h.s. is a function of θ only whereas the r.h.s. is a function of ϕ only. They can be equal only if each side is equal to a constant (say m^2) i.e.

$$\frac{1}{\Theta}\sin^2\theta\left[\frac{\partial^2\Theta}{\partial\theta^2}+\cot\theta\frac{\partial\Theta}{\partial\theta}\right]+\left(\sin^2\theta\right)\frac{2IE}{\hbar^2}=m^2$$

or

$$\left(\frac{\partial^2\Theta}{\partial\theta^2}+\cot\theta\frac{\partial\Theta}{\partial\theta}\right)+\left(\beta-\frac{m^2}{\sin^2\theta}\right)\Theta=0 \qquad (11i)$$

where

$$\beta=\left(\frac{2IE}{\hbar^2}\right)\text{ and }-\frac{1}{\Phi}\frac{\partial^2\Phi}{\partial\phi^2}=m^2$$

or

$$\frac{\partial^2\Phi}{\partial\phi^2}+m^2\Phi=0 \qquad (11ii)$$

Solution of the Φ-equation:

The general solution of (11) (ii) is
$$\Phi=a\,e^{im\phi}$$

where $m=0,\pm1,\pm2,\ldots$

The value of the constant a is obtained by normalization condition. Since ϕ varies from 0 to 2π, so

$$\int_0^{2\pi}\Phi\Phi^*d\phi=1\text{ or }a^2\int_0^{2\pi}e^{im\phi}e^{-im\phi}d\phi=1\text{ or }a^2\times2\pi=1$$

$$\therefore\qquad a=\frac{1}{\sqrt{2\pi}}$$

Hence $$\Phi = \frac{1}{\sqrt{2\pi}} e^{im\phi}, \quad m = 0, \pm 1, \pm 2, \ldots \qquad (12)$$

Solution of the Θ-equation:

Let $$x = \cos\theta \quad \therefore \quad \frac{\partial x}{\partial\theta} = -\sin\theta$$

$$\frac{\partial\Theta}{\partial\theta} = \frac{\partial\Theta}{\partial x} \cdot \frac{\partial x}{\partial\theta} = -\sin\theta\frac{\partial\Theta}{\partial x}$$

$$\Rightarrow \qquad \frac{\partial}{\partial\theta} = -\sin\theta\frac{\partial}{\partial x}$$

$$\Rightarrow \qquad \frac{\partial^2\Theta}{\partial\theta^2} = \frac{\partial}{\partial\theta}\left[-\sin\theta\frac{\partial\Theta}{\partial x}\right] = -\cos\theta\frac{\partial\Theta}{\partial x} - \sin\theta\frac{\partial}{\partial\theta}\left(\frac{\partial\Theta}{\partial x}\right)$$

$$= -x\frac{\partial\Theta}{\partial x} - \sin\theta\left(-\sin\theta\frac{\partial^2\Theta}{\partial x^2}\right)$$

or $$\frac{\partial^2\Theta}{\partial\theta^2} = -x\frac{\partial\Theta}{\partial x} + \sin^2\theta\frac{\partial^2\Theta}{\partial x^2}$$

$$= -x\frac{\partial\Theta}{\partial x} + \left(1 - x^2\right)\frac{\partial^2\Theta}{\partial x^2}$$

So, Eqn. (11) (*i*) becomes

$$\left(1 - x^2\right)\frac{\partial^2\Theta}{\partial x^2} - x\frac{\partial\Theta}{\partial x} + \frac{\cos\theta}{\sin\theta}\left(-\sin\theta\frac{\partial\Theta}{\partial x}\right) + \left(\beta - \frac{m^2}{1 - x^2}\right)\Theta = 0$$

or $$\left(1 - x^2\right)\frac{\partial^2\Theta}{\partial x^2} - 2x\frac{\partial\Theta}{\partial x} + \left(\beta - \frac{m^2}{1 - x^2}\right)\Theta = 0 \qquad (13i)$$

Since $x = \cos\theta$, this equation will have physical significance only for $-1 \le x \le +1$. Let the solution of Eqn. (13*i*), rewritten as

$$\frac{\partial}{\partial x}\left[\left(1 - x^2\right)\frac{\partial\Theta}{\partial x}\right] + \left(\beta - \frac{m^2}{1 - x^2}\right)\Theta = 0 \qquad (13ii)$$

be given by

$$\Theta(\theta) = \left(1 - x^2\right)^{m/2} X(x) \qquad (14)$$

where X is a function of x only.

Then

$$\frac{\partial\Theta}{\partial x} = \frac{m}{2}\left(1 - x^2\right)^{(m/2)-1}(-2x)X + \left(1 - x^2\right)^{m/2}\frac{\partial X}{\partial x}$$

So,

$$\frac{\partial}{\partial x}\left[\left(1 - x^2\right)\frac{\partial\Theta}{\partial x}\right] = \frac{\partial}{\partial x}\left[-mx\left(1 - x^2\right)^{m/2}X + \left(1 - x^2\right)^{(m/2)+1}\frac{\partial X}{\partial x}\right]$$

$$= \left[-m\left(1 - x^2\right)^{m/2}X - mx\left(\frac{m}{2}\right)\left(1 - x^2\right)^{(m/2)-1}(-2x)X - mx\left(1 - x^2\right)^{m/2}\frac{\partial X}{\partial x}\right.$$

$$+\left(\frac{m}{2}+1\right)\left(1-x^2\right)^{(m/2)}(-2x)\frac{\partial X}{\partial x}+\left(1-x^2\right)^{(m/2)+1}\frac{\partial^2 X}{\partial x^2}\Bigg]$$

$$=\left\{-m\left(1-x^2\right)^{m/2}+m^2x^2\left(1-x^2\right)^{(m/2)-1}\right\}X$$

$$-\left\{mx\left(1-x^2\right)^{m/2}+(m+2)(-x)\left(1-x^2\right)^{m/2}\right\}\frac{\partial X}{\partial x}+\left(1-x^2\right)^{(m/2)+1}\frac{\partial^2 X}{\partial x^2}$$

$$=\left\{-m\left(1-x^2\right)^{m/2}+m^2x^2\left(1-x^2\right)^{(m/2)-1}\right\}X$$

$$-\left\{2x(m+1)\left(1-x^2\right)^{m/2}\right\}\frac{\partial X}{\partial x}+\left(1-x^2\right)^{(m/2)+1}\frac{\partial^2 X}{\partial x^2}$$

Hence from Eqn. (13*ii*)

$$\left\{-m\left(1-x^2\right)^{m/2}+m^2x^2\left(1-x^2\right)^{(m/2)-1}\right\}X-\left\{2x(m+1)\left(1-x^2\right)^{m/2}\right\}\frac{\partial X}{\partial x}$$

$$+\left(1-x^2\right)^{(m/2)+1}\frac{\partial^2 X}{\partial x^2}+\left\{\beta-\frac{m^2}{1-x^2}\right\}\left(1-x^2\right)^{m/2}X=0$$

Dividing both sides by $\left(1-x^2\right)^{m/2}$, we have

$$\left(1-x^2\right)\frac{\partial^2 X}{\partial x^2}-2(m+1)x\frac{\partial X}{\partial x}+\left\{-m+\frac{m^2x^2}{1-x^2}+\beta-\frac{m^2}{1-x^2}\right\}X=0$$

or $\left(1-x^2\right)\frac{\partial^2 X}{\partial x^2}-2(m+1)x\frac{\partial X}{\partial x}+\left\{\beta+\frac{-m+mx^2+m^2x^2-m^2}{1-x^2}\right\}X=0$

or $\left(1-x^2\right)\frac{\partial^2 X}{\partial x^2}-2(m+1)x\frac{\partial X}{\partial x}+\left\{\beta+\frac{-m\left(1-x^2\right)-m^2\left(1-x^2\right)}{1-x^2}\right\}X=0$

or

$$\left(1-x^2\right)\frac{\partial^2 X}{\partial x^2}-2(m+1)x\frac{\partial X}{\partial x}+\left[\beta-m(1+m)\right]X=0 \qquad (15)$$

Let $\alpha=(m+1)$ and $\lambda=\beta-m(1+m)$, then

$$\left(1-x^2\right)\frac{\partial^2 X}{\partial x^2}-2\alpha x\frac{\partial X}{\partial x}+\lambda X=0 \qquad (16)$$

Now, let us represent X as a power series:

$$X=a_0+a_1x+a_2x^2+...$$

Then $\qquad \frac{\partial X}{\partial x}=a_1+2a_2x+3a_3x^2+...$

and $\qquad \frac{\partial^2 X}{\partial x^2}=2a_2+6a_3x+12a_4x^2+...$

Substituting these values in Eqn. (16), we get

$$\left(1-x^2\right)\left(2a_2 + 6a_3 x + 12a_4 x^2 + ...\right) - 2\alpha x\left(a_1 + 2a_2 x + 3a_3 x^2 + ...\right)$$

$$+\lambda\left(a_0 + a_1 x + a_2 x^2 + ...\right) = 0$$

or $\quad \left(2a_2 + \lambda a_0\right) + \left\{6a_3 x + (\lambda - 2\alpha)a_1\right\} x + \left\{12a_4 x^2 + (\lambda - 4\alpha - 2)a_2\right\} x^2$

$$....... + \left[(n+1)(n+2)a_{n+2} + \left\{\lambda - 2n\alpha - n(n-1)\right\} a_n\right] x_n + ... = 0$$

This series will be zero for all values of x only if coefficient of each power of x vanishes separately i.e.

$$(n+1)(n+2)a_{n+1}\left\{\lambda - 2n\lambda - n(n-1)\right\} a_n = 0 \quad \text{(where } n = 0, 1, 2, 3...)$$

or $\qquad a_{n+2} = \dfrac{2n\lambda + n(n-1) - \lambda}{(n+1)(n+2)} a_n$ (17i)

Substituting the values of α and λ, we get

$$a_{n+2} = \dfrac{(n+m)(n+m+1) - \beta}{(n+1)(n+2)} a_n \tag{17ii}$$

For the wave function to satisfy boundary conditions (i.e. it must be finite), so $X(x)$ should be finite i.e. it should be a polynomial which vanishes after a definite number of terms. This is possible only when numerator of the above equality is zero after n terms, i.e.

$$(n + m)(n + m + 1) - \beta = 0$$

or $\qquad\qquad \beta = (n + m)(n + m + 1)$

Since m is either zero or an integer and n is also either zero or an integer, therefore we can put $n + m = l$ (another constant), which is also either zero or an integer. Then

$$\beta = l(l + 1) \text{ where } l = 0, 1, 2, 3, ...$$

With this value of β, Eqn. (13) (*ii*) becomes

$$\frac{\partial}{\partial x}\left[\left(1-x^2\right)\frac{\partial\Theta}{\partial x}\right] + \left(l(l+1) - \frac{m^2}{1-x^2}\right)\Theta = 0 \tag{18}$$

The solution of this equation is

$$\Theta = BP_l^m(x) = BP_l^m(\cos\theta)$$

where B is normalization constant and $P_l^m(x)$ is the *associated Legendre function* defined by

$$P_l^m(x) = \left(1-x^2\right)^{m/2} \frac{d^m}{dx^m} P_l(x) \tag{19}$$

where $P_l(x)$ is the *Legendre polynomial* of the order l. The normalization condition gives

$$\int \Theta_{ml}\Theta_{ml}^* d\theta = 1 \text{ or } B^2 \int_{-1}^{+1} P_l^m(x)P_l^m(x)dx = 1$$

or

$$B^2 \int_{-1}^{+1} P_l^m(x) P_l^m(x) dx = 1$$

But according to the orthogonality property of the Legendre functions

$$\int_{-1}^{+1} P_k^m(x) P_l^m(x) dx = 0 \text{ if } k \neq l$$

$$= \frac{2}{(2l+1)} \frac{(l+m)!}{(l-m)!} \text{ if } k = l$$

Hence

$$B = \sqrt{\frac{(2l+1)}{2} \frac{(l-m)!}{(l+m)!}}$$

and the normalized Θ wave function becomes

$$\Theta = \sqrt{\frac{(2l+1)}{2} \frac{(l-m)!}{(l+m)!}} \ P_l^m(\cos\theta) \tag{20}$$

and $\psi \equiv \Theta \ \Phi$ gives eigen functions

$$\psi(\theta, \phi) = \frac{1}{\sqrt{2\pi}} \sqrt{\frac{(2l+1)}{2} \frac{(l-m)!}{(l+m)!}} \ P_l^m(\cos\theta).e^{im\phi} \tag{21}$$

where $l = 0, 1, 2, 3, \ldots$

and $m = 0, \pm1, \pm2, \pm3, \ldots$

The functions defined by

$$Y_l^m(\theta, \phi) = (-1)^m \sqrt{\left(\frac{2l+1}{4\pi}\right) \frac{(l-m)!}{(l+m)!}} \ P_l^m(\cos\theta).e^{im\phi} \tag{22}$$

are called *spherical harmonics*.

Eigenvalues

In the eigen functions, we have assumed

$$\beta = l \ (l + 1) \ (l = 0, 1, 2, 3, \ldots) \text{ where } \beta = \frac{2IE}{\hbar^2}$$

Therefore $\dfrac{2IE}{\hbar^2} = l \ (l + 1) \text{ or } E = \dfrac{l(l+1)\hbar^2}{2I}$ (23)

Here l is called the *rotational quantum number*. Expression (23) gives the energy eigenvalues of the rigid rotator. These are discrete and not continuous.

$$l = 0, \qquad E_0 = 0$$

$$l = 1, \qquad E_1 = 2\hbar^2 / 2I$$

$$l = 2, \qquad E_2 = 6\hbar^2 / 2I$$

$$l = 3, \qquad E_3 = 12\hbar^2 / 2I$$

$$\underline{} \qquad \underline{} \qquad \underline{}$$

$$\underline{} \qquad \underline{} \qquad \underline{}$$

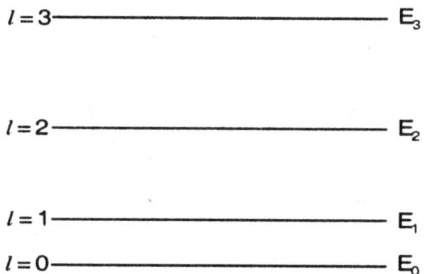

$l = 3$ —————————————————— E_3

$l = 2$ —————————————————— E_2

$l = 1$ —————————————————— E_1

$l = 0$ —————————————————— E_0

Fig. 6.2: Energy levels of a free axis rotator

The energy levels are not equidistant, as the value of l increases, the separation between two consecutive energy levels increases.

The angular momentum

$$J = \sqrt{2EI} = \sqrt{l(l+1)}\ \hbar \tag{24}$$

Since l assumes 0 and integral values, the angular momentum of the rigid rotator is quantised and assumes only discrete values.

Case II (Special case): Rotator in a fixed plane

If the two particles of a rigid rotator can rotate only in one plane, the direction of axis of rotation is fixed. It is then called rotator in fixed plane (e.g. a diatomic molecule). It can be assumed to be equivalent to a single particle of mass μ placed at the centre of mass. The time independent Schrödinger equation is

$$\nabla^2 \psi + \frac{2\mu}{\hbar^2}(E - V)\psi = 0$$

Since the distance r between the two particles is fixed, the potential energy (V) may be assumed to be zero. Then,

$$\nabla^2 \psi + \frac{2\mu}{\hbar^2} E\psi = 0 \tag{25}$$

In terms of spherical polar coordinates

$$\frac{1}{r^2}\frac{\partial}{\partial r}\left(r^2 \frac{\partial \psi}{\partial r}\right) + \frac{1}{r^2 \sin\theta}\frac{\partial}{\partial \theta}\left(\sin\theta \frac{\partial \psi}{\partial \theta}\right) + \frac{1}{r^2 \sin^2\theta}\frac{\partial^2 \psi}{\partial \phi^2} + \frac{2\mu E}{\hbar^2}\psi = 0 \tag{26}$$

Since the distance r is fixed, there will be no dependence of ψ on r. If we further assume that the rotator rotates only in X-Y plane, the value of $\theta = 90^0$ (constant), hence no θ–dependence. Thus, Schrödinger equation becomes

$$\frac{\partial^2 \psi}{\partial \phi^2} + \frac{2\mu r^2}{\hbar^2}E\psi = 0 \tag{27i}$$

$$\frac{\partial^2 \psi}{\partial \phi^2} + \frac{2IE}{\hbar^2}\psi = 0 \quad (I = \mu r^2) \tag{27ii}$$

Eigenfunctions:

In the above equation, the wave function ψ is function of ϕ only

$$\psi = \Phi(\phi)$$

$$\therefore \qquad \frac{\partial^2 \Phi}{\partial \phi^2} = -\frac{2IE}{\hbar^2}\Phi$$

or

$$\frac{\partial^2 \Phi}{\partial \phi^2} + m^2\Phi = 0 \quad \left(m^2 = \frac{2IE}{\hbar^2}\right)$$

The solution of this equation is

$$\Phi = a\, e^{im\phi} \text{ with } m = 0, \pm 1, \pm 2, \ldots \qquad (28i)$$

$$\int_0^{2\pi} \Phi\Phi^* d\phi = 1$$

or

$$a^2 \int_0^{2\pi} e^{im\phi} e^{-im\phi} d\phi = 1 \quad \therefore \quad a = \frac{1}{\sqrt{2\pi}} \qquad (28ii)$$

Hence

$$\psi = \Phi(\phi) = \frac{1}{\sqrt{2\pi}} e^{im\phi}$$

Eigenvalues:

We have

$$m^2 = \frac{2IE}{\hbar^2} \text{ or } E = \frac{m^2\hbar^2}{2I} \qquad (29)$$

where

$$m = 0, \pm 1, \pm 2, \ldots$$

Thus the energy levels of a fixed axis rotator are also discrete:

$$m = 0, \quad E_0 = 0$$
$$m = 1, \quad E_1 = \hbar^2/2I$$
$$m = 2, \quad E_2 = 4\hbar^2/2I$$
$$m = 3, \quad E_3 = 9\hbar^2/2I$$

The energy levels are not equidistant, as the value of m increases, the separation between two consecutive energy levels also increases.

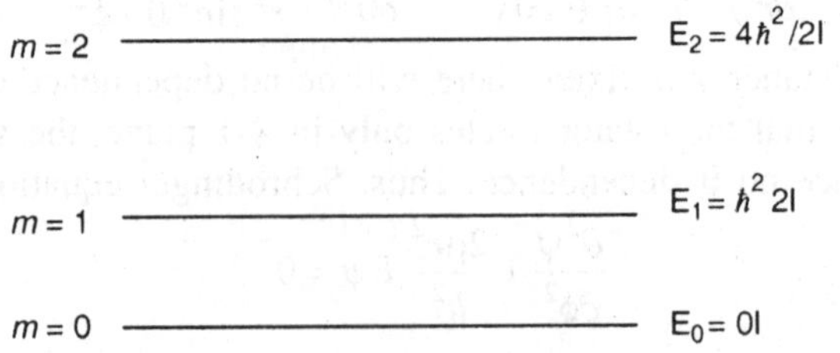

$$m = 2 \underline{\hspace{6cm}} E_2 = 4\hbar^2/2I$$

$$m = 1 \underline{\hspace{6cm}} E_1 = \hbar^2\, 2I$$

$$m = 0 \underline{\hspace{6cm}} E_0 = 0I$$

Fig. 6.3: Energy levels of a fixed axis rotator

6.2 Rotational Spectrum of a Diatomic Molecule

A diatomic molecule consists of two atoms at a distance r, which, to the first approximation, may be taken as constant. The two atoms rotate about an axis passing through their centre of mass and perpendicular to the line joining them. So, a diatomic molecule may be considered as a rigid rotator.

Angular momentum $\quad J_l = \sqrt{l(l+1)}\,\hbar$

and rotational energy $\quad E_t = \dfrac{l(l+1)\hbar^2}{2I}$

Obviously, the rotational energy levels are discrete and are not equispaced.

At very low temperatures (near absolute zero), generally all the molecules are in the ground state ($l = 0$). As the temperature increases, the molecules get excited to higher energy states. When there is a transition from higher energy state to the lower energy state, the difference in energy is emitted in the form of electromagnetic radiation, having frequency v, given by

$$h\nu = E_l - E_{l-1} = l(l+1)\frac{\hbar^2}{2I} - (l-1)l\frac{\hbar^2}{2I}$$

or $$h\nu = \frac{\hbar^2 l}{I} = \frac{h^2}{4\pi^2 I}\,l \qquad (30i)$$

or wave number $\bar{\nu} = \dfrac{1}{\lambda} = \dfrac{\nu}{c} = \dfrac{h^2}{4\pi^2 Ic}\,l \qquad (30ii)$

(where c = speed of light)

Since the rotational energy gap $E_l - E_{l-1}$ for a diatomic molecule is very small ($\sim 10^{-2}$eV), the radiations corresponding to rotational transitions are obtained in the far infra red or microwave region. From Eqn. (30i) it is evident that the difference in frequencies of radiations emitted corresponding to two consecutive transitions is the same i.e. common frequency interval

$$\Delta\nu = \nu_l - \nu_{l-1} = \frac{h}{4\pi^2 I}\,l - \frac{h}{4\pi^2 I}(l-1) = \frac{h}{4\pi^2 I} \qquad (31)$$

This fact has been experimentally verified.

6.3 Legendre Polynomials and Associated Legendre Functions

The associated Legendre functions are defined by Eqn. (19) as

$$P_l^m(x) = \left(1 - x^2\right)^{m/2} \frac{d^m}{dx^m} P_l(x) \qquad (32)$$

where Legendre polynomials $P_l(x)$ are given by Rodrigues' formula as

$$P_l(x) = \frac{1}{2^l\, l!}\left(\frac{d}{dx}\right)^l \left(x^2 - 1\right)^l \qquad (33)$$

For example,

$$P_2(x) = \frac{1}{2^2 \times 2!}\left(\frac{d}{dx}\right)^2 \left(x^2 - 1\right)^2$$

$$= \frac{1}{4 \times 2!}\frac{d}{dx}\left[2\left(x^2 - 1\right).2x\right]$$

$$= \frac{4}{8}\frac{d}{dx}\left(x^3 - x\right) = \frac{1}{2}\left(3x^2 - 1\right)$$

$$\therefore \qquad P_2^1(x) = \left(1 - x^2\right)^{1/2}\frac{d}{dx}\left[\frac{1}{2}\left(3x^2 - 1\right)\right]$$

$$= 3x\left(1 - x^2\right)^{1/2}$$

We may develop Table 6.1 for ($x = \cos\theta$). Fig. 6.4 gives graphical illustration of Legendre polynomials. Shapes of the associated Legendre polynomials as a function of the angle θ (the angle between the z-axis and the xy plane) are depicted in Fig. 6.5.

Table 6.1: Legendre and associated legendre functions

Legendre functions	Associated Legendre functions
$P_0(x) = 1$	$P_1^1(x) = \left(1 - x^2\right)^{1/2} = \sin\theta$
$P_1(x) = x = \cos\theta$	$P_2^1(x) = 3x\left(1 - x^2\right)^{1/2} = 3\cos\theta\sin\theta$
$P_2(x) = \frac{1}{2}\left(3x^2 - 1\right)$	$P_2^2(x) = 3\left(1 - x^2\right) = 3\sin^2\theta$
$\quad = \frac{1}{2}\left(3\cos^2\theta - 1\right)$	
$P_3(x) = \frac{1}{2}\left(5x^3 - 3x\right)$	$P_3^1(x) = \frac{3}{2}\left(5x^2 - 1\right)\left(1 - x^2\right)^{1/2} = \frac{3}{2}\left(5\cos^2\theta - 1\right)\sin\theta$
$\quad = \frac{1}{2}\left(5\cos^2\theta - 3\cos\theta\right)$	$P_3^2(x) = 15x\left(1 - x^2\right) = 15\cos\theta\sin^2\theta$
	$P_3^3(x) = 15\left(1 - x^2\right)^{3/2} = 15\sin^3\theta$

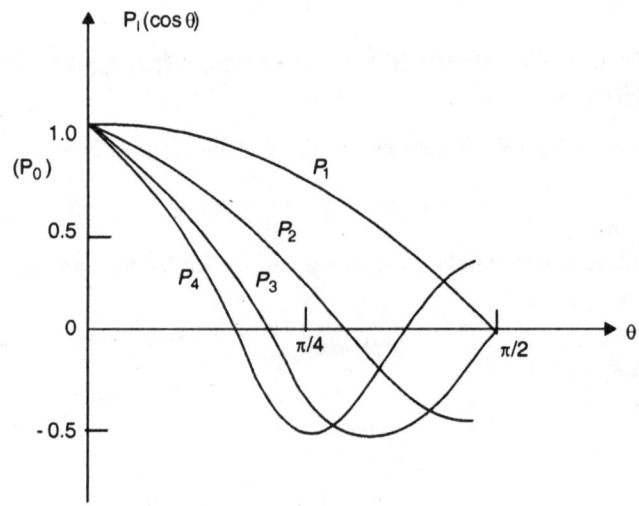

Fig. 6.4: Graphs of P_l (cos θ) for $0 \le \theta \le \pi/2$ and $l = 0, 1, 2, 3, 4$

Example 6.1: In a hydrogen molecule, the separation between two protons is 0.74 Å. Assuming it to be a rigid rotator, calculate the first three energy levels of the H_2 molecule. (Given, $m_p = 1.7 \times 10^{-27}$ kg, $\hbar = 1.05 \times 10^{-34}$ J-s.)

Solution: We have $\quad m_p = 1.7 \times 10^{-27}$ kg,

$$\hbar = 1.05 \times 10^{-34} \text{ J-s.}$$

$$r = 0.74 \text{ Å} = 0.74 \times 10^{-10} \text{ m}$$

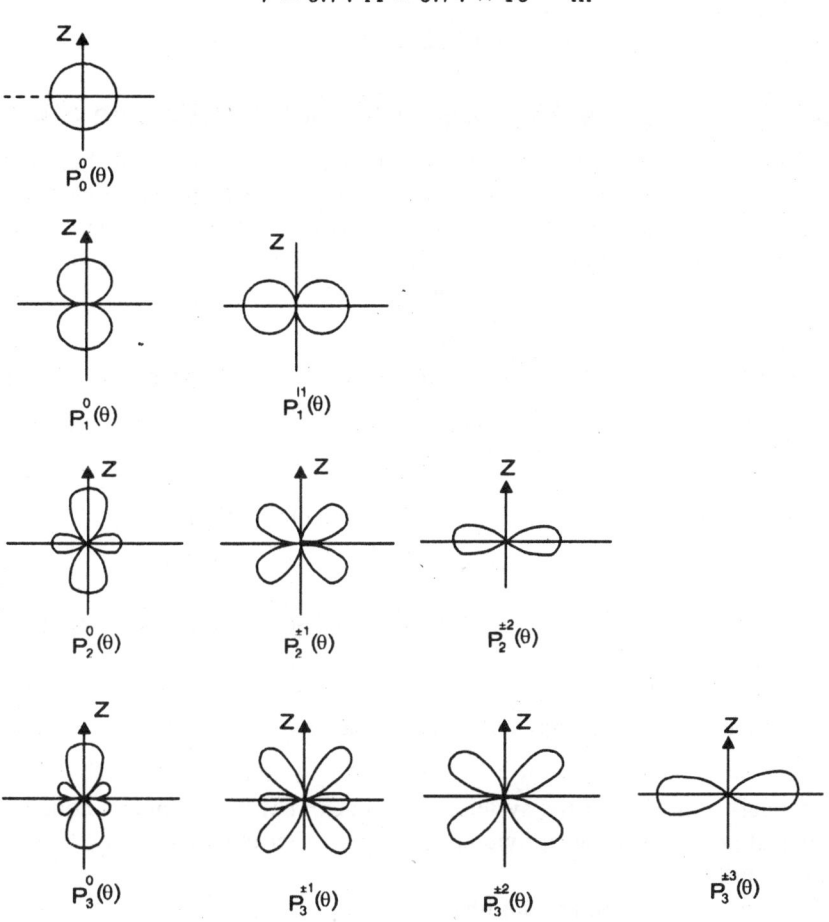

Fig. 6.5: Shapes of the associated Legendre polynomials $P_l^m(\theta)$

$$\text{Reduced mass} = \mu = \frac{m_p . m_p}{m_p + m_p} = \frac{m_p}{2} = 8.5 \times 10^{-28} \text{ kg}$$

Moment of inertia about an axis passing through the centre of mass is

$$I = \mu r^2$$
$$= 8.5 \times 10^{-28} \times (0.74 \times 10^{-10})^2$$
$$= 4.65 \times 10^{-48} \text{ kg m}^2$$

$$\text{Rotational kinetic energy } E = \frac{l(l+1)\hbar^2}{2I}$$

where $l = 0, 1, 2, \ldots$

For $l = 1$, $\quad E_1 = 2 \times \dfrac{\hbar^2}{2I} = \dfrac{2 \times \left(1.05 \times 10^{-34}\right)^2}{2 \times 4.65 \times 10^{-48}} = 2.37 \times 10^{-21}$ J

$l = 2$, $\quad E_2 = 6 \times \dfrac{\hbar^2}{2I} = 7.11 \times 10^{-21}$ J

$l = 3$, $\quad E_3 = 12 \times \dfrac{\hbar^2}{2I} = 14.22 \times 10^{-21}$ J

6.4 Schrödinger's Wave Equation for the Hydrogen Atom

A hydrogen atom consists of an electron moving around a proton which is about 1836 times heavier than the electron. The proton may be considered to be practically stationary, but, the correction for proton motion can be incorporated by replacing the electron mass m_e by the reduced mass of the system.

$$\mu = \frac{m_e m_p}{m_e + m_p}.$$

Schrödinger's equation for the electron motion in three dimensions in hydrogen atom is

$$\frac{\partial^2 \psi}{\partial x^2} + \frac{\partial^2 \psi}{\partial y^2} + \frac{\partial^2 \psi}{\partial z^2} + \frac{2\mu}{\hbar^2}(E - V)\psi = 0 \tag{34}$$

The potential V here is the electrostatic potential energy

$$V = -\frac{e^2}{4\pi\varepsilon_0 r} \tag{35}$$

because electronic charge is $-e$ and charge on the proton is $+e$, both separated by a distance r.

Since V is a function of r rather than of x, y, z, we cannot substitute Eqn. (35) directly into Eqn. (34). There are two alternatives, we may express V in terms of Cartesian coordinates x, y, z by replacing r by $\sqrt{x^2 + y^2 + z^2}$ or, owing to (spherical) symmetry of the system, we may express Schrödinger's equation in spherical polar coordinates. Symmetry suggests that latter alternative would be more appropriate.

In spherical polar coordinates

$x = r \sin \theta \cos \phi$
$y = r \sin \theta \sin \phi$
$z = r \cos \theta$

$$\theta = \cos^{-1}\left(\frac{z}{\sqrt{x^2 + y^2 + z^2}}\right)$$

and $\qquad\qquad \phi = \tan^{-1}(y/x).$

In spherical polar coordinates Schröndinger's equation becomes

$$\frac{1}{r^2}\frac{\partial}{\partial r}\left(r^2 \frac{\partial \psi}{\partial r}\right) + \frac{1}{r^2 \sin\theta}\frac{\partial}{\partial \theta}\left(\sin\theta \frac{\partial \psi}{\partial \theta}\right) + \frac{1}{r^2 \sin^2\theta}\frac{\partial^2 \psi}{\partial \phi^2}$$

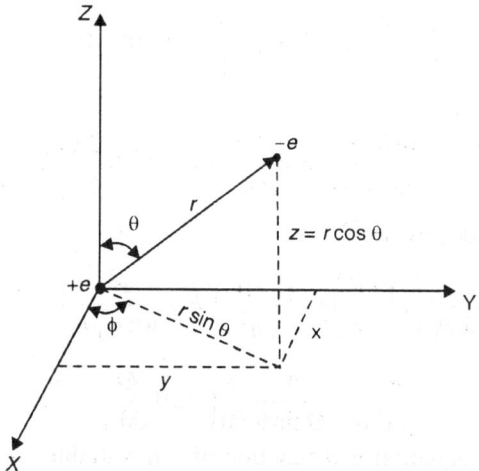

Fig. 6.6: Spherical polar coordinates

$$+\frac{2\mu}{\hbar^2}(E-V)\psi = 0 \qquad (36)$$

This is the partial differential equation for the wave function ψ of the electron in a hydrogen atom. Here it should be noted that in the Bohr's model, the electron's motion is one dimensional, and one quantum number is enough to specify the state of the electron. In the Schröndinger's equation (Eqn. 36), there are three variables (r, θ and ϕ) corresponding to three dimensions of spherical polar coordinates so three quantum numbers would govern the wave function.

We solve the Eqn. (36) by the method of separation of variables. Multiplying both sides of Eqn. (36) by $r^2 \sin^2 \theta$ and putting the value of $V(r)$, we get

$$\sin^2\theta \frac{\partial}{\partial r}\left(r^2 \frac{\partial \psi}{\partial r}\right) + \sin\theta \frac{\partial}{\partial \theta}\left(\sin\theta \frac{\partial \psi}{\partial \theta}\right) + \frac{\partial^2 \psi}{\partial \phi^2}$$

$$+\frac{2\mu}{\hbar^2}r^2 \sin^2\theta\left(E + \frac{e^2}{4\pi\varepsilon_0 r}\right)\psi = 0 \qquad (37)$$

We may write ψ as

$$\psi(r, \theta, \phi) = R(r)\, \Theta(\theta)\, \Phi(\phi) \qquad (38)$$

Substituting this in Eqn. (37) and dividing throughout by $R\,\Theta\,\Phi$, we obtain

$$\frac{\sin^2\theta}{R}\frac{\partial}{\partial r}\left(r^2 \frac{\partial R}{\partial r}\right) + \frac{\sin\theta}{\Theta}\frac{\partial}{\partial \theta}\left(\sin\theta \frac{\partial \Theta}{\partial \theta}\right)$$

$$+r^2 \sin^2\theta \frac{2\mu}{\hbar^2}\left(E + \frac{e^2}{4\pi\varepsilon_0 r}\right) = -\frac{1}{\Phi}\frac{\partial^2 \Phi}{\partial \phi^2} \qquad (39)$$

Here l.h.s is a function of r and θ and is independent of ϕ whereas r.h.s. is a function of ϕ only. Therefore this is possible only if the quantity on each side of this equation is a constant, say m^2:

$$-\frac{1}{\Phi}\frac{\partial^2 \Phi}{\partial \phi^2} = m^2 \quad \text{or} \quad \frac{\partial^2 \Phi}{\partial \phi^2} + m^2 \Phi = 0 \tag{40}$$

and

$$\frac{\sin^2 \theta}{R}\frac{\partial}{\partial r}\left(r^2 \frac{\partial R}{\partial r}\right) + \frac{\sin \theta}{\Theta}\frac{\partial}{\partial \theta}\left(\sin \theta \frac{\partial \Theta}{\partial \theta}\right) + r^2 \sin^2 \theta \frac{2\mu}{\hbar^2}\left(E + \frac{e^2}{4\pi \epsilon_0 r}\right) = m^2$$

Dividing both sides by $\sin^2 \theta$

$$\frac{1}{R}\frac{\partial}{\partial r}\left(r^2 \frac{\partial R}{\partial r}\right) + r^2 \left(\frac{2\mu}{\hbar^2}\right)\left(E + \frac{e^2}{4\pi \epsilon_0 r}\right)$$

$$= \frac{m^2}{\sin^2 \theta} - \frac{1}{\Theta \sin \theta}\frac{\partial}{\partial \theta}\left(\sin \theta \frac{\partial \Theta}{\partial \theta}\right) \tag{41}$$

The l.h.s of this equation is a function of one variable r only while r.h.s. is a function of θ only. Therefore, if this equation is to be satisfied, each side must be equal to the same constant (say β) separately. That is,

$$\frac{1}{R}\frac{\partial}{\partial r}\left(r^2 \frac{\partial R}{\partial r}\right) + r^2 \left(\frac{2\mu}{\hbar^2}\right)\left(E + \frac{e^2}{4\pi \epsilon_0 r}\right) = \beta$$

or

$$\frac{1}{r^2}\frac{\partial}{\partial r}\left(r^2 \frac{\partial R}{\partial r}\right) - \frac{\beta R}{r^2} + \frac{2\mu}{\hbar^2}\left(E + \frac{e^2}{4\pi \epsilon_0 r}\right)R = 0 \tag{42}$$

and

$$\frac{m^2}{\sin^2 \theta} - \frac{1}{\Theta \sin \theta}\frac{\partial}{\partial \theta}\left(\sin \theta \frac{\partial \Theta}{\partial \theta}\right) = \beta$$

or

$$\frac{1}{\sin \theta}\frac{\partial}{\partial \theta}\left(\sin \theta \frac{\partial \Theta}{\partial \theta}\right) + \left(\beta - \frac{m^2}{\sin^2 \theta}\right)\Theta = 0 \tag{43}$$

Eqns (40), (42) and (43) are now to be solved to determine the eigen functions.

Solution of the Φ-equation:

The general solution of Eqn. (40) is

$$\Phi_m = A e^{im\phi}$$

where

$$m = 0, \pm 1, \pm 2, \dots$$

A is normalization constant, determined by the condition

$$\int_0^{2\pi} \Phi^* \Phi \, d\phi = 1 \quad \text{or} \quad \int_0^{2\pi} A^* e^{-im\phi} A e^{im\phi} d\phi = 1$$

or

$$\int_0^{2\pi} A^2 d\phi = 1 \quad \text{or} \quad A^2 \times 2\pi = 1$$

$$\therefore \quad A = \frac{1}{\sqrt{2\pi}} \quad \text{or} \quad \Phi = \frac{1}{\sqrt{2\pi}} e^{im\phi} \tag{44}$$

The constant m is called the *magnetic quantum number* because the energy of a hydrogen atom in a magnetic field depends on the value of this quantum number.

Solution of the Θ-equation:

The Θ equation is exactly the same as that for a free axis rigid rotator. The solutions are

$$\Theta = \sqrt{\frac{(2l+1)}{2}\frac{(l-|m|)\ !}{(l+|m|)\ !}}\ P_l^m(\cos\theta) \tag{45}$$

where $\beta = l(l+1)$, $l = 0, 1, 2, 3, \ldots$ and P_l^m are associated Legendre polynomials. l is called *azimuthal quantum number*. The shapes of the associated Legendre polynomials are depicted in Fig. 6.5.

Solution of the Radial equation:

Substituting the value of β in Eqn. (42), we have

$$\frac{1}{r^2}\frac{\partial}{\partial r}\left(r^2\frac{\partial R}{\partial r}\right) - \frac{l(l+1)}{r^2}R + \frac{2\mu}{\hbar^2}\left(E - V(r)\right)R = 0$$

or
$$\frac{\partial^2 R}{\partial r^2} + \frac{2}{r}\frac{\partial R}{\partial r} + \left[-\frac{l(l+1)}{r^2} + \frac{2\mu E}{\hbar^2} + \frac{2\mu e^2}{4\pi\varepsilon_0\hbar^2 r}\right]R = 0 \tag{46}$$

According to classical theory, the orbit will be hyperbolic for $E > 0$ and elliptical for $E < 0$. Thus the electrons will be bound for the negative energy and the solution of Eqn. (46) will represent the eigen function for the normal hydrogen atom. Hence assuming $E < 0$, let

$$\alpha^2 = -\frac{2\mu E}{\hbar^2} \text{ and } \lambda = \frac{\mu e^2}{4\pi\varepsilon_0\hbar^2\alpha} \tag{47}$$

Then, from Eqn. (46), we get

$$\frac{\partial^2 R}{\partial r^2} + \frac{2}{r}\frac{\partial R}{\partial r} + \left[-\frac{l(l+1)}{r^2} - \alpha^2 + \frac{2\lambda\alpha}{r}\right]R = 0 \tag{48}$$

Now, if $\rho = 2\alpha r$, then

$$\frac{\partial R}{\partial r} = \frac{\partial R}{\partial \rho}\frac{\partial \rho}{\partial r} = 2\alpha\frac{\partial R}{\partial \rho} \tag{49i}$$

and
$$\frac{\partial^2 R}{\partial r^2} = \frac{\partial}{\partial r}\left(2\alpha\frac{\partial R}{\partial \rho}\right) = \frac{\partial}{\partial \rho}\left(2\alpha\frac{\partial R}{\partial \rho}\right)\cdot\frac{\partial \rho}{\partial r} = 4\alpha^2\frac{\partial^2 R}{\partial \rho^2} \tag{49ii}$$

Substituting these values of $\dfrac{\partial R}{\partial r}$ and $\dfrac{\partial^2 R}{\partial r^2}$ in Eqn. (48) and dividing by $4\alpha^2$,

we get

$$\frac{\partial^2 R}{\partial \rho^2} + \frac{2}{\rho}\frac{\partial R}{\partial \rho} + \left[-\frac{l(l+1)}{\rho^2} + \frac{\lambda}{\rho} - \frac{1}{4}\right]R = 0 \tag{50}$$

If $\rho \to \infty$, Eqn. (50) approaches the form

$$\frac{\partial^2 R(\rho)}{\partial \rho^2} - \frac{R(\rho)}{4} = 0$$

The solutions of this equation are
$$R(\rho) = e^{-\rho/2} \text{ and } R(\rho) = e^{\rho/2}$$
The value of ρ varies from 0 to ∞, therefore from Eqn. $R = e^{\rho/2}$, the value of R increases with increase in the value of ρ and from $R = e^{-\rho/2}$, the value of R decreases with increase in the value of ρ. Hence, only $R = e^{-\rho/2}$ will be the possible solution (an asymptotic solution). General solution of Eqn. (50) is
$$R = F(\rho)e^{-\rho/2} \tag{51}$$
Substituting this in Eqn. (50), we get
$$\frac{\partial^2 F}{\partial \rho^2} + \left[\frac{2}{\rho} - 1\right]\frac{\partial F}{\partial \rho} + \left[-\frac{l(l+1)}{\rho^2} + \frac{\lambda}{\rho} - \frac{1}{4}\right]F = 0 \tag{52}$$

We now try a solution of the form
$$F(\rho) = \rho^s \, G(\rho) \tag{53}$$

where
$$G(\rho) = \sum_{k=0}^{\infty} a_k \rho^k$$

so that
$$\frac{\partial F}{\partial \rho} = s\rho^{s-1}G + \rho^s \frac{\partial G}{\partial \rho}$$

and
$$\frac{\partial^2 F}{\partial \rho^2} = s(s-1)\rho^{s-2}G + 2s\rho^{s-1}\frac{\partial G}{\partial \rho} + \rho^s \frac{\partial^2 G}{\partial \rho^2}$$

Then from Eqn. (52),

$$\rho^{s+2}\frac{\partial^2 G}{\partial \rho^2} + 2s\rho^{s+1}\frac{\partial G}{\partial \rho} + s(s-1)\rho^s G + 2\rho^{s+1}\frac{\partial G}{\partial \rho} + 2s\rho^s G - \rho^{s+2}G$$

$$+ s\rho^{s+1}G + (\lambda - 1)\rho^{s+1}G - l(l+1)\rho^s G = 0$$

Dividing both sides by ρ^s and rearranging, we have

$$\rho^s \frac{\partial^2 G}{\partial \rho^2} + \rho[2(s+1) - \rho]\frac{\partial G}{\partial \rho} + [\rho(\lambda - s - 1) + s(s+1) - l(l+1)]G = 0$$

If $\qquad\qquad \rho = 0$ then $[s(s+1) - l(l+1)] = 0$

or $\qquad\qquad\qquad s(s+1) = l(l+1)$

or $\qquad\qquad\qquad s = l$ or $-(l+1)$

But according to the boundary condition $R(\rho)$ should be finite at $\rho = 0$. This requires that $s = l$

$\therefore \qquad\qquad \rho G'' + [2(l+1) - \rho]\, G' + (\lambda - l - 1)\, G = 0 \tag{55}$

Differentiating Eqn. (54) w.r.t. ρ, we get

$$G' = \sum_k a_k . k \, \rho^{k-1}$$

and
$$G'' = \sum_k a_k . k(k-1)\, \rho^{k-2}$$

Substituting values of G, G' and G'' in Eqn. (55), we get

$$\sum_k a_k.k(k-1)\,\rho^{k-1} + \left[2(l+1)-\rho\right]\sum_k a_k.k\,\rho^{k-1}$$

$$+(\lambda-l-1)\sum_k a_k\rho^k = 0 \tag{56}$$

The condition for the above equation to be satisfied is that the coefficients of each power of ρ should be zero. Collecting the coefficients of ρ^k and equating to zero, we get

$$a_{k+1}k(k+1) - a_k k + 2(l+1)a_{k+1}(k+1) + (\lambda-l-1)a_k = 0$$

or

$$a_{k+1}\left[2(l+1)(k+1) + k(k+1)\right] + (\lambda-l-1-k)a_k = 0$$

or

$$a_{k+1} = \frac{k++1-\lambda}{(k+1)(2l+k+2)}a_k \tag{57}$$

But, from $F(\rho) = \rho^s G(\rho) \equiv \rho^s \sum_{k=0}^{\infty} a_k \rho^k$, it is clear that as the value of ρ increases, the value of $F(\rho)$ (i.e. the value of R) also increases. We infer therefore, that the power series should be finite, then only the satisfactory wave function will be obtained. Let this series vanishes after ρ^k, then numerator in the recursion formula should be equal to zero, so

$$k + l + 1 - \lambda = 0$$

or

$$\lambda = k + l + 1 = n \text{ (say)} \tag{58}$$

Here k (= 1, 2, 3...) is called the *radial quantum number* and n (= 1, 2, 3...) is called the total or *principal quantum number*.

Energy Eigen values:

$$\alpha^2 = -\frac{2\mu E}{\hbar^2} \text{ and } \lambda^2 = \frac{1}{(4\pi\varepsilon_0)^2}\frac{\mu^2}{\alpha^2}\frac{e^2}{\hbar^4} = \frac{\mu^2 e^4 (4\pi\varepsilon_0)^{-2}}{\hbar^4\left(-\dfrac{2\mu E}{\hbar^2}\right)}$$

$$\therefore \qquad E_n = -\frac{\mu e^4}{(4\pi\varepsilon_0)^2 \times 2\hbar^2\,\lambda^2} = \frac{(1/2)\mu e^4}{(4\pi\varepsilon_0)^2\,\hbar^4 n^2} \tag{59}$$

This gives the energy eigen values for the hydrogen atom and is in agreement with the Bohr's theory.

Radial Wave Functions:

Substituting $\lambda = n$ in Eqn. (55), we get

$$\rho\frac{\partial^2 G}{\partial\rho^2} + \left[2(l+1)-\rho\right]\frac{\partial G}{\partial\rho} + (n-l-1)G = 0 \tag{60i}$$

If we put $2l + 1 = p$ and $n + l = n'$, then

$$\rho\frac{\partial^2 G}{\partial\rho^2} + (p+1-\rho)\frac{\partial G}{\partial\rho} + (n'-p)G = 0 \tag{60ii}$$

The solutions of this equation are

$$G(\rho) = C\ L_n^p{}'(\rho) = C\ L_{n+l}^{2l+1}(\rho) \tag{61}$$

where C is an arbitrary constant to be determined by normalization condition and $L_{n+l}^{2l+1}(\rho)$ are called *associated Leguerre polynomials* and are given by

$$L_{n+l}^{2l+1}(\rho) = \sum_{k=0}^{n-l-1} \frac{(-1)^{k+1}\left[(n+l)!\right]^2 \rho^k}{(n-l-1-k)!(2l+1+k)!k!} \tag{62}$$

Hence complete radial wave function is

$$R(r) = e^{-\rho/2}\rho^l G(\rho)$$

or

$$R_{nl}(r) = C\ e^{-\rho/2}\rho^l\ L_{n+l}^{2l+1}(\rho) \tag{63}$$

We have

$$\lambda = \frac{\mu e^2}{4\pi\varepsilon_0 \hbar^2 \alpha}$$

$$\Rightarrow \qquad \alpha = \frac{\mu e^2}{4\pi\varepsilon_0 \hbar^2 \lambda} = \frac{\mu e^2}{4\pi\varepsilon_0 \hbar^2 n} \tag{64}$$

$$\therefore \qquad \rho = 2\alpha r = 2\frac{\mu e^2}{4\pi\varepsilon_0 \hbar^2 n}r = \frac{2r}{n}\frac{\mu e^2}{4\pi\varepsilon_0 \hbar^2}$$

or

$$\rho = \frac{2r}{n} \times \frac{1}{a_0} \quad \text{where } a_0 = \frac{4\pi\varepsilon_0 \hbar^2}{\mu e^2} \tag{65}$$

is the Bohr radius (i.e. radius of the first orbit).

Applying the normalization condition

$$\int R_{nl}(r)\ R_{nl}^*(r)\ r^2 dr = 1$$

we get

$$C = \sqrt{\left(\frac{2}{na_0}\right)^3 \frac{(n-l-1)!}{2n\left[(n+l)!\right]^3}}$$

Thus the normalized radial eigenfunctions are

$$R_{nl}(r) = \left(\frac{2}{na_0}\right)^3 \frac{(n-l-1)\ !}{2n\left[(n+l)\ !\ \right]^3}\ e^{-\frac{r}{na_0}}\left(\frac{2r}{na_0}\right)^l L_{n+l}^{2l+1}\left(\frac{2r}{na_0}\right) \tag{66}$$

6.5 Hydrogen-Like Wave Functions

The complete wave functions for the discrete stationary states of a one electron (i.e. hydrogen-like) atom are given by

$$\psi_{nlm}\ (r,\ \theta,\ \phi) = R_{nl}(r)\ \Theta_{lm}\ (\theta)\ \Phi_m(\phi) \tag{67}$$

where

$$\Phi_m(\phi) = \frac{1}{\sqrt{2\pi}}e^{im\phi} \tag{68i}$$

$$\Theta_{lm}(\theta) = \left[\left(\frac{2l+1}{2}\right)\frac{(l-|m|)\ !}{(l+|m|)\ !}\right]^{1/2} P_l^{|m|}(\cos\theta) \tag{68ii}$$

and $\quad R_{nl}(r) = -\left[\left(\dfrac{2Z}{na_0}\right)^3 \dfrac{(n-l-1)\,!}{2n\left[(n+l)\,!\,\right]^3}\right]^{1/2} e^{\frac{-Zr}{na_0}} \left(\dfrac{2Zr}{na_0}\right)^l L_{n+l}^{2l+1}\left(\dfrac{2Zr}{na_0}\right)$ $\qquad$ (68iii)

The minus sign is introduced in Eqn. (68iii) (for convenience) to make the function positive for small values of r.

The wave functions of Eqn. (68) are normalized, so that

$$\int_0^\infty \int_0^\pi \int_0^{2\pi} \psi_{nlm}^*(r,\theta,\phi)\psi_{nlm}(r,\theta,\phi)r^2\,dr(\sin\theta)\,d\theta\,d\phi \qquad (69)$$

because an elemental volume in spherical polar coordinates as shown in Fig. 6.7 is

$$dv = (r\sin\theta\,d\phi)\,(r\,d\theta)\,(dr)$$

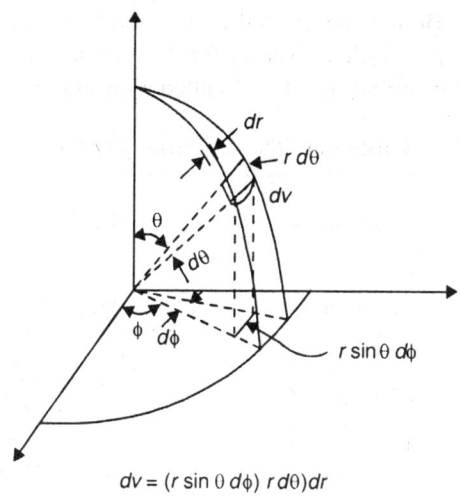

$$dv = (r\sin 0\,d\phi)\,r\,d\theta)dr$$

Fig. 6.7: A volume element dv in spherical polar coordinates

Moreover, the functions Φ_m, Θ_{lm} and R_{nl} are separately normalized:

$$\int_0^{2\pi} \Phi_m^*(\phi)\ \Phi_m(\phi)\ d\phi = 1 \qquad (70)$$

$$\int_0^\pi \{\Theta_{lm}(\theta)\}^2 \sin\theta\,d\theta = 1$$

and $\qquad \displaystyle\int_0^\infty \left[R_{nl}(r)\right]^2 r^2 dr = 1$

The eigen functions ψ_{nlm} are mutually orthogonal. That is, the integral

$$\int_0^\infty \int_0^\pi \int_0^{2\pi} \psi_{nlm}^*(r,\theta,\phi)\psi_{n'l'm'}(r,\theta,\phi)r^2(\sin\theta)\,d\theta\,d\phi\,dr$$

vanishes, except for $n = n'$, $l = l'$ and $m = m'$ in as much as if $m \neq m'$; the integral in ϕ vanishes; if $m = m'$ but $l \neq l'$, the integral in θ vanishes and if $m = m'$ and $l = l'$ but $n \neq n'$, the integral in r vanishes.

Expressions for normalized wave functions are given in Tables 6.2 to 6.4 and complete wave functions in Table 6.5.

Table 6.2: The functions $\Phi_m(\phi)$

$$\Phi_0(\phi) = \frac{1}{\sqrt{2\pi}} \; ; \qquad \Phi_1(\phi) = \frac{1}{\sqrt{2\pi}} e^{i\phi} \; ; \qquad \Phi_{-1}(\phi) = \frac{1}{\sqrt{2\pi}} e^{-i\phi}$$

$$\Phi_2(\phi) = \frac{1}{\sqrt{2\pi}} e^{i2\phi} \; ; \qquad \Phi_{-2}(\phi) = \frac{1}{\sqrt{2\pi}} e^{-i2\phi}$$

It is to be noted that the electron is no longer considered as moving in a fixed orbit as described in Bohr's theory, rather, its motion is described by the wave function ψ_{nlm} in the probabilistic sense, for a given n, with $l = 0$ called an s-orbital, $l = 1$ called p-orbital and $l = 2$ called d-orbital.

Table 6.3: The functions $\Theta_{lm}(\theta)$

s electrons	$l = 0, m = 0$	$\Theta_{0,0} = \dfrac{1}{\sqrt{2}}$
p electrons	$l = 1, m = 0$	$\Theta_{1,0} = \dfrac{\sqrt{6}}{2} \cos\theta$
	$l = 1, m = \pm 1$	$\Theta_{1,\pm1} = \dfrac{\sqrt{3}}{2} \sin\theta$
d electrons	$l = 2, m = 0$	$\Theta_{2,0} = \dfrac{\sqrt{10}}{4}\left(3\cos^2\theta - 1\right)$
	$l = 2, m = \pm 1$	$\Theta_{2,\pm1} = \dfrac{\sqrt{15}}{2} \sin\theta . \cos\theta$
	$l = 2, m = \pm 2$	$\Theta_{2,\pm2} = \dfrac{\sqrt{15}}{4} \sin^2\theta$

We may express the total wave function as a product of the radial and angular parts as

$$\psi_{nlm}(r, \theta, \phi) = R_{nl}(r)\, \psi_{lm}(\theta, \phi) \qquad (70a)$$

where
$$\psi_{lm}(\theta, \phi) = \Theta_{lm}(\theta)\, \Phi_m(\phi) \qquad (70b)$$

For $l = 0$, $m = 0$, we have an s-orbital. For $l = 1$, there are three values of m viz. 1, 0 and -1; the corresponding angular wave function ψ_{lm} may be designated as $\psi(p_{+1})$, $\psi(p_0)$ and $\psi(p_{-1})$:

$$\psi(p_{+1}) = \left(\frac{3}{8\pi}\right)^{1/2} \sin\theta . e^{i\phi}$$

$$\psi(p_0) = \left(\frac{3}{4\pi}\right)^{1/2} \cos\theta$$

$$\psi(p_{-1}) = \left(\frac{3}{8\pi}\right)^{1/2} \sin\theta . e^{-i\phi}$$

However, since for $m \neq 0$, the orbitals are imaginary functions, therefore an equivalent set of real functions is given by linear combinations as

$$\psi(p_x) = \frac{\psi(p_{+1}) + \psi(p_{-1})}{\sqrt{2}}$$

$$\psi(p_z) = \psi(p_0) \hspace{3cm} (71)$$

$$\psi(p_y) = -i\frac{\psi(p_{+1}) - \psi(p_{-1})}{\sqrt{2}}$$

Similarly for $l = 2$, there are five values of m, so there are five states: $\psi(d_{+2})$, $\psi(d_{+1})$, $\psi(d_0)$, $\psi(d_{-1})$, $\psi(d_{-2})$. The following five linear combinations give real d functions:

$$\psi(d_{z^2}) = \psi(d_0)$$

$$\psi(d_{xz}) = \frac{\psi(d_{+1}) + \psi(d_{-1})}{\sqrt{2}}$$

$$\psi(d_{yz}) = -\frac{i}{\sqrt{2}}\left(\psi(d_{+1}) - \psi(d_{-1})\right) \hspace{2cm} (72)$$

$$\psi\left(d_{x^2-y^2}\right) = \frac{\psi(d_{+2}) + \psi(d_{-2})}{\sqrt{2}}$$

$$\psi(d_{xy}) = -\frac{i}{\sqrt{2}}\left(\psi(d_{+2}) - \psi(d_{-2})\right)$$

Figure 6.8 illustrates polar representation of the angular wave functions ψ_{lm} for s, p and d orbitals. Their physical significance is that the value of ψ_{lm} in any direction (θ, ϕ) is proportional to the length of the straight line drawn from the origin to the surface in a specific direction (θ, ϕ). Their squares $|\psi_{lm}|^2$ are the probability distribution functions.

The radial part of the wave function i.e. $R_{nl}(r)$ determines the spatial extent of the orbital. It not only varies with r but does so in a different way for each set of quantum numbers n and l. Their values are given in Table 6.4.

6.6 The Ground State of the Hydrogen Atom

The eigenfunction of the hydrogen atom in its ground state ($1s$ with $n = 1$, $l = 0$, $m = 0$) is given by

$$\psi_{100} = \frac{1}{\sqrt{\pi} \, a_0^{3/2}} e^{-r/a_0} \hspace{3cm} (73)$$

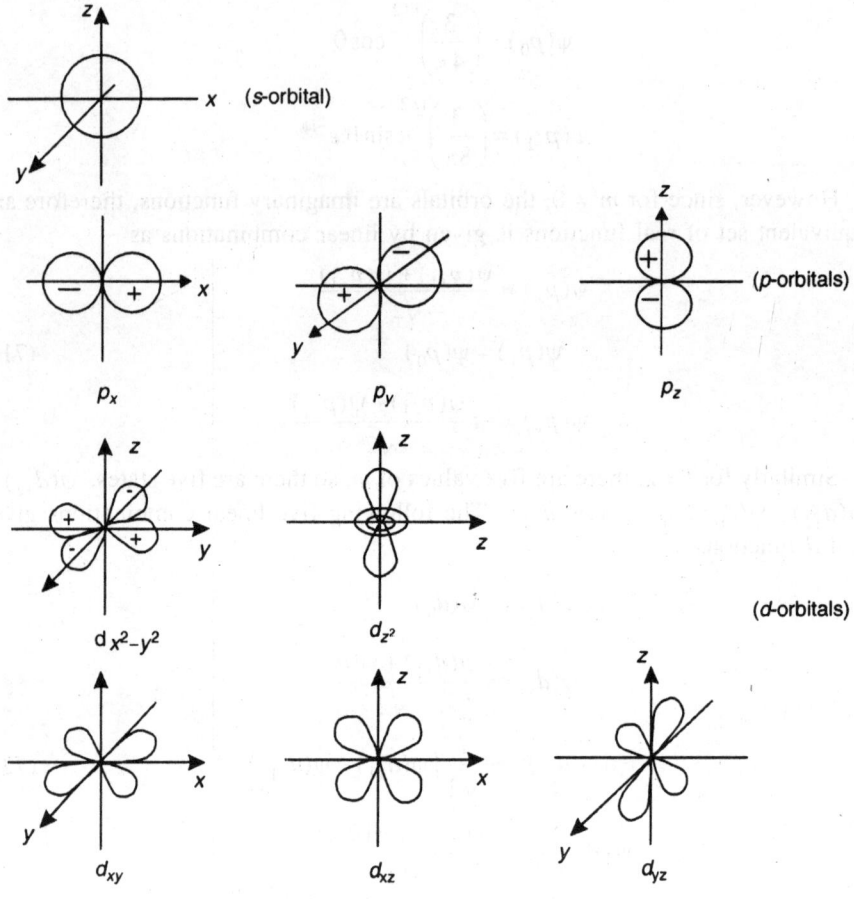

Fig. 6.8: Shapes of *s*-, *p*- and *d*- orbitals

Table 6.4: The functions $R_{nl}(r)$

($\rho = Zr/a_0$ where a_0 = Bohr radius of innermost orbit $\equiv$ 0.529 Å)

$n = 1, l = 0$	1s	$R_{1,0} = 2\left(\dfrac{Z}{a_0}\right)^{3/2} e^{-\rho}$
$n = 2, l = 0$	2s	$R_{2,0} = \dfrac{1}{2\sqrt{2}}\left(\dfrac{Z}{a_0}\right)^{3/2} (2-\rho)e^{-\rho/2}$
$n = 2, l = 1$	2p	$R_{2,1} = \dfrac{1}{2\sqrt{6}}\left(\dfrac{Z}{a_0}\right)^{3/2} \rho e^{-\rho/2}$

contd....

contd....

$n = 3, l = 0$	3s	$R_{3,0} = \dfrac{2}{81\sqrt{3}}\left(\dfrac{Z}{a_0}\right)^{3/2}(27 - 18\rho + 2\rho^2)e^{-\rho/3}$	
$n = 3, l = 1$	3p	$R_{3,1} = \dfrac{4}{81\sqrt{6}}\left(\dfrac{Z}{a_0}\right)^{3/2}(6\rho - \rho^2)e^{-\rho/3}$	
$n = 3, l = 2$	3d	$R_{3,2} = \dfrac{4}{81\sqrt{30}}\left(\dfrac{Z}{a_0}\right)^{3/2}\rho^2 e^{-\rho/3}$	

Table 6.5: The hydrogen like wave functions for $n = 1, 2$ ($\rho = Zr/a_0$)

n	l	m	$\psi_{nlm}(r, \theta, \phi)$
1	0	0	$\dfrac{1}{\sqrt{\pi}}\left(\dfrac{Z}{a_0}\right)^{3/2}e^{-\rho}$
2	0	0	$\dfrac{1}{4\sqrt{2\pi}}\left(\dfrac{Z}{a_0}\right)^{3/2}(2 - \rho)e^{-\rho/2}$
2	1	0	$\dfrac{1}{4\sqrt{2\pi}}\left(\dfrac{Z}{a_0}\right)^{3/2}\rho e^{-\rho/2}.\cos\theta$
2	1	± 1	$\dfrac{1}{8\sqrt{\pi}}\left(\dfrac{Z}{a_0}\right)^{3/2}\rho e^{-\rho/2}.\sin\theta.e^{\pm i\phi}$

Physically, $\psi * \psi = \dfrac{1}{\pi a_0^3}e^{-2r/a_0}$ represents probability distribution function for the electron relative to the nucleus. This expression being independent of θ and ϕ, we infer that the *H*-atom is *spherically symmetrical*. This property is not possessed by the normal (i.e. ground state of) Bohr atom, as the Bohr orbit was restricted to a circular orbit in a single plane.

The probability that the electron lies in the volume element $r^2 \, dr \sin\theta \, d\theta \, d\phi$ is $\psi * \psi \, r^2 \, dr \sin\theta \, d\theta \, d\phi$ and the probability that the electron lies between a spherical shell bounded by radii r and $r + dr$ is obtained by integrating over θ and ϕ to yield

$$D_{100}(r)dr = \frac{4}{a_0^3}r^2 e^{-2r/a_0}dr \tag{74}$$

$D_{100}(r)$ is known as the *radial distribution function*. It is indicated in Fig. 6.9.

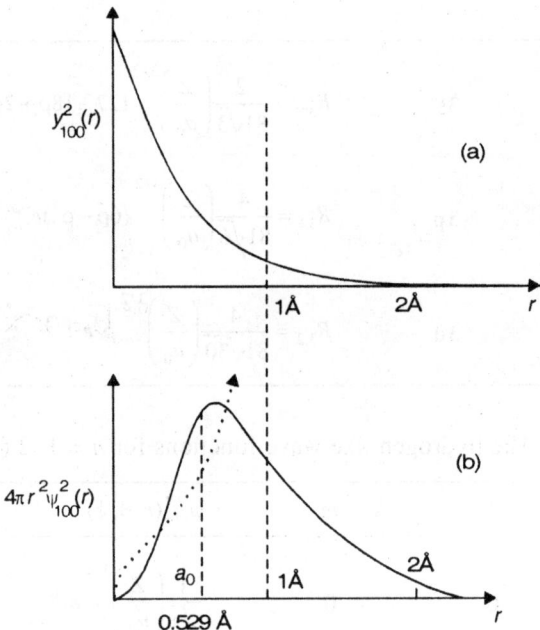

Fig. 6.9: (*a*) Probability density distribution function and (*b*) radial distribution function for the normal hydrogen atom. The dotted curve in Fig. (*b*) represents the probability distribution function for first Bohr orbit

We see that the probability that the electron remain within about 1Å of the nucleus is large and the "size" of the atom is about the same as given by Bohr theory. The most probable distance of the electron from the nucleus (i.e. the value of r for which $D(r)$ is maximum) is found to be $a_0 = 0.529$Å just the same as normal Bohr orbit radius. There is however a significant difference between the Bohr atom and the wave-mechanical atom. In the latter, the electron though most likely to be located at a distance a_0 from the nucleus, may also be elsewhere as indicated by the radial distribution function.

6.7 Discussion of the Hydrogen Like Radial Functions

The radial wave functions $R_{nl}(r)$ for $n = 1, 2, 3$ and $l = 0, 1$ are shown in the Fig. 6.10.

The x-axis represents the values of ρ with $R_{nl}(r)$ plotted along the y-axis. It is to be noted that for all *s*-states, R_{nl} is maximum at $r = 0$, i.e. at the nucleus itself, while for other states, the wave function is zero at $r = 0$. The wave function crosses the ρ-axis $n - l - 1$ times in the region between $\rho = 0$ and $\rho = \infty$.

We are more interested however, in the radial distribution function

$$D_{nl}(r) = 4\pi r^2 \left[R_{nl}(r)\right]^2 \qquad (75)$$

The plot of this function versus r gives the radial distribution curve.

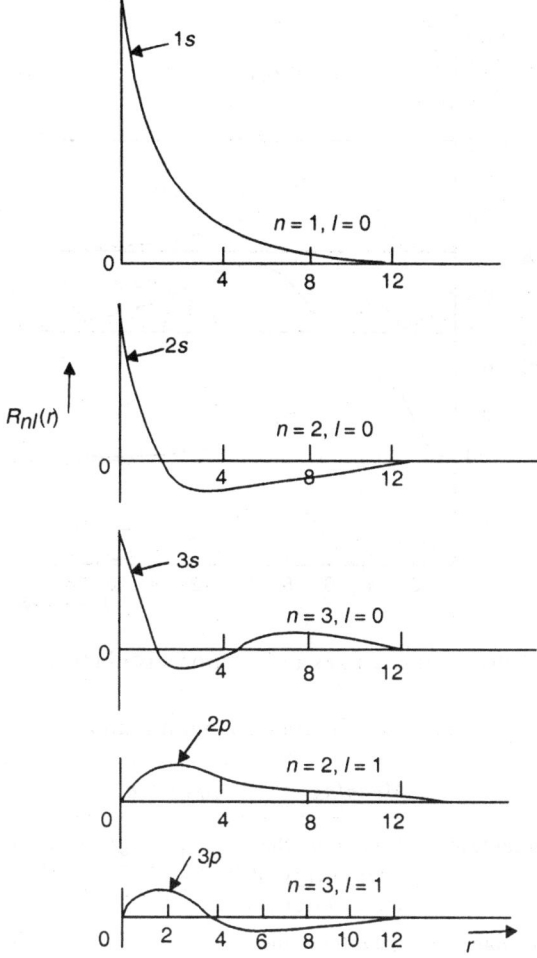

Fig. 6.10: Hydrogen atom radial wave functions

Fig. 6.11 gives all the radial distribution curves corresponding to the radial wave functions shown in the Fig. 6.9.

It is seen that though the electron density is maximum at the centre $r = 0$, the probability of finding the electron exactly there is zero. This is so because, mathematically the centre is a point of zero volume. It is also seen that there is a principal maximum in each of the plots (and subsidiary maxima in some of the plots) and, as n increases, the radius of the principal maximum increases. For example, for 1s and 2s, the principal maximum occurs at about a_0 and $5a_0$ respectively, i.e. the 1s electron is most likely to occur at 0.529Å but the 2s electron at a distance of 5×0.529Å from the nucleus. It is also noticed that for both 1s and 2s, the probability is zero at $r = 0$ and r = ∞, for the 2s state, the probability is zero also at $2 \times 0.529 = 1.058$Å. Table 6.6 summarized the four quantum numbers.

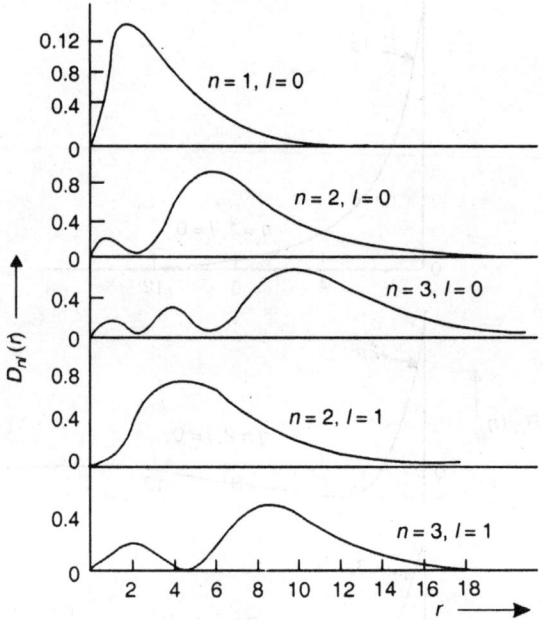

Fig. 6.11: Radial distribution curves for electron distribution in hydrogen atom.

Table 6.6: The four quantum numbers

Symbol	Name	Meaning	Equation	Values
n	Total quantum number	Quantizes the total energy of the electron	$E_n = \dfrac{1}{2} \dfrac{\mu Z^2 e^4}{(4\pi\varepsilon_0)^2 n^2 \hbar^2}$	$n = 1, 2, \ldots$
l	Orbital quantum number	Quantizes the magnitude of the total orbital angular momentum	$L = \sqrt{l(l+1)}\ \hbar$	$l = 0, 1, 2, ..(n-1)$ (n values of l)
m	Magnetic orbital quantum number	Quantizes the z-component of the the total orbital angular momentum. This is called spatial quantization	$L_z = m\hbar$	$m = -l.., 0, ..,+l$ ($2l+1$ values of m)
m_s	Magnetic spin quantum number	Quantizes the z-component of the spin	$s_z = m_s\hbar$	$m_s = -\tfrac{1}{2}, +\tfrac{1}{2}$

The formulae for average values of various powers of r are as follows:

$$\langle r \rangle = \frac{n^2 a_0}{Z} \left[1 + \frac{1}{2} \left\{ 1 - \frac{l(l+1)}{n^2} \right\} \right]$$

$$\langle r^2 \rangle = \frac{a_0^2 n^4}{Z^2} \left[1 + \frac{3}{2} \left\{ 1 - \frac{l(l+1) - \frac{1}{3}}{n^2} \right\} \right]$$

$$\left\langle \frac{1}{r} \right\rangle = \frac{Z}{a_0 n^2}$$ (76)

$$\left\langle \frac{1}{r^2} \right\rangle = \frac{Z^2}{a_0^2 n^3 (l + \frac{1}{2})}$$

$$\left\langle \frac{1}{r^3} \right\rangle = \frac{Z^3}{a_0^3 n^3 l (l + 1/2)(l + 1)}$$

6.8 Degeneracy

The energy of a nth atomic state of a hydrogen like atom (i.e. defined by the principal quantum number n) is given by

$$E_n = \frac{1}{2} \frac{\mu Z^2 e^4}{(4\pi\varepsilon_0)^2 n^2 \hbar^2}$$ (77)

Except for $n = 1$, each energy level n is degenerate with respect to both l and m and each level is therefore represented by more than one independent solution of the wave equation.

Total quantum number n assumes values $n = 0, 1, 2, 3, 4, \ldots$ For each value of n there are angular momentum quantum numbers $l = 0, 1, 2, 3, 4, \ldots (n - 1)$. For each of these l values, magnetic quantum number takes values $m = -l, -l +1, \ldots, -1, 0, +1, \ldots, (l - 1), l$ (These are $2l + 1$ values). The quantum number m specifies the relative distance of the electron from the nucleus. The orbital quantum number l describes the way in which the angular momentum of the electron is quantized (and has a small effect on the energy). It governs the strength with which an electron is attached with the nucleus, larger the value of l, smaller is the bond strength. The magnetic quantum number m is effective in presence of an externally applied magnetic field and is a measure of the angle between the electron's angular momentum vector and the applied magnetic field. Each different combination of n, l and m corresponds to a unique quantum state ψ_{nlm} called an *orbital*. According to Pauli's principle, each orbital can contain no more than two electrons and these two must have opposite spins (designated by a fourth quantum number m_s, which can have values of only $+\frac{1}{2}$ or $-\frac{1}{2}$).

Thus there are $(2l + 1)$ independent wave functions for a given value of n and l and the number of independent wave functions for a given value of n is given by

$$\sum_{l=0}^{n-1} (2l + 1) = 2\sum_{0}^{n-1} l + \sum_{l=0}^{n-1} 1 = 2\frac{(n-1)n}{2} + n = n^2$$

Table 6.7: Possible states of an electron

	K	L						M				
$n =$	1	2						3				
$l =$	0	0	1			0	1			2		
	s	s	p			s	p			d		
m	0	0	−1	0	+1	0	−1	0	+1	−2	−1	0	+1	+2
m_s	↑↓	↑↓	↑↓	↑↓	↑↓	↑↓	↑↓	↑↓	↑↓	↑↓	↑↓	↑↓	↑↓	↑↓

Table 6.7 shows, for the possible states of an electron in an atom, the division into groups and subgroups up to $n = 3$. Each of the before mentioned subgroup (with given n and l) is once more subdivided according to the values of m. All of the latter subgroups of states have the same energy for a given n and given l (in the absence of the magnetic field). Each of these ψ_{nlm} is once more subdivided into two subgroups (with $m_s = +\frac{1}{2}$ and $m_s = -\frac{1}{2}$) represented by an arrow with up direction and down direction (respectively).

The maximum number of electrons which can have the same n and l and yet not violate the Pauli's principle is given by the number of arrows in the table between the corresponding vertical lines and is obviously equal to $2(2l+1)$, since there are $(2l+1)$ possible values of m for a given l. Figure 6.12 depicts different eigen states in a hydrogen like atom.

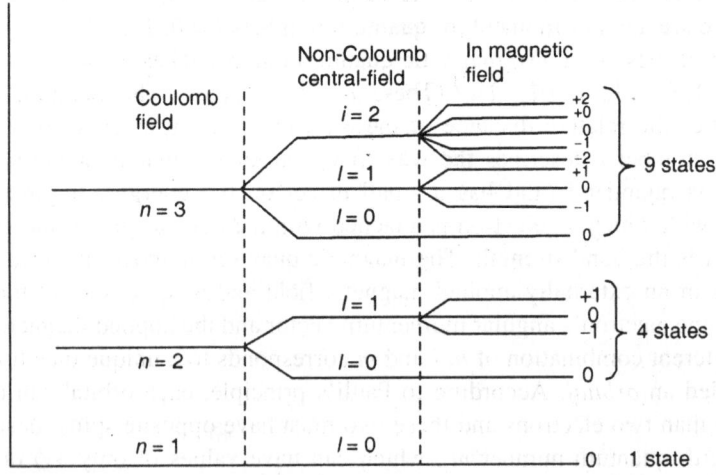

Fig. 6.12: Different eigenstates in hydrogen like atom

The degeneracy with respect to l and m is a characteristic of the central force field, i.e. the force field depending only on the radial distance r, while the $(2l+1)$ degeneracy with respect to m disappears in presence of some external field (e.g. a magnetic field) and the level is split up into n^2 different energy levels.

Example 6.2: Show that the three 2p-eigenfunctions of hydrogen atom are orthogonal to each other.

Solution: The three 2p-eigenfunctions are given by

$$\psi_{2,1,0} = c_1 \rho e^{-\rho/2} \cos \theta$$

and

$$\psi_{2,1,\pm1} = c_2 \rho e^{-\rho/2} \sin \theta \, e^{\pm i\phi}$$

where c_1 and c_2 are constants and $\rho = r/a_0$. In the integral $\int \psi^*_{2,1,1} \, \psi_{2,1,-1} d\tau$, the ϕ dependent term is $e^{-2i\phi}$. The corresponding integral gives

$$\int_0^{2\pi} e^{-2i\phi} d\phi = -\frac{1}{2i} \left[e^{-2i\phi} \right]_0^{2\pi} = 0$$

Similarly, ϕ integral terms of $\int \psi^*_{2,1,0} \, \psi_{2,1,1} d\tau$ and $\int \psi^*_{2,1,0} \, \psi_{2,1,-1} d\tau$ are respectively

$$\int_0^{2\pi} e^{i\phi} d\phi = 0$$

and

$$\int_0^{2\pi} e^{-i\phi} d\phi = 0 \quad .$$

Hence the result.

6.9 The Free Particle

In case of a free particle, the time independent Schrödinger equation is

$$-\frac{\hbar^2}{2m} \nabla^2 \psi(\vec{r}) = E \psi(\vec{r}) \tag{78}$$

Its plane wave solution is

$$\psi(\vec{r}) = A \exp(i \, \vec{p} \cdot \vec{r}) \tag{79}$$

where

$$\vec{p} = \hbar \vec{k} \quad \text{and} \quad |\vec{k}| = \frac{\sqrt{2mE}}{\hbar}$$

Spherical Wave Solution:

We can express $\nabla^2 \psi(\vec{r})$ in spherical coordinates and express ψ as

$$\psi(r, \theta, \phi) = R(\vec{r}) \; \Theta(\theta) \Phi(\phi) \tag{80}$$

The solution of the angular part are the spherical harmonics $Y_{lm}(\theta, \phi)$. The radial equation, with $V = 0$ assumes the form

$$\frac{1}{r^2} \frac{\partial}{\partial r} \left(r^2 \frac{\partial R_l}{\partial r} \right) + \left[\frac{2mE}{\hbar^2} - \frac{l(l+1)}{r^2} \right] R_l(r) = 0 \tag{81}$$

Letting

$$k^2 = \frac{2mE}{\hbar^2} \quad \text{and} \quad \rho \equiv kr$$

we have

$$\frac{\partial^2 R_l}{\partial \rho^2} + \frac{2}{\rho} \frac{\partial R_l}{\partial \rho} + \left[1 - \frac{l(l+1)}{\rho^2} \right] R_l(\rho) = 0 \tag{82}$$

Substituting

$$R(\rho) = P(\rho)\, \rho^{-1/2} \tag{83}$$

We can write Eqn. (82) as

$$\frac{\partial^2 P}{\partial \rho^2} + \frac{1}{\rho}\frac{\partial P(\rho)}{\partial \rho} + \left\{ 1 - \left[\frac{l+\frac{1}{2}}{\rho}\right]^2 \right\} P(\rho) = 0$$

or

$$\rho^2 \frac{\partial^2 P}{\partial \rho^2} + \rho \frac{\partial P}{\partial \rho} + \left[\rho^2 - (l+\tfrac{1}{2})^2 \right] P = 0 \tag{84}$$

This is Bessel's equation. Its general solution is

$$P(\rho) = A' J_{l+\frac{1}{2}}(\rho) + B' J_{-(l+\frac{1}{2})}(\rho) \tag{85}$$

which gives

$$R(\rho) = \frac{A'}{\rho^{1/2}} J_{l+\frac{1}{2}}(\rho) + \frac{B'}{\rho^{1/2}} J_{-(l+\frac{1}{2})}(\rho)$$

$$= A j_l(\rho) + B \eta_l(\rho) \tag{86}$$

where A and B are constants and $j_l(\rho)$ and $\eta_l(\rho)$ are the spherical Bessel functions and spherical Neumann functions respectively. These are defined as

$$j_l(\rho) = \sqrt{\frac{\pi}{2\rho}} \, J_{l+\frac{1}{2}}(\rho) \tag{87}$$

and

$$\eta_l(\rho) = (-1)^{l+1} \sqrt{\frac{\pi}{2\rho}} \, J_{-(l+\frac{1}{2})}(\rho) \tag{88}$$

Now, we know that Bessel functions J_l are given by

$$J_{(l+\frac{1}{2})}(\rho) = \sum_{s=0}^{\infty} \frac{(\rho/2)^{l+\frac{1}{2}+2s}}{s! \, \Gamma\left(l+s+\frac{3}{2}\right)} \tag{89}$$

Further

$$\Gamma\left(\frac{3}{2}\right) = \frac{1}{2}\Gamma\left(\frac{1}{2}\right) = \frac{\sqrt{\pi}}{2}$$

and

$$\frac{\sin\rho}{\rho} = \left[1 - \frac{\rho^2}{3!} + \frac{\rho^4}{5!} - \cdots \right] = \sum_{p=0}^{\infty} (-1)^p \frac{\rho^{2p}}{(2p+1)!} \tag{90}$$

Then from Eqn. (89), we get

$$J_{1/2}(\rho) = \frac{\rho^{1/2}}{\sqrt{2}\,\Gamma\left(\frac{3}{2}\right)} \left(1 - \frac{\rho^2}{3!} + \frac{\rho^4}{5!} - \cdots \right) = \left(\frac{2}{\pi\rho}\right)^{1/2} \sin\rho \tag{91}$$

and

$$J_{-1/2}(\rho) = \left(\frac{2}{\pi\rho}\right)^{1/2} \cos\rho \tag{92}$$

It is also known that the Bessel function of the first kind and of order n are given by

$$J_n(x) = \sum_{s=0}^{\infty} \frac{(-1)^s x^{2s+n}}{s! \; 2^{2s+n} \Gamma(n+s+1)} \tag{93}$$

This series functions give the recurrence relation

$$\left(\frac{2n}{\rho}\right) J_n = J_{n-1} + J_{n+1} \tag{94}$$

From Eqns (87), (88), (90), (91), (92) and (94), we get

$$\left. \begin{aligned} j_0 &= \frac{\sin\rho}{\rho}, & j_1 &= \frac{\sin\rho}{\rho^2} - \frac{\cos\rho}{\rho} \\[2mm] \eta_0 &= -\frac{\cos\rho}{\rho}, & \eta_1 &= -\frac{\cos\rho}{\rho^2} - \frac{\sin\rho}{\rho} \end{aligned} \right| \tag{95}$$

In closed form, these relations can be expressed as

$$j_l(\rho) = (-\rho)^l \left(\frac{1}{\rho}\frac{d}{d\rho}\right)^l \frac{\sin\rho}{\rho}$$

$$\left. \begin{aligned} j_l(\rho) &= (-\rho)^l \left(\frac{1}{\rho}\frac{d}{d\rho}\right)^l \frac{\sin\rho}{\rho} \\[2mm] \eta_l(\rho) &= (-\rho)^l \left(\frac{1}{\rho}\frac{d}{d\rho}\right)^l \left(-\frac{\cos\rho}{\rho}\right) \end{aligned} \right| \tag{96}$$

These functions are shown in Fig. 6.13. The solutions corresponding to a definite energy $E\left(=\dfrac{\hbar^2 k^2}{2m}\right)$ and a definite orbital angular momentum $\sqrt{l(l+1)}\,\hbar$ can be written as

$$\psi_{k,l}(r,\theta,\phi) = \left[A_l j_l(kr) + B_l \eta_l(kr) \right] Y_{lm}(\theta,\phi) \tag{97}$$

where $Y_{lm}(\theta,\phi)$ are spherical harmonics. The most general solution for a definite energy are

$$\psi_k(r,\theta,\phi) = \sum_{l=0}^{\infty} \left[A_l j_l(kr) + B_l \eta_l(kr) \right] Y_{lm}(\theta,\phi) \tag{98}$$

Since ψ must be finite everywhere we have $B_l = 0$ (because $\eta_l(kr)$ is not finite at the origin). Thus

$$\psi_k(r,\theta,\phi) = \sum_{l=0}^{\infty} A_l Y_{lm}(\theta,\phi) j_l(kr) = \sum_{l=0}^{\infty} R_{kl} Y_{lm}(\theta,\phi) \tag{99}$$

where

$$R_{kl} = A_l j_l(kr) = A_l(-1)^l \frac{r^l}{k^l}\left(\frac{1}{r}\frac{d}{dr}\right)^l \frac{\sin kr}{kr} \tag{100}$$

describe the stationary spherical waves.

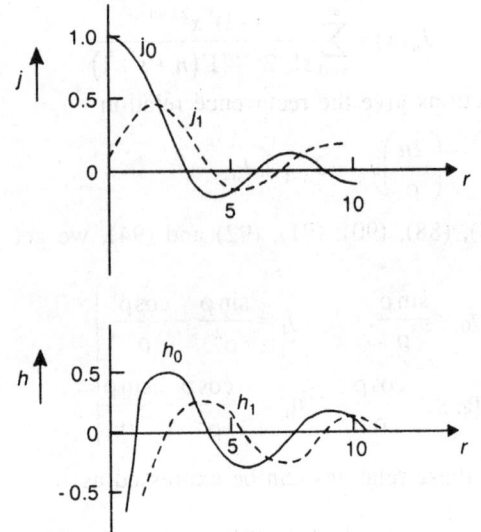

Fig. 6.13: (*a*) Spherical Bessel functions $j_l(\rho)$, $l = 0, 1$, and
(*b*) Spherical Neumann functions $\eta_l\ (\rho)$, $l = 0, 1$

6.10 Qualitative Discussion of Spherical Harmonics

6.10.1 Schrödinger Equation for Radial Symmetry

A central force acting on an object of mass μ (e.g. coulomb force on an electron in a hydrogen atom), depends only on the magnitude of the radial distance from the origin (centre of mass) and is directed along the line from the origin to the object.

In Classical physics, an important feature of a central force is that there is no torque relative to the origin on the object under the influence of the force, consequently, the angular momentum is conserved. We expect some thing similar in quantum mechanics.

Radial symmetry of central forces allows us to make use of spherical coordinates (*r*, θ, φ), as in Fig. 6.14.

Here, the potential *V* depends only on the radial variable *r*. The Schrödinger equation is in carterian coordinates:

$$ih\frac{\partial\psi(\vec{r},t)}{\partial t} \ = \ \frac{-\hbar^2}{2\mu}\Delta^2\psi(\vec{r},t)+V(\vec{r})\psi(\vec{r},t)$$

Where, $$\nabla^2 \ = \ \frac{\partial^2}{\partial x^2}+\frac{\partial^2}{\partial y^2}+\frac{\partial^2}{\partial z^2} \tag{101}$$

For incorporating change of variables from (*x*, *y*, *z*) to (*r*, θ, φ) we first remove the time dependence by writing eigenstates (of energy):

$$\psi(\vec{r},t) \ = \ \psi(\vec{r})e^{-iEt/\hbar} \tag{102}$$

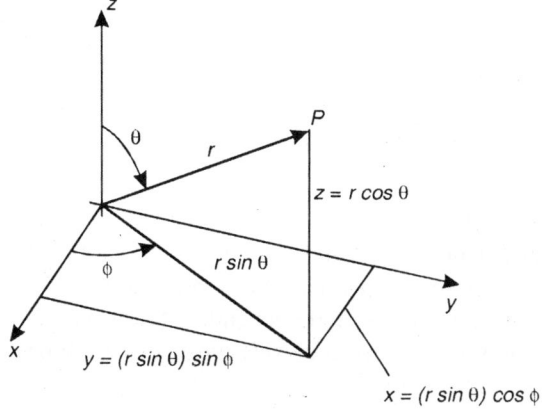

Fig. 6.14: Spherical (polar) coordinates

As a result, we are left with the equation

$$-\frac{\hbar^2}{2\mu}\left(\frac{\partial^2}{\partial x^2}+\frac{\partial^2}{\partial y^2}+\frac{\partial^2}{\partial z^2}\right)\psi(\vec{r})+V(r)\psi(\vec{r})=E\psi(\vec{r}) \qquad (103)$$

The vector $\vec{r}$ is the vector from the centre of force to the particle of mass μ under the influence of this force. Now,

$x = r\sin\theta\cos\phi,\ y = r\sin\theta\sin\phi,\ z = r\cos\theta$ (from Fig. 6.14)

Where θ is the polar angle and ϕ is the azimuthal angle.

In spherical coordinates, the time independent Schrödinger equation becomes

$$\frac{1}{r^2}\frac{\partial}{\partial r}\left(r^2\frac{\partial\psi}{\partial r}\right)+\frac{1}{r^2\sin\theta}\frac{\partial}{\partial\theta}\left(\sin\theta\frac{\partial\psi}{\partial\theta}\right)$$

$$+\frac{1}{r^2\sin\theta}\frac{\partial^2\psi}{\partial\phi^2}+\frac{2\mu}{\hbar^2}(E-V)\psi=0 \qquad (104)$$

or

$$+-\frac{\hbar^2}{2\mu}\left[\frac{\partial^2}{\partial r^2}+\frac{2}{r}\frac{\partial}{\partial r}+\frac{1}{r^2}\left(\frac{\partial^2}{\partial\theta^2}+\cot\theta\frac{\partial}{\partial\theta}+\frac{1}{\sin^2\theta}\frac{\partial^2}{\partial\phi^2}\right)\right]\psi(\vec{r})$$

$$+V(r)\ \psi(\vec{r})=E\ \psi(\vec{r}) \qquad (105)$$

This is an eigenvalue equation which is a partial differential equation which can be reduced to a set of one dimensional differential equations in r, or θ or ϕ that can be solved directly. For this, we employ the method of separation of variables and write a trial solution of the form:

$$y(\vec{r})=R(r)\ Y(\theta,\phi) \qquad (106)$$

Here, the function $R(r)$ is known as the *radial wave function* and $Y(\theta,\phi)$, which we can find explicitly, is known as a *spherical harmonic*.

When Eqn. (106) is substituted into Eqn. (105) the derivatives taken w.r.t.r act only on $R(r)$ and the derivatives taken with respect to θ and ϕ act only on $Y(\theta, \phi)$. Aside from these derivatives, functions R and Y are purely multiplicative. Then, if we next divide our equation by the product RY, we get

$$-\frac{\hbar^2}{2\mu}\frac{1}{R(r)}\left(\frac{\partial^2}{\partial r^2}+\frac{2}{r}\frac{\partial}{\partial r}\right)R(r)-\frac{\hbar^2}{2\mu}\frac{1}{r^2}\frac{1}{Y(\theta,\phi)}$$

$$\times\left(\frac{\partial^2}{\partial\theta^2}+\cot\theta\frac{\partial}{\partial\theta}+\frac{1}{\sin^2\theta}\frac{\partial}{\partial\phi^2}\right)Y(\theta,\phi)+V(r)=E \qquad (107)$$

If we now multiply by r^2, we see that the second term on the left is exclusively dependent on θ and ϕ and the 1st and 3rd terms on the left as well as the E term on the right are exclusively dependent on r. We can group all the r-dependent terms on the left side and all the angle dependent terms on the right (the method of seperation of variables). Then, each side must separately be a constant say λ. Then we get these two equations

$$\left(\frac{\partial^2}{\partial\theta^2}+\cot\theta\frac{\partial}{\partial\theta}+\frac{1}{\sin^2\theta}\frac{\partial^2}{\partial\phi^2}\right)Y(\theta,\phi)=\lambda Y(\theta,\phi) \qquad (108)$$

$$r^2\left\{-\frac{\hbar^2}{2\mu}\left[\frac{d^2}{dr^2}+\frac{2}{r}\frac{d}{dr}\right]+V(r)-E\right\}R(r)=\lambda R(r) \qquad (109)$$

or $\qquad -\frac{\hbar^2}{2\mu}\left(\frac{d^2R(r)}{dr^2}+\frac{2}{r}\frac{dR(r)}{dr}+\lambda\frac{R(r)}{r^2}\right)+V(r)R(r)=ER(r) \qquad (110)$

This Eqn. (110) for R is an ordinary differential equation and is an eigenvalue equation that has solutions only for certain values of energy E.

Eqn. (108) is more cumbersome which we know, can he solved by utilizing still another separation of variables. We write

$$Y(\theta,\phi)=F(\theta)\,\Phi\,(\phi) \qquad (111)$$

When this is substituted into Eqn. (108), by similar reasonning, we get

$$\frac{\partial^2\Phi(\phi)}{\partial\phi^2}=-m^2\Phi(\phi) \qquad (112)$$

and $\qquad \dfrac{d^2F(\theta)}{d\theta^2}+\cot\theta\dfrac{dF(\theta)}{d\theta}-\dfrac{m^2}{\sin^2\theta}\,F(\theta)=\lambda F(\theta) \qquad (113)$

where m^2 is our new "separation constant" for separation of Y into F and Φ.

Each of the above two equations is an ordinary differential equation, involving the single variables ϕ and θ respectively. We can solve these equations, even without knowledge of the potential, since the potential does not appear in either of the equations.

6.10.2 Probabilistic Interpretation of the Wave Function

We know that $|\psi(x)|^2\,dx$ represents the probability of finding the particle within a gap of width dx at the location dx. For three dimensions this is given by $|\psi(\vec{r})|^2$ $dxdydz$. It is the probability of finding the particle within a box of volume $d^3\vec{r}$ ($=dxdydz$) about the point $\vec{r}\,(x,y,z)$. In spherical coordinates the volume of the box is $r^2dr\sin\theta\,d\theta\,d\phi$ and probability of finding the particle within this box is

$$|\psi(\vec{r})|^2\,r^2\,dr\,\sin\theta\,d\theta\,d\phi \qquad (114)$$

Using Eqn (106), the normalisation condition is

$$1 = \int |R(r)|^2 |Y(\theta,\phi)|^2 \; r^2 \, dr \, \sin\theta \, d\theta \, d\phi$$

$$= \int_0^\infty |R(r)|^2 \, r^2 dr \times \int |Y(\theta,\phi)|^2 \, \sin\theta \, d\theta \, d\phi$$

There are now two separate normalisation conditions to apply, that the probability of finding the particle at some radius (r) among all possible radii ($r = 0$ to ∞) is unity and that the probability of finding at some angle (θ, ϕ) among all possible angles (θ from $-\pi$ to $+\pi$ and ϕ, 0 to 2π) is unity. Mathematically

$$\int_0^\infty |R(r)|^2 \, r^2 dr = 1 \tag{116}$$

and

$$\int_{\phi=0}^{2\pi} \int_{\theta=-\pi}^{+\pi} |Y(\theta,\phi)|^2 \sin\theta \, d\theta = 1 \tag{117}$$

6.10.3 Solving for Spherical Harmonics

We can start with the equation for $\Phi(\phi)$.

The solution is

$$\Phi(\phi) = Ce^{im\phi} \tag{118}$$

Where C is any contant. There is a constraint on m which follows from the fact that when we increase the angle ϕ by 2π, we are going around a circle about the origin in the xy plane, and under these circumstances, we do not want the solution to give us something new i.e. $\Phi(\phi)$ must be single valued, for this

$$\Phi(\phi + 2\pi) = \Phi(\phi) \tag{119}$$

or $\qquad C \exp [im (\phi + 2\pi)] = C \exp (im \, \phi)$

or $\qquad\qquad \exp (2\pi \, im) = 1 \tag{120}$

This last equation can be satisfied only if m is an integer, i.e.,

$$m = 0, \pm 1, \pm 2, \pm 3 \; \dots\dots \tag{121}$$

We have seen that the solution of Eqn. (113) for $F(\theta)$ is more difficult. We shall content ourselves here with a very brief discussion. Let us start with some conditions on its solution. We see that both m and λ appear in the equation. We have already found restrictions on m from the equation for Φ so, we can say that only λ appears as a parameter in the equation for F. But, in fact, this equation can be satisfied in a physically acceptable manner only if

$$\lambda = -l(l+1) \tag{122}$$

Where $\qquad \lambda = 0, 1, 2, 3 \dots. \tag{123}$

is a positive integer. Moreover, Eqn. (113) imposes a further restriction on m, namely, given l, (rather than being given all integer values), m can run only from $-l$ to $+l$ i.e. $m = -l, -l + 1, -l + 2, \dots -1, 0, +1 \dots. l - 1, l \tag{124}$

The structure of the Schrödinger equation has forced two parameters l and m to be integers. We know that these are the quantum numbers. In fact, these are associated with the quantisation of angular momentum.

Given a value of l (some positive integer), the function $F(\theta)$ is an lth order, polynomial in $\sin\theta$ and $\cos\theta$. Here, rather than presenting $F(\theta)$ separately, we write the combination $Y(\theta, \phi)$, the product of Φ and F. Here, F depends on both

l and m and Φ depends on m alone. Thus the form of the spherical harmonic $Y(\theta, \phi)$ depends on both l and m, and is accordingly labelled $Y_{lm}(\theta, \phi)$. Table 6.7 lists some spherical harmonics.

Table 6.7: Spherical harmonics

Y_{00}	$= \dfrac{1}{\sqrt{4\pi}}$		Y_{11}	$= \sqrt{\dfrac{3}{8\pi}}\sin\theta\exp(i\phi)$
Y_{10}	$= \sqrt{\dfrac{3}{4\pi}}\cos\theta$		Y_{22}	$= \sqrt{\dfrac{15}{32\pi}}\sin^2\theta\exp(2i\phi)$
Y_{21}	$= -\sqrt{\dfrac{15}{8\pi}}\sin\theta\cos\theta\exp(i\phi)$		Y_{20}	$= \sqrt{\dfrac{5}{16\pi}}\left(3\cos^2\theta-1\right)$

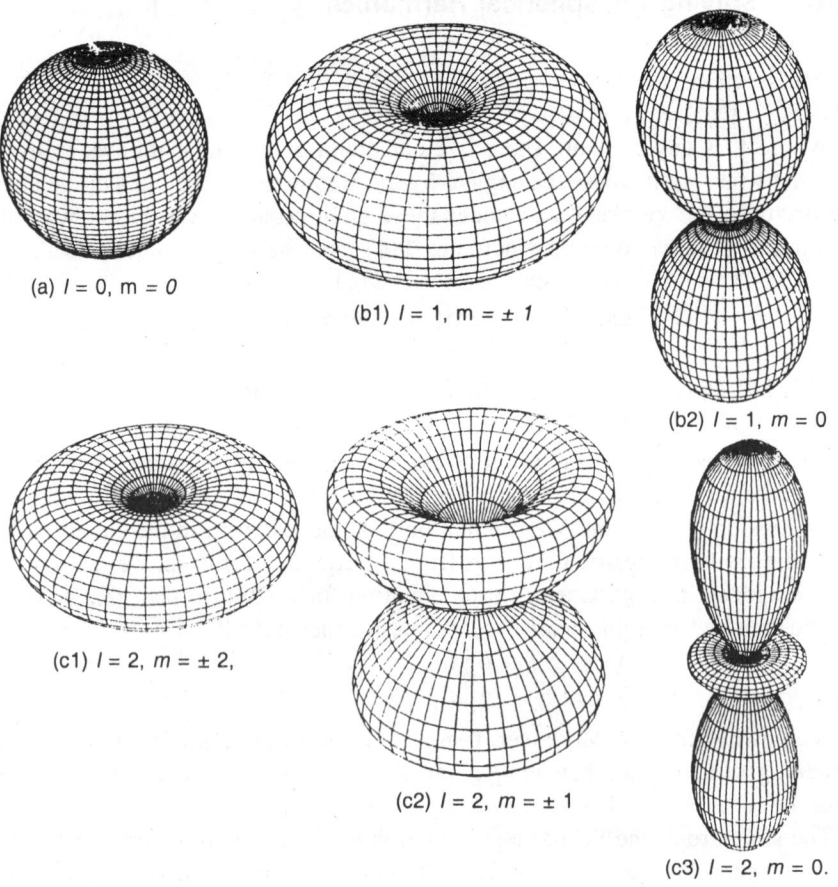

(a) $l = 0$, m = 0

(b1) $l = 1$, m = ± 1

(b2) $l = 1$, m = 0

(c1) $l = 2$, m = ± 2,

(c2) $l = 2$, m = ± 1

(c3) $l = 2$, m = 0.

Fig. 6.15: $|Y_{lm}|^2$ represents the probability of finding the electron in a solid angle $\sin\theta\, d\theta\, d\phi$, the distributions shown are for (a) $l = 0$, $m = 0$, (b) $l = 1$, $m = \pm 1$; $l = 1$, $m = 0$ and (c) $l = 2$, $m = \pm 2$, $l = 2$, $m = \pm 1$ and $l = 2$, $m = 0$

We can systematically verify by direct substitution that Y's are solutions of Eqn. (108) Let us, for expample, consider the case of Y_{00}.

$$\iint |Y_{00}(\theta,\phi)|^2 \sin\theta\, d\theta\, d\phi = \left(\frac{1}{\sqrt{4\pi}}\right)^2 \iint \sin\theta\, d\theta\, d\phi \qquad (125)$$

Now
$$\iint \sin\theta\, d\theta\, d\phi = \iint d(\cos\theta)\, d\phi = \int_{-1}^{+1} d(\cos\theta) \int_{0}^{2\pi} d\phi = 4\pi \qquad (126)$$

which is just the total solid angle about a sphere. Thus, the Y_{00} is correctly normalized.

Here, we can note generally that if the wave function of a particle under the influence of a central force (e.g. an electron in hydrogen atom)–is proportional to $Y_{lm}(\theta,\phi)$, Then the probability distribution (given by $|Y_{lm}|^2$ is independent of the azimuthal angle ϕ. This is, because $|Y_{lm}|^2$ contains $|\exp(im\phi)|^2$ and this equals unity. Figure 6.15 is a plot of the probability distribution for the l and m values of Table 6.7 as a function of θ.

Questions and Problems

1. Write the Schrödinger wave equation for a rigid rotator. Obtain its eigenfunctions and eigenvalues. How is the rotational spectrum of diatomic molecules explained by these results?

2. What is meant by a rigid rotator? Write Schrödinger wave equation for a free axis rigid rotator and solve it to obtain eigen functions and eigenvalues.

3. A diatomic molecule is free to rotate in a fixed plane. Obtain the Schrödinger equation to describe its motion and show that the eigenvalues of its rotational energy are given as $E_l = l(l+1)\hbar^2/(2I)$ where I is the moment of inertia of the molecule and l is the rotational quantum number.

4. Solve the wave equation for hydrogen atom and obtain an expression for eigenvalues and normalized wave function for the lowest state.

5. The normalized ground state wave function of a single electron in an atom with a nuclear charge Ze is

$$\psi(r) = 2\left(\frac{Z}{a_0}\right)^{3/2} e^{-Zr/a_0}$$

where a_0 is the Bohr radius. Show that the expectation value of the distance of the electron from the nucleus is

$$\langle r \rangle = \frac{3a_0}{2Z}$$

6. Explain how one can solve the problem of the hydrogen atom quantum mechanically. Solve the radial part of the Schrödinger equation for H-atom and obtain the energy eigenvalues. Explain the degeneracy in the spectrum.

7. Show that $E_n = \frac{1}{2} <V>$ in the stationary state of the hydrogen atom.

8. Show that the vector $\vec{A} = \frac{1}{2\mu}(\vec{L}\times\vec{p} - \vec{p}\times\vec{L}) - e^2\frac{\vec{x}}{r}$ is a constant of motion for the hydrogen atom.

9. Describe the four quantum numbers of an electron in an atom.

10. What are atomic orbitals. Give polar representations of angular part of s-, p- and d-orbitals.

11. Find the momentum distribution of the electron in the hydrogen atom in the 1s-, 2s- and 2p- states.

12. Evaluate the most probable distance of the electron of the hydrogen atom in its $2p$ state.

    [**Hint:** $P_{nl}(r) = r^2 |R_{nl}|^2$ and $d\rho_{21}/dr = 0$)   [**Ans.** $R = 4a_0$]

13. Show that the size of the hydrogen atom in its ground state (i.e. $\langle r^2 \rangle^{1/2}$) is $\sqrt{3}a_0$.

14. Set up and solve the wave equation for a rigid rotator that has no kinetic energy of rotation about a particular axis and has equal moments of inertia about the two perpendicular axes.

7

Matrix Formulation of Quantum Mechanics

7.1 Introduction

It has been already mentioned in chapter 2 that development of quantum mechanics took place according to two schools of thought, the wave mechanics and the matrix mechanics. The former one, developed by Schrödinger, stressed on the idea that position of continuity (waviness) is an essential element of nature and apparently, the observed discreteness of certain atomic phenomena followed from the requirement that the wave function be continuòus. On the other hand, Heisenberg favoured the idea of discrete quantum jumps as an essential feature of atomic phenomena, which was described by him by matrices. In this regard, he abandoned the search for an underlying picture and dealt directly with collection of discrete numbers representing the observed quantum transitions in the form of matrices.

Before dealing matrix-formulation of wave functions, it would be helpful to first discuss general properties of matrices.

7.2 Properties of Matrices

(1) Transpose Matrix: A^T:

The transpose of an arbitrary martix A is written as A^T and is obtained by interchanging corresponding rows and columns of A i.e. $A = a_{ij}$ and $A^T = a_{ji}$, for example

$$A = \begin{pmatrix} 1 & -2 & 3 \\ 2 & 0 & 1 \end{pmatrix} ; \quad A^T = \begin{pmatrix} 1 & 2 \\ -2 & 0 \\ 3 & 1 \end{pmatrix}$$

(2) Complex conjugate matrix: A^*:

The complex conjugate of an arbitrary matrix A is formed by taking the complex conjugate of each element i.e. $A^* = a_{ij}^*$ (for all i and j), e.g.

$$A = \begin{pmatrix} 3 + 2i & 4 - 2i \\ 3i & 4 \end{pmatrix} ; \quad A^* = \begin{pmatrix} 3 - 2i & 4 + 2i \\ -3i & 4 \end{pmatrix}$$

(3) Hermitian conjugate :A†

The Hermitian conjugate of an arbitrary matrix A is denoted by $A^\dagger$ and is obtained by taking the complex conjugate of the matrix and then the transpose of the complex conjugate matrix, for example

$$A = \begin{pmatrix} 2-3i & 4+5i \\ 3 & -4i \end{pmatrix}$$

$$A^\dagger = \begin{pmatrix} 2+3i & 4-5i \\ 3 & 4i \end{pmatrix}^T = \begin{pmatrix} 2+3i & 3 \\ 4-5i & 4i \end{pmatrix}$$

(4) Addition of Matrices:

Two matrices A and B can be added like ordinary numbers if they have the same number of rows and columns. $A + B = C$. In terms of components (elements) $c_{ij} = a_{ij} + b_{ij}$, e.g.

$$A = \begin{pmatrix} 4 & 3 \\ 5 & 7 \end{pmatrix} \text{ and } B = \begin{pmatrix} 3 & 2 \\ 4 & 1 \end{pmatrix}$$

$$C = A + B = \begin{pmatrix} 4+3 & 3+2 \\ 5+4 & 7+1 \end{pmatrix} = \begin{pmatrix} 7 & 5 \\ 9 & 8 \end{pmatrix}$$

Similarly, we can define subtraction of matrices.

(5) Multiplication of a matrix by a constant:

If $A = \begin{pmatrix} a_{11} & a_{12} \\ a_{21} & a_{22} \end{pmatrix}$ then if λ is some constant, scalar product

$$\lambda A = \lambda \begin{pmatrix} a_{11} & a_{12} \\ a_{21} & a_{22} \end{pmatrix} = \begin{pmatrix} \lambda a_{11} & \lambda a_{12} \\ \lambda a_{21} & \lambda a_{22} \end{pmatrix}$$

defines multiplication by a constant.

(6) Multiplication of Matrices:

A matrix A can be multiplied by the other matrix B if the number of columns in A is equal to the number of rows in B.

$$(A)_{mn}(B)_{nr} = (C)_{mr} \quad \text{or}$$

$$\begin{pmatrix} a_{11} & a_{12}\cdots & a_{1n} \\ a_{21} & a_{22}\cdots & a_{2n} \\ a_{m1} & a_{m2}\cdots & a_{mn} \end{pmatrix} \begin{pmatrix} b_{11} & b_{12}\cdots & b_{1r} \\ b_{21} & b_{22}\cdots & b_{2r} \\ b_{n1} & b_{n2}\cdots & b_{nr} \end{pmatrix} = \begin{pmatrix} c_{11} & c_{12}\cdots & c_{1r} \\ c_{21} & c_{22}\cdots & c_{2r} \\ c_{m1} & c_{m2}\cdots & c_{mr} \end{pmatrix}$$

where

$$c_{ij} = a_{i1}b_{1j} + a_{i2}b_{2j} + \ldots + a_{in}b_{nj}$$

For example,

$$A = \begin{pmatrix} 2 & 3 \\ 1 & 2 \\ -1 & 1 \end{pmatrix}; \quad B = \begin{pmatrix} 1 & 3 & 0 \\ -1 & 2 & 4 \end{pmatrix}$$

(A is a 3×2 matrix and B is 2×3 matrix)

The product of A and B is

$$C = AB = \begin{pmatrix} 2 & 3 \\ 1 & 2 \\ -1 & 1 \end{pmatrix}_{3\times 2} \begin{pmatrix} 1 & 3 & 0 \\ -1 & 2 & 4 \end{pmatrix}_{2\times 3}$$

$$C_{11} = a_{11}b_{11} + a_{12}b_{21}, C_{12} = a_{11}b_{12} + a_{12}b_{22}, C_{13} = a_{11}b_{13} + a_{12}b_{23}$$
$$C_{21} = a_{21}b_{11} + a_{22}b_{21}, C_{22} = a_{21}b_{12} + a_{22}b_{22}, C_{23} = a_{21}b_{12} + a_{22}b_{23}$$
$$C_{31} = a_{31}b_{11} + a_{32}b_{21}, C_{32} = a_{31}b_{12} + a_{32}b_{23}, C_{33} = a_{31}b_{13} + a_{32}b_{23}$$

$$C = \begin{pmatrix} 2 \times 1 + 3 \times -1 & 2 \times 3 + 3 \times 2 & 2 \times 0 + 3 \times 4 \\ 1 \times 1 + 2 \times -1 & 1 \times 3 + 2 \times 2 & 1 \times 0 + 2 \times 4 \\ -1 \times 1 + 1 \times -1 & -1 \times 3 + 1 \times 2 & -1 \times 0 + 1 \times 4 \end{pmatrix}$$

or

$$C = \begin{pmatrix} -1 & 12 & 12 \\ 0 & 7 & 8 \\ -2 & -1 & 4 \end{pmatrix}$$

It is to be noted that $AB \neq BA$.

(7) Matrices obey distributive law:

$$A(B + C) = AB + AC$$

(8) Matrices obey associated law of multiplication:

$$A(BC) = (AB)C$$

(9) Direct product of matrices:

Let

$$|A> = \begin{pmatrix} a_1 \\ a_2 \end{pmatrix}$$

be a ket vector and $<B| = (b_1, b_2)$ be a bra vector*. The product

$$|A><B| = \begin{pmatrix} a_1 \\ a_2 \end{pmatrix} \otimes (b_1, b_2)$$

is called the *direct product* of vectors $|A>$ and $<B|$. The result is

$$|A> \otimes <B| = \begin{pmatrix} a_1 \\ a_2 \end{pmatrix} \otimes (b_1, b_2) = \begin{pmatrix} a_1 b_1 & a_1 b_2 \\ a_2 b_1 & a_2 b_2 \end{pmatrix}$$

It must be noted here that this is different from the matrix multiplication described just above.

7.3 Determinant of a Matrix

Every matrix cannot have a determinant, only a square matrix can have a determinant. The determinant formed by the square array of elements constituting, a square matrix C is called the determinant of matrix C. It is denoted by $|C|$ or det C. i.e. if $C = (c_{ij})$ then det $C = |C| = |c_{ij}|$. The determinant of a square matrix having n rows and n columns is called a determinant of order n (In any determinant, the number of rows is equal to the number of columns). Consider the determinant

$$\Delta = \begin{vmatrix} a_{11} & a_{12} & a_{13} \\ a_{21} & a_{22} & a_{23} \\ a_{31} & a_{32} & a_{33} \end{vmatrix}$$

If we strike off any one row and any one column of the determinant, then the second order determinant so formed is called a minor of the given determinant, e.g.

(*) This notation has been discussed in detail in the next chapter.

$$\begin{vmatrix} a_{11} & a_{12} & a_{13} \\ a_{21} & a_{22} & a_{23} \\ a_{31} & a_{32} & a_{33} \end{vmatrix} \text{ or } \begin{vmatrix} a_{12} & a_{13} \\ a_{32} & a_{33} \end{vmatrix}$$

is the minor of the determinant or in particular minor of the element a_{21}. Thus, we will have $3^2 = 9$ minors of the determinant Δ.

Every determinant can be easily reduced to a single number e.g.

$$\Delta = \begin{vmatrix} a_{11} & a_{12} & a_{13} \\ a_{21} & a_{22} & a_{23} \\ a_{31} & a_{32} & a_{33} \end{vmatrix}$$

$$= a_{11} \begin{vmatrix} a_{22} & a_{23} \\ a_{32} & a_{33} \end{vmatrix} - a_{12} \begin{vmatrix} a_{21} & a_{23} \\ a_{31} & a_{33} \end{vmatrix} + a_{13} \begin{vmatrix} a_{21} & a_{22} \\ a_{31} & a_{32} \end{vmatrix}$$

$$= a_{11}(\text{minor of } a_{11}) - a_{12}(\text{minor of } a_{12}) + a_{13}(\text{minor of } a_{13})$$

7.4 Some Special Square Matrices

(1) Unit Matrix:

The unit matrix is defined by

$$I = \delta_{ij}$$

where

$$IA = AI = A$$

The 3×3 unit matrix is given by

$$I = \begin{pmatrix} 1 & 0 & 0 \\ 0 & 1 & 0 \\ 0 & 0 & 1 \end{pmatrix}$$

(2) Diagonal Matrix:

We write $D = D_{ij}\, \delta_{ij}$, δ_{ij} is Kronecker delta.

The following is an example of 3×3 diagonal matrix

$$D = \begin{pmatrix} 3 & 0 & 0 \\ 0 & 2 & 0 \\ 0 & 0 & -1 \end{pmatrix}$$

Thus for a diagonal matrix, $a_{ij} = 0$, when $i \neq j$.

(3) Singular matrix:

If $|A| = 0$ then A is said to be a singular matrix.

For example

$$A = \begin{pmatrix} 1 & 0 \\ 0 & 0 \end{pmatrix} \text{ or } |A| = 0$$

(4) Cofactor matrix:

Let there be a matrix

$$A = \begin{pmatrix} a_{11} & a_{12} & a_{13} \\ a_{21} & a_{22} & a_{23} \\ a_{31} & a_{32} & a_{33} \end{pmatrix}$$

Then, the cofactor matrix is defined as

$$A^c = \begin{pmatrix} A^{11} & A^{12} & A^{13} \\ A^{21} & A^{22} & A^{23} \\ A^{31} & A^{32} & A^{33} \end{pmatrix}$$

where

$$A^{11} = (-1)^{1+1}\begin{vmatrix} a_{22} & a_{23} \\ a_{32} & a_{33} \end{vmatrix}, \ A^{12} = (-1)^{1+2}\begin{vmatrix} a_{21} & a_{23} \\ a_{31} & a_{33} \end{vmatrix}, \ A^{13} = (-1)^{1+3}\begin{vmatrix} a_{21} & a_{22} \\ a_{32} & a_{32} \end{vmatrix}$$

$$A^{21} = (-1)^{2+1}\begin{vmatrix} a_{12} & a_{13} \\ a_{32} & a_{33} \end{vmatrix}, \ A^{22} = (-1)^{2+2}\begin{vmatrix} a_{11} & a_{13} \\ a_{31} & a_{33} \end{vmatrix}, \ A^{23} = (-1)^{2+3}\begin{vmatrix} a_{11} & a_{12} \\ a_{31} & a_{32} \end{vmatrix}$$

$$A^{31} = (-1)^{3+1}\begin{vmatrix} a_{12} & a_{13} \\ a_{22} & a_{23} \end{vmatrix}, \ A^{32} = (-1)^{3+2}\begin{vmatrix} a_{11} & a_{13} \\ a_{21} & a_{23} \end{vmatrix}, \ A^{33} = (-1)^{3+3}\begin{vmatrix} a_{11} & a_{12} \\ a_{21} & a_{22} \end{vmatrix}$$

(5) Adjoint of a matrix:

The adjoint of a matrix is written as adj A, it is defined as the cofactor transpose that is

$$\text{Adj } A = \left(A^c\right)^T$$

e.g.

$$A = \begin{pmatrix} 1 & 3 \\ 2 & 1 \end{pmatrix}; \ A^c = \begin{pmatrix} 1 & -2 \\ -3 & 1 \end{pmatrix}$$

∴

$$A = A^{CT} = \begin{pmatrix} 1 & -3 \\ -2 & 1 \end{pmatrix}$$

(6) Self adjoint matrix:

If adj $A = A$, A is said to be self-adjoint e.g.

$$A = \begin{pmatrix} -1 & 0 \\ 0 & -1 \end{pmatrix}; \ A^C = \begin{pmatrix} -1 & 0 \\ 0 & -1 \end{pmatrix}$$

$$A^{CT} = \text{adj } A = \begin{pmatrix} -1 & 0 \\ 0 & -1 \end{pmatrix} = A$$

(7) Symmetric matrix:

If $A^T = A$, A is said to be a symmetric matrix.

(8) Antisymmetric (skew) matrix:

If $A^T = -A$, A is said to be an anti-symmetric (skew) matrix e.g.

$$\sigma_2 = \begin{pmatrix} 0 & -i \\ i & 0 \end{pmatrix}$$

hence

$$\sigma_2^T = \begin{pmatrix} 0 & i \\ -i & 0 \end{pmatrix} = -\sigma_2$$

(9) Hermitian matrix:

If $A^\dagger = A$, i.e. if its hermitian conjugate (or conjugate of its transpose) equals itself, it is said to be a Hermitian matrix, e.g.

$$\sigma_2 = \begin{pmatrix} 0 & -i \\ i & 0 \end{pmatrix} \text{ or } \sigma_2^* = \begin{pmatrix} 0 & i \\ -i & 0 \end{pmatrix}$$

hence
$$\left(\sigma_2^*\right)^T = \begin{pmatrix} 0 & -i \\ i & 0 \end{pmatrix} = \sigma_2$$

(10) Unitary matrix:

If $AA^\dagger = I$, A is said to be a unitary matrix. In other words, a complex matrix A is called unitary if its complex conjugate transpose is the same as its inverse.

For example
$$\sigma_1 = \begin{pmatrix} 0 & 1 \\ 1 & 0 \end{pmatrix}, \ \sigma_1^* = \begin{pmatrix} 0 & 1 \\ 1 & 0 \end{pmatrix}$$

and
$$\left(\sigma_1^*\right)^T \equiv \sigma_1^\dagger = \begin{pmatrix} 0 & 1 \\ 1 & 0 \end{pmatrix}$$

and
$$\sigma_1 \sigma_1^\dagger = \begin{pmatrix} 0 & 1 \\ 1 & 0 \end{pmatrix}\begin{pmatrix} 0 & 1 \\ 1 & 0 \end{pmatrix} = \begin{pmatrix} 1 & 0 \\ 0 & 1 \end{pmatrix} = I$$

(11) Orthogonal matrix:

If $AA^T = I$, A is said to be an orthogonal matrix.

(12) The trace of a matrix:

The trace of a matrix is given by the sum of its diagonal elements, i.e.
$$Tr\, A = \sum_k a_{kk}$$

e.g.
$$A = \begin{pmatrix} 3 & 5 \\ 2 & 6 \end{pmatrix}; \ Tr\, A = 3 + 6 = 9$$

(13) The Inverse matrix:

If A is a matrix, then inverse matrix A^{-1} satisfies $AA^{-1} = 1$.

Example 7.1: Show that

$$A = \begin{pmatrix} \sqrt{2}/2 & -i\sqrt{2}/2 & 0 \\ i\sqrt{2}/2 & -\sqrt{2}/2 & 0 \\ 0 & 0 & 1 \end{pmatrix} \text{ is a unitary matrix.}$$

Solution: We have to show $A^\dagger A = I$

Now

$$A^\dagger A = (A^*)^T A = \begin{pmatrix} \sqrt{2}/2 & -i\sqrt{2}/2 & 0 \\ i\sqrt{2}/2 & -\sqrt{2}/2 & 0 \\ 0 & 0 & 1 \end{pmatrix}\begin{pmatrix} \sqrt{2}/2 & -i\sqrt{2}/2 & 0 \\ i\sqrt{2}/2 & -\sqrt{2}/2 & 0 \\ 0 & 0 & 1 \end{pmatrix}$$

$$= \begin{pmatrix} 1 & 0 & 0 \\ 0 & 1 & 0 \\ 0 & 0 & 1 \end{pmatrix} = I.$$

7.5 Relation between Adj *A* and *A*

There exists an important relation between A and Adj A. This is

$$A \, (\text{Adj } A) = (\text{Adj } A) \, A = |A| \, I$$

where I is a unit matrix. We can verify this as follows:

Let

$$A = \begin{bmatrix} a_1 & b_1 & c_1 \\ a_2 & b_2 & c_2 \\ a_3 & b_3 & c_3 \end{bmatrix} \text{ and } I = \begin{bmatrix} 1 & 0 & 0 \\ 0 & 1 & 0 \\ 0 & 0 & 1 \end{bmatrix}$$

$$\therefore \quad A \, (\text{adj } A) = \begin{bmatrix} a_1 & b_1 & c_1 \\ a_2 & b_2 & c_2 \\ a_3 & b_3 & c_3 \end{bmatrix} \begin{bmatrix} A_1 & A_2 & A_3 \\ B_1 & B_2 & B_3 \\ C_1 & C_2 & C_3 \end{bmatrix}$$

$$= \begin{bmatrix} a_1 A_1 + b_1 B_1 + c_1 C_1 & a_1 A_2 + b_1 B_2 + c_1 C_2 & a_1 A_3 + b_1 B_3 + c_1 C_3 \\ a_2 A_1 + b_2 B_1 + c_2 C_1 & a_2 A_2 + b_2 B_2 + c_2 C_2 & a_2 A_3 + b_2 B_3 + c_2 C_3 \\ a_3 A_1 + b_3 B_1 + c_3 C_1 & a_3 A_2 + b_3 B_2 + c_3 C_2 & a_3 A_3 + b_3 B_3 + c_3 C_3 \end{bmatrix}$$

$$= \begin{bmatrix} |A| & 0 & 0 \\ 0 & |A| & 0 \\ 0 & 0 & |A| \end{bmatrix}$$

$$\left(\because \ a_1 A_1 + b_1 B_1 + c_1 C_1 = |A| \ \text{and} \ a_1 A_2 + b_1 B_2 + c_1 C_2 = 0 \ \text{etc.} \right)$$

$$|A| = \begin{bmatrix} 1 & 0 & 0 \\ 0 & 1 & 0 \\ 0 & 0 & 1 \end{bmatrix} = |A| \, I$$

Also $(\text{Adj } A) \, A = |A| \, I$

$\therefore \quad A \, (\text{Adj } A) = (\text{Adj } A) \, A = |A| \, I$

$\therefore \quad$ If $|A| \neq 0$ then $A^{-1} = \dfrac{1}{|A|} (\text{Adj } A)$

Example 7.2: Determine the adjoint of the matrix

$$A = \begin{bmatrix} 2 & -3 & 1 \\ 0 & 3 & -2 \\ -1 & 4 & 5 \end{bmatrix}$$

Solution:

$$A^C = \begin{bmatrix} A_1 & B_1 & C_1 \\ A_2 & B_2 & C_2 \\ A_3 & B_3 & C_3 \end{bmatrix}$$

The cofactor elements for the first column are

$$A_1 = (-1)^{1+1} \begin{vmatrix} 3 & -2 \\ 4 & 5 \end{vmatrix} = (15 + 8) = 23$$

$$A_2 = (-1)^{1+2} \begin{vmatrix} -3 & 1 \\ 4 & 5 \end{vmatrix} = -1(-15 - 4) = 19$$

$$A_3 = (-1)^{1+3} \begin{vmatrix} -3 & 1 \\ 3 & -2 \end{vmatrix} = (6 - 3) = 3$$

Similarly,
$$B_1 = (-1)^{2+1} \begin{bmatrix} 0 & -2 \\ -1 & 5 \end{bmatrix} = -1(0-2) = 2$$

$$B_2 = (-1)^{2+2} \begin{bmatrix} 2 & 1 \\ -1 & 5 \end{bmatrix} = (10+1) = 11$$

$$B_3 = (-1)^{3+2} \begin{bmatrix} 2 & 1 \\ 0 & -2 \end{bmatrix} = -1(-4-0) = 4$$

and
$$C_1 = (-1)^{3+1} \begin{bmatrix} 0 & 3 \\ -1 & 4 \end{bmatrix} = (0+3) = 3$$

$$C_2 = (-1)^{3+2} \begin{bmatrix} 2 & -3 \\ -1 & 4 \end{bmatrix} = -1(8-3) = -5$$

$$C_3 = (-1)^{3+3} \begin{bmatrix} 2 & -3 \\ 0 & 3 \end{bmatrix} = (6+0) = 6$$

so that
$$A^C = \begin{bmatrix} 23 & 2 & 3 \\ 19 & 11 & -5 \\ 3 & 4 & 6 \end{bmatrix}$$

and
$$\text{Adj } A = A^{CT} = \begin{bmatrix} 23 & 19 & 3 \\ 2 & 11 & 4 \\ 3 & -5 & 6 \end{bmatrix}$$

7.6 Some Identities for Matrices

Following matrix identities can easily be verified:

$$(AB)^T = B^T A^T;$$

$$\text{or } (AB)' = B'A'$$

$$(AB)^\dagger = B^\dagger A^\dagger$$

$$(AB)^* = A^* B^*$$

$$|AB| = |A| \, |B|$$

$$|A^T| = |A|$$

$$|A^*| = |A|^*$$

$$|A^\dagger| = |A|^*$$

$$Tr \, (AB) = Tr \, (BA)$$

Example 7.3: Determine the matrix A^{-1}

where
$$A = \begin{bmatrix} 2 & -3 & 1 \\ 0 & 3 & -2 \\ -1 & 4 & 5 \end{bmatrix}$$

Solution: $A^{-1} = A \dfrac{1}{|A|} (\text{Adj } A)$

From example (7.1),

$$\text{Adj } A = \begin{bmatrix} 23 & 19 & 3 \\ 2 & 11 & 4 \\ 3 & -5 & 6 \end{bmatrix}$$

Now expanding $|A|$ in terms of first row,

$$|A| = 2[3 \times 5 - (-2 \times 4)] - (-3)[0 \times 5 - (-1 \times -2)] + 1[0 \times 4 - 3 \times (-1)]$$
$$= 2[15 + 8] + 3[-2] + 1[3] = 46 - 6 + 3 = 43$$

Hence $\qquad A^{-1} = \dfrac{1}{43} \begin{bmatrix} 23 & 19 & 3 \\ 2 & 11 & 4 \\ 3 & -5 & 6 \end{bmatrix}$

7.7 Matrix Form of Wave Functions and Operators

The multiplication law for two matrices is

$$(AB)_{mn} = \sum_i A_{mi} B_{in}$$

where m denotes the row and i denotes the column of matrix A.

We can associate with any operator A, a matrix, also denoted by A, whose matrix element is defined by

$$A\psi_m = \sum_{n=1}^{N} A_{nm} \psi_n = \sum_{n=1}^{N} \psi_n A_{nm}$$

$$A_{nm} = \int \psi_n^* A \psi_m \, dr \equiv \left(\psi_n, A\psi_m \right) \qquad (1)$$

with respect to a set of eigen functions ψ_n, which form an orthonormal basis of a vector space of N dimensions.

A matrix operating on a wave function (vector) can be represented as a linear transformation

$$A\chi = \phi \qquad (2)$$

where we can expand χ and ϕ in terms of the basis set ψ:

$$\chi = \sum_{j=1}^{N} b_j \psi_j = b_1 \psi_1 + b_2 \psi_2 + b_3 \psi_3 + \dots \qquad (3a)$$

$$\phi = \sum_{k=1}^{N} c_k \psi_k = c_1 \psi_1 + c_2 \psi_2 + c_3 \psi_3 + \dots \qquad (3b)$$

Each b_n (or c_n) corresponds to the (complex) component of an arbitrary vector in the direction of the axis corresponding to ψ_n.

From the orthonormality of ψ's, it follows that:

$$b_j = \int \psi_j^* \chi \, dq$$

and

$$c_k = \int \psi_k^* \phi \, dq = \int \psi_k^* A\chi \, dq$$

$$= \int \psi_k^* A\left(b_1\psi_1 + b_2\psi_2 + ...\right) dq$$

$$= A_{k1}b_1 + A_{k2}b_2 + ... \qquad \text{(by Eqn. 1)}$$

The set of equations (one equation for each k) giving c's in terms of the b's may be written:

$$\begin{pmatrix} c_1 \\ c_2 \\ ... \\ ... \end{pmatrix} = \begin{pmatrix} A_{11} & A_{12} & A_{13} & .... \\ A_{21} & A_{22} & A_{23} & .... \\ .... & .... & .... & .... \\ .... & .... & .... & .... \end{pmatrix} \begin{pmatrix} b_1 \\ b_2 \\ ... \\ ... \end{pmatrix}$$

Thus the operator A can be ascribed a matrix A whose elements are A_{ij}.

Comparison with Eqn. (2)$\phi = A\chi$ suggests that we can associate with any function χ a one column matrix (column-vector) b whose elements are the coefficients of ψ's in the expansion of χ. The matrix equation representing Eqn. (4) is

$$c = Ab, \; c_k = \sum_n A_{kn} b_n, \quad \left(\phi = \sum_k c_k \psi_k\right) \qquad (5)$$

Thus Eqn. (2) interprets that each linear operator corresponds to a process which changes each vector of a vector space $V_n(F)$ into some other (specified) vector in the same space F.

The complex conjugate of Eqn. (2) is

$$\phi^* = (A \chi)^*$$

$$c_k = \left(\int \psi_k^* \phi \, dq\right)^*$$

$$= \int \psi_k \phi^* \, dq = \int \psi_k \left\{b_1^* (A\psi_1)^* + b_2^* (A\psi_2)^* + ...\right\} dq$$

$$= b_1^* \int (A\psi_1)^* \psi_k dq + b_2^* \int (A\psi_2)^* \psi_k dq + ...$$

$$= b_1^* (A\psi_1, \psi_k) + b_2^* (A\psi_2, \psi_k) + ...$$

using the definition $(\psi_i, A^\dagger\psi_k) = (A\psi_i, \psi_k)$

The Eqn. becomes

$$c_k^* = b_1^* \int \psi_1^* A^\dagger \psi_k dq + b_2^* \int \psi_2^* A^\dagger \psi_k dq + ...$$

$$= b_1^* A_{1k}^\dagger + b_2^* A_{2k}^\dagger + ...$$

where $A^\dagger$ is Hermitian adjoint of A. Thus

$$c^\dagger = b^\dagger A^\dagger \qquad (6)$$

where $c^\dagger$ and $b^\dagger$ are the one-row matrices which are Hermitian conjugates of c and b. i.e.

$$c^\dagger = \widehat{c_1^* c_2^*} ...; \; b^\dagger = \widehat{b_1^* b_2^*} ... \qquad (7)$$

We therefore represent χ^* and ϕ^* by $b^\dagger$ and $c^\dagger$ respectively. Normalization of ϕ is expressed as $\int \phi^* \phi \, dq = c^\dagger c = 1$. It should be noted here that a row-matrix multiplies a square matrix from the left, and also, $c^\dagger c$ is a scalar (number), whereas $cc^\dagger$ is a matrix.

Matrices of certain special types have been given appropriate names. Those of special interest in quantum mechanics are:

Symmetric	:	$A = A^T$	real	:	$A = A*$
Skew symmetric	:	$A = -A^T$	imaginary	:	$A = -A*$ (8)
Hermitian	:	$A = A^\dagger$	unitary	:	$A^{-1} = A^\dagger$
Skew-Hermitian	:	$A = -A^\dagger$	diagonal	:	$a_{ij} = a_{ii}\,\delta_{ij}$

7.8 Adjoint of an Operator

The linear operator A' adjoint to the linear operator A is defined by the equation

$$(A^\dagger\psi,\phi) = (\psi, A\phi) \tag{9}$$

for all ψ and ϕ. Now, we know that

$$(\psi,\ \phi)* = (\phi,\ \psi)$$

If we take the complex conjugate of Eqn. (9), we find

$$(\psi,\ A\phi)* = (\phi,\ A^\dagger\psi) \tag{10}$$

A *self-adjoint operator* satisfies

$$A^\dagger = A$$

A self adjoint or Hermitian operator generates a Hermitian matrix

$$(\psi_n,\ A\psi_m) = (A^\dagger\psi_n,\ \psi_m) = \left(A\psi_n,\ \psi_m\right) \tag{11}$$

This means that

$$A_{nm} = A^*_{mn} = A^\dagger_{nm} \tag{12}$$

If A in Eqn. (10) is self adjoint

$$(\psi,\ A\phi)* = (\phi,\ A\psi) \tag{13}$$

If ϕ is replaced by ψ in this Eqn.

$$(\psi,\ A\psi)* = (\psi,\ A\psi) \tag{14}$$

i.e. $(\psi, A\psi)$ is real if A is self adjoint.

Example 7.4: Find the adjoint of the operator

$$\partial_x = \frac{\partial}{\partial\mathbf{x}}$$

Solution: We have

$$\psi*\frac{\partial\phi}{\partial x} = \frac{\partial}{\partial x}(\psi*\phi) - \frac{\partial\psi*}{\partial x}\phi$$

Integrating both sides

$$\int_{-\infty}^{+\infty}\psi*\frac{\partial\phi}{\partial x}\,dx = \int_{-\infty}^{+\infty}\frac{\partial}{\partial x}(\psi*\phi)\,dx - \int_{-\infty}^{+\infty}\frac{\partial\psi*}{\partial x}\phi\,dx$$

or

$$\int_{-\infty}^{+\infty}\psi*\frac{\partial\phi}{\partial x}\,dx = \psi*\phi\Big|_{-\infty}^{+\infty} - \int_{-\infty}^{+\infty}\frac{\partial\psi*}{\partial x}\phi\,dx$$

or

$$\int_{-\infty}^{+\infty}\psi*\frac{\partial\phi}{\partial x}\,dx = 0 - \int_{-\infty}^{+\infty}\frac{\partial\psi*}{\partial x}\phi\,dx$$

or $$\left(\psi, \partial_x \phi\right) = -\left(\partial_x \psi, \phi\right)$$

or $$\left(\partial_x \psi, \phi\right) = -\left(\psi, \partial_x \phi\right)$$

so adjoint defined in Eqn. (9) is $\partial_x^\dagger = -\partial_x$

7.9 The Matrix of an Operator and Secular or Characteristic Equation

We can find the eigenvalues and the eigenvectors of a linear operator if we are given the matrix representation for that operator.

Suppose, we have to find the eigenvalues of an operator A which satisfies the eigenvalue equation

$$A\psi = \lambda\psi \tag{15}$$

Let ψ be expressible as

$$\psi = \sum_j C_j \psi_j \tag{16}$$

so that using Eqn. (15)

$$\sum_j C_j A\psi_j = \lambda \sum_j C_j \psi_j \tag{17}$$

Multiplying both sides of this equation by ψ_i^* and integrating over all space

$$\sum_j C_j \int \psi_i^* A\psi_j d^3 r = \lambda \sum_j C_j \int \psi_i^* \psi_j \cdot d^3 r \tag{18}$$

or $$\sum_j C_j a_{ij} = \lambda \sum_j C_j \cdot \delta_{ij} \tag{19}$$

In an n-dimensional space, the above equation is a set of n equations

$$a_{11}C_1 + a_{12}C_2 + \ldots + a_{1n}C_n = \lambda C_1$$
$$a_{21}C_1 + a_{22}C_2 + \ldots + a_{2n}C_n = \lambda C_2$$
$$\cdots \quad \cdots \quad \cdots \quad \cdots \quad \cdots \quad \cdots$$
$$\cdots \quad \cdots \quad \cdots \quad \cdots \quad \cdots \quad \cdots$$
$$a_{n1}C_1 + a_{n2}C_2 + \ldots + a_{nn}C_n = \lambda C_n$$

or, $$(a_{11} - \lambda)C_1 + a_{12}C_2 + \ldots + a_{1n}C_n = 0$$
$$\cdots \quad\quad\quad \cdots \quad\quad\quad \cdots \tag{20}$$
$$\cdots \quad\quad\quad \cdots \quad\quad\quad \cdots$$
$$a_{n1}C_1 + a_{n2}C_2 + \ldots + (a_{nn} - \lambda)C_n = 0$$

So we have a system of n homogeneous linear equations in n unknowns $(C_1 \ldots C_n)$. There will be non zero solutions only if the determinant of the coefficients is zero, i.e.

$$\begin{vmatrix} a_{11} - \lambda & a_{12} & \cdots & a_{1n} \\ a_{21} & a_{22} - \lambda & \cdots & a_{2n} \\ \cdots & \cdots & \cdots & \cdots \\ \cdots & \cdots & \cdots & \cdots \\ a_{n1} & a_{n2} & \cdots & a_{nn} - \lambda \end{vmatrix} \equiv [A - \lambda I] = 0 \tag{21}$$

This determinant, when evaluated, gives a polynomial of degree n in λ called the characteristic polynomial of A (and Eqn. 21 is known as *characteristic equation*). The roots of this polynomial are the eigenvalues $(\lambda_1, \lambda_2, ..., \lambda_n)$ of the operator A. For every eigen value, the Eqns. (20) have a solution giving the corresponding eigenvector,

$$\psi \leftrightarrow \begin{pmatrix} C_1 \\ C_2 \\ ... \\ \\ ... \\ C_n \end{pmatrix} \tag{22}$$

However,

$$\int \psi * \psi d^3 r = 1 \tag{23}$$

So,

$$\int \sum_{ij} C_i^* C_j \cdot \psi_i \psi_j^* \, d^3 r = \sum_{ij} C_i^* C_j \delta_{ij}$$

$$= \sum_i C_i^* C_i = 1 \tag{24}$$

$$\left(C_1^* C_2^* .... C_n^* \right) \begin{bmatrix} C_1 \\ C_2 \\ ... \\ \\ C_n \end{bmatrix} = 1 \tag{25}$$

The above procedure diagonalizes the matrix A and the diagonalized matrix is given by

$$\text{diag } A = \begin{bmatrix} \lambda_1 & ... & ... & 0 \\ ... & \lambda_2 & ... & ... \\ ... & ... & ... & ... \\ 0 & ... & ... & \lambda_n \end{bmatrix} \tag{26}$$

Example 7.5: Find the eigenvalues and reduce to diagonal form the matrix

$$A = \begin{bmatrix} 1 & 2 & 3 \\ 0 & 3 & 4 \\ 0 & 0 & 2 \end{bmatrix}$$

Solution: The secular equation is $|A - \lambda I| = 0$

where

$$I = \begin{bmatrix} 1 & 0 & 0 \\ 0 & 1 & 0 \\ 0 & 0 & 1 \end{bmatrix}$$

or

$$\begin{bmatrix} 1 & 2 & 3 \\ 0 & 3 & 4 \\ 0 & 0 & 2 \end{bmatrix} - \begin{bmatrix} \lambda & 0 & 0 \\ 0 & \lambda & 0 \\ 0 & 0 & \lambda \end{bmatrix} = 0$$

or

$$\begin{bmatrix} 1-\lambda & 2 & 3 \\ 0 & 3-\lambda & 4 \\ 0 & 0 & 2-\lambda \end{bmatrix} = 0$$

On solving the above determinant, we obtain

$$(1-\lambda)\{(3-\lambda)(2-\lambda)-4\times0\} - 2\{0\times(2-\lambda)-0\} + 3\{0-(3-\lambda)\times0\} = 0$$

or

$$(1-\lambda)(3-\lambda)(2-\lambda) = 0$$

$\therefore$

$$\lambda = 1, 2 \text{ and } 3$$

The diagonal matrix can be written as

$$\begin{bmatrix} \lambda_1 & 0 & 0 \\ 0 & \lambda_2 & 0 \\ 0 & 0 & \lambda_3 \end{bmatrix} = \begin{bmatrix} 1 & 0 & 0 \\ 0 & 3 & 0 \\ 0 & 0 & 2 \end{bmatrix}$$

Example 7.6: Find the eigenvalues and eigenvectors of the matrix.

$$A = \begin{bmatrix} -1 & 0 \\ 0 & 1 \end{bmatrix}$$

Solution: We have

$$\begin{bmatrix} -1 & 0 \\ 0 & 1 \end{bmatrix} \begin{bmatrix} C_1 \\ C_2 \end{bmatrix} = \lambda \begin{bmatrix} C_1 \\ C_2 \end{bmatrix}$$

where λ is the eigenvalue and $\begin{bmatrix} C_1 \\ C_2 \end{bmatrix}$ the eigenvector of A. The eigenvectors of A are the solutions of the simultaneous equations

$$(-1-\lambda)\ C_1 = 0$$

$$(1-\lambda)\ C_2 = 0$$

These equations have a non zero solution only when

$$\begin{vmatrix} -1-\lambda & 0 \\ 0 & 1-\lambda \end{vmatrix} = -\left(-1-\lambda^2\right) = 0$$

i.e. when $\lambda = \pm1.$

The corresponding eigen vectors are

$$\alpha \equiv \begin{pmatrix} C_1 \\ C_2 \end{pmatrix} = \begin{pmatrix} 1 \\ 0 \end{pmatrix} \text{ for } \lambda = -1$$

and

$$\beta \equiv \begin{pmatrix} C_1' \\ C_2' \end{pmatrix} = \begin{pmatrix} 0 \\ 1 \end{pmatrix} \text{ for } \lambda = 1$$

7.10 Change of Basis

If the vectors ψ_i form an orthonormal basis for a space of N dimensions,

$$(\psi_i, \psi_j) = \int \psi_i^* \ \psi_j \cdot dr = \delta_{ij} \ (i, j = 1, 2, ..., N) \tag{27}$$

then a new basis can be constructed by forming N linear combinations of these vectors

$$\psi'_i = \sum_{j=0}^{N} u^*_{ij} \, \psi_j \tag{28}$$

where u_{ij} are suitably chosen complex numbers. This new basis will also be orthonormal provided that

$$(\psi'_i, \psi'_j) = \delta_{ij} \tag{29}$$

Using Eqn. (28), we have

$$\psi'^*_j = \sum_l u_{jl} \, \psi^*_l$$

$$\psi'^*_i = \sum_k u_{ik} \, \psi^*_k \tag{30}$$

so that Eqn. (29) becomes

$$(\psi'_i, \psi'_j) \equiv \sum_{k,l} u_{ik} \, u^*_{jl} \, (\psi_k, \psi_l) = \delta_{ij} \tag{31}$$

This, using Eqn. (27) becomes

$$\sum_{k,l} u_{ik} u^*_{jl} \, \delta_{kl} = \delta_{ij}$$

or

$$\sum_k u_{ik} u^*_{jk} = \delta_{ij} \tag{32}$$

In other words, the matrix $U = (u_{ij})$ must satisfy

$$UU^\dagger = 1 \tag{33}$$

It implies that U must be non singular and

$$U^{-1} = U^\dagger \tag{34}$$

This condition would also insure that the state vectors ψ'_i formed according to Eqn. (28) are orthonormal. Hence the necessary and sufficient condition for the vectors ψ'_i to form an orthonormal basis is that the matrix $U = (u_{ij})$ be unitary. Since

$$U^{-1}U = UU^{-1} = 1,$$

the condition of Eqn. (34) can be rewritten as

$$UU^\dagger = U^\dagger U = 1 \tag{35}$$

Such a unitary transformation has the effect of substituting the basis $\{\psi'_i\}$ for the basis $\{\psi_i\}$. If a state function ψ is represented with respect to the basis $\{\psi_i\}$ by

$$\psi = \sum_i b_i \psi_i \tag{36}$$

and with respect to the basis $\{\psi'_i\}$ by

$$\psi = \sum_i b'_i \psi'_i \tag{37}$$

then, since the two expressions (36 and 37) are representative of one and the same state, so

$$\sum_i b'_i \psi'_i = \sum_{i,j} b'_i u^*_{ij} \psi_j$$

or
$$\sum_j b_j \psi_j = \sum_{i,j} b'_i u^*_{ij} \psi_j$$

Hence, by orthogonality of the ψ_j

$$\sum_j b_j \psi_j \psi^*_j = \sum_{i,j} b'_i u^*_{ij} \psi_j \psi^*_j \left(\sum_j \psi_j \psi^*_j = \delta_{jj'} = 1 \right)$$

or
$$b_j = \sum_i u^*_{ij} b'_i \qquad (38)$$

This equation gives the connection between the components b_i and b'_i of the state vector ψ with respect to the bases $\{\psi_i\}$ and $\{\psi'_i\}$. Since U is unitary matrix, Eqn. (38) can be solved for the b'_j in terms of b_j. So, using Eqn. (32), we have

$$\sum_j u_{kj} b_j = \sum_{i,j} u_{kj} u^*_{ij} b'_i$$

$$= \sum_i \delta_{ki} b'_i$$

or
$$b'_k = \sum_j u_{kj} b_j \qquad (39)$$

It is to be noted here that the state vector has an identity independent of the basis chosen to represent it. In other words, (b_i) and (b'_i) column matrices represent the same vector with respect to different bases. Thus b, b', b'' etc are column matrices, while b is an abstract symbol denoting a definite vector in the vector space, keeping this in mind, we may rewrite Eqn. (39) as

$$b' = Ub, \ b = U^\dagger b'$$

i.e. b and b' are connected by a unitary matrix U defined in Eqns 30, 32, 33 and 34.

7.11 A Unitary Transformation

Let us now consider a linear transformation

$$c = Ab \quad \text{or} \quad c_i = \sum_j A_{ij} b_j \qquad (41)$$

This is a linear relation between the representatives of b and c relative to the basis $\{\psi\}$. Changing to the basis $\{\psi'\}$, we obtain

$$c' = Uc = UAb = UAU^\dagger b' \ \text{(using Eqn. 40)} \qquad (42)$$

and this can written as

$$c' = A'b' \qquad (43)$$

with the matrix A', given by

$$A' = UAU^\dagger = UAU^{-1} \qquad (44)$$

Here A' is the matrix which represents the linear transformation from basis $\{\psi\}$ to basis $\{\psi'\}$. In other words, Eqn. (44) represents a linear transformation from the matrix A to matrix A' which are represented in different bases.

We know that a matrix is said to be a square matrix if it has the same number of rows as columns. A square matrix is *diagonal* if it has non-vanishing elements only along the principal diagonal. We have seen that an operator can be represented by a matrix. The diagonal elements of a matrix operator are called *eigen values* of the matrix. The matrix A in Eqn. (44) is said to be *diagonalized* by the matrix U if the matrix A' (that results from the transformation) is diagonal i.e. when $A'_{ij} = A'_i \delta_{ij}$. In the equation (44), the transformation would be *unitary transformation* if the matrix U is *non singular* (i.e. $|U| \neq 0$).

Example 7.7: Prove that the eigenvalues of a matrix are not changed by a unitary transformation.

Solution: Suppose the operator A is represented by the matrix A in the original representation. The eigenvalue λ satisfies $AC = \lambda C$, where C is the representative of eigenfunction. If the basis functions are transformed, the operator matrix undergoes a similarity transformation:

$$A = UA'U^{-1}$$

and the eigenvalue equation may be written as

$$UA'U^{-1}C = \lambda C$$

Pre-multiplying both sides by U^{-1}, we get

$$A'(U^{-1}C) = \lambda(U^{-1}C) \quad \text{or} \quad A'C' = \lambda C'$$

where C' is the eigenfunction representative in the new basis. Hence the eigenvalues do not change.

Example 7.8: Show that the commutation relation $[x, p_x] = i\hbar$ remains unchanged under a unitary transformation.

Solution: Let U denote the operator effecting the transformation. Then

$$x' = UxU^\dagger \text{ and } p'_x = Up_xU^\dagger$$

$$[x', p'_x] = x'p'_x - p'_x x'$$

$$= \left(UxU^\dagger\right)\left(Up_xU^\dagger\right) - \left(Up_xU^\dagger\right)\left(UxU^\dagger\right)$$

$$= Uxp_xU^\dagger - Up_x xU^\dagger$$

$$= U\left(xp_x - p_x x\right)U^\dagger$$

$$= U[x, p_x]U^\dagger = i\hbar UU^\dagger = i\hbar$$

Example 7.9: Show that the vectors

$$A_1 = \begin{pmatrix} \cos\theta \\ \sin\theta \\ 0 \end{pmatrix}; \ A_1 = \begin{pmatrix} -\sin\theta \\ \cos\theta \\ 0 \end{pmatrix}; \ A_3 = \begin{pmatrix} 0 \\ 0 \\ 1 \end{pmatrix}$$

form an orthonormal set.

Solution: First, we have to show that

$$A_j^T A_k = \begin{cases} 1 & \text{if } j = k \\ 0 & \text{if } j \neq k \end{cases}$$

Hence, for $j = k = 1$, we have

$$A_1^T A_1 = (\cos\theta \ \sin\theta \ 0) \begin{pmatrix} \cos\theta \\ \sin\theta \\ 0 \end{pmatrix}$$

$$\cos^2\theta + \sin^2\theta + 0 = 1 \qquad\qquad (i)$$

Similarly, we find $A_2^T A_2 = 1$ and $A_3^T A_3 = 1$ i.e. A_1, A_2, A_3 are unit vectors. To show the orthogonality of any two, consider first (say) $j = 1$, $k = 2$. Then,

$$A_1^T A_2 = (\cos\theta \ \sin\theta \ 0) \begin{pmatrix} -\sin\theta \\ \cos\theta \\ 0 \end{pmatrix} = 0$$

Similarly, $A_1^T A_3 = 0$, $A_2^T A_3 = 0$. Hence the given vectors are mutually orthogonal (and from i), they form an orthonormal set of vectors.

Example 7.10: The base vectors of a representation are

$$\begin{pmatrix} 1 \\ 0 \end{pmatrix} \text{ and } \begin{pmatrix} 0 \\ 1 \end{pmatrix}$$

Construct a transformation matrix U for transformation to another representation having base vectors:

$$\begin{pmatrix} 1/\sqrt{2} \\ 1/\sqrt{2} \end{pmatrix} \text{ and } \begin{pmatrix} -1/\sqrt{2} \\ 1/\sqrt{2} \end{pmatrix}$$

Solution: The transformation matrix U must be such that

$$\begin{pmatrix} 1/\sqrt{2} \\ 1/\sqrt{2} \end{pmatrix} = \begin{pmatrix} U_{11} & U_{12} \\ U_{21} & U_{22} \end{pmatrix} \begin{pmatrix} 1 \\ 0 \end{pmatrix} \qquad\qquad (i)$$

and

$$\begin{pmatrix} -1/\sqrt{2} \\ 1/\sqrt{2} \end{pmatrix} = \begin{pmatrix} U_{11} & U_{12} \\ U_{21} & U_{22} \end{pmatrix} \begin{pmatrix} 0 \\ 1 \end{pmatrix} \qquad\qquad (ii)$$

From (*i*)

$$U_{11} \times 1 + U_{12} \times 0 = \frac{1}{\sqrt{2}} \Rightarrow U_{11} = 1/\sqrt{2}$$

$$U_{21} \times 1 + U_{22} \times 0 = \frac{1}{\sqrt{2}} \Rightarrow U_{21} = 1/\sqrt{2}$$

From (*ii*)

$$U_{11} \times 0 + U_{12} \times 1 = -\frac{1}{\sqrt{2}} \Rightarrow U_{12} = -1/\sqrt{2}$$

$$U_{21} \times 0 + U_{22} \times 1 = \frac{1}{\sqrt{2}} \Rightarrow U_{22} = 1/\sqrt{2}$$

So,

$$U = \begin{pmatrix} \dfrac{1}{\sqrt{2}} & -\dfrac{1}{\sqrt{2}} \\ \dfrac{1}{\sqrt{2}} & \dfrac{1}{\sqrt{2}} \end{pmatrix}$$

Example 7.11. (*i*) Find the eigenvalues and eigenvectors of the matrix

$$A = \begin{pmatrix} 5 & 7 & -5 \\ 0 & 4 & -1 \\ 2 & 8 & -3 \end{pmatrix}$$

and (*ii*) determine a set of unit eigenvectors.

Solution: (*i*) If $X = \begin{pmatrix} x_1 \\ x_2 \\ x_3 \end{pmatrix}$ we have to consider the equation $AX = \lambda X$ or

$$\begin{pmatrix} 5 & 7 & -5 \\ 0 & 4 & -1 \\ 2 & 8 & -3 \end{pmatrix} \begin{pmatrix} x_1 \\ x_2 \\ x_3 \end{pmatrix} = \lambda \begin{pmatrix} x_1 \\ x_2 \\ x_3 \end{pmatrix}$$

or

$$\begin{pmatrix} 5x_1 + 7x_2 - 5x_3 \\ 4x_2 - x_3 \\ 2x_1 + 8x_2 - 3x_3 \end{pmatrix} = \begin{pmatrix} \lambda x_1 \\ \lambda x_2 \\ \lambda x_3 \end{pmatrix}$$

Equating corresponding elements of these matrices, we get

$$\begin{vmatrix} (5 - \lambda)x_1 + 7x_2 - 5x_3 = 0 \\ (4 - \lambda)x_2 - x_3 = 0 \\ 2x_1 + 8x_2 - (3 + \lambda)x_3 = 0 \end{vmatrix} \qquad (i)$$

There will be non trivial solutions if

$$\begin{vmatrix} 5 - \lambda & 7 & -5 \\ 0 & 4 - \lambda & -1 \\ 2 & 8 & -3 - \lambda \end{vmatrix} = 0 \qquad (ii)$$

Expansion of this determinant yields

$$\lambda^3 - 6\lambda^2 + 11\lambda - 6 = 0$$

or

$$(\lambda - 1)(\lambda - 2)(\lambda - 3) = 0$$

Thus, the eigenvalues are $\lambda_1 = 1$, $\lambda_2 = 2$, $\lambda_3 = 3$.

To find eigenvectors:

Corresponding to $\lambda_1 = 1$, Eqn. (*i*) becomes

$$4x_1 + 7x_2 - 5x_3 = 0$$

$$3x_2 - x_3 = 0$$

$$2x_1 + 8x_2 - 4x_3 = 0$$

Solving for x_1 and x_3 in terms of x_2, we find $x_3 = 3x_2$ and $x_1 = 2x_2$. Then an eigenvector is

$$\begin{pmatrix} x_1 \\ x_2 \\ x_3 \end{pmatrix} = \begin{pmatrix} 2x_2 \\ x_2 \\ 3x_2 \end{pmatrix} = x_2 \begin{pmatrix} 2 \\ 1 \\ 3 \end{pmatrix} \text{ or simply } \begin{pmatrix} 2 \\ 1 \\ 3 \end{pmatrix}$$

since any eigenvector is a constant multiple of this. Similarly, corresponding to $\lambda_2 = 2$, Eqns. (*i*) lead to $x_3 = 2x_2$ and $x_1 = x_2$. This leads to the eigenvector

$$\begin{pmatrix} x_1 \\ x_2 \\ x_3 \end{pmatrix} = \begin{pmatrix} x_2 \\ x_2 \\ 2x_3 \end{pmatrix} = x_2 \begin{pmatrix} 1 \\ 1 \\ 2 \end{pmatrix} \quad \text{or simply} \quad \begin{pmatrix} 1 \\ 1 \\ 2 \end{pmatrix}$$

Again, for $\lambda_3 = 3$, we obtain $x_3 = x_2$ and $x_1 = -x_2$, which gives the eigenvector

$$\begin{pmatrix} x_1 \\ x_2 \\ x_3 \end{pmatrix} = \begin{pmatrix} -x_2 \\ x_2 \\ x_2 \end{pmatrix} = x_2 \begin{pmatrix} -1 \\ 1 \\ 1 \end{pmatrix} \quad \text{or simply} \quad \begin{pmatrix} -1 \\ 1 \\ 1 \end{pmatrix}$$

(*ii*) The unit eigenvectors have the property that the sum of the squares of their components = 1. For this, we divide each vector by the square root of the sum of the squares of the components. Thus, we get

$$\begin{pmatrix} 2/\sqrt{14} \\ 1/\sqrt{14} \\ 3/\sqrt{14} \end{pmatrix}, \begin{pmatrix} 1/\sqrt{5} \\ 1/\sqrt{5} \\ 2/\sqrt{5} \end{pmatrix}, \begin{pmatrix} -1/\sqrt{3} \\ 1/\sqrt{3} \\ 1/\sqrt{3} \end{pmatrix}.$$

Example 7.12: If $[A, B] = 0$ then using matrix method show that A and B have a common set of eigen functions for degenerate case.

Solution: Let the eigen value a be m-fold degenerate:

$$A\psi_j = a\psi_j \quad j = 1, ..., m \quad \text{(i)}$$

whence
$$(\psi_j, \psi_k) = \delta_{jk}$$

Then since $[A, B] = 0$ therefore

$$AB\psi_j = BA\psi_j$$

And the eigen value equation (*i*) implies

$$A(B)\psi_j = a(B\psi_j)$$

Thus, $B\psi_j$ is also an eigenfunction of A with the eigen value a and, a linear combination of the functions ψ_j.

$$B\psi_j = \sum_k C_{jk}\psi_k \quad \text{(ii)}$$

with coefficients

$$C_{jk} = (\psi_k, B\psi_j) = C_{kj}^*$$

The matrix (C_{jk}) is Hermitian and can be diagonalized by a unitary transformation U:

$$U^\dagger CU = C_D \quad \text{with} \quad U^\dagger U = UU^\dagger = \hat{1}$$

From this it follows that $CU = UC_D$ or, in components

$$\sum_k C_{ij}U_{jk} = U_{ik}C_{Dk}$$

and
$$\sum_i U_{ir}^* C_{ik} = C_{Dr} U_{kr}^* \quad \text{(iii)}$$

This means that the kth column vector of the matrix U, i.e.,

$$\begin{pmatrix} U_{1k} \\ \dots \\ \dots \\ U_{mk} \end{pmatrix}$$

is an eigen vector of the matrix C with eigen value C_{Dk}.

Multiplying Eqn. (*ii*) by U_{jr}^*, we find using Eqn. (*iii*)

$$\sum_j BU_{jr}^* \psi_j = \sum_{j,k} U_{jr}^* C_{jk} \psi_k = \sum_k C_{Dr} U_{kr}^* \psi_k = C_{Dr} \phi_r \qquad (iv)$$

Thus, we see that, the linear combinations

$$\phi_r = \sum_k U_{kr}^* \psi_k$$

of the degenerate eigenfunctions ψ_k represent eigenfunctions of both A and B. The eigenvalues of B are given by the diagonal elements C_{Dr} of the diagonal matrix C_D.

Example 7.13: Show that the matrix representation of a Hermitian operator is Hermitian.

Solution: Let A be a Hermitian operator. Then for any two given states ψ_i and ψ_j'.

$$\left(\psi_i, A\psi_j\right) = \left(A\psi_i, \psi_j\right) \qquad (i)$$

Now

$$a_{ij} = \left(\psi_i, A\psi_j\right)$$

$$\therefore \qquad A^\dagger = [a_{ij}]^\dagger = \left[a_{ji}^*\right] = \left[\left(\psi_j, A\psi_i\right)^*\right]$$

$$= \left[\left(A\psi_i, \psi_j\right)\right] = \left[\left(\psi_i, A\psi_j\right)\right] = \left[a_{ij}\right] = A \qquad \text{(using } (i)\text{)}$$

7.12 General Proof of Uncertainty Principle

For any operator F and its Hermitian adjoint $F^\dagger$ (defined by $F_{mn}^\dagger = F_{nm}^*$) we can easily find (as follows) that $FF^\dagger$ is Hermitian or self adjoint. With reference to an arbitrary basic set of eigenfunctions consider the matrix element

$$\left(FF^\dagger\right)_{mn} = \sum_i F_{mi} F_{in}^\dagger = \left(\sum_i F_{im}^\dagger F_{ni}\right)^* \qquad (45)$$

(since $F^\dagger$ is just the complex conjugate of F^T)

$$= \left(\sum_i F_{ni} F_{im}^\dagger\right)^* = (FF^\dagger)_{nm}^* \qquad (46)$$

(using $F_{mi} = (F_{mi}^\dagger)^\dagger = (F_{im}^*)^\dagger = (F_{im}^\dagger)^*$ and $F_{in}^\dagger = F_{ni}^*$)

Thus $FF^\dagger$ is Hermitian and

$$<FF^\dagger> = \int \psi^* \, FF^\dagger \, \psi \, dq$$
$$= \int (F^\dagger \psi)^* \, F^\dagger \psi \, dq = \int |F^\dagger \psi|^2 \, dq \geq 0 \qquad (47)$$

(using the definition of adjoint of an operator A viz.,

$$\int X^*A \; Y \; dq = \int (A^\dagger X)^* \; Y \; dq)$$

Now, we can express F as

$$F = C + i\lambda D \tag{48}$$

where C, D are two self adjoint operators and λ is a real number. It follows that

$$F_{mn}^\dagger = (C + i\lambda D)_{mn}^\dagger = C_{nm}^* - i\lambda D_{nm}^* = C_{mn}^\dagger - i\lambda D_{mn}^\dagger$$

so that

$$F^\dagger = C^\dagger - i\lambda D^\dagger = C - i\lambda D \text{ (by assumption } C = C^\dagger, D = D^\dagger) \tag{49}$$

Hence

$$<FF^\dagger> = <(C + i\lambda D) \; (C - i\lambda D)> \tag{50}$$

$$= <C^2> + \lambda^2 <D^2> - i\lambda <(CD - DC)> \;\; \geq 0$$

The minimum of this function occurs when $2\lambda <D^2> - i <(CD - DC)> = 0$. This gives the minimum value as

$$<C^2> + \frac{1}{4} \frac{<(CD - DC)>^2}{<D^2>} \geq 0$$

or

$$\left\langle C^2 \right\rangle \left\langle D^2 \right\rangle \geq -\frac{1}{4} \left\langle (CD - DC) \right\rangle^2 \tag{51}$$

Deviations from the mean value:

$$\delta C = C - <C>$$

and

$$\delta D = D - <D> \tag{52}$$

then

$$\delta C \delta D - \delta D \delta C = (C - <C>) \; (D - <D>) - (D - <D>)(C - <C>)$$

$$= CD - DC + <C> <D> - <D> <C> - <C><D> - <C>D + D<C> + <D>C$$

$$= CD - DC \tag{53}$$

because $<C><D>$ being constant numbers, they commute with every other quantity. Replacing C and D by δC and δD respectively in Eqn. (51), we obtain

$$\left\langle (\delta C)^2 \right\rangle \left\langle (\delta D)^2 \right\rangle \geq -\frac{1}{4} \left\langle (CD - DC) \right\rangle^2 \tag{54}$$

Since $<\delta C>$ is zero $\left(\left\langle (\delta C)^2 \right\rangle \neq 0 \right)$, we can estimate the magnitude of δC by the root mean square deviation ΔC defined by

$$\Delta C = +\sqrt{\left\langle (\delta C)^2 \right\rangle}$$

and

$$\Delta D = +\sqrt{\left\langle (\delta D)^2 \right\rangle} \tag{55}$$

Hence,

$$\Delta C \Delta D \geq \sqrt{-\frac{1}{4} \left\langle (CD - DC) \right\rangle^2} \tag{56}$$

which is the *uncertainty relation*. It means that it is not possible to make a precise simultaneous measurement of both C and D unless the average value of their commutator vanishes.

For a physical system described by coordinates q_m and its canonical conjugate momentum p_m, above equation becomes

$$\Delta q_m \Delta p_m \geq \sqrt{-\frac{1}{4}\langle q_m p_m - p_m q_m \rangle^2}$$

$$= \sqrt{-\frac{1}{4}(i\hbar)^2}$$

or $\qquad\qquad\qquad \Delta q_m \Delta p_m \geq \dfrac{\hbar}{2}$ (57)

In other words, in a simultaneous measurement of position and momentum, the product of their uncertainties is greater than or equal to $\hbar/2$. All other pairs of canonically conjugate quantities (as for example, energy and time) obey this uncertainty relation.

Questions and Problems

1. (a) How do you diagonalize a matrix?
   (b) What do you understand by the eigen functions and eigen values of a matrix?

2. By using the characteristic equation, diagonalize the matrix

$$A = \begin{bmatrix} 2 & \sqrt{2} & 0 \\ 0 & 2 & 0 \\ 0 & \sqrt{2} & 2 \end{bmatrix}$$

3. Given the matrix

$$A = \begin{bmatrix} 0 & \sqrt{2/3} & 0 \\ \sqrt{2/3} & 2 & \sqrt{1/3} \\ 0 & \sqrt{1/3} & 2 \end{bmatrix}$$

   Write the unitary matrix which diagonalizes A and prove by direct multiplication that UAU^{-1} is a diagonal matrix.

4. Is the matrix $A = \begin{bmatrix} 1 & i & 0 \\ -i & 0 & -2i \\ 0 & 2i & 2 \end{bmatrix}$ Hermitian?

5. Check that following matrices are unitary

   (i) $A = \begin{pmatrix} -1/2 & -\sqrt{3}/2 \\ \sqrt{3}/2 & -1/2 \end{pmatrix}$; (ii) $B = \begin{pmatrix} -1/2 & \sqrt{3}/2 \\ \sqrt{3}/2 & -1/2 \end{pmatrix}$

6. Find the eigenvalues and the eigenvectors of the matrix

$$A = \begin{bmatrix} 1 & 1 & 1 \\ 1 & 2 & 3 \\ 1 & 3 & 6 \end{bmatrix}$$

7. (*a*) How can one diagonalize a matrix by unitary transformation?

   (*b*) Show that under a unitary transformation, a Hermitian operator remains Hermitian and that the trace of operator remains unchanged.

8. A certain observable has a 3 × 3 matrix representation given by

$$\frac{1}{\sqrt{2}}\begin{pmatrix} 0 & 1 & 0 \\ 1 & 0 & 1 \\ 0 & 1 & 0 \end{pmatrix}$$

   Find the normalized eigenvectors of this observable and the corresponding eigenvalues. Is there any degeneracy?

9. Given

$$A = \begin{bmatrix} -\dfrac{1}{3} & \sqrt{\dfrac{2}{3}} & \dfrac{\sqrt{2}}{3} \\ \sqrt{\dfrac{2}{3}} & 0 & \dfrac{1}{\sqrt{3}} \\ \dfrac{\sqrt{2}}{3} & \dfrac{1}{\sqrt{3}} & -\dfrac{2}{3} \end{bmatrix}, \; B = \begin{bmatrix} \dfrac{5}{3} & \dfrac{1}{\sqrt{6}} & -\dfrac{1}{3\sqrt{2}} \\ \dfrac{1}{\sqrt{6}} & \dfrac{3}{2} & \dfrac{1}{2\sqrt{3}} \\ -\dfrac{1}{3\sqrt{2}} & \dfrac{1}{2\sqrt{3}} & \dfrac{11}{6} \end{bmatrix}$$

   Show that *A* and *B* commute. Find their eigenvalues and show that they determine a unitary transformation

$$U = \begin{bmatrix} \dfrac{1}{\sqrt{3}} & \dfrac{1}{\sqrt{2}} & \dfrac{1}{\sqrt{6}} \\ \dfrac{1}{\sqrt{3}} & 0 & -\dfrac{2}{\sqrt{6}} \\ \dfrac{1}{\sqrt{3}} & -\dfrac{1}{\sqrt{2}} & \dfrac{1}{\sqrt{6}} \end{bmatrix} \text{ which diagonalizes both } A \text{ and } B.$$

10. Find the eigenvalues and eigenvectors of

$$A = \begin{pmatrix} \cos\theta & -\sin\theta \\ \sin\theta & \cos\theta \end{pmatrix}$$

11. Using the recurrence relation of the Hermite polynomials

$$\frac{dH_n(y)}{dy} = 2nH_{n-1}(y)$$

   Show that the matrix elements of momentum for the harmonic oscillator are given by

$$p_{jk} = -i\hbar \int_{-\infty}^{+\infty} \psi_j^* \frac{\partial \psi_k}{\partial x} \, dx = i\left(\frac{m\hbar\omega}{2}\right)^{1/2} \left\{ \sqrt{j}\, \delta_{j,k+1} - \sqrt{j+1}\, \delta_{j,k-1} \right\}$$

$$\left[\textbf{Hint :} \quad \psi_n(y) = \left\{ \frac{\alpha}{2^n \sqrt{\pi}\, n!} \right\}^{1/2} H_n(y) e^{-y^2/2}, y = \left(\frac{m\omega}{\hbar}\right)^{1/2} x \right].$$

8

Dirac Representation Theory

8.1 An *n*-dimensional Vector Space

À set of *n*-vectors $\hat{e}_i$ is said to be linearly independent if there exists no relation of the form

$$\sum_{i=1}^{n} c_i \, \hat{e}_i = 0 \tag{1}$$

between them except for the trivial case $c_1 = c_2 = ... c_n = 0$. Existence of a relation would reduce the number of independent terms. In three dimensional space, the number of $\hat{e}_i$'s is three. In general, any set of n linearly independent vectors spans an *n*-dimensional linear space R_n.

8.2 The State Vector ψ in Dirac Representation and Hilbert Space

We know that for a description of a quantum state, we assign a wave function ψ_a called a *state vector* in a functional space e.g. $\psi(x)$ or $\psi(p)$. Here index '*a*' corresponds to a particular state and is known as a *state-index*. In Dirac notation a state-vector is represented by |ψ> called a *ket vector*. It is an abstract symbol that has nothing to do with geometric three dimensional vectors but stands for a particular physical state, i.e.

$$\text{wave function} \rightarrow |\psi> \quad \text{(Dirac notation)} \tag{2}$$

Just as a vector can be represented by a superposition in vector space $\left(\vec{v} = \sum_i v_i \hat{x}_i \right)$ we represent

$$\psi(x) = c_1\psi_1 + c_2\psi_2 + ... + c_n\psi_n + ...\infty = \sum_{n=0}^{\infty} c_n\psi_n(x) \tag{3}$$

That is, we can consider the wave function as a nondenumerable infinite expansion. In particular

$$\psi(x) = \sum_n c_n\psi_n(x) \quad \text{(energy-representation)}$$

$$= \int d\vec{r} c_r \psi_r(\vec{r}) \quad \text{(coordinate-representation)}$$

$$= \int dp \, c_p \, \psi_p(x) \quad \text{(momentum-representation)} \tag{4}$$

Here, expansion coefficients c_n, c_r and c_p etc. characterise the state ψ equally well.

In general any state $|\psi\rangle$ can be made up from a linear combination of base states as

$$|\psi\rangle = \sum_{i=1}^{n} c_i |\psi_i\rangle \tag{5}$$

where the c_i's are a set of ordinary complex numbers and $|\psi_1\rangle$, $|\psi_2\rangle$... etc. stand for base states in some base or representation. For time independent form a state function is denoted as $\psi_a(\xi)$, which, using Dirac's notation is represented as

$$\psi_a(\xi) = \langle \xi | a \rangle \tag{6}$$

where the notation $\langle \xi |$ represents a *bra-vector*. The $\langle \, | \, \rangle$ is termed as a bracket, [bra+ket = $\langle \, | \, \rangle$: bracket].

Thus according to Dirac, any state 'a' of a quantum system (as, for instance, electrons in a metal) may be represented by the ket-vectors. These ket vectors can be added together, and can be multiplied by complex scalar factors and so one obtains a new ket vector. The set of all possible ket vectors ψ_i's (which can be labelled by index i), forms an abstract complex vector space (since ψ, in general being complex) with an infinite (but countable) dimensions, called the *Hilbert-space*.

8.2.1 Properties of Hilbert Space

A Hilbert space H has the following properties:
  (1) The space is linear i.e. if c is a constant and ϕ is any element of the space then $c\phi$ is also an element of the space. Further the state vectors (wave functions) of the space satisfy following properties also:

$$|a\rangle + |b\rangle = |b\rangle + |a\rangle$$
$$(|a\rangle + |b\rangle) + |c\rangle = |a\rangle + (|b\rangle + |c\rangle)$$
$$1\,|a\rangle = |a\rangle \quad \text{(Identity operator)}$$
$$\alpha\,(\beta\,|a\rangle) = (\alpha\beta)\,|a\rangle \tag{7}$$
$$\alpha\,(|a\rangle + |b\rangle) = \alpha|a\rangle + \alpha\,|b\rangle$$
$$(\alpha + \beta)\,|a\rangle = \alpha\,|a\rangle + \beta\,|b\rangle$$

  (2) To each "ket" vector $|a\rangle$, there can be assigned a dual "bra" vector denoted by $\langle a|$ and connected with the ket vector through the simple relation $\langle a| \equiv |a\rangle^\dagger$ (adjoint of $|a\rangle$)
  (3) There is an inner-product $\langle \psi | \phi \rangle$ for any two elements in the space. For functions defined in the interval $a \leq x \leq b$ (in one dimension), we may take

$$\langle \phi | \psi \rangle = \int_a^b \phi^* \, \psi \, dx \tag{8}$$

(4) Any element of H has a *norm* ("length") that is related to the inner product as follows:

$$(\text{norm of } \psi)^2 \equiv \left|\psi^2\right| = \langle\psi|\psi\rangle \tag{9}$$

(5) H is complete. Every Cauchy sequence of function in H converges to an element of H. A Cauchy sequence $\{\psi_n\}$ is such that $\left|\psi_n - \psi_l\right| \rightarrow 0$ as n and l approach infinity. In other words, we may say that a Hilbert space contains all its limit points.

(6) If ψ_a and ψ_b are two state vectors, then $\psi_a{}^\dagger \psi_b = <a \mid b>$, where $<a|b>$ is a number).

8.2.2 Physical Significance of the Hilbert Space

When the systematics of a linear vector space of n-dimensions are carried over into a complex space with infinitely many ($n \rightarrow \infty$, but countable!) dimensions with state vectors satisfying above mentioned properties, we get a Hilbert space. Then we may visualize a geometric picture in which a state function such as ψ (which contains all the physically relevant information about physical state of a system) is regarded as a state vector in infinite dimensional Hilbert space. Each dimension then corresponds to one of the rows of the one column matrix

$$\psi \rightarrow \begin{pmatrix} c_1 \\ c_2 \\ ... \\ ... \\ c_n \end{pmatrix} \rightarrow |\psi> \tag{10}$$

that describes (see Eqn. 5) the state ($|\psi>$) and the component of the state vector ($|\psi_i>$) along that axis i of the Hilbert space is numerically equal to the corresponding element of the matrix c_i, for instance, an eigenfunction ψ_n can be represented geometrically by a vector along a certain "direction" (or "axis") 'n' in the Hilbert space.

Different choices for the orientation of the axes in the Hilbert space correspond to different choices for representation, e.g. energy-representation corresponds to choosing axes in such a way that a state vector oriented along one of the energy axes is an eigen state of the Hamiltonian.

The states specified by the different components ψ_i (forming basic set) are represented as equivalent to vectors which are orthogonal to each other. Any other vector ϕ in the Hilbert space can be represented as the geometric sum

$$\phi = \sum_{i=1}^{\infty} c_i \psi_i \tag{11}$$

8.2.3 Hilbert Space and Continuous Variable

In case, if we are concerned with a continuous variable x instead of the integer index i, we say that a function $f(x)$ is square-integrable in the interval (a, b) if the norm

$$(f,f) = \int_a^b |f(x)|^2 \, dx \tag{13}$$

exists. The class of square integrable functions defines a linear vector space in which we define the scalar product of two functions $f(x)$ and $g(x)$ as

$$(f,g) = \int_a^b f^*(x)g(x)dx \tag{14}$$

The Schwarz inequality

$$|(f,g)|^2 \leq (f,f)(g,g) \tag{15}$$

shows that (f, g) exists if $f(x)$ and $g(x)$ are square integrable. The vector space defined in the above way where the linear combination

$$h(x) = \alpha f(x) + \beta g(x) \tag{16}$$

is also square integrable defines a Hilbert space. In this space, the variable x (in place of i) varies continuously over the interval (a, b) and the vector $\vec{f}$ has infinitely many components $f(x)$ which are not countable. A linear transformation in Hilbert sapce i.e. $\vec{g} = \hat{A}\vec{f}$ is defined by the equation

$$g(x) = \int_a^b A(x,x')f(x')\, dx' \tag{17}$$

where the continuous matrix $A(x, x')$ is a function of the two variables x, x' and is a representative of the operator $\hat{A}$. The identity transformation is defined by

$$\vec{g} = \hat{1}\vec{f} = \vec{f}$$

The matrix $1(x, x')$ which represents the unit operator, (formally) must satisfy

$$g(x) = \int_a^b 1(x,x')f(x')dx' = f(x) \tag{18}$$

for every function $f(x)$. By the definition of the Dirac-delta function, this implies that

$$1(x,x') = \delta(x - x')$$

so that for the Hilbert space, $\delta(x - x')$ takes the place of the Kronecker symbol δ_{ij} in the finite dimensional space. Thus

$$f(x) = \int_{-\infty}^{\infty} \delta(x - x')f(x')dx' \tag{19}$$

$f(x')$ are the components of $f(x)$.

8.3 The Dual Vector Space

There exists a vector space dual to the "ket" vector Hilbert space which is assigned as

$$\psi^* \to (c_1^*, c_2^* .....) \to \langle\psi| \tag{20}$$

$\langle\psi|$ is a "bra" vector. (compare with Eqn. 10)

Ket and bra vectors have a different nature and thus cannot be added together. Both of these are neither purely real nor purely imaginary. The sum in the dual space results from the mapping.

$$\psi_c^* = \alpha^* \; \psi_a^* + \beta^* \psi_b^*$$

$$\rightarrow \langle c | = \alpha * \langle a | + \beta * \langle b |$$

Thus dual vectors also form a linear space. Hermitian oeprators $\hat{F} = \hat{F}^\dagger$ act upon ket-vectors from the left and upon the bra vectors from the right and transform them into other 'ket' and 'bra' vectors respectively, e.g.

$$|b\rangle = \hat{F} \; |a\rangle \tag{21}$$

and

$$\langle b | = \left(\hat{F} \; |a\rangle \right)^\dagger = < a | F^\dagger = < a | \hat{F} \tag{22}$$

The scalar product of two "ket" vectors |a> and |b> is denoted by the bracket <b|a>. The scalar product <b|a> is a normal complex number and satisfies the relation

$$< b \,|\, a > = < a \,|\, b > * \tag{23}$$

8.4 Dirac Representation of Operators

The product |b> <a|, in which a "ket" vector stands to the left of a "bra" vector is an operator. Any state vector |a> can be expanded in terms of the complete set of *orthonormal* vectors $|F_m\rangle$ corresponding to the operator $\hat{F}$:

$$|a\rangle = \sum_m |F_m\rangle \; \langle F_m |a\rangle \tag{24}$$

Similarly, we may expand any operator $\hat{A}$ in terms of the complete set of operators $|F_m\rangle \, \langle F_n|$ as:

$$\hat{A} = \sum_{m,n} A_{mn} |F_m\rangle \; \langle F_n | \tag{25}$$

If this equality is satisfied, we can (by using the orthonormality of the vectors $|F_m\rangle$), uniquely determine the matrix elements in the expansion

$$A_{mn} = \langle F_m | \hat{A} | F_n \rangle \tag{26}$$

In particular, the expansion of the unit operator $\hat{I}$ has the form

$$\hat{I} = \sum_m |F_m\rangle \; \langle F_m | \tag{27}$$

8.5 Observables

Any dynamical quantity (e.g. position momentum or angular-momentum etc.) that can be measured is known as an observable. The observables can be represented by linear operators and since the measured value must yield a real number, an observable is always represented by a real linear operator. We denote the *eigen kets* of the observable α by $|\alpha_n\rangle$ which are assumed to form an orthonormal set

$$< \alpha_m \,|\, \alpha_n > = \delta_{mn} \tag{28}$$

Since $|\alpha_n\rangle$ form a complete set, an arbitrary ket $|P\rangle$ can be expanded as

$$|P\rangle = \sum_n c_n |\alpha_n\rangle$$

If we now make a measurement of α, the probability of the outcome α_n is $|c_n|^2$, (we have assumed $|P>$ to be normalized). If we make a measurement, the system would yield the resulting state $|\alpha_n>$. For degenerate states, one can always choose an appropriate linear combination of eigen kets so that they form an orthonormal set (i.e. mutually orthogonal and linearly independent eigen kets).

8.6 The Completeness Condition

We have already known earlier that the eigen kets of an observable form a complete set. Let $|n>$, $n = 0, 1, 2, \ldots$ denote an arbitrary ket. Then

$$|P> = \sum_n c_n |n> \tag{29}$$

where $\sum$ denotes summation over the discrete states and an integration over the continuum states. Since the eigen kets can be assumed to form an orthonormal set

$$\langle m|n\rangle = \delta_{mn}$$

we have

$$\langle m|P\rangle = \sum_n c_n \langle m|n\rangle$$

$$= \sum_n c_n \delta_{mn} = c_m \tag{30}$$

Thus

$$|P\rangle = \sum_n |n\rangle\, c_n = \left\{\sum_n |n\rangle\, \langle n|\right\} |P\rangle \tag{31}$$

Since the above equation holds for an arbitrary ket $|P>$, the quantity inside the curly brackets must be a unit operator.

$$\sum_n |n\rangle\, \langle n| = 1 \tag{32}$$

This is known as the *completeness condition* (This may be compared with the completeness condition of Schrödinger wave functions).

Example 8.1: Show that $<n/m> = \sum <n/i> <i/m>$, where $|n>$ and $|m>$ are two arbitrary vectors and the vectors $|i>$ form a complete set.

Solution: Since the vectors $|i>$ form a complete set, so we can write

$$|n> = \sum_i a_i |i>, \quad a_i = <i|n>$$

and

$$|m> = \sum_j b_j |j>, \quad b_j = <j|m>$$

∴

$$<n|m> = \sum_{i,j} a_i^* <i| <j|m> |j> = \sum_{i,j} <n|i> <j|m> <i|j>$$

$$= \sum_{i,j} <n|i> <j|m> \delta_{ij} = \sum_i <n|i> <i|m>$$

8.7 Relation between Dirac's Bra and Kets and Matrices

Suppose an observable B has a complete set of eigenvectors $|n>$. Each vector $|\psi>$ is completely specified by the set of numbers (components) $<n|\psi>$.

The completeness relation can be used to write the inner product <φ|ψ> for any (dual) vector <φ| as

$$\langle \phi | \psi \rangle = \sum \langle \phi | n \rangle \; \langle n | \psi \rangle \tag{33}$$

where

$$|\phi\rangle = A|\psi\rangle \tag{34}$$

We can similarly specify the vector |φ> produced by the linear operator A by its components

$$<n|\phi> = <n|A|\psi>$$
$$= \sum_{m} < n | A | m >< m | \psi > \tag{35}$$

i.e. the observable A is completely specified by the numbers $<n|A|m> \equiv A_{nm}$. Thus the operator A corresponds to a matrix.

The vectors |ψ> and |φ> are represented by column vectors

$$\psi_m = < m | \psi >; \; \phi_n = < n | \phi > \tag{36}$$

and Eqn. (35) is the matrix product

$$\begin{pmatrix} \phi_1 \\ \phi_2 \\ \vdots \end{pmatrix} = \begin{pmatrix} A_{11} & A_{12} & \cdots \\ A_{21} & A_{22} & \cdots \\ \vdots & \vdots & \vdots \end{pmatrix} \begin{pmatrix} \psi_1 \\ \psi_2 \\ \vdots \end{pmatrix} \tag{37}$$

The inner product of a <φ| with a |ψ> in Eqn. (33) is the matrix product of a row vector and a column vector.

The matrix elements of the product of two operators A and B are

$$< n | AB | m >= \sum_{i} < n | A | i >< i | B | m > \tag{38}$$

This is the product of the matrix with elements A_{ni} with the matrix with elements B_{im}.

Example 8.2: Find three matrices A, B and C that satisfy the following equations

$$A^2 = B^2 = C^2 = 1$$
$$AB + BA = BC + CB = CA + AC = 0$$

where 1 is the unit matrix and 0, the null matrix. Obtain explicit expressions for all three matrices in a representation, in which A is diagonal, assuming that it is non-degenerate.

Solution:

$$A|m> = a_m |m>$$
$$A^2 |m> = 1 |m>$$

or $$a_m A |m> = 1 |m> \; \therefore a_m^2 = 1 \text{ or } a_m = \pm 1$$

Matrix element $A_{nm} = <n|A|m> = a_m <n|m> = a_m \delta_{nm}$

$$\therefore \qquad A = \begin{pmatrix} 1 & 0 \\ 0 & -1 \end{pmatrix}$$

Let $$B = \begin{pmatrix} b_1 & b_2 \\ b_3 & b_4 \end{pmatrix}$$

Now $$AB + BA = 0$$

$\therefore$
$$\begin{pmatrix} 1 & 0 \\ 0 & -1 \end{pmatrix}\begin{pmatrix} b_1 & b_2 \\ b_3 & b_4 \end{pmatrix} + \begin{pmatrix} b_1 & b_2 \\ b_3 & b_4 \end{pmatrix}\begin{pmatrix} 1 & 0 \\ 0 & -1 \end{pmatrix} = 0$$

or
$$\begin{pmatrix} b_1 & b_2 \\ -b_3 & -b_4 \end{pmatrix} + \begin{pmatrix} b_1 & -b_2 \\ b_3 & -b_4 \end{pmatrix} = 0$$

or
$$\begin{pmatrix} 2b_1 & 0 \\ 0 & -2b_4 \end{pmatrix} = 0 \Rightarrow \begin{matrix} b_1 = 0 \\ b_4 = 0 \end{matrix}$$

$\therefore$
$$B = \begin{pmatrix} 0 & b_2 \\ b_3 & 0 \end{pmatrix}$$

Now
$$B^2 = 1$$

or
$$\begin{pmatrix} 0 & b_2 \\ b_3 & 0 \end{pmatrix}\begin{pmatrix} 0 & b_2 \\ b_3 & 0 \end{pmatrix} = 1$$

or
$$\begin{pmatrix} b_2 b_3 & 0 \\ 0 & b_2 b_3 \end{pmatrix} = 1$$

$\therefore$
$$b_2 b_3 = 1, \ b_2 = \frac{1}{b_3} = b \ (\text{say})$$

$\therefore$
$$B = \begin{pmatrix} 0 & b \\ b^{-1} & 0 \end{pmatrix}$$

Similarly
$$C = \begin{pmatrix} 0 & c \\ c^{-1} & 0 \end{pmatrix}$$

Now
$$BC + CB = 0$$

or
$$\begin{pmatrix} 0 & b \\ b^{-1} & 0 \end{pmatrix}\begin{pmatrix} 0 & c \\ c^{-1} & 0 \end{pmatrix} + \begin{pmatrix} 0 & c \\ c^{-1} & 0 \end{pmatrix}\begin{pmatrix} 0 & b \\ b^{-1} & 0 \end{pmatrix} = 0$$

or
$$\begin{pmatrix} bc^{-1} & 0 \\ 0 & b^{-1}c \end{pmatrix} + \begin{pmatrix} cb^{-1} & 0 \\ 0 & c^{-1}b \end{pmatrix} = 0$$

$\therefore$
$$bc^{-1} + b^{-1}c = 0$$

or
$$b^2 + c^2 = 0$$

or
$$b^2 = -c^2 \ \text{or} \ b = \pm ic$$

Hence, we get two-parameter family of matrices

$$B = \begin{pmatrix} 0 & b \\ b^{-1} & 0 \end{pmatrix}; \ C = \begin{pmatrix} 0 & ib \\ -ib^{-1} & 0 \end{pmatrix}$$

Example 8.3: Given $H = \dfrac{p^2}{2m} + V(r)$, calculate $\sum (E_n - E_m)|x_{mn}|^2$.

Solution: $$[x, H] = \left[x, \frac{p^2}{2m} + V(r) \right]$$

$$= \left[x, \frac{p^2}{2m}\right] + [x, V(r)]$$

But $[x, V(r)] = 0$, since x and $V(r)$ commute. Therefore

$$[x, H] = \frac{1}{2m}[x, p^2] = \frac{1}{2m}[x, p_x^2]$$

because $p^2 = p_x^2 + p_y^2 + p_z^2$, and p_y and p_z commute with x.

$\therefore$
$$[x, H] = \frac{1}{2m}\{p_x[x, p_x] + [x, p_x]p_x\}$$

but
$$[x, p_x] = i\hbar$$

$\therefore$
$$[x, H] = \frac{2i\hbar}{2m}p_x = \frac{i\hbar}{m}p_x$$

Taking double commutator of this equation

$$\left[x, [x, H]\right] = \frac{i\hbar}{m}[x, p_x]$$

or
$$x[x, H] - [x, H]x = \frac{i\hbar}{m}i\hbar \equiv -\frac{\hbar^2}{m}$$

Now,

$$\langle n| \, x[x, H] - [x, H]x \, |m\rangle = -\frac{\hbar^2}{m} \, \langle n|m\rangle$$

Taking $n = m$

$$< n| \, x[x, H] - [x, H]x \, |n> = -\frac{\hbar^2}{m} \qquad (\because <n|n> = 1)$$

We can rewrite l.h.s. as

$$\text{L.H.S.} = \langle n| \, x[x, H] \, |n\rangle - \langle n| \, [x, H]x \, |n\rangle$$

$$= \sum_m \langle n| \, x \, |m\rangle \, \langle m| \, [x, H] \, |n\rangle - \sum_m \langle n| \, [x, H] \, |m\rangle \, \langle m| x \, |n\rangle$$

$$= \sum_m \langle n| \, x \, |m\rangle \, \langle m| \, (xH - Hx) \, |n\rangle - \sum_m \langle n| \, (xH - Hx) \, |m\rangle \, \langle m| \, x|n\rangle$$

$$= \sum_m \langle n| \, x \, |m\rangle \, \langle m| \, (xE_n - E_m x) \, |n\rangle$$

$$- \sum_m \langle n| \, (xE_m - E_n x) \, |m\rangle \, \langle m| \, x \, |n\rangle$$

$$= \sum_m 2(E_n - E_m) \, \langle n| \, x|m\rangle \, \langle m| \, x \, |n\rangle$$

$$= \sum_m 2(E_n - E_m) x_{nm} x_{mn}$$

But $x_{nm} = x_{mn}^* \ (= x_{mn})$, as x is Hermitian

$\therefore$ L.H.S. $= \sum_m 2(E_n - E_m)\left|x_{mn}\right|^2$

Hence $\sum_m 2(E_n - E_m)\left|x_{mn}\right|^2 = -\dfrac{\hbar^2}{m}$

or $\sum_m (E_n - E_m)\left|x_{mn}\right|^2 = -\dfrac{\hbar^2}{2m}$

Example 8.4: Using rules of bra and ket algebra prove that

(i) Trace $(XY) =$ Trace (YX) (ii) $(XY)^\dagger = Y^\dagger X^\dagger$

where X and Y are operators.

Solution: (*i*) Trace $(XY) = \sum_n < n\,|\,XY\,|\,n >$

Putting $\sum_m \left|m\right\rangle \left\langle m\right| = 1$ in between X and Y,

Trace $(XY) = \sum_m \sum_n \langle n|\,x\,|m\rangle \langle m|\,Y\,|m\rangle$

Since $<m|Y|n>$ or $<n|X|m>$ are matrix elements and they are just numbers, so we may interchange them

$\therefore$ Trace $(XY) = \sum_n \sum_m < m\,|\,Y\,|\,n ><n\,|\,x\,|\,m >$

$= \sum_m \langle m|\, YX\,|m\rangle =$ Trace (YX) $\left(\text{Using } \sum_n |n\rangle\langle n| = 1\right)$

(*ii*) Consider the matrix element $\{<P|XY|Q>\}^\dagger$, then it is equal to $<Q|(XY)^\dagger|P>$ because $(|A>)^\dagger = <A|$.

Let $Y|Q> = |Q'>$ and $<P|X = <P'|$, then

$\{<P|XY|Q>\}^\dagger = \{<P'|Q'>\}^\dagger = <Q'|P'>$

But $Y|Q> = |Q'>$ so $<Q'| = <Q|Y^\dagger$;

$<P|X = <P'|$, so $|P'> = X^\dagger|P>$.

Hence

$\{<P|XY|Q>\}^\dagger = <Q'|P'> = <Q|Y^\dagger X^\dagger\,|P>$.

But $\{<P|XY|Q>\}^\dagger = <Q|(XY)^\dagger|P>$

Therefore $(XY)^\dagger = Y^\dagger X^\dagger$

8.8 Coordinate and Momentum Representation

We often represent wave vectors as functions of either coordinates or momenta. We discuss now the relationship between this function space and the bra, ket notation.

We denote by $<r|u>$ the coordinate representation of ket $|u>$. Similarly $<p|u>$ is the momentum representation of $|u>$. Thus,

$$\vec{u}_j(\vec{r}) = <\vec{r}\,|\,u_j> \tag{39}$$

$$\vec{u}_j(\vec{p}) = <\vec{p}\,|\,u_j> \tag{40}$$

Also

$$\vec{u}_j^*(\vec{r}) = <u_j\,|\,\vec{r}> \tag{41}$$

and

$$\vec{u}_j^*(\vec{p}) = <u_j\,|\,\vec{p}> \tag{42}$$

The scalar product $<u_i\,|\,u_j>$ is given by

$$<u_i\,|\,u_j> \equiv \left(u_i, u_j\right)$$

$$= \int \vec{u}_i^*(\vec{r})\vec{u}_j(\vec{r})\,d\vec{r}$$

$$= \int <u_i\,|\,r><r\,|\,u_j>\,d\vec{r} \tag{43}$$

Hence,

$$\int |\vec{r}><\vec{r}\,|\,d\vec{r} = \hat{1} \tag{44}$$

Similarly

$$\int |\vec{p}><\vec{p}\,|\,d\vec{p} = \hat{1} \tag{45}$$

From Eqns. (40) and (44), we have

$$\vec{u}_j(\vec{p}) = <\vec{p}\,|\,u_j> = \int <\vec{p}\,|\,\vec{r}><\vec{r}\,|\,u_j>\,d\vec{r}$$

$$= \int u_j(\vec{r})<\vec{p}\,|\,\vec{r}>\,d\vec{r} \tag{46}$$

Similarly, from Eqns. (39) and (45), we have

$$\vec{u}_j(\vec{r}) = \int u_j(\vec{p})<\vec{r}\,|\,\vec{p}>\,d\vec{p}$$

$$= \int u_j(\vec{p})<p\,|\,r>^*\,d\vec{p} \tag{47}$$

Substituting p_x (x component of $\vec{p}$) for $\vec{p}$ and x for $\vec{r}$, in Eqns (46), we have

$$\vec{u}_j(p_x) = \int_{-\infty}^{+\infty} \vec{u}_j(x)<p_x\,|\,x>\,dx \tag{48}$$

Similarly, from Eqn. (47) we have

$$\vec{u}_j(x) = \int_{-\infty}^{+\infty} \vec{u}_j(p_x)<p_x\,|\,x>^*\,dp_x \tag{49}$$

Example 8.5: Using $f(\lambda) = e^{\lambda A}Be^{-\lambda A}$

show $e^A Be^{-A} = B + [A, B] + \dfrac{1}{2!}\left[A, [A, B]\right] + \ldots$

Solution:

$$\frac{\partial f}{\partial \lambda} = Ae^{\lambda A}Be^{-\lambda A} - e^{\lambda A}Be^{-\lambda A}A$$

$$= A f(\lambda) - f(\lambda) A = [A, f(\lambda)]$$

$$\frac{\partial^2 f}{\partial \lambda^2} = Af'(\lambda) - f'(\lambda)A = \left[A, f'(\lambda)\right]$$

$$= \left[A, [A, f(\lambda)]\,\right]$$

since

$$f(\lambda) = f(0) + \lambda f'(\lambda)\big|_{\lambda=0} + \frac{\lambda^2}{2!}f''(\lambda)\big|_{\lambda=0} + \ldots$$

$$= B + \frac{\lambda^2}{1!}[A, B] + \frac{\lambda^2}{2!}\Big[A, [A, B]\Big] + ...$$

Now, put $\lambda = 1$

$$\therefore \qquad e^A B e^{-A} = B + [A, B] + \frac{1}{2!}\Big[A, [A, B]\Big] + ...$$

Example 8.6: Show that $e^{xa}a * e^{-xa} = a * + x$, if $[a, a^\dagger] = 1$

Solution: From preceding problem

$$e^{xa}a * e^{-xa} = a * + [xa, a*] + \frac{1}{2!}\Big[xa, [xa, a*]\Big] + ...$$

$$= a * + x[a, a*] + \frac{1}{2!}[xa, x] + ...$$

$$= a * + x + 0 = a * + x$$

Example 8.7: Given $H = \frac{1}{2}(a^\dagger a + aa^\dagger)$ and $[a, H] = a$, then show $[a^2, a^\dagger] = 2a$. Don't use commutation relation.

Solution: $[a, H] = aH - Ha = a$

or $a \frac{1}{2}(a^\dagger a + aa^\dagger) - \frac{1}{2}(a^\dagger a + aa^\dagger) a = a$

or $\frac{1}{2}[aa^\dagger a + aaa^\dagger - a^\dagger aa - aa^\dagger a] = a$

or $a^2 a^\dagger - a^\dagger a^2 = 2a$

or $[a^2, a^\dagger] = 2a.$

Example 8.8: Given $[a, a^\dagger] = 1$; $[a, a] = [a^\dagger, a^\dagger] = 0$ and $a |0> = 0$. Then prove $<0|e^a e^{a\dagger}|0> = e$.

Solution: From example (3.33)

$$e^A e^B = e^{A+B+\frac{1}{2}[A,B]}$$

$$\therefore \qquad e^a e^{a\dagger} = e^{(a+a\dagger)} e^{\frac{1}{2}[a,a\dagger]}$$

$$e^{a\dagger} e^a = e^{(a\dagger+a)} e^{\frac{1}{2}[a\dagger,a]}$$

or $$e^{a\dagger} e^a = e^{(a+a\dagger)} e^{\frac{1}{2}[a\dagger,a]}$$

$$\therefore \qquad e^a e^{a\dagger} = e^{(a+a\dagger)} e^{\frac{1}{2}} \qquad\qquad (i)$$

$$e^{a\dagger} e^a = e^{(a+a\dagger)} e^{-\frac{1}{2}} \Rightarrow e^{(a+a\dagger)} = e^{\frac{1}{2}} \qquad\qquad (ii)$$

Substituting for $e^{(a+a\dagger)}$ from (ii) into (i), we get

$$e^a e^{a\dagger} = e^{\frac{1}{2}} e^{a\dagger} e^a e^{\frac{1}{2}}$$

$$= e^{a\dagger} e^a e$$

Hence

$$<0| \ e^a e^{a\dagger} |0> = <0| \ e^{a\dagger} e^a |0>e$$

$$= <0| \ (1 + a^\dagger + \frac{a^{\dagger \, 2}}{2!} + ...) \ (1 + a + \frac{a^2}{2!} + ...) \ |0 > e$$

Since $a \ |0> = 0$ etc, therefore only the first term (viz. 1) contributes and all the others contribute zero.

$$\therefore \qquad <0| \ e^a \ e^{a\dagger} \ |0> = <0|1|0> \ e = e.$$

8.9 Change of basis (Dirac Notation)

Consider the transformation of basis $|a_i\rangle$ to $|b_j\rangle$:

$$|b_j\rangle = I|b_j\rangle = \sum_i |a_i\rangle\langle a_i|b_j\rangle \qquad (50)$$

This Eqn. shows that $|a_i\rangle$ is transformed to $|b_j\rangle$ by a matrix whose elements are $\langle a_i|b_j\rangle$. Let us call this the (ji) element of a matrix U. Then Eqn. (50) is written as

$$|b_j\rangle = \sum_i (U)_{ji} |a_i\rangle \qquad (51)$$

where

$$(U)_{ji} = \langle a_i|b_j\rangle$$

(Note that it appears on the l.h.s. of $|a_i\rangle$ in Eqn. (51)). Therefore, we may write

$$(U^\dagger)_{ij'} = \langle b_{j'}|a_i\rangle \qquad (52)$$

From Eqns. (50) and (51), we find:

$$(UU^\dagger)_{jj'} = \sum_i (U_{ji}) \ (U^\dagger)_{ij'}$$

$$= \sum_i \langle a_i|b_j\rangle \ \langle b_{j'}|a_i\rangle$$

$$= \sum_i \langle b_{j'}|a_i\rangle \ \langle a_i|b_j\rangle \qquad (53)$$

(because complete bracket expressions are scalars, we may interchange them.)

Now $\sum_i |a_i\rangle \ \langle a_i| = 1$, therefore

$$(UU^\dagger)_{jj'} = \delta_{jj'} \text{ using } \langle b_{j'}|b_j\rangle = \delta_{jj'}$$

so

$$(UU^\dagger) = 1 \qquad (55)$$

Hence, the transformation matrix connecting the two bases is a unitary matrix. We can easily construct the matrix U as follows:

Let U be equal to $\sum_k |b_k\rangle \langle a_k|$. Then

$$U|a_i\rangle = \sum_k |b_k\rangle \langle a_k|a_i\rangle$$

$$= \sum_k |b_k\rangle \delta_{ki} = |b_i\rangle \tag{56}$$

From Eqn. (56), we note that U transforms $|a_i\rangle$ to $|b_i\rangle$.

We see $\qquad U = \sum_k |b_k\rangle \langle a_k|$ gives

$$UU^\dagger = \sum_k |b_k\rangle \langle a_k| \sum_m |a_m\rangle \langle b_m|$$

So,

$$U^\dagger = \sum_k \sum_m |b_k\rangle \langle a_k|a_m\rangle \langle b_m|$$

$$= \sum_k \sum_m |b_k\rangle \delta_{km} \langle b_m| = \sum_k |b_k\rangle \langle b_k| = 1 \tag{57}$$

Thus $\qquad U = \sum_k |b_k\rangle \langle a_k|$ is unitary and changes the basis $|a_i\rangle$ to $|b_i\rangle$.

Theorem: If a given physical system can be expanded in terms of complete orthonormal eigenvectors of two non-commutative operators, then these basic ket vectors are related by a unitary operator.

Proof: If $|\psi\rangle$ is any arbitrary state vector then

$$|\psi\rangle = \sum_k |a_k\rangle \langle a_k|\psi\rangle \tag{58}$$

or

$$\langle\psi| = \sum_k \{\langle\psi|a_k\rangle\} \langle a_k| \tag{59}$$

If $\langle\psi|a_k\rangle$'s are known then $\langle\psi|$ and equivalently $|\psi\rangle$ is also known. Similarly if $\langle\psi|b_k\rangle$ is known we get the representation of $|\psi\rangle$ in the new transformed base $|b_k\rangle$. One would like to know, how the given state vector $|\psi\rangle$ is related to the new $|b_k\rangle$ and old $|a_k\rangle$ basis.

From Eqn. (2)

$$|b_k\rangle = \sum_i (U)_{ki} |a_i\rangle$$

Applying $\langle\psi|$ on both sides, we get

$$\langle\psi|b_k\rangle = \sum_i (U)_{ki} \langle\psi|a_i\rangle$$

i.e. $\qquad\qquad$ <ψ|b> = U <ψ|a> $\tag{60}$

and $\qquad\qquad$ <b|ψ> = U† <a|ψ> $\tag{61}$

i.e. representation of $|\psi\rangle$ in the two bases are related by a unitary operator. Consider the matrix element of an observable X, namely $\langle b_k| X |b_l\rangle$

$$\langle b_k|X|b_l\rangle = \sum_l \sum_m \langle b_k|a_l\rangle \langle a_l| X |a_m\rangle \langle a_m|b_l\rangle \tag{62}$$

From $\quad U = \sum_l |b_l\rangle \langle a_l|$ and $U^\dagger = \sum_i |a_i\rangle \langle b_i|$ we find

$$<a_{k'}|U^\dagger|a_k> = \sum_i \langle a_{k'}|a_i\rangle \langle b_i|a_k\rangle$$

$$= \sum_i \delta_{k'i} \langle b_i|a_k\rangle = \langle b_{k'}|a_k\rangle \tag{63i}$$

Similarly, $\langle a_k| U |a_{k'}\rangle = \langle a_k|b_{k'}\rangle \tag{63ii}$

Using these results we get from Eqn. (62), using Eqns. (63 i) and (63 ii)

$$\langle b_k|X|b_l\rangle = \sum_l \sum_m \langle b_k|a_l\rangle \langle a_l| X |a_m\rangle \langle a_m|b_l\rangle \tag{64}$$

This equation gives the matrix element of X in $|b>$ basis in terms of its elements in the $|a>$ basis by means of the unitary matrix U defined as

$$U = \sum_l |b_l><a_l|$$

We could have defined $U = \sum_k |a_k\rangle \langle b_k|$. This is also unitary but then the role of U and $U^\dagger$ would have been interchanged. In that case

$$U|b_k> = \sum_l |a_l><b_l|b_k>$$

$$= \sum_l |a_l\rangle \delta_{lk} = |a_k\rangle.$$

So, $\quad U|b_k> = |a_k>$ and $|b_k> = U^\dagger |a_k> \tag{65}$

Questions and Problems

1. Prove that $|P> <Q| = |Q> <P|$. (**Hint:** Let $a = |P> <Q|$ and $|B> = a \, |A>$)
2. Let $|\psi>$ and $|\phi>$ be two vectors of finite norm. Show that
$$Tr \, (|\psi> <\phi|) = <\phi|\psi>$$

(**Hint:** $\mathrm{Tr}\, A = \sum_i <i|A|i>$)

3. Show that the eigenvalues of a Hermitian matrix are real.
(**Hint:** $A^\dagger = A$ and start with $A \, |\psi> = \lambda|\psi>$)
4. Prove the following properties of the projection operator P_i
(i) $P_i^2 = P_i$ (ii) $P_i P_j = P_i \delta_{ij}$ or $P_j \delta_{ij}$

(**Hint:** $P_i^2 |\psi> = P_i(P_i |\psi>)$ and $P_i = |\psi_i> <\psi_i|$)
5. Find the trace of $|A> <B|$.
6. Use bra-ket algebra to prove the following:
(i) Trace (AB) = Trace (BA)
(ii) $(AB)^\dagger = B^\dagger A^\dagger$
(iii) Trace (ABC) = Trace (CAB) = Trace (BCA)
where A, B and C are operators.

7. An operator is said to be unitary if it satisfies $UU^\dagger = U^\dagger U = 1$. Show that if $\langle\psi|\psi\rangle = 1$ then $\langle U\psi|U\psi\rangle = 1$.

8. Show that if operator A is Hermitian, then e^{iA} is unitary.

9. The Hamiltonian operator for a two state system is given by
$$H = a\,(|1\rangle\langle1| - |2\rangle\langle2| + |1\rangle\langle2| + |2\rangle\langle1|)$$
where a is a number with the dimension of energy. Find the energy eigen-values and the corresponding energy eigen kets (as linear combinations of $|1\rangle$ and $|2\rangle$).

9

Simple Harmonic Oscillator using Ladder Operators

9.1 Solution of Schrödinger's Equation using Creation and Annihilation Operators and their Matrices

The simple classical harmonic oscillator consists of the one dimensional motion of a point mass attracted to a fixed centre by a force that is proportional to the displacement from that centre. It forms one of the most important problems in quantum mechanics also. From a practical point of view it has applications in a variety of branches of modern physics-molecular spectroscopy, solid state physics, nuclear structure quantum field theory etc. such more complicated systems can be analysed in terms of normal modes of motion whenever the interparticle forces are linear functions of the relative displacements and these normal modes are formally equivalent to harmonic oscillators. The linearity of newtonian equations of motion here means that the potential energy is a bilinear function of the coordinates. We shall solve the oscillator problem using Dirac's operator method to obtain the energy eigenkets and energy eigenvalues. The basic Hamiltonian is:

$$H = \frac{p_x^2}{2m} + \frac{m\omega^2 x^2}{2} \tag{1}$$

where angular frequency ω and spring constant k are related by $\omega = \sqrt{k/m}$. The energy eigenvalue equation is

$$Hu_n = E_n u_n \tag{2}$$

The Schröndinger time dependent equation is

$$H\psi = i\hbar \frac{\partial \psi}{\partial t} \tag{3}$$

with the general solution

$$\psi = \sum_n \psi_n = \sum_n c_n u_n e^{-i(E_n/\hbar)t} \tag{4}$$

We have to find the possible eigenvalues of Eqn. (2) and the corresponding eigenfunctions. It is convenient to define two non-Hermitian ladder operators

$$a = \sqrt{\frac{m\omega}{2\hbar}}\left(x + \frac{ip}{m\omega}\right)$$

and
$$a^\dagger = \sqrt{\frac{m\omega}{2\hbar}}\left(x - \frac{ip}{m\omega}\right) \tag{5}$$

known as the *annihilation operator* and the *creation operator* respectively. Using the canonical commutation relation, we have,

$$[a, a^\dagger] = aa^\dagger - a^\dagger a \tag{6}$$

Now,
$$aa^\dagger = \frac{m\omega}{2\hbar}\left(x + \frac{ip}{m\omega}\right)\left(x - \frac{ip}{m\omega}\right)$$

$$= \frac{m\omega}{2\hbar}\left\{x^2 - \frac{ixp}{m\omega} + \frac{ipx}{m\omega} + \frac{p^2}{m^2\omega^2}\right\}$$

$$= \frac{1}{2\hbar}\frac{p^2}{m\omega} + \frac{m\omega^2 x^2}{2\hbar\omega} - \frac{i}{2\hbar}(xp - px)$$

$$= \frac{p^2}{2\hbar m\omega} + \frac{m\omega^2 x^2}{2\hbar\omega} - \frac{i}{2\hbar}[x, p]$$

$$= \frac{p^2}{2\hbar m\omega} + \frac{m\omega^2 x^2}{2\hbar\omega} + \frac{1}{2} \quad \text{(since } [x,p] = i\hbar) \tag{7}$$

Similarly,
$$a^\dagger a = \frac{p^2}{2\hbar m\omega} + \frac{m\omega^2 x^2}{2\hbar\omega} - \frac{1}{2} \tag{8}$$

Using (7) and (8) Eqn. (6) becomes
$$[a, a^\dagger] = 1 \tag{9}$$

We also define the *number operator*
$$N = a^\dagger a \tag{10}$$

which is obviously Hermitian.

From Eqn. (8) we have

or
$$a^\dagger a = \frac{H}{\hbar\omega} - \frac{1}{2} \tag{11}$$

Similarly
$$aa^\dagger = \frac{H}{\hbar\omega} + \frac{1}{2}$$

hence
$$N = \frac{H}{\hbar\omega} - \frac{1}{2}$$

or the Hamiltonian operator
$$H = \hbar\omega\left(N + \frac{1}{2}\right) \tag{12}$$

This indicates a relation between Hamiltonian and the number operator. Because H is a linear function of N, one can diagonalize N simultaneously with H. We denote an energy eigenket of N by its eigenvalue n so

$$N \,|n> = n \,| \,n> \tag{13}$$

Because of Eqn. (12)

$$H\,|\,n> = \left(n+\frac{1}{2}\right) \hbar\omega \,|n> \tag{14}$$

which means that the energy eigenvalues are given by

$$E_n = \left(n+\frac{1}{2}\right) \hbar\omega \tag{15}$$

The three operators $a^\dagger$, N and a are quantum mechanical analogues of three Gods, *Brahma* (the creator), *Vishnu* (the preserver) and *Shiva* (the destroyer) of Hindu mythology. To appreciate the physical significance of a, $a^\dagger$ and N, we first note that

$$[N,a] = [a^\dagger a,a] = a^\dagger \,[a,a] + [a^\dagger,a]a = -a \tag{16}$$

Similarly, we find

$$[N,a^\dagger] = a^\dagger \Rightarrow [N,a^\dagger] \,|a> = a^\dagger \,|n>$$

Consequently, we have

$$Na^\dagger \,|n> - a^\dagger N \,|n> = a^\dagger|n>$$
$$Na^\dagger|n> - a^\dagger N \,|n> = [N,a^\dagger] \,|n>$$

or
$$Na^\dagger|n> = ([N,a^\dagger] + a^\dagger N) \,|n>$$
$$= (n + 1) \,a^\dagger|n> \tag{17}$$

Similarly,

$$Na \,|n> = ([N,a] + aN) \,|n>$$
$$= (n - 1) \,a \,|n> \tag{18}$$

These relations imply that $a^\dagger \,|n>$ ($a \,|n>$) is also an eigenket of the number operator N with eigenvalue increased (decreased) by one. This amounts to the creation (annihilation) of one quantum unit energy $\hbar\omega$, hence the appropriateness of the name creation (and annihilation) operators for $a^\dagger$ and a respectively.

Eqn. (18) implies that $a \,|n>$ and $|n - 1>$ are the same up to a multiplicative constant:

$$a|\,n> = c \,| \,n{-}1> \tag{19}$$

Here, the multiplicative constant c is to be determined from the requirement that both $|n>$ and $|n{-}1>$ be normalized. First we note that

$$<n|a^\dagger a|n> = |c|^2 \text{ or } (a|n>) \,(a|n>)^*$$

or
$$(c|n{-}1>) \,(c|n{-}1>)^\dagger \text{ or } <n{-}1| \,c^*c \,|n{-}1> = |c|^2$$

We can evaluate the l.h.s. by noting that $a^\dagger a$ is just the number operator, so

$$n = |\,c\,|^2 \tag{20}$$

Taking c to be real and positive, we get

$$a \mid n >= \sqrt{n} \mid n-1 > \tag{21a}$$

Similarly, we get

$$a^\dagger \mid n >= \sqrt{n+1} \mid n+1 > \tag{21b}$$

Suppose, we keep on applying the annihilation operator a to both sides of Eqn. (19)

$$a^2 \mid n >= \sqrt{n(n-1)} \mid n-2 >$$

$$a^3 \mid n >= \sqrt{n(n-1)(n-2)} \mid n-3 > \tag{22}$$

In this way, we can obtain numerical operator eigenkets with smaller and smaller n until the sequence terminates. It is quite obvious since we start with a positive integer n, one may think that if we start with a non integer n, the sequnce will not terminate leading to eigen vectors with a negative value of n. But we also require norm of $a|n>$ to be positive:

$$n = <n|N|n> = (<n|a^\dagger) \cdot (a|n>) \geq 0$$

This implies that n can never be negative. So, the sequence must terminate with $n = 0$ and the allowed values of n are non-negative integers.

Since, the smallest possible value of n is zero, the ground state energy is

$$E_0 = \frac{1}{2} \hbar \omega \tag{23}$$

Now, we can apply successively the raising operator $a^\dagger$ to the ground state $|0>$

$$|1> = a^\dagger |0>$$

$$|2\rangle = \left(\frac{a^\dagger}{\sqrt{2}} \right) |1\rangle = \left[\frac{(a^\dagger)^2}{\sqrt{2}} \right] |0\rangle$$

$$|3\rangle = \left(\frac{a^\dagger}{\sqrt{3}} \right) |2\rangle = \left[\frac{(a^\dagger)^3}{\sqrt{3!}} \right] |0\rangle$$

$$\cdots \quad \cdots \quad \cdots \quad \cdots \quad \cdots$$

$$|n\rangle = \left[\frac{(a^\dagger)^n}{\sqrt{n!}} \right] |0\rangle \tag{24}$$

In this way, we get successive simultaneous eigenkets of N and H with energy eigenvalues

$$E_n = \left(n + \frac{1}{2} \right) \hbar \omega \quad (n = 0, 1, 2, 3, \dots) \tag{25}$$

From Eqn. (21) and the orthonormality requirement for $|n>$, we obtain the matrix elements

$$\langle n' | a | n \rangle = \sqrt{n}\, \delta_{n',n-1}$$

$$\langle n' | a^\dagger | n \rangle = \sqrt{n+1}\, \delta_{n',n+1} \tag{26}$$

We can find matrix representations of $a^\dagger$ and a explicitly, as follows with the help of equations (26). When $n = 0$, $n' = 1$, element is present and value of the element is $\sqrt{1}$. When $n = 1$, $n' = 2$, element is there and the value is $\sqrt{2}$ and so on. Hence

$$
n \rightarrow \quad
\begin{array}{cccc}
0 & 1 & 2 & \cdots
\end{array}
$$

$$
a^\dagger = n' \downarrow \begin{array}{c} 0 \\ 1 \\ 2 \\ 3 \\ \vdots \end{array}
\begin{pmatrix}
0 & 0 & 0 & \cdots & \cdots \\
\sqrt{1} & 0 & 0 & \cdots & \cdots \\
0 & \sqrt{2} & 0 & 0 & \cdots \\
0 & 0 & \sqrt{3} & 0 & \cdots \\
\vdots & \vdots & \vdots & \vdots & \ddots
\end{pmatrix}
$$

Similarly

$$
n \rightarrow \quad
\begin{array}{cccc}
0 & 1 & 2 & \cdots
\end{array}
$$

$$
a = n' \downarrow \begin{array}{c} 0 \\ 1 \\ 2 \\ \vdots \end{array}
\begin{pmatrix}
0 & \sqrt{1} & 0 & \cdots \\
0 & 0 & \sqrt{2} & \cdots \\
0 & 0 & 0 & \cdots \\
\vdots & \vdots & \vdots & \cdots
\end{pmatrix}
$$

Also,

$$
x = \sqrt{\frac{\hbar}{2m\omega}} \, (a^\dagger + a)
$$

and

$$
p = i\sqrt{\frac{m\hbar\omega}{2}} \, (a^\dagger - a) \tag{27}
$$

$\therefore$ From Eqns (26) and (27), the matrix elements of x and p operators are

$$
\langle n' | x | n \rangle = \sqrt{\frac{\hbar}{2m\omega}} \left(\sqrt{n} \, \delta_{n',n-1} + \sqrt{n+1} \, \delta_{n',n+1} \right)
$$

$$
\langle n' | p | n \rangle = i\sqrt{\frac{m\hbar\omega}{2}} \left(-\sqrt{n} \, \delta_{n',n-1} + \sqrt{n+1} \, \delta_{n',n+1} \right) \tag{28}
$$

Substituting matrix-values of $a^\dagger$ and a in Eqn. (27) or using Eqn. (28), we obtain

$$
x = \left(\frac{\hbar}{2m\omega} \right)^{1/2}
\begin{pmatrix}
0 & \sqrt{1} & 0 & \cdots \\
\sqrt{1} & 0 & \sqrt{2} & \cdots \\
0 & \sqrt{2} & 0 & \cdots \\
\vdots & \vdots & \vdots & \ddots
\end{pmatrix}
$$

and

$$
p = i\left(\frac{m\hbar\omega}{2} \right)^{1/2}
\begin{pmatrix}
0 & -\sqrt{1} & 0 & \cdots & \cdots \\
\sqrt{1} & 0 & -\sqrt{2} & \cdots & \cdots \\
0 & \sqrt{2} & 0 & -\sqrt{3} & \cdots \\
0 & 0 & \sqrt{3} & 0 & \cdots \\
\vdots & \vdots & \vdots & \vdots & \ddots
\end{pmatrix}
$$

Note that neither p nor x is diagonal in the N representation as they do not commute with N. Matrices for N and H are

$$N = a^\dagger a = \begin{pmatrix} 0 & 0 & 0 & \cdots & \cdots \\ 0 & 1 & 0 & \cdots & \cdots \\ 0 & 0 & 2 & \cdots & \cdots \\ 0 & 0 & 0 & 3 & \cdots \\ \vdots & \vdots & \vdots & \vdots & \vdots \end{pmatrix}$$

and

$$H = \hbar\omega \begin{pmatrix} 1/2 & 0 & 0 & \cdots \\ 0 & 3/2 & 0 & \cdots \\ 0 & 0 & 5/2 & \cdots \\ \vdots & \vdots & \vdots & \cdots \end{pmatrix}$$

Now from Eqn. (24), the eigen functions are

$$\psi_n(x) \equiv <x\,|\,n> = \left\langle x \left| \frac{\left(a^+\right)^n}{\sqrt{n!}} \right| 0 \right\rangle \tag{29}$$

9.2 Normalized Coherent States

We can write the eigenvalue equation using the annihilation operator a:

$$a|\beta> = \beta\,|\beta> \tag{30}$$

We can now solve this eigen value equation and obtain the eigenfunctions $|\beta>$. These $|\beta>$'s describe what are known as the coherent states which are in fact quantum mechanical analogues of the classical oscillator (see chapter 5). Since $|\beta>$'s form a complete set, we may write $|\beta>$ as an expression:

$$|\beta> = \sum_{n=0}^{\infty} c_n\,|n>$$

Now,

$$a|\beta> = \sum_{n=0}^{\infty} c_n a\,|n>$$

$$= \sum_{n=1}^{\infty} c_n \sqrt{n}\ |n-1> \tag{31}$$

Also

$$a|\beta> = \beta|\beta> = \beta \sum_{n=0}^{\infty} c_n\,|n> \tag{32}$$

Using Eqn. (30), from (31) and (32)

$$c_1\,|0> + c_2\sqrt{2}\,|1> + c_3\sqrt{3}\,|2> + ... = \beta\left(c_0\,|0> + c_1\,|1> + c_2\,|2> + ...\right)$$

or

$$c_1 = \beta c_0$$

$$c_2 = \beta\frac{c_1}{\sqrt{2}} = \frac{\beta^2}{\sqrt{2}}c_0$$

$$c_3 = \beta \frac{c_2}{\sqrt{3}} = \frac{\beta^3}{\sqrt{3!}} c_0$$

$$\vdots \qquad \vdots \qquad \vdots$$

$$\cdots \qquad \cdots \qquad \cdots$$

or

$$c_n = \frac{\beta^n}{\sqrt{n!}} c_0 \qquad (33)$$

Hence

$$|\beta> \; = c_0 \sum_{n=0}^{\infty} \frac{\beta^n}{\sqrt{n!}} |n> \qquad (34)$$

Normalizing $|\beta>$, we get

$$1 = <\beta|\beta> = |c_0|^2 \sum_n \sum_m \frac{\beta^n}{\sqrt{n!}} \frac{\beta^{*m}}{\sqrt{m!}} \delta_{nm}$$

$$= |c_0|^2 \sum_{n=0}^{\infty} \frac{\left(|\beta|^2\right)^n}{n!} = |c_0|^2 \, e^{|\beta|^2}$$

$$\Rightarrow \qquad c_0 = e^{-|\beta|^2/2} \qquad (35)$$

Thus

$$c_n = \frac{\beta^n}{\sqrt{n!}} e^{-\frac{1}{2}|\beta|^2} \qquad (36)$$

and

$$|\beta\rangle = e^{-\frac{1}{2}|\beta|^2} \sum_{n=0}^{\infty} \frac{\beta^n}{\sqrt{n!}} |n\rangle \qquad (37)$$

These represent the normalized coherent states.

9.3 The most General Solution

The most general solution of the time dependent Schrödinger equation for the harmonic oscillator can be written as

$$\psi(x,t) = \psi(x) e^{-iEt/\hbar}$$

Since the allowed values of E are given by

$$E \equiv E_n = \left(n + \tfrac{1}{2}\right) \hbar\omega , \qquad\qquad n = 0, 1, 2,\ldots$$

and the corresponding eigenfunctions by

$$\psi_n(y) = \left(\frac{\alpha}{2^n n! \sqrt{\pi}}\right)^{1/2} H_n(y) e^{-y^2/2} \qquad (38)$$

$$(y = \alpha x); \; \alpha = \sqrt{m\omega/\hbar} ; \qquad\qquad n = 0 , 1, 2, \ldots$$

Therefore, the most general solution of the equation

$$i\hbar \frac{\partial \psi(x,t)}{\partial t} = \left(-\frac{\hbar^2}{2\mu} \frac{\partial^2}{\partial x^2} + \frac{1}{2}\mu\omega^2 x^2\right) \psi(x,t)$$

would be given by

$$\psi(x,t) = \sum_{n=0}^{\infty} c_n \psi_n(x) e^{-iE_n t/\hbar}$$

$$= \sum_{n=0}^{\infty} c_n \psi_n(x) e^{-i\left(n+\frac{1}{2}\right)\omega t} \tag{39}$$

i.e. knowing $\psi(x, 0)$, $\psi(x, t)$ can be determined as follows:

$$\psi(x,t) = \sum_{n=0}^{\infty} c_n \psi_n(x)$$

$$\Rightarrow \qquad c_n = \int_{-\infty}^{+\infty} \psi_n^*(x)\psi(x,0)\, dx \tag{40}$$

These values of c_n can be substituted in Eqn. (39) and then we can sum the series to obtain $\psi(x, t)$. If now

$$\int_{-\infty}^{+\infty} |\psi(x,0)|^2\, dx = 1$$

then

$$\sum_n |c_n|^2 = 1$$

implying

$$\int_{-\infty}^{+\infty} |\psi(x,t)|^2\, dx = 1$$

$\psi(x,t)$ describes the time evolution of the wave packet and the quantity $|c_n|^2$ represents the probability of finding the oscillator in the nth state which is independent of time.

9.4 Time Development of Coherent States

We have

$$a|\beta> = \beta|\beta> \tag{41}$$

where a is the annihilation operator and eigenkets $|\beta>$ defined by this equation are the coherent states, given by

$$|\beta> = \exp\left(-\frac{1}{2}|\beta|^2\right)\sum_{n=0}^{\infty} \frac{\beta^n}{\sqrt{n!}}|n> \tag{42}$$

We are interested in finding out how a system described by the time independent Hamiltonian H will evolve with time, if the state of the system at $t = 0$ is described by the ket $|\psi(0)>$.

Now, the ket $|\psi(t)>$, which describes the evolution of the system with time, would satisfy the time dependent Schrödinger equation

$$i\hbar\frac{\partial}{\partial t}|\psi(t)> = H|\psi(t)> \tag{43}$$

Since the Hamiltonian is independent of time, we can integrate this expression to obtain

$$|\psi(t)\rangle = e^{-Ht/\hbar}|\psi(0)\rangle \tag{44}$$

That Eqn. (44) is a solution of Eqn. (43) can be checked by direct substitution of Eqn. (44) into Eqn. (43):

$$i\hbar \frac{\partial}{\partial t}\left[e^{-iHt/\hbar} \mid \psi(0) > \right] = i\hbar \left(-\frac{iH}{\hbar} \right) e^{-iHt/\hbar} \mid \psi(0) >$$

$$= He^{-iHt/\hbar} \mid \psi(0) >= H \mid \psi(t) > \qquad (45)$$

Now, using

$$\left| \psi \right\rangle = \left(\sum_n \left| n \right\rangle \left\langle n \right| \right) \left| \psi \right\rangle \qquad (46)$$

We obtain

$$\left| \psi(t) \right\rangle = \sum_n \left| n \right\rangle \left\langle n \middle| \psi(t) \right\rangle$$

$$= e^{-iHt/\hbar} \sum_n \left| n \right\rangle \left\langle n \middle| \psi(0) \right\rangle \qquad (47)$$

We know that an operator A is defined through the power series

$$e^A = 1 + \frac{1}{2!}AA + \frac{1}{3!}AAA + \dots \qquad (48i)$$

Using this, and $H \mid n> = E_n \mid n>$, we get

$$e^{-iHt/\hbar} = 1 + \left(-\frac{iHt}{\hbar} \right) + \frac{1}{2!}\left(-\frac{iHt}{\hbar} \right)\left(-\frac{iHt}{\hbar} \right) + \dots \qquad (48ii)$$

$$e^{-iHt/\hbar} \mid n >= 1 + \left(-\frac{iE_n t}{\hbar} \right) \mid n > + \frac{1}{2!}\left(-\frac{iE_n t}{\hbar} \right)\left(-\frac{iE_n t}{\hbar} \right) \mid n > + \dots$$

$$= \left[1 + \left(-\frac{iE_n t}{\hbar} \right) + \frac{1}{2!}\left(-\frac{iE_n t}{\hbar} \right)\left(-\frac{iE_n t}{\hbar} \right) + \dots \right] \mid n >$$

$$= e^{-iE_n t/\hbar} \mid n > \qquad (49)$$

So

$$\left| \psi(t) \right\rangle = \sum_n e^{-iHt/\hbar} \left| n \right\rangle \left\langle n \middle| \psi(0) \right\rangle$$

$$= \sum_n e^{-iE_n t/\hbar} \left| n \right\rangle \left\langle n \middle| \psi(0) \right\rangle \qquad (50)$$

We may use this equation to study the time development of coherent states. Thus,

$$\left| \psi(0) > \equiv \mid \beta > = \exp\left(-\frac{1}{2} \mid \beta \mid^2 \right) \sum_{n=0}^{\infty} \frac{\beta^n}{\sqrt{n!}} \mid n > \qquad (51)$$

so that

$$<n\mid\psi(0)> \equiv <n\mid\beta> = \exp\left(-\frac{1}{2} \mid \beta \mid^2 \right) \frac{\beta^n}{\sqrt{n!}} \qquad (52)$$

(because $<n\mid n> = 1$)

Hence from Eqns. (50) and (52) we have

$$|\psi(t)\rangle = \exp\left(-\frac{1}{2}|\beta|^2\right) \sum_n e^{-iE_n t/\hbar} \frac{\beta^n}{\sqrt{n!}} |n\rangle \qquad (53)$$

or

$$|\psi(t)\rangle = \exp\left(-\frac{1}{2}|\beta|^2\right) \sum_n \frac{\beta^n}{\sqrt{n!}} \exp\left[-i\left(n+\frac{1}{2}\right)\omega t\right] |n\rangle \qquad (54)$$

(because $E_n = \left(n+\frac{1}{2}\right)\hbar\omega$)

This equation gives the time development of the states.

9.5 Minimum Uncertainty Product for Coherent State

In terms of creation and annihilation operators,

$$x = \sqrt{\frac{\hbar}{2m\omega}} \ (a + a^\dagger) \qquad (55)$$

$$p = \sqrt{\frac{m\hbar\omega}{2}} \ i(a^\dagger - a) \qquad (56)$$

We have

$$a|\beta\rangle = \beta|\beta\rangle \qquad (57i)$$

so

$$\langle\beta|a^\dagger = \langle\beta|\beta^* \qquad (57ii)$$

(as a is not Hermitian)
We also know that

$$\begin{aligned}(\Delta x)^2 &= \langle x^2 \rangle - \langle x \rangle^2 \\ (\Delta p)^2 &= \langle p^2 \rangle - \langle p \rangle^2\end{aligned} \qquad (58)$$

Now let us calculate

$$\langle x \rangle = \langle\beta|x|\beta\rangle = \sqrt{\frac{\hbar}{2m\omega}} \ \langle\beta|a + a^\dagger|\beta\rangle$$

$$= \sqrt{\frac{\hbar}{2m\omega}} \ [\langle\beta|a|\beta\rangle + \langle\beta|a^\dagger|\beta\rangle]$$

Using Eqn. (57ii), we get

$$\langle x \rangle = \sqrt{\frac{\hbar}{2m\omega}} \ (\beta + \beta^*) \qquad (59)$$

(as $\langle\beta|\beta\rangle = 1$)
Now,

$$\langle x^2 \rangle = \langle\beta|x^2|\beta\rangle$$

$$= \frac{\hbar}{2m\omega} \ \langle\beta|a^2 + a^{\dagger 2} + aa^\dagger + a^\dagger a|\beta\rangle$$

$$= \frac{\hbar}{2m\omega} \ [\langle\beta|a^2|\beta\rangle + \langle\beta|a^{\dagger 2}|\beta\rangle + \langle\beta|2a^\dagger a|\beta\rangle + \langle\beta|\beta\rangle]$$

where we have used $aa^\dagger = 1 + a^\dagger a$ from $[a, a^\dagger] = 1$. Hence

$$\left\langle x^2 \right\rangle = \frac{\hbar}{2m\omega}\left[\beta^2 + \beta^{*2} + 2\beta^*\beta + 1\right] \tag{60}$$

Consequently,

$$(\Delta x)^2 = <x^2> - <x>^2$$

$$= \frac{\hbar}{2m\omega}\left[\beta^2 + \beta^{*2} + 2\beta^*\beta + 1 - \left(\beta + \beta^*\right)^2\right]$$

$$= \frac{\hbar}{2m\omega} \tag{61}$$

Proceeding similarly (using $p = \sqrt{\dfrac{\hbar m\omega}{2}}\ i(a^\dagger - a)$ in place of x), we get

$$(\Delta p)^2 = \frac{\hbar m\omega}{2} \tag{62}$$

Hence $\quad (\Delta p)^2 (\Delta x)^2 = \dfrac{\hbar}{2m\omega} \times \dfrac{\hbar m\omega}{2} = \dfrac{\hbar^2}{4}$

or $$\Delta p\, \Delta x = \frac{\hbar}{2}$$

It is the minimum uncertainty product for $|\beta>$ i.e. coherent state yields minimum uncertainty product close to classical situation.

Example 9.1: Show that the average value of the kinetic energy of the linear harmonic oscillator is equal to one half the total energy of the oscillator when it is in the state of definite energy.

Solution: We use ladder operators:

$$a = \sqrt{\frac{m\omega}{2\hbar}}\left(x + \frac{ip}{m\omega}\right) \tag{i}$$

$$a^\dagger = \sqrt{\frac{m\omega}{2\hbar}}\left(x - \frac{ip}{m\omega}\right)$$

and the relations $\qquad a^\dagger a = \dfrac{H}{\hbar\omega} - \dfrac{1}{2} \tag{ii}$

and $\qquad aa^\dagger = \dfrac{H}{\hbar\omega} + \dfrac{1}{2}$

Now, $\qquad a - a^\dagger = \sqrt{\dfrac{m\omega}{2\hbar}}\dfrac{2ip}{m\omega} = \sqrt{\dfrac{2}{\hbar\omega m}}\,ip$

$$(a - a^\dagger)^2 = -\frac{2}{\hbar\omega}\frac{p^2}{m}$$

or $\qquad a^2 + a^{\dagger 2} - aa^\dagger - a^\dagger a = -\left(\dfrac{4}{\hbar\omega}\right)\dfrac{p^2}{2m}$

or
$$\frac{1}{4}(a^2 + a^{\dagger 2}) - \frac{1}{4}(aa^\dagger + a^\dagger a) = -\frac{1}{\hbar\omega}\frac{p^2}{2m}$$

or
$$-\frac{1}{4}(a^2 + a^{\dagger 2}) + \frac{1}{4} \times \frac{2H}{\hbar\omega} = \frac{1}{\hbar\omega}\frac{p^2}{2m} \tag{iii}$$

$$(\because \text{ from Eqns } (ii) \; aa^\dagger \overset{\bullet}{+} a^\dagger a = \frac{2H}{\hbar\omega})$$

Now
$$\left\langle \frac{p^2}{2m} \right\rangle = \left(u_n, \frac{p^2}{2m} u_n \right)$$

$$= \left(u_n, \left\{ -\frac{\hbar\omega}{4}(a^2 + a^{\dagger 2}) + \frac{H}{2} \right\} u_n \right) \quad \text{(using Eqn. } (iii))$$

$$= -\frac{\hbar\omega}{4}(u_n, a^2 u_n) - \frac{\hbar\omega}{4}\left(u, a^{\dagger 2} u_n \right) + \frac{1}{2}(u_n, H u_n) \tag{iv}$$

But
$$u_n = c_n \left(a^\dagger \right)^n u$$

$\therefore$
$$(u_n, a^{\dagger 2} u_n) = (u_n, c_n (a^\dagger)^{n+2} u_0)^{(*)}$$

$$= \frac{c_n}{c_{n+2}} (u_n, c_{n+2} (a^\dagger)^{n+2} u_0)$$

$$= \frac{c_n}{c_{n+2}} (u_n, u_{n+2}) = 0 = (u_n, a^2 u_n)$$

Hence from Eqn. (iv)

$$\left\langle \frac{p^2}{2m} \right\rangle = \frac{1}{2}(u_n, H u_n)$$

$$\equiv \frac{1}{2} \int \bar{u}_n H u_n \, dr = \frac{1}{2} < H >_n$$

$$= \frac{1}{2} \int \bar{u}_n E_n u_n \, dr$$

$$= \frac{E_n}{2} \int \bar{u}_n u_n \, dr = \frac{E_n}{2}$$

(*)
$$\left(u_n, \frac{p^2}{2m} u_n \right) \equiv \int u_n^* \frac{p^2}{2m} u_n \, dr$$

$$(\phi, \psi) = \int \bar{\phi}\psi \, dx = \int \phi^* \psi \, dx$$

$$(\phi, A\psi) = \int \bar{\phi} A\psi \, dx \qquad \qquad (A \text{ is an operator})$$

$$(\phi, \alpha\psi) = \alpha \int \bar{\phi}\psi \, dx$$

Example 9.2: Show that $(a^\dagger). a^n = N(N-1) \ldots (N-n+1)$ where $N = a^\dagger a$, is the number operator.

Solution:
$$a^\dagger a = N$$
$$(a^\dagger)^2 a^2 = a^\dagger a^\dagger aa = a^\dagger Na = a^\dagger(aN - a)$$
$$= a^\dagger aN - a^\dagger a = N^2 - N = N(N-1)$$
$$(\because \; [N, a] = -a \text{ or } Na - aN = -a \text{ or } Na = aN - a)$$

Similarly,
$$a^{\dagger\,3}a^3 = a^\dagger a^{\dagger\,2}a^2 a = a^\dagger N(N-1)\, a$$
$$= a^\dagger NNa - a^\dagger Na$$
$$= a^\dagger N\,(aN - a) - N(N-1)$$
$$= a^\dagger NaN - a^\dagger Na - N\,(N-1)$$
$$= a^\dagger\,(aN - a)\,N - 2N\,(N-1)$$
$$= N^3 - N^2 - 2N(N-1)$$
$$= N(N^2 - N) - 2N(N-1)$$
$$= (N-2)(N-1)N$$

Hence
$$a^{\dagger\,n}\,a^n = N\,(N-1) \ldots (N-n+1)$$

Example 9.3: Show that for a simple harmonic oscillator

(a) $(E_l - E_k) < k\,|\,x\,|\,l > = \dfrac{i\hbar}{m} < k\,|\,p\,|\,l >$ and

(b) $< l\,|\,xp + px\,|\,l > = 0$

Solution: (a) For an oscillator, the Hamiltonian is
$$H = \frac{p^2}{2m} + \frac{m\omega^2 x^2}{2}$$

$\therefore$
$$\left[x, H\right] = i\hbar\, \dot{x} \qquad\qquad (i)$$

or
$$xH - Hx = \frac{i\hbar}{m}p \qquad\qquad (ii)$$

Taking the matrix element of both sides between the kth and l^{th} states, we have
$$< k\,|\,xH - Hx\,|\,l > = \frac{i\hbar}{m} < k\,|\,p\,|\,l >$$

or
$$< k\,|\,xH\,|\,l > - < k\,|\,Hx\,|\,l > = \frac{i\hbar}{m} < k\,|\,p\,|\,l >$$

or
$$E_l < k\,|\,x\,|\,l > - E_k < k\,|\,x\,|\,l > = \frac{i\hbar}{m} < k\,|\,p\,|\,l >$$

or
$$\left(E_l - E_k\right) < k\,|\,x\,|\,l > = \frac{i\hbar}{m} < k\,|\,p\,|\,l >$$

(b) From Eqn. (ii) we have
$$p = \frac{m}{i\hbar}(xH - Hx)$$

$$\therefore \qquad <k\,|\,xp+px\,|\,l> = \frac{m}{i\hbar}<k\,|\,x(xH-Hx)+(xH-Hx)x\,|\,l>$$

$$= \frac{m}{i\hbar}<k\,|\,x^2H-xHx+xHx-Hx^2\,|\,l>$$

$$= \frac{m}{i\hbar}<k\,|\,x^2H-Hx^2\,|\,l>$$

$$= \frac{m}{i\hbar}<k\,|\left(-Hx^2+x^2H\right)|\,l>$$

$$= \frac{m}{i\hbar}\left(E_l-E_k\right)<k\,|\,x^2\,|\,l>$$

Taking $k=l$, we get

$$<l\,|\,xp+px\,|\,l> = \frac{m}{i\hbar}\left(E_l-E_l\right)<l\,|\,x^2\,|\,l> = 0$$

Example 9.4: Show that for an oscillator transition from an initial state $|l>$ with energy E_l to a final state $|k>$ with energy E_k, we have

$$\sum_k \left(E_l-E_k\right)|<k\,|\,x\,|\,l>|^2 = \frac{\hbar^2}{2m}$$

Solution:

$$\sum_k \left(E_l-E_k\right)\,|\,<k\,|\,x\,|\,l>|^2$$

$$= \sum_k \left(E_l-E_k\right)<l\,|\,x\,|\,k><k\,|\,x\,|\,l>$$

$$= \sum_k <l\,|\,xH-Hx\,|\,k><k\,|\,x\,|\,l>$$

$$= \sum_k <l\,|\,[x,H]\,|\,k><k\,|\,x\,|\,l>$$

$$= \frac{i\hbar}{m}\sum_k <l\,|\,p\,|\,k><k\,|\,x\,|\,l>$$

$$= \frac{i\hbar}{m}<l\,|\,px\,|\,l> \qquad\qquad (i)$$

Now,

$$<l\,|\,px\,|\,l> = \frac{1}{2}<l\,|\,px-xp+xp+px\,|\,l>$$

$$= \frac{1}{2}<l\,|\,px-xp\,|\,l>+\frac{1}{2}<l\,|\,xp+px\,|\,l>$$

$$= \frac{1}{2}<l\,|\,[p,x]\,|\,l>+0$$

$$= -\frac{i\hbar}{2}<l\,|\,l> = -\frac{i\hbar}{2} \qquad\qquad (ii)$$

Therefore,

$$\sum_k \left(E_l-E_k\right)|<k\,|\,x\,|\,l>|^2 = \frac{i\hbar^2}{m}\left(-\frac{i\hbar}{2}\right) = \frac{\hbar^2}{2m}$$

Example 9.5: Show that $x_{nm} \neq 0$ unless $E_n - E_m = \pm\hbar\omega$ (selection rule) for harmonic oscillator whose hamiltonian is

$$H = \frac{p_x^2}{2\mu} + \frac{1}{2}kx^2 \text{ and } [x, p] = i\hbar$$

Solution: $[x, H] = \left[x, \frac{p_x^2}{2\mu} \right] + 0$

$$= \frac{xp_x^2 - p_x^2 x}{2\mu} = \frac{1}{2\mu}\left(xp_x^2 - p_x xp_x + p_x xp_x + p_x^2 x \right)$$

$$= \frac{\left(xp_x - p_x x \right) p_x + p_x \left(xp_x - p_x x \right)}{2\mu}$$

$$= \frac{2i\hbar}{2\mu} p_x = \left(\frac{i\hbar}{\mu} \right) p_x \qquad (i)$$

$$\left[[x, H], H \right] = \frac{i\hbar}{\mu}[p, H]$$

$$= \frac{i\hbar}{\mu}\left[p, \frac{p^2}{2\mu} + \frac{1}{2}kx^2 \right] ; \left(\left[p, \frac{p^2}{2\mu} \right] = 0 \right)$$

$$= \frac{i\hbar k}{2\mu}\left\{ [p, x]x - x[x, p] \right\}$$

$$= \frac{i\hbar k}{2\mu}\left\{ [p, x]x + x[p, x] \right\} = \frac{i\hbar k}{2\mu}\{-2i\hbar x\}$$

$$\therefore \quad \left[[x, H], H \right] = \frac{\hbar^2 k}{\mu} x \qquad (ii)$$

or

$$\langle n| \left[[x, H], H \right] |m \rangle = \frac{\hbar^2 k}{\mu} \langle n|x|m \rangle$$

or

$$\langle n|[x, H]H|m \rangle - \langle n|H[x, H]|m \rangle = \frac{\hbar^2 k}{\mu} \langle n|x|m \rangle$$

or

$$(E_m - E_n)\langle n|[x, H]|m \rangle = \frac{\hbar^2 k}{\mu} \langle n|x|m \rangle$$

or $(E_m - E_n)\langle n|xH|m \rangle - (E_m - E_n)\langle n|Hx|m \rangle = \frac{\hbar^2 k}{\mu} \langle n|x|m \rangle$

or

$$(E_m - E_n)^2 \langle n|x|m \rangle = \frac{\hbar^2 k}{\mu} \langle n|x|m \rangle$$

or

$$\left\{ (E_m - E_n)^2 - \hbar^2\omega^2 \right\}\langle n|x|m \rangle = 0$$

$$x_{nm} \text{ or } \langle n|x|m \rangle \neq 0$$

So,
$$\left(E_m - E_n\right)^2 = \hbar^2\omega^2$$

or
$$E_m - E_n = \pm\hbar\omega \quad \text{(selection rule)}$$

Example 9.6: Show that the raising and lowering operators defined by

$$R_+ = \frac{p_x}{\sqrt{2m}} + i\sqrt{\frac{k}{2}}\,x$$

and
$$R_- = \frac{p_x}{\sqrt{2m}} - i\sqrt{\frac{k}{2}}\,x$$

respectively are Hermitian adjoints of each other.

Solution: We have to show

$$\int \bar{\psi}_a R_- \psi_b\, dx = \int \overline{\left(R_+\psi_a\right)}\,\psi_b\, dx \quad \text{and} \quad \int \bar{\psi}_a R_+ \psi_b\, dx = \int \overline{\left(R_-\psi_a\right)}\,\psi_b\, dx$$

Now,
$$\overline{R_+\psi_a} = \overline{\left(\frac{p_x}{\sqrt{2m}} + i\sqrt{\frac{k}{2}}x\right)\psi_a}$$

$$= \frac{i\hbar}{\sqrt{2m}}\frac{\partial\bar{\psi}_a}{\partial x} - i\sqrt{\frac{k}{2}}\,x\,\bar{\psi}_a \qquad \left(\because p_x = -i\hbar\frac{\partial}{\partial x}\right)$$

$$= i\alpha\frac{\partial\bar{\psi}_a}{\partial x} - \beta x\bar{\psi}_a \qquad \left(\text{where } \alpha = \frac{\hbar}{\sqrt{2m}} \text{ and } \beta = i\sqrt{\frac{k}{2}}\right)$$

So that
$$\int\overline{\left(R_+\psi_a\right)}\,\psi_b\,dx = \int\left(i\alpha\frac{\partial\bar{\psi}_a}{\partial x} - \beta x\bar{\psi}_a\right)\psi_b\,dx$$

$$= i\alpha\int\frac{\partial\bar{\psi}_a}{\partial x}\psi_b\,dx - \beta\int x\bar{\psi}_a\psi_b\,dx \tag{i}$$

$$= i\alpha\left(\psi_b\bar{\psi}_a\Big|_{-\infty}^{+\infty} - \int\frac{\partial\psi_b}{\partial x}\bar{\psi}_a\,dx\right) - \beta\int x\bar{\psi}_a\psi_b\,dx \tag{ii}$$

(Integrating by parts)

Also,
$$\int\bar{\psi}_a R_-\psi_b\,dx = \int\bar{\psi}_a\left(-i\alpha\frac{\partial\psi_b}{\partial x} - \beta x\psi_b\right)dx$$

$$= -i\alpha\int\bar{\psi}_a\frac{\partial\psi_b}{\partial x}\,dx - \beta\int x\bar{\psi}_a\psi_b\,dx$$

$$= -i\alpha\left(\psi_b\bar{\psi}_a\Big|_{-\infty}^{+\infty} - \int\frac{\partial\psi_b}{\partial x}\bar{\psi}_a\,dx\right) - \beta\int x\bar{\psi}_a\psi_b\,dx$$

$$= 0 + i\alpha\int\frac{\partial\bar{\psi}_a}{\partial x}\psi_b\,dx - \beta\int x\bar{\psi}_a\psi_b\,dx \tag{iii}$$

∴ From (i) and (iii)

$$\int\bar{\psi}_a R_-\psi_b\,dx = \int\overline{\left(R_+\psi_a\right)}\,\psi_b\,dx$$

Questions and Problems

1. Consider a pair of operators a and $a^\dagger$ that satisfy commutation relation
$$[a,a^\dagger] = 1$$
   (a) Show that the operator $N = a^\dagger a$ has eigen values $n = 0, 1, 2,..$
   (b) Show that the harmonic oscillator Hamiltonian may be written in the form
$$H = \hbar\omega\left(N + \tfrac{1}{2}\right)$$

2. Show that for one dimensional harmonic oscillator
$$H = \frac{p^2}{2m} + \frac{1}{2}m\omega^2 x^2$$
   the energy eigenvalues can be obtained by diagonalization of an operator N given by
$$N = a^\dagger a$$
$$a^\dagger = \frac{1}{\sqrt{2m\omega\hbar}}(m\omega x - ip)$$
   and $a^\dagger$ is Hermitian adjoint of a. Explain the significance of $a, a^\dagger$. Give the matrices for the position and momentum operators.

3. Show that the average value of the kinetic energy of the linear harmonic oscillator is equal to one half the total energy of the oscillator when it is in a state of definite energy.

4. Show that the expectation values of the kinetic and potential energies are equal for a linear harmonic oscillator, i.e.
$$<n|\tfrac{1}{2}m\omega^2 x^2 |n> = <n|\frac{p^2}{2m}|n>$$

5. If $a, a^\dagger$ were required to satisfy the anticommutation rule $\left[a,a^\dagger\right]_+ = 1$, instead of $[a, a^\dagger] = 1$ and also $\left[a,a\right]_+ = \left[a^\dagger,a^\dagger\right]_+ = 0$, verify that equations
$$[a^\dagger a, a] = -a$$
   and $[a^\dagger a, a^\dagger] = a^\dagger$ still hold. Also verify that the operator $N = a^\dagger a$ have only 0 and 1 as eigen values.

6. Prove that, for a harmonic oscillator, the matrix elements of operators $\hat{x}$ and $\hat{p}$ are given respectively by
$$<n'|x|n> = \left(\frac{\hbar}{2m\omega}\right)^{1/2}\left[\sqrt{n}\,\delta_{n',n-1} + \sqrt{n+1}\,\delta_{n',n+1}\right]$$
$$<n'|p|n> = i\left(\frac{m\hbar\omega}{2}\right)^{1/2}\left[-\sqrt{n}\,\delta_{n',n-1} + \sqrt{n+1}\,\delta_{n',n+1}\right]$$

7. Show that for a harmonic oscillator
   (i) $\langle n|x^2|n\rangle = \hbar(2n+1)/(2m\omega)$
   (ii) $\langle n| \left(a+a^\dagger\right)^4 |n\rangle = 6n^2 + 6n + 3$

10

Introductory Quantum Dynamics (The Three Quantum Mechanical Pictures)

10.1 Introduction

In quantum mechanics, the state of a physical system is represented by a vector defined in an abstract Hilbert space. Therefore, the equation of motion for a quantum mechanical system could be a differential equation for the state vector. Since, the observable quantities are not the state vectors, the expectation values of a set of Hermitian operators therefore correspond to the dynamical variables. So the equations of motion are concerned with the evolution in time of these expectation values.

Now, the expectation value of an operator $\hat{A}$ in the state represented by the normalized state vector ψ is given by

$$<\hat{A}> = <\psi \,|\hat{A}|\, \psi> \tag{1}$$

The variation with time of $<\hat{A}>$ can, therefore, by viewed as arising in one of the following ways:

(a) The state vector ψ changes with time, but $\hat{A}$ remains unchanged

(b) $\hat{A}$ changes with time, ψ remaining constant

(c) Both $\hat{A}$ and ψ change with time. Correspondingly, we have the Schrödinger, the Heisenberg and the Interaction pictures of time development respectively.

The Schrödinger picture

Corresponding to every quantum mechanical system, there exists a family of linear operators $\hat{U}\,(t, t_0)$, defined on the infinite-dimensional Hilbert space of the system, which describes the evolution of the state vector from time t_0 to time t:

$$\psi(t) = \hat{U}\,(t, t_0)\,\psi(t_0) \tag{2}$$

$\hat{U}$ is called the *evolution* or *time development operator*.

The Heisenberg picture

In order to distinguish the state vectors and operators in this picture from those of the Schrödinger picture, we use the subscript H. Thus ψ_H and $\hat{A}_H$ respectively, denote a state vector and an operator in the Heisenberg picture. ψ_H is time independent but $\hat{A}_H$ depends on time. We note that if we define $\psi_H(t)$ by

$$\psi_H(t) = \hat{U}^{-1}(t, t_0)\, \psi\,(t) \tag{3}$$

where $\psi(t)$ is the state vector in the Schrödinger picture, then ψ_H is independent of time. From Eqn. (2)

$$\psi_H(t) = \hat{U}^{-1}(t,t_0)\, \psi(t) = \hat{U}^{-1}(t,t_0)\, \hat{U}(t,t_0)\, \psi(t_0^{\cdot}) = \psi(t_0) \tag{4}$$

The Interaction picture

This picture avails the advantages of both the Heisenberg and the Schrödinger pictures. The interaction approach is useful when the Hamiltonian can be split up into two parts, one part independent of time and the other part dependent on time. For example, when the system is in a external field. The time independent part would represent the Hamiltonian of the system in the absence of the external field and the time dependent part that arising from the presence of the external field

$$\hat{H}(t) = \hat{H}^{(0)} + \hat{H}^{(1)}(t) \tag{5}$$

10.2 Heisenberg's Matrix Equation of Motion

Regarding q's and p's we may write:

$$q_l q_m - q_m q_l = 0 \; ; \; p_l p_m - p_m p_l = 0$$

$$q_l p_m - p_m q_l = i\hbar \delta_{lm} I \tag{6}$$

where I is the unit matrix.

In terms of the elements, the last equation reads

$$\left(q_l p_m - p_m q_l\right)_{rs} = \int \psi_r^* \left\{ q_l \left(-i\hbar \frac{\partial}{\partial q_m}\right) - \left(-i\hbar \frac{\partial}{\partial q_m}\right) q_l \right\}$$

$$= -i\hbar \int \psi_r^* \left\{ q_l \frac{\partial \psi_s}{\partial q_m} - \frac{\partial}{\partial q_m}(q_l \psi_s) \right\} dq$$

$$= 0 \text{ if } l \neq m \tag{7}$$

$$= i\hbar \int \psi_r^* \psi_s \, dq = i\hbar \delta_{rs} I \text{ if } l = m$$

To obtain the equation of motion for any function A of q, p and possibly of t, we begin with the time dependent Schrödinger equation

$$i\hbar \frac{\partial \psi}{\partial t} = H\psi \; ; \; -i\hbar \frac{\partial \psi^*}{\partial t} = H * \psi * \tag{8}$$

where the Hamiltonian operator is Hermitian, i.e.

$$\int \psi * H\psi \, dq = \int (H * \psi^*)\psi \, dq \tag{9}$$

We define the matrix A_{rs} by

$$A_{rs} = \int \psi_r^* A \psi_s \, dq \tag{10}$$

Differentiating this with respect to time, we get

$$\frac{d}{dt} A_{rs} = \int \left\{ \psi_r^* A \dot{\psi}_s + \dot{\psi}_r A \psi_s + \psi_r \dot{A} \psi_s \right\} dq$$

$$= \frac{1}{i\hbar} \int \left\{ \psi_r^* \left(AH\psi_s\right) - \left(H^* \psi_r^*\right)(A\psi_s) \right\} dq + \int \psi_r^* \frac{\partial A}{\partial t} \psi_s \, dq$$

$$= \frac{1}{i\hbar} \int \left\{ \psi_r^* A H \psi_s - \left(\psi_r^* H \right)\left(A \psi_s \right) \right\} dq + \int \psi_r^* \frac{\partial A}{\partial t} \psi_s dq$$

$$= \frac{1}{i\hbar}(AH - HA)_{rs} + \left(\frac{\partial A}{\partial t} \right)_{rs} \qquad (11i)$$

it follows therefore

$$\frac{d}{dt} A = \frac{1}{i\hbar}[A, H] + \frac{\partial A}{\partial t} \qquad (11ii)$$

This is the Heisenberg's matrix form of the equation of motion of a dynamical variable. We say that A is a constant of motion if dA/dt is zero and this happens when (*i*) A is not explicitly a function of time ($\partial A/\partial t = 0$) and (*ii*) A commutes with H. When A is not explicitly function of time,

$$\frac{d}{dt} A = \frac{1}{i\hbar}[A, H], \quad \frac{dA_{rs}}{dt} = \frac{1}{i\hbar}[AH - HA]_{rs} \qquad (12)$$

10.3 Schrödinger's and Heisenberg's Representations

We know that in Schrödinger picture the state of the system at a given time t is represented by the state vector $\psi_s(t)$ and the evolution of the system is then described by the time dependence of $\psi_s(t)$ as

$$i\hbar \frac{\partial \psi_s(t)}{\partial t} = H_s \psi_s(t) \qquad (13)$$

H_s is the time independent Hamiltonian operator. As a matter of fact, in this picture, any operator A_s corresponding to a physical observable is time independent. However, its expectation value is

$$< A_s > = \int \psi_s^*(t) A_s \psi_s(t) dq \equiv \left(\psi_s, A_s \psi_s \right) \qquad (14)$$

This will in general be time dependent with

$$< \dot{A}_s > = \frac{d}{dt} < A_s > = \left(\psi_s(t), [A_s, H_s] \, \psi_s(t) \right)/i\hbar \qquad (15)$$

(using $\frac{d}{dt} < A > = < \dot{A} > = \int \psi^* \dot{A} \psi \, dq$ and Eqn. (12))

Let us consider now a time dependent unitary transformation $U(t)$ on $\psi_s(t)$ which transforms $\psi_s(t)$ to ϕ, that is

$$\phi = U(t) \, \psi_s(t)$$

$$U^{-1}\phi = \psi_s(t) \qquad (16)$$

$$U(t)U^\dagger(t) = U^\dagger(t) \, U(t) = I$$

$$U^\dagger(t) = U^{-1}(t) \qquad (17)$$

We can write also

$$UH_s \psi_s = UH_s U^{-1}(U\psi_s) = UH_s U^{-1}\phi$$

$$= i\hbar U \frac{\partial}{\partial t} \psi_s = i\hbar \left[\frac{\partial}{\partial t}(U\psi_s) - \frac{\partial U}{\partial t} \psi_s \right]$$

$$= i\hbar \left[\frac{\partial}{\partial t} \phi - \frac{\partial U}{\partial t} U^{-1} \phi \right] \tag{18}$$

Therefore ϕ obeys the equation

$$i\hbar \frac{\partial \phi}{\partial t} = \left(i\hbar \frac{\partial U}{\partial t} U^{-1} + UH_s U^{-1} \right) \phi \tag{19}$$

If we choose U to satisfy the equation

$$-i\hbar \frac{\partial U}{\partial t} = UH_s \tag{20}$$

or

$$-i\hbar \frac{\partial U(t)}{\partial t} U^{-1} = UH_s U^{-1} \tag{21}$$

then we will obtain from Eqn. (19)

$$\frac{\partial \phi}{\partial t} = 0 \tag{22}$$

that is, the transformed state vector ϕ_H (Heisenberg picture) is time independent. The time load in Heisenberg picture is carried by the operator $A_H(t)$. Thus we see how Schrödinger picture leads to Heisenberg picture which becomes further clear as follows:

The expectation value of the operator A_s can be written as

$$< A > = \int \psi_s^*(t) A_s \psi_s(t) \, dq$$

$$= \psi_s^*(t) U^{-1} U A_s U^{-1} U \psi_s(t) \, dq$$

$$= \int \phi_H^* U A_s U^{-1} \phi_H \, dq \tag{23}$$

where from Eqn. (16) we have written $\phi_H^* = \psi_s^* U^{-1}$, ψ_s^* being row matrix, appears on the left of U^{-1}. Defining $A_H(t)$ by

$$A_H(t) = U(t) A_s U^{-1}(t) \tag{24}$$

we can write in the abbreviated notation

$$< A > = \left(\psi_s(t), A_s \psi_s(t) \right) = \left(\phi_H, A_H(t) \phi_H \right) \tag{25}$$

Therefore the time depedent operator $A_H(t)$ in Heisenberg picture has the same expectation value in terms of ϕ_H, as A_s has in terms of $\psi_s(t)$.

Further, from Eqn. (21) we have (using $H_s^\dagger = H_s$)

$$-i\hbar \frac{\partial U}{\partial t} = UH_s$$

or

$$i\hbar \frac{\partial U^\dagger}{\partial t} = H_s^\dagger U^\dagger$$

$$\frac{d}{dt} A_H(t) = \frac{d}{dt} \left(UU^\dagger A_H(t) \right) = \frac{\partial U}{\partial t} U^\dagger A_H(t) + A_H U \frac{\partial U^\dagger}{\partial t} + UU^\dagger \frac{\partial A_H}{\partial t}$$

$$= -\frac{1}{i\hbar} UH_s U^\dagger A_H(t) + A_H U \left(\frac{1}{i\hbar} H_s U^\dagger \right) + \frac{\partial A_H}{\partial t} ,$$

$$\left(\because i\hbar\frac{\partial U}{dt} = \frac{-1}{i\hbar}UH_s \right)$$

$$\left(\because i\hbar\frac{\partial U^\dagger}{dt} = H_s^\dagger U^\dagger \right)$$

$$= \frac{1}{i\hbar}H_H A_H + \frac{1}{i\hbar}A_H H_H + \frac{\partial A_H}{\partial t} \qquad \text{(where } UH_s U^\dagger = H_H \text{)}$$

$$= \frac{1}{i\hbar}\left(A_H H_H - H_H A_H \right) + \frac{\partial A_H}{\partial t}$$

or
$$\frac{dA_H}{dt} = \frac{\partial A_H}{\partial t} + \frac{1}{i\hbar}[A_H, H_H] \tag{26i}$$

If A_H does not depend explicitly on time

$$\frac{dA_H}{dt} = \frac{1}{i\hbar}[A_H, H_H] \tag{26ii}$$

Equations (26i) and (26ii) are called *Heisenberg's equations of motion.*

Thus in both representations, the equation of motion has the same form. Equations (22) and (26) define the Heisenberg picture. From Eqn. (26) it is evident that $(d/dt)\,H_H = 0$ i.e. Hamiltonian is the same time independent operator in both the pictures, and in both cases, the constants of motion can be represented by observables which commute with the Hamiltonian. Particularly, if the dynamical state of a system is represented by an eigen vector of B in the Heisenberg picture:

$$B\phi_H = b\phi_H$$

the variable B maintains the same well defined value in the course of time, and we say that the eigenvalue b is a good quantum number.

10.4 Evolution of State Vector ψ and the "Action" (Change of representation)

From Eqn. (21) we may write

$$i\hbar\frac{\partial U(t)}{\partial t} = H\,U(t)$$

Applying this operator equation to the state vector $\psi(t_0)$, we get

$$i\hbar\frac{\partial \hat{U}(t,t_0)\psi(t_0)}{\partial t} = \hat{H}(t)\hat{U}(t,t_0)\psi(t_0) \tag{27}$$

i.e.
$$i\hbar\frac{\partial \psi(t)}{\partial t} = \hat{H}(t)\psi(t) \tag{28}$$

This is the equation of motion for the state vector (time dependent Schrödinger equaiton). We thus see that the evolution of the state vector ψ in time may be viewed as the continuous unfolding of a unitary transformation.

Equation (28) may be rewritten as

$$\frac{\partial}{\partial t}\left(\frac{\hbar}{i}\ln\psi \right) + \hat{H}(t) = 0 \tag{29}$$

This equation is analogous to the Hamilton-Jacobi equation in classical mechanics with the action S given by

$$S = \frac{\hbar}{i} \ln \psi$$

or
$$\psi = \exp[(i/\hbar)S] \tag{30}$$

In the coordinate representation, Eqn. (28) reads

$$i\hbar \frac{\partial \psi(\vec{r},t)}{\partial t} = \hat{H}(\vec{r},\nabla,t)\psi(\vec{r},t)$$

Often $\hat{H}$ does not depend explicitly on time. Then we have from Eqn. (27)

$$\int_{t_0}^{t} \frac{d\hat{U}(t',t_0)}{\hat{U}(t',t_0)} = -(i/\hbar)\hat{H} \int_{t_0}^{t} dt'$$

or
$$\hat{U}(t,t_0) = e^{-(i/\hbar)\hat{H}(t-t_0)} \tag{31}$$

so that
$$\psi(\vec{r},t) = e^{-(i/\hbar)\hat{H}(t-t_0)}\psi(\vec{r},t_0)$$

Choosing $t_0 = 0$ and writing $\psi(\vec{r},0) \equiv \psi_H(\vec{r})$ we get

$$\psi_s(\vec{r},t) = e^{-(i/\hbar)\hat{H}t}\psi_H(\vec{r}) \tag{32}$$

or
$$|\psi_H\rangle = e^{iHt/\hbar}|\psi_s(t)\rangle \tag{33}$$

Now from Eqns (24) and (31)
$$A_H(t) = e^{iHt/\hbar}A_s e^{-iHt/\hbar} \tag{34}$$

10.5 Dirac or Interaction Picture

In the interaction or Dirac picture the time evolution of quantum systems is intermediate between the Schrödinger and Heisenberg pictures and the evolution of the system with time is given as usual by

$$i\hbar \frac{\partial}{\partial t}|\psi_s,t\rangle = H_s|\psi_s,t\rangle \tag{35}$$

but the Hamiltonian is split up into two parts: time independent part $\hat{H}^{(0)}$ and the time dependent part $\hat{H}'$ when there is some perturbing force acting on the system:

$$\hat{H}(t) = \hat{H}^{(0)} + \hat{H}'(t) \tag{36}$$

such a picture is very suitable for the perturbation treatment of two interacting systems, e.g. mesons and nucleons. The Schrödinger equation becomes

$$i\hbar \frac{\partial}{\partial t}|\psi_s,t\rangle = (H^{(0)} + H')|\psi_s,t\rangle \tag{37}$$

Here $H^{(0)}$ is a constant operator and H' depends on time.

Let us introduce a state vector

$$|\psi,t\rangle_I = U|\psi_s,t\rangle$$

$$= e^{iH^{(0)}t/\hbar}|\psi_s,t\rangle \tag{38}$$

Obviously, the operator $U \equiv e^{iH^{(0)}t/\hbar}$ is a unitary operator since

$$UU^{-1} = \hat{1} \tag{39}$$

i.e.

$$|\psi_s,t\rangle = e^{-iH^{(0)}t/\hbar}|\psi,t\rangle_I \tag{40}$$

Using above equation, Eqn. (37) reduces to

$$i\hbar \frac{\partial}{\partial t}\left(e^{-iH^{(0)}t/\hbar}|\psi,t>_I \right) = (H^{(0)} + H')e^{-iH^{(0)}t/\hbar}|\psi,t>_I$$

or

$$= i\hbar e^{-iH^{(0)}t/\hbar}\frac{\partial}{\partial t}|\psi,t>_I + H^{(0)}e^{-iH^{(0)}t/\hbar}|\psi,t>_I$$

$$= H^{(0)}e^{-iH^{(0)}t/\hbar}|\psi,t>_I + H'e^{-iH^{(0)}t/\hbar}|\psi,t>_I$$

$$i\hbar e^{-iH^{(0)}t/\hbar}\frac{\partial}{\partial t}|\psi,t>_I = H'e^{-iH^{(0)}t/\hbar}|\psi,t>_I$$

or

$$i\hbar\frac{\partial}{\partial t}|\psi,t>_I = \left(e^{iH^{(0)}t/\hbar}H'e^{-iH^{(0)}t/\hbar} \right)|\psi,t>_I$$

The term on the r.h.s. within brackets we write as H'_I; then equation of motion in interaction picture may be written as

$$i\hbar\frac{\partial}{\partial t}|\psi,t>_I = H'_I|\psi,t>_I \tag{41}$$

In fact all operators A_s of Schröndinger picture may be transformed into interaction picture by the unitary transformation

$$A_I(t) = e^{iH^{(0)}t/\hbar}A_s(t)e^{-iH^{(0)}t/\hbar} \tag{42}$$

Also

$$\frac{dA_I}{dt} = \frac{i}{\hbar}\left[H^{(0)}, A_I \right] + \frac{\partial A_I}{dt} \tag{43}$$

Equation (42) for interaction picture operators has exactly the same form as the Heisenberg equation of motion but with one important difference: It is not the complete Hamiltonian H but the unperturbed part $H^{(0)}$ which enters the operators. Thus we may say that operators obey free equations of motion (as if $H \rightarrow H_0$). The effects of H' are manifested entirely through the time variation of the state vectors, $|\psi_I(t)\rangle$. Table 10.1 summarizes the difference in the Schrödinger picture, the Heisenberg picture and the interaction picture.

Since the operators, like the dynamical variables which they represent, change in time in the Heisenberg picture, the Heisenberg picture is more like classical dynamics than the Schrödinger picture.

Table 10.2 summarizes the equations of motion of the state vector $|\psi\rangle$ and of any observable A of the system in each of the three pictures.

Table 10.1

	State vectors	*Operators*
Schröndinger picture	Moving (changing)	Fixed
Heisenberg picture	Fixed	Moving (changing)
Interaction picture	Changing	changing

Since the operators, like the dynamical variables which they represent, change in time in the Heisenberg picture, the Heisenberg picture is more like classical dynamics than the Schrödinger picture.

Table 10.2 summarizes the equations of motion of the state vector |ψ> and of any observable A of the system in each of the three pictures.

Table 10.2

Schrödinger picture	$i\hbar\dfrac{d\mid \psi_s(t)>}{dt} = H\mid \psi_s(t)>$ $\dfrac{dA_s}{dt} = 0$
Heisenberg picture	$\dfrac{d\mid \psi_{II}>}{dt} = 0 \; ; \; \dfrac{dA_{II}(t)}{dt} = \dfrac{1}{i\hbar}\left[A_{II}, H_{II}\right]$ $\left\vert\psi_{II}\right\rangle = e^{iHt/\hbar}\mid \psi_s(t)>$ $A_{II}(t) = e^{iHt/\hbar}A_s e^{-iHt/\hbar}$
Interaction picture	$\left\vert\psi_I(t)\right\rangle = e^{iH^{(0)}t/\hbar}\left\vert\psi_s(t)\right\rangle$ $i\hbar\dfrac{\partial\mid \psi_I(t)>}{\partial t} = H'_I \mid \psi_I(t) > \; (*)$ $A_I(t) = e^{iH^{(0)}t/\hbar}A_s e^{-iH^{(0)}t/\hbar}$ $\dfrac{dA_I}{dt} = \dfrac{1}{i\hbar}\left[A_I, H^{(0)}\right]$ $H_I(t) = H^{(0)} + H'$

(*) Here $H'_I = e^{iH^{(0)}t/\hbar}H'e^{-iH^{(0)}t/\hbar}$

Example 10.1: Show that if the operator A, B and C satisfy the commutation relation

$$\{A,B] = iC$$

in the Schrödinger picture, then this relation is also valid in the Heisenberg picture.

Solution: In the Heisenberg picture

$$A_{II} = e^{iHt/\hbar}A e^{-iHt/\hbar}$$
$$B_{II} = e^{iHt/\hbar}B e^{-iHt/\hbar}$$

So,

$$\left[A_{II}, B_{II}\right] = A_{II}B_{II} - B_{II}A_{II}$$
$$= e^{iHt/\hbar}A e^{-iHt/\hbar} \cdot e^{iHt/\hbar}B e^{-iHt/\hbar}$$
$$\quad - e^{iHt/\hbar}B e^{-iHt/\hbar} \cdot e^{iHt/\hbar}A e^{-iHt/\hbar}$$
$$= e^{iHt/\hbar}ABe^{-iHt/\hbar} - e^{iHt/\hbar}BAe^{-iHt/\hbar}$$
$$= e^{iHt/\hbar}(AB - BA)e^{-iHt/\hbar}$$

$$= e^{iHt/\hbar}[A,B]e^{-iHt/\hbar}$$

$$= e^{iHt/\hbar}iCe^{-iHt/\hbar}$$

or $\qquad [A_H, B_H] = iC_H$.

10.6 Comparison of Heisenberg's Equations of Motion with Classical Hamilton's Equations of Motion

Hamilton's equations of motion are

$$\frac{dq_i}{dt} = \frac{\partial H}{\partial p_i} \; ; \; \frac{dp_i}{dt} = -\frac{\partial H}{\partial q_i} \qquad (44)$$

and

$$\frac{dA}{dt} = \frac{\partial A}{\partial t} + \{A, H\} \qquad (45)$$

where q_i and p_i are the generalised coordinates and momenta and H is the Hamiltonian. A is a general dynamical variable which is a function of q_i, p_i and t, $\{A, H\}$ is the *Poisson bracket* defined by

$$\{A, H\} = \sum_i \left(\frac{dA}{dq_i} \frac{\partial H}{\partial p_i} - \frac{\partial H}{\partial q_i} \frac{\partial A}{\partial p_i} \right) \qquad (46)$$

Now, from Heisenberg's equation of motion

$$\frac{dA_H}{dt} = \frac{1}{i\hbar}[A_H, H_H] \qquad (47)$$

We have, for the basic canonical operators $\hat{q}_j$ and $\hat{p}_j$:

$$\frac{d\hat{q}_j}{dt} = \frac{1}{i\hbar}[q_j, H_H] = \{q_j, H_H\} \qquad (48)$$

$$= \frac{\partial q_i}{\partial q_j}\frac{\partial H_H}{\partial p_j} - \frac{\partial H_H}{\partial q_j}\frac{\partial q_j}{\partial p_j}$$

$$= \frac{\partial H_H}{\partial p_j} \qquad \left(\because \; \frac{\partial q_j}{\partial p_j} = 0 \right) \qquad (49)$$

Similarly,

$$\frac{dp_j}{dt} = \frac{1}{i\hbar}[p_j, H_H]$$

$$= \{p_j, H_H\}$$

$$= \frac{\partial p_j}{\partial q_j}\frac{\partial H_H}{\partial p_j} - \frac{\partial H_H}{\partial q_j}\frac{\partial p_j}{\partial p_j}$$

$$= -\frac{\partial H_H}{\partial q_j} \qquad \left(\because \; \frac{\partial p_j}{\partial q_j} = 0 \right) \qquad (50)$$

Thus, we see that these equations are identical with the corresponding canonical equations of Hamilton in classical mechanics.

10.7 The Linear Harmonic Oscillator in Heisenberg Picture

Dropping the suffix H attached to Heisenberg operators, we have

$$\hat{H}(t) = \frac{\left[\hat{p}(t)\right]^2}{2m} + \frac{1}{2}K\left[\hat{x}(t)\right]^2$$

or
$$\hat{H}(t) = \frac{1}{2m}\left[\hat{p}^2(t) + m^2\omega^2\hat{x}^2(t)\right] \tag{51}$$

The equations of motion for $\hat{x}$ and $\hat{p}$ according to equations (49) and (50) are

$$\frac{d\hat{x}}{dt} = \frac{\partial\hat{H}}{\partial\hat{p}} = \frac{\hat{p}(t)}{m} \tag{52}$$

and
$$\frac{d\hat{p}}{dt} = -\frac{\partial\hat{H}}{\partial\hat{x}} = -m\omega^2\hat{x}(t) \tag{53}$$

Differentiating Eqn. (52) w.r.t. t once again,

$$\frac{d^2\hat{x}}{dt^2} = \frac{1}{m}\frac{d\hat{p}}{dt}$$

$$= -\omega^2\hat{x}(t) \qquad \text{(using Eqn. (53)}$$

or
$$\frac{d^2\hat{x}}{dt^2} + \omega^2\hat{x} = 0$$

This is of the same form as the classical equation of motion for the harmonic oscillator. The solution of this equation is

$$\hat{x}(t) = \hat{a}_1\cos\omega t + \hat{a}_2\sin\omega t$$

$$\hat{a}_1 = \hat{x}(0) \equiv \hat{x}_0 \qquad \text{(say)}$$

and
$$\hat{a}_2 \equiv \frac{1}{\omega}\frac{d\hat{x}}{dt}\Big|_{t=0} = \frac{1}{\omega}\frac{\hat{p}(0)}{m} \equiv \frac{\hat{p}_0}{m\omega}$$

Therefore
$$\hat{x}(t) = \hat{x}_0\cos\omega t + (\hat{p}_0/m\omega)\sin\omega t$$

and
$$\hat{p}(t) = m\frac{d\hat{x}}{dt} = -m\omega\hat{x}_0\sin\omega t + \hat{p}_0\cos\omega t$$

Substituting these value of $\hat{x}(t)$ and $\hat{p}(t)$ in Eqn. (51), we get

$$\hat{H} = \frac{1}{2m}\left(\hat{p}_0^2 + m^2\omega^2\hat{x}_0^2\right) \tag{54}$$

which is independent of time.

To obtain the energy levels, we have to calculate H_{nn}. Since the form of $\hat{H}$ is same as in Schrödinger representation, we may use the same eigen functions

$$u_n(x) = \frac{\sqrt{\alpha}}{\left(\sqrt{\pi}2^n n!\right)^{1/2}}\exp\left(-\frac{1}{2}\alpha^2 x^2\right)H_n(\alpha x)$$

$$H_{n'n} = \frac{\left(p_0^2\right)_{n'n}}{2m} + \frac{m\omega^2\left(x_0^2\right)_{n'n}}{2} \tag{55}$$

Here $\hat{p}_0$ and $\hat{x}_0$ are the Heisenberg operators at time $t = 0$, they must be identical with corresponding operators in Schröndinger's picture:

$$\hat{x}_0 \equiv x \text{ and } \hat{p}_0 = -i\hbar \frac{d}{dx}$$

Thus,

$$(p_0^2)_{n'n} = (-i\hbar)^2 \int_{-\infty}^{+\infty} u_{n'}^*(x) \frac{d^2}{dx^2} u_n(x) \, dx$$

$$= -\hbar^2 \alpha^2 \int_{-\infty}^{+\infty} \psi_n^*(y) \psi_n''(y) \, dy \tag{56}$$

where $y = \alpha x$, $\alpha = \sqrt{m\omega/\hbar}$ and ψ_n satisfies the differential equation

$$\psi_n'' + (1 + 2n - y^2)\psi_n = 0$$

so that

$$\psi_n'' = \left[y^2 - (1 + 2n) \right] \psi_n \tag{57}$$

$$\therefore \quad \frac{1}{2m}(p_0^2)_{n'n} = \frac{\hbar^2\alpha^2}{2m}\left[(2n+1)\delta_{nn'} - \alpha^2 (x_0^2)_{n'n} \right]$$

(using Eqn. (57) for ψ_n'' in Eqn. (56) and the orthonormality condition)

Now

$$\frac{\hbar^2\alpha^2}{2m} = \frac{1}{2}\hbar\omega \text{ and } \frac{\hbar^2\alpha^4}{2m} = \frac{1}{2}\hbar\omega \times \frac{m\omega}{\hbar} = \frac{1}{2}m\omega^2$$

Hence

$$\frac{1}{2m}(p_0^2)_{n'n} = \left(n + \tfrac{1}{2} \right) \hbar\omega\delta_{nn'} - \frac{1}{2}m\omega^2 (x_0^2)_{n'n}$$

or

$$\frac{1}{2m}(p_0^2)_{n'n} + \frac{1}{2}m\omega^2 (x_0^2)_{n'n} = \left(n + \tfrac{1}{2} \right)\hbar\omega\delta_{nn'} \tag{58}$$

Comparing Eqns (54) and (58), we get

$$H_{n'n} = \left(n + \tfrac{1}{2} \right)\hbar\omega\delta_{nn'} \tag{59}$$

$\therefore$ The eigen values of $\hat{H}$ are given by

$$H_{nn} \equiv E_n = \left(n + \tfrac{1}{2} \right)\hbar\omega, \quad n = 0, 1, 2, \ldots \tag{60}$$

Example 10.2: Prove that for any operator A in Heisenberg's picture

$$\frac{d}{dt}\langle A_H \rangle = \frac{1}{i\hbar}\langle [A_H, H_H] \rangle + \left\langle \frac{\partial A_H}{\partial t} \right\rangle$$

Solution:

$$\frac{d}{dt}\langle A_H \rangle = \frac{d}{dt}|\psi_H\rangle A_H(t) |\psi_H\rangle$$

$$= \langle \psi_H | \frac{dA_H}{dt} | \psi_H \rangle$$

$$= \left\langle \psi_H \left| \left(\frac{\partial A_H}{\partial t} + \frac{1}{i\hbar}[A_H, H_H] \right) \right| \psi_H \right\rangle$$

$$= \left\langle \frac{\partial A_H}{\partial t} \right\rangle + \left\langle \frac{1}{i\hbar}[A_H, H_H] \right\rangle$$

$$= \left\langle \frac{\partial A_H}{\partial t} \right\rangle + \frac{1}{i\hbar}\left\langle [A_H, H_H] \right\rangle$$

10.8 Time Evolution Operator and the Interaction Picture (The Scattering Matrix)

There are two special choices for the time development (i.e. time evolution) operator U, so that solution of the equation:

$$i\hbar \frac{\partial}{\partial t} | \psi_s > = H | \psi_s > \tag{61}$$

can be written as

$$| \psi_s(t) > = U(t,t_0) | \psi_s(t_0) > \tag{62a}$$

$$U(t,t_0) = e^{-iH(t-t_0)/\hbar} \tag{62b}$$

provided H is independent of time. Obviously

$$i\hbar \frac{\partial U(t,t_0)}{\partial t} = H \ U(t,t_0) \tag{63a}$$

and

$$\lim_{t \to t_0} U(t,t_0) = 1$$

and

$$U^{-1}(t,t_0) = U(t_0,t),$$

$$U(t,t_1) \ U(t_1,t_0) = U(t,t_0) \tag{63b}$$

If H is time dependent, Eqn. (62a) would still remain valid if U were defined through Eqns. (63a), though U would no longer have the explicit form of Eqn. (62b). Since U translates the state at time t_0 to that at time t, it is called the *time evolution operator*.

In the interaction picture, the unitary operator U is so chosen that the variation of operators is entirely due to H_0 and the states would not vary but for the existence of H' (see Eqns. (42)/(43) and (41) respectively).

The appropriate choice is

$$U = V_0^{-1}(t, t_0) \tag{64}$$

where V_0 satisfies Eqn. (63) with H replaced by H_0.

$$|\psi(t)\rangle_I = V_0^{-1}(t,t_0) \ |\psi_s(t)\rangle \tag{65}$$

$$A_I(t) = V_0^{-1}(t,t_0)A(t)V_0(t,t_0) \tag{66}$$

and

$$i\hbar \frac{dV_0(t,t_0)}{dt} = H_0 V_0(t,t_0) \tag{67}$$

Pre multiplying Eqn. (65) by V_0 and then differentiating, we get

$$H_0 V_0 | \psi(t) >_I + V_0 i\hbar \frac{\partial}{\partial t} | \psi(t) >_I = i\hbar \frac{\partial}{\partial t} | \psi_s(t) >$$

$$= H | \psi_s(t) > = HV_0 | \psi(t) >_I$$

Consequently,

$$V_0 i\hbar \frac{\partial}{\partial t} | \psi(t) >_I = H'V_0 | \psi(t) >_I \qquad \text{(using Eqn. 41)}$$

or

$$i\hbar \frac{\partial}{\partial t} | \psi(t) >_I = V_0^{-1} H'V_0 | \psi(t) >_I \ \equiv H'_I | \psi(t) >_I \tag{68}$$

with
$$H'_I \equiv V_0^{-1} H' V_0$$

We can obtain a perturbative solution of the form

$$\left|\psi(t)\right\rangle_I = \sum_n \left|\psi^{(n)}(t)\right\rangle_I \tag{69}$$

where $\left|\psi^{(n)}(t)\right\rangle_I$ is of nth order H'_I (i.e. contains n factors of H'_I). Substituting this in Eqn. (68) and equating terms of the same order on the two sides, we get

$$i\hbar \frac{\partial}{\partial t}\left|\psi^{(0)}(t)\right\rangle_I = 0$$

$$i\hbar \frac{\partial}{\partial t}\left|\psi^{(n)}(t)\right\rangle_I = H'_I \left|\psi^{(n-1)}(t)\right\rangle_I, \quad n = 1, 2, \ldots \tag{70}$$

At the initial moment, when the interaction has had no time to act, we take

$$\left|\psi^{(0)}(t_0)\right\rangle_I = \left|\psi^{(1)}(t_0)\right\rangle_I = \ldots = \left|\psi^{(n)}(t_0)\right\rangle_I = 0 \tag{71}$$

$$(n = 1, 2, \ldots)$$

The solutions of Eqn. (70) subject to these boundary conditions are

$$\left|\psi^{(0)}(t)\right\rangle_I = \left|\psi^{(0)}(t_0)\right\rangle_I ;$$

$$\left|\psi^{(n)}(t)\right\rangle_I = \frac{1}{i\hbar} \int_{t_0}^{t} H'_I(t') \left|\psi^{(n-1)}(t')\right\rangle_I dt' \tag{72}$$

The last equation enables us to yield recursive solutions:

$$\left|\psi^{(n)}(t)\right\rangle_I = \frac{1}{(i\hbar)^n} \left\{ \int_{t_0}^{t} dt_n \, H'_I(t_n) \int_{t_0}^{t_n} H'_I(t_{n-1}) dt_{n-1} \right.$$

$$\left. \ldots\ldots \int_{t_0}^{t_2} H'_I(t_1) dt_1 \right\} \left|\psi^{(0)}(t_0)\right\rangle_I \cdot \tag{73}$$

Substituting this in Eqn. (69), we obtain

$$\left|\psi(t)\right\rangle_I = V_I(t, t_0) \left|\psi(t_0)\right\rangle_I \tag{74}$$

where the operator $V_I(t, t_0)$ relating the interaction picture state at time t to that at time t_0 is given by

$$V_I(t, t_0) = 1 + \frac{1}{i\hbar} \int_{t_0}^{t} H'_I(t_1) dt_1 + \frac{1}{(i\hbar)^2} \int_{t_0}^{t} H'_I(t_2) dt_2 \times \int_{t_0}^{t_2} H'_I(t_1) dt_1$$

$$+ \frac{1}{(i\hbar)^3} \int_{t_0}^{t} H'_I(t_3) dt_3 \int_{t_0}^{t_3} H'_I(t_2) dt_2 \int_{t_0}^{t_2} H'_I(t_1) dt_1 + \ldots. \tag{75}$$

We might consider Eqn. (74) as the definition of $V_I(t, t_0)$ (independent of any perturbation expansion). Then by substituting this in Eqn. (68) and considering $\left|\psi(t_0)\right\rangle_I$ as quite arbitrary, we obtain

$$i\hbar \frac{\partial V_I(t,t_0)}{\partial t} = H_I' V_I(t,t_0) \tag{76}$$

and
$$V_I(t_0,t_0) = 1$$

Then the infinite series (Eqn. 75) is a perturbative solution of the Eqn. (76). We now proceed to express the state vector $|\psi(t)\rangle$ in the Schrödinger picture.

$$|\psi_s(t)\rangle = V_0(t,t_0) |\psi(t)\rangle_I$$

$$= V_0(t,t_0) V_I(t,t_0) |\psi(t_0)\rangle_I \quad \text{(using Eqn. 74)} \tag{77}$$

Since the states in both pictures coincide at t_0 (i.e. $|\psi(t_0)\rangle_I = |\psi_s(t_0)\rangle$), therefore

$$|\psi_s(t)\rangle = V(t,t_0) |\psi_s(t_0)\rangle \tag{78}$$

with
$$V(t,t_0) \equiv V_0(t,t_0) V_I(t,t_0)$$

giving the time evolution operator of the Schrödinger picture in terms of that of the interaction picture.

Using the expression (75) for V_I and the fact that

$$H_I'(t) = V_0^{-1}(t,t_0) H' V_0(t,t_0)$$

we get

$$V(t,t_0) = \sum_{n=0}^{\infty} V^{(n)}(t,t_0)$$

with

$$V^{(n)}(t,t_0) = \frac{1}{(i\hbar)^n} V_0(t,t_0) \int_{t_0}^{t} \int_{t_0}^{t_n} \cdots \int_{t_0}^{t_2}$$

$$\left[V_0^{-1}(t_n,t_0) H' V_0(t_n,t_0) \; V_0^{-1}(t_{n-1},t_0) H' V_0(t_{n-1},t_0) \cdots \right.$$

$$\left. V_0^{-1}(t_1,t_0) H' V_0(t_1,t_0) \right] dt_1 dt_2 \cdots dt_n \tag{79}$$

$$= (i\hbar)^{-n} \int_{t_0}^{t} \int_{t_0}^{t_n} \cdots \int_{t_0}^{t_2} \left[V_0(t,t_n) H' V_0(t_n,t_{n-1}) H' \right.$$

$$\left. \cdots V_0(t_2,t_1) H' V_0(t_1,t_0) \right] dt_1 dt_2 \cdots dt_n \tag{80}$$

Here we have used

$$V_0(t_a,t_b) V_0(t_b,t_c) = V_0(t_a,t_c)$$

(so that $V_0(t_2,t_1) V_0(t_1,t_0) = V_0(t_2,t_0)$ etc).

Since V_0 is the time evolution operator for the unperturbed system, the above picture of progress of the system is one of unperturbed propagation from t_0 to t_1, t_1 to t_2, ..., t_n to t, with interruptions (action of the perturbations at t_1, t_2,...., t_n manifested through the factors $H'/(i\hbar)$ occurring between pairs of V_0's.

The transition amplitude from the initial state $|i\rangle$ to the final $|f\rangle$ is $\langle f|\psi(t)\rangle$ under the condition that $|\psi(t_0)\rangle = |i\rangle$. This is simply $\langle f|V(t,t_0)|i\rangle$. The nth order contribution to it is $\langle f|V^{(n)}(t,t_0)|i\rangle$.

The Scattering Matrix

In scattering experiments, the time interval between the firing of the projectile and its observation after scattering event is extremely large compared to the time during which the colliding particles actually interact. So, practically, this interval may be considered infinite. Hence, in order to apply above treatment to scattering problems, it is customary to assume $t_0 \to -\infty$ and $t \to +\infty$. The operator

$$S = \lim_{\substack{t_0 \to -\infty \\ t \to +\infty}} V(t, t_0) \tag{81}$$

is known as the *scattering operator*. Its matrix elements $<f|S|i>$ between the unperturbed (free) states of the system of particles constitute the so called *scattering matrix* or *S-matrix*.

Questions and Problems

1. Discuss the Schrödinger, the Heisenberg and the interaction representations for describing the dynamical behaviour of a system. Obtain the equation of motion for a state function and for an operator in each representation.

2. Discuss the Schrödinger picture and Heisenberg picture of time evolution. Discuss various properties of time evolution operators. Prove that for any operator A associated with an observable,

$$\frac{d}{dt}\langle A_H \rangle = \frac{1}{i\hbar}\langle \, [A_H, H_H] \, \rangle + \left\langle \frac{\partial}{\partial t} A_H \right\rangle$$

in Heisenberg picture.

3. Show that for any operator which commutes with the Hamiltonian, the operators in Schrödinger and Heisenberg pictures are identical.

4. For Schrödinger operators A, B and C, let $[A, B] = C$. What is the commutation relation for the corresponding operators in the Heisenberg representation?

5. Discuss the problem of linear harmonic oscillator in Heisenberg picture.

6. Construct the position and momentum operators $\hat{x}(t)$ and $\hat{p}(t)$ respectively for a linear harmonic oscillator.

11

Angular-Momentum

11.1 Angular-Momentum Operator

In classical mechanics, the angular momentum of a particle is defined by

$$\vec{L} = \vec{r} \times \vec{p} \tag{1}$$

where $\vec{r}$ and $\vec{p}$ represent the position vector and linear momentum of the particle respectively. Thus

$$L_x = yp_z - zp_y \tag{2i}$$
$$L_y = zp_x - xp_z \tag{2ii}$$
$$L_z = xp_y - yp_x \tag{2iii}$$

In quantum mechanics, we replace p_x, p_y and p_z by the corresponding operators, so we obtain

$$L_x = -i\hbar \left(y\frac{\partial}{\partial z} - z\frac{\partial}{\partial y} \right) \tag{3i}$$

$$L_y = -i\hbar \left(z\frac{\partial}{\partial x} - x\frac{\partial}{\partial z} \right) \tag{3ii}$$

$$L_z = -i\hbar \left(x\frac{\partial}{\partial y} - y\frac{\partial}{\partial x} \right) \tag{3iii}$$

or using Einstein summation convention

$$L_i = \epsilon_{ijk}\, x_j\, p_k \tag{4}$$

where ϵ_{ijk} is the completely antisymmetric tensor of the third rank

$$\epsilon_{ijk} = \begin{cases} 1 & \text{for even permutations of (123),} \\ -1 & \text{for odd permutations of (123),} \\ 0 & \text{otherwise} \end{cases}$$

11.2 Commutation Relations

$$\left[L_x, L_y \right] = L_x L_y - L_y L_x$$

$$= \left(yp_z - zp_y \right)\left(zp_x - xp_z \right) - \left(zp_x - xp_z \right)\left(yp_z - zp_y \right)$$

$$= yp_z zp_x - yp_z xp_z - zp_y zp_x + zp_y xp_z$$
$$-zp_x yp_z + zp_x zp_y + xp_z yp_z - xp_z zp_y \qquad (\because \ [p_x, p_y] = 0)$$
$$= -yp_x(-p_z z + zp_z) + p_y x(zp_z - p_z z)$$
$$= i\hbar(xp_y - yp_x) = i\hbar L_z \qquad \qquad (5i)$$

Similarly
$$\left[L_y, L_z\right] = i\hbar L_x \qquad \qquad (5ii)$$

and
$$\left[L_x, L_z\right] = i\hbar L_y \qquad \qquad (5iii)$$

In summation notation
$$\left[L_i, L_j\right] = i\hbar \in_{ijk} L_k \qquad \qquad (6)$$

Now,
$$= \left[yp_z, y\right] - \left[zp_y, y\right]$$
$$= y\left[p_z, y\right] + [y, y]p_z - z\left[p_y, y\right] - [z, y]p_y$$
$$= y.0 + 0.p_z - z(-i\hbar) - 0.p_y$$

or
$$\left[L_x, y\right] = i\hbar z \qquad \qquad (7i)$$

Similarly
$$\left[L_x, x\right] = 0 \qquad \qquad (7ii)$$

and
$$\left[L_z, x\right] = -i\hbar y \qquad \qquad (7iii)$$

Hence
$$\left[L_i, x_j\right] = i\hbar \in_{ijk} x_k \qquad \qquad (8)$$

Further
$$\left[L_x, p_y\right] = \left[yp_z - zp_y, p_y\right] = \left[yp_z, p_y\right] - \left[zp_y, p_y\right]$$
$$= y\left[p_z, p_y\right] + [y, p_y]p_z - z\left[p_y, p_y\right] - [z, p_y]p_y$$
$$= y.0 + i\hbar p_z - z.0 - 0.p_y$$
$$= i\hbar p_z \qquad \qquad (9i)$$

Similarly
$$\left[L_x, p_x\right] = 0 \qquad \qquad (9ii)$$

and
$$\left[L_x, p_z\right] = -i\hbar p_y \qquad \qquad (9iii)$$

In summation convention
$$\left[L_i, p_j\right] = i\hbar \in_{ijk} p_k \qquad \qquad (10)$$

Further,
$$\left[L^2, L_x\right] = \left[L_x^2 + L_y^2 + L_z^2, L_x\right]$$
$$= \left[L_x^2, L_x\right] + \left[L_y^2, L_x\right] + \left[L_z^2, L_x\right]$$
$$= \left[L_y^2, L_x\right] + \left[L_z^2, L_x\right] \qquad \left(\text{because } \left[L_x^2, L_x\right] = 0\right)$$

Now

$$\left[L_y^2, L_x\right] = L_y^2 L_x - L_x L_y^2$$

$$= L_y L_y L_x - L_x L_y L_y$$

$$= L_y\left(L_y L_x - L_x L_y\right) + \left(L_y L_x - L_x L_y\right)L_y$$

$$= L_y\left[L_y, L_x\right] + \left[L_y, L_x\right]L_y$$

$$= L_y(-i\hbar L_z) + (-i\hbar L_z)L_y$$

$$= -i\hbar(L_y L_z + L_z L_y)$$

Similarly

$$\left[L_z^2, L_x\right] = i\hbar(L_y L_z + L_z L_y)$$

Hence

$$\left[L^2, L_x\right] = -i\hbar(L_y L_z + L_z L_y) + i\hbar(L_y L_z + L_z L_y) \qquad (11i)$$

$$= 0$$

Similarly

$$\left[L^2, L_y\right] = 0 \text{ and } \left[L^2, L_z\right] = 0$$

Hence

$$\left[L^2, L\right] = 0 \qquad (11ii)$$

i.e. L^2 commutes with angular momentum operator L. So L^2 and L have simultaneous eigenfunctions and are simultaneously measurable.

11.3 Representation of Angular Momentum in Spherical Polar Coordinates

$$x = r \sin \theta \cos \phi \qquad (12i)$$
$$y = r \sin \theta \sin \phi \qquad (12ii)$$
$$z = r \cos \theta \qquad (12iii)$$

$$r = \left(x^2 + y^2 + z^2\right)^{1/2}$$

$$\tan \theta = \frac{\sqrt{x^2 + y^2}}{z}$$

$$r^2 = x^2 + y^2 + z^2 \qquad (13i)$$

$$\tan^2 \theta = \frac{(x^2 + y^2)}{z^2} \qquad (13ii)$$

$$\tan \phi = \frac{y}{x} \qquad (13iii)$$

$$2r\left(\frac{\partial r}{\partial x}\right) = 2x \qquad \text{(from Eqn. (13i))}$$

$$\frac{\partial r}{\partial x} = \left(\frac{x}{r}\right) = \sin \theta \cos \phi \qquad (14)$$

Differentiating Eqn. (13*ii*) w.r.t. x

$$2\tan\theta\sec^2\theta = \left(\frac{\partial\theta}{\partial x}\right) = \frac{2x}{z^2} = \frac{2r\sin\theta\cos\phi}{r^2\cos^2\theta} = \frac{2\tan\theta\cos\phi}{r\cos\theta}$$

$$\therefore \qquad \frac{\partial\theta}{\partial x} = \left(\frac{1}{r}\right)\cos\theta\,\cos\phi \qquad (15)$$

From Eqn. (13*iii*)

$$\tan\phi = \frac{y}{x} = yx^{-1}$$

Differentiating

$$\sec^2\phi\,d\phi = y(-1)x^{-2}dx$$

or

$$\frac{\partial\phi}{\partial x} = -\frac{y}{x^2}\cos^2\phi = -\frac{r\sin\theta\sin\phi}{r^2\sin^2\theta\cos^2\phi}\cos^2\phi$$

$$= -\left(\frac{1}{r}\right)\frac{\sin\phi}{\sin\theta} \qquad (16)$$

Similar to Eqns (14), (15) and (16), we get

$$\frac{\partial r}{\partial y} = \sin\theta\sin\phi \qquad (17)$$

$$\frac{\partial\theta}{\partial y} = \left(\frac{1}{r}\right)\cos\theta\sin\phi \qquad (18)$$

$$\frac{\partial\phi}{\partial y} = \left(\frac{1}{r}\right)\frac{\cos\phi}{\sin\theta} \qquad (19)$$

Now, we have

$$\frac{\partial\psi}{\partial x} = \frac{\partial\psi}{\partial r}\left(\frac{\partial r}{\partial x}\right) + \frac{\partial\psi}{\partial\theta}\left(\frac{\partial\theta}{\partial x}\right) + \frac{\partial\psi}{\partial\phi}\left(\frac{\partial\phi}{\partial x}\right)$$

$$= \sin\theta\cos\phi\left(\frac{\partial\psi}{\partial r}\right) + \frac{\cos\theta\cos\phi}{r}\left(\frac{\partial\psi}{\partial\theta}\right) - \frac{\sin\phi}{r\sin\theta}\left(\frac{\partial\psi}{\partial\phi}\right)$$

Since ψ is arbitrary, therefore

$$\frac{\partial}{\partial x} = \sin\theta\cos\phi\frac{\partial}{\partial r} + \frac{\cos\theta\cos\phi}{r}\frac{\partial}{\partial\theta} - \frac{\sin\phi}{r\sin\theta}\frac{\partial}{\partial\phi} \qquad (20i)$$

Similarly,

$$\frac{\partial}{\partial y} = \sin\theta\,\sin\phi\frac{\partial}{\partial r} + \frac{\cos\theta\,\sin\phi}{r}\frac{\partial}{\partial\theta} + \frac{\cos\phi}{r\sin\theta}\frac{\partial}{\partial\phi} \qquad (20ii)$$

$$\frac{\partial}{\partial z} = \cos\theta\frac{\partial}{\partial r} - \frac{\sin\theta}{r}\frac{\partial}{\partial\theta} \qquad (20iii)$$

Using Eqn. (12) and Eqns (14) to (20), we obtain

$$L_z\psi(r,\theta,\phi) = -i\hbar\left(x\frac{\partial\psi}{\partial y} - y\frac{\partial\psi}{\partial x}\right)$$

$$= -i\hbar\left\{x\left(\frac{\partial\psi}{\partial r}\frac{\partial r}{\partial y} + \frac{\partial\psi}{\partial\theta}\frac{\partial\theta}{\partial y} + \frac{\partial\psi}{\partial\phi}\frac{\partial\phi}{\partial y}\right)\right.$$

$$-y\left(\frac{\partial\psi}{\partial r}\frac{\partial r}{\partial x} + \frac{\partial\psi}{\partial\theta}\frac{\partial\theta}{\partial x} + \frac{\partial\psi}{\partial\phi}\frac{\partial\phi}{\partial x}\right)\Bigg\}\quad\quad (21)$$

$$=-i\hbar\Bigg[r\sin\theta\cos\phi\Bigg\{\frac{\partial\psi}{\partial r}\sin\theta\sin\phi + \frac{\partial\psi}{\partial\theta}\left(\frac{1}{r}\cos\theta\sin\phi\right)$$

$$+\frac{\partial\psi}{\partial\phi}\left(\frac{1}{r}\cos\phi\cos ec\,\theta\right)\Bigg\}$$

$$-r\sin\theta\sin\phi\Bigg\{\frac{\partial\psi}{\partial r}(\sin\theta\cos\phi) + \frac{\partial\psi}{\partial\theta}\left(\frac{1}{r}\cos\theta\cos\phi\right)$$

$$+\frac{\partial\psi}{\partial\phi}\left(-\frac{1}{r}\sin\phi\cos ec\,\theta\right)\Bigg\}\Bigg]$$

$$=-i\hbar\Bigg[r\sin^2\theta\cos\phi\sin\phi\frac{\partial\psi}{\partial r} + \sin\theta\cos\theta\sin\phi\cos\phi\frac{\partial\psi}{\partial\theta}$$

$$+\cos^2\phi\frac{\partial\psi}{\partial\phi} - r\sin^2\theta\cos\phi\sin\phi\frac{\partial\psi}{\partial r}$$

$$-\sin\theta\cos\theta\sin\phi\cos\phi\frac{\partial\psi}{\partial\theta} + \sin^2\phi\frac{\partial}{\partial\phi}\Bigg]$$

or $\quad\quad L_z = -i\hbar\dfrac{\partial}{\partial\phi}$ $\quad\quad\quad\quad\quad\quad (22)$

Similarly, $\quad L_x = i\hbar\left(\sin\phi\dfrac{\partial}{\partial\theta} + \cot\theta\cos\phi\dfrac{\partial}{\partial\phi}\right)$ $\quad\quad (23)$

$$L_y = i\hbar\left(-\cos\phi\frac{\partial}{\partial\theta} + \cot\theta\sin\phi\frac{\partial}{\partial\phi}\right)\quad\quad (24)$$

Now,

$$L^2 = L_x^2 + L_y^2 + L_z^2 \quad\quad\quad\quad\quad\quad (25)$$

$$=-\hbar^2\Bigg[\left(\sin\phi\frac{\partial}{\partial\theta} + \cot\theta\cos\phi\frac{\partial}{\partial\phi}\right)\left(\sin\phi\frac{\partial}{\partial\theta} + \cot\theta\cos\phi\frac{\partial}{\partial\phi}\right)\Bigg]$$

$$-\hbar^2\Bigg[\left(-\cos\phi\frac{\partial}{\partial\theta} + \cot\theta\sin\phi\frac{\partial}{\partial\phi}\right)\left(-\cos\phi\frac{\partial}{\partial\theta} + \cot\theta\sin\phi\frac{\partial}{\partial\phi}\right)\Bigg]$$

$$-\hbar^2\Bigg[\left(\frac{\partial}{\partial\phi}\right)\left(\frac{\partial}{\partial\phi}\right)\Bigg]$$

or $\quad L^2 = -\hbar^2\Bigg[\sin^2\phi\dfrac{\partial^2}{\partial\theta^2} + \sin\phi\cos\phi\dfrac{\partial}{\partial\theta}\left(\cot\theta\dfrac{\partial}{\partial\phi}\right)$

$$+\cot\theta\cos\phi\frac{\partial}{\partial\phi}\left(\sin\phi\frac{\partial}{\partial\theta}\right) + \cot^2\theta\cos\phi\frac{\partial}{\partial\phi}\left(\cos\phi\frac{\partial}{\partial\phi}\right)$$

$$+\cos^2\phi\frac{\partial^2}{\partial\theta^2}-\cos\phi\sin\phi\frac{\partial}{\partial\theta}\left(\cot\theta\frac{\partial}{\partial\phi}\right)$$

$$\left.-\cot\theta\sin\phi\frac{\partial}{\partial\phi}\left(\cos\phi\frac{\partial}{\partial\theta}\right)+\cot^2\theta\sin^2\phi\frac{\partial^2}{\partial\phi^2}+\frac{\partial^2}{\partial\phi^2}\right]$$

$$=-\hbar^2\left[\frac{\partial^2}{\partial\theta^2}+\cot\theta\cos\phi\left(\cos\phi\frac{\partial}{\partial\theta}+\sin\phi\frac{\partial^2}{\partial\phi\partial\theta}\right)\right.$$

$$+\cot^2\theta\cos\phi\left(-\sin\phi\frac{\partial}{\partial\phi}+\cos\phi\frac{\partial^2}{\partial\phi^2}\right)$$

$$\left.-\cot\theta\sin\phi\left(-\sin\phi\frac{\partial}{\partial\theta}+\cos\phi\frac{\partial^2}{\partial\phi\partial\theta}\right)+\cot^2\theta\sin^2\phi\frac{\partial^2}{\partial\phi^2}+\frac{\partial^2}{\partial\phi^2}\right]$$

$$=-\hbar^2\left[\frac{\partial^2}{\partial\theta^2}+\cot\theta\cos^2\phi\frac{\partial}{\partial\theta}+\cot\theta\cos\phi\sin\phi\frac{\partial^2}{\partial\phi\partial\theta}\right.$$

$$-\cot^2\theta\cos\phi\sin\phi\frac{\partial}{\partial\phi}+\cot^2\theta\cos^2\phi\frac{\partial^2}{\partial\phi^2}$$

$$\left.+\cot\theta\sin^2\phi\frac{\partial}{\partial\theta}-\cot\theta\sin\phi\cos\phi\frac{\partial^2}{\partial\phi\partial\theta}+\cot^2\theta\sin^2\phi\frac{\partial^2}{\partial\phi^2}+\frac{\partial^2}{\partial\phi^2}\right]$$

$$=-\hbar^2\left[\frac{\partial^2}{\partial\theta^2}+\cot\theta\frac{\partial}{\partial\theta}+\cot^2\theta\frac{\partial^2}{\partial\phi^2}+\frac{\partial^2}{\partial\phi^2}\right]$$

$$=-\hbar^2\left[\frac{\partial^2}{\partial\theta^2}+\frac{\cos\theta}{\sin\theta}\frac{\partial}{\partial\theta}+\mathrm{cosec}^2\theta\frac{\partial^2}{\partial\phi^2}\right]$$

or $\qquad L^2=-\hbar^2\left[\frac{1}{\sin\theta}\frac{\partial}{\partial\theta}\left(\sin\theta\frac{\partial}{\partial\theta}\right)+\frac{1}{\sin^2\theta}\frac{\partial^2}{\partial\phi^2}\right]$ $\qquad\qquad$ (26)

Example 11.1: Show that for a particle moving in a conservative potential $V(\vec{r})$,

$$[H,L_z]=i\hbar\frac{\partial V(\vec{r})}{\partial\phi}\ and\ \left[H,L^2\right]=\left[V,L^2\right]$$

and hence prove that the orbital angular momentum is a constant of motion for the central field.

Solution: For a particle moving in a conservative field

$$H=\frac{p^2}{2m}+V(\vec{r})$$

$$\Rightarrow\qquad [H,L_z]=\left[\frac{p^2}{2m}+V(\vec{r}),L_z\right]=\frac{1}{2m}\left[p^2,L_z\right]+\left[V(\vec{r}),L_z\right]\qquad (i)$$

Now, $\left[p^2, L_z\right] = \left[p_x^2 + p_y^2 + p_z^2, L_z\right]$

$$= \left[p_x^2, L_z\right] + \left[p_y^2, L_z\right] \qquad\qquad \left(\because \; \left[p_z^2, L_z\right] = 0\right)$$

$$= p_x\left[p_x, L_z\right] + \left[p_x, L_z\right]p_x + p_y\left[p_y, L_z\right] + \left[p_y, L_z\right]p_y$$

$$= -i\hbar p_x p_y - i\hbar p_y p_x + i\hbar p_y p_x + i\hbar p_x p_y = 0 \qquad (ii)$$

and $\left[V(\vec{r}), L_z\right] = \left[V(\vec{r}), -i\hbar\dfrac{\partial}{\partial\phi}\right]$

$$= -i\hbar V(\vec{r})\frac{\partial}{\partial\phi} - (-i\hbar)V(\vec{r})\frac{\partial}{\partial\phi} + i\hbar\frac{\partial V(\vec{r})}{\partial\phi}$$

$$= i\hbar\frac{\partial V(\vec{r})}{\partial\phi} \qquad (iii)$$

From (*i*), (*ii*) and (*iii*) $\qquad \left[H, L_z\right] = i\hbar\dfrac{\partial V(\vec{r})}{\partial\phi}$

Now, $$\left[H, L^2\right] = \frac{1}{2m}\left[p^2, L^2\right] + \left[V(\vec{r}), L^2\right] \qquad (iv)$$

But p^2 commutes with L_z. Similarly we can show p^2 commutes with L_x and L_y. Thus p^2 will commute with L^2. Hence from (*iv*), we get

$$\left[H, L^2\right] = \left[V(\vec{r}), L^2\right]$$

For central field, V will be a function of r only, independent of θ and ϕ. Then $\dfrac{\partial V(\vec{r})}{\partial\phi} = 0$. Hence L_z commutes with H. Similarly L_x and L_y also commute with H. Hence $\vec{L}$ will commute with the Hamiltonian or it will be a constant of motion for central field.

11.4 Eigenvalues and Eigenfunctions for L²

The eigenvalue equation for L^2 can be written as:

$$L^2\psi = \lambda\psi \qquad (27)$$

where ψ is the eigenfunction of L^2 with an eigenvalue λ. We know that

$$L^2 = -\hbar^2\left[\frac{1}{\sin\theta}\frac{\partial}{\partial\theta}\left(\sin\theta\frac{\partial}{\partial\theta}\right) + \frac{1}{\sin^2\theta}\frac{\partial^2}{\partial\phi^2}\right]$$

so, we have

$$\frac{1}{\sin\theta}\frac{\partial}{\partial\theta}\left(\sin\theta\frac{\partial\psi}{\partial\theta}\right) + \frac{1}{\sin^2\theta}\frac{\partial^2\psi}{\partial\phi^2} = -\frac{\lambda}{\hbar^2}\psi \qquad (28)$$

This differential equation may be solved by the method of separation of variables. Let

$$\psi = \Theta(\theta)\,\Phi(\phi) \qquad (29)$$

Substituting Eqn. (29) in Eqn. (28) and dividing throughout by ψ, we get

$$\frac{1}{\Theta}\frac{1}{\sin\theta}\frac{\partial}{\partial\theta}\left(\sin\theta\frac{\partial\Theta}{\partial\theta}\right) + \frac{1}{\Phi}\frac{1}{\sin^2\theta}\frac{\partial^2\Phi}{\partial\phi^2} = -\frac{\lambda}{\hbar^2}$$

or

$$-\frac{1}{\Phi}\frac{d^2\Phi}{d\phi^2} = \frac{\sin^2\theta}{\Theta}\left\{\frac{1}{\sin\theta}\frac{d}{d\theta}\left(\sin\theta\frac{d\Theta}{d\theta}\right) + \frac{\lambda\Theta}{\hbar^2}\right\} \tag{30}$$

The l.h.s. of this equation is independent of θ and the r.h.s. is independent of φ. Thus the derivative of the l.h.s. w.r.t. θ vanishes and hence l.h.s. is equal to a constant independent of θ. Similarly, the r.h.s. is a constant independent of φ. Since the two sides are equal, we denote the (same) constant value by m^2.

$$\frac{d^2\Phi}{d\phi^2} + m^2\Phi = 0 \tag{31}$$

and

$$\frac{1}{\sin\theta}\frac{d}{d\theta}\left(\sin\theta\frac{d\Theta}{d\theta}\right) - \frac{m^2\Theta}{\sin^2\theta} + \frac{\lambda}{\hbar^2}\Theta = 0 \tag{32}$$

Solution of Eqn. (31) is

$$\Phi(\phi) = c_1 e^{\pm im\phi} \tag{33}$$

where c_1 is the constant of integration.

Now, physically meaningful wavefunctions are those which are finite, single valued and whose derivatives are continuous. Since values of φ differing by integral multiples of 2π correspond to the same point in space, so, solutions (Eqn. 33) will satisfy the condition of single valuedness only if

$$e^{\pm im\phi} = e^{\pm im(\phi+2\pi)}$$

i.e.

$$e^{2\pi im} = 1 \Rightarrow m = 0, \pm 1, \pm 2, \ldots$$

Thus m will be an integer, and the physical solutions of Eqn. (31) are

$$\Phi(\phi) = c_1 e^{im\phi}, \ m = 0, \pm 1, \pm 2, \ldots \tag{34i}$$

(Note: Since both signs of m are included, it is not necessary to mention $e^{-im\phi}$ separately.)

To evaluate c_1:

$$\int_0^{2\pi} \Phi*\Phi \, d\phi = 1$$

or

$$|c_1|^2 \int_0^{2\pi} d\phi = 1 \Rightarrow c_1 = \frac{1}{\sqrt{2\pi}}$$

Hence

$$\Phi(\phi) = \frac{1}{\sqrt{2\pi}}e^{im\phi}, \ m = 0, \pm 1, \pm 2, \ldots \tag{34ii}$$

To solve Eqn. (32), we put $\mu = \cos\theta$ in it, we get:

$$\frac{d}{d\mu}\left\{(1-\mu^2)\frac{d\Theta}{d\mu}\right\} + \left(\frac{\lambda}{\hbar^2} - \frac{m^2}{1-\mu^2}\right)\Theta = 0$$

or

$$(1-\mu^2)\frac{d^2\Theta}{d\mu^2} - 2\mu\frac{d\Theta}{d\mu} + \left(\frac{\lambda}{\hbar^2} - \frac{m^2}{1-\mu^2}\right)\Theta = 0 \tag{35}$$

For the special case $m = 0$, Eqn. (35) assumes the form:

$$(1 - \mu^2)\frac{d^2\Theta}{d\mu^2} - 2\mu\frac{d\Theta}{d\mu} + \frac{\lambda}{\hbar^2}\Theta = 0 \tag{36}$$

To find the solution of Eqn. (36), we write a power series for Θ:

$$\Theta = \sum_{k=0}^{\infty} c_k \mu^k \tag{37}$$

Substituting it in Eqn. (36) and comparing the coefficients of μ^k, we get the following recurrence relation for the coefficients:

$$c_{k+2} = \frac{k(k+1) - \lambda/\hbar^2}{(k+1)(k+2)} c_k \tag{38}$$

For very large values of k, we have from Eqn. (38),

$$\frac{c_{k+2}}{c_k} \to \frac{k}{k+2}$$

This ratio permits the conclusion that for large k, Θ behaves like the series $\sum_{0}^{\infty} \mu^k / k$, beçause the ratio of consecutive coefficients in this series is also $k/(k+2)$.

For $\mu = \pm 1$ i.e. $\theta = 0$, π, the series $\sum_{0}^{\infty} \mu^k / k$ diverges. Thus, Θ and hence ψ is not finite and cannot be admitted as an eigenfunction of L^2. The only way to avoid this situation is to choose λ in such a way that the power series for Θ is cut off after some finite number of terms (making Θ a polynomial). Suppose that the power series terminates at some finite value of $k = l$ (say), where l is a positive integer (and all higher powers vanish). According to Eqn. (38), it will be so if λ has the value given by

$$\lambda = \hbar^2 \, l(l+1) \tag{39}$$

These are the eigenvalues of L^2 for $l = 0, 1, 2, \ldots$ (l is known as the *orbital angular momentum quantum number*).

Using Eqn. (39), we can write Eqn. (36) as

$$(1 - \mu^2)\frac{d^2\Theta}{d\mu^2} - 2\mu\frac{d\Theta}{d\mu} + l(l+1)\Theta = 0 \tag{40}$$

This is the well known Legendre's equation having its solutions the *Legendre's polynomial* of degree l:

$$\Theta = P_l(\mu) = \frac{1}{l! \; 2^l}\frac{d^l(\mu^2 - 1)^l}{d\mu^l} \tag{41}$$

Thus we have solved Eqn. (35) for the special case $m = 0$. Now, we can find the solution for $m \neq 0$ by using the solution for $m = 0$. For this, we differentiate equation (36) m times (using Lebniz's rule) and write $\dfrac{d^m\Theta}{d\mu^m} = v$, we obtain

$$(1 - \mu^2)\frac{d^2 v}{d\mu^2} - 2\mu(m+1)\frac{dv}{d\mu} + (l - m)(l + m + 1)v = 0 \tag{42}$$

Since $P_l(\mu)$ is a solution of Eqn. (36); Eqn. (42) is satisfied by

$$v = \frac{d^m P_l(\mu)}{d\mu^m} \tag{43}$$

If we substitute $w = v(1-\mu^2)^{m/2}$ in Eqn. (42), we get

$$(1-\mu^2)\frac{d^2 w}{d\mu^2} - 2\mu\frac{dw}{d\mu} + \left(l(l+1) - \frac{m^2}{1-\mu^2}\right)w = 0 \tag{44}$$

This Eqn. is identical to Eqn. (35) for $\lambda = \hbar^2 \, l(l+1)$ and is known as the *Associated Legendre's equation*. From it, we see that solution of Eqn. (35) is given by

$$\Theta(\theta) = w = P_l^m(\mu) = (1-\mu^2)^{m/2}\frac{d^m P_l(\mu)}{d\mu^m} \tag{45}$$

Here $w = P_l^m(\mu)$ are called the associated Legendre's polynomials and these are defined for positive integers $m \le l$. These are the only physically acceptable solutions of Eqn. (35). For normalization:

$$|c|^2 \int_{-1}^{+1} P_l^m(\cos\theta) \, P_l^m(\cos\theta) \; d(\cos\theta) = 1 \tag{46}$$

or $$|c|^2 \frac{2}{2l+1}\cdot\frac{(l+m)!}{(l-m)!} = 1 \quad \text{(using the orthogonality relation for } P_l^m\text{)}$$

$$\Rightarrow c = \sqrt{\frac{(2l+1)}{2}\cdot\frac{(l-m)!}{(l+m)!}}$$

Therefore, normalized Θ are given by:

$$\Theta(\theta) = \sqrt{\frac{(2l+1)}{2}\cdot\frac{(l-m)!}{(l+m)!}} \; P_l^m(\cos\theta) \tag{47}$$

and the normalized eigenfunctions of L^2 are:

$$\psi_{lm} = Y_{l,m}(\theta,\phi) = \Theta(\theta)\,\Phi(\phi) = \Theta_{lm}(\theta).\frac{e^{im\phi}}{\sqrt{2\pi}}$$

or $$\psi_{lm} = \epsilon\left\{\frac{(2l+1)}{4\pi}\cdot\frac{(l-m)!}{(l+m)!}\right\}^{1/2} P_l^m(\cos\theta)\,.e^{im\phi} \tag{48}$$

$Y_{l,m}$ are spherical harmonics and ϵ is a phase factor ($= (-1)^{l+m}$). $\epsilon = (-1)^m$ for $m \ge 0$ and for negative values of m, the spherical harmonics are determined from the condition:

$$Y_{l,-m}(\theta,\phi) = (-1)^m Y_{l,m}^*(\theta,\phi)$$

The first few spherical harmonics are

$$Y_{00} = \left(\frac{1}{4\pi}\right)^{1/2}; \qquad\qquad Y_{10} = \left(\frac{3}{4\pi}\right)^{1/2}\cos\theta$$

$$Y_{11} = -\left(\frac{3}{8\pi}\right)^{1/2}\sin\theta \; e^{i\phi}; \qquad\qquad Y_{1,-1} = \left(\frac{3}{8\pi}\right)^{1/2}\sin\theta \, . \, e^{-i\phi}$$

$$Y_{20} = \left(\frac{5}{16\pi}\right)^{1/2}(3\cos^2\theta - 1); \qquad Y_{21} = -\left(\frac{15}{8\pi}\right)^{1/2}\sin\theta.\cos\theta.\,e^{i\phi}$$

$$Y_{2,-1} = \left(\frac{15}{8\pi}\right)^{1/2}\sin\theta.\cos\theta.\,e^{-i\phi}; \qquad Y_{22} = \left(\frac{15}{32\pi}\right)^{1/2}\sin^2\theta.\,e^{2i\phi}$$

$$Y_{2,-2} = \left(\frac{15}{32\pi}\right)^{1/2}\sin^2\theta.\,e^{-2i\phi} \tag{49}$$

The dependence of the spherical harmonics on θ is shown in the polar diagrams given in Fig. (11.1). The functions Y_{lm} form an orthonormal set:

$$\int Y_{lm}^* Y_{l'm'} d\Omega = \int_0^{2\pi}\int_0^{\pi} Y_{l,m}^*(\theta,\phi) Y_{l'm'}(\theta,\phi)\sin\theta\, d\theta\, d\phi = \delta_{ll'}\delta_{mm'}$$

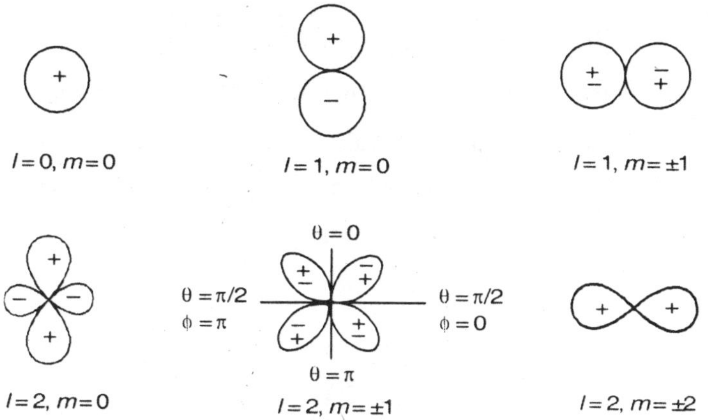

$l=0, m=0$ $\qquad\qquad$ $l=1, m=0$ $\qquad\qquad$ $l=1, m=\pm1$

$l=2, m=0$ $\qquad\qquad$ $l=2, m=\pm1$ $\qquad\qquad$ $l=2, m=\pm2$

Fig. 11.1: Polar diagrams for the $Y_{lm}(\theta,\phi)$ for points in the x-z plane. The distance from the centre to the curve (at an angle θ to the vertical) gives the magnitude of $Y_{lm}(\theta,\phi)$

11.5 Eigenvalues and Eigenfunctions for the Components of Angular Momentum $\vec{L}$ (Spherical Harmonics)

We have seen that L^2 commutes with all the three components of $\vec{L}$. Therefore it is possible to construct the simultaneous eigenfunctions for L^2 and a component of $\vec{L}$.

We note $\qquad\qquad L_z = -i\hbar\dfrac{\partial}{\partial\phi}$ and

$$L_z Y_{l,m}(\theta,\phi) = -i\hbar\frac{\partial}{\partial\phi}Y_{l,m}(\theta,\phi) = -i\hbar\frac{\partial}{\partial\phi}\left[\Theta_l(\theta)\frac{e^{im\phi}}{\sqrt{2\pi}}\right]$$

or $\qquad\qquad L_z Y_{l,m}(\theta,\phi) = \hbar m Y_{l,m}(\theta,\phi)$

so $Y_{l,m}(\theta,\phi)$ are also eigenfunctions of L_z with eigenvalues $m\hbar$ ($m = 0,\pm1,\pm2, \ldots$). In fact, for the hydrogen atom

$$H\psi_{nlm} = E_n\psi_{nlm}\,;\quad \psi_{nlm} = R_{nl}(r)Y_{lm}(\theta,\phi) \tag{50}$$

$$L^2 \psi_{nlm} = l(l+1)\hbar^2 \psi_{nlm} \; ; \; (l = 0, 1, ..., n-1)$$

$$L_z \psi_{nlm} = m\hbar \psi_{nlm} ; \; (m = -l, -l+1, ..., l-1, l) \tag{51}$$

Since the complete set (ψ_{nlm}) are eigenfunctions of H, L^2, L_z, these three operators commute among themselves. Both $\bar{L}$ and L_z are constants of motion for the hydrogen atom ($\because$ H commutes with L^2 and L_z). As L^2 and L_z commute, their simultaneous measurement is possible. The angular momentum $|\bar{L}|$ has the exact value $\hbar\sqrt{l(l+1)}$ and L_z has the exact value $\hbar m$ ($m = -l, ..., 0, ..., +l$). But we do not know anything about L_x or L_y precisely (unlike classical mechanics where all the three components could be specified simultaneously).

However, the component

$$\sqrt{L_x^2 + L_y^2} = \sqrt{L^2 - L_z^2} \equiv \sqrt{\hbar^2(l^2 + l) - \hbar^2 m^2}$$

lies in the *xy* plane, and there is complete uncertainty about the angle ϕ. Also if $m = l$; $\sqrt{L_x^2 + L_y^2}$ has the minimum value $\hbar\sqrt{l}$ and L_z has its corresponding maximum value $\hbar l$ and due to the uncertainty principal the state is unable to attain the full value $\hbar\sqrt{l(l+1)}$ in the *z*-direction.

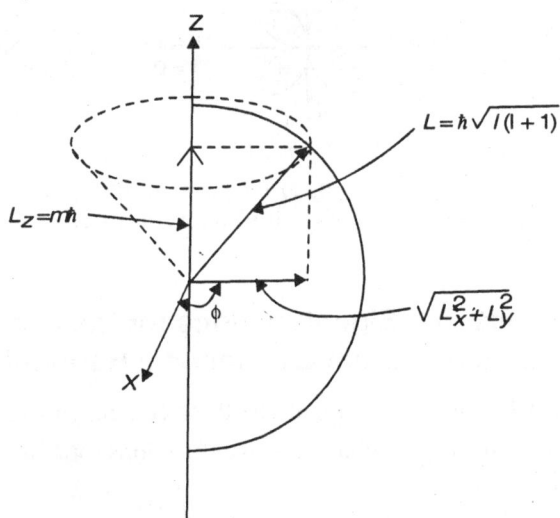

Fig. 11.2: Angular momentum vectors

The oeprators L_x, L_y, L_z do not commute among themselves.

Example 11.2: Does there exist any state ϕ which is an eigenstate of both L_x and L_y.

Solution:
$$L_x \phi = m_x \phi; \; L_y \phi = m_y \phi \tag{i}$$
$$L_x L_y \phi = L_x m_y \phi = m_x m_y \phi = L_y L_x \phi$$
$$(\because \; m_x\phi = L_x \phi \text{ and } m_x \text{ and } m_y \text{ are just numbers})$$

hence
$$L_z \phi = \frac{1}{i\hbar}\left(L_x L_y - L_y L_x\right)\,\phi = 0$$

i.e. if equation (*i*) exists, ϕ belongs to the zero eigenvalue of L_z ($m_z = 0$).

Again, starting with $L_y\,\phi = m_y\,\phi$ and $L_z\,\phi = m_z\,\phi$, we get $m_x = 0$. And with $L_x\,\phi = m_x\,\phi$ and $L_z\,\phi = m_z\,\phi$, we get $m_y = 0$.

So, if more than one (i.e. two) component of angular momentum is to have a definite value, it must satisfy
$$L_x\,\phi = L_y\,\phi = L_z\,\phi = 0$$
i.e. such a state does not exist.

11.6 Addition of Orbital Angular Momenta (Clebsch Gordan Coefficients)

In classical mechanics we combine angular momenta by vector addition. We wish to know how to combine these in quantum mechanics. Consider two angular momenta $\vec{L}_1$ and $\vec{L}_2$ which are independent, i.e.

$$\left[L_{1x}, L_{2x}\right] = \left[L_{1y}, L_{2y}\right] = \ldots = \left[L_{1z}, L_{2z}\right] = 0 \tag{52}$$

The angular momenta under consideration may be of two electrons in different atoms or in the same atom if the mutual interaction of such electrons is disregarded. We write

$$\vec{L} = \vec{L}_1 + \vec{L}_2 \tag{53}$$

so that
$$\begin{aligned}
\left[L_y, L_z\right] &= \left[L_{1y} + L_{2y}, L_{1z} + L_{2z}\right]\\
&= \left[L_{1y}, L_{1z}\right] + \left[L_{2y}, L_{2z}\right]\\
&= i\hbar L_{1x} + i\hbar L_{2x} = i\hbar L_x \text{ etc.}
\end{aligned} \tag{54}$$

which establishes that $\vec{L}$ is an angular momentum. The quantum number associated with this is (say) *l*.

Leaving out the uninterested radial part of the wave function, we can write the wave function of the combined system as

$$Y_{l_1}^{(m_1)}(1)Y_{l_2}^{(m_2)}(2), \quad \{m_1 = -l_1,\ldots,0,\ldots,l_1;\ m_2 = -l_2,\ldots,0,\ldots,l_2\} \tag{55}$$

for which the total angular momentum L^2 assumes each of the values $l(l+1)\hbar^2$ with

$$l = l_1 + l_2, l_1 + l_2 - 1, \ldots, l_1 - l_2 + 1, l_1 - l_2. \tag{56}$$

because classically $\vec{L}_1$ and $\vec{L}_2$ have a resultant $\vec{L}$ whose magnitude lies in the range (see Fig. 11.3),

$$\left|L_1 - L_2\right| \le L \le \left|L_1 + L_2\right| \text{ say } L_1 > L_2 \tag{57}$$

Each eigenvalue $\hbar^2 l(l+1)$ is $(2l+1)$ fold degenerate, each eigenvalue of the independent electron being $(2l_i + 1)$, $i = 1, 2$, fold degenerate.

Then there are $(2l+1)$ eigenfunctions $Y_l^{(m)}(1,2)$ of L_z with m varying from $-l$ to $+l$. Actually,

$$\left.\begin{aligned}
L^2 Y_l^{(m)}(1,2) &= \hbar^2 l(l+1)\,Y_l^{(m)}(1,2)\\
L_z Y_l^{(m)}(1,2) &= \hbar^2 m Y_l^{(m)}(1,2)
\end{aligned}\right\} = (m = -l, \ldots, l) \tag{58}$$

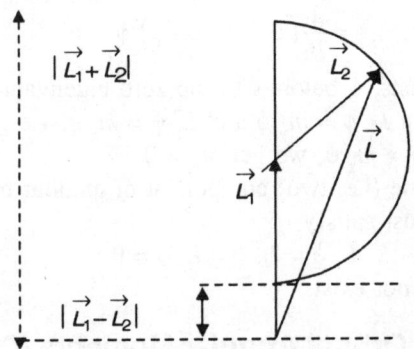

Fig. 11.3: Addition of angular momenta

Since all functions $Y_l^{(m)}(1,2)$ are linear combinations of the product functions $Y_{l_1}^{(m_1)}(1)$, $Y_{l_2}^{(m_2)}(2)$, we can write

$$Y_l^{(m)}(1,2) = \sum_{m_1=-l_1}^{l_1} \sum_{m_2=-l_2}^{l_2} C_{l_1 m_1 l_2 m_2}^{l,m} Y_{l_1}^{(m_1)}(1) Y_{l_2}^{(m_2)}(2) \tag{59}$$

Since $L_z = L_{1z} + L_{2z}$, it follows that

$$L_z Y(1,2) = \hbar m Y(1,2) = \hbar(m_1 + m_2) Y(1) Y(2) \tag{60}$$

and so only those terms occur in Eqn. (59) for which

$$m = m_1 + m_2 \tag{61}$$

We can now simplify Eqn. (59) to

$$Y_l^{(m)}(1,2) = \sum_{\substack{m_2+m_2 \\ =m}} C_{l_1 m_1 l_2 m_2}^{l,m} Y_{l_1}^{(m_1)}(1) \, Y_{l_2}^{(m_2)}(2) \tag{62}$$

The coefficients $C_{l_1 m_1 l_2 m_2}^{l,m}$, called the *Clebsch-Gordan coefficients*, are determined by the condition that $Y_l^{(m)}(1,2)$ should be a normed eigenfunction of L^2 and L_z satisfying Eqn. (58). The calculation of these coefficients using the properties of angular momentum is tedius, but in practice one can obtain them from the tables (Theory of Atomic Spectra by E.U. Condon, G.H. Shortley, Cambridge U.P., 1951, pp. 76-79). In Table 11.1, we give the coefficients for $l_2 = 1$ i.e. for $C_{l_1 m_1 l_2 m_2}^{l,m}$.

As an example we write the eigenfunctions $Y_l^{(m)}$ of two particles, each of which is in the p state ($l_1 = 1$, $l_2 = 1$) (possible values of l are 0, 1, 2). Then, using Eqn. (62) and table (11.1) we have, for $l = 0$, and $m = 0$,

$$Y_l^{(m)}(1,2) = \frac{1}{\sqrt{3}} Y_{l_1=1}^{(1)}(1) Y_1^{(m_2=-1)}(2) + \frac{1}{\sqrt{3}} Y_1^{(-1)}(1) Y_1^{(1)}(2)$$

$$-\frac{1}{\sqrt{3}} Y_1^{(0)}(1) Y_1^{(0)}(2) \tag{63}$$

Similarly, we can obtain eigenfunctions corresponding to $l = 1$ and $l = 2$.

We have seen in article (11.5) that the eigen states of orbital angular momentum have a coordinate representation having only spherical symmetry (no radial

Table 11.1: The C.G. coefficients

m_1	$m_2 = 1$	$m_2 = 0$	$m_2 = -1$.
$= l_1+1$	$\left[\dfrac{(l_1+m)(l_1+m+1)}{(2l_1+1)(2l_1+2)}\right]^{1/2}$	$\left[\dfrac{(l_1-m+1)(l_1+m+1)}{(2l_1+1)(l_1+1)}\right]^{1/2}$	$\left[\dfrac{(l_1-m)(l_1-m+1)}{(2l_1+1)(2l_1+2)}\right]^{1/2}$
$= l_1$	$-\left[\dfrac{(l_1+m)(l_1-m+1)}{2l_1(l_1+1)}\right]^{1/2}$	$\dfrac{m}{[l_1(l_1+1)]^{1/2}}$	$\left[\dfrac{(l_1-m)(l_1+m+1)}{2l_1(l_1+1)}\right]^{1/2}$
$= l_1-1$	$\left[\dfrac{(l_1-m)(l_1-m+1)}{2l_1(2l_1+1)}\right]^{1/2}$	$-\left[\dfrac{(l_1-m)(l_1+m)}{l_1(2l_1+1)}\right]^{1/2}$	$\left[\dfrac{(l_1+m+1)(l_1+m)}{2l_1(2l_1+1)}\right]^{1/2}$

dependence) and so they are represented by $|lm>$. The coordinate space wave functions for a single particle are denoted by $<\theta\phi|lm>$ and are the spherical harmonics. The C.G. coefficients have been discussed in detail in chapter 13.

11.7 Ladder Operators

We define

$$\left(L_+\right) = L_x + iL_y \;;\; \left(L_-\right) = L_x - iL_y \tag{64}$$

$$L_x = +i\hbar\left(\sin\phi\,\frac{\partial}{\partial\theta} + \cot\theta\cos\phi\,\frac{\partial}{\partial\phi}\right)$$

$$L_y = i\hbar\left(-\cos\phi\,\frac{\partial}{\partial\theta} + \cot\theta\sin\phi\,\frac{\partial}{\partial\phi}\right)$$

$$L_x + iL_y = i\hbar\sin\phi\,\frac{\partial}{\partial\theta} + i\hbar\cot\theta\cos\phi\,\frac{\partial}{\partial\phi} + \hbar\cos\phi\,\frac{\partial}{\partial\theta} - \hbar\cot\theta\sin\phi\,\frac{\partial}{\partial\phi}$$

$$= \hbar e^{i\phi}\,\frac{\partial}{\partial\theta} + i\hbar\cot\theta(\cos\phi + i\sin\phi)\,\frac{\partial}{\partial\phi}$$

or

$$L_+ = L_x + iL_y = \hbar e^{i\phi}\left(\frac{\partial}{\partial\theta} + i\cot\theta\,\frac{\partial}{\partial\phi}\right)$$

Similarly,

$$L_- = L_x - iL_y = \hbar e^{-i\phi}\left(-\frac{\partial}{\partial\theta} + i\cot\theta\,\frac{\partial}{\partial\phi}\right)$$

Using the properties of spherical harmonics,

$$Y_l^{(m)}(\theta,\phi) = \left\{\frac{(2l+1)}{4\pi}\cdot\frac{(l-m)!}{(l+m)!}\right\}^{1/2}(-1)^m P_l^m(\cos\phi)\,.e^{im\phi}$$

$$\left[Y_l^{(m)}(\theta,\phi)\right]^* = (-1)^m Y_l^{-m}(\theta,\phi)$$

one can show that

$$L_+ Y_{lm}(\theta,\phi) = \{l(l+1) - m(m+1)\}^{1/2}\,\hbar Y_{l,m+1}(\theta,\phi)$$

$$L_- Y_{lm}(\theta,\phi) = \{l(l+1) - m(m-1)\}^{1/2}\,\hbar Y_{l,m-1}(\theta,\phi)$$

$$\left(L_\pm\right)^\dagger = L_\mp$$

$$\left[L_z, L_\pm\right] = i\hbar L_y \pm \hbar L_x = \pm \hbar L_\pm \tag{65}$$

For example,

$$\left[L_z, L_+\right] = \left[L_z, L_x + iL_y\right] = \left[L_z, L_x\right] + \left[L_z, iL_y\right]$$

$$= i\hbar L_y + i\left[L_z, L_y\right] = i\hbar L_y - i^2 \hbar L_x = i\hbar L_y + \hbar L_x$$

$$= \hbar\left(L_x + iL_y\right) = \hbar L_+$$

$$\left[L_+, L_-\right] = -2i\left[L_x, L_y\right] = 2\hbar L_z \tag{66}$$

$$\left[L^2, L_\pm\right] = 0 \tag{67}$$

$$L_+ L_- = \left(L_x + iL_y\right)\left(L_x - iL_y\right)$$

$$= L_x\left(L_x - iL_y\right) + iL_y\left(L_x - iL_y\right)$$

$$= L_x^2 - iL_x L_y + iL_y L_x - i^2 L_y^2$$

$$= L_x^2 + L_y^2 - i\left[L_x, L_y\right] = L_x^2 + L_y^2 + \hbar L_z$$

so,

$$L^2 = L_x^2 + L_y^2 + L_z^2 = L_+ L_- - \hbar L_z + L_z^2$$

$$= L_- L_+ + \hbar L_z + L_z^2 \tag{68}$$

Now, let ψ_{lz} be an eigenfunction of L_z

$$\therefore \qquad L_z \psi_{lz} = l_z \psi_{lz}$$

With Eqn. (65) we find

$$L_z L_\pm - L_\pm L_z = \pm \hbar L_z$$

$$\therefore \qquad L_z L_\pm \psi_{lz} = L_\pm L_z \psi_{lz} \pm \hbar L_\pm L_z$$

and thus

$$L_z \left(L_\pm \psi_{lz}\right) = (l_z \pm \hbar) L_\pm \psi_{lz}$$

Thus $L_{+\atop(-)}$ raises (lowers) the eigenvlaue l_z by $\hbar$.

$$L^2 \psi_{lm} = \hbar^2 l(l+1)\psi_{lm} \quad l \geq 0$$

$$L_z \psi_{lm} = \hbar m \psi_{lm} \tag{69}$$

From Eqn. (67) it further follows that

$$L^2 \left(L_\pm \psi_{lm}\right) = L_\pm L^2 \psi_{lm} = \hbar^2 l(l+1)\left(L_\pm \psi_{lm}\right)$$

$\left(L_\pm \psi_{lm}\right)$ is thus an eigenfunction of L^2 with the same eigenvalue as that of ψ_{lm}.

Further information can be obtained from the normalization using Eqns. (64), (68) and (69).

$$\left(L_\pm \psi_{lm}, L_\pm \psi_{lm}\right) = \left(\psi_{lm}, L_\mp L_\pm \psi_{lm}\right)$$

$$= \left(\psi_{lm}, \left(L^2 - L_z^2 \mp \hbar L_z\right)\psi_{lm}\right)$$

$$= \hbar^2 \left(l(l+1) - m^2 \mp m\right)$$

where we assume that ψ_{lm} is normalized. Hence, it follows that

$$L_{\pm}\psi_{lm} = \hbar\sqrt{l(l+1) - m(m\pm 1)}\ \psi_{l,m\pm 1} \qquad (70)$$

11.8 Angular Momentum and Infinitesimal Rotations

We shall now see that the angular momentum $\vec{L}$ is related to the rotations of system. In fact each component of $\vec{L}$ is related to an infinitesimal rotation about the corresponding axis.

Let us consider the effect of a small rotation $\delta\phi$ about the z-axis as shown in the Fig. 11.4.

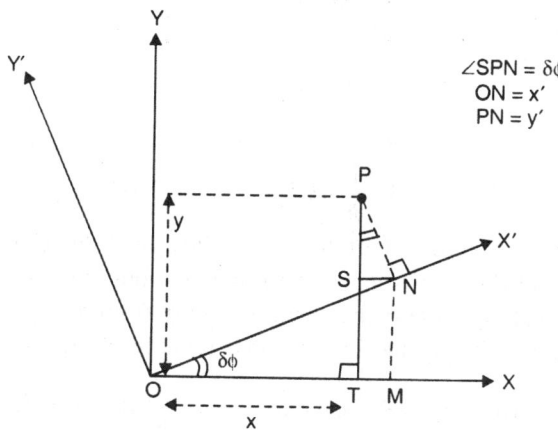

Fig. 11.4: A small rotation by $\delta\phi$ about the z-axis

$$SN = y'\sin(\delta\phi) = TM$$
$$PS = y'\cos(\delta\phi)$$
$$OM = x'\cos(\delta\phi) = OT + TM = x + y'\sin(\delta\phi)$$
$$NM = x'\sin(\delta\phi)$$
$$PT = y = PS + ST = y'\cos(\delta\phi) + x'\sin(\delta\phi)$$

Hence,

$$x = x'\cos\delta\phi - y'\sin\delta\phi \qquad (71)$$

and
$$y = x'\sin\delta\phi + y'\cos\delta\phi \qquad (72)$$

Multiply Eqn. (71) by $\cos\delta\phi$ and (72) by $\sin\delta\phi$ and add, we get

$$x' = x\cos\delta\phi + y\sin\delta\phi \qquad (73)$$

Similarly multiply Eqn. (71) by $-\sin\delta\phi$ and (72) by $\cos\delta\phi$ and add, we get

$$y' = -x\sin\delta\phi + y\cos\delta\phi \qquad (74)$$

If we assume $\delta\phi$ to be infinitesimal, we write

$$x' = x + y.\delta\phi$$
$$y' = y - x.\delta\phi \qquad (75)$$

Thus if $R(\hat{z},\delta\phi)$ represents the rotation operator, then its effect on an arbitrary wave function ψ is given by

$$R(\hat{z},\delta\phi)\ \psi(x,y,z) = \psi(x',y',z')$$

$$= \psi(x + y\delta\phi, y - x\delta\phi, z)$$

$$= \psi(x, y, z) + y.\delta\phi\frac{\partial\psi}{\partial x} - x.\delta\phi\frac{\partial\psi}{\partial y} \tag{76}$$

where we have retained the first term only in the Taylor expansion because $\delta\phi$ is infinitesimal. Thus

$$R(\hat{z}, \delta\phi) \; \psi(x, y, z) = \left[1 + \delta\phi\left(y.\frac{\partial}{\partial x} - x.\frac{\partial}{\partial y}\right)\right] \psi(x, y, z)$$

$$= \left[1 + \frac{\delta\phi}{i\hbar} L_z\right] \psi(x, y, z)$$

$$= \left[1 - \frac{i}{\hbar}\delta\phi L_z\right] \psi(x, y, z) \tag{77}$$

Since $\psi(x, y, z)$ is arbitrary

$$R(\hat{z}, \delta\phi) = 1 - \frac{i}{\hbar} \delta\phi \, L_z \tag{78}$$

Thus operator L_z is generator of the infinitesimal rotation $\delta\phi$ about the $\hat{z}$-axis.

In order to find $R_z(\phi)$ for some finite rotation about the z-axis, we note that the effect of a small increase in the magnitude of ϕ, from ϕ to $\phi + \delta\phi$ is to follow the finite rotation $R_z(\phi)$ by the infinitesimal rotation $R_z(\delta\phi)$ to give

$$R_z(\phi + \delta\phi) = R_z(\delta\phi).R_z(\phi) = \left\{1 - \frac{i}{\hbar} \delta\phi \, L_z\right\}.R_z(\phi) \tag{79}$$

Thus
$$\frac{dR_z}{d\phi} = \lim_{\Delta\phi \to 0} \frac{R_z(\phi + \delta\phi) - R_z(\phi)}{\Delta\phi} = -\frac{i}{\hbar}L_z$$

or
$$dR_z = -\frac{i}{\hbar}L_z.d\phi$$

Integrating, we get

$$R_z(\phi) = \exp\left(-\frac{i}{\hbar}\phi \, L_z\right)$$

or
$$R(\hat{n}, \phi) = \exp\left(-\frac{i}{\hbar}\phi\vec{L}.\hat{n}\right) \tag{80}$$

Thus operator $\hat{R}$ defined by this equation ($\hat{n}$ is a unit vector) acts as a generator of rotation about the respective axis. Now if there exists any operator Q which commutes with R

$$[Q,R] = 0$$

then the operator Q is said to be rotationally invariant. This equation is satisfied if operator Q satisfies the following commutation relations:

$$[Q,L_z] = 0, \; [Q,L_y] = 0 \text{ and } [Q,L_z] = 0$$

As the components of L are hermitian, ϕ is real, therefore $R(\hat{n},\phi)$ is unitary.

11.9 Representation of Orbital Angular Momentum Operators by Matrices

We have seen in the preceding article that operators L_x, L_y and L_z are generators of infinitesimal rotation about the respective axes.

There are two kinds of transformations. The first kind are unitary transformations from one representaion to another that rotate axes in Hilbert space without change in the state vector. In the second kind involve dynamical variables (of the sort $e^{-iHt/\hbar}$) that produce generalised rotations of the state vectors in Hilbert space without change in the axes. Transformations involving angular momentum are of the second kind in which a state $\psi_\alpha(r)$, (represented by the ket $|\alpha>$) changes to the state $\psi_{\alpha'}(r)$ (represented by ket $|\alpha'>$), by a linear operator R which is so defined that any vector $\vec{r}$ is rotated into the vector $\vec{r}_R \equiv R\vec{r}$. For 3-dimensional Euclidean rotation, if the rotation is of infinitesimal amount $\delta\phi$ around an arbitrary axis, $\vec{r}_R$ and $\vec{r}$ are related by

$$\vec{r}_R \approx \vec{r} + \delta\vec{\phi} \times \vec{r} \tag{81}$$

(ϕ is of infinitesimal length and retaining only quantities of the first order in $\delta\phi$). The norm of the ket $|\alpha>$ is not affected by the displacement.

We use the column vector representation for $\vec{r}_R$ and $\vec{r}$ namely

$$\vec{r}_R = \begin{pmatrix} x_R \\ y_R \\ z_R \end{pmatrix} \equiv |r_R\rangle \tag{82i}$$

and

$$|r\rangle = \begin{pmatrix} x \\ y \\ z \end{pmatrix} \tag{82ii}$$

For $\delta\phi$, we use $\delta\phi = (\delta\phi_x, \delta\phi_y, \delta\phi_z)$ where $\delta\phi_x$, $\delta\phi_y$, $\delta\phi_z$ are the infinitesimal rotation around x, y, z axes respectively. Since

$$\delta\vec{\phi} \times \vec{r} = \begin{pmatrix} \delta\phi_y z - \delta\phi_z y \\ \delta\phi_z x - \delta\phi_x z \\ \delta\phi_x y - \delta\phi_y x \end{pmatrix} = \begin{pmatrix} 0 & -\delta\phi_z & \delta\phi_y \\ \delta\phi_z & 0 & -\delta\phi_x \\ -\delta\phi_y & \delta\phi_x & 0 \end{pmatrix} \begin{pmatrix} x \\ y \\ z \end{pmatrix} \tag{83}$$

and

$$\begin{pmatrix} 0 & -\delta\phi_z & \delta\phi_y \\ \delta\phi_z & 0 & -\delta\phi_x \\ -\delta\phi_y & \delta\phi_x & 0 \end{pmatrix} = \delta\phi_x \begin{pmatrix} 0 & 0 & 0 \\ 0 & 0 & -1 \\ 0 & 1 & 0 \end{pmatrix} + \delta\phi_y \begin{pmatrix} 0 & 0 & 1 \\ 0 & 0 & 0 \\ -1 & 0 & 0 \end{pmatrix}$$

$$+\delta\phi_z \begin{pmatrix} 0 & -1 & 0 \\ 1 & 0 & 0 \\ 0 & 0 & 0 \end{pmatrix} \tag{84i}$$

So, we can write from Eqn. (83)

$$\delta\vec{\phi} \times \vec{r} = \left(i\delta\vec{\phi} \cdot \frac{\vec{L}}{\hbar} \right) |r\rangle \tag{84ii}$$

with

$$L_x = -i\hbar \begin{pmatrix} 0 & 0 & 0 \\ 0 & 0 & -1 \\ 0 & 1 & 0 \end{pmatrix} \tag{85i}$$

$$L_y = -i\hbar \begin{pmatrix} 0 & 0 & 1 \\ 0 & 0 & 0 \\ -1 & 0 & 0 \end{pmatrix} \tag{85ii}$$

and

$$L_z = -i\hbar \begin{pmatrix} 0 & -1 & 0 \\ 1 & 0 & 0 \\ 0 & 0 & 0 \end{pmatrix} \tag{85iii}$$

One can check that L_x, L_y and L_z given by equations (85) are hermitian. Further

$$\left[L_i, L_j \right] = i\hbar \,\epsilon_{ijk} \, L_k \tag{86}$$

i, j, k run for 1, 2, 3 representing $L_1 = L_x$, $L_2 = L_y$, $L_3 = L_z$ respectively. So, the matrices for L_x, L_y and L_z satisfy the commutation relation. Hence

$$|r_R\rangle = \left(1 + i\delta\vec{\phi}.\vec{L}/\hbar\right) |r\rangle \tag{87}$$

Example 11.3: Using vector methods, prove that

$$L^2\psi = -\hbar^2 r^2 \nabla^2 \psi + \hbar^2 \frac{\partial}{\partial r}\left(r^2 \frac{\partial \psi}{\partial r} \right)$$

Solution:
$$L = \vec{r} \times \vec{p} = -i\hbar\vec{r} \times \vec{\nabla}$$

$$\therefore \qquad L^2\psi = \left(-i\hbar\vec{r} \times \nabla\right)^2 \psi$$

$$= -\hbar^2 \left(\vec{r} \times \nabla\right) \cdot \left(\vec{r} \times \nabla\psi\right)$$

$$= -\hbar^2 \vec{r} \cdot \left[\nabla \times \left(\vec{r} \times \nabla\psi\right)\right]$$

(by interchanging the dot and cross in the triple product)

Now, we may use the vector identity

$$\nabla \times \left(\vec{A} \times \vec{B}\right) = \left(\vec{B} \cdot \nabla\right)A - \vec{B}\left(\nabla.\vec{A}\right) - \left(\vec{A}.\nabla\right)\vec{B} + \vec{A}\left(\nabla.\vec{B}\right)$$

so that
$$\nabla \times \left(\vec{r} \times \nabla\psi\right) = \left(\nabla\psi \cdot \nabla\right)\vec{r} - \nabla\psi\left(\nabla.\vec{r}\right) - \left(\vec{r}.\nabla\right)\nabla\psi + \vec{r}\left(\nabla.\nabla\psi\right)$$

Thus,
$$L^2\psi = -\hbar^2\vec{r} \cdot \left[\left(\nabla\psi \cdot \nabla\right)\vec{r} - \nabla\psi\left(\nabla.\vec{r}\right) - \left(\vec{r}.\nabla\right)\nabla\psi + \vec{r}\left(\nabla^2\psi\right) \right]$$

$$= -\hbar^2 \left[r\frac{\partial\psi}{\partial r} - r\frac{\partial\psi}{\partial r}3 - \vec{r} \cdot \left(r\frac{\partial}{\partial r}\nabla\psi\right) + r^2\nabla^2\psi \right]$$

$$= -\hbar^2 \left[-2r\frac{\partial\psi}{\partial r} - r^2\frac{\partial^2\psi}{\partial r^2} + r^2\nabla^2\psi \right]$$

$$= -\hbar^2 \left[r^2\nabla^2\psi - \frac{\partial}{\partial r}\left(r^2\frac{\partial\psi}{\partial r}\right) \right]$$

$$= -\hbar^2 r^2 \left[\nabla^2\psi - \frac{1}{r^2}\frac{\partial}{\partial r}\left(r^2\frac{\partial\psi}{\partial r}\right) \right]$$

Example 11.4: Using the relation

$$p_r = -i\hbar \left[\frac{\partial}{\partial r} + \frac{1}{r} \right] \text{ show that } \frac{p^2}{2m} = \frac{p_r^2}{2m} + \frac{L^2}{2mr^2}.$$

Solution: From example (11.3), we have

$$L^2 = r^2 p^2 + \hbar^2 \frac{\partial}{\partial r} \left(r^2 \frac{\partial}{\partial r} \right) \qquad (i)$$

Now,

$$\frac{p_r^2}{2m} \psi = \frac{(-i\hbar)^2}{2m} \left(\frac{\partial}{\partial r} + \frac{1}{r} \right) \left(\frac{\partial}{\partial r} + \frac{1}{r} \right) \psi$$

$$= -\frac{\hbar^2}{2m} \left(\frac{\partial^2 \psi}{\partial r^2} - \frac{\psi}{r^2} + \frac{2}{r} \frac{\partial \psi}{\partial r} + \frac{1}{r^2} \psi \right)$$

$$= -\frac{\hbar^2}{2m} \left[\frac{1}{r^2} \frac{\partial}{\partial r} \left(r^2 \frac{\partial}{\partial r} \right) \right] \psi \qquad (ii)$$

Hence, from Eqn. (i) and (ii)

$$L^2 = r^2 p^2 - r^2 p_r^2$$

or $$\frac{L^2}{2mr^2} = \frac{p^2}{2m} - \frac{p_r^2}{2m}$$

or $$\frac{p^2}{2m} = \frac{p_r^2}{2m} + \frac{L^2}{2mr^2}.$$

Example 11.5: Verifiy that the spherical harmonic $Y_{11}(\theta, \phi)$ is an eigen function of L^2 and of L_z with quantum numbers $l = 1$ and $m = 1$ by using the explicit form of Y_{11}

Solution: We have

$$Y_{11}(\theta, \phi) = \sqrt{\frac{3}{8\pi}} \sin \theta e^{i\phi}$$

We want to act on this function with the operators L^2 and L_z. The L^2 equation is

$$L^2 Y_{11}(\theta, \phi) = -\hbar^2 \left(\frac{\partial^2}{\partial \theta^2} + \cot \theta \frac{\partial}{\partial \theta} + \frac{1}{\sin^2 \theta} \frac{\partial^2}{\partial \phi^2} \right) \sqrt{\frac{3}{8\pi}} \sin \theta e^{i\phi}$$

$$= -\hbar^2 \sqrt{\frac{3}{8\pi}} \left[-\sin \theta + \cot \theta (\cos \theta) + \frac{1}{\sin^2 \theta} (\sin \theta) i^2 \right] e^{i\phi}$$

The product $\cot \theta (\cos \theta) = (\cos \theta / \sin \theta) \cos \theta = \cos^2 \theta / \sin \theta$ Further, $i^2 = -1$, then the quantity in the square bracket is

$$-\sin \theta + \frac{\cos^2 \theta}{\sin \theta} - \frac{1}{\sin \theta} = -\sin \theta + \frac{1}{\sin \theta} (\cos^2 \theta - 1)$$

$$= -\sin\theta + \frac{1}{\sin\theta}\left(-\sin^2\theta\right) = -2\sin\theta$$

Thus

$$L^2 Y_{11} = 2\hbar^2 \sqrt{\frac{3}{8\pi}} \sin\theta \cdot e^{i\phi}$$

We can recognise Y_{11} on the right, verifying that it is an eigen function of L^2. The eigenvalue is the factor multiplying Y_{11} (on r.h.s.) namely $2\hbar^2$. If we compare this factor with $\hbar^2 l(l+1)$ we find $l=1$.

Similarly for L_2

$$L_2 Y_{11}(\theta,\phi) = -i\hbar \frac{\partial}{\partial\phi} \sqrt{\frac{3}{8\pi}} \sin\theta \cdot e^{i\phi}$$

$$= -i\hbar \cdot i \cdot \sqrt{\frac{3}{8\pi}} \sin\theta \exp(i\phi)$$

$$= \hbar \sqrt{\frac{3}{8\pi}} \sin\theta \cdot e^{i\phi}$$

Again Y_{11} reappears on the right side, showing that it is an eigenfunction of L_z and the eigenvalue is $\hbar$, comparing this with $\hbar m$, we find $m=1$.

Questions and Problems

1. (*i*) From the classical definition $\vec{L} = \vec{r} \times \vec{p}$ of angular momentum and the commutation rules for $\vec{r}$ and $\vec{p}$ operators, derive the commutation rules for the operators L_x, L_y and L_z.
   (*ii*) Calculate the eigenvalue spectrum of raising and lowering operators $L_\pm = L_x \pm iL_y$ in a representation in which both L^2 and L_z are diagonal.

2. Define angular momentum vector $\vec{L}$. Express the commutation relations of components of $\vec{L}$ vector with linear momenta. Show that $(L^2, \vec{L}) = 0$. What is the consequence of this relation?

3. Express the operators L_x, L_y, L_z and L^2 in polar coordinates.

4. Obtain expressions for L_+, L_- and L^2 in spherical polar coordinates.

5. Deduce the commutation relation for the components L_x, L_y, L_z of the orbital angular momentum and show that each of these commutes with L^2. Derive eigenvalues of L^2 and L_z.

6. Write short notes on

   (*i*) Angular momentum commutations relations.
   (*ii*) Coupling of orbital angular momenta and C.G. coefficients.
   (*iii*) Matrix representation of operators L_x, L_y, L_z.

7. (*i*) Prove that $L_+ = L_x + iL_y = \hbar e^{i\phi}\left(\frac{\partial}{\partial\theta} + i\cot\theta \frac{\partial}{\partial\phi}\right)$

$$L_- = L_x - iL_y = \hbar e^{-i\phi}\left(-\frac{\partial}{\partial\theta} + i\cot\theta \frac{\partial}{\partial\phi}\right)$$

(*ii*) Using expression in (*i*) and $L_z = -i\hbar \dfrac{\partial}{\partial \phi}$, prove that

$$\frac{1}{2}\left(L_+ L_- + L_- L_+\right) + L_z^2 = L^2$$

8. Prove that $\vec{L} \times \vec{L} = i\hbar\vec{L}$ if $\left[L_i, L_j\right] = i\hbar \,\epsilon_{ijk}\, L_k$
9. Given that,

$$Y_{0,0} = \frac{1}{\sqrt{4\pi}}; \qquad Y_{1,0} = \left(\frac{3}{4\pi}\right)^{1/2} \cos\theta$$

$$Y_{1,1} = -\left(\frac{3}{8\pi}\right)^{1/2} \sin\theta\; e^{i\phi}; \qquad Y_{1,-1} = \left(\frac{3}{8\pi}\right)^{1/2} \sin\theta\,.\,e^{-i\phi}$$

Verify the following equations using operator representations of L_x and L_y:

$$L_x Y_{1,1} = \frac{1}{\sqrt{2}}\hbar Y_{1,0}\;; \qquad\qquad L_y Y_{1,1} = \frac{i}{\sqrt{2}}\hbar Y_{1,0}$$

$$L_x Y_{1,0} = \frac{1}{\sqrt{2}}\hbar\left(Y_{1,1} + Y_{1,-1}\right); \qquad L_y Y_{1,0} = -\frac{i}{\sqrt{2}}\hbar\left(Y_{1,1} - Y_{1,-1}\right)$$

10. Prove the operator identity $\vec{L}^2 + \left(\vec{r}.\vec{p}\right)^2 = r^2\vec{p}^2 + i\hbar\vec{r}.\vec{p}$.

11. Using $L_x \pm iL_y = \hbar e^{\pm i\phi}\left(\pm\dfrac{\partial}{\partial\theta} + i\;\cot\theta\dfrac{\partial}{\partial\phi}\right)$ derive

$$\left(L_x \pm iL_y\right)Y_l^m = \hbar\sqrt{(l \mp m)(l \pm m + 1)}\; Y_l^{m\pm1}$$

12. Prove that the orbital angular momentum operator commutes with the Klein-Gordon Hamiltonian.

    [**Hint:** see chapter (21) for K.G. Hamiltonian.]

13. Prove that $\vec{L} \times \vec{r} - i\hbar\vec{r} = i\hbar\vec{r} - \vec{r} \times \vec{L}\;(\equiv \vec{K})$ and show that this operator is Hermitian. Also show that $\left[\vec{L}^2, \vec{r}\right] = -2i\hbar\vec{K}$.

14. Show that for the hydrogen atom

$$\begin{pmatrix} \psi_{211} \\ \psi_{210} \\ \psi_{21-1} \end{pmatrix} = \frac{r\left[e^{-r/2a_0}\right]}{\sqrt{3}\,a_0\left(2a_0\right)^{3/2}} \begin{pmatrix} Y_{1,1} \\ Y_{1,0} \\ Y_{1,-1} \end{pmatrix}.$$

12

The Internal
Angular Momentum

12.1 Experimental Evidences and Spin Angular Momentum Commutation Relations

1. According to the definition of angular momentum $\vec{L} = -i\hbar(\vec{r} \times \nabla) \equiv \vec{r} \times \vec{p}$ we find in case of hydrogen atom that given the quantum numbers n and l there are $(2l + 1)$ degenerate states of the atom corresponding to $(2l + 1)$ values of m_l (Since m_l can take values from $-l$ through 0 to $+l$). The Normal Zeeman effect is a direct experimental evidence of this fact. In presence of an external magnetic field $\vec{B}$, the interaction hamiltonian becomes

$$H_{int} = -\frac{e}{2mc}\vec{B} \cdot \vec{L} \tag{1}$$

where magnetic moment $\quad \vec{\mu} = \frac{e}{2mc}\vec{L} \tag{2}$

$(e/(2mc)$ is known as the *gyromagnetic ratio*). m_l's denote the eigenvalues of L_z. The contribution to H_{int} is according to

$$\mu_B B m_l \tag{3}$$

A remarkable fact which is experimentally observed is that in atoms with odd atomic number, the splitting is as if m_1 were half integer.

2. As early as 1921, Compton deduced from the properties of ferromagnetic materials that the electron must possess intrinsic magnetic moment. During experiments it is found that the factor for angular momentum is $e/(m_0 c)$ and not $e/(2m_0 c)$.

3. In the case of alkali spectra, (instead of uneven $2l+1$) a doublet structure, i.e. even number of lines are observed (due to further splitting of each line). In the Stern—Gerlach experiment, silver has a spherically symmetric charge distribution plus one 5-s electron ($l = 0$) and no splitting should occur. If this external electron were in the 5-p state, one might expect a splitting into three beams, but experiment gives a splitting into two beams.

The anomalies were surmounted by Uhlenbeck and Goudsmit by suggesting that besides the orbital angulat momentum due to orbital motion, the electron also has an intrinsic (spin) angular momentum. In their own words:

"The electron rotates about its own axis with angular momentum $\hbar/2$. For this value of the angular momentum, there are only two orientations of the angular momentum vector. The gyromagnetic ratio is twice as large for the rotation about its own axis as for the orbital motion".

Thus in addition to orbital angular momentum each electron has some intrinsic angular momentum $\vec{S}$ (spin angular momentum). There is associated a spin magnetic moment $\vec{\mu} = -\dfrac{e}{mc}\vec{S}$ with the spin angular momentum. The value of the spin angular momentum for the electron was assigned to be ($\frac{1}{2}$) $\hbar$, which has two components ($\pm\frac{1}{2}$)$\hbar$ in any direction, and in particular

$$S_z = \pm\frac{1}{2}\hbar \tag{4}$$

The two projections in the Stern Gerlach experiment can now be associated with the two components of the spin magnetic moment. Thus, now, the quantity L^2 is not in general a constant of the motion and the quantum number l cannot be used to characterize a particle. The *spin angular momentum* $\vec{S}$ satisfies the commutation relation such that S^2 commutes with all dynamical variables because for $\vec{S}$ a necessary condition is that it cannot be expressed in terms of $\vec{r}$ and $\vec{p}$. Then S^2 is strictly a constant of the motion and can be replaced by the number $(s\,(s+1)\,\hbar^2)$ where s may be an integer (for bosons) or half (an odd) integer (for fermions).

The spin angular momentum is not an attribute of the electrons only; all quantum mechanical systems are assigned with a value of it. Electrons, protons, neutrons, neutrinos and muons each have a value ($\frac{1}{2}$)$\hbar$, photons have $\hbar$ and pions (π-mesons) have zero value of the spin angular momentum. It has no classical analogue. Hence the spin angular momentum is an observable represented by an operator $\vec{S}$ with three components S_x, S_y and S_z which satisfy the following commutation relations:

$$\left[\hat{S}_x, \hat{S}_y\right] = \hat{S}_x\hat{S}_y - \hat{S}_y\hat{S}_x = i\hbar\hat{S}_z$$

$$\left[\hat{S}_y, \hat{S}_z\right] = \hat{S}_y\hat{S}_z - \hat{S}_z\hat{S}_y = i\hbar\hat{S}_x \tag{5}$$

$$\left[\hat{S}_z, \hat{S}_x\right] = \hat{S}_z\hat{S}_x - \hat{S}_x\hat{S}_z = i\hbar\hat{S}_y$$

Since the characteristic values of spin (for electron) are $\pm\frac{1}{2}$, the operators should also satisfy

$$\hat{S}_x^2 = \hat{S}_y^2 = \hat{S}_z^2 = \frac{\hbar^2}{4}\hat{I} \tag{6}$$

where $\hat{I}$ is identity operator.

Defining

$$S_+ = S_x + iS_y$$

$$S_- = S_x - iS_y \tag{7}$$

and
$$S^2 = S_x^2 + S_y^2 + S_z^2$$

We can easily derive the followign commutation relations:

$$\left[S_+, S_z\right] = -\hbar S_+$$

$$[S_-, S_z] = \hbar S_-$$

$$[S_+, S_-] = 2\hbar S_z \tag{8}$$

$$[S^2, S_x] = [S^2, S_y] = [S^2, S_z] = 0$$

The operators S_x, S_y and S_z may be expressed as 2×2 matrices in the following form:

$$\hat{S}_x = \begin{bmatrix} a_x & b_x \\ b_x^* & c_x \end{bmatrix}; \; \hat{S}_y = \begin{bmatrix} a_y & b_y \\ b_y^* & c_y \end{bmatrix}; \; \hat{S}_z = \begin{bmatrix} \hbar/2 & 0 \\ 0 & -\hbar/2 \end{bmatrix} \tag{9}$$

It is known that operators expressed as matrices are Hermitian and therefore the corresponding coefficients on opposite sides of the diagonal are complex conjugate of each other. Now if Eqn. (9) is inserted in the 2nd equation of (5), we get

$$\hat{S}_y \hat{S}_z - \hat{S}_z \hat{S}_y = \begin{bmatrix} a_y & b_y \\ b_y^* & c_y \end{bmatrix} \begin{bmatrix} \hbar/2 & 0 \\ 0 & -\hbar/2 \end{bmatrix} - \begin{bmatrix} \hbar/2 & 0 \\ 0 & -\hbar/2 \end{bmatrix} \begin{bmatrix} a_y & b_y \\ b_y^* & c_y \end{bmatrix}$$

$$= \frac{\hbar}{2} \begin{bmatrix} a_y & -b_y \\ b_y^* & -c_y \end{bmatrix} - \frac{\hbar}{2} \begin{bmatrix} a_y & b_y \\ -b_y^* & -c_y \end{bmatrix}$$

$$= \begin{bmatrix} 0 & -\hbar b_y \\ \hbar b_y^* & 0 \end{bmatrix}$$

or $\quad \hat{S}_y \hat{S}_z - \hat{S}_z \hat{S}_y = \begin{bmatrix} 0 & -\hbar b_y \\ \hbar b_y^* & 0 \end{bmatrix} = i\hbar \begin{bmatrix} a_x & b_x \\ b_x^* & c_x \end{bmatrix} \tag{10}$

From Eqn. (10), we get

$$a_x = c_x = 0 \text{ and } i\hbar b_x = -\hbar b_y \text{ or } b_y = -ib_x \tag{11}$$

Also from the last of Eqn. (5) and Eqn. (9), we get

$$\hat{S}_z \hat{S}_x - \hat{S}_x \hat{S}_z = \begin{bmatrix} \hbar/2 & 0 \\ 0 & -\hbar/2 \end{bmatrix} \begin{bmatrix} a_x & b_x \\ b_x^* & c_x \end{bmatrix} - \begin{bmatrix} a_x & b_x \\ b_x^* & c_x \end{bmatrix} \begin{bmatrix} \hbar/2 & 0 \\ 0 & -\hbar/2 \end{bmatrix}$$

$$= \frac{\hbar}{2} \begin{bmatrix} a_x & b_x \\ -b_x^* & -c_x \end{bmatrix} - \frac{\hbar}{2} \begin{bmatrix} a_x & -b_x \\ b_x^* & -c_x \end{bmatrix}$$

$$= \begin{bmatrix} 0 & \hbar b_x \\ -\hbar b_x^* & 0 \end{bmatrix}$$

or $\quad \hat{S}_z \hat{S}_x - \hat{S}_x \hat{S}_z = \begin{bmatrix} 0 & \hbar b_x \\ -\hbar b_x^* & 0 \end{bmatrix} = i\hbar \begin{bmatrix} a_y & b_y \\ b_y^* & c_y \end{bmatrix} \tag{12}$

From Eqn. (12), we get

$$a_y = c_y = 0 \text{ and } b_x = -ib_y \tag{13}$$

From Eqn. (6), we get

$$\hat{S}_z^2 = \hat{S}_x^2 = \frac{\hbar^2}{4}\hat{I}$$

$$\hat{S}_z^2 = \begin{pmatrix} a_x^2 & b_x^2 \\ b_x^{*2} & c_x^2 \end{pmatrix} = \begin{pmatrix} \frac{1}{4}\hbar^2 & 0 \\ 0 & \frac{1}{4}\hbar^2 \end{pmatrix}$$

or $\qquad \hat{S}_z^2 = \begin{pmatrix} 0 & |b_x|^2 \\ |b_x|^2 & 0 \end{pmatrix} = \begin{pmatrix} \frac{1}{4}\hbar^2 & 0 \\ 0 & \frac{1}{4}\hbar^2 \end{pmatrix}$ $(\because a_x = 0 = c_x)$ $\qquad$ (14)

For Eqn. (14) to be true, we should have

$$|b_x|^2 = \tfrac{1}{4}\hbar^2 \text{ or } b_x = \tfrac{1}{2}\hbar \qquad (15)$$

On substituting Eqns (11), (13) and (15) in Eqn. (9), we get

$$\hat{S}_x = \begin{pmatrix} 0 & \frac{1}{2}\hbar \\ \frac{1}{2}\hbar & 0 \end{pmatrix}; \ \hat{S}_y = \begin{pmatrix} 0 & -\frac{i\hbar}{2} \\ +\frac{i\hbar}{2} & 0 \end{pmatrix}; \ \hat{S}_z = \begin{pmatrix} \frac{1}{2}\hbar & 0 \\ 0 & -\frac{1}{2}\hbar \end{pmatrix}$$

or $\qquad \hat{S}_x = \frac{1}{2}\hbar\begin{pmatrix} 0 & 1 \\ 1 & 0 \end{pmatrix}; \ \hat{S}_y = \frac{\hbar}{2}\begin{pmatrix} 0 & -i \\ +i & 0 \end{pmatrix}; \ \hat{S}_z = \frac{\hbar}{2}\begin{pmatrix} 1 & 0 \\ 0 & -1 \end{pmatrix}$ $\qquad$ (16)

12.2 Pauli Spin Matrices

If we consider the case with spin ½ (e.g. electron), then according to Uhlanbeck and Goudsmit hypothesis, each of the operators $\hat{S}_x$, $\hat{S}_y$ and $\hat{S}_z$ must have just two eigenvalues $(\tfrac{1}{2})\hbar$ and $(-\tfrac{1}{2})\hbar$ (i.e. the electron rotates about its axis with an invariable angular momentum ½$\hbar$). We introduce three new auxiliary operators σ_x, σ_y, σ_z defined by

$$\hat{S}_x = \frac{1}{2}\hbar\hat{\sigma}_x, \ \hat{S}_y = \frac{1}{2}\hbar\hat{\sigma}_y, \ \hat{S}_z = \frac{1}{2}\hbar\hat{\sigma}_z \qquad (17)$$

where σ_x, σ_y and σ_z are called *Pauli spin matrices*.

Properties of Pauli Spin Matrices (Commutation Relations)

Since the eigenvalues of each S are to be just $(\tfrac{1}{2})\hbar$ and $(-\tfrac{1}{2})\hbar$, the eigenvalues of the each σ must be 1 and -1. Each of the operators $\hat{\sigma}_x^2$, $\hat{\sigma}_y^2$ and $\hat{\sigma}_z^2$ must therefore have only the eigenvalues $+1$ and such operator is only unit operator:

$$\hat{\sigma}_x^2 = \hat{\sigma}_y^2 = \hat{\sigma}_z^2 = \hat{1} \qquad (18)$$

According to Eqns (5), (17) and (18), the commutation relations satisfied by σ's must be

$$\left[\sigma_x, \sigma_y\right] = \sigma_x\sigma_y - \sigma_y\sigma_x$$

$$= \left(\frac{2}{\hbar}\right)^2 \left[S_x, S_y\right] = \frac{4}{\hbar^2}i\hbar S_z = \frac{4}{\hbar}iS_z$$

or $\qquad \left[\sigma_x, \sigma_y\right] = \frac{4i}{\hbar} \times \frac{1}{2}\hbar\sigma_z = 2i\sigma_z$

Similarly, $\left[\sigma_y, \sigma_z\right] = 2i\sigma_x$ (19)

$$\left[\sigma_z, \sigma_x\right] = 2i\sigma_y$$

Now, $2i\left(\sigma_x\sigma_y + \sigma_y\sigma_x\right) = \left(2i\sigma_x\right)\sigma_y + \sigma_y\left(2i\sigma_x\right)$

$$= \left(\sigma_y\sigma_z - \sigma_z\sigma_y\right)\sigma_y + \sigma_y\left(\sigma_y\sigma_z - \sigma_z\sigma_y\right)$$

$$= -\sigma_z\sigma_y^2 + \sigma_y^2\sigma_z = -\sigma_z + \sigma_z = 0 \quad \left(\because \sigma_y^2 = 1\right)$$

So, σ_x and σ_y anticommute. Similarly any two of the σ's anticommute:

$$\sigma_x\sigma_y + \sigma_y\sigma_x = 0 \; ; \; \sigma_y\sigma_z + \sigma_z\sigma_y = 0 \; ; \; \sigma_z\sigma_x + \sigma_x\sigma_z = 0 \quad (20)$$

$$\sigma_x\sigma_y - \sigma_y\sigma_x = 2i\sigma_z \text{ or } 2\sigma_x\sigma_y = 2i\sigma_z \quad \text{(using Eqns 19 and 20)}$$

or $\sigma_x\sigma_y = i\sigma_z$

Similarly $\sigma_y\sigma_z = i\sigma_x$ (21)

$$\sigma_z\sigma_x = i\sigma_y$$

Since each σ has two eigenvalues, so (2×2) matrix may fulfil the purpose and we may begin by associating with σ_z, the simplest (2×2) matrix

$$\sigma_z = \begin{pmatrix} 1 & 0 \\ 0 & -1 \end{pmatrix}$$

Let $\sigma_x = \begin{pmatrix} a & b \\ c & d \end{pmatrix}$ so that $\sigma_x\sigma_z = \begin{pmatrix} a & b \\ c & d \end{pmatrix}\begin{pmatrix} 1 & 0 \\ 0 & -1 \end{pmatrix} = \begin{pmatrix} a & -b \\ c & -d \end{pmatrix}$

But σ_x and σ_z anticommute, so we must have

$$\sigma_x\sigma_z + \sigma_z\sigma_x = \begin{pmatrix} a & -b \\ c & -d \end{pmatrix} + \begin{pmatrix} a & b \\ -c & -d \end{pmatrix} = 0$$

or $\begin{pmatrix} 2a & 0 \\ 0 & -2d \end{pmatrix} = 0 \Rightarrow a = d = 0$

i.e., every matrix that anticommutes with σ_z must have the form $\begin{pmatrix} 0 & b \\ c & 0 \end{pmatrix}$ whose eigenvalues are $\pm\sqrt{bc}$. Note that the eigenvalues λ are given by

$$\begin{pmatrix} 0 - \lambda & b \\ c & 0 - \lambda \end{pmatrix} = 0$$

so that if they are to be 1 and −1, we must set $bc = 1$ and the simplest probability is to take $b = c = 1$

$$\therefore \qquad \sigma_x = \begin{pmatrix} 0 & 1 \\ 1 & 0 \end{pmatrix}$$

$$\sigma_z\sigma_x = i\sigma_y \Rightarrow \begin{pmatrix} 1 & 0 \\ 0 & -1 \end{pmatrix}\begin{pmatrix} 0 & 1 \\ 1 & 0 \end{pmatrix} = i\sigma_y$$

or $\begin{pmatrix} 1\times 0 + 0\times 1 & 1\times 1 + 0\times 0 \\ 0\times 0 + (-1)\times 1 & 0\times 1 + (-1)\times 0 \end{pmatrix}$ or $\begin{pmatrix} 0 & 1 \\ -1 & 0 \end{pmatrix} = i\sigma_y$

(row × column multiplication)

or $\qquad \sigma_y = -i\begin{pmatrix} 0 & 1 \\ -1 & 0 \end{pmatrix}$ or $\sigma_y = \begin{pmatrix} 0 & -i \\ i & 0 \end{pmatrix}$ (22)

Hence,

$$\sigma_x = \begin{pmatrix} 0 & 1 \\ 1 & 0 \end{pmatrix} ; \sigma_y = \begin{pmatrix} 0 & -i \\ i & 0 \end{pmatrix} ; \sigma_z = \begin{pmatrix} 1 & 0 \\ 0 & -1 \end{pmatrix}$$

These are the *Pauli spin matrices*.

Commutation relation between s^2 and its components σ_x , σ_y and σ_z:

We have

$$\left[\sigma^2 , \sigma_x \right] = \left[\sigma_x^2 + \sigma_y^2 + \sigma_z^2 , \sigma_x \right] \tag{23}$$

$$= \left[\sigma_x^2 , \sigma_x \right] + \left[\sigma_y^2 , \sigma_x \right] + \left[\sigma_z^2 , \sigma_x \right]$$

$$= \sigma_x^3 - \sigma_x^3 + \left[\sigma_y^2 , \sigma_x \right] + \left[\sigma_z^2 , \sigma_x \right]$$

$$= \left[\sigma_y \sigma_y , \sigma_x \right] + \left[\sigma_z \sigma_z , \sigma_x \right]$$

Since $[ab,c] = a[b,c] + [a,c]b$, therefore, $\tag{24}$

$$\left[\sigma^2 , \sigma_x \right] = \sigma_y \left[\sigma_y , \sigma_x \right] + \left[\sigma_y , \sigma_x \right] \sigma_y + \sigma_z \left[\sigma_z , \sigma_x \right] + \left[\sigma_z , \sigma_x \right] \sigma_z$$

$$= \sigma_y \left(-2i\sigma_z \right) + \left(-2i\sigma_z \sigma_y \right) + \sigma_z \left(2i\sigma_y \right) + \left(2i\sigma_y \right) \sigma_z$$

$$= -2i\sigma_y \sigma_z - 2i\sigma_z \sigma_y + 2i\sigma_z \sigma_y + 2i\sigma_y \sigma_z = 0$$

Hence,

$$\left[\sigma^2 , \sigma_x \right] = 0 ; \left[\sigma^2 , \sigma_y \right] = 0 ; \left[\sigma^2 , \sigma_z \right] = 0 \tag{25}$$

So, σ^2 commutes with each of σ_x, σ_y and σ_z.

12.3 Representation for Spin Angular Momentum Wave Function

Consider a particle having an intrinsic angular momentum (spin) which is unrelated to its orbital motion i.e.

$$\left[\vec{S}, \vec{L} \right] = 0 \tag{26}$$

We know that $\vec{S}^2$ and S_z commute, therefore there exist simultaneous eigenstates $|sm_s\rangle$ of $\vec{S}^2$ and S_z.

$$S^2 |sm_s\rangle = s(s+1)\hbar^2 |sm_s\rangle$$

$$S_z |sm_s\rangle = m_s \hbar |sm_s\rangle \tag{27}$$

$$S_\pm |sm_s\rangle = \hbar \sqrt{s(s+1) - m_s(m_s \pm 1)} \; |s, m_s \pm 1\rangle \tag{28}$$

where $m_s = s, s-1, ..., 0, ..., s$, $(2s+1)$ values). Here it is worthwhile to note the difference between orbital and spin angular momenta: any particle can have arbitrary value $l\hbar$ ($l = 0, 1, 2, ...$) for the orbital angular momentum but the spin (s) of a given kind of particle has a fixed value and only m_s can vary (for electron $s = \frac{1}{2}$, therefore, $m_s = \pm \frac{1}{2}$), it can take any of the $(2s + 1)$ values. Thus the $(2s + 1)$ states $|sm_s\rangle$ form a complete set for the spin states of the particle. Any arbitrary spin state is represented by

$$|\chi\rangle = \sum_{m_s=-s}^{+s} |sm_s\rangle\langle sm_s|\chi\rangle \qquad (29)$$

All such states constitute the *spin-space* of a particle of spin s. The set of $(2s + 1)$ coefficients $\langle sm_s|\chi\rangle$ (considered as a function of m_s) constitute the *spin wave function* representing the spin state $|\chi\rangle$. It is written as a column whose elements are labelled by the values of m_s in decreasing order. These eigenfunctions are $(2s+1)$ – row, one column matrices that have zeros in all positions except one. For example, in the case of the electron ($s = \frac{1}{2}$), the wave functions for the basis states $|\frac{1}{2},\frac{1}{2}\rangle$ and $|\frac{1}{2},-\frac{1}{2}\rangle$ are

$$\left|\frac{1}{2},\frac{1}{2}\right\rangle \leftrightarrow \alpha \equiv \begin{pmatrix} 1 \\ 0 \end{pmatrix}$$

and

$$\left|\frac{1}{2},-\frac{1}{2}\right\rangle \leftrightarrow \beta \equiv \begin{pmatrix} 0 \\ 1 \end{pmatrix} \qquad (30)$$

respectively. (The α and β are standard abbreviations, α is the wave function for $m_s = \frac{1}{2}$ or *spin up state* while β is for the $m_s = -\frac{1}{2}$ or *spin down state*.)

12.4 Vector Representation of Spin Angular Momentum

The square of the total spin angular momentum is

$$S^2 = (\hbar/2)^2 \left(\sigma_x^2 + \sigma_y^2 + \sigma_z^2\right) = \frac{3}{4}\hbar^2 \begin{pmatrix} 1 & 0 \\ 0 & 1 \end{pmatrix} \qquad (31)$$

Taking the positive sign, the eigenvalue of S is $\left(\sqrt{3/4}\right)\hbar$, which can be obtained from $\sqrt{s(s+1)}\,\hbar$ by taking $s = \frac{1}{2}$. In vector model $\left(\sqrt{3/4}\right)\hbar$ will be the length of the spin vector. Its projection in the (say) positive z-direction will be $\hbar/2$.

When S_z is specified, it is not possible to specify x- or y-components as

$$\langle\sigma_x\rangle = \left(S_{\pm\frac{1}{2}}, \sigma_x S_{\pm\frac{1}{2}}\right) = \langle\sigma_y\rangle = 0 \qquad (32)$$

Since $\sigma_x^2 = \sigma_y^2 = 1$, $\left(S_x\right)^2$ and $\left(S_y\right)^2$ will have the value $\hbar^2/4$.

Hydrogen-like wave function including spin can be written as

$$\psi_{nlm_l m_s} = \psi_{nlsm_l} S_{m_s}$$

$$= C_{nlm_l m_s} R_{nl}(r)P_l^{m_l}(\cos\,\theta)\, e^{im_l\phi} S_{m_s}\left(\sigma_z\right) \qquad (33)$$

where the quantum number s is equal to $\frac{1}{2}$, m_s can take up either of the values $+\frac{1}{2}$ or $-\frac{1}{2}$. Here ψ_{nlsm_l} is to be regarded as a complex number. This form of the wave function is obtained by solving for a central field problem (e.g. H-atom). The central forces are characterised by invariance under separate rotations of space and spin coordinates.

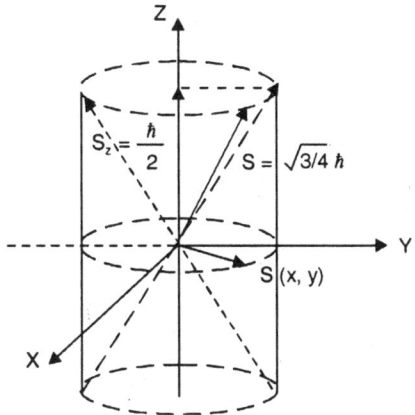

Fig. 12.1: Spin angular momentum vector. If $S_z = \hbar/2$, then $< S(x, y) > = 0$

Example 12.1: If σ_x, σ_y and σ_z are Pauli spin matrices and $\vec{A}$ and $\vec{B}$ any constant vectors, show that

$$\left(\vec{\sigma}\cdot\vec{A}\right)\left(\vec{\sigma}\cdot\vec{B}\right) = \vec{A}\cdot\vec{B} + i\vec{\sigma}\cdot\left(\vec{A}\times\vec{B}\right).$$

Solution: We have

$$\left(\vec{\sigma}\cdot\vec{A}\right)\left(\vec{\sigma}\cdot\vec{B}\right) = \left(\sigma_x A_x + \sigma_y A_y + \sigma_z A_z\right)\left(\sigma_x B_x + \sigma_y B_y + \sigma_z B_z\right)$$

$$= \sigma_x^2 A_x B_x + \sigma_y^2 A_y B_y + \sigma_z^2 A_z B_z$$

$$+ \sigma_x\sigma_y A_x B_y + \sigma_y\sigma_x A_y B_x + \sigma_y\sigma_z A_y B_z + \sigma_z\sigma_y A_z B_y$$

$$+ \sigma_x\sigma_z A_x B_z + \sigma_z\sigma_x A_z B_x$$

Keeping in view the following properties of Pauli spin matrices

$$\sigma_x^2 = \sigma_y^2 = \sigma_z^2 = 1$$

and

$$\sigma_x\sigma_y = -\sigma_y\sigma_x = i\sigma_z \qquad \text{(i.e. they anti-commute in pairs)}$$

$$\sigma_y\sigma_z = -\sigma_z\sigma_y = i\sigma_x$$

$$\sigma_z\sigma_x = -\sigma_x\sigma_z = i\sigma_y$$

we get

$$\left(\vec{\sigma}\cdot\vec{A}\right)\left(\vec{\sigma}\cdot\vec{B}\right) = A_x B_x + A_y B_y + A_z B_z + i\sigma_z\left(A_x B_y - A_y B_x\right)$$

$$+ i\sigma_x\left(A_y B_z - A_z B_y\right) + i\sigma_y\left(A_z B_x - A_x B_z\right)$$

$$= \vec{A}\cdot\vec{B} + i\vec{\sigma}\cdot\left(\vec{A}\times\vec{B}\right).$$

12.5 Representation of an Operator in Spin Space

In general, we represent the common set of eigenfunctions of S^2 and S_z by

$$|\chi\rangle = \begin{pmatrix} a_+ \\ a_- \end{pmatrix} \text{ and } S_z|\chi\rangle = \frac{1}{2}\hbar\sigma_z|\chi\rangle \qquad (34)$$

Now,

$$|\chi\rangle = \begin{pmatrix} a_+ \\ a_- \end{pmatrix} = a_+ \begin{pmatrix} 1 \\ 0 \end{pmatrix} + a_- \begin{pmatrix} 0 \\ 1 \end{pmatrix} \equiv a_+ \left| \tfrac{1}{2}, \tfrac{1}{2} \right\rangle + a_- \left| \tfrac{1}{2}, -\tfrac{1}{2} \right\rangle \tag{35}$$

$$= a_+ \chi_+ + a_- \chi_- \equiv \sum_{m_s=-1/2}^{+1/2} a_{m_s} \chi_{m_s} \tag{36}$$

$$= \sum_{m_s=-1/2}^{+1/2} |sm_s\rangle \langle sm_s|\chi\rangle \tag{37}$$

Here the two component column vectors

$$\chi_+ = \begin{pmatrix} 1 \\ 0 \end{pmatrix} \equiv \alpha \text{ and } \chi_- = \begin{pmatrix} 0 \\ 1 \end{pmatrix} \equiv \beta \tag{38}$$

are called *spinors*. χ_+ and χ_- are orthonormal. The a_{m_s} must satisfy the normalization condition

$$\langle\chi|\chi\rangle = \left(a_+^* a_-^* \right) \begin{pmatrix} a_+ \\ a_- \end{pmatrix} = |a_+|^2 + |a_-|^2 = 1 \tag{39}$$

$|a_{m_s}|^2$ represent the probability of obtaining the eigenvalue $m_s \hbar$ for S_z as a result of measurement.

Any operator in the spin space will be represented by a 2×2 matrix with complex elements, using the orthonormal spinors χ_+ and χ_- as basis:

$$(A) = \begin{pmatrix} \langle \chi_+|A|\chi_+\rangle & \langle \chi_+|A|\chi_-\rangle \\ \langle \chi_-|A|\chi_+\rangle & \langle \chi_-|A|\chi_-\rangle \end{pmatrix} \tag{40}$$

and this can be expressed as a linear combination of the unit matrix I and the Pauli matrices:

$$(A) = A_0 I + A_x \sigma_x + A_y \sigma_y + A_z \sigma_z \tag{41}$$

$$= A_0 \begin{pmatrix} 1 & 0 \\ 0 & 1 \end{pmatrix} + A_x \begin{pmatrix} 0 & 1 \\ 1 & 0 \end{pmatrix} + A_y \begin{pmatrix} 0 & -i \\ i & 0 \end{pmatrix} + A_z \begin{pmatrix} 1 & 0 \\ 0 & -1 \end{pmatrix}$$

$$= \begin{pmatrix} A_0 + A_z & A_x - iA_y \\ A_x + iA_y & A_0 - A_z \end{pmatrix} \tag{42}$$

where A_0, A_x, A_y and A_z are operators.

Example 12.2: The common set of eigenfunctions of J^2 and J_z are represented by ψ_{jm_j}

$$J^2 \psi_{jm_j} = j(j+1)\hbar^2 \psi_{jm_j}$$

and

$$J_z \psi_{jm_j} = m_j \hbar \psi_{jm_j}$$

Obtain the Pauli operators associated with spin $-1/2$ particles for J^2 and J_z.

Solution: $\vec{J} = \vec{L} + \vec{S}$

and $\qquad J_z = L_s + S_z$ (as we shall see in next chapter)

$$J^2 = \left(\vec{L} + \vec{S}\right) \cdot \left(\vec{L} + \vec{S}\right)$$

$$= L^2 + S^2 + 2\vec{L} \cdot \vec{S}$$

$$= L^2 + S^2 + 2\left(L_x S_x + L_y S_y + L_z S_z\right)$$

$$\equiv A_0 I + A_x \sigma_x + A_y \sigma_y + A_z \sigma_z$$

$$\equiv \begin{pmatrix} A_0 + A_z & A_x - iA_y \\ A_x + iA_y & A_0 - A_z \end{pmatrix} \qquad \text{(from Eqn. 42)}$$

where $\quad A_0 = L^2 + S^2$

and $\quad A_x - iA_y = \left(L_x - iL_y\right)\hbar = \hbar L_-$

$$A_x + iA_y = \left(L_x + iL_y\right)\hbar = \hbar L_+$$

$$A_z = \hbar L_z$$

$$J^2 \psi_{jm_j} \equiv \begin{pmatrix} A_0 + A_z & A_x - iA_y \\ A_x + iA_y & A_0 - A_z \end{pmatrix} \psi_{jm_j}$$

$$\Rightarrow \qquad J^2 = \begin{pmatrix} L^2 + \frac{3}{4}\hbar^2 + \hbar L_z & \hbar L_- \\ \hbar L_+ & L^2 + \frac{3}{4}\hbar^2 - \hbar L_z \end{pmatrix}$$

$$J_z = L_z + S_z = \begin{pmatrix} L_z + S_z & 0 \\ 0 & L_z - S_z \end{pmatrix} = \begin{pmatrix} L_z + \frac{1}{2}\hbar & 0 \\ 0 & L_z - \frac{1}{2}\hbar \end{pmatrix}$$

12.6 Eigenvalues and Eigenfunctions of the Spin Operators S_x, S_y and S_z

The eigenvalues of S_x, S_y and S_z are $\pm \hbar/2$. The normalized eigenfunctions can be obtained as follows:

The eigenvalue equation

$$\hat{S}_x |\chi\rangle = \frac{1}{2}\hbar\sigma_x |\chi\rangle = m_s \hbar |\chi\rangle \tag{43}$$

gives

$$\hat{S}_x |\chi\rangle = \frac{\hbar}{2}\begin{pmatrix} 0 & 1 \\ 1 & 0 \end{pmatrix}\begin{pmatrix} a_+ \\ a_- \end{pmatrix} = \pm\frac{\hbar}{2}\begin{pmatrix} a_+ \\ a_- \end{pmatrix}$$

or

$$\begin{pmatrix} a_- \\ a_+ \end{pmatrix} = \pm\begin{pmatrix} a_+ \\ a_- \end{pmatrix} \tag{44}$$

$$\therefore \qquad a_+ = \pm a_- .$$

From normalization condition $a_+ = a_- = 1/\sqrt{2}$ or $a_+ = -a_- = 1/\sqrt{2}$.

Hence,

$$\chi_+^x = \frac{1}{\sqrt{2}}\left(\chi_+ + \chi_-\right) = \frac{1}{\sqrt{2}}\begin{pmatrix} 1 \\ 1 \end{pmatrix} \equiv \alpha_x$$

$$\chi_-^x = \frac{1}{\sqrt{2}}(\chi_+ - \chi_-) = \frac{1}{\sqrt{2}}\begin{pmatrix} 1 \\ -1 \end{pmatrix} \equiv \beta_x \tag{45}$$

Similarly,

$$\chi_+^y = \frac{1}{\sqrt{2}}(\chi_+ + i\chi_-) = \frac{1}{\sqrt{2}}\begin{pmatrix} 1 \\ i \end{pmatrix} \equiv \alpha_y$$

$$\chi_-^y = \frac{1}{\sqrt{2}}(\chi_+ - i\chi_-) = \frac{1}{\sqrt{2}}\begin{pmatrix} 1 \\ -i \end{pmatrix} \equiv \beta_y \tag{46}$$

and

$$\chi_+^z \equiv \alpha_z = \begin{pmatrix} 1 \\ 0 \end{pmatrix} ; \chi_-^z \equiv \beta_z = \begin{pmatrix} 0 \\ 1 \end{pmatrix} \tag{47}$$

The fact that $\chi_\pm^x, \chi_\pm^y$ and $\chi_\pm^z$ are all different means that no pair of S_x, S_y and S_z observables are compatible. So any of the three pairs of eigenspinors $\{\alpha_x, \beta_x\}, \{\alpha_y, \beta_y\}, \{\alpha_z, \beta_z\}$ may be taken as a basis. We can expand an arbitrary spinor $|\chi\rangle$ as

$$|\chi\rangle = \begin{pmatrix} a_+ \\ a_- \end{pmatrix} = a_+ \alpha_z + a_- \beta_z$$

$$= \frac{1}{\sqrt{2}}\left[(a_+ + a_-)\alpha_x + (a_+ - a_-)\beta_x \right]$$

$$= \frac{1}{\sqrt{2}}\left[(a_+ - ia_-)\alpha_y + (a_+ + ia_-)\beta_y \right] \tag{48}$$

We have, for an arbitrary spinor $|\chi\rangle$

$$\langle S_x \rangle = \langle \chi | S_x | \chi \rangle = \langle \chi | \frac{\hbar}{2}\sigma_x | \chi \rangle \tag{49i}$$

or

$$\langle S_x \rangle = \begin{pmatrix} a_+^* & a_-^* \end{pmatrix} \frac{\hbar}{2}\begin{pmatrix} 0 & 1 \\ 1 & 0 \end{pmatrix}\begin{pmatrix} a_+ \\ a_- \end{pmatrix} = \frac{1}{2}\hbar\left(a_+^* a_- + a_-^* a_+ \right) \tag{49ii}$$

Similarly,

$$\langle S_y \rangle = -\frac{i}{2}\hbar\left(a_+^* a_- - a_-^* a_+ \right) \tag{50}$$

If the electron is known to be in the spin state α_z, $\langle S_z \rangle = \hbar/2$, then $a_- = 0$ and therefore $\langle S_x \rangle = 0 = \langle S_y \rangle$. This is depicted geometrically in fig. (12.1). Thus, due to the commutation relation $\left[S_x^-, S_y^- \right] = i\hbar S_z$, only one of the components of $\vec{S}$ can be accurately measured.

Example 12.3: Using the result of example (12.1) show that

$$e^{i\theta \vec{\sigma} \cdot \hat{n}} = \cos\theta + i\vec{\sigma} \cdot \hat{n}\sin\theta$$

Solution: $\left(\vec{\sigma} \cdot \vec{A}\right)\left(\vec{\sigma} \cdot \vec{B}\right) = \vec{A} \cdot \vec{B} + i\vec{\sigma} \cdot \left(\vec{A} \times \vec{B}\right)$

Putting $\vec{A} = \vec{B} = \hat{n}$, we get

$$\left(\vec{\sigma} \cdot \hat{n}\right)\left(\vec{\sigma} \cdot \hat{n}\right) = 1 + i\vec{\sigma} \cdot \left(\hat{n} \times \hat{n}\right) \text{ (because } \hat{n} \cdot \hat{n} = 1) \qquad (i)$$

Now, we may expand

$$e^{i\theta\vec{\sigma}\cdot\hat{n}} = 1 + i\theta\frac{\vec{\sigma} \cdot \hat{n}}{1!} + \frac{i^2\theta^2}{2!}(\vec{\sigma} \cdot \hat{n})(\vec{\sigma} \cdot \hat{n}) + \frac{i^3\theta^3}{3!}(\vec{\sigma} \cdot \hat{n})(\vec{\sigma} \cdot \hat{n})(\vec{\sigma} \cdot \hat{n}) + \ldots$$

Using (i), this expression becomes

$$e^{i\theta\vec{\sigma}\cdot\hat{n}} = 1 + \frac{i\theta \cdot \vec{\sigma} \cdot \hat{n}}{1!} - \frac{\theta^2}{2!}\left(1 + i\vec{\sigma} \cdot \left(\hat{n} \times \hat{n}\right)\right) - \frac{i\theta^3}{3!}(\vec{\sigma} \cdot \hat{n})\left(1 + i\vec{\sigma} \cdot \left(\hat{n} \times \hat{n}\right)\right) + \ldots$$

$$= \left(1 - \frac{\theta^2}{2!} + \ldots\right) + i\vec{\sigma} \cdot \hat{n}\left(\frac{\theta}{1!} - \frac{\theta^3}{3!} + \ldots\right) \qquad \left(\because \hat{n} \times \hat{n} = 0\right)$$

$$= \cos\theta + i\vec{\sigma} \cdot \hat{n}\sin\theta \cdot$$

Example 12.4: The Hamiltonian for a spin 1/2 particle with charge +e in an external magnetic field $\vec{B}$ is

$$H = -\frac{ge}{2mc}\vec{s} \cdot \vec{B}$$

Calculate the operator $\dfrac{d\vec{s}}{dt}$ if $\vec{B} = B\hat{y}$.

Solution: $\dfrac{d\vec{s}}{dt} = \dfrac{1}{i\hbar}[\vec{s}, H] = -\dfrac{ge}{i2m\hbar c}\left[\vec{s}, \vec{s} \cdot \vec{B}\right].$

Now $\left[\vec{s} \, \vec{s} \cdot \vec{B}\right] = \left[\vec{s}_x, \vec{s} \cdot \vec{B}\right]\hat{i} + \left[\vec{s}_y, \vec{s} \cdot \vec{B}\right]\hat{j} + \left[\vec{s}_z, \vec{s} \cdot \vec{B}\right]\hat{k}$

where $\hat{i}, \hat{j}, \hat{k}$ are unit vectors along the x, y and z axes respectively.

Now, $\left[s, \vec{s} \cdot \vec{B}\right] = \left[s_x, s_x\right]B_x + \left[s_x, s_y\right]B_y + \left[s_x, s_z\right]B_z$

$$= i\hbar\left(s_z B_y - s_y B_z\right)$$

$$= i\hbar\left(\vec{B} \times \vec{s}\right)_x$$

(we have used $\left[s_x, s_y\right] = i\hbar s_z$ and $\left[s_x, s_z\right] = -i\hbar s_y$)

So, $\left[\vec{s}, \vec{s} \cdot \vec{B}\right] = -i\hbar\left(\vec{s} \times \vec{B}\right)$

Hence $\dfrac{d\vec{s}}{dt} = +\dfrac{ge}{i2m\hbar c}i\hbar\left(\vec{s} \times \vec{B}\right)$

$$= \frac{ge}{2mc}\left(\vec{s} \times \vec{B}\right)$$

Example 12.5: Using the Hamiltonian $H = \omega S_z$, write the Heisenberg's equation of motion for $S_{x,y,z}(t)$ where S is the electron spin $\vec{S} = \frac{1}{2}\sigma$. Solve them to obtain $S_{x,y,z}$ as a function of time.

Solution: Heisenberg's equations of motions are

$$\dot{S}_x = \frac{1}{i\hbar}[S_x, H] = \frac{\omega}{i\hbar}[S_x, S_z] = -\frac{i\hbar\omega}{i\hbar}S_y = -\omega S_y$$

Similarly,

$$\dot{S}_y = \frac{\omega}{i\hbar}[S_y, S_z] = i\hbar\frac{\omega}{i\hbar}S_x = \omega S_x$$

$$\dot{S}_z = \frac{\omega}{i\hbar}[S_z, S_z] = 0$$

i.e. S_z is constant in time, $\dot{S}_x = -\omega S_y$, $\dot{S}_y = \omega S_x$. Since, $S_\pm = S_x \pm iS_y$, so we have $\dot{S}_+ = i\omega S_+$ and $\dot{S}_- = i\omega S_-$ giving $S_\pm(t) = Ce^{\pm i\omega t}$, where C is the value of $S_\pm$ at $t = 0$.

12.7 Isospin

Irrespective of their charge, neutron and proton are very much alike and have similar properties e.g. both exist inside the nucleus, have same spin (= ½) and almost same mass. Just as in presence of a magnetic field, a spin ½ electron comes in two states (spin up and spin down), similarly, according to Heisenburg, a nucleon in an electromagnetic field has an isotopic spin (I = ½) and comes in two states in the Isospin space[*]:

Isotopic spin-up (I_z = ½ a proton) and isotopic spin down (I_z = – ½ a neutron). Just as we cannot distinguish between two electronic states in absence of a magnetic field, we also cannot distinguish between a proton and a neutron in absence of an electromagnetic field. They are two different states of a nucleon in the isospin space. Like $\hat{S} = \frac{1}{2}\hat{\sigma}$ for spin, we have matrix representation for *isospin*:

$$\hat{I} = \frac{1}{2}\hat{\tau}$$

where

$$\tau_1 = \begin{pmatrix} 0 & 1 \\ 1 & 0 \end{pmatrix}; \ \tau_2 = \begin{pmatrix} 0 & -i \\ i & 0 \end{pmatrix}; \ \tau_3 = \begin{pmatrix} 1 & 0 \\ 0 & -1 \end{pmatrix}$$

and

$$[\tau_i, \tau_j] = 2i\tau_k \text{ for } i, j, k \text{ cyclic } (= 1,2,3).$$

---

[*] For a nucleon $\psi = \psi_{space}\psi_{spin}\psi_{charge}$ where ψ_{charge} is equivalent to $\psi_{isospin}$ and refers to an abstract two-dimensional space called isospin space in which the charge operator is denoted by $\hat{I}$. In isospin space, vector attached to nucleon is called a spinor. The isospin symmetry in strong interactions is a consequence of charge independence of nucleon-nucleon force.

The eigenkets can be represented by $|\text{Im}_I\rangle$ such that

$$\hat{I}^2|\text{Im}_I\rangle = I(I+1)|\text{Im}_I\rangle$$

and $\quad I_z|\text{Im}_I\rangle = m_I|\text{Im}_I\rangle$

where $m_I = \pm 1/2$. Thus $\left|\dfrac{1}{2}\dfrac{1}{2}\right\rangle \equiv \begin{pmatrix} 1 \\ 0 \end{pmatrix}$ refers to a proton (p) and $\left|\dfrac{1}{2}-\dfrac{1}{2}\right\rangle \equiv \begin{pmatrix} 0 \\ 1 \end{pmatrix}$ to a neutron (n). A vector in the isospin space is known as an *isospinor* and has a general representation.:

$$|\text{nucleon>} = a\begin{pmatrix} 1 \\ 0 \end{pmatrix} + b\begin{pmatrix} 0 \\ 1 \end{pmatrix}$$

We define $\quad I_\pm \equiv I_1 \pm iI_2$ as operators which satisfy

$$I_\pm|\text{Im}_I\rangle = \sqrt{(I \mp m_I)(I \pm m_I + 1)}\ |\text{Im}_I \pm 1\rangle$$

$$[I_3, I_\pm] = \pm I_\pm, \quad [I_\pm, I_\mp] = \pm 2I_3$$

and $\quad I_+|-\rangle = |+\rangle,\ I_-|+\rangle = |-\rangle,\ I_\pm|\pm\rangle = 0$

where $|+\rangle$ and $|-\rangle$ are respectively the proton ket and neutron ket. We also define:

$$\tau_+ = \begin{pmatrix} 0 & 1 \\ 1 & 0 \end{pmatrix} = \frac{1}{2}(\tau_1 + i\tau_2)\ ;\ \tau_- = \begin{pmatrix} 0 & 0 \\ 1 & 0 \end{pmatrix} = \frac{1}{2}(\tau_1 - i\tau_2)$$

$$\begin{cases} \tau_+ n = p & ;\ \tau_+ p = 0 \\ \tau_- p = n & ;\ \tau_- n = 0 \end{cases}$$

The charge operator Q (has eigenvalues 1 and 0 in the proton and neutron states respectively) and is defined as

$$Q = \begin{pmatrix} 1 & 0 \\ 0 & 1 \end{pmatrix} = \frac{1}{2}(1 + \tau_3)$$

Similar to spin, isospin is an intrinsic property of the particles. The spinor-kets $|+\rangle$ and $|-\rangle$ (having m_I +½ and –½ respectively, can be regarded as vectors in the abstract isospin space. The idea of isospin was introduced to explain Yukawa's meson-theory of nuclear forces. The pi-mesons were thought to be exchanged between neutrons and protons which were responsible for strong nuclear binding force. Thus in case of pi-mesons we have isospin $I = 1$; $I_3 = 1$ for π^+; $I_3 = 0$ for π^0 and $I_3 = -1$ for π^-. Later, the concept of isospin was extended to other hadrons.

12.8 Magnetic Moment of an Electron and Gyromagnetic Ratio

The interaction Hamiltonian for a charged particle in a constant uniform magnetic field $\vec{B}$ is

$$H_{\text{int}} = -\vec{\mu}_L \cdot \vec{B}$$

where
$$\vec{\mu}_L = \frac{e}{2mc}\vec{L} = -\frac{|e|}{2mc}\vec{L} \tag{51}$$

$\vec{\mu}_L$ is the magnetic moment due to the orbital motion. If spin is taken into account, for an electron, then the additional interaction energy is

$$H_s = -\vec{\mu}_s \cdot \vec{B}, \ \vec{\mu}_s = \frac{e}{mc}\vec{S} = -\frac{|e|}{mc}\vec{S} \tag{52}$$

The ratio of magnetic momentum to the angular momentum is known as the *gyromagnetic ratio*. The gyromagnetic ratios are written in terms of the Bohr magneton, $\mu_B = \dfrac{|e|\hbar}{2mc}$. That is

$$\left|\frac{\vec{\mu}_L}{\vec{L}}\right| \equiv \gamma_l = \frac{ge}{2mc} = \frac{\mu_B}{\hbar} \tag{53}$$

and
$$\left|\frac{\vec{\mu}_s}{\vec{S}}\right| \equiv \gamma_s = \frac{ge}{2mc}\left(\cong \frac{e}{mc}\right) \tag{54}$$

where g is the spectroscopic splitting factor ($g = 1$ for γ_l and $g = 2.0023$ for γ_s). Here l is the electronic charge, m, the mass of the electron and c, the velocity of light ($\hbar = h/2\pi$, h being Planck's constant).

Questions and Problems

1. Explain why Pauli introduced a set of 2×2 spin matrices and obtain the commutation relations satisfied by the three components of the spin vector.
2. What are Pauli spin operators? Express Pauli spin operators in the form of 2×2 matrices.
3. Prove that Pauli matrices are unitary and any two different Pauli matrices anticommute.
4. Prove that for Pauli spin matrices
    (a) $\vec{\sigma} \times \vec{\sigma} = 2i\vec{\sigma}$
    (b) $\left(\vec{\sigma}\cdot\vec{A}\right)\left(\vec{\sigma}\cdot\vec{B}\right) = \vec{A}\cdot\vec{B} + i\vec{\sigma}\cdot\left(\vec{A}\times\vec{B}\right)$
        where $\vec{A}$ and $\vec{B}$ are any constant vectors.
    (c) $e^{i\theta\vec{\sigma}\cdot\hat{n}} = \cos\theta + i\vec{\sigma}\cdot\hat{n}\sin\theta$
5. Show that
    (i) $\exp\left(i\sigma_y\phi\right) = \cos\phi + i\sigma_y\sin\phi$
    (ii) $\sin\left(\sigma_x\phi\right) = \sigma_x\sin\phi$
    (iii) $\cos\left(\sigma_z\phi\right) = \cos\phi$
6. For a spatial rotation through angle ϕ about an axis of rotation along a unit vector $\hat{n}$, transformations in spinor space are represented by
$$U = e^{i\phi\hat{n}\cdot\vec{S}/\hbar}$$

(*i*) Show using $\left(\vec{\sigma}\cdot\vec{A}\right)\left(\vec{\sigma}\cdot\vec{B}\right) = \vec{A}\cdot\vec{B} + i\vec{\sigma}\cdot\left(\vec{A}\times\vec{B}\right)$

that $\qquad U = \cos\dfrac{\phi}{2} + i\left(\vec{\sigma}\cdot\hat{n}\right)\sin\dfrac{\phi}{2}$

(*ii*) Using this formula for U and using the identities for the Pauli matrices show that

$$U\sigma U^{\dagger} = \hat{n}\left(\hat{n}\cdot\vec{\sigma}\right) - \hat{n}\times\left(\hat{n}\times\vec{\sigma}\right)\cos\phi + \left(\hat{n}\times\vec{\sigma}\right)\sin\phi$$

7. Show that the Pauli matrices have the properties:

   (*i*) $\quad \sigma_x\sigma_y\sigma_z = i$

   (*ii*) $\left[\sigma_x, \sigma_y\right] = 2i\sigma_z$ (for $x\,y\,z$ cyclic)

8. Using Pauli's spin matrix representation, reduce the following operators

   (*i*) $S_x^2 S_y S_z^2$ $\qquad\qquad$ (*ii*) $S_x^2 S_y^2 S_z^2$

   $$\left[\textbf{Ans. } (i)\;\left(\frac{\hbar}{2}\right)^5\sigma_y\,;\;(ii)\;\left(\frac{\hbar}{2}\right)^6\right]$$

9. Consider a spin ½ system represented by the normalized state vector

   $$\begin{pmatrix}\cos\theta\\ \sin\theta\cdot e^{i\phi}\end{pmatrix}$$

   What is the probability that a measurement of S_y yields $-\hbar/2$.

13

Total Angular Momentum

13.1 Total Angular Momentum of Atoms

Each electron in an atom has a definite orbital angular momentum $\vec{L}$ and a definite spin angular momentum $\vec{S}$ both of which contribute to give the total angular momentum $\vec{J}$ of the atom. $\vec{J}$ is quantized with a magnitude given by

$$J = \sqrt{j(j+1)}\ \hbar \qquad (1)$$

and a component J_z in the z-direction given by

$$J_z = m_j \hbar \qquad (2)$$

where j and m_j are quantum numbers governing $|\vec{J}|$ and J_z.

Let us consider an atom whose total angular momentum is provided by a single electron (e.g. hydrogen, Li, Na, He$^+$, Be$^+$, Mg$^+$). The exclusion principle insures that the total angular momentum and so magnetic moment of closed shells (in these atoms) are zero.

The magnitude L of the orbital angular momentum $\vec{L}$ of an atomic electron is determined by its orbital quantum number l according to the relation:

$$L = \sqrt{l(l+1)}\ \hbar \qquad (3)$$

while the component L_z of $\vec{L}$ along the z-axis is determined by

$$L_z = m_l \hbar \qquad (4)$$

where m_l is the magnetic quantum number. Similarly, the magnitude S of the spin angular momentum $\vec{S}$ is determined by the spin quantum number s (which possesses the value $+\frac{1}{2}$ only) according to the relation

$$S = \sqrt{s(s+1)}\ \hbar \qquad (5)$$

while the component S_z of $\vec{S}$ along the z-axis is determined by the magnetic spin quantum number m_s according to

$$S_z = m_s \hbar \quad (m_s = \pm\tfrac{1}{2}\, h) \qquad (6)$$

Vectors $\vec{L}$ and $\vec{S}$ add (vectorially) to yield the total angular momentum $\vec{j}$:

$$\vec{J} = \vec{L} + \vec{S} \tag{7}$$

The symbols j and m_j are used to denote the quantum numbers that describe J and J_z for a single electron:

$$J = \sqrt{j(j+1)}\ \hbar \tag{8}$$

$$J_z = m_j \hbar \tag{9}$$

Let us consider the z components of $\vec{J}$, $\vec{L}$ and $\vec{S}$.

$$J_z = L_z \pm S_z$$

$$m_j \hbar = m_l \hbar \pm m_s \hbar \tag{10}$$

so
$$m_j = m_l \pm m_s$$

The possible values of m_l range from $-l$ through 0 to $+l$ and those of m_s are $\pm s$. The quantum number l is always an integer or zero while $s = \frac{1}{2}$. Thus, m_j must be half integral. The possible values of m_j range from $+j$ through 0 to $-j$ (in integral steps). Therefore

$$j = l \pm s$$

and like m_j, j is also always half integral.

Because of the simultaneous quantization of $\vec{J}, \vec{L}$ and $\vec{S}$, they can have only certain specific relative orientations. For example, for a one electron atom, there are only two relative orientation possible: one corresponding to $j = l + s$ (i.e. $J > L$) and the other corresponding to $j = l - s$ ($J < L$).

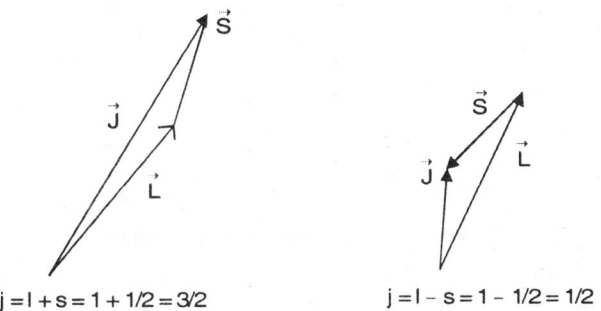

$$j = l + s = 1 + 1/2 = 3/2 \qquad\qquad j = l - s = 1 - 1/2 = 1/2$$

Fig. 13.1: Two possible orientations of $\vec{L}$ and $\vec{S}$
(for $l = 1$ and $s = \frac{1}{2}$) corresponding to $\vec{J} = \vec{L} + \vec{S}$

The angular momenta $\vec{L}$ and $\vec{S}$ interact magnetically and so, exert torques on each other. When there is no external magnetic field $\vec{j}$ is conserved in magnitude and direction and due to mutual interaction vectors $\vec{L}$ and $\vec{S}$ precess around the direction of their resultant $\vec{J}$. However, if an external magnetic field $\vec{B}$ is present, $\vec{j}$ precesss about the direction of $\vec{B}$ while $\vec{L}$ and $\vec{S}$ continue precessing about $\vec{j}$. The possibility of different orientations of $\vec{j}$ relative to B gives rise to the anomalous Zeeman effect (as different relative orientations invole slightly different energies).

13.2 L S Coupling (Russel Saunders Coupling)

When there are more than one electron contributing in an atom, the orbital angular momenta L_i of the various electrons are coupled together (vectorially) into a single resultant $\vec{L}$ and the spin angular momenta $\vec{S}_i$ are coupled together independently into resultant $\vec{S}$ (electrons in the closed shells donot contribute). Then, $\vec{L}$ and $\vec{S}$ interact magnetically (via spin orbit effect) to form a total angular momentum $\vec{J}$:

$$\vec{L} = \sum_i \vec{L}_i \;;\quad \vec{S} = \sum_i \vec{S}_i \;;\quad \vec{J} = \vec{L} + \vec{S} \tag{11}$$

(Here L_i's couple according to as mentioned in article 12.6 and S_i's couple in the same way as L_i's.)

The LS scheme owes its existence to the relative strengths of the electrostatic forces that couple the individual orbital angular momenta into a resultant $\vec{L}$ and the individual spin angular momenta into a resultatnt $\vec{S}$. The coupling between the various $\vec{L}_i$ is usually such that the configuration of lowest energy is the one for which L is a maximum and also $\vec{S}_i$ always combine into a ground state configuration in which S is a maximum. Then $\vec{L}$ and $\vec{S}$ combine such that J is a minimum so as to result in the lowest energy.

Example 13.1: Two p electrons in an atom

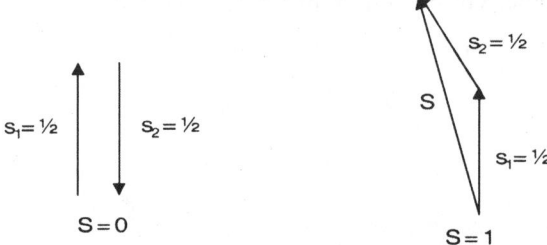

Fig. 13.2a: Vector combinations of spin angular momenta of two electrons

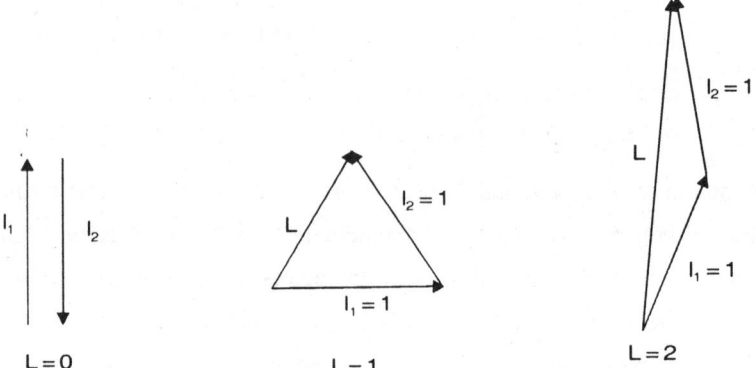

Fig. 13.2b: Vector combinations of orbital angular momenta for two p electrons

(In general $L = l_1 + l_2, l_1 + l_2 - 1, l_1 + l_2 - 2, ..., l_1 - l_2$)
And in general possible values of J are

$$J = L + S, L + S - 1, L + S - 2, ... , |L - S|: (2S + 1 \text{ values})$$

$|L - S|$ indicates only positive (or zero) values of J are significant when $S >$ L.

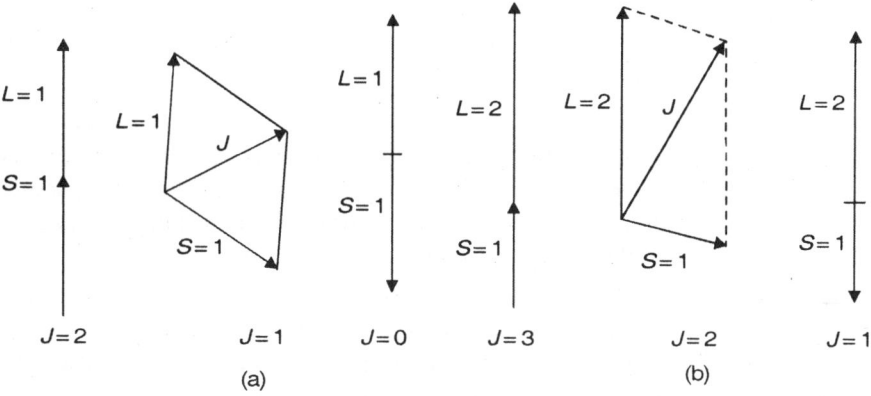

Fig. 13.3: Combination of vectors $\vec{S}$ and $\vec{L}$ (a) $S = 1, L = 1$ (b) $S = 1, L = 2$
(the magnitudes are in units of $\hbar$)

13.3 JJ Coupling

The electrostatic forces that couple the $\vec{L}_i$ into a single vector $\vec{L}$ and the $\vec{S}_i$ into a vector $\vec{S}$ are stronger than the magnetic spin orbit forces that couple $\vec{L}$ and $\vec{S}$ to form $\vec{J}$ (in light atoms), and dominate even when a moderate external magnetic field is applied. In that case $\vec{J}$ precesses around $\vec{B}$ slowly as compared to precession of $\vec{L}$ and $\vec{S}$ around $\vec{J}$.

However, in heavy atoms, the nuclear charge produces spin orbit interactions of magnitude comparable to electrons' coupling between $\vec{L}_i$ and $\vec{S}_i$. Consequently, the $L\,S$ coupling begins to break down in heavier atoms. In these cases the following scheme known as the $j\,j$ *coupling* holds true:

$$\vec{L}_i + \vec{S}_i = \vec{J}_i$$

and
$$\vec{J} = \sum_i \vec{J}_i \tag{12}$$

where $\vec{J}_i$ are the total angular momenta of the individual electrons and $\vec{J}$ is the angular momentum of the entire atom. A similar break down gives rise to Paschen-Back effect in atomic spectra in presence of strong external magnetic fields.

13.4 Spectroscopic Notation

We have the following spectroscopic notation for energy levels of an atom:

$$^{2S+1}L_j \tag{13}$$

where S is the total spin quantum number, L is the total orbital angular momentum quantum number and J is the total angular momentum quantum number of that particular level or state. Here we have used J, L and S without the vector notation by which we mean the quantum number instead of the actual magnitude of angular momentum. In analogy with the notation for electronic orbital angular momentum, we have the following desgination for atomic states:

L	:	0	1	2	3	4	5	...
Term designation	:	S	P	D	F	G	H	...

Consider the example of a two electron system such as helium ($1s^2$). For this system $l = 0$ for both electrons and the only way the spins can align is antiparallel (according to exclusion principle). Thus $S = 0$ for the combined electrons and $L = 0 + 0 = 0$ for the orbital angular momentum. This also makes $J = 0$ ($\because J = L + S$). Therefore, the term designation for the ground state of helium is $^1S_0$.

In case of excited states, the designations become a bit complicated since there are many possible ways of combining the electrons. Suppose that one of the electrons is excited to the 2s level, yielding a configuration of 1s 2s. This state (energy level of atom) also has $l = 0$ for both the electrons so $L = 0 + 0 = 0$. But, the spins can be either parallel or antiparallel (the exclusion principle doesn't apply because either electron has a different value of n). We therefore have two possible states for the spin $S = \frac{1}{2} + \frac{1}{2} = 1$ and $S = \frac{1}{2} - \frac{1}{2} = 0$. In the first case, the multiplicity (number of levels) is 1 even though $2S + 1 = 3$, since the multiplicity is determined by whichever is the smaller of $2S + 1$ and $2L + 1$. The second case ($S = 0$) leads to a multiplicity of 1. We thus have two possible states $^3S_1$ and $^1S_0$ for the electron configuration 1s 2s (the first is referred to as a *triplet* and the latter, as *singlet*).

For the excited state 1s 2p of helium, $l = 0$ for one electron and $l = 1$ for the other electron. This leads to only one possible value for L viz. $L = 0 + 1 = 1$ and so we have an angular momentum designation of P. The spins combine with two possibilities, leading to $S = 1$ and $S = 0$. Thus for $S = 1$, $L = 1$ we have $J = 1 + 1 = 2$. $J = 1 + 0 = 1$ and $J = 1 - 1 = 0$ (The first one corresponds to spin and orbital momentum aligned, the second, spin perpendicular to orbital and the third, spin antiparallel to orbital).

The possible values of J vary by integral values. Thus, we start with the maximum value of J ($= L + S$), decrease it by 1 for each state until we reach $J = (L - S)$:

$$1s\,(l_1 = 0,\ s_1 = \tfrac{1}{2}) \qquad\qquad 2p\,(l_2 = 1,\ s_2 = \tfrac{1}{2})$$

$$L = l_1 + l_2 = 0 + 1 = 1 \qquad S = s_1 + s_2 = \tfrac{1}{2} + \tfrac{1}{2} = 1,\ \ S = s_1 - s_2 = 0$$

$$S = 1: \vec{J} = \vec{L} + \vec{S} = 1 + 1 = 2 \Rightarrow\ ^3P_2$$

$$= 1 + 0 = 1 \Rightarrow\ ^3P_1$$

$$= 1 - 1 = 0 \Rightarrow\ ^3P_0$$

$$S = 0: \vec{J} = \vec{L} + \vec{S} = 1 + 0 = 1 \Rightarrow\ ^1P_1$$

Table 13.1: Ground state terms of some elements

Element	Atomic number	Electronic configuration (outer)	Term
H	1	1s	$^2S_{1/2}$
He	2	$1s^2$	$^1S_0$
Li	3	(He) 2s	$^2S_{1/2}$
Be	4	(He) $2s^2$	$^1S_0$
B	5	(He) $2s^2$ 2p	$^2P_{1/2}$
C	6	(He) $2s^2$ $2p^2$	$^3P_0$
N	7	(He) $2s^2$ $2p^3$	$^4S_{3/2}$
O	8	(He) $2s^2$ $2p^4$	$^3P_2$
F	9	(He) $2s^2$ $2p^5$	$^2P_{3/2}$

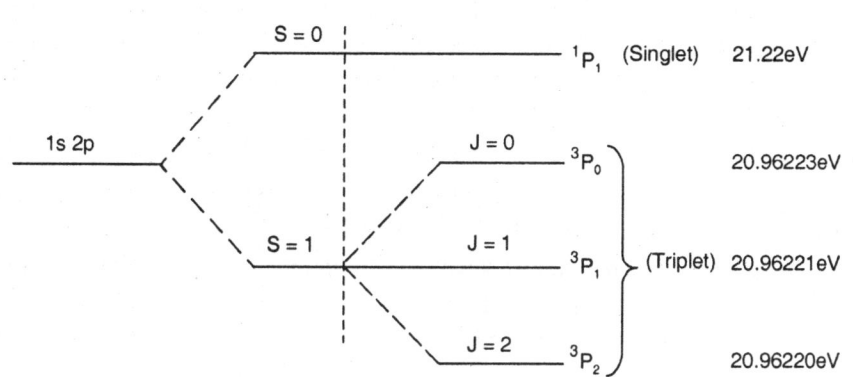

Fig. 13.4: Singlet and triplet energy levels of the excited state 1s 2p of the helium atom

Example 13.2: The Fig. 13.5 illustrates how vectors $\vec{L}$ and $\vec{S}$ combine vectorially to give specific spectrospic states.

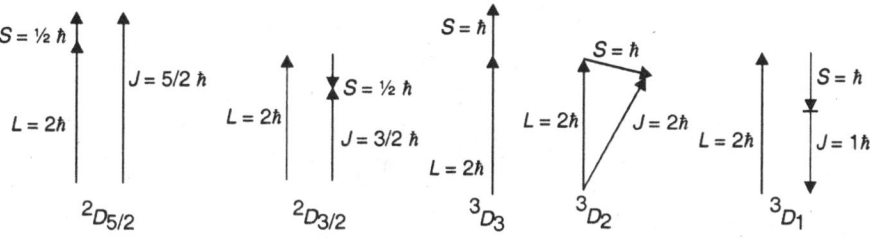

Fig. 13.5: Vectorial addition of $\vec{L}$ and $\vec{S}$ to produce resultant states $^{2S+1}L_J$ for (a) $S = \frac{1}{2}$, $L = 2$ and (b) $S = 1$, $L = 2$

13.5 The Vector Model of Angular Momentum

The total angular momentum eigenstates can be represented by two different sets of good quantum numbers in accordance with two different schemes of coupling of orbital and spin angular momenta in a vector model of atom. If a one electron atom is placed in a magnetic field, which is parallel to the z-axis, then the precession of $\vec{L}$ and $\vec{S}$ and $\vec{J}$ around the z-axis results in a constant value of their z-component, which must be one of the possible eigenvalues $(m_l + m_s)\hbar$ or $m_j\hbar$.

In the ψ_{jm_j} representation, the operators H, L^2, S^2, J^2 and J_z commute and so, there is a set of common eigenfunctions which satisfy:

$$H\psi_{jm_j} = E_n\psi_{jm_j}$$
$$L^2\psi_{jm_j} = l(l+1)\hbar^2\psi_{jm_j}$$
$$S^2\psi_{jm_j} = s(s+1)\hbar^2\psi_{jm_j} \qquad (14a)$$
$$J^2\psi_{jm_j} = j(j+1)\hbar^2\psi_{jm_j}$$
$$J_z\psi_{jm_j} = m_j\hbar\psi_{jm_j}$$

This *description corresponds to L S coupling* in which an applied magnetic field is weak compared with the extremely strong fields inside the atom so that $\vec{J} = (\vec{L} + \vec{S})$ precesses as a whole aroung the z-axis adjusting its orientation such such that its z-component J_z has one of the allowed values $m_j\hbar$. The good quantum numbers in this case are l, s, j and m_j (since $m_l\hbar$ and $m_s\hbar$ are varying in time due to precession) and ψ_{jm_j} are the suitable eigenfunctions.

For an applied field stronger than the internal fields of the atom, $j\,j$ coupling predominates and $\vec{L}$ and $\vec{S}$ adjust separately so that their z-components become (separately) constants of motion. In this case $\vec{L}$ and $\vec{S}$ precess independently about the z-axis. Thus L^2, S^2, L_z and S_z only are quantized so that m_l and m_s are good quantum numbers.

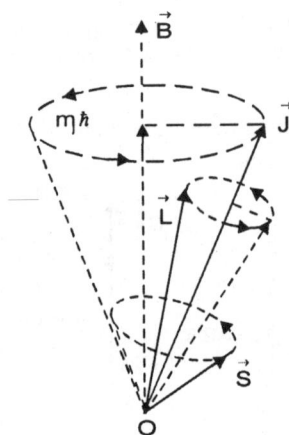

Fig. 13.6: Precession of $\vec{J}$ around the field direction with $J_z = m_j\hbar$ as a constant of motion

The eigenstates of the angular momentum can be described by the set of common eigenfunctions $\Psi_{m_l m_s}$ of the operators H, L^2, S^2, L_z and S_z which satisfy:

$$H\Psi_{m_l m_s} = E_n \Psi_{m_l m_s}$$

$$L^2\Psi_{m_l m_s} = l(l+1)\hbar^2 \Psi_{m_l m_s}$$

$$S^2\Psi_{m_l m_s} = s(s+1)\hbar^2 \Psi_{m_l m_s} \qquad (14b)$$

$$L_z\Psi_{m_l m_s} = m_l \hbar \Psi_{m_l m_s}$$

$$S_z\Psi_{m_l m_s} = m_s \hbar \Psi_{m_l m_s}$$

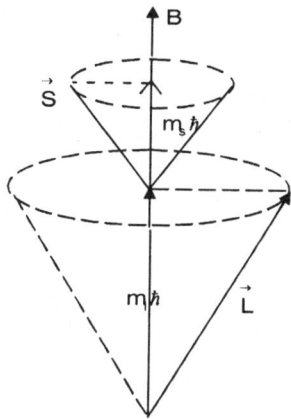

Fig. 13.7: Independent precession of $\vec{L}$ and $\vec{S}$ around the field direction with $m_l \hbar$ and $m_s \hbar$ as constants of motion

Here H denotes the central force Hamiltonian defined by

$$H = -\frac{\hbar^2}{2m}\left[\frac{1}{r^2}\frac{\partial}{\partial r}\left(r^2\frac{\partial}{\partial r}\right) + \frac{1}{r^2\sin\theta}\frac{\partial}{\partial\theta}\left(\sin\theta\frac{\partial}{\partial\theta}\right) + \frac{1}{r^2\sin^2\theta}\frac{\partial^2}{\partial\phi^2}\right] + V(r)$$

Example 13.3: In case of L–S coupling, an atom is placed in a magnetic field $\vec{B}$. Show that

$$\vec{\mu} = G\vec{J} = \frac{e}{2Mc}\left(\vec{J} + \vec{S}\right)$$

and $\quad \left\langle njl'm_j'|\mu_z B|njlm_j\right\rangle = m_j g_L \dfrac{e\hbar}{2Mc} B\delta_{m_j m_j'}\delta_{ll'}$

where $\quad g_L = 1 + \dfrac{j(j+1) + s(s+1) - l(l+1)}{2j(j+1)}$

Solution: Refer to Figs 13.6 and 13.8

$$\left|\vec{\mu}_l\right| = \frac{e}{2Mc}\left|\vec{L}\right| ; \quad \left|\vec{\mu}_s\right| = 2\left(\frac{e}{2Mc}\right)\left|\vec{S}\right|$$

$$\vec{J} = \vec{L} + \vec{S}$$

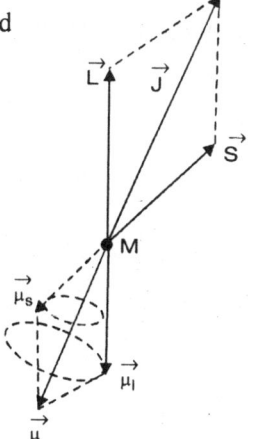

Fig. 13.8

In external magnetic field, $\vec{L}$ and $\vec{S}$ precess round $\vec{J}$ and $\vec{J}$ precesses round the field direction.

$$\vec{\mu} = \vec{\mu}_l + \vec{\mu}_s$$

or $\quad \mu =$ component $\vec{\mu}_l$ along $\vec{J}$ + component of $\vec{\mu}_s$ along $\vec{J}$

or

$$\mu = \left[L \cos(\vec{L}, \vec{J}) + 2S \cos(\vec{S}, \vec{J}) \right] \frac{e}{2Mc}$$

From cosine law,

$$S^2 = L^2 + J^2 - 2LJ \cos(\vec{L}, \vec{J})$$

$\therefore \qquad L \cos(\vec{L}, \vec{J}) = \dfrac{J^2 + L^2 - S^2}{2J}$

Similarly

$$S \cos(\vec{S}, \vec{J}) = \frac{J^2 + S^2 - L^2}{2J}$$

$\therefore \qquad \mu = \left[\dfrac{J^2 + L^2 - S^2}{2J} + 2 \dfrac{J^2 + S^2 - L^2}{2J} \right] \dfrac{e}{2Mc}$

or $\qquad \mu = \left[1 + \dfrac{J^2 + S^2 - L^2}{2J^2} \right] J \dfrac{e}{2Mc}$

or $\qquad |\vec{\mu}| = \left[1 + \dfrac{j(j+1) + s(s+1) - l(l+1)}{2j(j+1)} \right] J \dfrac{e}{2Mc}$ $\qquad$ (i)

$$\equiv g_L J \left(\frac{e}{2Mc} \right) \equiv GJ \qquad (ii)$$

(g_L is called Lande's splitting factor).

Now $\qquad \hat{L}^2 = (\hat{J} - \hat{S})^2 = (\hat{J} - \hat{S}) \cdot (\hat{J} - \hat{S})$

or $\qquad \hat{S} \cdot \hat{J} = \dfrac{1}{2}(\hat{J}^2 + \hat{S}^2 - \hat{L}^2)$ $\qquad$ (iii)

Comparing (i) and (ii) with (iii), we get

$$g_L = 1 + \left\langle \frac{\hat{S} \cdot \hat{J}}{\hat{J}^2} \right\rangle \qquad (iv)$$

Now, the vector $\vec{S}$ is a non-constant vector $(\vec{J} = \vec{L} + \vec{S})$ and precesses about the constant vector $\vec{J}$, therefore only the components of $\vec{S}$ parallel to $\vec{J}$ is relevant (other components average out). So, we can represent $\vec{S}$ as

$$\langle \vec{S} \rangle = \left(\vec{S} \cdot \frac{\vec{J}}{|\vec{J}|} \right) \frac{\vec{J}}{|\vec{J}|} = \left(\frac{\vec{S} \cdot \vec{J}}{J^2} \right) \vec{J}$$

This gives

$$\langle S_z \rangle = \left(\frac{\vec{S} \cdot \vec{J}}{J^2} \right) J_z = (g_L - 1) J_z \qquad \text{(from Eqn. iv)} \qquad (v)$$

or
$$\vec{S}_z = (g_L - 1)\vec{J}_z$$

or
$$\left(\vec{S}_z + \vec{J}_z\right) = g_L \vec{J}_z \qquad (vi)$$

From Eqn. (*iii*), we get

$$\mu_z = g_L \left(\frac{e}{2Mc}\right) J_z$$

Eqn. (*vi*) gives

$$\left(\vec{J} + \vec{S}\right) = g_L \vec{J}$$

Hence
$$\vec{\mu} = \frac{e}{2Mc} g_L \vec{J} = \frac{e}{2Mc}\left(\vec{J} + \vec{S}\right)$$

and from Eqn. (*ii*)

$$\langle \mu_z \rangle = \frac{e}{2Mc} g_L \langle \hat{J}_z \rangle$$

(Suppose $\vec{B}$ is directed along the z-axis)

Then,

$$\left\langle njl'm_j' \middle| \mu_z B \middle| njlm_j \right\rangle \equiv \frac{e}{2Mc} g_L B \left\langle njl'm_j' \middle| \hat{J}_z \middle| njlm_j \right\rangle$$

$$= \frac{e}{2Mc} g_L B m_j \hbar \delta_{m_j m_j'} \delta_{ll'}$$

$$= m_j g_L \left(\frac{e\hbar}{2Mc}\right) B \delta_{m_j m_j'} \delta_{ll'}$$

13.5.1 Ordering of Atomic Energy-Levels in L S Coupling (Hund's Rules)

In $L\,S$ coupling, first spin orbit interaction is ignored and ordering is determined qualitatively. Then spin orbit interaction is introduced as a small perturbation to deduce the final multiplicity of energy levels.

In determining ordering of energy eigenstates, the Pauli's principle does not permit more than two electrons in a single state and for the state with the largest total spin, the spins are essentially parallel to each other. There is thus a tendency of the electrons to avoid each other, so as to minimize the mutual repulsion. Consequently, the lowest energy state is generally the state with the largest total spin S. The filling of the electron in energy levels and consequently ordering of levels in the atomic or ionic shells is governed by Hund's rules.

Hund's first rule: Out of the many states one can form by placing N electrons into the $2(2l + 1)$ levels* of the partially filled shell, those that lie lowest in energy have the largest total spin S** that is consistent with the exclusion principle. Thus if $N \le 2l + 1$, all electrons can have parallel spins without multiple occupation.

---

* There are $2(2l + 1)$ one electron levels.

** The largest value S can have is equal to the largest magnitude S_z can have.

Then $S = N/2$ when $N \leq 2l + 1$. When $N = 2l + 1$, S has its maximum $l + \frac{1}{2}$. Since electrons after the $(2l + 1)^{\text{th}}$ are required by the exclusion principle to have their spins opposite to the spins of the first $2l + 1$ electrons, S is reduced from its maximum value by half a unit for each electron filling after the $(2l + 1)^{\text{th}}$.

Hund's second rule: The total angular momentum L of the lowest lying states has the largest value that is consistent with the Hund's first rule. In other words the ground state would have the largest possible S value. The largest possible L value is equal to the largest magnitude that L_z can have. Thus the first electron in the shell will go into a level with $|l_z|$ equal to its maximum value l. The second (according to Hund's first rule), must have the same spin as the first and so, is forbidden by the exclusion principle from having the same value of l_z. It would have $|l_z| = l - 1$, leading to a total L of $l + (l - 1) = 2l - 1$. Continuing in this way, we will have $L = l + (l - 1) + ... + (l - (n - 1))]$, for a shell less than half filled. When the shell is exactly half filled, all values of l_z are to be taken, then $L = 0$. The second half of the shell is filled with electrons with spin opposite to those in the first half and then the exclusion principle allows us again to permit L to have the same series of values which we traversed in filling the first half. Thus, due to Hund's second rule, out of terms of a given multiplicity, the lowest one will have greatest L. Further, for a given S, the energy of the states will increase as L decreases.

Hund's third rule: The first two rules determine the values of L and S assumed by the states of lowest energy. However, there are $(2L + 1)(2S + 1)$ possible states, which can take on all integral J values between $|L - S|$ and $L + S$. Thus, finally, the degeneracy of the $(2L + 1)(2S + 1)$ states is removed by the consideration of spin orbit interaction. There is one more rule known as the *Multiplet Rule* which governs the ground state configuration in $L S$ coupling. It relies on the sign of the constant C in the perturbation

$$H' = C\vec{L} \cdot \vec{S} \tag{15}$$

where C is a constant for a given L and S. This gives a perturbative contribution to energy

$$\Delta E = \frac{1}{2}C\left[J(J+1) - L(L+1) - S(S+1)\right] \tag{16i}$$

It is found that *the constant C is positive for multiplets formed from a subshell that is half or less than half filled and negative for multiplets formed from a subshell which is more than half filled.* As a result, for a subshell which is half filled or less, the energy within the multiplet increases as J increases, and for a subshell which is more than half filled, the energy within the multiplet decreases as J increases (*Multiplet rule*). This means that *the ground state of an atom has the smallest J value subject to Hund's rule if the subshell is half filled or less, and the largest J value if the subshell is more than half filled.*

i.e.
$$J = |L - S|; \ N \leq 2l + 1$$
$$J = L + S; \ N \geq 2l + 1 \tag{16ii}$$

The separation between the levels with values $(J + 1)$ and J (but the same L and S) is obtained as

$$E_{J+1} - E_J = C(J+1) \qquad (17)$$

This is known as the *Lande's interval rule*. It states that *the spacing between consecutive levels of a fine structure multiplet is proportional to the larger of the two J-values of the levels.*

The Hund's rule may appear to be difficult to understand but are easier to apply in principle. The Table 13.2 illustrates the application of Hund's rules in constructing ground state of ions with partially filled *d*-shells.

In terms of term values Hund's rule is as follows:

Of terms for equivalent electrons,

(1) the one with the greatest multiplicity $2S + 1$ lies lowest and term energies increase as the multiplicity decreases, and

(2) of terms of a given multiplicity, the lowest is that with greatest *L*. Then, energy of the states (of given *S*) increases as *L* decreases.

The degeneracy of the states having same *L* and same *S* values but different *J* values is lifted due to spin orbit interaction and ordering of the levels is governed by the multiplet rule (see the last column of the Table 13.2).

Table 13.2: Application of Hund's rules to ground states of ions with partially filled *d* shells ($l = 2$)
(After "Solid State Physics", Ashcroft and Mermin,
Harcourt Brace College Publishers, U.S.A.)

N	$l_z=$ 2,	1,	0,	−1,	−2	S	$L=\|\Sigma l_z\|$	J	$^{2S+1}L_J$
1	↓					½	2	3⁄2	$^2D_{3/2}$
2	↓	↓				1	3	2	$^3F_2$
3	↓	↓	↓			3⁄2	3	3⁄2	$^4F_{3/2}$
4	↓	↓	↓	↓		2	2	0	$^5D_0$
5	↓	↓	↓	↓	↓	5⁄2	0	5⁄2	$^6S_{5/2}$
6	↓↑	↑	↑	↑	↑	2	2	4	$^5D_4$
7	↓↑	↓↑	↑	↑	↑	3⁄2	3	9⁄2	$^4F_{9/2}$
8	↓↑	↓↑	↓↑	↑	↑	1	3	4	$^3F_4$
9	↓↑	↓↑	↓↑	↓↑	↑	½	2	5⁄2	$^2D_{5/2}$
10	↓↑	↓↑	↓↑	↓↑	↓↑	0	0	0	$^1S_0$

For N = 1 to 4: $J = |L - S|$

For N = 6 to 9: $J = L + S$

13.6 Spin Orbit Interaction and Total Angular Momentum

The electron is a charged particle and associated with its intrinsic rotation (spin), there is a spin magnetic moment

$$\vec{\mu}_s = \frac{e\vec{S}}{m_e c} \qquad (18)$$

In an atom, this interacts with the magnetic field created by its orbital motion, which is proportional to orbital angular momentum $\vec{L}$.

'Without spin consideration' the unperturbed Hamiltonian is

$$H_0 = \frac{\vec{p}^2}{2m_e} + V_c(r) \tag{19}$$

where $V_c(r) = -Ze^2/r$ is the central (spin independent) potential of the nucleus.

Due to the interaction of spin with the orbital angular momentum, there arises a perturbation term which is obtained from the Dirac's relativistic equation, and is given by

$$H' = \frac{1}{2m_e^2 c^2} \vec{S} \cdot \vec{L} \frac{1}{r} \frac{d}{dr} V_c(r) \tag{20}$$

In order to understand this equation, we note that an electron moving with velocity $\vec{v}$ through an electric field $\vec{E}$ (of the atom) will experience a magnetic field equal to

$$\vec{B}_{eff} = -\frac{\vec{v}}{c} \times \vec{E}$$

This leads to an interaction energy

$$H' = -\vec{\mu}_s \cdot \vec{B}_{eff} \quad 22$$

$$= -\vec{\mu}_s \cdot \left(-\frac{\vec{v}}{c} \times \vec{E} \right) = \frac{e}{m_e c} \vec{S} \cdot \frac{\vec{v}}{c} \times \vec{E}$$

$$= -\frac{e}{m_e c^2} \vec{S} \cdot \vec{v} \times \vec{E}$$

$$= -\frac{e}{m_e^2 c^2} \left(\vec{E} \times \vec{p} \right) \cdot \vec{S} \tag{21}$$

Now, field

$$e\vec{E} = -\nabla V_c = -\hat{r} \frac{\partial V_c}{\partial r} = -\vec{r} \left(\frac{1}{r} \frac{dV_c}{dr} \right)$$

$$\therefore \quad H' = \frac{1}{m_e^2 c^2} \frac{1}{r} \frac{dV_c}{dr} \vec{r} \times \vec{p} \cdot \vec{S} \text{ or } H' = \frac{1}{m_e^2 c^2} \frac{1}{r} \frac{dV_c}{dr} \left(\vec{L} \cdot \vec{S} \right) \tag{22}$$

The magnitude of interaction energy calculated in this way turns out to be larger by a factor of 2. This discrepancy arises because, here we have considered relative orbital motion of the nucleus about the electron, and the frame of electron is not an inertial frame. There is a classical explanation for this due to spin precession (known as *Thomas-precession* after L.H. Thomas)(*). Due to this, the correct result, after taking into account relativistic kinematic considerations is

$$(H = H_0 + H') \tag{23}$$

_____

(*) See: J.D. Jackson, Classical Electrodynamics (Wiley, New York, 1975).

where, for the hydrogen atom, we have

$$H' = \xi(r)\vec{L} \cdot \vec{S} = \xi(r)\left(L_x S_x + L_y S_y + L_z S_z\right) \tag{24}$$

with

$$\xi(r) = \frac{1}{2m_e^2 c^2}\frac{Ze^2}{r^3} \qquad \left(\because \ dV_c \ / \ dr = Ze^2 \ / \ r^2\right)$$

It can be seen that L_z and S_z commute with H_0 but they do not commute with H'. In fact,

$$\left[L_z, H\right] = \left[L_z, H_0\right] + \left[L_z, H'\right] \qquad \text{(using Eqn. 23)}$$

$$= \xi\left[L_z, \vec{L} \cdot \vec{S}\right]$$

$$= \xi\left(\left[L_z, L_x\right]S_x + \left[L_z, L_y\right]S_y + \left[L_z, L_z\right]S_z\right)$$

$$= i\hbar\xi\left(L_y S_x - L_x S_y\right) \qquad \left(\because \ \left[L_z, L_x\right] = i\hbar L_y \ \text{etc}\right)$$

i.e. L_z does not commute with H.

Similarly

$$\left[S_z, H\right] = i\hbar\xi\left(L_x S_y - L_y S_x\right) = -\left[L_z, H\right]$$

Thus,

$$\left[L_z + S_z, H\right] = 0$$

i.e. $L_z + S_z = J_z$ the z-component of the total angular momentum operator (and similarly J_x and J_y) commute with H.

$$\left[J_x, H\right] = \left[J_y, H\right] = \left[J_z, H\right] = 0 \tag{26}$$

Hence if we now apply perturbation theory to hydrogenic atoms *using H' as the perturbation*, then (with just H_0) we are free to choose either of the two sets as the base kets:

Set – 1: The eigenkets of $\vec{L}^2, \vec{S}^2, L_z, S_z$

Set –2: The eigenkets of $\vec{L}^2, \vec{S}^2, \vec{J}^2, J_z$ $\qquad$ (27)

Without H', either set is satisfactory in the sense that the base kets are also energy eigenkets. With H' added, it is far better to use set-2 because $\vec{L} \cdot \vec{S}$ (which does not commute with L_z and S_z) commutes with $\vec{J}^2$ and J_z (we have to choose those unperturbed kets that diagonalize the perturbation).

We note that

$$\vec{S} + \vec{L} = \vec{J} \tag{28}$$

$\Rightarrow$

$$\vec{S}^2 + 2\vec{S} \cdot \vec{L} + \vec{L}^2 = \vec{J}^2$$

or

$$\vec{S} \cdot \vec{L} = \frac{1}{2}\left(\vec{J}^2 - \vec{L}^2 - \vec{S}^2\right) \tag{29}$$

(for a given n and l, there are $2(2l + 1)$ degenerate eigenstates of H_0). If we combine the degenerate eigenfunctions into linear combinations that are eigenfunctions of $\vec{J}^2$ (they already are eigenfunctions of J_z) then these linear combinations will diagonalize H'.

The states $\left| j, m_j, l \right\rangle$ or $\left| l \pm \frac{1}{2}, m_j, l \right\rangle$ diagonalize the operator $\vec{S} \cdot \vec{L}$, where j, m_j, l and s are defined as

$$J^2 = \hbar^2 j(j+1)$$
$$j = \left(l \pm \frac{1}{2} \right)$$
$$S^2 = \hbar^2 s(s+1)$$
$$L^2 = \hbar^2 l(l+1)$$
$$J_z = m_j \hbar$$

$$\vec{S} \cdot \vec{L} \left| l \pm \tfrac{1}{2}, m_j, l \right\rangle = \frac{\hbar^2}{2} \left(j(j+1) - l(l+1) - s(s+1) \right) \left| l \pm \tfrac{1}{2}, m_j, l \right\rangle$$

$$= \frac{\hbar^2}{2} \left(\begin{array}{c} \left(l + \tfrac{1}{2} \right) \left(l + \tfrac{3}{2} \right) - l(l+1) - \tfrac{1}{2} \times \tfrac{3}{2} \\ \left(l - \tfrac{1}{2} \right) \left(l + \tfrac{1}{2} \right) - l(l+1) - \tfrac{1}{2} \times \tfrac{3}{2} \end{array} \right) \left| l \pm \tfrac{1}{2}, m_j, l \right\rangle$$

$$= \frac{\hbar^2}{2} \left(\begin{array}{c} l^2 + 2l + \left(\tfrac{3}{4} \right) - l^2 - l - \left(\tfrac{3}{4} \right) \\ l^2 - \left(\tfrac{1}{4} \right) - l^2 - l - \left(\tfrac{3}{4} \right) \end{array} \right) \left| l \pm \tfrac{1}{2}, m_j, l \right\rangle$$

$$= \frac{\hbar^2}{2} \left(\begin{array}{c} l \\ -l-1 \end{array} \right) \left| l \pm \tfrac{1}{2}, m_j, l \right\rangle \tag{30}$$

The operators H, $\vec{J}^2$, J_z, $\vec{L}^2$, $\vec{S}^2$ now form a complete set and the eigenfunctions can be characterized by their eigenvalues.

The correct unperturbed states (for perturbation theory) are therefore

$$\left\langle \vec{x} \middle| n, j = l \pm \tfrac{1}{2}, m_j, l \right\rangle$$

$$= R_{nl}(r) \left(\alpha_\pm Y_{lm_j - \frac{1}{2}} (\theta, \phi) \middle| \uparrow \right\rangle + \beta_\pm Y_{lm_j + \frac{1}{2}} (\theta, \phi) \middle| \downarrow \right\rangle \right) \tag{31}$$

where the coefficients $\alpha_\pm$ and $\beta_\pm$ are defined as

$$\alpha_\pm = \pm \sqrt{\frac{l \pm m_j + \frac{1}{2}}{2l + 1}} = \pm \beta_\mp \tag{32}$$

(These are known as *C.G. coefficients* and derived in section 13.18. The states (Eqn. 31) are eigenstates of H_0 with energy eigenvalues E_n swhich are $2n^2$ fold degenerate (and diagonalize the perturbation).

The correction from first order perturbation theory yields

$$\left\langle H' \right\rangle_{n, j = l \pm \frac{1}{2}, l, m_j} = \frac{1}{2m_e^2 c^2} \frac{\hbar^2}{2} \left(\begin{array}{c} l \\ -l-1 \end{array} \right) Ze^2 \left\langle \frac{1}{r^3} \right\rangle_{nl} \tag{33}$$

Substituting

$$\left\langle \frac{1}{r^3} \right\rangle = \int_0^\infty \frac{1}{r^3} |R_{nl}|^2 r^2 dr = \frac{1}{n^3 l\left(l + \frac{1}{2}\right)(l+1)} \left(\frac{Z}{a_0}\right)^3 .$$

in Eqn. (33), we get

$$\langle H' \rangle = \frac{e^2 \hbar^2 Z^4}{4m_e^2 c^2 a_0^3 n^3 l\left(l + \frac{1}{2}\right)(l+1)} \begin{pmatrix} l \\ -l-1 \end{pmatrix} \qquad (34)$$

Thus,

$$E_{nl} = \begin{cases} E_n + \frac{1}{2} l \zeta_{nl} & ; \ j = l + \frac{1}{2} \\ E_n - \frac{1}{2}(l+1)\zeta_{nl} & ; \ j = l - \frac{1}{2} \end{cases} \qquad (35)$$

$$\zeta_{nl} = \frac{e^2 \hbar^2 Z^4}{2m_e^2 c^2 a_0^3 n^3 l\left(l + \frac{1}{2}\right)(l+1)} \qquad (36)$$

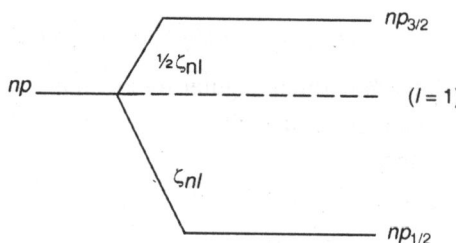

Fig. 13.9: Splitting of a np-state due to spin-orbit coupling

The s-electron state (with $l = 0$) are unaffected, and the splitting of all other states is proportional to $(2l + 1)$. The level corresponding to $j = l + \frac{1}{2}$ will be higher than that for $j = l - \frac{1}{2}$ (and $E'_n = E_n \pm \langle H' \rangle$).

Example 13.4: Using the vector model of angular momentum, determine the allowed values for the angle between vector observables $\vec{L}$ and $\vec{S}$ of an electron in a one electron atom.

Solution: Vectors $\vec{L}$ and $\vec{S}$ precess around $\vec{J}$ and $\hat{J}^2 = \hat{L}^2 + \hat{S}^2 + 2\hat{L} \cdot \hat{S}$

i.e.

$$\hat{L} \cdot \hat{S} = \frac{1}{2}\left(\hat{J}^2 - \hat{L}^2 - \hat{S}^2\right)$$

or, in the eigenstates $|jm_j\rangle$ of the total angular momentum, we have

$$\langle \hat{L} \cdot \hat{S} \rangle = \frac{\hbar^2}{2}\{j(j+1) - l(l+1) - s(s+1)\}$$

The angle between the two vectors $\vec{L}$ and $\vec{S}$ is given by

$$\left\langle \hat{L} \cdot \hat{S} \right\rangle = \left| \vec{L} \right| \, \left| \vec{S} \right| \, \cos\left(\vec{L}, \vec{S} \right)$$

or

$$\cos\left(\vec{L}, \vec{S} \right) = \frac{\left\langle \hat{L} \cdot \hat{S} \right\rangle}{\left| \vec{L} \right| \, \left| \vec{S} \right|} = \frac{\left\langle \hat{L} \cdot \hat{S} \right\rangle}{\sqrt{\left\langle \hat{L}^2 \right\rangle \left\langle \hat{S}^2 \right\rangle}} = \frac{j(j+1) - l(l+1) - s(s+1)}{2\sqrt{l(l+1)\, s(s+1)}}$$

where

$$j = l \pm 1/2 \text{ and } s = 1/2 \, .$$

13.7 Generalized Angular Momentum

We have seen that the definition $\vec{L} = \vec{r} \times \vec{p}$ of the angular momentum cannot admit $\hbar/2$ as a possible value rather $\vec{J} = \vec{L} + \vec{S}$ commutes with the Hamiltonian, we therefore define a general angular momentum vector $\vec{J}$ (with the components J_x, J_y and J_z) satisfying the commutation relations

$$\left[J_x, J_y \right] = i\hbar J_z \, ; \left[J_y, J_z \right] = i\hbar J_x \, ; \left[J_z, J_x \right] = i\hbar J_y \tag{37}$$

$\vec{J}$ may be in particular the orbital angular momentum $\vec{L}$, or the spin angular momentum $\vec{S}$ or a combination of both.

We define the operators,

$$J_+ = J_x + iJ_y \text{ and } J_- = J_x - iJ_y$$

and

$$J^2 = J_x^2 + J_y^2 + J_z^2 \tag{38i}$$

Now,

$$\left[J^2, J_x \right] = \left[J_x^2 + J_y^2 + J_z^2, J_x \right]$$

$$= \left[J_x^2, J_x \right] + \left[J_y^2, J_x \right] + \left[J_z^2, J_x \right]$$

$$= \left[J_y^2, J_x \right] + \left[J_z^2, J_x \right] \qquad \left(\because \left[J_x^2, J_x \right] = 0 \right) \tag{38ii}$$

$$\left[J_y^2, J_x \right] = J_y^2 J_x - J_x J_y^2$$

$$= J_y J_y J_x - J_x J_y J_y$$

$$= J_y \left(J_y J_x - J_x J_y \right) + \left(J_y J_x - J_x J_y \right) J_y$$

$$= J_y \left(-i\hbar J_z \right) + \left(-i\hbar J_z \right) J_y$$

$$= -i\hbar \left(J_y J_z + J_z J_y \right) \tag{38iii}$$

Similarly,

$$\left[J_z^2, J_x \right] = i\hbar \left(J_y J_z + J_z J_y \right) \tag{38iv}$$

So, from Eqn. (38ii, iii and iv)

$$\left[J^2, J_x \right] = 0 \tag{39}$$

Likewise,

$$\left[J^2, J_y\right] = 0 \text{ and } \left[J^2, J_z\right] = 0$$

i.e., J^2 and J_z commute, so they will have simultaneous eigenfunctions.

13.7.1 Relations between J^2, J_z, J_+ and J_-

$$\left[J_z, J_+\right] = \left[J_z, J_x + iJ_y\right]$$
$$= \left[J_z, J_x\right] + i\left[J_z, J_y\right]$$
$$= i\hbar J_y + i\left(-i\hbar J_x\right)$$
$$= \hbar\left(iJ_y + J_x\right) = \hbar J_+$$

So, $$\left[J_z, J_+\right] = \hbar J_+ \qquad (40i)$$

Similarly, $$\left[J_z, J_-\right] = -\hbar J_- \qquad (40ii)$$

Now, $$J_- J_+ = \left(J_x - iJ_y\right)\left(J_x + iJ_y\right)$$
$$= J_x\left(J_x + iJ_y\right) - iJ_y\left(J_x + iJ_y\right)$$
$$= J_x^2 + i\left[J_x, J_y\right] + J_y^2$$
$$\Rightarrow \qquad J_x^2 + J_y^2 = J_- J_+ - i\left[J_x, J_y\right]$$
$$= J_- J_+ - i\left(i\hbar J_z\right)$$
$$= J_- J_+ + \hbar J_z$$

$$\Rightarrow \qquad J^2 = J_x^2 + J_y^2 + J_z^2 \text{ or } J^2 = J_- J_+ + J_z^2 + \hbar J_z \qquad (41i)$$

Similarly,

$$J_+ J_- = \left(J_x + iJ_y\right)\left(J_x - iJ_y\right)$$
$$= J_x^2 + iJ_y J_x - iJ_x J_y - i^2 J_y^2$$
$$= J_x^2 + J_y^2 - i\left[J_x, J_y\right]$$
$$= J_x^2 + J_y^2 - i\left(i\hbar J_z\right) = J_x^2 + J_y^2 + \hbar J_z$$

or $$J_x^2 + J_y^2 = J_+ J_- - \hbar J_z \quad \therefore \quad J^2 = J_+ J_- + J_z^2 - \hbar J_z \qquad (41ii)$$

Adding Eqns (41i and ii), we get

$$2J^2 = J_- J_+ + J_+ J_- + 2J_z^2 \text{ or } J^2 = \frac{1}{2}\left(J_+ J_- + J_- J_+\right) + J_z^2 \qquad (42)$$

It can also be proved that

$$\left[J_+, J_-\right] = 2\hbar J_z \qquad (43)$$

Example 13.5: Prove that (a) $\left[J_z, J_+^n\right] = n\hbar J_+^n$ and (b) $\left[J_z, J_-^n\right] = -n\hbar J_-^n$.

Solution: Since J^2 and J_z commute, they possess simultaneous eigenfunctions. Let a and b be eigenvalues of J^2 and J_z respectively for the (simultaneous) state $|\psi>$ i.e.

$$J^2|\psi> = a|\psi> \text{ and } J_z|\psi> = b|\psi>$$

Now,

$$\left[J_z, J_+\right] = J_z J_+ - J_+ J_z$$

∴
$$J_z J_+ = \left[J_z, J_+\right] + J_+ J_z \qquad (i)$$

$$J_z J_+ |\psi\rangle = \left[J_z, J_+\right]|\psi\rangle + J_+ J_z |\psi\rangle$$

$$= \hbar J_+ |\psi\rangle + J_+ b |\psi\rangle \quad \left(\because \left[J_z, J_+\right] = \hbar J_+\right)$$

Therefore,

$$J_z J_+ |\psi\rangle = (b + \hbar) J_+ |\psi\rangle \qquad (ii)$$

Further,

$$J_z J_+^2 |\psi\rangle = J_z J_+ \left(J_+ |\psi\rangle\right)$$

$$= \left\{ \left[J_z, J_+\right] + J_+ J_z \right\} J_+ |\psi\rangle \qquad \text{(using Eqn. (i) for } J_z J_+ \text{)}$$

$$= \left\{ \hbar J_+ + J_+ J_z \right\} J_+ |\psi\rangle$$

$$= \hbar J_+^2 |\psi\rangle + J_+ J_z J_+ |\psi\rangle$$

$$= \hbar J_+^2 |\psi\rangle + J_+ (b + \hbar) J_+ |\psi\rangle \qquad \text{(using ii)}$$

$$= \hbar J_+^2 |\psi\rangle + (b + \hbar) J_+^2 |\psi\rangle$$

$$= (b + 2\hbar) J_+^2 |\psi\rangle$$

By induction

$$J_z J_+^n |\psi\rangle = (b + n\hbar) J_+^n |\psi\rangle$$

Similarly,

$$J_z J_-^n |\psi\rangle = (b - n\hbar) J_-^n |\psi\rangle$$

$$\left[J_z, J_+^n\right] = J_z J_+^n - J_+^n J_z = (b + n\hbar)J_+^n - J_+^n J_z \qquad (iii)$$

But
$$J_+^n J_z |\psi\rangle = b J_+^n |\psi\rangle \qquad (iv)$$

∴   From (iii) and (iv) $\left[J_z, J_+^n\right] = n\hbar J_+^n$. Similarly, we can prove (b).

13.8 The Eigenvalue Spectrum for $\vec{J}^2$ and J_z and Matrix Elements for J_+ and J_-

Let us denote the eigenvalue of J_z by $m\hbar$ and that of J^2 by $\lambda\hbar^2$. Then the simultaneous eigenkets of J^2 and J_z can be written as $|\lambda m\rangle$ and the eigenvalue equations as:

$$J^2 \left| \lambda m \right\rangle = \lambda \hbar^2 \left| \lambda m \right\rangle \tag{44}$$

$$J_z \left| \lambda m \right\rangle = m \hbar \left| \lambda m \right\rangle \tag{45}$$

Now, we will see that the eigenvalues λ and m satisfy $\lambda \geq m^2$. We have ladder operators J_+ and J_- which are adjoint to each other.

$$J_+ = J_-^\dagger , J_- = J_+^\dagger$$

and since J_x, J_y and J_z are observables, they are represented by real operators

$$J_x^* = J_x , \text{etc.}$$

so that

$$\left\langle \lambda m \right| J^2 - J_z^2 \left| \lambda m \right\rangle = \left\langle \lambda m \right| J_x^2 + J_y^2 \left| \lambda m \right\rangle$$

representing $\qquad \left\langle \lambda m \right| \equiv \left\langle P \right|$ and $\left| \lambda m \right\rangle \equiv \left| P \right\rangle$

We have

$$\left\langle \lambda m \right| J^2 - J_z^2 \left| \lambda m \right\rangle = \left\langle P \right| J_x J_x \left| P \right\rangle + \left\langle P \right| J_y J_y \left| P \right\rangle$$

or $\qquad \left(\lambda - m^2 \right) \hbar^2 \left\langle \lambda m \middle| \lambda m \right\rangle = \left\langle Q \middle| Q \right\rangle + \left\langle R \middle| R \right\rangle \geq 0 \qquad (46)$

where $\left| Q \right\rangle = J_x \left| P \right\rangle$ and $\left| R \right\rangle = J_y \left| P \right\rangle$ and because J_x and J_y are real operators

$\therefore \qquad\qquad \left\langle Q \right| = \left\langle P \right| J_x^* = \left\langle P \right| J_x$

and $\qquad\qquad \left\langle R \right| = \left\langle P \right| J_y^* = \left\langle P \right| J_y$

Thus, from Eqn. (46), we get

$$\lambda - m^2 \geq 0 \text{ or } \lambda \geq m^2 \tag{47}$$

Now, we consider the operation of $J_z J_+$ on $\left| \lambda m \right\rangle$. We have

$$\left[J_z , J_+ \right] = \hbar J_+ \text{ or } J_z J_+ - J_+ J_z = \hbar J_+ \text{ or } J_z J_+ = \hbar J_+ + J_+ J_z$$

$\Rightarrow \qquad J_z J_+ \left| \lambda m \right\rangle = \left(\hbar J_+ + J_+ J_z \right) \left| \lambda m \right\rangle$

$$= \hbar J_+ \left| \lambda m \right\rangle + J_+ J_z \left| \lambda m \right\rangle$$

$$= \hbar J_+ \left| \lambda m \right\rangle + J_+ (\hbar m) \left| \lambda m \right\rangle$$

$$= \hbar (m+1) J_+ \left| \lambda m \right\rangle \tag{48i}$$

Similarly, $\quad J_z J_- \left| \lambda m \right\rangle = \hbar (m-1) J_- \left| \lambda m \right\rangle \tag{48ii}$

Eqns (48) show that $J_\pm \left| \lambda m \right\rangle$ are also eigenkets of J_z with eigenvlaues $(m \pm 1)\hbar$. We may therefore write

$$J_+ \left| \lambda m \right\rangle = C_+ \left| \lambda , m+1 \right\rangle$$

and $\qquad\qquad J_- \left| \lambda m \right\rangle = C_- \left| \lambda , m-1 \right\rangle \tag{49}$

where C_+ and C_- are proportionality constants and can be determined by the normalization condition of $J_\pm \left| \lambda m \right\rangle$. For a given vlaue of λ, the inequality (47) limits the magnitude of m. If j is the greatest value of m for any given λ, application of the raising operator J_+ to the eigenket $\left| \lambda j \right\rangle$ should not lead to a new eigenket, therefore

$$J_+ \left| \lambda j \right\rangle = 0$$

operating on the left by J_-, we obtain

$$J_- J_+ |\lambda j\rangle = 0$$

or $$\left(J^2 - J_z^2 - \hbar J_z \right) |\lambda j\rangle = 0 \qquad \text{(using Eqn. (41}i\text{)}$$

or $$\left(\lambda - j^2 - j \right) \hbar |\lambda m\rangle = 0$$

Since, $|\lambda j\rangle \neq 0$, therefore $\lambda - j^2 - j = 0$

or $$\lambda = j(j+1) \qquad (50)$$

Similarly, if j' is the lowest value of m, then

$$J_- |\lambda j'\rangle = 0$$

operating on the left by J_+, we obtain

$$\left(J^2 - J_z^2 + \hbar J_z \right) |\lambda j'\rangle = 0$$

$\therefore$ $$\lambda - j'^2 + j' = 0 \text{ or } \lambda = j'(j'-1) \qquad (51)$$

Using Eqn. (50) and (51), we obtain

$$j'(j'-1) = j(j+1)$$

$\Rightarrow$ either $$j' = -j \text{ or } j' = j+1$$

The 2nd solution is meaningless because j is the greatest value of m, hence

$$j' = -j \qquad (52)$$

Thus, we have a family of kets which correspond to the same eigenvalue $\lambda = j(j+1)$ of J^2 but different eigenvalues of J_z. These are denoted by

$$|\lambda, j\rangle, |\lambda, j-1\rangle, ..., |\lambda, -j\rangle \qquad (53)$$

Successive application of the ladder operator J_- on $|\lambda, j\rangle$ ultimately leads to $|\lambda, -j\rangle$. Also successive application of J_+ on $|\lambda, -j\rangle$ leads to $|\lambda, j\rangle$. This will happen only if $j - j' \equiv 2j$ is an integer, so, the allowed values of j are

$$j = 0, 1/2, 1, 3/2, 2, ...$$

The corresponding values of $\lambda = j(j+1)$ are 0, 3/4, 2, 15/4, 6, ... For each value of j, the m-values go from $-j$ to $+j$ in steps of unity. Hence for a given value of j, the eigenvalues of J_z are

$$m\hbar \Rightarrow j\hbar, (j-1)\hbar, ..., -(j-1)\hbar, -j\hbar \qquad (54)$$

These are $(2j + 1)$ in number. Consequently, there are $(2j + 1)$ eigenkets of J^2 each having the same eigenvalue $\lambda = j(j+1)$. We label the eigenkets as

$$|j, m\rangle; m = -j, -j+1, ..., j-1, j \; (j = 0, 1/2, 1, 3/2, 2, ...) \qquad (55)$$

The kets satisfy the orthonormality condition

$$\langle j', m' | j, m \rangle = \delta_{jj'} \delta_{mm'} \qquad (56)$$

We now determine the constants $C_\pm$ of Eqn. (49). We have

$$\langle \lambda m | J_+^\dagger = \langle \lambda, m+1 | C_+^*$$

or $$\langle \lambda m| \, J_- = \langle \lambda, m+1| \, C_+^*$$

∴ $$\langle \lambda m| J_- J_+ |\lambda m\rangle = |C_+|^2 \langle \lambda, m+1|\lambda, m+1\rangle \tag{57}$$

Now assuming that the eigenvectors |λm> are normalized to unity, we can write this Eqn. as

$$\langle \lambda m| J^2 - J_z^2 - \hbar J_z |\lambda m\rangle = |C_+|^2 \tag{58}$$

or $$\left[j(j+1) - m^2 - m \right] \hbar^2 \ \langle \lambda m|\lambda m\rangle = |C_+|^2$$

or $$C_+ = \sqrt{j(j+1) - m(m+1)} \ \hbar$$

Hence,

$$J_+ |j,m\rangle = \left[j(j+1) - m(m+1) \right]^{1/2} \hbar \ |j,m+1\rangle \tag{59i}$$

and similarly,

$$C_- = \sqrt{j(j+1) - m(m-1)} \ \hbar$$

$$J_- |j,m\rangle = \left[j(j+1) - m(m-1) \right]^{1/2} \hbar \ |j,m-1\rangle \tag{59ii}$$

13.9 Eigenvalues of J_x and J_y and $J_\pm$

$$J_x = \frac{1}{2}(J_+ + J_-) \ \text{ and } \ J_y = \frac{1}{2i}(J_+ - J_-) = -\frac{i}{2}(J_+ - J_-)$$

We know that

$$\langle j'm'| J_x |jm\rangle = \langle j'm'| \frac{1}{2}(J_+ + J_-) |jm\rangle$$

$$= \frac{1}{2}\langle j'm'| J_+ |jm\rangle + \frac{1}{2}\langle j'm'| J_- |jm\rangle$$

$$= \frac{1}{2}\hbar \left[(j-m)(j+m+1) \right]^{1/2} \delta_{j',j}\delta_{m',m+1}$$

$$+ \frac{1}{2}\hbar \left[(j+m)(j-m+1) \right]^{1/2} \delta_{j',j}\delta_{m',m-1} \ \text{(from Eqns 59)} \tag{60}$$

Similarly matrix elements of J_y will be given by

$$\langle j'm'| J_y |jm\rangle = \langle j'm'| \left(-\frac{i}{2} \right) (J_+ - J_-) |jm\rangle$$

$$= -\frac{i}{2}\langle j'm'| J_+ |jm\rangle + \frac{i}{2}\langle j'm'| J_- |jm\rangle$$

$$= -\frac{i\hbar}{2}\left[(j-m)(j+m+1) \right]^{1/2} \delta_{j',j}\delta_{m',m+1}$$

$$+ \frac{i\hbar}{2}\left[(j+m)(j-m+1) \right]^{1/2} \delta_{j',j}\delta_{m',m-1} \tag{61}$$

In case $j' = j$, then $\delta_{j',j} = 1$ and Eqns (60) and (61) become

$$\langle j'm'| J_x |jm\rangle = \frac{\hbar}{2}\sqrt{(j-m)(j+m+1)} \ \delta_{m',m+1}$$

$$+\frac{\hbar}{2}\sqrt{(j+m)(j-m+1)}\ \delta_{m',m-1} \tag{62}$$

In case $\quad m' = m +1,\quad \delta_{m',m+1} = 1$

and $\quad m' = m - 1,\quad \delta_{m',m-1} = 1$

then matrix element of J_x will be non zero.

In case $\quad m' \neq m + 1,\quad \delta_{m',m+1} = 0$

and $\quad m' \neq m - 1,\quad \delta_{m',m-1} = 0$

then matrix element of J_x will be zero.

Similarly,

$$\langle j'm' | J_y | jm \rangle = -\frac{i\hbar}{2}\sqrt{(j-m)(j+m+1)}\ \delta_{m',m+1}$$

$$+\frac{i\hbar}{2}\sqrt{(j+m)(j-m+1)}\ \delta_{m',m-1} \tag{63}$$

In case $m' \neq m \pm 1$, the matrix element of J_y will be zero and when $m' = m$ ± 1, elements will be finite.

Thus, if $m' = (m + 1)$, we obtain

$$\langle j,m+1 | J_x | jm \rangle = \frac{1}{2}\hbar\sqrt{(j-m)(j+m+1)} \tag{64}$$

In case $m' = (m - 1)$, we have

$$\langle j,m-1 | J_x | jm \rangle = \frac{1}{2}\hbar\sqrt{(j+m)(j-m+1)} \tag{65}$$

other elements for which $m' \neq m \pm 1$ will be zero. Similarly for J_y, if $m' = m$ $+ 1$, we obtain

$$\langle j,m+1 | J_y | jm \rangle = -\frac{i\hbar}{2}\sqrt{(j-m)(j+m+1)} \tag{66}$$

and for $m' = m - 1$, we get

$$\langle j,m-1 | J_y | jm \rangle = +\frac{i\hbar}{2}\sqrt{(j+m)(j-m+1)} \tag{67}$$

Eigenvalues of $J_\pm$:

If $\bar{J}$ is orbital in nature (no spin dependence), $j = l$ is an integer and the spherical harmonics $Y_{lm}(\theta,\phi)$ are simultaneous eigenvalues for J^2 and J_z (diagonal matrices) then, since

$$J_x = -\frac{\hbar}{i}\left(\sin\phi\ \frac{\partial}{\partial\theta} + \cot\theta\cos\theta\ \frac{\partial}{\partial\phi}\right)$$

$$J_y = \frac{\hbar}{i}\left(\cos\phi\ \frac{\partial}{\partial\theta} - \cot\theta\sin\theta\ \frac{\partial}{\partial\phi}\right) \tag{68i}$$

$$J_z = \frac{\hbar}{i} \frac{\partial}{\partial \phi},$$

we can represent $J_\pm$ as

$$J_+ = \hbar e^{i\phi} \left(\frac{\partial}{\partial \theta} + i \cot \theta \frac{\partial}{\partial \phi} \right) \qquad (68ii)$$

$$J_- = \hbar e^{-i\phi} \left(-\frac{\partial}{\partial \theta} + i \cot \theta \frac{\partial}{\partial \phi} \right)$$

then, using the properties of the spherical harmonics, we can show that

$$J_+ Y_{lm}(\theta,\phi) = \left[l(l+1) - m(m+1) \right]^{1/2} \hbar Y_{l,m+1}(\theta,\phi)$$

$$J_- Y_{lm}(\theta,\phi) = \left[l(l+1) - m(m-1) \right]^{1/2} \hbar Y_{l,m-1}(\theta,\phi) \qquad (69)$$

as expected from Eqns (59i) and (59ii).

13.10 Form of the Matrices for Total Angular Momentum (J^2 and J_z)

It is known that the matrices for J^2 and J_z are diagonal and

$$J_z = m\hbar$$

and

$$J^2 = j(j+1)\hbar^2 \qquad (70)$$

The value of magnetic quantum number m varies from $-j$ to $+j$ and total values of m are given by $(2j+1)$. Thus, the dimensions of these (diagonal) matrices will be $(2j+1)$, i.e. *there will be $(2j+1)$ rows and $(2j+1)$ columns.* For a given value of j, J_z-matrix will be of the form:

$$J_z = \hbar \begin{pmatrix} j & 0 & 0 & \cdots & 0 \\ 0 & (j-1) & 0 & \cdots & 0 \\ 0 & 0 & (j-2) & \cdots & 0 \\ \vdots & \vdots & \vdots & \vdots & \vdots \\ 0 & \cdots & \cdots & \cdots & -j \end{pmatrix} \quad (2j+1) \times (2j+1) \text{ (diagonal matrix)}$$

J^2 matrix for a given j-value will be of the form

$$J^2 = \hbar^2 \begin{pmatrix} j(j+1) & 0 & \cdots & 0 \\ 0 & j(j+1) & \cdots & 0 \\ \vdots & \vdots & \vdots & \vdots \\ 0 & \cdots & \cdots & j(j+1) \end{pmatrix} \quad (2j+1) \times (2j+1) \text{ (diagonal matrix)}$$

i.e. J^2 and J_z matrices will contain elements only along the principal diagonal. Each of matrices J_z and J^2 can be represented by the form:

		j'	0	½		1		...
		m'	0	½	- ½	1	0	-1	...
j 0	m 0		(E.P) $(2j+1) \times (2j+1)$	(0)		(0)		...
½	$\begin{cases} ½ \\ -½ \end{cases}$		(0)	(E.P) $(2j+1) \times (2j+1)$		(0)		...
1	$\begin{cases} 1 \\ 0 \\ -1 \end{cases}$		(0)	(0)		(E.P) $(2j+1) \times (2j+1)$		...
⋮	⋮		⋮	⋮		⋮		⋮

where E.P. represents elements present (see Tables 13.3 and 13.4).

Matrices for J_+ and J_-:

From Eqn. (59i)

$$\langle j,m+1| J_+ |j,m\rangle = \hbar\left[(j-m)(j+m+1)\right]^{1/2}$$

$$(\text{since, } \delta_{jj} = 1 = \delta_{m+1,m+1})$$

(other elements will be zero)

i.e. matrix for J_+ will be of the following type:

		m'	0	1	2	3
	m					
	0		0	(E.P.)	0	...
J_+	1		0	0	(E.P.)	...
	2		0	0	0	(E.P.)
	3		⋮	⋮	⋮	⋮
	⋮					

From Eqn. (59ii)

$$\langle j,m-1| J_- |j,m\rangle = \hbar\left[(j+m)(j-m+1)\right]^{1/2}$$

$$(\text{since, } \delta_{jj} = 1 = \delta_{m-1,m-1})$$

(other elements will be zero)

The matrix for J_- will be of the following type:

m' \ m	0	1	2	...
0	0	0	0	...
J_- 1	(E.P.)	0	0	...
2	0	(E.P.)	0	...
:	:	:	:	...

J_+ matrix contains only upper diagonal.

J_- matrix contains only lower diagonal.

Example 13.6: Let $j = 1/2$, $m = 1/2, -1/2$

$$J^2 = \hbar^2 j(j+1) = \frac{1}{2}\hbar^2\left(\frac{1}{2}+1\right) = \frac{3}{4}\hbar^2$$

The dimensions of J^2 matrix $= (2j + 1) = (2 \times 1/2 + 1) = 2$

(there will be two rows and two columns in J^2 matrix)

$$J^2 = \hbar^2 \begin{bmatrix} \frac{3}{4} & 0 \\ 0 & \frac{3}{4} \end{bmatrix}$$

$J_z = m\,\hbar = 1/2\,\hbar$ for $m = 1/2$ or $= -1/2\,\hbar$ for $m = -1/2$

$$\therefore \quad J_z = \hbar \begin{pmatrix} \frac{1}{2} & 0 \\ 0 & -\frac{1}{2} \end{pmatrix}$$

Now,

$$J_+ = \hbar\sqrt{(j-m)(j+m+1)}$$

$$= \hbar\sqrt{\left(\frac{1}{2}-\frac{1}{2}\right)\left(\frac{1}{2}+\frac{1}{2}+1\right)} = 0 \text{ for } j = 1/2, \; m = 1/2$$

$$J^+ = \hbar\sqrt{\left(\frac{1}{2}+\frac{1}{2}\right)\left(\frac{1}{2}-\frac{1}{2}+1\right)} = \hbar \text{ for } j = 1/2, \; m = -1/2$$

$$\therefore \quad J^+ = \hbar \begin{pmatrix} 0 & 1 \\ 0 & 0 \end{pmatrix}$$

Table 13.3: The form of the matrices for $J_z/\hbar$

j		0	½	½	1	1	1	3/2	3/2	3/2	3/2	...
	m	0	½	−½	1	0	−1	3/2	½	−½	−3/2	...
0	0	0	0	0								
½	½	0	½	0								
½	−½	0	0	−½ (2 × 2)								
1	1				1	0	0					
1	0				0	0	0					
1	−1				0	0	−1 (3 × 3)					
3/2	3/2							3/2	0	0	0	
3/2	½							0	½	0	0	
3/2	−½							0	0	−½	0	
3/2	−3/2							0	0	0	−3/2 (4 × 4)	
⋮	⋮											

Table 13.4: The form of the matrices for $J^2/\hbar^2$

j		0	½	½	1	1	1	3/2	3/2	3/2	3/2	...
	m	0	½	−½	1	0	−1	3/2	½	−½	−3/2	...
0	0											
½	½		¾	0								
½	−½		0	¾								
1	1				2	0	0					
1	0				0	2	0					
1	−1				0	0	2					
3/2	3/2							15/4	0	0	0	
3/2	½							0	15/4	0	0	
3/2	−½							0	0	15/4	0	
3/2	−3/2							0	0	0	15/4	
$j(j+1)=$ 15/4												
⋮	⋮											

$$J_- = \hbar\sqrt{(j+m)(j-m+1)}$$
$$= \hbar \quad \text{for } j = ½,\ m = ½$$
$$= 0 \quad \text{for } j = ½,\ m = -½$$

$$\therefore \qquad J_- = \hbar \begin{pmatrix} 0 & 0 \\ 1 & 0 \end{pmatrix}$$

Example 13.7: For an eigenstate of J^2 and J_z show that $\langle J_x \rangle = \langle J_y \rangle = 0$

and $\langle J_x^2 \rangle = \langle J_y^2 \rangle = \dfrac{1}{2}\hbar^2 \left[j\,(j+1) - m^2 \right]$

From these relations develop a classical model for angular momentum.

Solution: $J_x = \dfrac{1}{2}(J_+ + J_-)$

$$\therefore \qquad \langle J_x \rangle = \langle j,m | J_x | j,m \rangle$$

$$= \frac{1}{2}\langle j,m | J_+ | j,m \rangle + \frac{1}{2}\langle j,m | J_- | j,m \rangle$$

$$= \frac{1}{2}\langle j,m | \{(j-m)(j+m+1)\}^{1/2}\, \hbar\, | j,m+1 \rangle$$

$$+ \frac{1}{2}\langle j,m | \{(j+m)(j-m+1)\}^{1/2}\, \hbar\, | j,m-1 \rangle$$

$$= \frac{\hbar}{2}\{(j-m)(j+m+1)\}^{1/2}\, \langle j,m | j,m+1 \rangle$$

$$+ \frac{\hbar}{2}\{(j+m)(j-m+1)\}^{1/2}\, \langle j,m | j,m-1 \rangle = 0$$

$$(\because\ \langle j,m | j,m+1 \rangle = \delta_{jj}\delta_{m,m+1} = 0\,;\ \langle j,m | j,m-1 \rangle = \delta_{jj}\delta_{m,m-1} = 0)$$

$$J_x^2 = \frac{1}{2}(J_+ + J_-)\frac{1}{2}(J_+ + J_-)$$

$$= \frac{1}{4}\left[J_+ J_+ + J_+ J_- + J_- J_+ + J_- J_- \right]$$

$$\therefore \quad \langle j,m | J_x^2 | j,m \rangle = \frac{1}{4}\Big[\langle j,m | J_+ J_+ | j,m \rangle + \langle j,m | J_+ J_- | j,m \rangle$$

$$+ \langle j,m | J_- J_+ | j,m \rangle + \langle j,m | J_- J_- | j,m \rangle \Big]$$

$$= \frac{1}{4}\Big[0 + \langle j,m | J_+ \{(j+m)(j-m+1)\}^{1/2}\, \hbar | j,m-1 \rangle$$

$$+ \langle j,m | J_- \{(j-m)(j+m+1)\}^{1/2}\, \hbar | j,m+1 \rangle + 0 \Big]$$

$$= \frac{1}{4}\Big[\hbar\{(j+m)(j-m+1)\}^{1/2}\, \langle j,m | J_+ | j,m-1 \rangle$$

$$+ \hbar\{(j-m)(j+m+1)\}^{1/2}\, \langle j,m | J_- | j,m+1 \rangle \Big]$$

$$= \frac{\hbar^2}{4}\Big[\{(j+m)(j-m+1)\}^{1/2}\, \langle j,m | \{(j+m)(j-m+1)\}^{1/2} | j,m \rangle$$

$$+ \{(j-m)(j+m+1)\}^{1/2}\, \langle j,m | \{(j-m)(j+m+1)\}^{1/2} | j,m \rangle \Big]$$

$$= \frac{\hbar^2}{4}\left[(j+m)(j-m+1)+(j-m)(j+m+1)\right]$$

$$= \frac{\hbar^2}{4}\left[\left(j^2-m^2\right)+j+m+j^2-m^2+j-m\right]$$

$$= \frac{\hbar^2}{4}\left[2j^2+2j-2m^2\right]$$

$$= \frac{\hbar^2}{2}\left[j(j+1)-m^2\right] \qquad (i)$$

By symmetry $\qquad \left\langle J_y^2 \right\rangle = \frac{\hbar^2}{2}\left[j(j+1)-m^2\right] \qquad (ii)$

Obviously $\qquad \left\langle J_z^2 \right\rangle = (m\hbar)^2 = m^2\hbar^2 \qquad (iii)$

Adding (i), (ii) and (iii)

$$\left\langle J^2 \right\rangle = \left\langle J_x^2 \right\rangle + \left\langle J_y^2 \right\rangle + \left\langle J_z^2 \right\rangle$$

$$= \hbar^2\left[j(j+1)-m^2\right]+m^2\hbar^2 = j(j+1)\hbar^2$$

We can picturize the above angular momentum relations as shown in Fig. 13.10.

Thus, the vector $\vec{J}$ precesses about the z-axis and $AB = \sqrt{J^2 - J_z^2}$

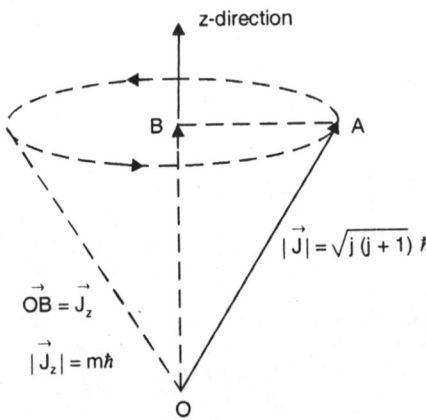

Fig. 13.10: Classical model for angular momentum vector

Example 13.8: Let $j = 1$, $m = 1, 0, -1$

$$J^2 = \hbar^2\begin{pmatrix} 2 & 0 & 0 \\ 0 & 2 & 0 \\ 0 & 0 & 2 \end{pmatrix} \quad ; \quad J_z = \hbar\begin{pmatrix} 1 & 0 & 0 \\ 0 & 0 & 0 \\ 0 & 0 & -1 \end{pmatrix}$$

$$J_+ = \hbar \begin{pmatrix} 0 & \sqrt{2} & 0 \\ 0 & 0 & \sqrt{2} \\ 0 & 0 & 0 \end{pmatrix} \; ; \; J_- = \hbar \begin{pmatrix} 0 & 0 & 0 \\ \sqrt{2} & 0 & 0 \\ 0 & \sqrt{2} & 0 \end{pmatrix}$$

$$J_x = \frac{J_+ + J_-}{2} \Rightarrow J_x = \frac{\hbar}{\sqrt{2}} \begin{pmatrix} 0 & 1 & 0 \\ 1 & 0 & 1 \\ 0 & 1 & 0 \end{pmatrix}$$

$$J_y = \frac{J_+ - J_-}{2i} \Rightarrow J_y = \frac{\hbar}{2i} \begin{pmatrix} 0 & \sqrt{2} & 0 \\ -\sqrt{2} & 0 & \sqrt{2} \\ 0 & -\sqrt{2} & 0 \end{pmatrix}$$

or

$$J_y = \frac{i\hbar}{\sqrt{2}} \begin{pmatrix} 0 & -1 & 0 \\ 1 & 0 & -1 \\ 0 & 1 & 0 \end{pmatrix}$$

13.11 Addition of (Total) Angular Momenta (Clebsch-Gordan coefficients)

We often have to deal with systems in which the total angular momentum is composed of two (or more) parts that are (to some approximation) independent of each other, e.g. the particles with spin (in non-relativistic limit), systems containing two or more particles (such as many electron atoms). The problem is to find how the total angular momentum is related to its component parts.

Let $\vec{J}_1$ and $\vec{J}_2$ be two independent angular momentum vectors in different subspaces and the components of $\vec{J}_1 \left(\vec{J}_2 \right)$ satisfy the usual commutation relation

$$\left[J_{1i}, J_{1j} \right] = i\hbar \varepsilon_{ijk} J_{1k} \tag{71i}$$

and

$$\left[J_{2i}, J_{2j} \right] = i\hbar \varepsilon_{ijk} J_{2k} \tag{71ii}$$

But

$$\left[J_{1k}, J_{2l} \right] = 0 \tag{72}$$

i.e. between any pair of operators from different two subspaces (i.e. J_1 and J_2 commute with each other)

If two distinct physical systems (or sets) of dynamical variables of one system (which are described in two different vector spaces) are merged, then the states of the composite system are vectors in the direct product space of the two previously separate vector spaces.

We designate the simultaneous eigenkets of J^2 and J_z by $|jm\rangle$ when

$$J^2 |jm\rangle = j(j+1)\hbar^2 |jm\rangle \tag{73}$$

and

$$J_z |jm\rangle = m\hbar |jm\rangle \tag{74}$$

Here, the orthonormal set of eigenkets $\left| j_1, m_1 \right\rangle$ (of J_1^2, J_{1z}) form the basis for a $(2j_1+1)$ dimensional vector space. Similarly the eigenkets $\left| j_2, m_2 \right\rangle$ (of J_2^2 and J_{2z}) are representative of a different vector space of dimension $(2j_2+1)$.

The infinitesimal rotation operator that affects both subspace 1 and subspace 2 is written (according to Eqn. (78) of chapter 11) as

$$\left(1 - \frac{i \vec{J}_1 \cdot \hat{n}\delta\phi}{\hbar} \right) \otimes \left(1 - \frac{i \vec{J}_2 \cdot \hat{n}\delta\phi}{\hbar} \right) = 1 - \frac{i \left(\vec{J}_1 \otimes 1 + 1 \otimes \vec{J}_2 \right) \cdot \hat{n}\delta\phi}{\hbar} \tag{75}$$

We define the total angular momentum by

$$\vec{J} = \vec{J}_1 \otimes 1 + 1 \otimes \vec{J}_2 \tag{76}$$

or more simply

$$\vec{J} = \vec{J}_1 + \vec{J}_2 \tag{77}$$

We note that

$$\left[J_y, J_z \right] = J_y J_z - J_z J_y$$

$$= \left(J_{1y} + J_{2y} \right) \left(J_{1z} + J_{2z} \right) - \left(J_{1z} + J_{2z} \right) \left(J_{1y} + J_{2y} \right)$$

$$= \left(J_{1y} J_{1z} - J_{1z} J_{1y} \right) + \left(J_{2y} J_{2z} - J_{2z} J_{2y} \right)$$

$$= i \left(J_{1x} + J_{2x} \right) = i J_x \tag{78}$$

(because of Eqns (72) and (71))

Hence the total $\vec{J}$ satisfies the angular momentum commutation relation:

$$\left[J_i, J_j \right] = i\hbar \varepsilon_{ijk} J_k \tag{79}$$

In other words $\vec{J}$ is also an angular momentum and the generator (of rotation) for the entire system.

Now

$$\left(\vec{J} \times \vec{J} \right)_x = \left(J_y J_z - J_z J_y \right)$$

$$= \left(J_{1y} + J_{2y} \right) \left(J_{1z} + J_{2z} \right) - \left(J_{1z} + J_{2z} \right) \left(J_{1y} + J_{2y} \right)$$

$$= \left(J_{1y} J_{1z} - J_{1z} J_{1y} \right) + \left(J_{2y} J_{2z} - J_{2z} J_{2y} \right)$$

$$= i \left(J_{1x} + J_{2x} \right) = i J_x$$

(as J_1 commutes with J_2, all other terms cancel out) and in general,

$$\vec{J} \times \vec{J} = i\vec{J} \tag{80}$$

i.e. $\vec{J}$ has the required properties of angular momentum operator.

Our problem is to obtain the eigenvalues of J^2, J_z and their eigenvectors in terms of the direct product of the eigenvectors of $\left(J_1^2, J_{1z} \right)$ and $\left(J_2^2, J_{2z} \right)$. The normalized simultaneous eigenvectors of the four operators $J_1^2, J_2^2, J_{1z}, J_{2z}$ can be symbolized by the direct product kets:

$$\left| j_1 m_1 \right\rangle \otimes \left| j_2 m_2 \right\rangle \equiv \left| j_1 j_2 m_1 m_2 \right\rangle \tag{81}$$

These constitute a basis for the product space. We have to construct from this basis the eigenvectors of $\left(J^2, J_z \right)$. We write the eigenkets |jm> $\left(\text{of } J^2, J_z \right)$ as a linear combination of the eigenvectors (Eqn. 81) as

$$\left| jm \right\rangle = \sum_{m_1, m_2} C\left(j_1 j_2 m_1 m_2, jm \right) \left| j_1 j_2 m_1 m_2 \right\rangle \tag{82i}$$

using the orthonormality condition, the coefficients of expansion C's are given by

$$C^{jm}_{j_1 m_1 j_2 m_2} \equiv C\left(j_1 j_2 m_1 m_2, jm \right) = \left\langle j_1 j_2 m_1 m_2 \middle| jm \right\rangle \tag{82ii}$$

$$\therefore \quad \left| jm \right\rangle = \sum_{m_1, m_2} \left\langle j_1 j_2 m_1 m_2 \middle| jm \right\rangle \left| j_1 j_2 m_1 m_2 \right\rangle \tag{83}$$

Summation in Eqn. (83) over m_1 is from $-j_1$ to j_1 and over m_2 is from $-j_2$ to j_2. The coefficients of expansion in Eqn. (83) are known as *Clebsch Gordan* or *Vector Coupling Coefficients.*

Now the question is: given the momenta (j_1, m_1) and (j_2, m_2), what are the possible values of j and m of the composite system. For this, we operate $J_z = J_{1z} + J_{2z}$ on Eqn. (83)

$$J_z \left| jm \right\rangle = \sum_{m'_1, m'_2} \left(J_{1z} + J_{2z} \right) \left| j_1 j_2 m'_1 m'_2 \right\rangle \left\langle j_1 j_2 m'_1 m'_2 \middle| jm \right\rangle$$

or $\quad m\hbar \left| jm \right\rangle = \sum_{m'_1, m'_2} \left(m'_1 \hbar + m'_2 \hbar \right) \left| j_1 j_2 m'_1 m'_2 \right\rangle \left\langle j_1 j_2 m'_1 m'_2 \middle| jm \right\rangle$

(using Eqn. (81) and $\left[J_{1z}, J_{2z} \right] = 0$)

or $\quad m \left| jm \right\rangle = \sum_{m'_1, m'_2} \left(m'_1 + m'_2 \right) \left| j_1 j_2 m'_1 m'_2 \right\rangle \left\langle j_1 j_2 m'_1 m'_2 \middle| jm \right\rangle$

or $\quad m \left\langle j_1 j_2 m_1 m_2 \middle| jm \right\rangle = \sum_{m'_1, m'_2} \left(m'_1 + m'_2 \right) \left\langle j_1 j_2 m_1 m_2 \right|$

$$\left| j_1 j_2 m'_1 m'_2 \right\rangle \left\langle j_1 j_2 m'_1 m'_2 \middle| jm \right\rangle$$

or $\quad m \left\langle j_1 j_2 m_1 m_2 \middle| jm \right\rangle = \sum_{m'_1, m'_2} \left(m'_1 + m'_2 \right) \delta_{m_1 m'_1} \delta_{m_2 m'_2} \left\langle j_1 j_2 m'_1 m'_2 \middle| jm \right\rangle$

or $\quad m \left\langle j_1 j_2 m_1 m_2 \middle| jm \right\rangle = \left(m_1 + m_2 \right) \left\langle j_1 j_2 m_1 m_2 \middle| jm \right\rangle$

or $$m = m_1 + m_2 \tag{84}$$

Thus we have a rule for the addition of angular momenta: *the sum of the z-components of the angular momenta equals their resultant and only those basis vectors will contribute for which this rule is satisfied.*

Now, we know that the largest value of m is equal to j and this would occur once only when m_1 and m_2 have their largest values viz. j_1 and j_2. Thus the largest value of $j = j_1 + j_2$ and there is only one such state for which this value is possible. The next (second) largest value of m is $(j_1 + j_2 - 1)$ and it can occur for two cases:

$$\begin{cases} m_1 = j_1 \\ m_2 = j_2 - 1 \end{cases} \text{ and } \begin{cases} m_1 = j_1 - 1 \\ m_2 = j_2 \end{cases}$$

There exist two linearly independent orthogonal combinations of these states which correspond to $j = j_1 + j_2$ and $(j_1 + j_2 - 1)$ and both with the projection $(j_1 + j_2 - 1)$. Next, $m = j_1 + j_2 - 2$ can be realized in three ways:

$$\begin{cases} m_1 = j_1 \\ m_2 = j_2 - 2 \end{cases} ; \begin{cases} m_1 = j_1 - 1 \\ m_2 = j_2 - 1 \end{cases} ; \begin{cases} m_1 = j_1 - 2 \\ m_2 = j_2 \end{cases}$$

There exist three orthogonal linear combination of these states which correspond to $j = j_1 + j_2$, $(j_1 + j_2 - 1)$ and $(j_1 + j_2 - 2)$ all with projection equal to $(j_1 + j_2 - 2)$. Clearly, this sequence ends, if either m_1 or m_2 reaches its lowest value i.e. $m_1 = -j_1$ or $m_2 = -j_2$. The lowest value of j in that case will be $|j_1 - j_2|$. Thus j can have the values:

$$j_1 + j_2 \geq j \geq |j_1 - j_2| \tag{85}$$

So we have another rule for the addition of angular momenta:

The magnitude of the sum of two angular momentum vectors can have any value ranging from the sum of their magnitude to the difference of their magnitude by integral steps.

Thus, we have two important properties of C.G. coefficients:

(*i*) the coefficients vanish unless $m = m_1 + m_2$

(*ii*) the coefficients vanish unless $j_1 + j_2 \geq j \geq |j_1 - j_2|$

The second property appears obvious from the vector model of angular momentum addition. Further, it becomes obvious too from the point of view of dimensionality. We may show that if inequality (85) holds then the dimensionality of space spanned by $\{|j_1 j_2 m_1 m_2\rangle\}$ is the same as that of the space spanned by $|jm\rangle$. This we check as follows.

For the (m_1, m_2) way of counting we obtain dimensionality

$$N = (2j_1 + 1)(2j_2 + 1) \tag{86}$$

because for given j_1 there are $(2j_1 + 1)$ possible values of m_1 and for given j_2 there are $(2j_2 + 1)$ values of m_2.

For the (j, m) way of counting, we note that for each j, there are $2j + 1$ states and according to Eqn. (85), j itself runs from $j_1 - j_2$ to $j_1 + j_2$ (assuming $j_1 \geq j_2$). We then obtain

$$N = \sum_{j=j_1-j_2}^{j_1+j_2} (2j + 1)$$

$$= \frac{1}{2}\Big[\{2(j_1 - j_2) + 1 + 2(j_1 + j_2) + 1\} (2j_2 + 1)\Big]$$

$$\left(\because \text{ Sum } S_n = \frac{n}{2}(a+l) \text{ and } n = \left(j_1 + j_2 \right) - \left(j_1 - j_2 \right) + 1 \right)$$

$$= \left(2j_1 + 1 \right) \left(2j_2 + 1 \right)$$

Because both ways of counting give the same value of N, therefore we see that Eqn. (85) is quite consistent. The C.G. coefficients of Eqn. (83) are just the elements of the transformation from one basis to another in the $\left(2j_1 + 1 \right) \left(2j_2 + 1 \right)$ dimensional Hilbert space. Since the set of initial states $\{ | j_1 m_1, j_2 m_2 \rangle \}$ and the set of final states $\{ | jm \rangle \}$ are assumed to be orthonormal, the matrix of C.G. coefficients is a $\left(2j_1 + 1 \right) \left(2j_2 + 1 \right)$ dimensional unitary matrix.

13.12 Recursion Relation for C.G. Coefficients

To simplify the notations, we use $| j_1 m_1, j_2 m_2 \rangle = | m_1 m_2 \rangle$ keeping in mind that j_1 and j_2 are the maximum values of m_1 and m_2. We can write Eqn. (83) as

$$| jm \rangle = \sum_{m_1, m_2} | j_1 m_1, j_2 m_2 \rangle \langle j_1 m_1, j_2 m_2 | jm \rangle$$

$$\equiv \sum_{m_1, m_2} | m_1 m_2 \rangle \langle m_1 m_2 | jm \rangle$$

Applying the raising operator J_+ to the l.h.s. and the equal operator $\left(J_{1+} + J_{2+} \right)$ to the r.h.s. of this equation, we get

$$J_+ | jm \rangle = \sum_{m_1, m_2} \left(J_{1+} + J_{2+} \right) | m_1 m_2 \rangle \langle m_1 m_2 | jm \rangle$$

or

$$\hbar \sqrt{(j-m)(j+m+1)} \ | j, m+1 \rangle$$

$$= \sum_{m_1, m_2} \left[\hbar \sqrt{(j_1 - m_1)(j_1 + m_1 + 1)} \ | m_1 + 1, m_2 \rangle \right.$$

$$\left. + \hbar \sqrt{(j_2 - m_2)(j_2 + m_2 + 1)} \ | m_1, m_2 + 1 \rangle \right] \langle m_1 m_2 | jm \rangle$$

or

$$\sqrt{(j-m)(j+m+1)} \ \langle m_1' m_2' | j, m+1 \rangle$$

$$= \sum_{m_1, m_2} \left[\sqrt{(j_1 - m_1)(j_1 + m_1 + 1)} \ \langle m_1' m_2' | m_1 + 1, m_2 \rangle \right.$$

$$\left. + \sqrt{(j_2 - m_2)(j_2 + m_2 + 1)} \ \langle m_1' m_2' | m_1, m_2 + 1 \rangle \right] \langle m_1 m_2 | jm \rangle$$

or

$$\sqrt{(j-m)(j+m+1)} \ \langle m_1' m_2' | j, m+1 \rangle$$

$$= \sqrt{(j_1 - m_1' + 1)(j_1 + m_1')} \ \langle m_1' - 1, m_2' | jm \rangle$$

$$+ \sqrt{(j_2 - m_2' + 1)(j_2 + m_2')} \ \langle m_1', m_2' - 1 | jm \rangle \tag{87i}$$

Because

$$\langle m_1' m_2' | m_1 + 1, m_2 \rangle = \delta_{m_1', m_1 + 1} \delta_{m_2', m_2}$$

$$\left(m_1 \to m_1' - 1, m_2 \to m_2' \right)$$

and

$$\langle m_1' m_2' | m_1, m_2 + 1 \rangle = \delta_{m_1', m_1} \delta_{m_2', m_2 + 1}$$

$$\left(m_1 \to m_1', m_2 \to m_2' - 1 \right)$$

Similarly, applying the lowering operator J_-, we get,

$$\langle m_1' m_2' | j, m - 1 \rangle \sqrt{(j + m)(j - m + 1)}$$

$$= \sqrt{\left(j_1 + m_1' + 1 \right) \left(j_1 - m_1' \right)} \, \langle m_1' + 1, m_2' | jm \rangle$$

$$+ \sqrt{\left(j_2 + m_2' + 1 \right) \left(j_2 - m_2' \right)} \, \langle m_1', m_2' + 1 | jm \rangle \qquad (87ii)$$

Equations ($87i$) and ($87ii$) are the desired recursion relations.

It is important to note here that because the $J_\pm$ operators have shifted the m values, the non-vanishing condition (84) for the C.G. coefficients has now become

$$m_1 + m_2 = m \pm 1 \qquad (88)$$

13.13 Two Options for Representation

As for the choice of base kets, we have two options:

Let $\psi_1 \left(j_1, m_1 \right)$ be the simultaneous eigenvector of J_1^2 and J_{1z} and $\psi_2 \left(j_2, m_2 \right)$. represent the simultaneous eigenvector of J_2^2 and J_{2z}. The product function

$$\psi_1 \left(j_1, j_2, m_1, m_2 \right) = \psi_1 \left(j_1, m_1 \right) \psi_2 \left(j_2, m_2 \right) \qquad (89)$$

represent a set of simultaneous eigenvectors of J_1^2, J_2^2, J_{1z} and J_{2z}. In other words (since these four operators commute with each other) we have following two options:

Option A: Simultaneous eigenkets of $\vec{J}_1^2, \vec{J}_2^2, J_{1z}$ and J_{2z} denoted by $| j_1 j_2, m_1 m_2 \rangle$ or $\psi \left(j_1 j_2, m_1 m_2 \right)$.

Option B: Simultaneous eigenkets of $\vec{J}_1^2, \vec{J}_2^2, \vec{J}^2$ and J_z . (This set of operators also mutually commute) Hence one can construct vectors $\phi \left(j_1, j_2, j, m \right)$ which are simultaneous eigenvectors of the operators $\vec{J}_1^2, \vec{J}_2^2, \vec{J}^2$ and J_z belonging to the eigenvalues $j_1 \left(j_1 + 1 \right), j_2 \left(j_2 + 1 \right), j \left(j + 1 \right)$ and m respectively.

We can choose the ψ functions as the base states and we can represent ϕ functions as a linear combination in terms of the C.G. coefficients:

$$\phi \left(j_1, j_2, j, m \right) = \sum_{m_1 m_2} C \left(j_1, j_2; j, m, m_1, m_2 \right) \psi \left(j_1, j_2, m_1, m_2 \right) \qquad (90)$$

where

$$C = \langle j_1, j_2, m_1, m_2 | j_1, j_2, j, m \rangle \qquad (91)$$

This can be rewritten in brief as

$$\phi(j,m) = \sum_{m_1 m_2} \langle m_1 m_2 | jm \rangle \; \psi_1(j_1, m_1) \; \psi_2(j_2, m_2) \qquad (92)$$

13.14 Construction of C.G. Coefficients

Now we describe the procedure for construction of C.G. coefficients $\langle m_1 m_2 | jm \rangle$. The total number of $|jm\rangle$ states will be equal to $(2j_1 + 1)(2j_2 + 1)$. Thus the matrix $\langle m_1 m_2 | jm \rangle$ has $(2j_1 + 1)(2j_2 + 1)$ number of rows and the same number of columns, but breaks up into disconnected submatrices in accordance with the value of m ($= m_1 + m_2$).

The highest value of $m = j$ ($= j_1 + j_2$) occurs only once (for $m_1 = j_1$ and $m_2 = j_2$), hence there will be a 1×1 submatrix. The next highest value of $m = j - 1$ ($= j_1 + j_2 - 1$) and it occurs twice ($m_1 = j_1$, $m_2 = j_2 - 1$ and $m_1 = j_1 - 1$ and $m_2 = j_2$), hence there will be then a 2×2 submatrix. Counting like this, we shall get a 3×3 matrix and so on until a maximum order is reached and maintained for one or more submatrices. Further, there occurs a decrease in unit steps till the last submatrix has $m = -j$ ($= -(j_1 + j_2)$). To find the maximum order of the possible submatrix we note that:

for $\quad m = j_1 + j_2 - 0$ we get 1×1 submatrix

for $\quad m = j_1 + j_2 - 1$ we get 2×2 submatrix

for $\quad m = j_1 + j_2 - 2$ we get 3×3 submatrix

$\quad \vdots \qquad \vdots \qquad \vdots \qquad \vdots \qquad \vdots$

for $\quad m = j_1 + j_2 - k$ we get $(k + 1) \times (k + 1)$ submatrix

Now, $m = j_1 + j_2 - k$ can occur for $m_1 = j_1$, $m_2 = j_2 - k$ and $m_1 = j_1 - k$ and $m_2 = j_2$. If $j_2 - k \le -j_2$, then $(k + 1) \times (k + 1)$ submatrix is not possible because minimum value of m_2 is j_2. Similarly, $j_1 - k \le -j_1$ is not possible. Thus, the extreme value for k should be such that $j_2 - k = -j_2$ or $j_1 - k = -j_1$, i.e. $k_{max} = 2j_2$ or $2j_1$ whichever is smaller. Hence the maximum order of submatrix is $(k + 1) = (2j_1 + 1)$ or $(2j_2 + 1)$ whichever is smaller. Consequently, the general appearance of the C.G. coefficient matrix will be as in Table 13.5.

For $m = j_1 + j_2 - 0$, there will be only one ψ vector and only one ϕ vector which will be equal (i.e. $\phi(j_1 + j_2, j_1 + j_2) = \psi(j_1, j_1) \; \psi(j_2, j_2)$.

For $m = j_1 + j_2 - 1$, there will be two ψ vectors $\psi(j_1, j_2 - 1)$ and $\psi(j_1 - 1, j_2)$ and two ϕ vectors

$$\phi(j_1 + j_2, j_1 + j_2 - 1)$$

and $\qquad\qquad\qquad\qquad \phi(j_1 + j_2 - 1, j_1 + j_2 - 1)$

For $m = j_1 + j_2 - k$, there will be $(k + 1)$, ψ vectors:

$$\psi(j_1, j_2 - k) ; \psi(j_1 - 1, j_2 - k + 1) .... \psi(j_1 - k + 1, j_2 - 1)$$

and the $(k + 1)$, ϕ vectors

$$\phi(j_1 + j_2, j_1 + j_2 - k) ; \phi(j_1 + j_2 - 1, j_1 + j_2 - k) ... \phi(j_1 + j_2 - k, j_1 + j_2 - k)$$

Table 13.5: General appearance of C.G. coefficient matrix

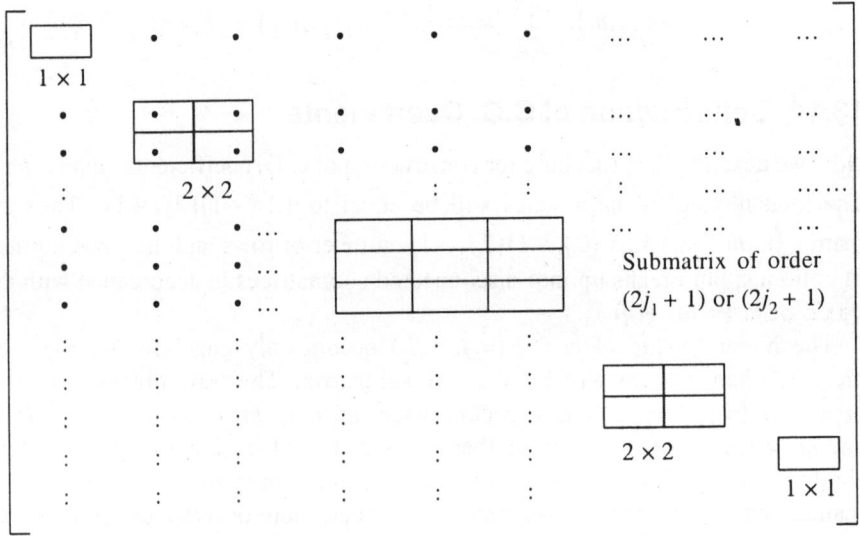

We assume $j_1 \geq j_2$, so $k_{max} = 2j_2$ and $j_{min} = j_1 + j_2 - k_{max} = j_1 - j_2$. Each ϕ vector, then can be expressed as a linear combination of $(k + 1)$, ψ vectors and the C.G. coefficients.

13.15 Evaluation of C.G. Coefficients

We have

$$\phi(j,m) = \sum_{m_1 m_2} \langle m_1 m_2 | jm \rangle \, \psi_1(j_1,m_1) \, \psi_2(j_2,m_2) \tag{93}$$

operating by J_z

$$J_z \phi(j,m) = (J_{1z} + J_{2z}) \langle m_1 m_2 | jm \rangle \, \psi_1(j_1,m_1) \, \psi_2(j_2,m_2)$$

$$m\phi(j,m) = (m_1 + m_2) \langle m_1 m_2 | jm \rangle \, \psi_1(j_1,m_1) \, \psi_2(j_2,m_2)$$

Substituting the value of $\phi(j, m)$ (from Eqn. (93) on l.h.s. of above equation and then transposing to the right, we get

$$\sum_{m_1 m_2} (m_1 + m_2 - m) \langle m_1 m_2 | jm \rangle \, \psi_1(j_1,m_1) \, \psi_2(j_2,m_2) = 0 \tag{94}$$

Since $\psi_1(j_1,m_1) \, \psi_2(j_2,m_2)$ are linearly independent and orthogonal therefore each coefficient in the sum must vanish. This gives either

$$\langle m_1 m_2 | jm \rangle = 0$$

or

$$m = m_1 + m_2$$

Hence C.G. coefficients may be expressed as

$$\langle m_1 m_2 | jm \rangle = f(j_1 j_2 j; m_1 m_2 m) \delta_{m, m_1 + m_2}$$

i.e. C.G. coefficients vanish if $m \neq m_1 + m_2$. The numbers j_1 and j_2 are the largest values of m_1 and m_2 respectively. Therefore the largest possible value of $m = j_1 + j_2$. It $(j_1 + j_2)$ is also the maximum possible value of j. For this case,

$$\phi(j,m) \to \phi\left(j_1 + j_2, j_1 + j_2\right) \equiv \phi\langle j_1 j_2 \mid jm \rangle$$

$$= \langle j_1 j_2 \mid j_1 + j_2, j_1 + j_2 \rangle \psi_1\left(j_1, m_1\right) \, \psi_2\left(j_2, m_2\right) \tag{95}$$

For ϕ to be normalized,

$$\left| \langle j_1 j_2 \mid j_1 + j_2, j_1 + j_2 \rangle \right|^2 = 1 \tag{96}$$

This implies that the C.G. coefficient for $j = j_1 + j_2$ and $m = j_1 + j_2$ is given by

$$\langle j_1 j_2 \mid j_1 + j_2, j_1 + j_2 \rangle = 1$$

so that

$$\phi\left(j_1 + j_2, j_1 + j_2\right) = \psi_1\left(j_1, j_1\right) \, \psi_2\left(j_2, j_2\right) \tag{97}$$

Operating equation (97) with $J_- = J_{1-} + J_{2-}$, we get

$$J_- \phi\left(j_1 + j_2, j_1 + j_2\right) = \left(J_{1-} + J_{2-}\right) \psi_1\left(j_1, j_1\right) \, \psi_2\left(j_2, j_2\right) \tag{98}$$

This Eqn. is same as

$$J_- \phi(j,m) = \left(J_{1-} + J_{2-}\right) \psi_1\left(j_1, m_1\right) \, \psi_2\left(j_2, m_2\right)$$

Thus, remembering that eigenvalue of J_- is $\sqrt{(j+m)(j-m+1)}$ Eqn. (98) becomes

$$J_- \phi(j,m) = \left(J_{1-} + J_{2-}\right) \psi\left(m_1, m_2\right) \tag{99i}$$

or

$$\sqrt{2\left(j_1 + j_2\right)} \, \phi\left(j_1 + j_2, j_1 + j_2 - 1\right) = \sqrt{2j_1} \, \psi_1\left(j_1, j_1 - 1\right) \psi_2\left(j_2, j_2\right)$$

$$+ \sqrt{2j_2} \, \psi_1\left(j_1, j_1\right) \psi_2\left(j_2, j_2 - 1\right) \tag{99ii}$$

(because operation by J_- lowers the value of m by one).

So, equivalently

$$\phi\left(j_1 + j_2, j_1 + j_2 - 1\right) = \left(\sqrt{\frac{j_1}{j_1 + j_2}}\right) \psi_1\left(j_1, j_1 - 1\right) \psi_2\left(j_2, j_2\right)$$

$$+ \left(\sqrt{\frac{j_2}{j_1 + j_2}}\right) \psi_1\left(j_1, j_1\right) \psi_2\left(j_2, j_2 - 1\right) \tag{100}$$

This case corresponds to

$$j = j_1 + j_2 \text{ and } m = j_1 + j_2 - 1$$

Corresponding to this, we get from Eqn. (92)

$$\phi\left(j_1 + j_2, j_1 + j_2 - 1\right) = \langle j_1 - 1, j_2 \mid j_1 + j_2, j_1 + j_2 - 1 \rangle \psi_1\left(j_1, j_1 - 1\right) \psi_2\left(j_2, j_2\right)$$

$$+ \langle j_1, j_2 - 1 \mid j_1 + j_2, j_1 + j_2 - 1 \rangle \psi_1\left(j_1, j_1\right) \psi_2\left(j_2, j_2 - 1\right) \tag{101}$$

Comparing equations (100) and (101) we get two C.G. coefficients,

$$\langle j_1 - 1, j_2 | j_1 + j_2, j_1 + j_2 - 1 \rangle = \sqrt{\frac{j_1}{j_1 + j_2}}$$

and $\qquad\qquad\qquad\qquad\qquad\qquad\qquad\qquad\qquad\qquad\qquad$ (102)

$$\langle j_1, j_2 - 1 | j_1 + j_2, j_1 + j_2 - 1 \rangle = \sqrt{\frac{j_2}{j_1 + j_2}}$$

A continuation of this process would give all the coefficients for $j = j_1 + j_2$. Equation (100) represents $\phi(j_1 + j_2, j_1 + j_2 - 1)$ as a linear combination of

$$\psi_1(j_1, j_1 - 1) \; \psi_2(j_2, j_2) \text{ and } \psi_1(j_1, j_1) \; \psi_2(j_2, j_2 - 1)$$

Now, consider two linear combinations for two given orthonormal vectors α and β viz.

$$\chi_1 = c_1 \alpha + c_2 \beta \qquad\qquad\qquad (103)$$

$$\chi_2 = c_3 \alpha + c_4 \beta \qquad\qquad\qquad (104)$$

χ_1 is normalized if $|c_1|^2 + |c_2|^2 = 1$.

χ_2 is normalized if $|c_3|^2 + |c_4|^2 = 1$.

Now, we try to find the condition that χ_1 and χ_2 are mutually orthogonal. This means

$$(\chi_1, \chi_2) = 0 \text{ or } \int \chi_1^* \chi_2 d\tau = 0$$

or $\qquad \int \left(c_1^* \alpha^* + c_2^* \beta^* \right) \left(c_3 \alpha + c_4 \beta \right) d\tau = 0$

or $\qquad \int \left(c_1^* c_3 \alpha^* \alpha + c_1^* c_4 \alpha^* \beta + c_2^* c_3 \beta^* \alpha + c_2^* c_4 \beta^* \beta \right) d\tau = 0$

or $\qquad \int c_1^* c_3 \alpha^* \alpha \, d\tau + \int c_1^* c_4 \alpha^* \beta \, d\tau + \int c_2^* c_3 \beta^* \alpha \, d\tau + \int c_2^* c_4 \beta^* \beta \, d\tau = 0$

Since α and β are orthonormal, we have

$$\int \alpha^* \beta \, d\tau = 0 = \int \beta^* \alpha \, d\tau \,; \; \int \alpha^* \alpha \, d\tau = 1 = \int \beta^* \beta \, d\tau \text{ so, } c_1^* c_3 + c_2^* c_4 = 0$$

or $\qquad\qquad\qquad c_4 = -c_3 \left(\frac{c_1^*}{c_2^*} \right) \qquad\qquad\qquad (105)$

Since $|c_3|^2 + |c_4|^2 = 1$, therefore

$$|c_3|^2 + |c_3|^2 \frac{|c_1|^2}{|c_2|^2} = 1$$

or $\qquad\qquad\qquad \frac{|c_3|^2}{|c_2|^2} \left(|c_2|^2 + |c_1|^2 \right) = 1$

Comparing this with $|c_1|^2 + |c_2|^2 = 1$, we get

$$|c_3|^2 = |c_2|^2$$

or

$$|c_3|^2 = c_2^* c_2$$

Applying this condition, we find that linear combination orthogonal to Eqn. (100) is

$$\phi' = -\sqrt{\frac{j_2}{j_1 + j_2}} \ \psi_1(j_1, j_1 - 1) \ \psi_2(j_2, j_2) + \sqrt{\frac{j_1}{j_1 + j_2}} \ \psi_1(j_1, j_1) \ \psi_2(j_2, j_2 - 1) \quad (106)$$

It is an eigenvector of J_z belonging to eigenvalue $j_1 + j_2 - 1$. It is also an eigenvector of J^2 belonging to the eigenvalue $j = j_1 + j_2 - 1$, i.e.

$$\phi' = \phi(j_1 + j_2 - 1, j_1 + j_2 - 1) \tag{107}$$

But from Eqn. (93), we get for this

$$\phi(j_1 + j_2 - 1, j_1 + j_2 - 1) = \sum_{m_1 m_2} \langle m_1 m_2 | jm \rangle \ \psi_1(j_1, m_1) \ \psi_2(j_2, m_2)$$

$$(m_1 = j_1 - 1, m_2 = j_2 - 1)$$

$$(j = j_1 + j_2 - 1, m = j_1 + j_2 - 1)$$

or $\quad \phi(j_1 + j_2, j_1 + j_2 - 1) = \langle j_1 - 1, j_2 | j_1 + j_2 - 1, j_1 + j_2 - 1 \rangle \times$

$$\psi_1(j_1, j_1 - 1) \ \psi_2(j_2, j_2) + \langle j_1, j_2 - 1 | j_1 + j_2 - 1, j_1 + j_2 - 1 \rangle \times$$

$$\psi_1(j_1, j_1) \ \psi_2(j_2, j_2 - 1) \tag{108}$$

Using Eqn. (107) and comparing equations (106) and (108), we get two more C.G. coefficients

$$\langle j_1 - 1, j_2 | j_1 + j_2 - 1, j_1 + j_2 - 1 \rangle = -\sqrt{\frac{j_2}{j_1 + j_2}}$$

and

$$\langle j_1, j_2 - 1 | j_1 + j_2 - 1, j_1 + j_2 - 1 \rangle = +\sqrt{\frac{j_1}{j_1 + j_2}} \tag{109}$$

Thus, from Eqns. (102) and (109), we obtain the Table 13.4. Further application of operator J_- to equations (100) and (108) would lead to $\phi(j_1 + j_2, j_1 + j_2 - 2)$ and $\phi(j_1 + j_2 - 1, j_1 + j_2 - 2)$ which are linear combinations of $\psi_1(j_1, j_1)$ $\psi_2(j_2, j_2 - 2)$, $\psi_1(j_1, j_1 - 1) \ \psi_2(j_2, j_2)$ and $\psi_1(j_1, j_1 - 2) \ \psi_2(j_2, j_2)$. Similar to above arguments, we can find a third linear combination of these three which is orthogonal to other two combinations. This third linear combination would belong to a new eigenvalue of J^2 corresponding to $j = j_1 + j_2 - 2$. In continuation of this process, a new value of J^2 appears each time for every new value of m, ultimately successive application of J_- annihilates one of the ψ's. The minimum value of j obtained in this way is $|j_1 - j_2|$.

Table 13.6: Clebsch-Gordan coefficients,

$$C^{jm}_{m_1 m_2} \text{ for } |m_1 m_2\rangle = |j_1 - 1, j_2\rangle \text{ and } |j_1, j_2 - 1\rangle.$$

m_1	m_2	$|jm\rangle$ (*)	
		$|j_1 + j_2, j_1 + j_2 - 1\rangle$	$|j_1 + j_2 - 1, j_1 + j_2 - 1\rangle$
$j_1 - 1$	j_2	$\left(\dfrac{j_1}{j_1 + j_2}\right)^{1/2}$	$-\left(\dfrac{j_2}{j_1 + j_2}\right)^{1/2}$
j_1	$j_2 - 1$	$\left(\dfrac{j_2}{j_1 + j_2}\right)^{1/2}$	$\left(\dfrac{j_1}{j_1 + j_2}\right)^{1/2}$

(*) Here, $|jm\rangle = \sum\limits_{m_1, m_2} C^{jm}_{m_1 m_2} |m_1 m_2\rangle$ where coefficients $C^{jm}_{m_1 m_2} = \langle m_1 m_2 | jm \rangle$.

13.16 Calculation of C.G. Coefficients

(i) *Calculation for* $j_1 = j_2 = 1/2$:

For $j_1 = j_2 = 1/2$, $\quad m_1 = \pm 1/2$, $\quad m_2 = \pm 1/2$

Thus there are four ψ functions

$$\psi(m_1, m_2); \; \psi\left(\tfrac{1}{2}, \tfrac{1}{2}\right), \psi\left(\tfrac{1}{2}, -\tfrac{1}{2}\right), \psi\left(-\tfrac{1}{2}, \tfrac{1}{2}\right), \psi\left(-\tfrac{1}{2}, -\tfrac{1}{2}\right)$$

The allowed values of j are $|j_1 - j_2| \le j \le j_1 + j_2$ or 0 and 1. Hence there are four ϕ functions $(\phi(j, m))$:

$$\phi(1, 1), \; \phi(1, 0), \; \phi(1, -1) \text{ and } \phi(0, 0)$$

(Note that there is no $\phi(0, -1)$, see article 13.9)

Obviously

$$\phi(1, 1) = \psi\left(\tfrac{1}{2}, \tfrac{1}{2}\right)$$

because $\psi\left(\tfrac{1}{2}, \tfrac{1}{2}\right)$ is the state for which $m_1 + m_2 = 1$ and

$$\phi(1, -1) = \psi\left(-\tfrac{1}{2}, -\tfrac{1}{2}\right) \tag{110}$$

corresponding to last value $m = -j$. $k_{max} = 2j_2 (= 1)$. Therefore $\phi(1, 0)$ and $\phi(0, 0)$ will be expressed as linear combinations

$$\phi(1, 0) = c_1 \psi\left(\tfrac{1}{2}, -\tfrac{1}{2}\right) + c_2 \psi\left(-\tfrac{1}{2}, \tfrac{1}{2}\right) \tag{111}$$

and

$$\phi(0, 0) = c_3 \psi\left(\tfrac{1}{2}, -\tfrac{1}{2}\right) + c_4 \psi\left(-\tfrac{1}{2}, \tfrac{1}{2}\right) \tag{112}$$

Now

$$J_- \phi(1, 1) = \left(J_{1-} + J_{2-}\right) \psi\left(\tfrac{1}{2}, \tfrac{1}{2}\right)$$

Using $J_- |j, m\rangle = \sqrt{(j + m)(j - m + 1)} \; |j, m - 1\rangle$ on l.h.s. of the above equation, we get

$$\sqrt{2}\ \phi(1,0) = \psi\left(-\tfrac{1}{2}, \tfrac{1}{2}\right) + \psi\left(\tfrac{1}{2}, -\tfrac{1}{2}\right)$$

$$\left(\because\ J_{1-}\ \psi(m_1, m_2) = \psi(m_1 - 1, m_2)\ etc.\right)$$

or $$\phi(1,0) = \frac{1}{\sqrt{2}} \psi\left(\tfrac{1}{2}, -\tfrac{1}{2}\right) + \frac{1}{\sqrt{2}} \psi\left(-\tfrac{1}{2}, \tfrac{1}{2}\right) \tag{113}$$

Comparing equations (111) and (113)

$$c_1 = \frac{1}{\sqrt{2}} = c_2$$

For $\phi(1, 0)$ and $\phi(0, 0)$ to be mutually orthogonal,

$$c_3 = c_2^* = \frac{1}{\sqrt{2}} \text{ and } c_4 = -c_1^* = -\frac{1}{\sqrt{2}}$$

So, from Eqn. (112), we get

$$\phi(0,0) = \frac{1}{\sqrt{2}} \psi\left(\tfrac{1}{2}, -\tfrac{1}{2}\right) - \frac{1}{\sqrt{2}} \psi\left(-\tfrac{1}{2}, \tfrac{1}{2}\right) \tag{114}$$

Hence from Eqns (109), (110), (113) and (114), we may represent the relation amongst ϕ's and ψ's by the matrix equation

$$\begin{pmatrix} \phi(1,1) \\ \phi(1,0) \\ \phi(0,0) \\ \phi(1,-1) \end{pmatrix} = \begin{pmatrix} 1 & 0 & 0 & 0 \\ 0 & \frac{1}{\sqrt{2}} & \frac{1}{\sqrt{2}} & 0 \\ 0 & \frac{1}{\sqrt{2}} & -\frac{1}{\sqrt{2}} & 0 \\ 0 & 0 & 0 & 1 \end{pmatrix} \begin{pmatrix} \psi\left(\tfrac{1}{2}, \tfrac{1}{2}\right) \\ \psi\left(\tfrac{1}{2}, -\tfrac{1}{2}\right) \\ \psi\left(-\tfrac{1}{2}, \tfrac{1}{2}\right) \\ \psi\left(-\tfrac{1}{2}, \tfrac{1}{2}\right) \end{pmatrix} \tag{115}$$

Example 13.9: Calculation of the C.G. coefficients for $j_1 = 1$ and $j_2 = 1/2$ (e.g. p state of electron)

m_1 takes $(2j_1 + 1)$ values: $-j_1$ to $+j_1$: $\quad m_1 \rightarrow 1, 0, -1$
m_2 can take $(2j_2 + 1)$ values : $\quad m_2 \rightarrow \tfrac{1}{2}, -\tfrac{1}{2}$

Consequently there are six $\psi(m_1, m_2)$ functions (*)

$$\psi\left(1, \tfrac{1}{2}\right)\ ;\ \psi\left(1, -\tfrac{1}{2}\right)\ ;\ \psi\left(0, \tfrac{1}{2}\right)\ ;\ \psi\left(0, -\tfrac{1}{2}\right)\ ;\ \psi\left(-1, \tfrac{1}{2}\right)\ ;\ \psi\left(-1, -\tfrac{1}{2}\right)$$

Thus, the values of $m\ (= m_1 + m_2)$ are

$$\frac{3}{2}, \frac{1}{2}, -\frac{1}{2}, \left(\frac{1}{2}\right), \left(-\frac{1}{2}\right) \text{ and } -\frac{3}{2}$$

The allowed values of j are

$$j_1 + j_2 \geq j \geq |j_1 - j_2| \text{ or } \frac{3}{2} \text{ and } \frac{1}{2}$$

(*) Expressing $\psi(j_1 j_2 m_1 m_2)$ as $\psi(m_1 m_2)$

Therefore, the corresponding $\phi(j, m)$ vectors are[**]

$$\phi\left(\tfrac{3}{2}, \tfrac{3}{2}\right), \ \phi\left(\tfrac{1}{2}, \tfrac{1}{2}\right), \ \phi\left(\tfrac{3}{2}, \tfrac{1}{2}\right), \ \phi\left(\tfrac{3}{2}, -\tfrac{1}{2}\right), \ \phi\left(\tfrac{1}{2}, -\tfrac{1}{2}\right), \ \phi\left(\tfrac{3}{2}, -\tfrac{3}{2}\right)$$

$\phi\left(j_{max}, m_{max}\right)$ corresponds to $\psi\left(m_{1\,max}, m_{2\,max}\right)$ or

$$\phi\left(\tfrac{3}{2}, \tfrac{3}{2}\right), = \psi\left(1, \tfrac{1}{2}\right) \tag{i}$$

because $\psi\left(1, \tfrac{1}{2}\right)$ is the state for which $m_1 + m_2 \,(= m) = \tfrac{3}{2}$.

Obviously

$$\phi\left(\tfrac{3}{2}, -\tfrac{3}{2}\right) = \psi\left(-1, -\tfrac{1}{2}\right) \tag{ii}$$

because only this $\psi\left(m_1, m_2\right)$ satisfies $m = \left(m_1 + m_2\right) = -\dfrac{3}{2}$ (as this ψ corresponds to the last ψ for which $m_1 = -j_1$ and $m_2 = -j_2$)

Other ϕ's will be linear combinations, obtained as follows:

$$\phi\left(\tfrac{3}{2}, \tfrac{1}{2}\right) = c_1 \psi\left(1, -\tfrac{1}{2}\right) + c_2 \psi\left(0, \tfrac{1}{2}\right)$$

$$\phi\left(\tfrac{1}{2}, \tfrac{1}{2}\right) = c_3 \psi\left(1, -\tfrac{1}{2}\right) + c_4 \psi\left(0, \tfrac{1}{2}\right)$$

$$\phi\left(\tfrac{3}{2}, -\tfrac{1}{2}\right) = c_5 \psi\left(0, -\tfrac{1}{2}\right) + c_6 \psi\left(-1, \tfrac{1}{2}\right)$$

$$\phi\left(\tfrac{1}{2}, -\tfrac{1}{2}\right) = c_7 \psi\left(0, -\tfrac{1}{2}\right) + c_8 \psi\left(-1, \tfrac{1}{2}\right)$$

Using $J_- |j, m\rangle = \sqrt{(j+m)(j-m+1)} \ |j, m-1\rangle$ on Eqn. (i)

$$J_- \phi\left(\tfrac{3}{2}, \tfrac{3}{2}\right) = \left[(J_1)_- + (J_2)_-\right] \psi\left(1, \tfrac{1}{2}\right)$$

or $\quad \sqrt{3}\ \phi\left(\tfrac{3}{2}, \tfrac{1}{2}\right) = \sqrt{2}\ \psi\left(0, \tfrac{1}{2}\right) + \psi\left(1, -\tfrac{1}{2}\right)$ [*]

or $\quad \phi\left(\tfrac{3}{2}, \tfrac{1}{2}\right) = \sqrt{\tfrac{2}{3}}\ \psi\left(0, \tfrac{1}{2}\right) + \sqrt{\tfrac{1}{3}}\ \psi\left(1, -\tfrac{1}{2}\right) \tag{iii}$

$$\therefore \qquad c_1 = \frac{1}{\sqrt{3}} \text{ and } c_2 = \sqrt{\frac{2}{3}}$$

---

[*] j's maximum value $= j_1 + j_2$

j's minimum value $= j_1 - j_2$

Then

$$\phi\left(j_1 + j_2, j_1 + j_2\right) = \psi\left(j_1, j_2\right) \left(= \psi\left(m_1, m_2\right)\right)$$

$$J_- \phi\left(j_1 + j_2, j_1 + j_2\right) = \left[(J_1)_- + (J_2)_-\right] \psi\left(m_1, m_2\right)$$

$$\sqrt{\left(\tfrac{3}{2} + \tfrac{3}{2}\right)\left(\tfrac{3}{2} - \tfrac{3}{2} + 1\right)} \times \phi\left(j_1 + j_2, j_1 + j_2 - 1\right)$$

$$= \sqrt{2j_1}\ \psi\left(m_1 - 1, m_2\right) + \sqrt{2j_2}\ \psi\left(m_1, m_2 - 1\right)$$

($\because$ for the 1st term on r.h.s. $j = j_1 = m_1 = 1; \ m_2 = \tfrac{1}{2}$ and for the 2nd term $j = j_2 \,(= m_2) = \tfrac{1}{2}$ and $j_1 = m_1 = 1$))

[**] Expressing $\phi\left(j_1\, j_2\, j\ m\right)$ as $\phi\,(j\ m)$.

By symmetry

$$\phi\left(\tfrac{3}{2}, -\tfrac{1}{2}\right) = \sqrt{\tfrac{2}{3}}\, \psi\left(0, -\tfrac{1}{2}\right) + \sqrt{\tfrac{1}{3}}\, \psi\left(-1, \tfrac{1}{2}\right) \qquad (iv)$$

$$\text{(i.e. } c_5 = \sqrt{\tfrac{2}{3}} \text{ and } c_6 = \sqrt{\tfrac{1}{3}} \text{)}$$

Now, $\phi\left(\tfrac{1}{2}, \tfrac{1}{2}\right)$, which is a linear combination of $\psi\left(1, -\tfrac{1}{2}\right)$ and $\psi\left(0, \tfrac{1}{2}\right)$ has to be orthogonal to $\phi\left(\tfrac{3}{2}, \tfrac{1}{2}\right)$. So,

$$c_1^* c_3 + c_2^* c_4 = 0$$

or

$$\frac{c_4}{c_3} = -\frac{c_1^*}{c_2^*} = -\frac{1/\sqrt{3}}{\sqrt{2/3}} = -\frac{1}{\sqrt{2}}$$

$$\therefore \qquad \phi\left(\tfrac{1}{2}, \tfrac{1}{2}\right) = c_3\left[\psi\left(1, -\tfrac{1}{2}\right) - \frac{1}{\sqrt{2}}\psi\left(0, \tfrac{1}{2}\right)\right]$$

Normalization of this wave function gives

$$|c_3|^2 + \frac{|c_3|^2}{2} = 1 \text{ or } c_3 = \sqrt{\tfrac{2}{3}}$$

$$\therefore \qquad \phi\left(\tfrac{1}{2}, \tfrac{1}{2}\right) = \sqrt{\tfrac{2}{3}}\, \psi\left(1, -\tfrac{1}{2}\right) - \sqrt{\tfrac{1}{3}}\, \psi\left(0, \tfrac{1}{2}\right) \qquad (v)$$

$$\text{(i.e. } c_3 = \sqrt{\tfrac{2}{3}} \text{ and } c_4 = -\sqrt{\tfrac{1}{3}} \text{)}$$

Application of J_- to Eqn. (v) gives

$$\phi\left(\tfrac{1}{2}, -\tfrac{1}{2}\right) = -\sqrt{\tfrac{2}{3}}\, \psi\left(-1, \tfrac{1}{2}\right) + \sqrt{\tfrac{1}{3}}\, \psi\left(0, -\tfrac{1}{2}\right) \qquad (vi)$$

$$\text{(i.e. } c_8 = -\sqrt{\tfrac{2}{3}} \text{ and } c_7 = \sqrt{\tfrac{1}{3}} \text{)}$$

Hence, we may represent the transformation from ψ's to ϕ's with the help of matrix of C.G. coefficients as

$$
\begin{bmatrix}
\phi\left(\tfrac{3}{2}, \tfrac{3}{2}\right) \\
\phi\left(\tfrac{3}{2}, \tfrac{1}{2}\right) \\
\phi\left(\tfrac{1}{2}, \tfrac{1}{2}\right) \\
\phi\left(\tfrac{3}{2}, -\tfrac{1}{2}\right) \\
\phi\left(\tfrac{1}{2}, -\tfrac{1}{2}\right) \\
\phi\left(\tfrac{3}{2}, -\tfrac{3}{2}\right)
\end{bmatrix}
=
\begin{bmatrix}
1 & 0 & 0 & 0 & 0 & 0 \\
0 & \sqrt{\tfrac{1}{3}} & \sqrt{\tfrac{2}{3}} & 0 & 0 & 0 \\
0 & \sqrt{\tfrac{2}{3}} & -\sqrt{\tfrac{1}{3}} & 0 & 0 & 0 \\
0 & 0 & 0 & \sqrt{\tfrac{2}{3}} & \sqrt{\tfrac{1}{3}} & 0 \\
0 & 0 & 0 & \sqrt{\tfrac{1}{3}} & -\sqrt{\tfrac{2}{3}} & 0 \\
0 & 0 & 0 & 0 & 0 & 1
\end{bmatrix}
\begin{bmatrix}
\psi\left(1, \tfrac{1}{2}\right) \\
\psi\left(1, -\tfrac{1}{2}\right) \\
\psi\left(0, \tfrac{1}{2}\right) \\
\psi\left(0, -\tfrac{1}{2}\right) \\
\psi\left(-1, \tfrac{1}{2}\right) \\
\psi\left(-1, -\tfrac{1}{2}\right)
\end{bmatrix}
$$

13.17 How to use Recursion Relations with the help of m_1m_2-plane

The Recursion relations $(87i)$ and $(87ii)$ can be put together in the form

$$\sqrt{(j \mp m)(j \pm m + 1)} \left\langle m_1\, m_2 \middle| j, m \pm 1 \right\rangle$$

$$= \sqrt{(j_1 \mp m_1 + 1)(j_1 \pm m_1)} \left\langle m_1 \mp 1, m_2 \middle| jm \right\rangle$$

$$+ \sqrt{(j_2 \mp m_2 + 1)(j_2 \pm m_2)} \left\langle m_1, m_2 \mp 1 \middle| jm \right\rangle \qquad (116)$$

These recursion relations together with normalization condition

$$\sum_{m_1\, m_2} \left| \left\langle m_1\, m_2 \middle| jm \right\rangle \right|^2 = 1 \qquad (117)$$

almost uniquely determine all C.G. coefficients. We assert "almost uniquely" because certain sign conventions are yet to be specified. We proceed as follows:

We consider $m_1\, m_2$-plane (see Fig. 13.11). The J_+ recursion relation (upper sign) tells us that the coefficient at $(m_1\, m_2)$ is related to the coefficients at $(m_1 - 1, m_2)$ and as shown in Fig. 13.11a. Similarly, the J_- recursion relation (lower sign) relates the three coefficients whose m_1, m_2 values are given in Fig. 13.11b.

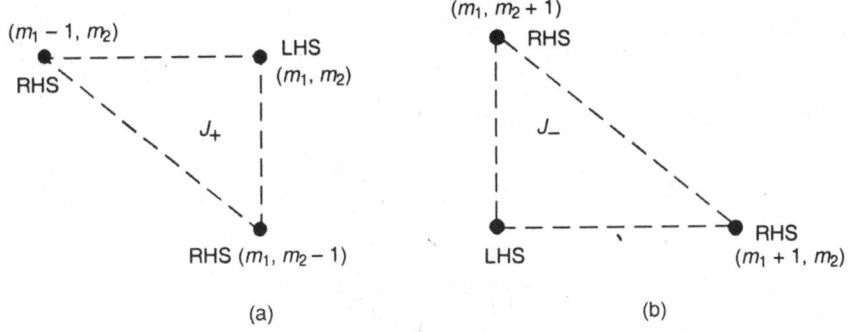

(a) (b)

Fig. 13.11: (a and b) m_1m_2-plane showing the Clebsch-Gordan coefficients related by the recursion relations (116)

Figure 13.12a shows a plot of the boundary of the allowed region corresponding to the conditions

$$|m_1| \le j_1;\ |m_2| \le j_2 \text{ and } -j \le m_1 + m_2 \le j \qquad (118)$$

Suppose we start with the upper right hand corner at A. Fig. 13.12b is a more detailed one. We apply the J_- recursion relation Eqn. (116) (lower sign) with $(m_1, m_2 + 1)$ corresponding to A. Observe now that the recursion relation connects A and B because the site corresponding to $(m_1 + 1, m_2)$ is forbidden by $|m_1| \le j_1$. Then we can obtain the C.G. coefficients of B in terms of the coefficients of A, e.g. from Eqn. (116):

$$\sqrt{(j+m)(j-m+1)}\ \langle m_1, m_2 | j, m-1 \rangle$$
$$= \sqrt{(j_2 + m_2 + 1)(j_2 - m_2)}\ \langle m_1, m_2 + 1 | jm \rangle$$

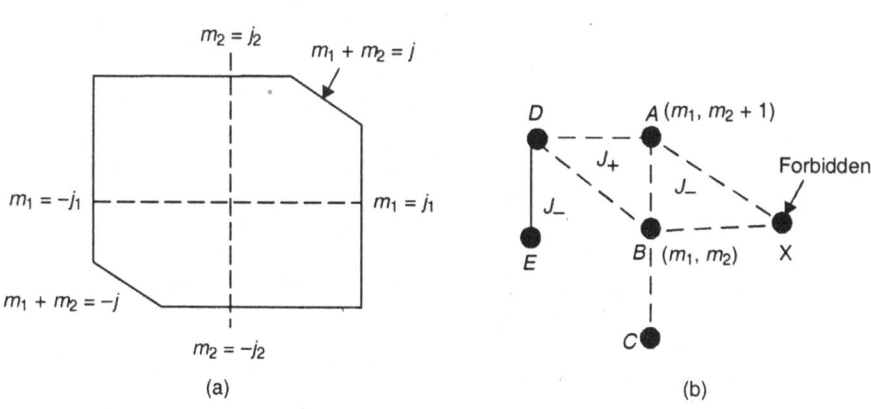

Fig. 13.12: (a and b) use of the recursion relations to obtain the C.G. coefficients

Next, we form a J_+ triangle made up of A, B and D. This enables us to obtain the coefficients of D once the coefficient of A is specified.

Knowing B and D we can get the coefficients of E, and so on. In this way, we can get the C.G. coefficients of every site of points in $m_1 m_2$ - plane in terms of the coefficient of starting site A. For overall normalization Eqn. (117) is used.

13.18 Addition of Orbital and Spin-angular Momentum of a Spin 1/2 Particle

We have

$$j_1 = l \quad \text{(integer)} \quad m_1 = m_l$$
$$j_2 = s = \frac{1}{2} \quad m_2 = m_s = \pm \frac{1}{2} \tag{119}$$

The allowed values of j are given by

$$j_1 + j_2 \geq j \geq \left| j_1 - j_2 \right| \text{ or } j = l \pm \tfrac{1}{2}, \quad l > 0$$

$$j = \tfrac{1}{2}, \quad l = 0 \tag{120}$$

For each l there are two possible j values, for example for $l = 1$ (p state) we have (in spectroscopic notation) $p_{3/2}$ and $p_{1/2}$ (the subscripts refer to j-value).

The $m_1 m_2$ - plane, in other words $m_l m_s$ - plane of this problem is as shown in Fig. 13.13. The allowed sites form only two rows: the upper row for $m_s = 1/2$ and the lower row for $m_s = -1/2$. Let us consider the case $j = l + 1/2$. Because m_s cannot exceed 1/2, we can use the J_- recursion relation in such a way that we always stay in the upper row ($m_2 = m_s = 1/2$) while the m_l value changes by one unit each time we consider a new J_- triangle.

($m_2 + 1$ is forbidden)

$$m \to m+1 \begin{cases} m_1 + m_2 = m \\ \text{or} \quad m_1 + \frac{1}{2} = m \\ \therefore \quad m_1 = \left(m - \frac{1}{2}\right)(=m_l) \; ; \quad m_2 = \frac{1}{2} \end{cases}$$

$$j_1 = l, j = l + 1/2$$

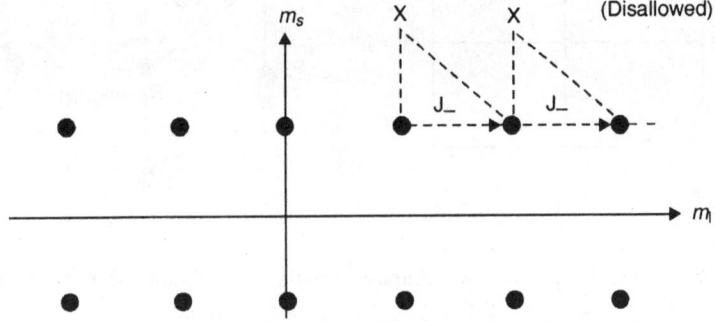

Fig. 13.13: Use of recursion relations to obtain the C.G. coefficients for
$$j_1 = l \text{ and } j_2 = s = \frac{1}{2}$$

From Eqn. (116) (lower sign)

$$\sqrt{(l + \tfrac{1}{2} + m + 1)(l + \tfrac{1}{2} - m - 1 + 1)} \; \left\langle m - \tfrac{1}{2}, \tfrac{1}{2} \middle| l + \tfrac{1}{2}, m \right\rangle$$

$$= \sqrt{(l + m - \tfrac{1}{2} + 1)(l - m + \tfrac{1}{2})} \; \left\langle m + \tfrac{1}{2}, \tfrac{1}{2} \middle| l + \tfrac{1}{2}, m + 1 \right\rangle \tag{121}$$

(We have used: $\sqrt{(j + m)(j - m + 1)} \; \left\langle m_1, m_2 \middle| j, m - 1 \right\rangle$

$$= \sqrt{(j_1 + m_1 + 1)(j_1 - m_1)} \; \left\langle m_1 + 1, m_2 \middle| j, m \right\rangle)$$

In other words,

$$\left\langle m - \tfrac{1}{2}, \tfrac{1}{2} \middle| l + \tfrac{1}{2}, m \right\rangle = \sqrt{\frac{l + m + \tfrac{1}{2}}{l + m + \tfrac{3}{2}}} \; \left\langle m + \tfrac{1}{2}, \tfrac{1}{2} \middle| l + \tfrac{1}{2}, m + 1 \right\rangle \tag{122}$$

Thus we move in the Fig. 13.13 horizontally by one unit.

We can in turn express $\left\langle m + \tfrac{1}{2}, \tfrac{1}{2} \middle| l + \tfrac{1}{2}, m + 1 \right\rangle$ in terms of $\left\langle m + \tfrac{3}{2}, \tfrac{1}{2} \middle| l + \tfrac{1}{2}, m + 2 \right\rangle$ and so on. Evidently, this procedure can be continued until m_l (or m_1) reaches l, the maximum possible value.

$$\left\langle m - \tfrac{1}{2}, \tfrac{1}{2} \middle| l + \tfrac{1}{2}, m \right\rangle = \sqrt{\frac{l + m + \tfrac{1}{2}}{l + m + \tfrac{3}{2}}} \sqrt{\frac{l + m + \tfrac{3}{2}}{l + m + \tfrac{5}{2}}} \; \left\langle m + \tfrac{3}{2}, \tfrac{1}{2} \middle| l + \tfrac{1}{2}, m + 2 \right\rangle$$

$$= \sqrt{\frac{l+m+\frac{1}{2}}{l+m+\frac{3}{2}}} \sqrt{\frac{l+m+\frac{3}{2}}{l+m+\frac{5}{2}}} \sqrt{\frac{l+m+\frac{5}{2}}{l+m+\frac{7}{2}}} \times \left\langle m+\tfrac{5}{2}, \tfrac{1}{2} \middle| l+\tfrac{1}{2}, m+3 \right\rangle.$$

$$\vdots \qquad \vdots \qquad \vdots \qquad \vdots$$

$$= \sqrt{\frac{l+m+\frac{1}{2}}{2l+1}} \left\langle l, \tfrac{1}{2} \middle| l+\tfrac{1}{2}, l+\tfrac{1}{2} \right\rangle \tag{123}$$

Let us now consider the angular momentum configuration in which m_l and m_s are both maximal i.e. $m_l = l$ and $m_s = 1/2$.

The total $m = m_l + m_s$ is $l + \frac{1}{2}$, which is possible only for $j = l + \frac{1}{2}$ and not for $j = l - \frac{1}{2}$. So, $\left| m_l = l, \; m_s = \tfrac{1}{2} \right\rangle$ must be equal to $\left| j = l + \tfrac{1}{2}, \; m = l + \tfrac{1}{2} \right\rangle$ upto a phase factor (real and positive by convention). Then we have

$$\left\langle l, \tfrac{1}{2} \middle| l + \tfrac{1}{2}, l + \tfrac{1}{2} \right\rangle = 1$$

Consequently,

$$\left\langle m - \tfrac{1}{2}, \tfrac{1}{2} \middle| l + \tfrac{1}{2}, m \right\rangle = \sqrt{\frac{l+m+\frac{1}{2}}{2l+1}} \tag{124}$$

However, we are interested in determing the values of the coefficients córresponding to question marks in the following expression:

$$\left| j = l + \tfrac{1}{2}, m \right\rangle = \sqrt{\frac{l+m+\frac{1}{2}}{2l+1}} \; \left| m_l = m - \tfrac{1}{2}, m_s = \tfrac{1}{2} \right\rangle$$

$$+ ? \; \left| m_l = m + \tfrac{1}{2}, m_s = -\tfrac{1}{2} \right\rangle \tag{125}$$

and
$$\left| j = l - \tfrac{1}{2}, m \right\rangle = ? \; \left| m_l = m - \tfrac{1}{2}, m_s = \tfrac{1}{2} \right\rangle$$

$$+ ? \; \left| m_l = m + \tfrac{1}{2}, m_s = -\tfrac{1}{2} \right\rangle \tag{126}$$

We note that the transformation matrix with fixed m from the (m_l, m_s) basis to the (j, m) basis (because of orthogonality) is expected to be of the form

$$\begin{pmatrix} \cos\alpha & \sin\alpha \\ -\sin\alpha & \cos\alpha \end{pmatrix} \tag{127}$$

Comparison of this with Eqn. (125) shows that cos α is itself (124), so we can readily determine sin α except for the sign:

$$\sin^2 \alpha = 1 - \frac{\left(l + m + \tfrac{1}{2} \right)}{(2l+1)} = \frac{\left(l - m + \tfrac{1}{2} \right)}{(2l+1)}$$

We can say that $\left\langle m_l = m + \tfrac{1}{2}, m_s = -\tfrac{1}{2} \middle| j = l + \tfrac{1}{2}, m \right\rangle$ must be positive because all $j = l + \frac{1}{2}$ states are reachable by applying the J_- operator successively to $\left| j = l + \tfrac{1}{2}, \; m = l + \tfrac{1}{2} \right\rangle$ and the matrix elements of J_- are always +ive by convention. So the 2×2 transformation matrix for C.G. coefficients in present case is

$$\begin{pmatrix} \sqrt{\dfrac{l+m+\frac{1}{2}}{(2l+1)}} & \sqrt{\dfrac{l-m+\frac{1}{2}}{(2l+1)}} \\[3ex] -\sqrt{\dfrac{l-m+\frac{1}{2}}{(2l+1)}} & \sqrt{\dfrac{l+m+\frac{1}{2}}{(2l+1)}} \end{pmatrix}$$

(128i)

Consequently, we have from Eqns. (125) and (126)

$$\left| j = l + \tfrac{1}{2}, m \right\rangle = \sqrt{\dfrac{l+m+\frac{1}{2}}{2l+1}} \ \left| m_l = m - \tfrac{1}{2}, m_s = \tfrac{1}{2} \right\rangle$$

$$+ \sqrt{\dfrac{l-m+\frac{1}{2}}{2l+1}} \ \left| m_l = m + \tfrac{1}{2}, m_s = -\tfrac{1}{2} \right\rangle$$

(128ii)

and

$$\left| j = l - \tfrac{1}{2}, m \right\rangle = - \sqrt{\dfrac{l-m+\frac{1}{2}}{2l+1}} \ \left| m_l = m - \tfrac{1}{2}, m_s = \tfrac{1}{2} \right\rangle$$

$$+ \sqrt{\dfrac{l+m+\frac{1}{2}}{2l+1}} \ \left| m_l = m + \tfrac{1}{2}, m_s = -\tfrac{1}{2} \right\rangle$$

(128iii)

Here the coefficients of the eigenvectors on the right hand side are the C.G. coefficients defined by matrix (128i).

In the subsequent article, we are going to discuss an important theorem pertaining to angular momentum operators related with spherical tensors. Therefore, it would be worthwhile to first discuss reducible and irreducible vector spaces and spherical tensors.

13.19 Reducible and Irreducible Vector Space

Consider the stationary Schrödinger equation

$$H\psi = Ey$$

(129)

Suppose that this equation remains invariant under a group (G) of transformations R in G (for example coordinate rotations for a central potential $V(r)$ in the Hamiltonian H) i.e.

$$H_R = RHR^{-1} = H$$

(130)

Now, let a solution ψ of Eqn. (129) be rotated so that $\psi \rightarrow R\psi$. Then $R\psi$ has the same energy E because

$$RH\psi = E (R\psi)$$

$$RHR^{-1} (R\psi)$$

$$= H (R\psi) \quad \text{(using Eqn. 130)}$$

(131)

i.e. all rotated solutions $R\psi$ are degenerate in energy. Let this vector space of transformed solutions has a finite dimension n such that $\psi_1, \psi_2, ..., \psi_n$ form a basis. Because $R\psi_i$ is one of the degenerate states, we can expand it in terms of the basis as

$$R\psi_i = \sum_{j1}^{n} r_{ij}\psi_j \tag{132}$$

i.e. with each R in G, we can associate a matrix (r_{ij}) called a representation. If we take any element of vector·space V_n (e.g. ψ_j) such that by rotating with all elements R (of G) transform the element into all other elements of V_n, then the vector space V_n is called *irreducible*. If, however, all elements of V_n are not reached, then V_n splits into a direct sum of two or more vector subspaces

$$V_n = V_1 \oplus V_2 \oplus ...$$

which map into themselves under a rotation. In this case, the vector space is called *reducible*. Then a basis can be found in V_n such that

$$U\left(r_{ij}\right)U^\dagger = \begin{pmatrix} \vec{r}_1 & 0 & \cdots \\ 0 & \vec{r}_2 & \cdots \\ \vdots & \vdots & \end{pmatrix}$$

(where U is a unitary matrix), for all R of G and all matrices (r_{ij}). Here $\vec{r}_1, \vec{r}_2,...$ etc are matrices of lower dimension than (r_{ij}).

13.20 Spherical Tensors

Consider the second rank tensor $T_{ij} = x_i y_j$ for $i, j = 1, 2, 3$. It contains the symmetric tensor $S_{ij} = \left(x_i y_j + x_j y_i\right)/2$ and the antisymmetric tensor $A_{ij} = \left(x_i y_j - x_j y_i\right)/2$ so that $T_{ij} = S_{ij} + A_{ij}$. Under rotations, the scalar product $\vec{x} \cdot \vec{y}$ is invariant, so we may form SO(3) irreducible tensor

$$S'_{ij} = \frac{1}{2}\left(x_i y_j + x_j y_i\right) - \frac{1}{3}\vec{x} \cdot \vec{y}\, \delta_{ij}$$

A tensor of a given rank that is irreducible with respect to the full group may well become reducible for the rotation group SO(3) (i.e. the group of rotations in 3-dimensions with determinant +1). For instance, when we form products of the components of the coordinate vector $\vec{r}$, then in polar coordinates (matching SO(3) symmetry), we end up with spherical harmonics:

$$r_{+1} = -\frac{1}{\sqrt{2}}\left(r_x + ir_y\right) = -\frac{1}{\sqrt{2}}r\,\sin\theta\,e^{i\phi} = r\sqrt{\frac{4\pi}{3}}\,Y_{11}$$

$$r_{-1} = \frac{1}{\sqrt{2}}\left(r_x - ir_y\right) = \frac{1}{\sqrt{2}}r\,\sin\theta\,e^{-i\phi} = r\sqrt{\frac{4}{3}}\,Y_{1,-1}$$

$$r_0 = r_z = r\sqrt{\frac{4}{3}}\,Y_{1,0}$$

where $Y_{lm}(\theta,\phi)$ are the spherical harmonics. Similarly, the ladder operators J_+ and J_- for SO(3) lead us to work with the spherical components of the vector $\vec{j}$, defined by:

$$J_{+1} = -\frac{1}{\sqrt{2}}\left(J_x + iJ_y\right) = -\frac{J_+}{\sqrt{2}}$$

$$J_{-1} = \frac{1}{\sqrt{2}}\left(J_x - iJ_y\right) = \frac{J_-}{\sqrt{2}}$$

$$J_0 = J_z$$

13.21 Irreducible Tensor Operators

A tensor is irreducible if the space V_n in which it is defined is irreducible.

The jth order *irreducible tensor operator* is a tensor T_{jm} with $(2j+1)$ components $(-j \le m \le +j)$ that are transformed by rotation of the coordinate system into a linear combination of the same components:

$$T'_{jm} = RT_{jm}R^{-1} = \sum_{m'}D^j_{m'm}\left(\alpha\right)T_{jm'} \tag{133}$$

under a rotation $R(\alpha)$, the Y_{lm} transform as

$$Y_{lm}\left(\hat{r}'\right) = \sum_{m'}Y_{lm'}\left(\hat{r}\right)D^l_{m'm}(R)$$

where $\hat{r}' = \left(\theta',\phi'\right)$ are obtained from $\hat{r} = (\theta,\phi)$ by the rotation R and are the angles of the same point, and the rotation matrices $D^j_{m'm}$ are defined by $D^j_{m'm}\left(\alpha\right) = \left\langle Jm\left|e^{i\alpha J_z}e^{i\alpha J_y}e^{i\alpha J_z}\right|Jm'\right\rangle$ (i.e. the components of the spherical tensor operator T are like spherical harmonics). From the expression of the infinitesimal rotation operator $R_v\left(\varepsilon\right)$ through an angle ε around the Ov axis, the physicist G. Racah in 1942 was able to give an equivalent definition (to Eqn. 133) of irreducible tensor operators using the following procedure.

We use the generator of infinitesimal rotations through the ε angle around Ov axis

$$R_v\left(\varepsilon\right) = 1 - i\varepsilon J_v \tag{134}$$

to write the elements of the rotation matrix

$$\begin{aligned}D^j_{m'm}\left(\varepsilon\right) &= \left\langle jm'\left|R_v\left(\varepsilon\right)\right|jm\right\rangle \\ &= \left\langle jm'\left|1 - i\varepsilon J_v\right|jm\right\rangle \\ &= \left\langle jm'\left|1\right|jm\right\rangle - i\varepsilon\left\langle jm'\left|J_v\right|jm\right\rangle \\ &= \delta_{m'm} - i\varepsilon\left\langle jm'\left|J_v\right|jm\right\rangle \end{aligned} \tag{135}$$

Substituting this in Eqn. (133)

$$\left(1 - i\varepsilon J_v\right)T_{jm}\left(1 + i\varepsilon J_v\right) = \sum_{m'}\left[\delta_{m'm} - i\varepsilon\left\langle jm'\left|J_v\right|jm\right\rangle\right]T_{jm'}$$

By making a first order approximation in ε, we get

$$T_{jm} - i\varepsilon J_v T_{jm} + i\varepsilon T_{jm}J_v = \sum_{m'}\delta_{m'm}T_{jm'} - i\varepsilon\sum_{m'}\left\langle jm'\left|J_v\right|jm\right\rangle T_{jm'}$$

or $\qquad T_{jm} - i\varepsilon\left[J_v, T_{jm}\right] = T_{jm} - i\varepsilon\sum_{m'}\left\langle jm'\left|J_v\right|jm\right\rangle T_{jm'}$

or
$$\left[J_v, T_{jm}\right] = \sum_{m'} \langle jm' | J_v | jm \rangle \, T_{jm'} \tag{136}$$

The choice of the Ov axis as OZ defines the matrix element

$$\langle jm' | J_z | jm \rangle = m\hbar\delta_{m'm} \tag{137}$$

From Eqns (136) and (137), we have

$$\left[J_z, T_{jm}\right] = m\hbar T_{jm} \tag{138}$$

By choosing the Ov axis to be OX and then OY and by forming the linear combinations $J_x \pm iJ_y = J_\pm$,

(since $J_+ | j,m \rangle = \left[(j-m)(j+m+1)\right]^{1/2} \, | j,m+1 \rangle$

and $J_- | j,m \rangle = \left[(j+m)(j-m+1)\right]^{1/2} \, | j,m-1 \rangle$)

so, the matrix element of Eqn. (136) becomes

$$\langle jm' | J_\pm | jm \rangle = \left[(j \mp m)(j \pm m + 1)\right]^{1/2} \delta_{m'm\pm 1} \tag{139}$$

This transforms Eqn. (136) into

$$\left[J_\pm, T_{jm}\right] = \left[(j \pm m + 1)(j \mp m)\right]^{1/2} \hbar T_{j,m\pm 1} \tag{140}$$

Relations (138) and (140) are called the *Racah definitions of irreducible tensor operators.*

13.22 The Wigner-Eckart Theorem

The matrix elements of the tensor operators between the eigenstates are of great physical importance. The Wigner-Eckart theorem is based on the application of spherical tensor operator to angular momentum. The theorem gives the ratio of the matrix element of a tensor operator between angular momentum eigenstates in terms of C.G. coefficients. Thus all the matrix elements of a tensor operator can be related to one particular element which may be chosen to be the simplest one.

The theorems states that *a matrix element of a spherical tensor operator $T_q^{(k)}$ of the rank k between states of angular momentum j and j' factorizes into a C.G. coefficient and a so called reduced matrix element (denoted by double bars) that no longer has any dependence on the projection quantum numbers m, m' and q,* i.e.

$$\langle \alpha' j' m' | \, T_q^{(k)} \, | \alpha j m \rangle = \langle j' k m' q | jm \rangle \frac{\langle \alpha, j | \, \left| T^{(k)} \right| \, | \alpha' j' \rangle}{\sqrt{2j+1}} \tag{141}$$

In other words such a matrix element splits into a dynamic part (the reduced matrix element) and a geometric part (the C.G. coefficient) that contains the rotational properties expressed by the projection quantum numbers q, m and m'. The factor $\sqrt{2j+1}$ has been introduced for convenience. The state symbol $|ajm\rangle$ means that the rotational properties of the state are described by J^2 and J_z having eigenvalues $j(j + 1)\hbar^2$ and $m\hbar$ respectively. The other properties of the state are lumped into symbol α.

Proof: Consider $(2k + 1)(2j + 1)$ eigenvectors

$$T_q^{(k)} \left|\alpha'j'm'\right\rangle \begin{cases} q = -k,-(k-1),...,(k-1),k \\ m' = -j',-(j'-1),...,(j'-1),j' \end{cases} \tag{142}$$

By taking the linear combination of these states, we may construct the new states defined by

$$\left|\lambda j''m''\right\rangle = \sum_{m'q}\left\langle j'km'q\,|\,j''m''\right\rangle T_q^{(k)}\left|\alpha'j'm'\right\rangle \tag{143}$$

Orthogonality of C.G. coefficients gives

$$T_q^{(k)}\left|\alpha'j'm'\right\rangle = \sum_{j''m''}\left\langle j'km'q\,|\,j''m''\right\rangle \left|\lambda j''m''\right\rangle \tag{144}$$

We have the commutation relations

$$\left[J_z, T_q^{(k)}\right] = q\hbar T_q^{(k)}$$

$$\left[J_+, T_q^{(k)}\right] = \sqrt{(k-q)(k+q+1)}\ \hbar T_{q+1}^{(k)} \tag{145}$$

$$\left[J_-, T_q^{(k)}\right] = \sqrt{(k+q)(k-q+1)}\ \hbar T_{q-1}^{(k)}$$

Also,

$$J_+\left|\alpha jm\right\rangle = \sqrt{(j-m)(j+m+1)}\ \hbar\left|\alpha,j,m+1\right\rangle \tag{146}$$

Operating Eqn. (142) by J_+, we get

$$J_+ T_q^{(k)}\left|\alpha'j'm'\right\rangle = \left[J_+, T_q^{(k)}\right]\left|\alpha'j'm'\right\rangle + T_q^{(k)}J_+\left|\alpha'j'm'\right\rangle$$

$$= \hbar\sqrt{(k-q)(k+q+1)}\ T_{q+1}^{(k)}\left|\alpha'j'm'\right\rangle$$

$$+\hbar\sqrt{(j'-m')(j'+m'+1)}\ T_q^{(k)}\left|\alpha',j',m'+1\right\rangle \tag{147}$$

Now, operating both sides of Eqn. (143) by J_+ and using Eqn. (147), we get

$$J_+\left|\lambda j''m''\right\rangle = \sum_{m',q}\left\langle j'km'q\,|\,j''m''\right\rangle \times \hbar\Big\{\sqrt{(k-q)(k+q+1)}\ T_{q+1}^{(k)}\left|\alpha'j'm'\right\rangle$$

$$+\sqrt{(j'-m')(j'+m'+1)}\ T_q^{(k)}\left|\alpha',j',m'+1\right\rangle\Big\}$$

Changing the summation index from q to $q - 1$ in first term and m' to $m' - 1$ in second term on r.h.s. in the sum, we get

$$J_+\left|\lambda j''m''\right\rangle = \sum_{m',q}\Big[\ \left\langle j'km'q-1\,|\,j''m''\right\rangle$$

$$\times\hbar\Big\{\sqrt{(k-q+1)(k+q)}\ T_q^{(k)}\left|\alpha'j'm'\right\rangle\ \Big\}$$

$$+\hbar\left\langle j'km'-1,q\,|\,j''m''\right\rangle\sqrt{(j'-m'+1)(j'+m')}\ T_q^{(k)}\left|\alpha',j',m'\right\rangle\Big]$$

or
$$J_+\left|\lambda j''m''\right\rangle = \sum_{m',q}\Big\{\ \left\langle j'km'q-1\big|j''m''\right\rangle \sqrt{(k+q)(k-q+1)}$$

$$+\left\langle j'km'-1,q\big|j''m''\right\rangle\sqrt{(j'+m')(j'-m'+1)}\ \Big\}\ T_q^{(k)}\left|\alpha',j',m'\right\rangle \qquad (148)$$

Using recursion relation for C.G. coefficients viz.

$$\sqrt{(j-m)(j+m+1)}\ \left\langle j_1 j_2 m_1 m_2\big|j,m+1\right\rangle$$

$$=\sqrt{\left(j_1-m_1+1\right)\left(j_1+m_1\right)}\ \left\langle j_1,j_2,m_1-1,m_2\big|jm\right\rangle$$

$$+\sqrt{\left(j_2-m_2+1\right)\left(j_2+m_2\right)}\ \left\langle j_1,j_2,m_1,m_2-1\big|jm\right\rangle$$

(Comparing the r.h.s. of this relation with r.h.s. of Eqn. (148), $j_1 \leftrightarrow j'$, $j_2 \leftrightarrow k$, $m_1 \leftrightarrow m'$, $m_2 \leftrightarrow q$. Hence

$$J_+\left|\lambda j''m''\right\rangle = \sum_{m',q}\hbar\sqrt{\left(j''-m''\right)\left(j''+m''+1\right)}$$

$$\left\langle j'km'q\big|j'',m''+1\right\rangle T_q^{(k)}\left|\alpha'j'm'\right\rangle \qquad (149)$$

Comparing the r.h.s. of this Eqn. with r.h.s. of Eqn. (143), we get

$$J_+\left|\lambda j''m''\right\rangle = \hbar\sqrt{\left(j''-m''\right)\left(j''+m''+1\right)}\left|\lambda j''m''+1\right\rangle \qquad (150)$$

Similarly, we can show that

$$J_-\left|\lambda j''m''\right\rangle = \hbar\sqrt{\left(j''+m''\right)\left(j''-m''+1\right)}\left|\lambda j''m''-1\right\rangle \qquad (151)$$

and
$$J_z\left|\lambda j''m''\right\rangle = m''\hbar\ \left|\lambda j''m''\right\rangle \qquad (152)$$

From relations (150), (151) and (152), we conclude that

(1) The states $\left|\lambda j''m''\right\rangle$ defined by Eqn. (143) behave like states of angular momentum $\left|j''m''\right\rangle$.

(2) Some of the states $\left|\lambda j''m''\right\rangle$ may vanish because all $T_q^{(k)}\left|\alpha'j'm'\right\rangle$ states may not be linearly independent.

(3) If some of the states $\left|\lambda j''m''\right\rangle$ vanish, the states $\left|\lambda j''m''\right\rangle$ will be unnormalized.

Therefore, the scalar product $\left\langle \alpha jm\big|\lambda j''m''\right\rangle$ are all zero, except those for which $j'' = j$ and $m'' = m$ i.e. $(2j+1)$ products and obviously, these are independent of m.

Now, the matrix element of $T_q^{(k)}$ between states $\left\langle \alpha jm\right|$ and $\left|\alpha'j'm'\right\rangle$ may be expressed as

$$\left\langle \alpha jm\big|T_q^{(k)}\big|\alpha'j'm'\right\rangle \qquad (153)$$

Using Eqn. (144), we get

$$\left\langle \alpha jm\big|T_q^{(k)}\big|\alpha'j'm'\right\rangle = \sum_{j''m''}\left\langle j'km'q\big|j''m''\right\rangle\ \left\langle \alpha jm\big|\lambda j''m''\right\rangle \qquad (154)$$

Since the operators J^2 and J_z are hermitian and their values exist only for $j'' = j$ and $m'' = m$, so that the sum in Eqn. (154) reduces to a single term, giving

$$\langle \alpha jm | T_q^{(k)} | \alpha' j' m' \rangle = \langle j' km' q | jm \rangle \, \langle \alpha jm | \lambda jm \rangle \qquad (155)$$

We define

$$\langle \alpha jm | \lambda jm \rangle = \frac{\langle \alpha j| \, \left\| T^{(k)} \right\| \, |\alpha' j' \rangle}{\sqrt{2j+1}}$$

then Eqn. (155) leads to

$$\langle \alpha jm | T_q^{(k)} | \alpha' j' m' \rangle = \langle j' km' q | jm \rangle \, \frac{\langle \alpha j| \, \left\| T^{(k)} \right\| \, |\alpha' j' \rangle}{\sqrt{2j+1}}$$

where $1/\sqrt{2j+1}$ is a constant of proportionality.

This proves the Wigner-Eckart theorem. This theorem offers the advantage that in many problems, the purely geometrical aspects (pertaining to C.G. coefficients) can be evaluated without detailed knowledge of the physical behaviour of the tensor operator.

Questions and Problems

1. Two p-electrons ($l = 1$) can have $L = 0, 1$ or 2 and $S = 0$ or 1 in the Russel Saunders coupling. Are all combinations of L and S permitted if the n-values of the two electrons are different? Are they all permitted if the n-values are the same?

2. Use Hund's rules to find the spectroscopic description of the ground states of the following atoms: N ($Z = 7$), K ($Z = 19$), Sc ($Z = 21$)
   Also write their electronic configurations .

3. Use Hund's rules to check the (S, L, J) quantum numbers of the elements with $Z = 14, 15, 24$.

4. Show by direct computation that $\vec{J} = \vec{L} + \vec{S}$ commutes with $\vec{L} \cdot \vec{S}$ and hence commutes with the Hamiltonian

$$H = -\frac{\hbar^2}{2m} \nabla^2 + V(r) + \xi(r) \vec{L} \cdot \vec{S}$$

5. How many angular momentum states arise for a system with two angular momentum $j_1 = 1$ and $j_2 = \frac{1}{2}$. Specify the states.

6. What do you understand by Clebsch-Gordan coefficients in relation to addition of angular momenta?

7. Obtain Clebsch-Gordan coefficients for the addition of orbital and spin angular momentum for electron in p-state.

8. For a state having well defined values for J_1^2, J_2^2, J_{1z} and J_{2z} viz. $j_1(j_1+1)\hbar^2$, $j_2(j_2+1)\hbar$, $m_1\hbar$ and $m_2\hbar$ respectively, show that

$$\left\langle J^2 \right\rangle = \left[j_1 \left(j_1 + 1 \right) + j_2 \left(j_2 + 1 \right) + 2m_1 m_2 \right] \hbar^2$$

[Hint: $J^2 = J_x^2 + J_y^2 + J_z^2 = \displaystyle\sum_{i=x,y,z} \left(J_{1i} + J_{2i} \right)^2 = J_1^2 + J_2^2 + 2\vec{J}_1 \cdot \vec{J}_2$)]

9. If J_x, J_y and J_z are angular momentum operators, show that

$$\left[J^2, J_\pm\right] = 0; \left[J_z, J_\pm\right] = \pm\hbar J_\pm ; \left[J_+, J_-\right] = 2\hbar J_z$$

$$J_+ J_- = J^2 - J_z\left(J_z + \hbar\right); J_- J_+ = J^2 - J_z\left(J_z - \hbar\right)$$

and $$J^2 = \frac{1}{2}\left(J_+ J_- + J_- J_+\right) + J_z^2$$

where $J_+ = J_x + iJ_y$; $J_- = J_x - iJ_y$; $J^2 = J_x^2 + J_y^2 + J_z^2$

10. Show that $J_\pm |jm\rangle = \left[j(j+1) - m(m \pm 1)\right]^{1/2} \hbar |j, m \pm 1\rangle$

11. Deduce the eigenvalues of J^2 and J_z.

12. Obtain the non-vanishing martix elements for the non-Hermitian operators $J_+ = J_x + iJ_y$ and $J_- = J_x - iJ_y$.

13. Obtain Clebsch-Gordan coefficients whentwo angular momenta $j_1 = \frac{1}{2}$ and $j_2 = \frac{1}{2}$ are coupled.

14. What do you mean by irreducible tensor operators? State and prove Wigner-Eckart theorem. What is its importance?

15. Calculate the matrix representation of the angular momentum operators L_x, L_y, L_z and L^2 for the values $l = 1/2, 1$ and $3/2$ by using the formulae:

$$\langle l', m'|L^2|l, m\rangle = \hbar^2 \delta_{ll'}\delta_{mm'}\, l(l+1)$$

$$\langle l', m'|L_z|l, m\rangle = \hbar\delta_{ll'}\delta_{mm'}\, m$$

$$\langle l', m'|L_-|l, m\rangle = \hbar\sqrt{(l+m)(l-m+1)}\; \delta_{ll'}\delta_{m-1,m'}$$

$$\langle l', m'|L_+|l, m\rangle = \hbar\sqrt{(l-m)(l+m+1)}\; \delta_{ll'}\delta_{m+1,m'}$$

where $-l \le m \le l$.

16. What is spin orbit interaction? Why it is zero for s-electron and very large in heavy elements?

17. Prove the identities: $\left[J_x^2, J_y^2\right] = \left[J_y^2, J_z^2\right] = \left[J_z^2, J_x^2\right]$

and show that these commutators are all zero in states for which $j = 0, \frac{1}{2}$ or 1.

18. In case of an atom placed in a magnetic field B, show that

$$\vec{\mu} = \frac{g_L e}{2Mc}\vec{J} = \frac{e}{2Mc}\left(\vec{J} + \vec{S}\right)$$

and $$\langle njl'm_j'|\mu_z B|njlm_j\rangle = m_j g_L \frac{e\hbar}{2Mc} B\delta_{m_j m_j'}\delta_{ll'}$$

where $$g_L = 1 + \frac{j(j+1) + s(s+1) - l(l+1)}{2j(j+1)}$$

14

Symmetries, Invariance Principles and Conservation Laws

14.1 Symmetries: Motivation from Classical Physics

The equation of motion of a non-relativistic test particle of mass m (and charge q) in the Coulomb field of a particle of charge q' is given by

$$m\frac{d^2\vec{x}}{dt^2} = \frac{qq'}{4\pi\varepsilon_0 r^3}\vec{x} \tag{1}$$

`Fig. 14.1

Here $|\vec{x}| = r$ and origin is assumed to be at the position of q'. Now, if we make the substitution

$$\vec{x} \rightarrow -\vec{x} \tag{2}$$

the equation (1) remains invariant. This is called parity operation, and so, *parity is a symmetry.*

Another symmetry which keeps the Eqn. (1) invariant is *time-reversal*, which is defined by the transformation

$$t \rightarrow -t \tag{3}$$

Since Eqn. (1) involves a double derivative w.r.t. time on the l.h.s., it is unaffected by the change of sign of time.

Apart from parity and time reversal symmetries, Eqn. (1) is also invariant if we change the sign of the charges of each particle

$$q \rightarrow -q, q' \rightarrow -q'$$

This symmetry is called *charge conjugation symmetry* (This charge-conjugation has been discussed further in the chapter on Relativistic Theory).

14.2 Symmetry Transformations in Quantum Mechanics

In due course of dynamical development of a physical system, there might be one or more physical observables which do not change with time. Such observables are called constants of motion and the principle embodying their constancy are referred to as conservation laws. For an observable A, we have

$$\frac{d < \hat{A} >}{dt} = \left\langle \frac{\partial \hat{A}}{\partial t} \right\rangle + \frac{1}{i\hbar} \left\langle \left[\hat{A}, \hat{H} \right] \right\rangle \tag{4}$$

where $\hat{A}$ is operator corresponding to the observable A and $\hat{H}$ is the Hamiltonian operator.

$$\therefore \qquad \frac{d < \hat{A} >}{dt} = 0 \text{ if } \left[\hat{A}, \hat{H} \right] = 0 \tag{5}$$

that is, *an observable is a constant of motion if the corresponding operator commutes with the Hamiltonian.*

Suppose, operator $\hat{A}$ transforms to $\hat{A}'$ under a unitary transformation:

$$\hat{A}' = U \hat{A} U^\dagger \tag{6}$$

We can define the unitary operator U by

$$U = \hat{I} + i\varepsilon \hat{G} \tag{7}$$

where I is unit operator, ε is a real (arbitrary) small parameter and operator G is called the *generator* of the infinitesimal unitary transformation.

$$U^\dagger U = (\hat{I} - i\varepsilon \hat{G}^\dagger)(\hat{I} + i\varepsilon \hat{G})$$
$$\cong \hat{I} + i\varepsilon (\hat{G} - \hat{G}^\dagger) \text{ (}\varepsilon^2 \text{ term is neglected) } \tag{8}$$

For U to be unitary

$$\hat{G} - \hat{G}^\dagger = 0 \text{ or } \hat{G} = \hat{G}^\dagger \tag{9}$$

i.e. $\hat{G}$ is Hermitian.

Now,
$$\hat{A}' = U\hat{A}U^\dagger = (I + i\varepsilon \hat{G}) \hat{A} (I - i\varepsilon \hat{G})$$
$$\cong \hat{A} + i\varepsilon (\hat{G} \hat{A} - \hat{A} \hat{G})$$
$$= \hat{A} + i\varepsilon [\hat{G}, \hat{A}] \tag{10}$$

Thus, for A to be invariant (i.e. A' = A) we require

$$[\hat{G}, \hat{A}] = 0 \tag{11}$$

Therefore, *a dynamical observable whose operator commutes with the generator G of the infinitesimal unitary transformation remains invariant under a unitary transformation* (from Eqns 10 and 11).

In particular if $\hat{A} \equiv \hat{H}$, then

$$[\hat{G}, \hat{H}] = 0 \tag{12}$$

i.e. $\hat{G}$ commutes with the Hamiltonian, therefore, $\hat{G}$ *is a constant of motion.* Again

$$\hat{H}' = U \hat{H} U^\dagger \tag{13}$$

Using Eqns (4) to (7), we get

$$\hat{H}' = \hat{H} + i\varepsilon [\hat{G}, \hat{H}]$$

But

$$[\hat{G}, \hat{H}] = 0 \tag{14}$$

Hence

$$\hat{H}' = \hat{H}.$$

Hence, if the observable corresponding to an operator $\hat{A}$ is conserved during the motion of the system, then the Hamiltonian of the system is invariant under unitary transformation generated by $\hat{A}$. Such transformations that leave the Hamiltonian invariant are called *symmetry transformations*. Now, if Hamiltonian remains invariant ($H' = H$) and if Eqn. (13) is true, then

$$\hat{H} = U \hat{H} U^\dagger$$

or

$$\hat{H} U = U \hat{H} U^\dagger U$$

or

$$\hat{H} U = U \hat{H}$$

$$\Rightarrow \qquad U \hat{H} |\psi> = \hat{H} U |\psi> \tag{15}$$

This equation is a mathematical statement of symmetry. We may therefore define symmetry as follows:

A physical system is symmetric with respect to the operation U when U commutes with H.

Any definite symmetry transformation is associated with a particular conservation law.

Among the various symmetries, there are the *geometrical symmetries* and the *dynamical symmetries*. The geometrical symmetries are the symmetries arising from the homogeneity of space and time. The geometrical symmetries are associated with geometrical operations like, translation in space (leading to conservation of . linear momentum), translation in time (leading to conservation of energy), rotation (associated with isotropy of space-leading to conservation of angular momentum) and inversion in space and time. Space inversion invariance leads to parity conservation and invariance under time reversal leads to applicability of Schrödinger equation with reversed (inverted) time. Out of these geometrical symmtries, translations (in space and time) and rotations in space come under *continuous symmetries* and those associated with space and time inversion come under *discrete symmetries*.

The dynamical symmetries are those which are associated with particular features of the interaction involved, for example degeneracies associated with the energy levels of the hydrogen atom and the isotropic harmonic oscillator.

Now we will discuss the conservation laws associated with geometrical symmetries.

14.3 Spatial Translation (*Conservation of Linear Momentum*)

Consider a particle described by wave function which is a funciton of x coordinate only. Let there be two reference frames S and S' in which the coordinates of the particle are measured x and x'. Let S' be shifted by R with respect to S. Let ψ and ψ' denote the wave functions in S and S'. It is assumed that the form of the wavefunction will not change by a translation through an arbitrary amount R.

Hence

$$\psi'(x') = \psi(x), \ (x' = x - R)$$

or

$$\psi'(x - R) = \psi(x) \text{ or } \psi'(x) = \psi(x + R)$$

where R is an infinitesimal translation. Expanding this equation, we get

$$\psi'(\vec{x}) = \psi(x) + \vec{R} \cdot \frac{\partial \psi(x)}{\partial x} + \dots$$

$$= \left[1 + \frac{i}{\hbar} \vec{R} \cdot \hat{p}_x + \frac{1}{2} \left(\frac{i}{\hbar} \vec{R} \cdot \hat{p}_x \right)^2 + \dots \right] \psi(x) = \hat{T}_R \psi(x)$$

where $\hat{T}_R \equiv e^{(i/\hbar)\vec{R} \cdot \hat{p}_x}$ is the operator of finite displacement.

or $\psi'(x) \cong (1 + iRG_x) \psi(x)$ (neglecting higher order terms) $\qquad (16)$

where

$$G_x = -i \frac{\partial}{\partial x} = -\frac{i\hbar}{\hbar} \frac{\partial}{\partial x} = \frac{p_x}{\hbar} \qquad (17)$$

The wavefunction $\psi(x)$ is transformed to $\psi'(x)$ by the action of the operator iRG_x on $\psi(x)$, therefore G_x is called the *generator of infinitesimal translation* in x direction. Thus the momentum operator is the generator of infinitesimal translation in space. Let the position eigenstates for the particle be denoted by $|x>$ and $|x'>$ at the coordinates x and x' measured from O and O' respectively.

$$\psi(x) = <x|\psi> \text{ and } \psi'(x) = <x'|\psi> \qquad (18)$$

where

$$<x'|\psi> = \left(1 + \frac{iRp_x}{\hbar} \right) \psi(x) \text{ (using Eqn. 16)} \qquad (19)$$

or

$$<x'|\psi> = \left(1 + \frac{iRp_x}{\hbar} \right) <x|\psi>$$

$$= <x| \left(1 + \frac{iRp_x}{\hbar} \right) |\psi>$$

Therefore

$$<x'| = <x| \left(1 + \frac{iRp_x}{\hbar} \right)$$

Taking complex conjugate

$$|x'> = \left(1 - \frac{iRp_x}{\hbar} \right) |x> \qquad (20)$$

From a generalization of this equation, the unitary infinitesimal translation operator is given by

$$U_T = \left(1 - i \frac{\vec{R} \cdot \vec{p}}{\hbar} \right) \qquad (21)$$

If follows that

$$H' = U_T H U_T^\dagger = \left(I - \frac{i\vec{R}\cdot\vec{p}}{\hbar}\right) H \left(I + \frac{i\vec{R}\cdot\vec{p}}{\hbar}\right)$$

$$\cong H - \frac{i\vec{R}}{\hbar}\cdot[\vec{p}, H] \qquad \qquad (22)$$

Hence invariance of the Hamiltonian under translation requires that $\vec{p}$ must commute with H, i.e. then the linear momentum is conserved. Hence conservation of linear momentum of an isolated system is a consequence of the translational invariance of the system.

14.4 Translation in Time (*Conservation of Energy*)

For an infinitesimal translation in time by τ,

$$\psi'(x,t) = \psi(x,t+\tau)$$

$$= \psi(x,t) + \tau\frac{\partial\psi}{\partial t}(x,t)$$

$$= \psi(x,t) + \frac{i}{\hbar}\tau\left(-i\hbar\frac{\partial}{\partial t}\right)\psi(x,t)$$

$$= \left[1 + i\tau\left(-\frac{H}{\hbar}\right)\right]\psi(x,t) \qquad \qquad (23)$$

Thus $(-H/\hbar)$ is the generator of time translations. The corresponing unitary operator is

$$U = \left(I - \frac{i\tau}{\hbar}H\right) \qquad \qquad (24)$$

where H is the Hamiltonian which is independent of time. The invariance of the Hamiltonian now requires $H' = UHU^\dagger = H$ or $UH = HU$. Since H is independent of time so U commutes with H. Consequently, H is a constant of the motion or total energy of the system is conserved.

Example 14.1: Show that the operator

$$\hat{S}(\varepsilon) = e^{-i\varepsilon\hat{p}/\hbar}$$

is the generator of a finite translation of origin through ε along the x_i-axis where x_i stands for x, y or z.

Solution: A finite translation of origin transforms an arbitrary wave function as

$$\hat{S}(\varepsilon)\psi(x_i) = \psi(x_i - \varepsilon)$$

$\psi(x_i - \varepsilon)$ may be expanded in a power series

$$\psi(x_i - \varepsilon) = \psi(x_i) - \frac{\varepsilon}{1!}\frac{d\psi(x_i)}{dx_i} + \frac{\varepsilon^2}{2!}\frac{d^2\psi(x_i)}{dx_i^2} + \dots$$

$$= \psi(x_i) + \sum_{n=1}^{\infty}(-1)^n\frac{\varepsilon^n}{n!}\frac{d^n\psi(x_i)}{dx_i^n}$$

$$= \left[1 + \sum_{n=1}^{\infty} \frac{1}{n!} \left(\frac{i}{\hbar} (-\varepsilon) p_i \right)^n \right] \psi(x_i) \quad \left(\because \; p_i = -i\hbar \frac{\partial}{\partial x_i} \right)$$

$$= e^{-i\varepsilon p_i / \hbar} \psi(x_i)$$

Hence the result.

14.5 Conservation of Angular Momentum (Rotational Invariance of the Hamiltonian)

Under 3-dimensional Euclidean rotation let a position vector $\vec{r}$ be rotated around an arbitrary axis through an angle of infinitesimal amount $\delta\phi$. Then $\vec{r}$ is transformed as

$$\vec{r}' = \vec{r} + \delta\vec{r} = r + \delta\vec{\phi} \times \vec{r} \tag{25}$$

where $\delta\phi$ is along the axis of rotation. Similarly, for the momentum vector $\vec{p}$, the change $\delta\vec{p}$ is

$$\delta\vec{p} = \delta\vec{\phi} \times \vec{p} \tag{26}$$

Consider a classical Hamiltonian $H(\vec{r}, \vec{p})$. Let under 3-dimensional rotation, H is transformed to H'. The infinitesimal change δH in H is

$$\delta H = \frac{\partial H}{\partial r_i} \delta r_i + \frac{\partial H}{\partial p_i} \delta p_i \qquad (i = 1, 2, 3)$$

$$= \frac{\partial H}{\partial r_i} \left(\delta\vec{\phi} \times \vec{r} \right)_i + \frac{\partial H}{\partial p_i} \left(\delta\vec{\phi} \times \vec{p} \right)_i$$

Thus

$$\delta H = \frac{\partial H}{\partial r_i} \varepsilon_{ijk} \left(\delta\phi \right)_j r_k + \frac{\partial H}{\partial p_i} \varepsilon_{ijk} \left(\delta\phi \right)_j p_k \tag{27}$$

It is to be noted here that we may write the second term as

$$-\varepsilon_{jik} \left(\delta\phi \right)_j \left(\frac{\partial H}{\partial p_i} \right) p_k \tag{28}$$

Using Poisson bracket $[r_i, H]_{PB}$ for $\left(\frac{\partial H}{\partial p_i} \right)$, we can write Eqn. (28) as[*]

$$-\varepsilon_{jik} \left(\delta\phi \right)_j \left[r_i, H \right]_{PB} p_k \tag{29}$$

$$\left(\because \left[r_i, H \right]_{PB} = \sum_i \frac{\partial r_i}{\partial r_i} \frac{\partial H}{\partial p_i} - 0 \right)$$

[*] Poisson bracket is defined as

$$[f, g]_{PB} = \sum_i \left(\frac{\partial f}{\partial q_i} \frac{\partial g}{\partial p_i} - \frac{\partial f}{\partial p_i} \frac{\partial g}{\partial q_i} \right)$$

[*] For a particle of spin s traveling at the speed of light, only the $+s$ and $-s$ substates of the possible set of $(2s + 1)$ such states, need be considered.

Similarly, the first term of Eqn. (27) can be written as

$$-\varepsilon_{jik}\left(\delta\phi\right)_j\left[p_k, H\right]_{PB} r_i \tag{30}$$

From Eqns (29) and (30), Eqn. (27) reduces to

$$\delta H = -\left(\delta\phi\right)_j \varepsilon_{jik} r_i \left[p_k, H\right]_{PB} - \left(\delta\phi\right)_j \varepsilon_{jik}\left[r_i, H\right]_{PB} p_k$$

$$= -\left(\delta\phi\right)_j \left[\varepsilon_{jik} r_i p_k, H\right] \tag{31}$$

But $\varepsilon_{jik} r_i p_k = L_j$ the jth component of angular momentum vector. Hence

$$\delta H = -\delta\phi\left[\vec{L}, H\right]_{PB} \tag{32}$$

If H is invariant under 3-dimensional rotation, then for classical system

$$\left[\vec{L}, H\right]_{PB} = 0 \tag{33}$$

since $\delta\phi$ is any arbitrary rotation to go over to quantum mechanics, we replace classical Poisson Bracket by $\left(\dfrac{1}{i\hbar}\right) \times$ Commutator Bracket. So, if a quantum Hamiltonian is invariant under 3-dimensional rotation then

$$\frac{1}{i\hbar}\left[\vec{L}, H\right] \to 0$$

i.e. $\vec{L}$ commutes with Hamiltonian. Then by Heisenberg's equation, we have,

$$\dot{L}_i = \frac{1}{i\hbar}\left[L_i, H\right] = 0 \tag{34}$$

This implies $\vec{L}_i$ is a constant of motion

14.6 Space Inversion (Parity operator)

The parity operator (reflection operator) $\hat{P}$ is defined by

$$\hat{P} f(x) = f(-x) \tag{35}$$

It is to be noted here that $\hat{P}$ is not just a mirror reflection, because that requires defining the plane in which we put the mirror. There is a special kind of a reflection that does not require the specification of a plane. We define the $\overset{\bullet}{\hat{P}}$ this way:

First you reflect in a mirror in the z-plane so that z goes to –z, x stays x and y stays y; then you turn the system by 180° about the z-axis so that x is made to go to –x and y to –y. The whole process corresponds to what is called an *inversion* (Every point is projected through the origin to the diametrically opposite position and all the coordinates are reversed).

Suppose, now that we have a state $\left|\psi_0\right\rangle$ which under inversion operation goes to $e^{i\delta}\left|\psi_0\right\rangle$, that is

$$\left|\psi_0'\right\rangle = \hat{P}\left|\psi_0\right\rangle = e^{i\delta}\left|\psi_0\right\rangle \tag{36}$$

Then suppose that we invert again. After two inversions, we are right back to the starting (initial) position. We must have that

$$\hat{P} \left| \psi_0' \right\rangle = \hat{P} \cdot \hat{P} \left| \psi_0 \right\rangle = \left| \psi_0 \right\rangle$$

But

$$\hat{P} \cdot \hat{P} \left| \psi_0 \right\rangle = \hat{P} e^{i\delta} \left| \psi_0 \right\rangle = \left(e^{i\delta} \right)^2 \left| \psi_0 \right\rangle$$

It follows that

$$\left(e^{i\delta} \right)^2 = 1$$

So if the inversion operator is a symmetry operation of a state, there are only two possibilities for δ:

$$e^{i\delta} = \pm 1$$

which means that

$$\hat{P} \left| \psi_0 \right\rangle = \left| \psi_0 \right\rangle \ \text{or} \ \hat{P} \left| \psi_0 \right\rangle = -\left| \psi_0 \right\rangle$$

Classically, if a state is symmetric under an inversion, we get back the same state. However, in quantum mechanics, when we get the same state or minus the same state. When we get the same state, the state is said to have *even-parity* when, the sign is reversed, the state is said to have *odd-parity*.

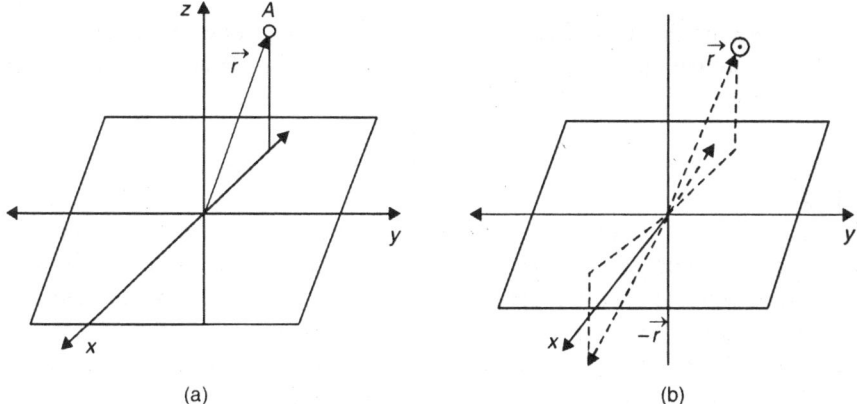

(a)   (b)

Fig. 14.2: The operation of inversion

A physical state needs to have a definite parity (i.e. an eigenvector of P) only if the Hamiltonian (interaction) is invariant under the parity operation. It is now well established that parity is conserved in strong interactions, and so far only the weak interaction is known to violate parity-conservation i.e. phenomenon mediated by weak interaction such as beta decay.

A state with a definite angular momentum has a definite parity. In fact a state with orbital angular momentum quantum number l has the parity $\pi_l = (-1)^l$. However, the intrinsic parity of the particle is not given by a simple formula involving the spin quantum number s, but has to be determined relative to a particle whose intrinsic parity is already fixed either by convention or by some other means. For example, the intrinsic parity of the nucleons (spin ½) is conventionally fixed as positive. Then, the parity of pions can be determined relative to that of the nucleons, for example, from the reaction

$$\pi^+ + d \rightarrow p + p$$

where d is deuteron (consisting of a proton and a neutron), p represents a proton and π^+ a positively charged pion. The parity of the pion turns out to be negative.

14.7 Time-reversal Invariance and Antilinear Operators

In quantum mechanics, the basic wave equation is the Schrödinger equation:

$$i\hbar \frac{\partial \psi}{\partial t} = \left(-\frac{\hbar^2}{2m} \nabla^2 + V \right) \psi \tag{37}$$

Suppose $\psi(\vec{x}, t)$ is a solution. It can be easily verified that $\psi(\vec{x}, -t)$ is not a solution, because of the appearance of the first order time derivative. However $\psi^*(\vec{x}, -t)$ is a solution, as may be verified by complex conjugation of Eqn. (37). We can convince ourselves of this point for an energy eigenstate by substituting

$$\psi(\vec{x}, t) = u_n(\vec{x}) e^{-iE_n t / \hbar}$$

and

$$\psi^*(\vec{x}, -t) = u_n^*(\vec{x}) e^{-iE_n t / \hbar}.$$

into the Schrödinger's equation (37). Thus we conjecture that time reversal must have something to do with complex conjugation. If at $t = 0$, the wave function is given by

$$\psi = <x|\alpha>$$

then the wave function for the corresponding time-reversed state is given by $<x|\alpha>^*$.

The reversal in time of a state represented by the ket $|\alpha>$ (or the wave function ψ_a) changes it into the ket $|\alpha'>$ (or the wave function $\psi_{a'}$) that develops in accordance with the opposite sense of progression of time. For this new state, the signs of all linear and angular momenta are reversed but other quantities are unchanged. Time reversal is effected by a time independent operator T such that

$$T|\alpha> = |\alpha'> \text{ or } T\psi_a = \psi_{a'} \tag{38i}$$

It is assumed that T is a symmetry operation for closed isolated physical systems. This means that if $|k>$ or U_k represents an eigenstate of the Hamiltonian (which is constant in time) with energy eigenvalue E_k, then $T|k>$ or Tu_k also represents an eigenstate with the same eigenvalue. At some particular time, say $t = 0$, the wave function ψ_a can be expanded in terms of the energy eigenfunction u_k in accordance with Eqn.

$$\psi_\alpha(\vec{r}) = \sum_k a_{\alpha k} u_k(\vec{r}) \tag{38ii}$$

where coefficients $a_{\alpha k}$ are given by

$$a_{\alpha k} = \int u_k^*(\vec{r}) \psi_\alpha(\vec{r}) d^3 r \tag{39}$$

We now consider two pairs of operations that can be performed on ψ_α that are expected to lead to the same physical state. In the first case we allow the state to propagate to time t and then reverse it. In the second case we reverse it at $t = 0$ and then allow the reversed state to propagate with the opposite sense of progression of time (i.e. to time $-t$). With the first pair of operations, propagation to time t takes the wave function Eqn. (38ii) into

$$\sum_k a_{\alpha k} e^{-iE_k t/\hbar} u_k \tag{40}$$

If now we assume that T is a linear operator (which we shall find leads to a contradiction), time reversal gives

$$\sum_k a_{\alpha k} e^{-iE_k t/\hbar} T u_k \tag{41}$$

with the second pair of operations, time reversal at $t = 0$ takes the wave function (38) into

$$\sum_k a_{\alpha k} T u_k$$

Since the energy eigenvalue of $T u_k$ is E_k, propagation to time $-t$ then gives

$$\sum_k a_{\alpha k} e^{iE_k t/\hbar} T u_k \tag{42}$$

It is apparent that, in general, the wave functions Eqn. (41) and Eqn. (42) are not multiples of each other, as they must be if they are to represent the same physical state. This contradiction is clearly connected somehow with the operation of complex conjugation. This is obvious, since, a change in the sign of t in the Schrödinger equation for ψ_α changes it into the equation for ψ_α^*, provided that H is real and independent of the time. Therefore, instead of assuming that T is a linear operator with the property

$$T\left(a_1\psi_1 + a_2\psi_2\right) = a_1 T\psi_1 + a_2 T\psi_2$$

we try the assumption that T has the property

$$T\left(a_1\psi_1 + a_2\psi_2\right) = a_1^* T\psi_1 + a_2^* T\psi_2 \tag{43}$$

Such an operator is said to be *antilinear* with the assumption Eqn. (43) for T, the two states (Eqn. 41 and Eqn. 42) both become

$$\sum_k a_{\alpha k}^* e^{iE_k t/\hbar} T u_k$$

and the contradiction disappears.

Application of T to both sides of the Schrödinger equation

$$i\hbar \frac{\partial \psi_\alpha}{\partial t} = H\psi_\alpha$$

gives
$$-i\hbar \frac{\partial\left(T\psi_\alpha\right)}{\partial t} = TH\psi_\alpha$$

Thus if T commutes with the Hamiltonian, so that

$$[T, H] = 0 \tag{44}$$

then $T\psi_\alpha$ satisfies the Schrödinger equation with t replaced by $-t$. It also follows from Eqn. (44) that $T u_k$ is an energy eigenfunction with eigenvalue E_k. Thus the condition that T be a symmetry operation is that Eqn. (44) be satisfied. Then it leads to the time reversal invariance of the Schrödinger equation. It is to be noted here that it may be apparent that all Hamiltonians would satisfy the time reversal invariance condition. But Hamiltonians corresponding to weak nuclear interactions are not time reversal invariant.

Example 14.1: Show that

(*i*) Translation and parity commute.

(*ii*) Using result of (*i*), show that momentum $\vec{p}$ must change sign under parity.

Solution: Let P and T denote parity and translation operators respectively.

(*i*)   under $P, \; \vec{x} \rightarrow -\vec{x}$

under $T, \; \vec{x} \rightarrow \vec{x} + \vec{a}$

Hence $T\psi(\vec{x}) \rightarrow \psi(\vec{x} + \vec{a})$

$$PT\psi(\vec{x}) \rightarrow P\psi(\vec{x} + \vec{a}) = \psi(-\vec{x} - \vec{a})$$

$$P\psi(\vec{x}) \rightarrow \psi(-\vec{x})$$

$$TP\psi(\vec{x}) \rightarrow T\psi(-\vec{x}) = \psi\left[-(\vec{x} + \vec{a})\right] = \psi(-\vec{x} - \vec{a})$$

Hence $PT\psi(\vec{x}) = TP\psi(\vec{x})$

or $[P, \; T] = 0.$

(*ii*)   Since $PT = TP$

$\therefore$ $PTP^{-1} = TPP^{-1} = T$

Writing unitary representation for operator T,

$$T = e^{i\vec{a} \cdot \vec{p} / \hbar}, \text{ we get}$$

$$P e^{i\vec{a} \cdot \vec{p} / \hbar} P^{-1} = e^{i\vec{a} \cdot \vec{p} / \hbar}$$

or $e^{-i\vec{a} \cdot \vec{p} / \hbar} PP^{-1} = e^{i\vec{a} \cdot \vec{p} / \hbar}$ (since P changes $\vec{a} \rightarrow -\vec{a}$)

Now, since $e^{-i\vec{a} \cdot \vec{p} / \hbar}$ cannot be equal to $e^{i\vec{a} \cdot \vec{p} / \hbar}$, we must also change sign of $\vec{p}$, i.e.

$$P e^{i\vec{a} \cdot \vec{p} / \hbar} P^{-1} = e^{i(-\vec{a}) \cdot (-\vec{p}) / \hbar} PP^{-1}$$

or under $P, \; \vec{p} \rightarrow -\vec{p}$, i.e. $\vec{p}$ must change sign under parity.

Example 14.2: Prove that the parity of spherical harmonics $Y_{l,m}(\theta, \phi)$ is $(-1)^l$.

Solution: In spherical polar coordinates, a change of $\vec{r}$ by $-\vec{r}$ is equivalent to a change:

$$\theta \rightarrow (\pi - \theta)$$

and $$\phi \rightarrow (\phi + \pi)$$

We have $Y_{l,m}(\theta, \phi) = CP_l^m(\cos \theta) e^{im\phi}$ (*C* is a constant)

Thus,

$$Y_{l,m}(\pi - \theta, \phi + \pi) = CP_l^m\left[\cos(\pi - \theta)\right] \exp\left[im(\phi + \pi)\right]$$

$$= CP_l^m(-\cos\theta) \exp(im\phi) \exp(im\pi)$$

$$= CP_l^m(\cos\theta)(-1)^{l+m} \exp(im\phi)(-1)^m$$

$$=(-1)^l CP_l^m (\cos\theta)\, e^{im\phi}$$

$$= (-1)^l Y_{l,m} (\theta,\phi)$$

Here, we have used $P_l^m (-x) = (-1)^{l+m} P_l^m (x)$. Hence parity of spherical harmonics $Y_{l,m} (\theta,\phi)$ is given by $(-1)^l$.

14.8 Identical Particles and Symmetry of Wave Functions

Let us consider N identical particles (say, electrons or π-mesons). Their Hamiltonian

$$H = H (1, 2, ..., N) \tag{45}$$

is symmetric in the variables 1, 2, ... i.e.

$$H (1, 2, ..., N) = H (2, 1, ..., N) \tag{46}$$

because the particels are identical.

Here $1 \equiv (x_1, \chi_1)$ includes position and spin. Similarly, we can write, the wavefunction

$$\psi = \psi (1, 2, ... N) \tag{47}$$

Now, we define the permutation operator or particle exchange operator P_{ij} as an operator which interchanges $i \leftrightarrow j$ i.e. for an arbitrary N-particle wave function $\psi (..., i, j, ...)$

$$P_{ij} \psi (..., i, j, ...) = \psi (..., j, i, ...) \tag{48}$$

In particular, for a system consisting of two identical particles

$$\hat{P}_{12} \psi (1,2) = \psi (2,1) \tag{49}$$

i.e. operator $\hat{P}_{12}$ interchanges all coordinates of particles 1, and 2. If identical particles are indistinguishable, the interchange should not produce any observable effect. This could be so if the operation by $\hat{P}_{12}$ produces a change only at most by a phase factor δ which leaves $\psi^*\psi$ unchanged

$$\hat{P}_{12} \psi (1, 2) = \psi(2, 1) = e^{i\delta} \psi(1, 2)$$

Now,
$$\hat{P}_{12}^2 \psi(1, 2) = \hat{P}_{12} \psi(2, 1)$$

$$= \hat{P}_{12}\, e^{i\delta} \psi(1, 2)$$

$$= e^{i\delta}\, \hat{P}_{12} \psi(1, 2)$$

$$= e^{i\delta} \psi(2, 1)$$

$$= e^{2i\delta} \psi(1, 2) \tag{50}$$

Since, the repeated interchange must yield the same (original) wave function, so

$$\hat{P}_{12}^2 \psi(1,2) = \psi(1,2) \tag{51}$$

From Eqn. (50) and (51)

$$e^{2i\delta} = 1 \qquad \text{or } \left(e^{i\delta}\right)^2 = 1 \text{ or } e^{i\delta} = \pm 1$$

Thus, either

$$\psi(1, 2) = \psi(2, 1) \tag{52}$$

or
$$\psi(1, 2) = -\psi(2, 1) \tag{53}$$

In Eqn. (52), the wave function does not changes sign and is said to be *symmetric* under exchange of two particles, whereas in Eqn. (53), the wave function changes sign and is said to be *antisymmetric*. It is a law of nature that the symmetry or antisymmetry under an exchange of two particles is a property of the particles themselves. Thus, indistinguishability of similar particles demands imposition of symmtry requirements with respect to interchange of two particles and so, we cannot digress ourselves from appreciating the beauty of nature in case of identical indistinguishable (i.e. quantum) particles that there are exactly either completely symmtric states or completely antisymmetric states.

14.9 Bose-particles and Fermi-particles

Let us consider scattering of two identical particles labelled a and b. The scattering may take place in two ways as shown in the Fig. 14.3.

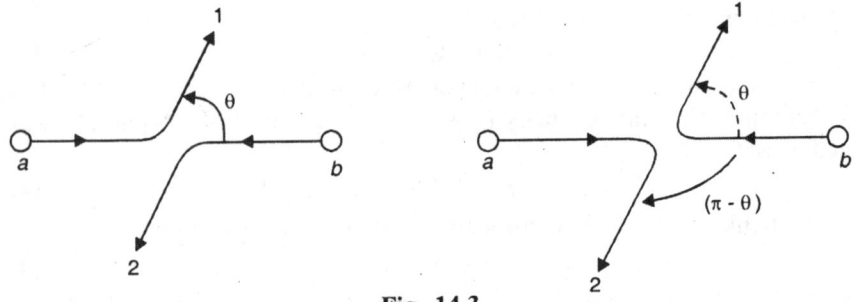

Fig. 14.3

In Fig. (a), particle 'a' scatters in the direction 1 and particle 'b' scatters in the direction 2. Let $f(\theta)$ denote the amplitude for this process and the probability of observing this event is $P_1 = |f(\theta)|^2$. It is quite obvious that particle 'b' would have scattered in direction 1 and particle 'a' in direction 2. If we assume that there are no special directions defined by spins etc., the probability P_2 of happening the process in second way is $P_2 = |f(\pi - \theta)|^2$ (i.e. what was observed by placing a counter at location 1 of Fig. (a) would be observed by placing the counter at 2 in Fig. (b) for which $\theta \to \pi - \theta$).

Now, the particles are identical, so the two different processes shown in Fig. (a) and (b) cannot be distinguished. There is an amplitude that either 'a' or 'b' goes into counter 1, and the other goes into counter 2. This amplitude is the sum of the amplitudes of the two processes. If we call the first one as $f(\theta)$, then the second one is $e^{i\delta} f(\pi-\theta)$. (Here either $e^{i\delta} = +1$ or $e^{i\delta} = -1$). In quantum mechanics, there is no way of distinguishing the two final states, so the two amplitudes can interfere. The exchanged case (Fig. b) either contributes with the same sign or it contributes with the opposite sign. Particles which interfere with a positive sign are called *Bose-particles* and those which interfere with a negative sign are called *Fermi-particles* i.e.:

The amplitude for scattering of identical particles is either

(Amplitude direct) + (Amplitude exchanged)   ...Bose-particles

(Amplitude direct) – (Amplitude exchanged)   ...Fermi-particles

It was conjectured by Pauli and observed experimentally that:

Systems consisting of idenitcal particles of integral spin 0, $1\hbar$, $2\hbar$, ... are described by symmetric wave function ψ_S.

Systems consisting of identical particles of half-odd-integral spin $\frac{1}{2}\hbar$, $\frac{3}{2}\hbar$, $\frac{5}{2}\hbar$, ... are described by antisymmetric wave functions ψ_S.

Particles of type (1) obey Bose Einstein statistics and are *Bose particles*. Particles of type (2) are *Fermi particles* as they obey Fermi Dirac statistics. Examples of *Bosons* are photon, π-meson, ^{4}He atom, gravitons, Cooper pairs in a superconductor. Examples of *Fermions* are the electron, the proton, the neutron, the muon, baryons, the neutrinos and ^{3}He atom.

14.10 Exchange Degeneracy

For a state described by two particles 1 and 2, the product wave function

$$\psi(1, 2) = \psi_a(\vec{r}_1)\, \psi_b(\vec{r}_2) \tag{54}$$

is a solution of the Schrödinger equation with energy $E = E_a + E_b$ (in independent particle model). Since the Hamiltonian (and therefore the Schrödinger equation) is completely symmetric under an exchange of two electrons in states a and b, therefore, the wave function

$$\psi(2, 1) = \psi_b(\vec{r}_1)\, \psi_a(\vec{r}_2) \tag{55}$$

is also a solution of the Schrödinger equation corresponding to the same eigenvalue $E = E_a + E_b$. Thus above wave functions $\psi(1, 2)$ and $\psi(2, 1)$ are degenerate and this degeneracy is known as the *exchange degeneracy* (arising because of possibility of exchange of two electrons).

The above simple product wave functions must now be properly symmetrized in view of quantum statistics. We could write linear combinations as symmetric and antisymmetric wave functions

$$\psi_s(1, 2) = \frac{1}{\sqrt{2}}\left[\psi_a(\vec{r}_1)\, \psi_b(\vec{r}_2) + \psi_a(\vec{r}_2)\, \psi_b(\vec{r}_1)\right] \tag{56}$$

$$\psi_A(1, 2) = \frac{1}{\sqrt{2}}\left[\psi_a(\vec{r}_1)\, \psi_b(\vec{r}_2) - \psi_a(\vec{r}_2)\, \psi_b(\vec{r}_1)\right] \tag{57}$$

which are also the solutions of the Schrödinger equation belonging to the same eigenvalue $E = E_a + E_b$. The factor $1/\sqrt{2}$ is the normalization constant. The wave functions ψ_S and ψ_A are symmetric and antisymmetric respectively because under an exchange of indices 1 and 2 ψ_S remains the same but ψ_A changes sign:

$$\psi_S(1, 2) = \psi_S(2, 1) \qquad \text{(Bosons)}$$

$$\psi_A(1, 2) = -\psi_A(2, 1) \qquad \text{(Fermions)}$$

If we set $a = b$, we find $\psi_A = 0$, that is two fermions cannot be in the same quantum state. The *Pauli's exclusion principle* states that no two electrons in an atom can be in the same quantum state. Thus comparison of the two statements leads to the conclusion that systems of electrons are described by wave functions that reverse sign upon exchange of any two of them.

Slater Determinant

We have

$$\psi_A(1, 2) = \frac{1}{\sqrt{2}} \begin{vmatrix} \psi_a(1) & \psi_a(2) \\ \psi_b(1) & \psi_b(2) \end{vmatrix} = -\psi_A(2, 1)$$

This holds for a two-electron system. This is also true for a system of N fermions. Then the above equation has the form

$$\psi_A(1, 2, ..., N) = \frac{1}{\sqrt{N!}} \begin{vmatrix} \psi_a(1) & \psi_a(2) & \cdots & \psi_a(N) \\ \psi_b(1) & \psi_b(2) & \cdots & \psi_b(N) \\ \vdots & \vdots & & \vdots \\ \psi_k(1) & \psi_k(2) & \cdots & \psi_k(N) \end{vmatrix}$$

known as Slater's determinant ($1/\sqrt{N!}$ is the normalization constant). This has an interesting property that the determinant vanishes if two or more ψ_i's are same.

Now we have an interesting Symmetrization Postulate according to which the possible states of an N number of identical particles are those which are either symmetrical or antisymmetrical with respect to interchange of any two parciles.

The interchange of two particles involves the interchange of two columns in the determinant. If we attempt to put two fermions in the same state ($\psi_a(1) = \psi_a(2)$), then the determinant will vanish. Thus there will be no solution of wave equation

$$(H - E)\,\psi_A = 0$$

if there is more than one electron in any one state.

The statement that no two fermions can be in the same quantum state is called · the *Pauli exclusion principle*.

$$\psi\,(..., x\sigma, ..., x\sigma, ...) = 0$$

In other words, two fermions having same spin cannot occupy the same state.

14.11 Importance of Pauli's Exclusion Principle

The exclusion principle explains the classification of elements in the Periodic Table. According, two electrons in one atomic orbit (same n, l, m_l) cannot have the same spin orientation (i.e. same spin quantum number m_s). In other words, two electron having same n, l, m_l have oppositely directed spins. So, in each m_l state, there can be at most two electron with antiparallel spins. Then each member of the pair will have different spin quantum number). This, distribution of electrons in various orbits will be as shown in the Table 14.1.

Table 14.1: Arrangement of electrons in first three orbits of an atom

Orbit	State	Quantum numbers			No. of electrons	With opposite spins	Total no. of electrons
		n	l	m_l			
1	1s	1	0	0	2	2	2
2	2s	2	0	0	2	2	

(Contd.)

Orbit	State	Quantum numbers			No. of electrons	With opposite spins	Total no. of electrons
		n	l	m_l			
	2p	2	1	+1	2	6	
				0	2		8
				−1	2		
3	3s	3	0	0	2	2	
	3p	3	1	+1	2		
				0	2	6	
				−1	2		
	3d	3	2	2	2		18
				1	2		
				0	2	10	
				−1	2		
				−2	2		

Proceeding similarly, we find that in general the maximum number of electrons in an orbit of total quantum number n is $2n^2$.

14.12 Physical Significance of Pauli's Exclusion Principle

Pauli's exclusion principle governs the group behaviour of identical elementary particles having half odd integral spin e.g. electrons, protons, etc. One most important consequence of the principle is that in the interaction of such particles, (e.g. two electron in an assembly, each of spin $\hbar/2$), new forces of non-dynamical character occur. This becomes further clear by following example. Consider two electrons moving along the same line, having coordinates (x_1, p_1) and (x_2, p_2). The de Broglie wave associated with them are given by $\exp(ip_1x_1/\hbar)$ and $\exp(ip_2x_2/\hbar)$. In case, both are present, they are indistinguishable and it is impossible to label them as 1 or 2. So, we can interchange them without affecting any physical observable. Then, wave function for this system of two electrons (according to Pauli) must be antisymmetric. We can construct antisymmetric wave function (space part) for this system as

$$\psi_A = \frac{1}{\sqrt{2}}\left[\exp\left\{\frac{i}{\hbar}(p_1x_1 + p_2x_2)\right\} - \exp\left\{\frac{i}{\hbar}(p_1x_2 + p_2x_1)\right\}\right]$$

$$\equiv \frac{1}{\sqrt{2}}\left[\cos C + i\sin C - \cos D - i\sin D\right] \tag{58i}$$

$$\psi_A^* = \frac{1}{\sqrt{2}}\left[\cos C - i\sin C - \cos D + i\sin D\right] \tag{58ii}$$

$$\psi_A^*\psi_A = \frac{1}{2}\left[(\cos C - \cos D)^2 + (\sin C - \sin D)^2\right]$$

$$= \left[1 - \{\cos C \cos D + \sin C \sin D\}\right]$$

$$= 1 - \cos (C - D)$$

$$= 1 - \cos \frac{1}{\hbar} \left(p_1 x_1 - p_1 x_2 + p_2 x_2 - p_2 x_1 \right)$$

$$= 1 - \cos \left\{ \frac{1}{\hbar} \left[\left(p_1 - p_2 \right) \left(x_1 - x_2 \right) \right] \right\} \tag{59}$$

This represents the probability density for the two electrons to occupy the coordinates x_1 and x_2. If p_1 and p_2 are fixed, $\psi_A^* \psi_A \to 0$, as $x_1 \to x_2$, i.e. the probability density tends to zero if the two electrons were to occupy the same position. But since $\psi_A^* \psi_A \neq 0$, so the two electrons cannot occupy the same position coordinate. Thus the requirement of antisymmetry introduces an effective interaction between two fermions. Irrespective of their charges, there is brought into action a repulsive interaction between two fermions due to which two fermions in the same state tend to stay away from each other. This repulsive force is of non dynamical character and is different from coulumb repulsion. Due to this reason the properties of an electron gas are quite different from that of ordinary gas. This also accounts for very low compressibility of solids and liquids as in that case, the electronic structures of atoms begin to overlap, and very strong repulsive forces are come into play.

We owe reason for this to the special nature of fermions which is quite contrary to the nature of bosons:

Fermions in an ensemble exhibit marked individualistic tendencies whereas bosons strive for unification (they can settle in the same state in any number).

Since the wave function for a particle depends on the degrees of freedom associated with the particle, the spin endows the electron with an additional degree of freedom. This was discovered by Ulhenbeck and Goudsmit and later, the existence of the spin degree of freedom emerged as a natural mathematical consequence of Dirac's relativistic quantum theory.

For a particle of spin s, there are $(2s + 1)$ substates provided the particle is not traveling at the speed of light.[*] Since $|\vec{s}| = \frac{1}{2}$ for electrons, therefore a state of given energy, angular momentum, parity etc. can be occupied by two electron (of spin $\pm\frac{1}{2}$), but by not more than two electrons. This is a restricted version of Pauli's exclusion principle. It is to be emphasized here that Fermi particles obey the exclusion principle only when they are in the same system, i.e. they move in a common force field. Then only each member of the system would be in a different quantum state.

14.13 Two Electron Spin Wave Functions

Let us consider a spin state χ for two electrons coupled by their spins only (i.e. $l = 0$ and $j_1 = j_2 = \frac{1}{2}$) i.e. $j_1 = s_1$ and $j_2 = s_2$ the spins of the two electrons. The total spin s of the two electron system can take the values $(j_1 + j_2$ to $|j_1 - j_2|)$ i.e.

---

[*] For a particle of spin s travelling at the speed of light, only the $+s$ and $-s$ substates of the possible set of $(2s + 1)$ such states, need be considered.

1 and 0. Let us first consider $s = 1$. The three wave functions for $s = 1 (2s + 1 = 2 + 1 = 3)$ (correspondingly $m_s = 1, 0, -1$) are given in terms of the spin functions $\alpha(1)$, $\beta(1)$ of the 1st electron and $\alpha(2)$, $\beta(2)$ of the 2nd electron as follows

(i) $\chi(1, 1) = \alpha(1) \, \alpha(2)$ (both spin up)

(ii) $\chi(1, 0) = \dfrac{1}{\sqrt{2}} \left[\alpha(1)\beta(2) + \beta(1)\alpha(2) \right]$ (60)

(iii) $\chi(1, -1) = \beta(1) \, \beta(2)$ (both spin down)

For (i) $m = m_1 + m_2 = \frac{1}{2} + \frac{1}{2} = 1$ and for (iii) $m = -1$. All the three functions are symmetric under interchange of the two electrons (i.e. labels 1 and 2). The wave function (ii) only differs from either (i) and (iii) in that it is having a different direction of the angular momentum vector. It can be easily verified that $\chi(1, 0)$ satisfies

$$\vec{S}^2 \chi(1, 0) = 1(1 + 1)\hbar^2 \, \chi(1, 0)$$

$$S_z \chi(1, 0) = 0 \tag{61}$$

where $$\vec{S} = \frac{1}{2} \hbar \left(\vec{\sigma}_1 + \vec{\sigma}_2 \right)$$

Here $\vec{\sigma}_1$ (corresponding to $\sigma_{1x}, \sigma_{1y}, \sigma_{1z}$) act on $\alpha(1)$, $\beta(1)$ and $\vec{\sigma}_2$ (corresponding to $\sigma_{2x}, \sigma_{2y}, \sigma_{2z}$) act on $\alpha(2)$, $\beta(2)$ only.

$$\vec{S}^2 = \frac{1}{4} \hbar^2 \left(\vec{\sigma}_1 + \vec{\sigma}_2 \right)^2$$

$$= \frac{1}{4} \hbar^2 \left(\vec{\sigma}_1^2 + \vec{\sigma}_2^2 + 2\vec{\sigma}_1 \cdot \vec{\sigma}_2 \right)$$

$$= \frac{1}{4} \hbar^2 \left(\vec{\sigma}_{1x}^2 + \vec{\sigma}_{1y}^2 + \vec{\sigma}_{1z}^2 + \vec{\sigma}_{2x}^2 + \vec{\sigma}_{2y}^2 + \vec{\sigma}_{2z}^2 + 2\vec{\sigma}_1 \cdot \vec{\sigma}_2 \right)$$

$$= \frac{3}{2} \hbar^2 + \frac{1}{2} \hbar^2 \vec{\sigma}_1 \cdot \vec{\sigma}_2 \tag{62}$$

Now, we define σ_+ and σ_- by

$$S_+ = S_x + iS_y = \frac{1}{2} \hbar \sigma_+ \tag{63i}$$

and $$S_- = S_x - iS_y = \frac{1}{2} \hbar \sigma_- \tag{63ii}$$

$$S_- \alpha = \hbar \beta, \ S_+ \beta = \hbar \alpha \tag{63iii}$$

(verify this using matrix forms)

(i.e. S_- flips the spin from up to down and S_+ flips the down spin to up.)
From (63iii)

$$\frac{1}{2} \hbar \sigma_+ \beta = \hbar \alpha \ \text{or} \ \sigma_+ \beta = 2\alpha$$

Similarly $\dfrac{1}{2}\hbar\sigma_-\alpha = \hbar\beta$ or $\sigma_-\alpha = 2\beta$ $\hspace{4cm}$ (64)

Hence

$$\vec{S}^2 = \frac{3}{2}\hbar^2 + \frac{1}{2}\hbar^2\left(\frac{1}{2}\sigma_{1+}\sigma_{2-} + \frac{1}{2}\sigma_{1-}\sigma_{2+} + \sigma_{1z}\sigma_{2z}\right) \hspace{1cm} (65)$$

$$S^2\alpha(1)\beta(2) = \frac{3}{2}\hbar^2\alpha(1)\beta(2) + \frac{1}{2}\hbar^2\left[\frac{1}{2}\sigma_{1+}\alpha(1)\sigma_{2-}\beta(2) + \frac{1}{2}\sigma_{1-}\alpha(1)\sigma_{2+}\beta(2)\right.$$
$$\left. +\sigma_{1z}\alpha(1)\sigma_{2z}\beta(2)\right]$$

$$= \frac{3}{2}\hbar^2\alpha(1)\beta(2) + \frac{1}{2}\hbar^2\left[0 + \frac{1}{2}2\beta(1)2\alpha(2) - \alpha(1)\beta(2)\right]$$

$$(\because\ \sigma_{1z}\alpha(1) = \alpha(1)\ \text{and}\ \sigma_{2z}\beta(2) = -\beta(2))$$

$$= \hbar^2\left[\alpha(1)\beta(2) + \alpha(2)\beta(1)\right] \hspace{2cm} (66)$$

Similarly

$$S^2\beta(1)\alpha(2) = \hbar^2\left[\beta(1)\alpha(2) + \beta(2)\alpha(1)\right] \hspace{2cm} (67)$$

From Eqn. (66) and (67)

$$S^2\chi(1,0) = \frac{1}{\sqrt{2}}S^2\left[\alpha(1)\beta(2) + \beta(1)\alpha(2)\right]$$

$$= \frac{1}{\sqrt{2}}\hbar^2\left[2\alpha(1)\beta(2) + 2\beta(1)\alpha(2)\right]$$

$$= 2\hbar^2\chi(1,0) = 1(1+1)\hbar^2\chi(1,0) \hspace{2cm} (68)$$

Thus identification for $\chi(1, 0)$ with linear combination of states (with $m_1 = $ ½ and $m_2 = $ –½ and $m_1 = $ –½, $m_2 = $ ½) is correct.

It may similarly be verified that

$$\chi(0,0) = \frac{1}{\sqrt{2}}[\alpha(1)\beta(2) - \beta(1)\alpha(2)] \hspace{2cm} (69)$$

satisfies the equations

$$\vec{S}^2\chi(0,0) = s(s+1)\hbar^2\chi(0,0) \hspace{2cm} (70)$$

The wave function $\chi(0, 0)$ is orthogonal to $\chi(1, 0)$. The first set of symmetric wave functions for which the total spin (s) is 1 (viz. $\chi(1, 1)$ $\chi(1, 0)$ and $\chi(1,-1)$) is collectively known as the *triplet states* (denoted by $^3\chi$) and the antisymmetric wave function $\chi(0, 0)$ is known as the *singlet state* $^1\chi$ ($s = 0$). Due to Eqn. (62), the triplet and singlet states are eigenstates of $\vec{\sigma}_1 \cdot \vec{\sigma}_2$ with the eigenvalues 1 and –3 respectively (because S^2 has the values $2\hbar^2$ and 0 for $^3\chi$ and $^1\chi$ respectively i.e. $\vec{\sigma}_1 \cdot \vec{\sigma}_2 = 1$ and –3 respectively)

Thus,

$$P_t \equiv \frac{1}{4}(3 + \vec{\sigma}_1 \cdot \vec{\sigma}_2)$$

and
$$P_s \equiv \frac{1}{4}\left(1 - \vec{\sigma}_1 \cdot \vec{\sigma}_2\right)$$
(71)

are projection operators to the triplet and singlet states respectively (because these satisfy $S^2 = \frac{3}{4}\hbar^2$).C

14.14 When the Two Electron Wave Function is to be Antisymmetrized?

No two electrons in an atom can be in the same quantum state. Is it true for any system containing two electrons? The previous statement demands anti-symmetrization of the combined wave function with reference to the coordinates of the two electrons, i.e. the uncorrelated wave function for the two electrons viz.

$$\psi_a\left(x_1\right)\psi_b\left(x_2\right)$$

must be replaced by the antisymmetrized wave function

$$\psi\left(x_1, x_2\right) = \frac{1}{N}\left[\psi_a\left(x_1\right)\psi_b\left(x_2\right) - \psi_a\left(x_2\right)\psi_b\left(x_1\right)\right]$$
(72)

The normalization factor N is obtained by the condition

$$\int \psi^*\left(x_1, x_2\right)\, \psi\left(x_1, x_2\right)\, dx_1 dx_2 = 1$$

or
$$\frac{1}{N^2}\int dx_1 \int dx_2 \left[\psi_a^*\left(x_1\right)\,\psi_b^*\left(x_2\right) - \psi_a^*\left(x_2\right)\psi_b^*\left(x_1\right)\right]$$
$$\times\left[\psi_a\left(x_1\right)\,\psi_b\left(x_2\right) - \psi_a\left(x_2\right)\psi_b\left(x_1\right)\right] = 1$$
(73)

Now,
$$\int dx\left|\psi_a\left(x\right)\right|^2 = \int dx\left|\psi_b\left(x\right)\right|^2 = 1$$

So, Eqn. (73) for $x_1 = x_2 = x$ becomes

$$N^2 = 2\left(1 + \left|\int dx\,\psi_a^*(x)\,\psi_b(x)\right|^2\right)$$
(74a)

(because
$$\int \psi_a^*(x_1)\,\psi_a(x_2)\,\psi_b^*(x_2)\,\psi_b(x_1)\, dx_1 dx_2$$
$$= \int \psi_a^*(x)\psi_b(x)dx \int \psi_a(x)\psi_b^*(x)dx$$
$$\left|\int dx\,\psi_a^*(x)\,\psi_a(x)\right|^2)$$

We can rewrite (74a) as

$$N^2 = 2\left(1 + \left|O_{ab}\right|^2\right)$$
(74b)

where $\quad O_{ab} \equiv \int dx\,\psi_a^*(x)\,\psi_b(x)$ is the overlap integral.

Now suppose that in our two electron system, the electron with label a is confined to some spatial region S and the electron with label b can be anywhere. Then the uncorrelated wave function is

$$\psi(x,y) = \psi_a(x)\,\psi_b(y)$$
(75)

and the corresponding probability density

$$P(S) = \int \psi^*(x,y)\psi(x,y)dxdy = \int_S dx \int_{-\infty}^{+\infty} dy |\psi_a(x)|^2 |\psi_b(y)|^2$$

$$= \int_S dx |\psi_a(x)|^2 \int_{-\infty}^{+\infty} |\psi_b(y)|^2 dy$$

$$= \int_S dx |\psi_a(x)|^2 \qquad\qquad (76)$$

The correlated (antisymmetrized) wave function is

$$\psi_A = \frac{1}{N}\left[\psi_a(x)\psi_b(y) - \psi_b(x)\psi_a(y)\right] \qquad\qquad (77)$$

The probability density is

$$\psi_A^* \cdot \psi_A = \frac{1}{N}\left[\psi_a^*(x)\psi_b^*(y) - \psi_b^*(x)\psi_a^*(y)\right]$$

$$\times\left[\psi_a(x)\psi_b(y) - \psi_b(x)\psi_a(y)\right] \qquad\qquad (78)$$

The probability that the electron with label a will be found in region S is given by

$$P_A(S) = \int_S dx \int_{-\infty}^{+\infty} \psi_A^*(x,y)\,\psi_A(x,y)\,dy \qquad\qquad (79)$$

Now the integration corresponding to the label 'a' will have to be done over the domain S and over whole range of coordinates for label 'b'.

$$P_A(S) = \frac{1}{N^2}\int_S dx |\psi_a(x)|^2 \int_{-\infty}^{+\infty} dy |\psi_b(y)|^2$$

$$+ \frac{1}{N^2}\int_S dy |\psi_a(y)|^2 \int_{-\infty}^{+\infty} |\psi_b(x)|^2 dx$$

$$- \frac{1}{N^2}\int_S dx \int_{-\infty}^{+\infty} dy \left[\psi_a^*(x)\,\psi_b(x)\,\psi_b^*(y)\,\psi_a(y)\right.$$

$$\left. + \psi_b^*(x)\,\psi_a(x)\psi_a^*(y)\,\psi_b(y)\right] \qquad\qquad (80)$$

The third term is the interference term and both integrals are over the region S. This minus term will be significant only if the overlap (or exchange) integral (i.e. $\int dx\,\psi_a^*(x)\,\psi_b(x)$ or $\int dx\,\psi_b^*(x)\,\psi_a(x)$ or $\int dy\,\psi_b^*(y)\,\psi_a(y)$ or $\int dy\,\psi_a^*(y)\,\psi_b(y)$ is dominant in the region S. Since wave functions fall off exponentially with distance for bound states, the overlap integral will be significant only for atoms or molecules and insignificant for atoms separated by significant distances. Thus Pauli's principle is important for electrons in an atom or a molecule but not for electrons of different atoms in a crystal lattice. So antisymmetrization is necessary in the former case but not necessary for crystal lattice atoms (core electrons will satisfy Pauli principle and the valence electrons will be free in metals).

If the wave functions of two electrons labelled 'a' and 'b' do not overlap, then interference term is zero and it is as good as to treat the two electrons as uncorrelated and described by only the product wave function $\psi_a(x)\,\psi_b(y)$.

In case of free electrons in a metal, suppose we consider exchange interaction for an electron pair. Then, although exchange interaction for an electron pair in a bond is negative, it is positive for free electrons. This can be seen as follows by considering two free electrons i and j and their pair wave function ψ_{ij}. For electrons with the same spin, the pair wave function must be antisymmetric in space coordinates. From this requirement, we obtain

$$\psi_{ij} = \frac{1}{\sqrt{2}\,V}\left(e^{i\vec{k}_i\cdot\vec{r}_i} e^{i\vec{k}_j\cdot\vec{r}_j} - e^{i\vec{k}_i\cdot\vec{r}_j} e^{i\vec{k}_j\cdot\vec{r}_i} \right)$$

$$= \frac{1}{\sqrt{2}\,V}\, e^{i\left(\vec{k}_i\cdot\vec{r}_i + \vec{k}_j\cdot\vec{r}_j\right)}\left[1 - e^{-i\left(\vec{k}_i - \vec{k}_j\right)\cdot\left(\vec{r}_i - \vec{r}_j\right)} \right] \tag{81}$$

The probability that electron i is to be found in volume element $d\vec{r}_i$ and that electron j is to be found in volume element $d\vec{r}_j$ is then equal to $\left|\psi_{ij}\right|^2 d\vec{r}_i\, d\vec{r}_j$

$$= \frac{1}{\sqrt{2}}\left[1 - \cos\left(\vec{k}_i - \vec{k}_j\right)\cdot\left(\vec{r}_i - \vec{r}_j\right) \right] d\vec{r}_i\, d\vec{r}_j \tag{82}$$

This expression shows that the probability of finding two electrons (with the same spin) at the same place vanishes for every k_i and k_j.

In case of a superconductor in the superconducting state, the free electrons form pairs of electrons of equal momenta but opposite spins. Each pair becomes a boson and the total wave function, which is the product of all pair wave functions is symmetric under exchange of any two pairs. All the pair wave functions have the same form and same phase, so their motions are correlated. These pairs short out the normal electrons and experience no resistance, endowing the superconductor with the zero resistivity (Bose-Einstein condensation).

14.15 The H₂ Molecule (Heitler and London Theory)

Heitler and London in 1927 first gave a quantum mechanical treatment of H_2 molecule. In this approach, atoms are assumed to maintain their individual identity in a molecule and the bond arises due to the interaction of the valence electrons when the atoms come closer.

Consider two hydrogen atoms far apart so that there is no interaction between them. Labelling the electrons as 1 and 2, the nuclei as a and b and the electron-nucleus distances by r_{a_1} and r_{a_2} (see Fig. 14.4).

The Schrödinger equations are

$$H_a(1)\psi_a(1) = E_a\psi_a(1)$$

$$H_b(2)\psi_b(2) = E_b\psi_b(2) \tag{83}$$

where $E_a = E_b = E_H$, the ground state energy of the hydrogen atom and $\psi_a(1)$ and $\psi_b(2)$ are the $1s$ hydrogenic wave functions

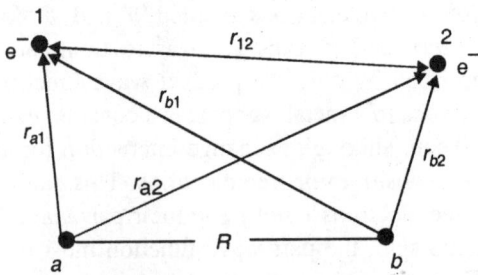

Fig. 14.4: H_2 molecule

$$\psi_a(1) = \psi_b(2) = \frac{1}{\sqrt{\pi}\, a_0^{3/2}}\, e^{-r/a_0} \tag{84}$$

($a_0 = 0.53$ Å is the Bohr radius)

$$H_a(1) = -\frac{\hbar^2}{2m}\nabla_1^2 - \frac{e^2}{r_{a_1}}\; ; \; H_b(2) = -\frac{\hbar^2}{2m}\nabla_2^2 - \frac{e^2}{r_{b_2}} \tag{85}$$

When the two atoms are brought closer, so long as there is no interaction between the two, the total Hamiltonian of the system is

$$H = H_a(1) + H_b(2) \tag{86}$$

It is equivalent to the situation that electron 1 is moving about proton a and electron 2 is moving about proton b without any interaction between the two atoms. This system is described by the wave function

$$\psi_1(1, 2) = \psi_a(1)\psi_b(2) \tag{87}$$

with energy eigenvalue $E_a + E_b$, since

$$H\left[\psi_a(1)\psi_b(2)\right] = \left[H_a(1) + H_b(2)\right]\psi_a(1)\psi_b(2)$$

$$= H_a(1)\psi_a(1)\psi_b(2) + H_b(2)\,\psi_a(1)\psi_b(2)$$

$$= E_a\psi_a(1)\psi_b(2) + E_b\psi_a(1)\psi_b(2)$$

$$= \left(E_a + E_b\right)\psi_a(1)\psi_b(2) \tag{88}$$

The two electrons being indistinguishable, this system can equally well be described with the same energy with structure in which electron 1 is associated with atom 'b' and electron 2 associated with atom 'a', this is known as *exchange degeneracy*, the wave function being given by

$$\psi_2(2, 1) = \psi_a(2)\psi_b(1) \tag{89}$$

In other words, $\psi_1(1, 2)$ and $\psi_2(2, 1)$ are eigen functions of the Hamiltonian $(H_a + H_b)$ with eigenvalue $(E_a + E_b)$. The two wave functions differ only in the interchange of the two electrons between the orbitals. In general case, the wave functions of this two electron system would be a linear combination of $\psi_1(1, 2)$ and $\psi_2(2, 1)$.

When two atoms are brought still closer, there will be interaction between the two atoms, if R denotes the equilibrium interatomic distance in H_2 molecule, its Hamiltonian is given by

$$H = H_a(1) + H_b(2) + H' + \frac{e^2}{R} \tag{90}$$

where $\qquad H' = -\frac{e^2}{r_{b_1}} - \frac{e^2}{r_{a_2}} + \frac{e^2}{r_{12}}$ is the perturbation term. $\qquad$ (91)

The problem can be solved by either the perturbation method or the variation method. We adopt here perturbation method as proposed by Heitler and London.

Since it is a two electron system and since electrons are indistinguishable, hence due to exchange degeneracy, the wave function would be either symmetric (ψ_S) or antisymmetric (ψ_a)

$$\psi_S = N_S\left[\psi_a(1)\psi_b(2) + \psi_a(2)\psi_b(1)\right] \tag{92a}$$

$$\psi_a = N_a\left[\psi_a(1)\psi_b(2) - \psi_a(2)\psi_b(1)\right] \tag{92b}$$

where N_s and N_a are normalization constants which can be obtained by following condition

$$\langle\psi_s|\psi_s\rangle = \iint N_S^2\left\{\psi_1(1,2) + \psi_2(2,1)\right\}^2 d\tau_1 d\tau_2$$

or $\qquad \langle\psi_s|\psi_s\rangle = N_S^2\left\{\langle\psi_1|\psi_1\rangle + \langle\psi_2|\psi_2\rangle + 2\langle\psi_a|\psi_b\rangle\langle\psi_b|\psi_a\rangle\right\}$

But $\qquad \langle\psi_1|\psi_1\rangle = \langle\psi_2|\psi_2\rangle = 1$

and $\qquad \langle\psi_a|\psi_b\rangle = \langle\psi_b|\psi_a\rangle \equiv \Delta$

Δ represents the orthogonality integral.

$\therefore \qquad N_S^2\left(1 + 1 + 2\Delta^2\right) = 1$

or $\qquad N_S^2 = \dfrac{1}{2\left(1 + \Delta^2\right)}$

Similarly $\qquad N_a^2 = \dfrac{1}{2\left(1 - \Delta^2\right)} \qquad$ (93)

Now the total wave function would be product of space and spin wave functions. Since, for electrons, the total wave function must be antisymmetric hence we must assign an antisymmetric spin function to the symmetric space wave function and symmetric spin function to the antisymmetric space wave function. Consequently, the Heitler-London wave functions are:

(i) $\quad N_S\left[\psi_a(1)\psi_b(2) + \psi_a(2)\psi_b(1)\right]\dfrac{1}{\sqrt{2}}\left[\alpha(1)\beta(2) - \beta(1)\alpha(2)\right]$ and

(ii) $\quad N_a\left[\psi_a(1)\psi_b(2)-\psi_a(2)\psi_b(1)\right] \begin{cases} \alpha(1)\alpha(2) \\ \dfrac{1}{\sqrt{2}}\left[\alpha(1)\beta(2)+\alpha(2)\beta(1)\right] \\ \beta(1)\beta(2) \end{cases}$ (94)

The first one corresponds to $S = 0$; antisymmetric spin function, singlet state and the second one to $S = 1$, symmetric spin function, hence triplet states.

Since the Hamiltonian does not contain spin terms, spin terms do not contribute and the space part alone can be taken as unperturbed wave functions for evaluation of energy using H'. The first order correction to the energy E' is the diagonal matrix element given by

$$E_1' = N_S^2 \left\langle \psi_a(1)\psi_b(2)+\psi_a(2)\psi_b(1)\middle|H'\middle|\psi_a(1)\psi_b(2)+\psi_a(2)\psi_b(1)\right\rangle$$

$$= \frac{2J+2K}{2\left(1+\Delta^2\right)} = \frac{J+K}{1+\Delta^2} \tag{95}$$

where

$$J = \left\langle \psi_a(1)\psi_b(2)\middle|H'\middle|\psi_a(1)\psi_b(2)\right\rangle \tag{96}$$

is called the *Coulomb integral* and

$$K = \left\langle \psi_a(1)\psi_b(2)\middle|H'\middle|\psi_a(2)\psi_b(1)\right\rangle \tag{97}$$

is called the *exchange integral*.

The energy of the singlet state corrected to the first order is

$$E_S = 2E_H + E_1' + \frac{e^2}{R}$$

$$= 2E_H + \frac{J+K}{1+\Delta^2} + \frac{e^2}{R} \tag{98}$$

The first order correction to the triplet state is

$$E_2' = \frac{J-K}{1-\Delta^2} \tag{99}$$

and so, the energy of the triplet state

$$E_a = 2E_H + E_2' + \frac{e^2}{\cdot R}$$

or $\qquad E_a = 2E_H + \dfrac{J-K}{1-\Delta^2} + \dfrac{e^2}{R}$ (100)

The Coulomb integral represents the classical coulomb interaction of electron charge cloud about one nucleus with the charge in the other nucleus and the interaction of the two charge clouds with one another neglecting correlation caused by symmetry (or anisymmetry) of the wave functions (92a) or (92b). *Exchange integral represents a non-classical interaction and is a consequence of exchange degeneracy.* This arises from correlation in the motion of the two electron arising from antisymmetrization of the wave functions in accordance with Pauli principle. Nevertheless, it is a quantum mechanical manifestation of the basic Coulomb interaction. This integral shows that electron 1 is partly in state a and partly in state b. Similarly electron 2 is partly in state b and partly in state a.

The overlap integral Δ is zero when the two atoms are far apart and is equal to 1 when they are in contact. For large R, both K and J are zero. For intermediate values of R both are negative and K is several times larger than J and lies between 0 and 1. Hence E_s has a value less than $2E_H$ while E_a is always larger than $2E_H$.

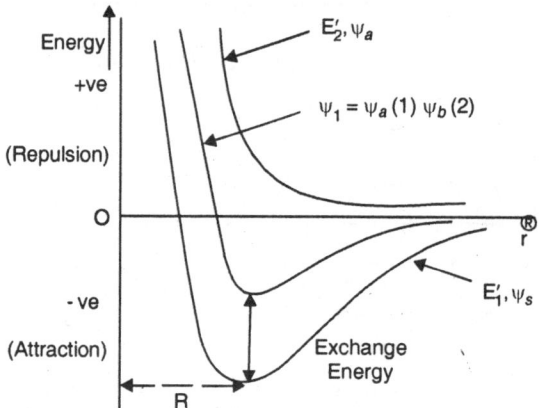

Fig. 14.5: Plot of energies E_1' and E_2' as a function of internuclear distance R

The state characterized by ψ_s shows a minimum (at $r = R$) and leads to formation of a stable molecule. The state characterized by ψ_a corresponds to replusion for all values of R and is unstable consequently, does not exist.

The theoretically evaluated value of R is 0.80 Å which is in close agreement with the experimental value 0.74 Å. The binding energy corresponding to the theoretical value is 72 k cal/mole. If exchange degeneracy is not considered, one would have got a binding energy of about 6 k cal/mole. The additional 66 k cal/mole of binding energy is termed as *exchange energy*, arising due to exchange integral.

Thus the two participating electrons in a covalent bond pair up with antiparallel spins and the bonding is through an exchange interaction, and, the molecule cannot be formed in an antisymmetric state. When two H atoms come closer, their chances of forming H_2 molecule are only 1:4 because there is only one wave function for the singlet state and three wave functions for triplet state.

Saturation property

Binding of two neutral atoms to form a H_2 molecule shows saturation i.e. in the simplest case of formation of the H_2 molecule, a H-atom binds just one other H-atom, but not two or three. This cannot be explained classically and the reason for this is that formation of H_2 molecule is a phenomenon involving exchange interaction which is exclusively a quantum mechanical artefact.

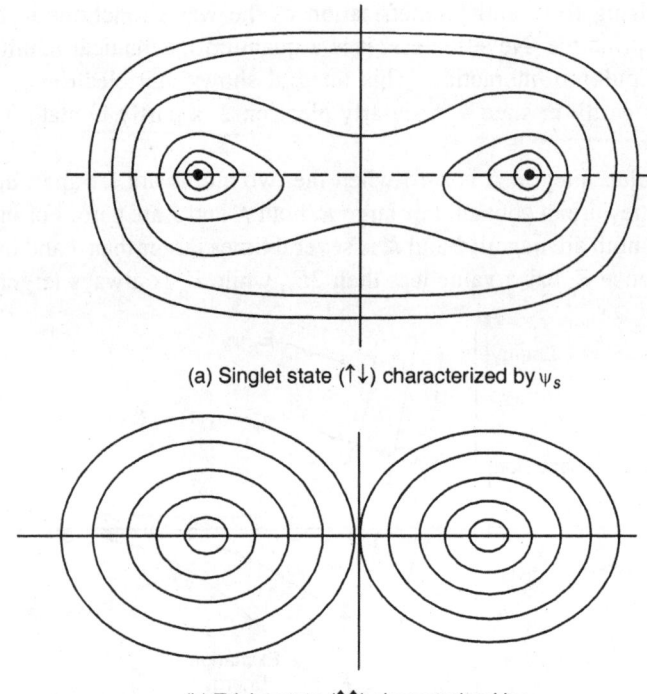

(a) Singlet state (↑↓) characterized by ψ_s

(b) Triplet state (↑↑) characterized by ψ_a

Fig. 14.6: Distribution of charge density of electron clouds in: (*a*) Symmetric and (*b*) antisymmetric states of H_2 molecule (corresponding to respectively the singlet, and triplet states)

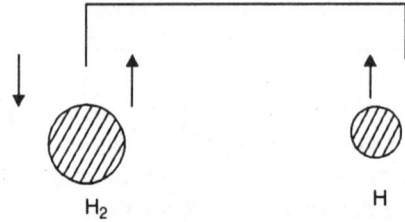

H_2 H

Fig. 14.7

H_2 is in a singlet state with antiparallel spins. If a H-atom comes closer, its electron can be exchanged with one of H_2 which has the same spin direction. Such an exchange leads to repulsion. Therefore H atom would be repelled by the H_2 molecule. The moment spin of the electron of the added H-atom becomes antiparallel to one of the spins of the electrons of H_2, the other electrons spin (of H_2) will try to oppose this.

14.16 Scattering of Identical Particles

In the scattering of identical particles (Fig. 14.2), there are two indistinguishable situations for every scattering angle in which the incident particle 1 and scatterer 2 have equal and opposite velocities in the centre of mass frame of reference.

If we suppose that the cross-section for the process (a) is given classically by $\sigma\left(\vec{k},\vec{k}'\right)$ then the cross-section for the process (b) is $\sigma\left(\vec{k},-\vec{k}'\right)$ since for scattered particle direction $\vec{k}'$, a recoil particle is observed in the direction $-\vec{k}'$. Consequently, the classical cross-section is

$$\sigma_c = \sigma\left(\vec{k},\vec{k}'\right) + \sigma\left(\vec{k},-\vec{k}'\right) \tag{101}$$

In quantum mechanics, the cross-section for the process (a) is $\left|f\left(\vec{k},\vec{k}'\right)\right|^2$ where $f\left(\vec{k},\vec{k}'\right)$ is the amplitude of the outgoing scattered wave. The asymptotic form of wave function is

$$\psi\left(\vec{k},\vec{k}',\vec{r}\right) \sim e^{i\vec{k}\cdot\vec{r}} + f\left(\vec{k},\vec{k}'\right)\frac{e^{ikr}}{r} \tag{102}$$

The correct wave function is obtained by symmetrization with respect to interchange of 1 and 2. Since $\vec{r} = \vec{r}_1 - \vec{r}_2$ and since $\vec{k}'$ has the same direction as $\vec{r}$, so

$$P_{12}\psi\left(\vec{k},\vec{k}',\vec{r}\right) = \psi\left(\vec{k},-\vec{k}',-\vec{r}\right) \tag{103}$$

Hence the degenerate wave functions are

$$\psi_s = \psi\left(\vec{k},\vec{k}',\vec{r}\right) + \psi\left(\vec{k},-\vec{k}',-\vec{r}\right)$$

$$\psi_A = \psi\left(\vec{k},\vec{k}',\vec{r}\right) - \psi\left(\vec{k},-\vec{k}',-\vec{r}\right) \tag{104}$$

The asympotic forms of these wave functions are

$$\psi \sim \left[e^{i\vec{k}\cdot\vec{r}} \pm e^{-i\vec{k}\cdot\vec{r}}\right] + f\left(\vec{k},\vec{k}'\right) \pm f\left(\vec{k},-\vec{k}'\right)\frac{e^{ikr}}{r}$$

For a pair of spin zero-particles, the scattering is described by the wave function ψ_s. The first term in the above expression represents unscattered (unit) intensities. The probability for observing an outgoing particle is given by the second term as

$$\left|f\left(\vec{k},\vec{k}'\right) + f\left(\vec{k},-\vec{k}'\right)\right|^2 \frac{d\vec{r}\,d\vec{R}}{r^2} \tag{106}$$

where $d\vec{r}\,d\vec{R}$ is the volume element in terms of the relative coordinate $\vec{r}$ and centre of mass coordinate $\vec{R}$. Thus, the cross-section is

$$\sigma_0\left(\vec{k},\vec{k}'\right) = \left|f\left(\vec{k},\vec{k}'\right) + f\left(\vec{k},-\vec{k}'\right)\right|^2$$

$$= \left|f\left(\vec{k},\vec{k}'\right)\right|^2 + \left|f\left(\vec{k},-\vec{k}'\right)\right|^2 + 2\,\mathrm{Re}\left[f*\left(\vec{k},\vec{k}'\right)f\left(\vec{k},-\vec{k}'\right)\right] \tag{107}$$

So, in addition to the classical cross section (given by the first two terms of this expression), we have an interference term. If $f\left(\vec{k},\vec{k}'\right)$ is independent of the angle ϕ, then

$$\sigma_0(\theta) = \left|f(\theta) + f(\pi - \theta)\right|^2 \tag{108}$$

$$\Rightarrow \qquad \sigma_0\left(\frac{\pi}{2}\right) = \left|2f\left(\frac{\pi}{2}\right)\right|^2 = 4\left|f\left(\frac{\pi}{2}\right)\right|^2 \tag{109}$$

i.e. at $\theta = \pi/2$, the cross-section is twice the classical cross-section at the same angle.

In case of spin $-$ ½ particles, since the singlet spin state is antisymmetric it must be combined with ψ_s, giving the cross section

$$^s\sigma_{\frac{1}{2}}\left(\vec{k},\vec{k}'\right)=\left|f\left(\vec{k},\vec{k}'\right)+f\left(\vec{k},-\vec{k}'\right)\right|^2 \tag{110}$$

The triplet state is symmetric and has to be combined with ψ_A leading to

$$^t\sigma_{\frac{1}{2}}\left(\vec{k},\vec{k}'\right)=\left|f\left(\vec{k},\vec{k}'\right)-f\left(\vec{k},-\vec{k}'\right)\right|^2 \tag{111}$$

In the absence of external forces, the four spin states are occupied equally, giving total cross section

$$\sigma_{\frac{1}{2}}\left(\vec{k},\vec{k}'\right)=\frac{1}{4}\,{}^s\sigma_{\frac{1}{2}}\left(\vec{k},\vec{k}'\right)+\frac{3}{4}\,{}^t\sigma_{\frac{1}{2}}\left(\vec{k},\vec{k}'\right)$$

$$=\left|f\left(\vec{k},\vec{k}'\right)\right|^2+\left|f\left(\vec{k},-\vec{k}'\right)\right|^2-\left(\tfrac{1}{2}-\tfrac{3}{2}\right)\operatorname{Re}f*\left(\vec{k},\vec{k}'\right)f\left(\vec{k},-\vec{k}'\right)$$

$$=\left|f\left(\vec{k},\vec{k}'\right)\right|^2+\left|f\left(\vec{k},-\vec{k}'\right)\right|^2-\operatorname{Re}f*\left(\vec{k},\vec{k}'\right)f\left(\vec{k},-\vec{k}'\right) \tag{112}$$

The factors ¼ and ¾ are statistically weights of the singlet and triplet states respectively. Thus,

$$\sigma_{\frac{1}{2}}\left(\frac{\pi}{2}\right)=\left|f\left(\frac{\pi}{2}\right)\right|^2 \tag{113}$$

which is one half of the classical cross-section. Hence the three cross sections at $\theta = \pi/2$ are in the ratio:

$$\sigma_0\left(\frac{\pi}{2}\right):\sigma_c\left(\frac{\pi}{2}\right):\sigma_{\frac{1}{2}}\left(\frac{\pi}{2}\right)=4:2:1 \tag{114}$$

Thus, knowing the classical cross-section with the help of Rutherford's formula (e.g. for low energy protons), we can determine the spin of the particles (with the help of the last relation). Similar experiments have revealed that the alpha particles have zero spin.

Questions and Problems

1. Comment: Every symmetry transformation is associated with a definite conservation law.

2. Prove that if an operator corresponding to an observable commutes with the generator of an infinitesimal transformation, then that operator remains invariant under the unitary transformation.

3. Prove that if an observable corresponding to an operator $\hat{A}$ is conserved during motion of a system then the Hamiltonian of the system remains invariant under the unitary transformation generate by $\hat{A}$. What is meant by symmetry of a physical system under a unitary transformation?

4. Explain how invariance of Hamiltonian under spatial translation leads to conservation of linear momentum.

5. Explain how rotational invariance of Hamiltonian leads to conservation of angular momentum.

6. Explain why time reversal operator T must be antilinear for $T\psi$ to satisfy the Schrödinger equation.

7. Prove that the parity operator is Hermitian and unitary.

8. Show that in case of idential indistinguishable particles, the application of exchange operator on the two particle wave function leads to either symmetric or antisymmetric states.

9. What is exchange degeneracy. Show that in case of fermions, two fermions cannot occupy the same quantum state (exclusion principle).

10. Discuss physical significance of Pauli's exclusion principle.

11. With P_t and P_s defined by $P_t = \frac{1}{4}(3 + \vec{\sigma}_1 \cdot \vec{\sigma}_2)_t$; $P_s = \frac{1}{4}(1 - \vec{\sigma}_1 \cdot \vec{\sigma}_2)$ show that $(P_t - P_s)$ has the same effect on $^3\chi$ and $^1\chi$ as the interchange of the two particles. The operator $(P_t - P_s) \equiv \frac{1}{2}(1 + \vec{\sigma}_1 \cdot \vec{\sigma}_2)$ is called the spin-exchange operator.

12. Discuss Heitler and London theory of bonding in H_2 molecule.

13. Show that if $\psi_{\alpha'} = T\psi_\alpha$ and $\psi_{\beta'} = T\psi_\beta$ then $(\psi_{\alpha'}, \psi_{\beta'}) = (\psi_\alpha, \psi_\beta)^* = (\psi_\beta, \psi_\alpha)$. From this show that the norm of a state vector is unchanged by time reversal.

15

Time Independent
Perturbation Theory

15.1 Introduction

The exact solution of Schrödinger's equation can be found in a comparatively small number of the simplest cases e.g. hydrogen atom, harmonic oscillator, rigid rotator etc. The majority of problems in quantum mechanics lead to equations which are too complex to be solved exactly. Often, however, quantities of different orders of magnitude appear in the conditions of the problem; among them there may be small quantities such that, when they are neglected, the problem is so much simplified that its exact solution becomes possible. In such cases, the first step in solving \the physical problem concerned is to solve exactly the simplified problem, and the second step is to calculate approximately the errors due to the small terms that have been neglected in the simplified problem. There is a general method of calculating these errors; it is called *perturbation theory*. It was developed by Schrödinger in 1926.

If the problem at hand is sufficiently similar to one that can be solved exactly, then the Hamiltonian can be broken up into two parts, one of which is large and characterizes the system for which the Schrödinger equation can be solved exactly, while the other part is small and can be treated as a perturbation. Perturbation theories are of two kinds:

(*a*) Time independent perturbation theory and

(*b*) Time dependent perturbation theory.

In the time independent method, Hamiltonian does not depend on time, while in the time dependent method Hamiltonian depends explicitly on time.

15.2 Non-degenerate Case

Let us suppose that the discrete set of all the eigenvalues $E_n^{(0)}$ and the orthonormal set of eigenfunctions $\psi_n^{(0)}$ of the unperturbed Hamiltonian $H^{(0)}$ of a system are known:

$$H^{(0)}\psi_n^{(0)} = E_n^{(0)}\psi_n^{(0)} \tag{1}$$

in which $E_n^{(0)}$ is the energy of the nth state of the system with the corresponding state function $\psi_n^{(0)}$. Now, if a small perturbation is added to the system so that the Hamiltonian changes to H then the energy levels and the stationary states of the are described by

$$H\psi = E\psi \tag{2}$$

Usually this equation is not exactly solvable. The perturbation theory provides a systematic method of successive approximations to the eigenfunction ψ and the eigenvalue E in terms of the unperturbed eigenfunctions $\psi_n^{(0)}$ and eigenvalues $E_n^{(0)}$. The basic idea is to write

$$H = H^{(0)} + \lambda H' \tag{3}$$

where H' is the perturbation and λ characterises the strength of the perturbation.

We now develop the perturbation theory assuming that E and ψ can be expressed as power series in λ such that the zero, the first etc. powers of λ correspond to the zero, the first etc. orders of approximation.

$$\psi = \psi_0 + \lambda\psi_1 + \lambda^2\psi_2 + \dots \tag{4}$$

$$E = E_0 + \lambda E_1 + \lambda^2 E_2 + \dots \tag{5}$$

Substituting Eqns (3), (4) and (5) in Eqn. (2) and grouping the terms of the same order together, we have

$$\left(H^{(0)} - E_0\right)\psi_0 + \left[\left(H^{(0)} - E_0\right)\psi_1 + \left(H' - E_1\right)\psi_0\right]\lambda$$
$$+ \left[\left(H^{(0)} - E_0\right)\psi_2 + \left(H' - E_1\right)\psi_1 - E_2\psi_0\right]\lambda^2 + \dots = 0 \tag{6}$$

We want this equation to be valid for any arbitrary (but small) strength of the perturbation λ. This can happen only if the coefficient of each power of λ vanishes. Thus, we should have

$$\left(H^{(0)} - E_0\right)\psi_0 = 0 \tag{7a}$$

$$\left(H^{(0)} - E_0\right)\psi_1 = \left(E_1 - H'\right)\psi_0 \tag{7b}$$

$$\left(H^{(0)} - E_0\right)\psi_2 = \left(E_1 - H'\right)\psi_1 + E_2\psi_0 \tag{7c}$$

etc. These are the zeroth order, first order, second order, ..., equations of the perturbation theory. The zeroth order equation (7a) means that ψ_0 is any one of the unperturbed eigenfunctions. Let us take

$$\psi_0 = \psi_m^{(0)}$$

and

$$E_0 = E_m^{(0)} \tag{8}$$

and suppose that $\psi_m^{(0)}$ is non degenerate.

15.2.1 Evaluation of First Order Energy E$_1$

Using Eqn. (8), we can write the first order perturbation equation (7b) as

$$\left(H^{(0)} - E_m^{(0)}\right)\psi_1 = \left(E_1 - H'\right)\psi_m^{(0)} \tag{9}$$

For the solution of this equation, we expand ψ_1 in terms of the complete set of unperturbed wave functions $\psi_n^{(0)}$

$$\psi_1 = \sum_n a_n^{(1)} \psi_n^{(0)} \tag{10}$$

$$\therefore \quad H^{(0)}\psi = \sum_n a_n^{(1)} H^{(0)} \psi_n^{(0)}$$

$$= \sum_n a_n^{(1)} E_n^{(0)} \psi_n^{(0)} \tag{11}$$

Substituting Eqns (10) and (11) into (9), we get

$$\sum_n a_n^{(1)} E_n^{(0)} \psi_n^{(0)} - E_m^{(0)} \sum_n a_n^{(1)} \psi_n^{(0)} = \left(E_1 - H'\right)\psi_m^{(0)}$$

or

$$\sum_n a_n^{(1)} \left(E_n^{(0)} - E_m^{(0)}\right)\psi_n^{(0)} = \left(E_1 - H'\right)\psi_m^{(0)} \tag{12}$$

Multiplying Eqn. (12) by $\psi_m^{(0)*}$ and integrating over the space variables, we have

$$\int \sum_n \psi_m^{(0)*} a_n^{(1)} \left(E_n^{(0)} - E_m^{(0)}\right)\psi_n^{(0)} d^3r = \int \psi_m^{(0)*} \left(E_1 - H'\right)\psi_m^{(0)} d^3r$$

Using the orthonormality condition

$$\int \psi_m^{(0)} \psi_n^{(0)} d^3r = 0 \text{ if } m \neq n$$

$$= 1 \text{ if } m = n$$

We have from the above equation

$$0 = \int \psi_m^{(0)*} \left(E_1 - H'\right)\psi_m^{(0)} d^3r$$

Since E_1 is first order energy correction, it can be taken outside the integral sign treating as constant.

$$\therefore \quad E_1 \int \psi_m^{(0)*} \psi_m^{(0)} d^3r = \int \psi_m^{(0)*} H' \psi_m^{(0)} d^3r$$

or

$$E_1 = \int \psi_m^{(0)*} H' \psi_m^{(0)} d^3r \quad \text{(using the normality of } \psi_m^{(0)}\text{)}$$

Denoting the $\psi_m^{(0)}$ by the ket $|m\rangle$, we can write it in the bra and ket notations as

$$E_1 = \langle m|H'|m\rangle \tag{13}$$

15.2.2 Evaluation of First Order Wave Function ψ_1

In order to find out the wave function ψ_1, we multiply Eqn. (12) by $\psi_k^{(0)*}$ and integrate it over all space:

$$\int \sum_n a_n^{(1)} \left(E_n^{(0)} - E_m^{(0)}\right)\psi_k^{(0)*} \psi_n^{(0)} d^3r = \int \psi_k^{(0)*} \left(E_1 - H'\right) \psi_m^{(0)} d^3r$$

Using the orthonormality of the unperturbed functions $\psi_n^{(0)}$, we get

$$a_k^{(1)}\left(E_k^{(0)} - E_m^{(0)}\right) = -\int \psi_k^{(0)} {}^* H' \, \psi_m^{(0)} d^3 r$$

or

$$a_k^{(1)}\left(E_k^{(0)} - E_m^{(0)}\right) = -\langle k|H'|m\rangle \Rightarrow a_k^{(1)} = \frac{-\langle k|H'|m\rangle}{\left(E_k^{(0)} - E_m^{(0)}\right)}$$

or

$$a_n^{(1)} = \frac{-\langle n|H'|m\rangle}{\left(E_n^{(0)} - E_m^{(0)}\right)} \; ; \; n \neq m \tag{14}$$

Using Eqn. (14) into Eqn. (10) we get the first order wave function

$$\psi_1 = \sum_{\substack{n \\ n \neq m}} \frac{\langle n|H'|m\rangle \psi_n^{(0)}}{E_m^{(0)} - E_n^{(0)}} \tag{15}$$

The summation is for $n \neq m$ only.

If we take $\lambda = 1$ or include the parameter λ into H', then in the first order perturbation.

$$H = H^{(0)} + H'$$

$$E = E_0 + E_1 = E_m^{(0)} + \langle m|H'|m\rangle$$

and

$$\psi = \psi_0 + \psi_1 = \psi_m^{(0)} + \sum_{n \neq m} \frac{\langle n|H'|m\rangle \psi_n^{(0)}}{\left(E_m^{(0)} - E_n^{(0)}\right)}$$

15.2.3 Evaluation of Second Order Energy E_2

Using Eqn. (8), we can write the second order perturbation Eqn. (7c) as

$$\left(H^{(0)} - E_m^{(0)}\right)\psi_2 = \left(E_1 - H'\right)\psi_1 + E_2\psi_m^{(0)} \tag{16}$$

For the solution of this equation we expand ψ_1 and ψ_2 in terms of the complete set of unperturbed wave functions $\psi_n^{(0)}$:

$$\psi_1 = \sum_n a_n^{(1)}\psi_n^{(0)} \text{ and } \psi_2 = \sum_n a_n^{(2)}\psi_n^{(0)} \tag{17}$$

Substituting Eqns. (17) into (16), we get

$$\sum_n a_n^{(2)}\left(H^{(0)} - E_m^{(0)}\right)\psi_n^{(0)} = \sum_n a_n^{(1)}\left(E_1 - H'\right)\psi_n^{(0)} + E_2\psi_m^{(0)}$$

or

$$\sum_n a_n^{(2)}\left(E_n^{(0)} - E_m^{(0)}\right)\psi_n^{(0)} = \sum_n a_n^{(1)}\left(E_1 - H'\right)\psi_n^{(0)} + E_2\psi_m^{(0)} \tag{18}$$

Multiplying Eqn. (18) by $\psi_m^{(0)}{}^*$ and integrating over the space variables:

$$\int \psi_m^{(0)}{}^* \sum_n a_n^{(2)}\left(E_n^{(0)} - E_m^{(0)}\right)\psi_n^{(0)}d^3 r$$

$$= \int \psi_m^{(0)}{}^* \sum_n a_n^{(1)}\left(E_1 - H'\right)\psi_n^{(0)}d^3 r + \int \psi_m^{(0)}{}^* E_2\psi_m^{(0)}d^3 r \tag{19}$$

For $m = n$, $\left(E_n^{(0)} - E_m^{(0)} \right)$ vanishes and for $m \neq n$, $\int \psi_m^{(0)} \psi_n^{(0)} d^3r$ vanishes, therefore the integral on the left hand side of the above equation vanishes for all values of n. Also, for $n \neq m$, the first part of the first integral on the right hand side vanishes due to the orthonormality of functions $\psi_n^{(0)}$. Thus, using the ortho-normality of functions $\psi_n^{(0)}$, we can write Eqn. (19) as

$$0 = - \sum_{n \neq m} a_n^{(1)} \int \psi_m^{(0)} {}^* H' \psi_n^{(0)} d^3r + E_2$$

or

$$E_2 = \sum_n a_n^{(1)} \langle m | H' | n \rangle$$

Susbtituting the value of $a_n^{(1)}$ from Eqn. (19) into the above Eqn., we get

$$E_2 = \sum_{\substack{n \\ n \neq m}} \frac{\langle n | H' | m \rangle \langle m | H' | n \rangle}{\left(E_m^{(0)} - E_n^{(0)} \right)} \tag{20}$$

The term $n = m$ has to be omitted.

15.2.4 Evaluation of Second Order Function ψ_2

In order to evaluate the wave function, we multiply Eqn. (18) by $\psi_k^{(0)}{}^*$ and integrate over all space:

$$\int \psi_k^{(0)}{}^* \sum_n a_n^{(2)} \left(E_n^{(0)} - E_m^{(0)} \right) \psi_n^{(0)} d^3r$$

$$= \int \psi_k^{(0)}{}^* \sum_n a_n^{(1)} \left(E_1 - H' \right) \psi_n^{(0)} d^3r + \int \psi_k^{(0)}{}^* E_2 \psi_n^{(0)} d^3r$$

Using the orthonormality of $\psi_n^{(0)}$'s, we can write

$$a_k^{(2)} \left(E_k^{(0)} - E_m^{(0)} \right) = a_k^{(1)} E_1 - \sum_n a_n^{(1)} \langle k | H' | n \rangle + 0$$

Using Eqns. (13) and (14) in this equation, we write

$$a_k^{(2)} \left(E_k^{(0)} - E_m^{(0)} \right) = - \frac{\langle k | H' | m \rangle \langle m | H' | m \rangle}{\left(E_k^{(0)} - E_m^{(0)} \right)} + \sum_n \frac{\langle k | H' | n \rangle \langle n | H' | m \rangle}{\left(E_k^{(0)} - E_m^{(0)} \right)}$$

or

$$a_k^{(2)} = \sum_n \frac{\langle k | H' | n \rangle \langle n | H' | m \rangle}{\left(E_m^{(0)} - E_n^{(0)} \right) \left(E_m^{(0)} - E_k^{(0)} \right)} - \frac{\langle k | H' | m \rangle \langle m | H' | m \rangle}{\left(E_k^{(0)} - E_m^{(0)} \right)^2} \tag{21}$$

Using Eqn. (21) into (17), we get the 2nd order wave function:

$$\psi_2 = \sum_k a_k^{(2)} \psi_k^{(0)} = \sum_k \psi_k^{(0)} \left[\sum_n \frac{\langle k | H' | n \rangle \langle n | H' | m \rangle}{\left(E_m^{(0)} - E_n^{(0)} \right) \left(E_m^{(0)} - E_k^{(0)} \right)} \right.$$

$$\left. - \frac{\langle k | H' | m \rangle \langle m | H' | m \rangle}{\left(E_k^{(0)} - E_m^{(0)} \right)^2} \right] \tag{22a}$$

Hence, the energy and wave function for second order perturbation are given by

$$E = E_0 + E_1 + E_2 = E_m^{(0)} + \langle m|H'|m\rangle + \sum' \frac{\left|\langle m|H'|n\rangle\right|^2}{\left(E_m^{(0)} - E_n^{(0)}\right)}$$

and $\quad \psi = \psi_0 + \psi_1 + \psi_2$

$$= \psi_m^{(0)} + \sum_k' \psi_k^{(0)} \left\{ \frac{\langle k|H'|m\rangle}{\left(E_m^{(0)} - E_k^{(0)}\right)} \left(1 - \frac{\langle m|H'|m\rangle}{\left(E_m^{(0)} - E_k^{(0)}\right)}\right) \right.$$

$$\left. + \sum_n' \frac{\langle k|H'|n\rangle \langle n|H'|m\rangle}{\left(E_m^{(0)} - E_k^{(0)}\right)\left(E_m^{(0)} - E_n^{(0)}\right)} \right\} \qquad (22b)$$

Here the prime over the summation indicates that $k = m$ or $k = n$ are to be avoided or $n \neq k \,(\neq m)$.

Example 15.1: Consider an electron inside an infinitely deep one dimensional potential well

$$V(x) = 0, \; 0 < x < L$$
$$= \infty \text{ for } x < 0 \text{ and } x > L$$

The normalized wave functions are

$$\psi_n = \sqrt{\frac{2}{L}} \sin\frac{n\pi x}{L}$$

Assuming a perturbation of the form $H' = -eFx$, show that in the first order perturbation theory, each level gets shifted by $-1/2 \, eFL$.

Solution: $\qquad H' = -eFx$

$$E_1 = \int_0^L \psi_n^* H' \psi_n \, dx$$

$$= \int_0^L \frac{2}{L}(-eFx) \sin^2\frac{n\pi x}{L} \, dx$$

$$= -\frac{2eF}{L} \int_0^L x \sin^2\frac{n\pi x}{L} \, dx$$

$$= -\frac{2eF}{L} \frac{L^2}{n^2\pi^2} \int_0^{n\pi} \theta \sin^2\theta \, d\theta \qquad \left(\theta = \frac{n\pi x}{L}\right)$$

$$= -\frac{2eEL}{n^2\pi^2} \int_0^{n\pi} \theta \,(1 - \cos 2\theta) \, d\theta$$

Now, $\quad \int_0^{n\pi} \theta \, d\theta = \left[\frac{\theta^2}{2}\right]_0^{n\pi} = \frac{n^2\pi^2}{2}$

and $\quad \int_0^{n\pi} \theta \cos 2\theta \, d\theta = \theta \int_0^{n\pi} \cos 2\theta \, d\theta \Big|_0^{n\pi} + \int_0^{n\pi} \left(\int \cos 2\theta \, d\theta\right) d\theta = 0$

Hence , $E_1 = -\dfrac{1}{2}eFL$

Example 15.2: The 2s wave function of the H atom in the standard notation $\psi_{nlm}(r,\theta,\phi)$ is

$$\psi_{200}(r,\theta,\phi) = \frac{1}{\sqrt{4\pi}}\left(\frac{1}{2a_0}\right)^{3/2}\left(2-\frac{r}{a_0}\right)e^{-r/2a_0}$$

This is perturbed by a potential of the form $V(r) = \dfrac{\lambda}{r^2}$. Calculate the first order shift in energy.

Solution: We have

$$\psi_{200}(r,\theta,\phi) = \frac{1}{\sqrt{4\pi}}\left(\frac{1}{2a_0}\right)^{3/2}\left(2-\frac{r}{a_0}\right)e^{-r/2a_0}$$

$$H' = \frac{\lambda}{r^2}$$

First order energy shift

$$E_1 = \int \psi^* H'\psi \, d\tau$$

$$= \frac{1}{4\pi}4\pi\int_0^\infty\left(\frac{1}{2a_0}\right)^3\left(2-\frac{r}{a_0}\right)^2 e^{-r/a_0}\frac{\lambda}{r^2}\cdot r^2 dr$$

$$= \frac{\lambda}{(2a_0)^3}\int_0^\infty\left(4+\frac{r^2}{a_0^2}-4\frac{r}{a_0}\right)e^{-r/a_0}dr$$

Now,

$$\int_0^\infty e^{-r/a_0}dr = \frac{1}{(1/a_0)} = a_0$$

$$\int_0^\infty r^2 e^{-r/a_0}dr = \frac{2}{(1/a_0)^3} = 2a_0^3$$

$$\int_0^\infty r e^{-r/a_0}dr = \frac{1}{(1/a_0)^2} = a_0^2$$

$\therefore$

$$E_1 = \frac{\lambda}{2a_0^3}\left[4a_0 + \frac{1}{a_0^2}2a_0^3 - \frac{4}{a_0}a_0^2\right] = \frac{\lambda}{a_0^2}$$

Example 15.3: A system has three unperturbed states, can be represented by the perturbed Hamiltonian matrix

$$\begin{pmatrix} E_1 & 0 & a \\ 0 & E_1 & b \\ a* & b* & E_2 \end{pmatrix} \text{ where } E_2 > E_1$$

The quantities a and b are to be regarded as perturbations that are of the same order and are small compared with $E_2 - E_1$. Use the second order non-degenerate perturbation theory to calculate the perturbed eigenvalues. Then diagonalise the matrix to find the exact eigenvalues.

Solution:
$$\begin{pmatrix} E_1 & 0 & a \\ 0 & E_1 & b \\ a* & b* & E_2 \end{pmatrix} = \begin{pmatrix} E_1 & 0 & 0 \\ 0 & E_1 & 0 \\ 0 & 0 & E_2 \end{pmatrix} + \begin{pmatrix} 0 & 0 & a \\ 0 & 0 & b \\ a* & b* & 0 \end{pmatrix}$$

or
$$H \equiv H_0 + H'$$

The eigenvalues of H_0 are E_1, E_1 and E_2. The corresponding eigenfunctions are obtained as follows.

For eigenvalue E_1: Any column matrix $\begin{pmatrix} x_1 \\ x_2 \\ 0 \end{pmatrix}$ is an eigenvector because

$$\begin{pmatrix} E_1 & 0 & 0 \\ 0 & E_1 & 0 \\ 0 & 0 & E_2 \end{pmatrix} \begin{pmatrix} x_1 \\ x_2 \\ 0 \end{pmatrix} = E_1 \begin{pmatrix} x_1 \\ x_2 \\ 0 \end{pmatrix}$$

We may choose $\quad u_1 = \begin{pmatrix} 1 \\ 0 \\ 0 \end{pmatrix}; u_2 = \begin{pmatrix} 0 \\ 1 \\ 0 \end{pmatrix}$

For eigenvalue E_2:

$$\begin{pmatrix} E_1 - E_2 & 0 & 0 \\ 0 & E_1 - E_2 & 0 \\ 0 & 0 & 0 \end{pmatrix} \begin{pmatrix} x_1 \\ x_2 \\ x_3 \end{pmatrix} = 0$$

Since $E_1 \neq E_2$, therefore $\quad x_1 = 0 = x_2$

$\therefore \quad u_3 = \begin{pmatrix} 0 \\ 0 \\ 1 \end{pmatrix}$

Note that first order energy correction for non-degenerate as well as degenerate case is zero.

We now apply the second order non degenerate perturbation theory to find the perturbed energy eigenvalue. We can use this procedure if we take u_3 as u_0 (one of the u's which is fixed). If there were only the two degenerate states in the system then we could not apply this procedure.

$\therefore \quad u_0 = \begin{pmatrix} 0 \\ 0 \\ 1 \end{pmatrix}$

$$H'_{10} = u_1^* \begin{pmatrix} 0 & 0 & a \\ 0 & 0 & b \\ a* & b* & 0 \end{pmatrix} u_0$$

$$= (1\ 0\ 0) \begin{pmatrix} 0 & 0 & a \\ 0 & 0 & b \\ a* & b* & 0 \end{pmatrix} \begin{pmatrix} 0 \\ 0 \\ 1 \end{pmatrix}$$

$$= (1\ 0\ 0) \begin{pmatrix} a \\ b \\ 0 \end{pmatrix} = a$$

Similarly, $$H'_{20} = (0\ 1\ 0) \begin{pmatrix} 0 & 0 & a \\ 0 & 0 & b \\ a* & b* & 0 \end{pmatrix} \begin{pmatrix} 0 \\ 0 \\ 1 \end{pmatrix}$$

$$= (0\ 1\ 0) \begin{pmatrix} a \\ b \\ 0 \end{pmatrix} = b$$

$$\therefore \qquad E^{(2)} = \sum_{n\neq 0} \frac{|H'_{n0}|^2}{E_0 - E_n} = \frac{a^2 + b^2}{E_2 - E_1} \qquad (\because E_0 = E_2,\ E_n = E_1)$$

The exact eigenvalues λ of H are solutions of

$$\begin{pmatrix} E_1 - \lambda & 0 & a \\ 0 & E_1 - \lambda & b \\ a* & b* & E_2 - \lambda \end{pmatrix} = 0$$

Expanding the determinant of the matrix, we get

$$-a*\left[a(E_1 - \lambda) \right] - b*\left[b(E_1 - \lambda) \right] + (E_2 - \lambda)\,(E_1 - \lambda)^2 = 0$$

or $$\left(-a^2 - b^2 \right)(E_1 - \lambda) + (E_2 - \lambda)\,(E_1 - \lambda)^2 = 0$$

$\therefore$ $\lambda_1 = E_1$ otherwise

$$(E_2 - \lambda)\,(E_1 - \lambda) - a^2 - b^2 = 0$$

or $$\lambda^2 - (E_1 + E_2)\lambda + \left(E_1 E_2 - a^2 - b^2 \right) = 0$$

or $$\lambda = \frac{(E_1 + E_2) \pm \sqrt{(E_1 + E_2)^2 - 4\left(E_1 E_2 - a^2 - b^2 \right)}}{2}$$

$$= \left[(E_1 + E_2) \pm \sqrt{(E_1 - E_2)^2 + 4\left(a^2 + b^2 \right)} \right] \times \frac{1}{2}$$

$$= \frac{E_1 + E_2}{2} \pm \frac{1}{2}(E_1 - E_2) \left[1 + \frac{4\left(a^2 + b^2 \right)}{(E_1 - E_2)^2} \right]^{1/2}$$

$$\cong \frac{E_1 + E_2}{2} \pm \frac{1}{2}(E_1 - E_2) \left[1 + \frac{2\left(a^2 + b^2 \right)}{(E_1 - E_2)^2} + \dots \right]$$

$$\cong E_1 \pm \frac{a^2 + b^2}{\left(E_1 - E_2\right)} + \dots \text{ or } E_2 \pm \frac{a^2 + b^2}{\left(E_1 - E_2\right)}$$

15.3 Time Independent Perturbation Theory for Degenerate Case (Doubly Degenerate)

Now, the non-degenerate perturbation method is not applicable because there it has been assumed that the perturbed wave functions differ slightly from one function $\psi_m^{(0)}$ which was the solution of the unperturbed wave equation for a given energy value. In case of degenerate level there are several such functions which correspond to the same energy.

For simplicity, we first consider a doubly degenerate level. Let $\psi_m^{(0)}$ and $\psi_l^{(0)}$ be two orthonormal eigenfunctions belonging to the level $E_m^{(0)}$. Since $\psi_m^{(0)}$ and $\psi_l^{(0)}$ are states corresponding to unperturbed energy, therefore their linear combination will also be a state corresponding to this energy, i.e.

$$\psi_0 = a_m \psi_m^{(0)} + a_l \psi_l^{(0)} \tag{23}$$

Substituting this value in the first order perturbation equation (7b), we have

$$\left(H^{(0)} - E_0\right)\psi_1 = \left(E_1 - H'\right)\left(a_m \psi_m^{(0)} + a_l \psi_l^{(0)}\right) \tag{24}$$

Multiplying Eqn. (24) by $\psi_m^{(0)*}$ and integrating over the entire space, we get

$$\int \psi_m^{(0)*} \left(H^{(0)} - E_0\right)\psi_1 d^3r = \int \psi_m^{(0)*} \left(E_1 - H'\right)a_m \psi_m^{(0)} d^3r$$

$$+ \int \psi_m^{(0)*} \left(E_1 - H'\right) a_l \psi_l^{(0)} d^3r \tag{25}$$

We have

$$\int \psi_m^{(0)*} H^{(0)}\psi_1 d^3r = \int \left[H^{(0)}\psi_m^{(0)}\right]^* \psi_1 d^3r \quad (\because H^{(0)} \text{ is Hermitian})$$

$$= \int \left[E_m^{(0)}\psi_m^{(0)}\right]^* \psi_1 d^3r$$

$$= E_m^{(0)} \int \psi_m^{(0)}\psi_1 d^3r \quad (\because E_m^{(0)} \text{ is real}) \tag{26}$$

Also $E_0 = E_m^{(0)}$. Therefore, using Eqn. (26) we find that the l.h.s. of Eqn. (25) vanishes. Hence from Eqn. (25), we get

$$\int \psi_m^{(0)*} \left(E_1 - H'\right)a_m \psi_m^{(0)} d^3r + \int \psi_m^{(0)*} \left(E_1 - H'\right) a_l \psi_l^{(0)} d^3r = 0 \tag{27}$$

or

$$a_m E_1 \int \psi_m^{(0)*} \psi_m^{(0)} d^3r - a_m \int \psi_m^{(0)*} H'\psi_m^{(0)} d^3r$$

$$+ a_l E_1 \int \psi_m^{(0)*} \psi_l^{(0)} d^3r - a_l \int \psi_m^{(0)*} H'\psi_l^{(0)} d^3r = 0 \tag{28}$$

Using the orthonormality of $\psi_m^{(0)}$ and $\psi_l^{(0)}$, this equation reduces to

$$a_m E_1 - a_m \langle m|H'|m\rangle - a_l \langle m|H'|l\rangle = 0 \tag{29}$$

or
$$\left(\langle m | H' | m \rangle - E_1 \right) a_m + \langle m | H' | l \rangle a_l = 0 \tag{30}$$

or
$$\left(H'_{mm} - E_1 \right) a_m + H'_{ml} a_l = 0$$

Similarly, multiplying Eqn. (24) by $\psi_l^{(0)*}$ and integrating over the entire space,
$$H'_{lm} a_m + \left(H'_{ll} - E_1 \right) a_l = 0 \tag{31}$$

Eqns. (30) and (31) are two homogenous equations for a_m and a_l and they have a non-zero solution only if the determinant of coefficients vanishes i.e.

$$\begin{vmatrix} \left(H'_{mm} - E_1 \right) & H'_{ml} \\ H'_{lm} & \left(H'_{ll} - E_1 \right) \end{vmatrix} = 0$$

This is called the secular equation. It is a second degree algebraic equation for E_1. Explicitly,

$$E_1^2 - \left(H'_{mm} + H'_{ll} \right) E_1 + \left(H'_{mm} H'_{ll} - H'_{ml} H'_{lm} \right) = 0$$

$$\therefore \quad E_1 = \frac{1}{2} \left[H'_{mm} + H'_{ll} \right] \pm \frac{1}{2} \left[\left\{ H'_{mm} + H'_{ll} \right\}^2 + 4 H'_{ml} H'_{lm} \right]^{1/2} \tag{32}$$

Thus there are two possible values of E_1, say $E_1^{(1)}$ and $E_1^{(2)}$ (for $\pm$ signs). For Hermitian operator H', both of these values are real. Hence the original energy level splits into two: $\left(E_m^{(0)} + E_1^{(1)} \right)$ and $\left(E_m^{(0)} + E_1^{(2)} \right)$ as a result of perturbation. We may say that the perturbation lifts the degeneracy. However, if for some perturbation $E_1^{(1)} = E_1^{(2)}$, then we have to go for higher order perturbation calculations to remove the degeneracy.

15.4 Physical Application of Perturbation Theory (First order)

15.4.1 Normal Helium Atom (without spin consideration)

A normal He atom consists of a nucleus of charge Ze (Z = 2) and two orbital electrons. The potential energy for the system is

$$V = -\frac{Ze^2}{r_1} - \frac{Ze^2}{r_2} + \frac{e^2}{r_{12}} \tag{33}$$

The Hamiltonian of the system is

$$H = -\frac{\hbar^2}{2m_0} \left(\nabla_1^2 + \nabla_2^2 \right) - \frac{Ze^2}{r_1} - \frac{Ze^2}{r_2} + \frac{e^2}{r_{12}} \tag{34}$$

The wave Eqn. for unperturbed Hamiltonian
$$H^0 \, \psi^0 = E^0 \, \psi^0 \tag{35}$$

can be solved easily, taking as sum of two H-atoms.

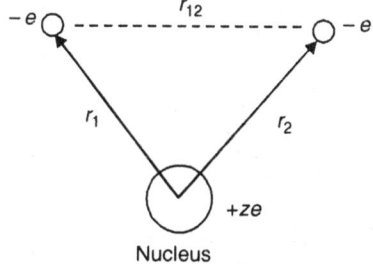

Fig. 15.1: Normal He atom

The wave-equation for two electrons in He atom is

$$H\psi = \left\{ -\frac{\hbar^2}{2m_0}\left(\nabla_1^2 + \nabla_2^2\right) + \left[-\frac{Ze^2}{r_1} - \frac{Ze^2}{r_2} + \frac{e^2}{r_{12}}\right]\right\}\psi = E\psi \qquad (36)$$

This can be rewritten as

$$H\psi = \left\{ -\frac{\hbar^2}{2m_0}\left(\nabla_1^2 + \nabla_2^2\right) - \frac{Ze^2}{r_1} - \frac{Ze^2}{r_2}\right\}\psi + \frac{e^2}{r_{12}}\psi = E\psi$$

If we write the Hamiltonian in the form

$$H = H^0 + H'$$

we note that unperturbed Hamiltonian

$$H^0 = -\frac{\hbar^2}{2m_0}\left(\nabla_1^2 + \nabla_2^2\right) - \frac{Ze^2}{r_1} - \frac{Ze^2}{r_2}$$

and perturbation $H' = \dfrac{e^2}{r_{12}}$.

The wave equation for unperturbed state would be

$$\left\{ -\frac{\hbar^2}{2m_0}\left(\nabla_1^2 + \nabla_2^2\right) - \frac{Ze^2}{r_1} - \frac{Ze^2}{r_2}\right\}\psi^0 = E^0\psi^0 \qquad (37)$$

If we substitute $\psi^0 = \psi_1^0 \psi_2^0$ and energy $E^0 = E_1^0 + E_2^0$, then Eqn. (37) may be split into two component equations one for each hydrogen-like atom:

$$\nabla_1^2 \psi_1^0 + \frac{2m_0}{\hbar^2}\left(E_1^0 + \frac{Ze^2}{r_1}\right)\psi_1^0 = 0 \qquad (38)$$

and

$$\nabla_2^2 \psi_2^0 + \frac{2m_0}{\hbar^2}\left(E_2^0 + \frac{Ze^2}{r_2}\right)\psi_2^0 = 0 \qquad (39)$$

The solutions for these two H-like wave equations are

$$E_1^0 = E_2^0 = -\frac{Z^2 m_0 e^4}{2\hbar^2 n^2} \equiv -\frac{Z^2 W_H}{n^2} \qquad (40)$$

where

$$W_H = \frac{m_0 e^4}{2\hbar^2} = \frac{e^2}{2a_0} \quad \text{(with } a_0 = \frac{\hbar}{m_0 e^2}\text{)} \qquad (41)$$

and
$$\psi_1^0 = \sqrt{\frac{Z^3}{\pi a_0^3}} e^{-Zr_1/a_0}$$

$$\psi_2^0 = \sqrt{\frac{Z^3}{\pi a_0^3}} e^{-Zr_2/a_0}$$

Total unperturbed energy is
$$E^0 = E_1^0 + E_2^0 = -2Z^2 W_H \qquad \text{(for } n = 1, \text{ the ground state)}$$

The first order perturbation energy E' is the average value of the perturbation function $H' = e^2/r_{12}$ over the unperturbed state of the system

$$E' = \int \psi_0^* H' \psi_0 \, dv$$

where
$$\psi_0 = \psi_1^0 \psi_2^0 = \left(\frac{Z^3}{\pi a_0^3}\right) e^{-\rho_1/2} e^{-\rho_2/2} \text{ and } \rho_1 = \frac{2Zr_1}{a_0}, \rho_2 = \frac{2Zr_2}{a_0}$$

or
$$\psi^0 = \frac{Z^3}{\pi a_0^3} \exp\left(\frac{-\rho_1 - \rho_2}{2}\right)$$

volume element for two electrons in spherical polar coordinates (r, θ, ϕ) is
$$dv = r_1^2 dr_1 \sin\theta_1 d\theta_1 d\phi_1 \, r_2^2 dr_2 \sin\theta_2 d\theta_2 d\phi_2$$

$$\therefore \qquad E' = \frac{Ze^2}{32\pi^2 a_0} \int_0^{2\pi} \int_0^{\pi} \int_0^{\infty} \int_0^{2\pi} \int_0^{\pi} \int_0^{\infty} \frac{e^{-\rho_1} - e^{-\rho_2}}{\rho_{12}} \rho_1^2 d\rho_1 \sin\theta_1 d\theta_1 d\phi_1$$
$$\times \rho_2^2 d\rho_2 \sin\theta_2 d\theta_2 d\phi_2$$

where
$$\rho_{12} = \frac{2Zr_{12}}{a_0}$$

This may be expressed as
$$E' = \frac{Ze^2}{32\pi^2 a_0} \int \int \frac{e^{-\rho_1} - e^{-\rho_2}}{\rho_{12}} dv_1 dv_2 \qquad (42)$$

where
$$dv_1 = \rho_1^2 d\rho_1 \sin\theta_1 d\theta_1 d\phi_1$$

and
$$dv_2 = \rho_2^2 d\rho_2 \sin\theta_2 d\theta_2 d\phi_2$$

The integrands (beside the factor $\frac{Ze^2}{32\pi^2 a_0}$) represent electrostatic distributions with density functions $e^{-\rho_1}$ and $e^{-\rho_2}$. The integral may be evaluated by calculating first the potential due to first of these by integrating over dv_1 and then performing the 2nd integral.

The potential at a point r due to a spherical shell of radius ρ_1 and thickness $d\rho_1$, i.e. of total charge $4\pi\rho_1^2 e^{-\rho_1} d\rho_1$ falls off with distance outside the shell as if the total charge were concentrated at the centre, so that for $r > \rho_1$, the potential

$$\Phi(r > \rho_1) = \frac{q_1}{r} = \frac{4\pi\rho_1^2 e^{-\rho_1} d\rho_1}{r}$$

Within the shell the potential is constant and has value equal to that at shell's surface

$$\Phi(r \le \rho_1) = \frac{4\pi\rho_1^2 e^{-\rho_1} d\rho_1}{\rho_1} = 4\pi\rho_1 e^{-\rho_1} d\rho_1$$

The potential due to the whole charge distribution is therefore given by

$$\Phi(r) = \frac{4\pi}{r} \int_0^r e^{-\rho_1} \rho_1^2 d\rho_1 + 4\pi \int_r^\infty e^{-\rho_1} \rho_1 d\rho_1$$

$$= \frac{4\pi}{r}\left[\left(-r^2 - 2r - 2\right) e^{-r} + 2\right] + 4\pi\left[r + 1\right] e^{-r}$$

$$= \frac{4\pi}{r}\left[\left(-r^2 - 2r - 2 + r + r^2\right) e^{-r} + 2\right]$$

$$= \frac{4\pi}{r}\left[2 - e^{-r}\left(r + 2\right)\right]$$

The energy of the second charge distribution in this potential is then

$$\int \Phi(\rho_2) e^{-\rho_2} dv_2 = \int \frac{4\pi}{\rho_2}\left[2 - e^{-\rho_2}\left(\rho_2 + 2\right)\right] e^{-\rho_2} dv_2$$

$$= \int_0^\infty \frac{4\pi}{\rho_2}\left[2 - e^{-\rho_2}\left(\rho_2 + 2\right)\right] e^{-\rho_2} \cdot 4\pi\rho_2^2 \, d\rho_2$$

$$= 16\pi^2 \int_0^\infty \left[2 - e^{-\rho_2}\left(\rho_2 + 2\right)\right] e^{-\rho_2} \rho_2 d\rho_2$$

$$= 16\pi^2 \left\{ \int_0^\infty 2\rho_2 e^{-\rho_2} d\rho_2 - \int_0^\infty \rho_2^2 e^{-2\rho_2} d\rho_2 - 2 \int_0^\infty \rho_2 e^{-2\rho_2} d\rho_2 \right\}$$

$$= 16\pi^2 \left\{ \frac{2}{1} - \frac{2}{8} - \frac{2}{4} \right\} = 16\pi^2 \times \frac{5}{4} = 20\pi^2 \tag{43}$$

that is, the value of the integral in Eqn. (42) is $20\pi^2$.

Hence

$$E' = \frac{Ze^2}{32\pi^2 a_0} \times 20\pi^2 = \frac{5Ze^2}{8a_0} = \frac{5}{8} \frac{Ze^2}{\hbar^2 / (m_0 e^2)}$$

$$= \frac{5}{8} \frac{Zm_0 e^4}{\hbar^2} = \frac{5}{4} ZW_H$$

Total energy $E = E_0 + E' = -2Z^2 W_H + \frac{5}{4} ZW_H$

$$= -\left(2Z^2 - \frac{5}{4} Z\right) W_H \tag{44}$$

For He-atom, $Z = 2$,

$$\therefore \quad E = -\left(2 \times 4 - \frac{5}{4} \times 2\right) W_H = -\frac{11}{2} W_H$$

$$= -\frac{11}{2} \times \frac{m_0 e^4}{2\hbar^2} = -\frac{11}{4} \frac{e^2}{a_0} = -2.75 \frac{e^2}{a_0}$$

The energy of He atom in groundstate without perturbation is $-2Z^2 W_H = -8W_H = -4e^2/a_0$ which is less than the corrected energy by perturbation theory.

15.4.2 Zeeman Effect (without electron-spin)

The change in the energy levels of an atom when it is placed in a uniform external magnetic field, is known as the Zeeman effect.

Let us consider that the field strength $\vec{B}$ is applied on a H-atom, so that an electron (reduced mass μ) carrying a charge $-e$ is moving in the field of vector potential $\vec{A}$:

$$\vec{B} = \nabla \times \vec{A}$$

then

$$\vec{B} \times \vec{r} = \nabla \times \vec{A} \times \vec{r} = (\nabla \cdot \vec{r})\vec{A} - (\vec{A} \cdot \nabla)\vec{r} = 3\vec{A} - \vec{A} = 2\vec{A}$$

$$\therefore \quad \vec{A} = \frac{1}{2}(\vec{B} \times \vec{r}) \tag{45}$$

The classical Hamiltonian of a particle of mass μ carrying $-e$ and moving in field of vector potential $\vec{A}$ may be expressed as

$$H(\vec{p}, \vec{r}) = \frac{1}{2\mu}\left(\vec{p} - \frac{e}{c}\vec{A}\right)^2 + V(\vec{r})$$

$$= \frac{p^2}{2\mu} + V(\vec{r}) - \frac{e}{2\mu c}\left(\vec{p} \cdot \vec{A} + \vec{A} \cdot \vec{p}\right) + \frac{e^2}{2\mu}\vec{A}^2$$

$$= H^0 + H' + H''$$

where

$$H^0 = \frac{p^2}{2\mu} + V(\vec{r}) \; ; \; H' = -\frac{e}{2\mu c}\left(\vec{p} \cdot \vec{A} + \vec{A} \cdot \vec{p}\right) \text{ and } H'' = \frac{e^2}{2\mu c^2}\vec{A}^2$$

First order Zeeman effect

For weak fields the second order perturbation term H' containing A^2 may be neglected and hence

$$H = H^0 + H'$$

$$= \frac{p^2}{2\mu} + V(\vec{r}) - \frac{e}{2\mu c}\left(\vec{p} \cdot \vec{A} + \vec{A} \cdot \vec{p}\right) \tag{46}$$

writing

$$\vec{p} \to \hat{p} = \frac{\hbar}{i}\nabla$$

and since

$$\nabla \cdot (\vec{A}\psi) = (\nabla \cdot \vec{A})\psi + \vec{A} \cdot \nabla\psi$$

we get

$$\left(\vec{p} \cdot \vec{A} + \vec{A} \cdot \vec{p}\right)\psi = \left(\frac{\hbar}{i}\nabla \cdot \vec{A} + \vec{A} \cdot \frac{\hbar}{i}\nabla\right)\psi$$

$$= \frac{\hbar}{i}\nabla \cdot (A\psi) + \frac{\hbar}{i}\vec{A} \cdot \nabla\psi$$

$$= \frac{\hbar}{i}\left\{(\nabla \cdot \vec{A})\psi + \vec{A} \cdot \nabla\psi\right\} + \frac{\hbar}{i}\vec{A} \cdot \nabla\psi$$

$$= \frac{\hbar}{i}\left\{(\nabla \cdot \vec{A})\psi + 2\vec{A} \cdot \nabla\psi\right\} \tag{47}$$

Also

$$\nabla \cdot \vec{A} = \nabla \cdot \left\{\frac{1}{2}(\vec{B} \times \vec{r})\right\} = \frac{1}{2}\nabla \cdot (\vec{B} \times \vec{r})$$

Using the vector identity

$$\nabla \cdot \left(\vec{a} \times \vec{b}\right) = \vec{b} \cdot \text{curl } \vec{a} - \vec{a} \cdot \text{curl } \vec{b}$$

we get

$$\frac{1}{2} \nabla \cdot \left(\vec{B} \times \vec{r}\right) = \frac{1}{2} \left\{\vec{r} \cdot \text{curl } \vec{B} - \vec{B} \cdot \text{curl } \vec{r}\right\}$$

$$= 0 \quad (\because \nabla \times B = 0 \text{ and } \nabla \times \vec{r} = 0) \qquad (48)$$

Also

$$\vec{A} \cdot \nabla \psi = \frac{1}{2} \left(\vec{B} \times \vec{r}\right) \cdot \nabla \psi = \frac{1}{2} \vec{B} \cdot \vec{r} \times \nabla \psi \qquad (49)$$

Using Eqns (48) and (49), Eqn. (47) gives

$$\left(\vec{p} \cdot \vec{A} + \vec{A} \cdot \vec{p}\right) \psi = \frac{\hbar}{i} \vec{B} \cdot \left(\vec{r} \times \nabla\right) \psi$$

$$= \vec{B} \cdot \left(\vec{r} \times \frac{\hbar}{i} \nabla\right) \psi = \vec{B} \cdot \left(\vec{r} \times \vec{p}\right) \psi = B \cdot \left(\vec{r} \times \vec{p}\right) \psi$$

$$\therefore \qquad \vec{p} \cdot \vec{A} + \vec{A} \cdot \vec{p} = \vec{B} \cdot \vec{L}$$

It is customary to choose magnetic field along the Z-axis then $\vec{B} \cdot \vec{L} = BL_Z$. The energy eigenfunctions are usually eigenstates of L_Z with eigenvalues $m\hbar$, m being magnetic quantum number. First order energy correction

$$E' = \langle n|H'|n\rangle$$

$$= \langle n|\frac{e}{2\mu c} BL_Z|n\rangle$$

$$= \frac{e}{2\mu c} Bm\hbar\langle n|n\rangle = m\frac{e\hbar}{2\mu c} B \qquad (50)$$

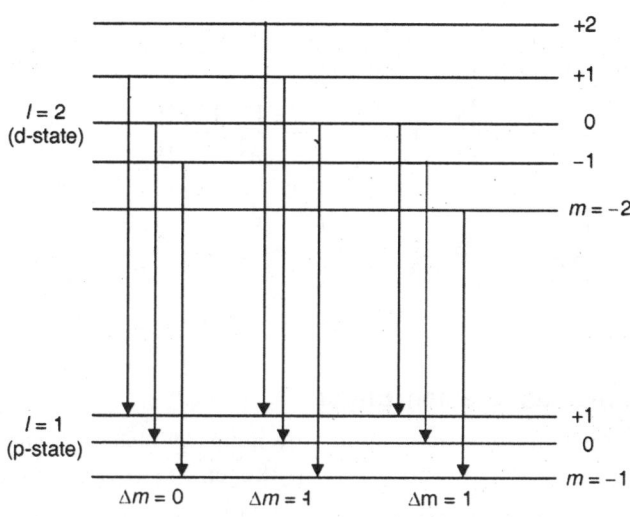

Fig. 15.2: Normal Zeeman splitting resulting due to superposition of several transitions

As we know, m can take values $-l$, ..., 0, ..., $+l$ (i.e. $2l + 1$) values. The selection rule permits only those transitions in which magnetic quantum number m changes by 0 or ± 1. This coupled with Eqn. (50) suffices to explain the normal Zeeman splitting. As an example we may consider the transitions between two states with $l = 2$ and $l = 1$ (i.e. between d and p states).

The second order Zeeman effect

If the applied magnetic field is large, then the term $e^2 A^2/(2\mu c^2)$ cannot be neglected. If again we consider the field $\vec{B}$ along z-axis, then according to expression $\vec{A} = \frac{1}{2}\left(\vec{B} \times \vec{r}\right)$, the components of $\vec{A}$ are

$$A_x = -\frac{1}{2}B \cdot \vec{y} \; ; \; A_y = -\frac{1}{2}B \cdot \vec{x} \; ; \; A_z = 0 \tag{51}$$

so that 2nd order perturbation term in Hamiltonian

$$H'' = \frac{e^2 A^2}{2\mu c^2} = \frac{e^2}{2\mu c^2}\left(A_x^2 + A_y^2 + A_z^2\right)$$

$$= \frac{e^2}{8\mu c^2}B^2\left(x^2 + y^2\right)_n \tag{52}$$

where suffix n indicates the energy in the nth state. If there is spherical symmetry, then

$$x^2 = y^2 = z^2 = \frac{r^2}{3} \tag{53}$$

and the average value of perturbation energy

$$\langle H'' \rangle = \frac{e^2 B^2}{8\mu c^2}\left\langle x^2 + y^2 \right\rangle_n = \frac{e^2 B^2}{8\mu c^2} \cdot \frac{2}{3}\langle r^2 \rangle_n \tag{54}$$

Now, it can be shown that for a state with quantum numbers n, l, m in a hydrogen like atom with nuclear charge Z:

$$\left\langle r^2 \right\rangle_{nlm} = \frac{a_0^4 n^4}{Z^2}\left\{1 + \frac{3}{2}\left[1 - \frac{l(l+1)-1/3}{n^2}\right]\right\} \tag{55}$$

Therefore

$$\langle H'' \rangle = \frac{e^2 B^2}{8\mu c^2} \cdot \frac{2}{3} \cdot \frac{a_0^4 n^4}{Z^2}\left\{1 + \frac{3}{2}\left[1 - \frac{l(l+1)-1/3}{n^2}\right]\right\} \tag{56}$$

This equation gives second order Zeeman effect under the selection rule $\Delta l = \pm 1$.

15.4.3 Anomalous Zeeman Effect

The normal Zeeman effect is exhibited by atoms in states in which the total electronic spin is zero. Now, consider a hydrogen like (i.e. one electron) atom in a uniform magnetic field $\vec{B}$. The Zeeman effect, i.e. change in energy levels in a uniform externally applied magnetic field, with the effect of electron-spin considered is called the *anomalous Zeeman effect*.

The interaction Hamiltonian is given by

$$H = H_0 + H_{LS} + H_B \tag{57}$$

where

$$H_0 = \frac{p^2}{2\mu} + V(r) \tag{58}$$

The spin orbit interaction term,

$$H_{LS} = \frac{1}{2\mu^2 c^2} \frac{1}{r} \frac{dV(r)}{dr} \vec{L} \cdot \vec{S} \tag{59}$$

$$H_B = -\frac{e}{2\mu c} \left| \vec{B} \right| L_z + \frac{e^2}{8\mu c^2} \left| \vec{B} \right|^2 \left(x^2 + y^2 \right) - \frac{e}{\mu c} \left| \vec{B} \right| S_z \tag{60}$$

In the equation (60), the quadratic term $\left| \vec{B} \right|^2 \left(x^2 + y^2 \right)$ is unimportant for a one electron atom. The third term (in Eqn. 60) is because of spin-magnetic moment interaction. We may write

$$H_B = -\frac{e \left| \vec{B} \right|}{2\mu c} (L_z + 2S_z) \tag{61}$$

The factor 2 with S_z indicates that the g-factor of the electron is 2.

Suppose H_B is treated as a small perturbation. We can study the effect of H_B using the eigenkets of $H_0 + H_{LS}$ – the $\vec{J}^2$, J_z eigenkets as base kets. We have

$$L_z + 2S_z = J_z + S_z \tag{62}$$

so, the first order shift can be written as

$$-\frac{e \left| \vec{B} \right|}{2\mu c} \left\langle J_z + S_z \right\rangle_{j=l\pm\frac{1}{2},m_j} \tag{63}$$

The expectation value of J_z gives $m\hbar$. For $<S_z>$, we know that

$$\left| j = l \pm \tfrac{1}{2}, m_j \right\rangle = \pm \sqrt{\frac{l \pm m_j + \frac{1}{2}}{(2l+1)}} \left| m_l = m_j - \tfrac{1}{2}, m_s = \tfrac{1}{2} \right\rangle$$

$$+ \sqrt{\frac{l \mp m + \frac{1}{2}}{(2l+1)}} \left| m_l = m_j + \tfrac{1}{2}, m_s = -\tfrac{1}{2} \right\rangle \tag{64}$$

which gives

$$\left\langle S_z \right\rangle_{j=l\pm\frac{1}{2},m_j} = \frac{\hbar}{2} \left(\left| a_+ \right|^2 - \left| a_- \right|^2 \right)$$

$$= \frac{\hbar}{2} \frac{1}{(2l+1)} \left[\left(l \pm m_j + \tfrac{1}{2} \right) - \left(l \mp m_j + \tfrac{1}{2} \right) \right] = \pm \frac{m_j \hbar}{(2l+1)} \tag{65}$$

Hence, energy shift due to $\vec{B}$ field is

$$\Delta E_B = -\frac{e \hbar B}{2\mu c} m_j \left[1 \pm \frac{1}{(2l+1)} \right]$$

or
$$\Delta E_B = -\frac{e\hbar B}{2\mu c}m_j\frac{2l+2}{2l+1} \text{ or } -\frac{e\hbar B}{2\mu c}m_j\frac{2l}{2l+1}$$ (66)

The splitting is depicted in Fig. 15.3.

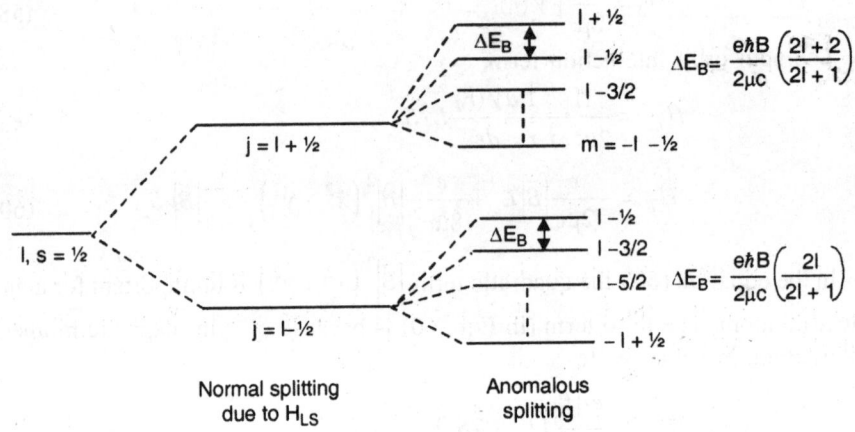

Fig. 15.3: Anomalous Zeeman effect (a general representation). The selection rule for the transitions is $\Delta m_j = \pm 1, 0$

So far we have treated the magnetic field as a small perturbation. But if the magnetic field is very intense, then effect of H_B becomes more important than that of H_{LS} and we have to consider $(H_0 + H_B)$ only (with effect of H_{LS} as only a small perturbation). In this case, the good quantum numbers are L_z and S_z. J^2 is now no good because spherical symmetry is completely destroyed, we are left with cylindrical symmetry only – i.e. invariance under rotation about the z-axis. So, L_z, S_z eigenkets $|l,s=\frac{1}{2},m_l,m_s\rangle$ are to be used as base kets.

$$\langle H_B\rangle_{m_l m_s} = -\frac{e|\vec{B}|\hbar}{2\mu c}(m_l + 2m_s)$$ (67)

The $2(2l+1)$ degeneracy in m_l and m_s existing originally with H_0 is now reduced by H_B to states with the same $(m_l) + (2m_s)$ value. Therefore, we have to evaluate the expectation value of $\vec{L}\cdot\vec{S}$ with respect to $|m_l,m_s\rangle$

$$\langle\vec{L}\cdot\vec{S}\rangle = \langle L_z S_z + \frac{1}{2}(L_+S_- + L_-S_+)\rangle_{m_l m_s}$$

$$= \hbar m_l m_s$$ (68)

because $\quad \langle L_\pm\rangle_{m_l} = 0$ and $\langle S_\pm\rangle_{m_s} = 0$

so, from Eqn. (57)

$$\langle H_{LS}\rangle_{m_l m_s} = \frac{\hbar^2 m_l m_s}{2\mu^2 c^2}\langle\frac{1}{r}\frac{dV}{dr}\rangle$$ (69)

Table 15.1: Relative importance of different angular momentum operators

External field	Dominant interaction	Almost*) good	No good	Always good
Weak $\vec{B}$	H_{LS}	$\vec{J}^2$ (or $\vec{L}\cdot\vec{S}$)	L_z, S_z	$\vec{L}^2, \vec{S}^2, J_z$
Strong $\vec{B}$	H_B	L_z, S_z	$\vec{J}^2$ (or $\vec{L}\cdot\vec{S}$)	

(*) In the table, almost good means good to the extent that the less dominant interaction can be ignored.

The expression (67) (i.e. in case of a strong magnetic field) is responsible for the *Paschen-Back effect*. This can be understood as follows:

With the help of equations (67) and (68) we can write from Eqn. (57), the expression for energy:

$$E = E_0 + \mu_B |\vec{B}| \left(m_l + 2m_s\right) + m_l m_s \zeta_{nl} \tag{70}$$

where $\quad \mu_B = -\dfrac{e\hbar}{2\mu c}$ and $\zeta_{nl} = \dfrac{\hbar^2}{2\mu^2 c^2}\left\langle \dfrac{1}{r}\dfrac{dV}{dr}\right\rangle$.

The Fig. 15.4 illustrates the transition from weak field Zeeman levels to the strong field Paschen-Back levels for a 2p state of a one electron atom.

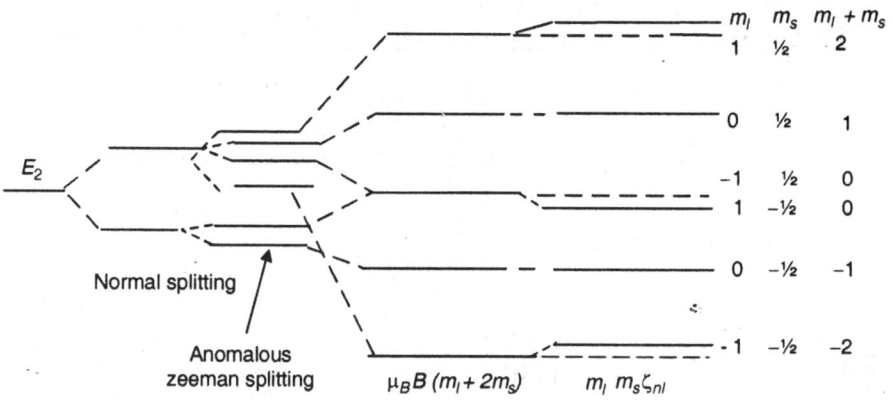

Fig. 15.4. Paschen-Back effect for 2p-state of a one electron atom

The effect of the strong magnetic field is to split a state of given quantum numbers n and l and energy E_0 into $2l + 3$ states (for $l \neq 0$) with equal separation (corresponding to $\mu_B B(m_l + 2m_s)$). The small correction due to the spin orbit interaction is specified in terms of n, l, m_l and m_s instead of n, l, j and m_j. This for the states with $j = (l + 1/2)$ corresponding to the highest ($m_l = 1$, $m_s = 1/2$) and the lowest ($m_l = -1$, $m_s = -1/2$) levels are depicted in the figure.

Example 15.4: Apply the perturbation theory to evaluate the first order energy shift in the ground state of a linear harmonic oscillator by a small perturbing potential Cx^4 in the hamiltonian.

Solution: The hamiltonian

$$H = \frac{p^2}{2m} + \frac{1}{2}kx^2 + Cx^4 \equiv H_0 + H'$$

where $H_0 = \frac{p^2}{2m} + \frac{1}{2}kx^2$ and small perturbation $H' = Cx^4$.

The unperturbed normalised wavefunctions of harmonic oscillators are given by

$$\psi_n(x) = \left\{ \frac{\alpha}{2^n n! \sqrt{\pi}} \right\}^{1/2} e^{-\alpha^2 x^2 / 2} H_n(\alpha x)$$

where $\alpha = \sqrt{\frac{m\omega}{\hbar}}$ so that

$$\psi_n(x) = \left(\frac{m\omega}{\hbar} \right)^{1/4} \frac{1}{\left(2^n n! \sqrt{\pi} \right)^{1/2}} e^{-\left(\frac{m\omega}{2\hbar} \right)x^2} H_n\left(\sqrt{\frac{m\omega}{\hbar}} x \right)$$

For ground state $n = 0$, therefore the ground state wavefunction of the harmonic oscillator becomes

$$\psi_0(x) = \left(\frac{m\omega}{\pi\hbar} \right)^{1/4} e^{-m\omega x^2 / (2\hbar)}$$

The first order energy correction, according to perturbation theory is given as

$$E' = \langle \psi_0 | H' | \psi_0 \rangle = \int_{-\infty}^{+\infty} \psi_0^* H' \psi_0 dx$$

$$= \left(\frac{m\omega}{\pi\hbar} \right)^{1/2} C \int_{-\infty}^{+\infty} e^{-m\omega x^2 / (2\hbar)} x^4 e^{-m\omega x^2 / (2\hbar)} dx$$

or

$$E' = \left(\frac{m\omega}{\pi\hbar} \right)^{1/2} C \int_{-\infty}^{+\infty} x^4 e^{-m\omega x^2 / \hbar} dx$$

Since

$$\int_{-\infty}^{+\infty} x^4 e^{-\beta x^2} dx = \frac{3}{4} \sqrt{\frac{\pi}{\beta^5}}$$

Therefore

$$E' = \left(\frac{m\omega}{\pi\hbar} \right)^{1/2} C \cdot \frac{3}{4} \sqrt{\frac{\pi}{(m\omega/\hbar)^5}}$$

or

$$E' = \frac{3}{4} C \frac{\hbar^2}{(m\omega)^2}; \ \omega = \frac{\sqrt{k}}{m^{1/2}}; \text{ or } \omega^2 = \frac{k}{m}$$

or

$$E' = \frac{3}{4} C \frac{\hbar^2}{m^2 k / m} \text{ or } E' = \frac{3}{4} C \frac{\hbar^2}{mk}$$

Example 15.5: A linear harmonic oscillator of charge e is perturbed by an electric field F in the positive x-direction. Show that the first order change in the energy level is zero. Why it is so? Also calculate the energy upto second order.
Solution: The Schrödinger equation is

$$\frac{d^2\psi}{dx^2} + \frac{2m}{\hbar^2}\left(E - \frac{1}{2}kx^2 + eFx\right)\psi = 0$$

The additional term $-eFx$ can be regarded as a perturbation.

$$\lambda H' = -eFx$$

Putting $\qquad \lambda = -eF, \; H' = x$

The unperturbed wavefunctions are given by

$$\psi_n(y) = N_n H_n(y)\exp\left(-\frac{y^2}{2}\right)$$

where $\qquad N_n = \left\{\frac{\alpha}{\sqrt{\pi}\,2^n n!}\right\}^{1/2}$; $\alpha = \sqrt{\frac{m\omega}{\hbar}}$; $\omega = \sqrt{\frac{k}{m}}$

When perturbation is switched on, the matrix elements H'_{mn} are given as

$$H'_{mn} = \int_{-\infty}^{+\infty}\left\{\frac{\alpha}{\sqrt{\pi}2^m m!}\right\}^{1/2} H_m(y)\exp\left(-y^2/2\right)x\left\{\frac{\alpha}{\sqrt{\pi}2^n n!}\right\}^{1/2}$$

$$\times H_n(y)\exp\left(-y^2/2\right)dx \qquad\qquad \left(x = \frac{y}{\alpha}\right)$$

$$= \int_{-\infty}^{+\infty}\frac{1}{\alpha}\left\{\frac{1}{\pi 2^{m+n} n! m!}\right\}^{1/2} H_m(y) y H_n(y)\exp\left(-y^2\right)dy$$

Now, we know that

$$y H_n(y) = \frac{1}{2}H_{n+1}(y) + n H_{n-1}(y)$$

$$\therefore \qquad H'_{mn} = \frac{1}{\alpha}\left\{\frac{1}{\pi 2^{m+n} n! m!}\right\}^{1/2}\int_{-\infty}^{+\infty} H_m(y)$$

$$\times \left[\frac{1}{2}H_{n+1}(y) + n H_{n-1}(y)\right]\exp\left(-y^2\right)dy$$

The r.h.s. is non zero only if $m = n + 1$ or $m = n - 1$. Also

$$\int_{-\infty}^{+\infty}\left[H_{n+1}(y)\right]^2\exp\left(-y^2\right)dy = \left(\sqrt{\pi}\right)2^{n+1}(n+1)!$$

$$\therefore \qquad H'_{n+1,n} = \frac{1}{\alpha}\left\{\frac{1}{\pi 2^{2n+1}(n+1)! n!}\right\}^{1/2}\frac{\sqrt{\pi}}{2}2^{n+1}(n+1)!$$

$$= \frac{1}{\alpha}\sqrt{\frac{n+1}{2}}$$

For $m = n - 1$, we get

$$H'_{n-1,n} = \frac{1}{\alpha}\left\{\frac{1}{\pi 2^{2n-1}(n-1)! n!}\right\}^{1/2} n\sqrt{\pi}2^{n-1}(n-1)! = \frac{1}{\alpha}\sqrt{\frac{n}{2}}$$

So,

$$
H'_{mn} = \begin{cases}
\dfrac{1}{\alpha}\sqrt{(n+1)/2} & \text{for } m = n+1 \\[2mm]
0 & \text{for } m \neq n+1 \\[2mm]
\dfrac{1}{\alpha}\sqrt{n/2} & \text{for } m = n-1
\end{cases}
$$

Since $E_1 = \langle m|H'|m \rangle$, therefore $E_1 = 0$. This comes to be zero because H' is an odd function of x and depends only on x. Therefore first order energy correction is zero.

Now,

$$
E_2 = \sum_n{}' \frac{\langle n|H'|m \rangle \, \langle m|H'|n \rangle}{E_m^{(0)} - E_n^{(0)}}
$$

$$
= \sum_n{}' \frac{\int \psi_n^{(0)*} H' \psi_m^{(0)} d\tau \, \int \psi_m^{(0)*} H' \psi_n^{(0)} d\tau}{E_m^{(0)} - E_n^{(0)}}
$$

$$
= \frac{-\dfrac{1}{\alpha^2}\left(\dfrac{n+1}{2}\right) + \dfrac{1}{\alpha^2}\left(\dfrac{n}{2}\right)}{\hbar\omega} = -\frac{1}{2\alpha^2\hbar\omega} = -\frac{1}{2m\omega^2}
$$

But

$$
E_n^{(0)} = \left(n + \frac{1}{2}\right)\hbar\omega
$$

So,

$$
E_n = E_n^{(0)} + \lambda^2 E_2
$$

$$
= \left(n + \frac{1}{2}\right)\hbar\omega - \frac{e^2 F^2}{2m\omega^2}
$$

15.4.4 The Perturbed Harmonic Oscillator

As an example, we consider the approximate energy levels of the system whose wave equation is

$$
\frac{d^2\psi}{dx^2} + \frac{2m}{\hbar^2}\left[E - \frac{1}{2}Kx^2 - ax^3 - bx^4\right]\psi = 0 \tag{71}
$$

where K is related to frequency ω by $\omega = \sqrt{K/m}$.

In Eqn. (70) if a and b are zero, this reduces to a wave equation for harmonic oscillator. In case a and b are small, we can treat these terms as perturbation

$$
H' = ax^3 + bx^4 \tag{72}
$$

The first order perturbed energy is given by

$$
E_1 = \int \psi_n^{(0)*} H' \psi_n^{(0)} d\tau
$$

$$
= \int_{-\infty}^{+\infty} \psi_n^{(0)*} ax^3 \psi_n^{(0)} dx + \int_{-\infty}^{+\infty} \psi_n^{(0)*} bx^4 \psi_n^{(0)} dx
$$

$$= a \int_{-\infty}^{+\infty} \psi_n^{(0)*} x^3 \psi_n^{(0)} dx + b \int_{-\infty}^{+\infty} \psi_n^{(0)*} x^4 \psi_n^{(0)} dx \qquad (73i)$$

Since x^3 is an odd function $\psi_n^{(0)*} \psi_n^{(0)}$ is an even function, therefore the value of the first integral is zero, i.e. first order perturbation due to ax^3 is zero. Thus

$$E_1 = b \int_{-\infty}^{+\infty} \psi_n^{(0)*} x^4 \psi_n^{(0)} dx \qquad (73ii)$$

To solve it, let

$$I = \int_{-\infty}^{+\infty} \psi_n^{(0)*} x^4 \psi_n^{(0)} dx \qquad (74)$$

We know that in case of linear harmonic oscillator:

$$\psi_n(x) = N_n H_n(y) \exp\left[-y^2/2\right] \qquad (75)$$

where $\qquad\qquad y = \alpha x \text{ and } \alpha = \sqrt{m\omega/\hbar}$.

$H_n(y)$ are Hermite polynomials.

Substituting Eqn. (75) in Eqn. (74)

$$I = \int_{-\infty}^{+\infty} N_n^2 H_n^2(y) e^{-y^2} \left(\frac{y^4}{\alpha^4}\right) \frac{dy}{\alpha}$$

$$= \frac{N_n^2}{\alpha^5} \int_{-\infty}^{+\infty} H_n^2(y) e^{-y^2} \cdot y^4 \, dy \qquad (76)$$

We also know that

$$y H_n(y) = \frac{1}{2} H_{n+1}(y) + n H_{n-1}(y) \quad \text{(see Appendices)} \qquad (77)$$

so $\qquad\qquad y^2 H_n(y) = \frac{1}{2} y H_{n+1}(y) + ny H_{n-1}(y) \qquad (78)$

From Eqn. (77) substituting $n \to (n + 1)$ and $n \to (n - 1)$,

$$y H_{n+1}(y) = \frac{1}{2} H_{n+2}(y) + (n+1) H_n(y)$$

$$y H_{n-1}(y) = \frac{1}{2} H_n(y) + (n-1) H_{n-2}(y)$$

Substituting these values in Eqn. (78), we get

$$y^2 H_n(y) = \frac{1}{2} \left\{ \frac{1}{2} H_{n+2}(y) + (n+1) H_n(y) \right\}$$

$$+ n \left\{ \frac{1}{2} H_n(y) + (n-1) H_{n-2}(y) \right\}$$

or $\qquad y^2 H_n(y) = \frac{1}{4} H_{n+2}(y) + \left(n + \frac{1}{2}\right) H_n(y) + n(n-1) H_{n-2}(y) \qquad (79)$

Substituting Eqn. (79) in Eqn. (76), we get

$$I = \frac{N_n^2}{\alpha^5} \int_{-\infty}^{+\infty} e^{-y^2} \left\{ \frac{1}{4} H_{n+2}(y) + \left(n + \frac{1}{2}\right) H_n(y) + n(n-1) H_{n-2}(y) \right\}^2 dy \qquad (80)$$

We now know that

$$\int_{-\infty}^{+\infty} e^{-y^2} H_n(y) H_m(y)\, dy = 0 \text{ if } m \neq n$$

$$= 2^n\, n!\, (\pi)^{1/2} \text{ if } m = n \tag{81}$$

Using Eqn. (81), Eqn. (80) becomes

$$I = \frac{N_n^2}{\alpha^5}\sqrt{\pi}\left\{\frac{1}{16}(n+2)!\,2^{(n+2)} + \left(n+\frac{1}{2}\right)^2 2^n n! + n^2(n-1)^2 2^{n-2}(n-2)!\right\}$$

$$= \left\{\frac{\alpha}{\pi^{1/2}\,2^n\,n!}\right\}\frac{1}{\alpha^5}\sqrt{\pi}\left\{\frac{1}{16}(n+2)!\,2^{(n+2)} + \left(n+\frac{1}{2}\right)^2 2^n \cdot n!\right.$$

$$\left. + n^2(n-1)^2 2^{(n-2)}(n-2)!\right\} \tag{82}$$

$$N_n = \left[\frac{\alpha}{\pi^{1/2}\,2^n\,n!}\right]^{1/2} \qquad\qquad \therefore I = \frac{3}{4\alpha^4}\left(2n^2 + 2n + 1\right)$$

From Eqn. (82) and (73)

$$E_1 = b\frac{3}{4\alpha^4}\left(2n^2 + 2n + 1\right)$$

so, $$E = E^{(0)} + E_1 = \left(n+\frac{1}{2}\right)\hbar\omega + \frac{3b}{4}\frac{\hbar^2}{mK}\left(2n^2 + 2n + 1\right)$$

where $\omega = \sqrt{K/m}$ and $\alpha^4 = mK/\hbar^2$

15.5 First Order Perturbation Theory for a Degenerate Level (α-fold degenerate)

Let $\psi_m^{(0)}$ represent the unperturbed eigenfunction corresponding to the energy eigenvalue E_m. Let E_m be α-fold degenerate, i.e. there are α linearly independent wavefunctions $\psi_{m1}, \psi_{m2}, ..., \psi_{m\alpha}$. For the perturbed state, the wave Eqn. is

$$H\psi - E\psi = 0 \tag{83}$$

where $$H = H^{(0)} + \lambda H^{(1)} + \lambda^2 H^{(2)} + ... \tag{84}$$

so that for the unperturbed system, we have

$$H^{(0)}\psi^{(0)} = E^{(0)}\psi^{(0)} \tag{85}$$

The solutions of the above equations are as follows:

$\psi_{01}^{(0)}, \psi_{02}^{(0)}, \psi_{03}^{(0)}, ..., \psi_{0\alpha}^{(0)}$ corresponding to energy $E_0^{(0)}$

$\psi_{11}^{(0)}, \psi_{12}^{(0)}, \psi_{13}^{(0)}, ..., \psi_{1\alpha}^{(0)}$ ... ... $E_1^{(0)}$

... ...

... ...

$\psi_{m1}^{(0)}, \psi_{m2}^{(0)}, \psi_{m3}^{(0)}, ..., \psi_{m\alpha}^{(0)}$ ... ... $E_m^{(0)}$

The problem now is to find the set of unperturbed wavefunctions to which the perturbed functions reduce when the perturbation vanishes. In other words, we have to determine coefficients involving linear transformations converting the initially chosen wavefunctions into the "correct zeroth order wavefuntions". The correct linear combination is represented as:

$$\chi_{ml}^{(0)} = k_{l1}\psi_{m1}^{(0)} + k_{l2}\psi_{m2}^{(0)} + \ldots + k_{l\alpha}\psi_{m\alpha}^{(0)}$$

$$= \sum_{l'=1}^{\alpha} k_{ll'}\psi_{ml'}^{(0)} \tag{86}$$

where $l = 1, 2, 3, \ldots, \alpha$

or

$$\chi_{m1}^{(0)} = k_{11}\psi_{m1}^{(0)} + k_{12}\psi_{m2}^{(0)} + \ldots + k_{1\alpha}\psi_{m\alpha}^{(0)}$$

$$\chi_{m2}^{(0)} = k_{21}\psi_{m1}^{(0)} + k_{22}\psi_{m2}^{(0)} + \ldots + k_{2\alpha}\psi_{m\alpha}^{(0)}$$

the function ψ_{ml} is represented as

$$\psi_{ml} = \chi_{ml}^{(0)} + \lambda\psi_{ml}^{(1)} + \lambda^2\psi_{ml}^{(2)} + \ldots \tag{87}$$

$$E_{ml} = E_m^{(0)} + \lambda E_{ml}^{(1)} + \lambda^2 E_{ml}^{(2)} + \ldots \tag{88}$$

The perturbed wave equation can be expressed as

$$H\psi_{ml} - E_{ml}\psi_{ml} = 0 \tag{89}$$

On substituting the values of H, ψ_{ml} and E_{ml} from Eqns (84) (87) and (88) in Eqn. (89), we get

$$\left(H^{(0)} + \lambda H^{(1)} + \lambda^2 H^{(2)} + \ldots\right)\left(\chi_{ml}^{(0)} + \lambda\psi_{ml}^{(1)} + \lambda^2\psi_{ml}^{(2)} + \ldots\right)$$

$$-\left(E_m^{(0)} + \lambda E_{ml}^{(1)} + \lambda^2 E_{ml}^{(2)} + \ldots\right)\left(\chi_{ml}^{(0)} + \lambda\psi_{ml}^{(1)} + \lambda^2\psi_{ml}^{(2)} + \ldots\right) = 0 \tag{90}$$

On solving this equation, we get

$$H^{(0)}\chi_{ml}^{(0)} - E_m^{(0)}\chi_{ml}^{(0)} + \lambda H^{(0)}\psi_{ml}^{(1)} + H^{(1)}\chi_{ml}^{(0)}$$

$$-E_m^{(0)}\psi_{ml}^{(1)} - E_{ml}^{(1)}\chi_{ml}^{(0)} + \ldots = 0 \tag{91}$$

Now, we can get the first order perturbation equation by equating the coefficients of λ equal to zero:

$$H^{(0)}\psi_{ml}^{(1)} + H^{(1)}\chi_{ml}^{(0)} - E_m^{(0)}\psi_{ml}^{(1)} - E_{ml}^{(1)}\chi_{ml}^{(0)} = 0 \tag{92}$$

On expanding $\psi_{ml}^{(1)}$ as follows

$$\psi_{ml}^{(1)} = \sum_{m'l'} a_{mlm'l'}\psi_{m'l'}^{(0)} \tag{93}$$

Substituting for $\psi_{ml}^{(1)}$ and $\chi_{ml}^{(0)}$ from Eqns (93) and (86) respectively into Eqn. (92), we get

$$\sum_{m',l'} a_{mlm'l'}H^{(0)}\psi_{m'l'}^{(0)} + \sum_{l'=1}^{\alpha} k_{ll'}H^{(1)}\psi_{ml'}^{(0)}$$

$$-\sum_{m',l'} a_{mlm'l'}E_m^{(0)}\psi_{m'l'}^{(0)} - \sum_{l'=1}^{\alpha} k_{ll'}E_{ml}^{(1)}\psi_{ml'}^{(0)} = 0$$

or
$$\sum a_{mlm'l'} \left(E_{m'}^{(0)} - E_m^{(0)} \right) \psi_{m'l'}^{(0)} = \sum_{l'=1}^{\alpha} k_{ll'} \left(E_{ml}^{(1)} - H^{(1)} \right) \psi_{ml'}^{(0)} \qquad (94)$$

Multiplying both sides by $\psi_{mj}^{(0)}{}^*$ and integrating over all space, we get

$$\sum_{m'l'} a_{mlm'l'} \left(E_{m'}^{(0)} - E_m^{(0)} \right) \int \psi_{mj}^{(0)}{}^* \psi_{m'l'}^{(0)} d\tau$$

$$= \sum_{l'=1}^{\alpha} k_{ll'} \left[E_{ml}^{(1)} \int \psi_{mj}^{(0)}{}^* \psi_{ml'}^{(0)} d\tau - \int \psi_{mj}^{(0)}{}^* H^{(1)} \psi_{ml'}^{(0)} d\tau \right] \qquad (95)$$

The l.h.s. of this Eqn. is equal to zero since $\psi_{mj}^{(0)}$ and $\psi_{m'l'}^{(0)}$ are orthogonal if $m \ne m'$ and $E_{m'}^{(0)} - E_m^{(0)}$ is zero in case $m = m'$ Now put

$$\int \psi_{mj}^{(0)}{}^* H^{(1)} \psi_{ml'}^{(0)} d\tau = H_{jl'}^{(1)} \qquad (96)$$

and
$$\int \psi_{mj}^{(0)}{}^* \psi_{ml'}^{(0)} d\tau = \Delta_{jl'} \qquad (97)$$

So, Eqn. (95) becomes

$$= \sum_{l'=1}^{\alpha} k_{ll'} \left[H_{jl'}^{(1)} - \Delta_{jl'} E_{ml}^{(1)} \right] = 0 \; j = 1, 2, 3, ..., \alpha \qquad (98)$$

Eqn. (98) represents a system of α homogenous linear simultaneous equations in the α-unknown quantities

$$k_{l1}, k_{l2}, ..., k_{l\alpha}$$

These equations can be written as

$$\left(H_{11}^{(1)} - \Delta_{11} E_{ml}^{(1)} \right) k_{l1} + \left(H_{12}^{(1)} - \Delta_{12} E_{ml}^{(1)} \right) k_{l2} + ... + \left(H_{1\alpha}^{(1)} - \Delta_{1\alpha} E_{ml}^{(1)} \right) k_{l\alpha} = 0$$

$$\left(H_{21}^{(1)} - \Delta_{21} E_{ml}^{(1)} \right) k_{l1} + \left(H_{22}^{(1)} - \Delta_{22} E_{ml}^{(1)} \right) k_{l2} + ... + \left(H_{2\alpha}^{(1)} - \Delta_{2\alpha} E_{ml}^{(1)} \right) k_{l\alpha} = 0$$

$$\cdots \quad \cdots \quad \cdots \quad \cdots \quad \cdots \quad \cdots \quad \cdots$$

$$\left(H_{\alpha 1}^{(1)} - \Delta_{\alpha 1} E_{ml}^{(1)} \right) k_{l1} + \left(H_{\alpha 2}^{(1)} - \Delta_{\alpha 2} E_{ml}^{(1)} \right) k_{l2} + ... + \left(H_{\alpha\alpha}^{(1)} - \Delta_{\alpha\alpha} E_{ml}^{(1)} \right) k_{\alpha\alpha} = 0 \quad (99)$$

The necessary condition which has to be satisfied by such a set of equations to have non zero solution is that the determinant of the coefficients should vanish i.e.

$$\begin{vmatrix} \left(H_{11}^{(1)} - \Delta_{11} E_{ml}^{(1)} \right) & \left(H_{12}^{(1)} - \Delta_{12} E_{ml}^{(1)} \right) & \cdots & \left(H_{1\alpha}^{(1)} - \Delta_{1\alpha} E_{ml}^{(1)} \right) \\ \left(H_{21}^{(1)} - \Delta_{21} E_{ml}^{(1)} \right) & \left(H_{22}^{(1)} - \Delta_{22} E_{ml}^{(1)} \right) & \cdots & \left(H_{2\alpha}^{(1)} - \Delta_{2\alpha} E_{ml}^{(1)} \right) \\ \cdots & \cdots & \cdots & \cdots \\ \cdots & \cdots & \cdots & \cdots \\ \left(H_{\alpha 1}^{(1)} - \Delta_{\alpha 1} E_{ml}^{(1)} \right) & \left(H_{\alpha 2}^{(1)} - \Delta_{\alpha 2} E_{ml}^{(1)} \right) & \cdots & \left(H_{\alpha\alpha}^{(1)} - \Delta_{\alpha\alpha} E_{ml}^{(1)} \right) \end{vmatrix} = 0 \qquad (100)$$

On using the condition
$$\Delta_{jl'} = 1 \; \text{if} \; j = l'$$
$$= 0 \quad \text{if} \; j \ne l'$$

we get

$$
\begin{vmatrix}
\left(H_{11}^{(1)} - E_{ml}^{(1)}\right) & H_{12}^{(1)} & \cdots & H_{1\alpha}^{(1)} \\
H_{21}^{(1)} & \left(H_{22}^{(1)} - E_{ml}^{(1)}\right) & \cdots & H_{2\alpha}^{(1)} \\
\cdots & \cdots & \cdots & \cdots \\
\cdots & \cdots & \cdots & \cdots \\
H_{\alpha 1}^{(1)} & H_{\alpha 2}^{(1)} & \cdots & \left(H_{\alpha\alpha}^{(1)} - E_{ml}^{(1)}\right)
\end{vmatrix} = 0 \qquad (101)
$$

The Eqns. (100) and (101) are known as secular equations.

It is to be noted here that in case the secular equation has the form

$$
\begin{vmatrix}
\left(H_{11}^{(1)} - E_{ml}^{(1)}\right) & 0 & \cdots & 0 \\
0 & \left(H_{22}^{(1)} - E_{ml}^{(1)}\right) & \cdots & 0 \\
\cdots & \cdots & \cdots & \cdots \\
\cdots & \cdots & \cdots & \cdots \\
0 & 0 & \cdots & \left(H_{\alpha\alpha}^{(1)} - E_{ml}^{(1)}\right)
\end{vmatrix} = 0 \qquad (102)
$$

then the functions which were assumed initially $\psi_{ml}^{(0)}, \psi_{m2}^{(0)}, ..., \psi_{ma}^{(0)}$ are the correct zero order functions for the perturbation $H^{(1)}$. The secular equation (of the form Eqn. 102) in which all the elements (except principal diagonal) are zero, is known as the diagonal form. The roots $E_{ml}^{(1)}$ can be found from an equation in this form as the algebraic equation equivalent to it can be written as

$$
\left(H_{11}^{(1)} - E_{ml}^{(1)}\right)\left(H_{22}^{(1)} - E_{ml}^{(1)}\right) \cdots\cdots \left(H_{\alpha\alpha}^{(1)} - E_{ml}^{(1)}\right) = 0
$$

with the roots

$$
E_{ml}^{(1)} = H_{11}^{(1)}, H_{22}^{(1)}, ..., H_{\alpha\alpha}^{(1)}
$$

15.6 First Order Stark Effect in Hydrogen Atom

The splitting of energy levels of an atom by a uniform external electric field is known as Stark effect. As a direct application of degenerate perturbation theory, we now consider the splitting of the first excited state of the H-atom by an electric field.

The unperturbed Hamiltonian for the H-atom is

$$
H^{(0)} = \frac{\hbar^2}{2\mu}\nabla^2 - \frac{e^2}{r} \qquad (103)
$$

The perturbation is

$$
\lambda H^{(1)} = -Q\vec{F}\cdot\vec{r} = eFr \cos\theta \quad (Q = -e) \qquad (104)
$$

e is the electronic charge, F is the strength of the applied electrostatic field and $z = r \cos\theta$ in polar coordinates.

The ground state of H-atom ($n = 1$, $l = 0$, $m = 0$)

$$
\psi_{100} = R_{10}(r)Y_{00}(\theta,\phi) \qquad (105)
$$

is spherically symmetric and is non-degenerate. Also, for the ground state (we find $\lambda H_{100,100}^{(1)} = 0$) i.e. there is no first order Stark effect.

The first excited level ($n = 2$ state) is 4-fold degenerate. The eigenfunctions are

$$\psi_{200} = R_{20}Y_{00} = \frac{1}{\sqrt{4\pi}} R_{20}(r) \qquad\qquad (n = 2, l = 0, m = 0)$$

$$\psi_{210} = R_{21}Y_{10} = \sqrt{\frac{3}{4\pi}} R_{21}(r) \cos\theta \qquad (n = 2, l = 1, m = 0)$$

$$\psi_{211} = R_{21}Y_{11} = -\sqrt{\frac{3}{8\pi}} R_{21}(r) \sin\theta \cdot e^{i\phi} \quad (n = 2, l = 1, m = 1)$$

$$\psi_{2,1,-1} = R_{21}Y_{1,-1} = \sqrt{\frac{3}{8\pi}} R_{21}(r) \sin\theta \cdot e^{-i\phi} \quad (n = 2, l = 1, m = -1) \quad (106)$$

where

$$R_{20}(r) = \frac{1}{\sqrt{2}} \frac{1}{a_0^{3/2}} \left(1 - \frac{r}{2a_0} \right) e^{-r/2a_0} \tag{107}$$

and

$$R_{21}(r) = \frac{1}{2\sqrt{6}} \frac{1}{a_0^{3/2}} \frac{r}{a_0} e^{-r/2a_0} \tag{108}$$

The corresponding (unperturbed) energy eigenvalue is $E^{(0)} = -E_H/4$ (where $-E_H \cong -13.6$ eV). Let us denote $\psi_{200} \equiv u_1$, $\psi_{210} \equiv u_2$, $\psi_{211} \equiv u_3$ and $\psi_{2,1,-1} \equiv u_4$. We rewrite (from Eqn. 92)

$$H^{(0)}\psi^{(1)} + H^{(1)}\chi^{(0)} = E^{(0)}\psi^{(1)} + E^{(1)}\chi^{(0)} \tag{109}$$

where

$$\chi^{(0)} = c_1 \cdot u_1 + c_2 \cdot u_2 + c_3 \cdot u_3 + c_4 \cdot u_4 \tag{110}$$

and

$$\psi^{(1)} = \sum_m a_m^{(1)} u_m \tag{111}$$

so, we get from Eqn. (109)

$$H^{(0)} \sum_m a_m^{(1)} u_m + H^{(1)} \left[c_1 \cdot u_1 + c_2 \cdot u_2 + c_3 \cdot u_3 + c_4 \cdot u_4 \right]$$

$$= E^{(0)} \sum_m a_m^{(1)} u_m + E^{(1)} \left[c_1 \cdot u_1 + c_2 \cdot u_2 + c_3 \cdot u_3 + c_4 \cdot u_4 \right] \tag{112}$$

Thus

$$\sum_m a_m^{(1)} E_m u_m + H^{(1)} \left[c_1 \cdot u_1 + c_2 \cdot u_2 + c_3 \cdot u_3 + c_4 \cdot u_4 \right]$$

$$= E^{(0)} \sum_m a_m^{(1)} u_m + E^{(1)} \left[c_1 \cdot u_1 + c_2 \cdot u_2 + c_3 \cdot u_3 + c_4 \cdot u_4 \right]$$

Multiplying both sides by u_1^* and integrating, we obtain (using $E_m = E^{(0)}$):

$$a_1^{(1)} E^{(0)} + \left[c_1 H_{11}^{(1)} + c_2 H_{12}^{(1)} + c_3 H_{13}^{(1)} + c_4 H_{14}^{(1)} \right]$$

$$= E^{(0)} a_1^{(1)} + E^{(1)} c_1 \tag{113}$$

where
$$H_{ij}^{(1)} = \int u_i^* H^{(1)} u_j d\tau$$

Rearranging, we obtain from Eqn. (112)
$$c_1\left(H_{11}^{(1)} - E^{(1)}\right) + c_2 H_{12}^{(1)} + c_3 H_{13}^{(1)} + c_4 H_{14}^{(1)} = 0 \qquad (114)$$

Similarly, if we multiply Eqn. (112) by u_2^*, u_3^*, u_4^* respectively and integrate, we obtain following respective equations

$$c_1 H_{21}^{(1)} + c_2\left(H_{22}^{(1)} - E^{(1)}\right) + c_3 H_{23}^{(1)} + c_4 H_{24}^{(1)} = 0 \qquad (115)$$

$$c_1 H_{31}^{(1)} + c_2 H_{32}^{(1)} + c_3\left(H_{33}^{(1)} - E^{(1)}\right) + c_4 H_{34}^{(1)} = 0 \qquad (116)$$

$$c_1 H_{41}^{(1)} + c_2 H_{42}^{(1)} + c_3 H_{43}^{(1)} + c_4\left(H_{44}^{(1)} - E^{(1)}\right) = 0 \qquad (117)$$

For a nontrivial solution, the determinant of the coefficients should vanish:

$$\begin{vmatrix} H_{11}^{(1)} - E^{(1)} & H_{12}^{(1)} & H_{13}^{(1)} & H_{14}^{(1)} \\ H_{21}^{(1)} & H_{22}^{(1)} - E^{(1)} & H_{23}^{(1)} & H_{24}^{(1)} \\ H_{31}^{(1)} & H_{32}^{(1)} & H_{33}^{(1)} - E^{(1)} & H_{34}^{(1)} \\ H_{41}^{(1)} & H_{42}^{(1)} & H_{43}^{(1)} & H_{44}^{(1)} - E^{(1)} \end{vmatrix} = 0 \qquad (118)$$

This is the secular equation for 4-fold degeneracy. In present case
$$H^{(1)} = -Q\vec{F}\cdot\vec{r} = +eFr\,\cos\theta \qquad (119)$$

Now, $H_{23}^{(1)} = \int u_2^* H^{(1)} u_3 d\tau = 0$ because $\int_0^{2\pi} e^{i\phi} d\phi = 0$. Due to the same reason, the matrix elements $H_{32}^{(1)}$, $H_{13}^{(1)}$, $H_{31}^{(1)}$, $H_{41}^{(1)}$ and $H_{23}^{(1)}$ vanish. By carrying out the integration over θ, it can be shown that $H_{11}^{(1)}$, $H_{22}^{(1)}$, $H_{33}^{(1)}$ and $H_{44}^{(1)}$ would also vanish. The only non-vanishing matrix elements are $H_{12}^{(1)}$ and $H_{21}^{(1)}$.

Now,
$$u_1 = \left(\frac{1}{\sqrt{4\pi}}\right)\left(\frac{1}{2a_0}\right)^{3/2}\left(2 - \frac{r}{a_0}\right)e^{-r/2a_0}$$

and
$$u_2 = \sqrt{\frac{3}{4\pi}}\left(\frac{1}{2a_0}\right)^{3/2}\frac{r}{a_0\sqrt{3}}e^{-r/2a_0}\cos\theta$$

$\therefore$
$$\left(u_1, H^{(1)} u_2\right) = \int u_1^* H^{(1)} u_2\, d\tau$$

$$= +eF\left(\frac{1}{2a_0}\right)^3 \frac{1}{\sqrt{4\pi}}\sqrt{\frac{3}{4\pi}}\frac{1}{a_0\sqrt{3}}\int\left(2 - \frac{r}{a_0}\right)r^2\cos^2\theta\ e^{-r/a_0}d\tau$$

$$(d\tau = r^2 \sin\theta\, dr\, d\theta\, d\phi)$$

$$= \frac{+eF}{32\pi}\times\frac{1}{a_0^4}2\pi\int_{r=0}^{\infty}\int_{\mu=-1}^{+1}r^4\left(2 - \frac{r}{a_0}\right)e^{-r/a_0}dr\,\mu^2\,d\mu \qquad (\mu = \cos\theta)$$

Now,
$$\int_{-1}^{+1} \mu^2 d\mu = \left[\frac{\mu^3}{3}\right]_{-1}^{+1} = \frac{2}{3}$$

$$\therefore \quad \left(u_1, H^{(1)} u_2\right) = +\frac{eF}{32} \times \frac{1}{a_0^4} \times \frac{2}{3} \times 2 \int_{r=0}^{\infty} r^4 \left(2 - \frac{r}{a_0}\right) e^{-r/a_0} dr$$

$$= +\frac{eF}{24 a_0^4} \left\{ 2 \cdot \frac{4 \times 3 \times 2}{\left(\dfrac{1}{a_0}\right)^5} - \frac{1}{a_0} \frac{5 \times 4 \times 3 \times 2}{\left(\dfrac{1}{a_0}\right)^6} \right\}$$

$$= -3eFa_0 \equiv -g \quad \text{(using } \int_0^{\infty} r^n e^{-\alpha r} dr = \frac{n!}{\alpha^{n+1}} \text{)} \qquad (120)$$

So, the secular equation becomes

$$\begin{vmatrix} -E^{(1)} & -g & 0 & 0 \\ -g & -E^{(1)} & 0 & 0 \\ 0 & 0 & -E^{(1)} & 0 \\ 0 & 0 & 0 & -E^{(1)} \end{vmatrix} = 0 \qquad (121)$$

The roots are $E^{(1)} = +g, -g, 0, 0$.

Now substituting the values of the matrix elements, Eqn. (114) to (117) become

$$-c_1 E^{(1)} - c_2 g = 0 \; ; \; -c_1 g - c_2 E^{(1)} = 0 \; ; \; c_3 E^{(1)} = 0 \; ; \; c_4 E^{(1)} = 0$$

Clearly, for $E^{(1)} = g$, $c_1 = -c_2$, $c_3 = c_4 = 0$ and the required linear combination
is
$$\phi_1 = \frac{1}{\sqrt{2}} \left(u_1 - u_2\right) \qquad (122)$$

Similarly, for $E^{(1)} = -g$, $c_1 = c_2$, $c_3 = c_4 = 0$ and the required linear combination is

$$\phi_2 = \frac{1}{\sqrt{2}} \left(u_1 + u_2\right) \qquad (123)$$

For the roots $E^{(1)} = g$, $c_1 = c_2 = 0$, c_3 and c_4 are indeterminate. Thus,

$$\phi_3 \text{ and } \phi_4 \; \left(= c_3 u_3 + c_4 u_4\right) \qquad (124)$$

Thus a H-atom in its first excited state behaves as though it has a permanent electric dipole moment $3ea_0$ that can be oriented in three different ways as in Fig. 15.4. The splitting of the level is shown in Fig. 15.5. As can be seen, the levels characterised by u_3 and u_4 remain degenerate and so, degeneracy is partially lifted.

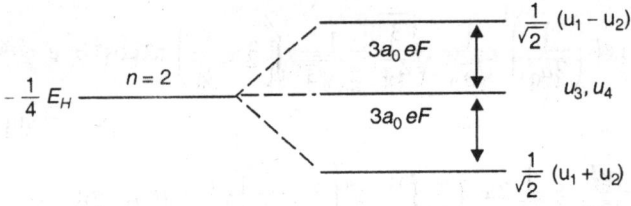

Fig. 15.5: Stark effect (splitting of the $n = 2$ level) in H-atom.
F is the field strength of the applied electric field

Questions and Problems

1. Give the first order perturbation theory for a non-degenerate case. Show that the first order perturbation energy for a non-degenerate state of a system is just the perturbation function averaged over the corresponding unperturbed state of the system. Derive an expression for the eigenfunction of the perturbed system.

2. Evaluate the first order and second order corrections to the energy of the $n = 1$ state of an oscillator subjected to a potential $V(x) = \frac{1}{2}m\omega^2 x^2 + bx$, where b is independent of x and $bx << \frac{1}{2}m\omega^2 x^2$. [**Ans.** 0, $-\dfrac{b^2}{2m\omega^2}$]

3. (*a*) Develop the stationary perturbation theory for a non-degenerate case up to second order. Show that the second order correction to the energy of the normal state is always negative.

   (*b*) Apply the first order perturbation result to calculate the ground state energy of He-atom.

4. Give the theory of first order Stark effect on the basis of quantum mechanics and discuss the splitting of energy levels.

5. Give the perturbation theory for a degenerate case and discuss the normal Zeeman effect.

6. Discuss the effect of a weak magnetic field on the energy states of an atom using perturbation theory. How is this effect modified if the magnetic field is strong?

7. Apply the perturbation theory to evaluate the first order energy shift in the ground state of a linear harmonic oscillator by a small perturbing potential Cx^2 in the Hamiltonian. [**Ans.** $\dfrac{C\hbar}{m\omega}\left(n+\dfrac{1}{2}\right)$]

8. Why the hydrogen atom in the ground state does not show a first order Stark effect?

9. For the harmonic oscillator in the ground state, find the probability of the transition to the nth eigenstate of eigenvalue $E_n = \hbar\omega (n + \frac{1}{2})$ if a constant perturbation $V = -\alpha x$ is abruptly turned on.

$$\left[\textbf{Ans. } P = \left(\frac{\alpha^2}{m\hbar\omega^3}\right)^n \frac{1}{2^n n!} e^{-\alpha^2/2m\hbar\omega^3} \right]$$

10. According to the special theory of relativity, for a free particle, the energy and momentum are related by

$$E^2 = c^2 p^2 + m^2 c^4$$

where c is the speed of the light. In presence of potential $V(r)$, the above equation modifies to

$$\left[E - V(r)\right]^2 = c^2 p^2 + m^2 c^4$$

Replacing $\hat{p}$ by $-i\hbar\nabla$, we obtain

$$\left[-\hbar^2 c^2 \nabla^2 + m^2 c^4\right]\psi = \left[E - V(r)\right]^2 \psi$$

(It is Schrödinger's relativistic equation)

If we write $E = E' + mc^2$, the above equation is equivalent to

$$\left(H_0 + H'\right)\psi = E'\psi$$

where

$$H_0 = -\frac{\hbar^2}{2m}\nabla^2 + V(r)$$

and

$$H' = -\frac{\left(E' - V(r)\right)^2}{2mc^2}$$

If we neglect H' we get non relativistic Schrödinger wave equation with $E' = E_n$. Treating H' as perturbation, calculate its first order effect on the energy levels of hydrogen like atoms (replace E' by E_n in the expression for H' and $V(r) = \dfrac{Ze^2}{r}$).

[**Hint:** $\displaystyle\iiint \psi_{nl'm'} H' \psi_{nlm}\,d\tau = -\frac{1}{2mc^2}\left[\int\left\{E_n + \frac{Ze^2}{r}\right\}^2 |R_{nl}|^2 r^2 dr\right]\delta_{ll'}\delta_{mm'}$

where E_n are the energy levels of hydrogen atom (non relativistic) with

$$E_n = -\frac{Z^2 E_H}{n^2}.]$$

[**Ans.** $-\dfrac{Z^2 E_H}{n^2}\left(\dfrac{\alpha Z}{n}\right)^2\left[\dfrac{n}{l+1/2} - \dfrac{3}{4}\right]$ where $\alpha = e^2/\hbar c$]

16

The Variation Method

There are many problems of wave mechanics which cannot be conveniently treated either by direct solution or by the use of perturbation theory e.g. helium atom is such a system:

$$H_0 = \frac{p_1^2}{2m} + \frac{p_2^2}{2m} - \frac{Ze^2}{r_1} - \frac{Ze^2}{r_2}, \ H' = \frac{e^2}{r_{12}}$$

We can calculate the first order energy correction E_1 by perturbation theory but energy obtained in this way is not accurate enough. We can go to the second order, it will improve the accuracy but at a tremendous calculational cost. For such systems variation method can be used.

16.1 Mathematical Principle

$$H = H_0 + H'$$

Let u_n denote the exact set of orthonormal eigenfunctions of the Hamiltonian H:

$$H u_n = E_n u_n \quad \int u_n^* u_{n'} d\tau = \delta_{nn'}$$

These eigenfunctions constitute a complete set, therefore an arbitrary bounded quadratically integrable function ϕ can be expanded in terms of these eigenfunctions:

$$\phi = \sum_n a_n u_n$$

Then, by definition

$$\langle H \rangle = \frac{\int \phi^* H \phi \, d\tau}{\int \phi^* \phi \, d\tau} = \frac{\int \left(\sum_n a_n^* u_n^* H \sum_{n'} a_{n'} u_{n'} \right) d\tau}{\int \left(\sum_n a_n^* u_n^* \right) \left(\sum_{n'} a_{n'} u_{n'} \right) d\tau}$$

(The integration is extended over the entire range of all the coordinates of the system.)

$$\langle H \rangle = \frac{\displaystyle\sum_{n}\sum_{n'} a_n^* a_{n'} E_{n'} \int u_n^* u_{n'}\, d\tau}{\displaystyle\sum_{n}\sum_{n'} a_n^* a_{n'} \int u_n^* u_{n'}\, d\tau}$$

or

$$= \frac{\displaystyle\sum_{n}\sum_{n'} a_n^* a_{n'} E_{n'} \delta_{nn'}}{\displaystyle\sum_{n}\sum_{n'} a_n^* a_{n'} \delta_{nn'}} = \frac{\displaystyle\sum_{n} a_n^* a_n E_n}{\displaystyle\sum_{n} a_n^* a_n}$$

Subtracting E_0, the energy of the ground state, from both sides of the equation, we have

$$\langle H \rangle - E_0 = \frac{\displaystyle\sum_{n} |a_n|^2 (E_n - E_0)}{\displaystyle\sum_{n} |a_n|^2}$$

Now, $E_n \geq E_0$ because by definition E_0 is the ground state energy, and since $|a_n|^2 > 0$ for all n, therefore we must have

$$<H> - E_0 \geq 0$$

or

$$<H> \geq E_0$$

Now ϕ is not known. We choose a suitable function ϕ known as the trial function which depends on some free parameters α_i ($i = 1, 2, ...$). Then we calculate $<H>$ which will be a function of α_i.

$$\langle H \rangle = \frac{\int \phi^* H \phi\, d\tau}{\int \phi^* \phi\, d\tau} = f(\alpha_1, \alpha_2, ...)$$

To make $<H>$ as close as possible to E_0, we minimize $f(\alpha_i)$ according to the condition

$$\frac{\partial f}{\partial \alpha_i} = 0 \quad (i = 1, 2, 3, ...)$$

The minimum value of $<H>$ determined in this way provides an upper bound to E_0, which will be close to the actual value E_0 if the trial function ϕ has a form closely resembling that of the actual ground state function u_0, so that more wisely is ϕ chosen, the more closely will $<H>$ approach E_0.

16.2 Linear Variation Function (Rayleigh Ritz Method)

It is convenient to assume

$$\phi = a_1 u_1 + a_2 u_2 + ... a_n u_n \tag{1}$$

where $u_1, u_2, ...$ etc. are the known linearly independent functions, which are not necessarily orthonormal, and the unknown coefficients $a_1, a_2, ...$ etc. are the free parameters which are to be determined to give the lowest value of $<H>$.

$$\int u_i^* u_j\, d\tau \equiv \Delta_{ij} \tag{2}$$

$$\int u_i^* H \, u_j \, d\tau \equiv H_{ij} \tag{3}$$

$$\langle H \rangle = \frac{\int \phi^* H \phi \, d\tau}{\int \phi^* \phi \, d\tau} = \frac{\int \left(\sum_i a_i u_i^* \right) H \left(\sum_j a_j u_j \right) d\tau}{\int \left(\sum_i a_i^* u_i^* \right) \left(\sum_j a_j u_j \right) d\tau} \tag{4}$$

or

$$\langle H \rangle = \frac{\sum_i \sum_j a_i \int u_i^* H \, u_j \, d\tau \, a_j}{\sum_i \sum_j a_i \int u_i^* u_j \, d\tau \, a_j}$$

$$= \frac{\sum_i \sum_j a_i H_{ij} a_j}{\sum_i \sum_j a_i \Delta_{ij} a_j} = \frac{\sum_{i,j} a_i H_{ij} a_j}{D} \quad \text{where } D = \sum_{i,j} a_i \Delta_{ij} a_j$$

In matrix notation $D = \tilde{a} \Delta a$

$$= \begin{pmatrix} a_1 & a_2 & \cdots & a_n \end{pmatrix} \begin{pmatrix} \Delta_{11} & \Delta_{12} & \cdots & \Delta_{1n} \\ \Delta_{21} & \Delta_{22} & \cdots & \Delta_{2n} \\ \vdots & \vdots & \vdots & \vdots \\ \Delta_{n1} & \Delta_{n2} & \cdots & \Delta_{nn} \end{pmatrix} \begin{pmatrix} a_1 \\ a_2 \\ \vdots \\ a_n \end{pmatrix}$$

$$D \langle H \rangle = \sum_{i,j} a_i H_{ij} a_j \tag{5}$$

To find the values of $a_1, a_2, \ldots$ etc. which make $<H>$ minimum, we differentiate both sides of this equation with respect to each a_k and put $\dfrac{\partial \langle H \rangle}{\partial a_k} = 0$

$$\frac{\partial D}{\partial a_1} \langle H \rangle + D \frac{\partial \langle H \rangle}{\partial a_1} = \frac{\partial}{\partial a_1} \sum_{i,j} a_i H_{ij} a_j$$

$$\frac{\partial D}{\partial a_2} \langle H \rangle + D \frac{\partial \langle H \rangle}{\partial a_2} = \frac{\partial}{\partial a_2} \sum_{i,j} a_i H_{ij} a_j$$

$$\cdots \quad \cdots \quad \cdots \quad \cdots \quad \cdots$$
$$\cdots \quad \cdots \quad \cdots \quad \cdots \quad \cdots$$

$$\frac{\partial D}{\partial a_n} \langle H \rangle + D \frac{\partial \langle H \rangle}{\partial a_n} = \frac{\partial}{\partial a_n} \sum_{i,j} a_i H_{ij} a_j$$

Now,

$$\frac{\partial D}{\partial a_k} = \frac{\partial}{\partial a_k} \sum_{i,j} a_i \Delta_{ij} a_j$$

$$= \sum_{i,j} \Delta_{ij} \left(a_i \frac{\partial a_j}{\partial a_k} + \frac{\partial a_i}{\partial a_k} a_j \right)$$

$$= \sum_{i,j} \Delta_{ij} \left(a_i + a_j \right) = 2 \sum_{j} \Delta_{ij} a_j$$

$$\therefore \qquad 2 \sum_{j} \Delta_{ij} a_j \langle H \rangle + 0 = 2 \sum_{j} H_{ij} a_j$$

or $\qquad \sum_{j} a_j \left(H_{ij} - \Delta_{ij} \langle H \rangle \right) = 0 \qquad (6)$

or $\qquad a_1 \left(H_{i1} - \Delta_{i1} \langle H \rangle \right) + a_2 \left(H_{i2} - \Delta_{i2} \langle H \rangle \right) + \ldots = 0$

Now, since $i = 1, 2, \ldots, n$

Therefore we have

$$a_1 \left(H_{11} - \Delta_{11} \langle H \rangle \right) + a_2 \left(H_{12} - \Delta_{12} \langle H \rangle \right) + \ldots = 0$$

$$a_1 \left(H_{21} - \Delta_{21} \langle H \rangle \right) + a_2 \left(H_{22} - \Delta_{22} \langle H \rangle \right) + \ldots = 0$$

$$\cdots \quad \cdots \quad \cdots \quad \cdots \quad \cdots \quad \cdots \quad \cdots$$

$$a_1 \left(H_{n1} - \Delta_{n1} \langle H \rangle \right) + a_2 \left(H_{n2} - \Delta_{n2} \langle H \rangle \right) + \ldots + a_n \left(H_{nn} - \Delta_{nn} \langle H \rangle \right) = 0 \qquad (7)$$

This set of n simultaneous equations in n variables $a_1, a_2, \ldots, a_n$ will have a non-trivial solution only when the determinant of the coefficients vanishes.

$$\begin{vmatrix} H_{11} - \Delta_{11} \langle H \rangle & H_{12} - \Delta_{12} \langle H \rangle & \cdots & H_{1n} - \Delta_{1n} \langle H \rangle \\ H_{21} - \Delta_{21} \langle H \rangle & H_{22} - \Delta_{22} \langle H \rangle & \cdots & H_{2n} - \Delta_{2n} \langle H \rangle \\ \cdots & \cdots & \cdots & \cdots \\ H_{n1} - \Delta_{n1} \langle H \rangle & H_{n2} - \Delta_{n2} \langle H \rangle & \cdots & H_{nn} - \Delta_{nn} \langle H \rangle \end{vmatrix} = 0 \qquad (8)$$

This equation may be solved for $\langle H \rangle$ and the lowest root $\langle H \rangle_0$ gives an upper limit to the energy E_0. On substituting for $\langle H \rangle_0$ in equation (7), we can solve them for $a_1, a_2, \ldots$ etc. which will give us the ground state (variation) function ϕ_0 corresponding to the energy E_0.

16.3 Applications of the Variation Method

16.3.1 Ground State Energy of the Hydrogen Atom

The Hamiltonian is given by

$$H = -\frac{\hbar^2}{2\mu} \left(\nabla^2 + \frac{A}{r} \right) \text{ where } A = \frac{2\mu e^2}{\hbar^2} \qquad (9)$$

For the trial function, we choose $\psi = e^{-\alpha r}$ with α as the variational parameter, to estimate the ground state energy of the atom. As the energy occurs in the radial equation, we consider only the radial part of ∇^2,

$$\nabla^2 = \frac{1}{r^2} \frac{\partial}{\partial r} \left(r^2 \frac{\partial}{\partial r} \right) \qquad (10)$$

$$\therefore \qquad \nabla^2 \psi = \left[\frac{1}{r^2} \frac{\partial}{\partial r} \left(r^2 \frac{\partial}{\partial r} \right) \right] \psi = \left[\frac{1}{r^2} \frac{\partial}{\partial r} \left(r^2 \frac{\partial}{\partial r} \right) \right] e^{-\alpha r}$$

$$= \frac{1}{r^2}\left(2r\frac{\partial}{\partial r} + r^2\frac{\partial^2}{\partial r^2}\right)e^{-\alpha r}$$

$$= \frac{1}{r^2}\left(2r(-\alpha)e^{-\alpha r} + r^2\alpha^2 e^{-\alpha r}\right)$$

$$= e^{-\alpha r}\left[\alpha^2 - \frac{2\alpha}{r}\right] \tag{11}$$

$$\langle H \rangle = \frac{\int \psi^* H\psi \, d\tau}{\int \psi^* \psi \, d\tau} \quad (d\tau = 4\pi r^2 \, dr)$$

$$= \frac{\int_0^\infty e^{-\alpha r}\left(-\frac{\hbar^2}{2\mu}\right)\left[e^{-\alpha r}\left(\alpha^2 - \frac{2\alpha}{r}\right) + \frac{Ae^{-\alpha r}}{r}\right]d\tau}{\int e^{-\alpha r}e^{-\alpha r}\,d\tau}$$

$$= \left(-\frac{\hbar^2}{2\mu}\right)\frac{\int_0^\infty \left\{e^{-2\alpha r}\left(\alpha^2 - \frac{2\alpha}{r}\right) + \frac{Ae^{-2\alpha r}}{r}\right\}(4\pi)r^2\,dr}{\int_0^\infty (4\pi)r^2 e^{-2\alpha r}\,d\tau}$$

$$= \left(-\frac{\hbar^2}{2\mu}\right)\frac{\int_0^\infty \alpha^2 r^2 e^{-2\alpha r}\,dr - \int_0^\infty 2\alpha r e^{-2\alpha r}\,dr + \int_0^\infty Ar e^{-2\alpha r}\,dr}{\int_0^\infty r^2 e^{-2\alpha r}\,d\tau}$$

Applying the relation (7), we get

$$\langle H \rangle = \left(-\frac{\hbar^2}{2\mu}\right)\frac{\alpha^2 \dfrac{2\,!}{(2\alpha)^3} - 2\alpha\dfrac{1}{(2\alpha)^2} + A\dfrac{1}{(2\alpha)^2}}{\dfrac{2}{(2\alpha)^3}}$$

$$= -\frac{\hbar^2}{2\mu}\left(\alpha^2 - 2\alpha^2 + A\alpha\right)$$

or $\qquad \langle H \rangle = +\dfrac{\hbar^2}{2\mu}\left(\alpha^2 - A\alpha\right) \tag{12}$

For <H> to be minimum $\dfrac{\partial \langle H \rangle}{\partial \alpha} = 0$

$$\frac{\hbar^2}{2\mu}\frac{\partial}{\partial \alpha}\left(\alpha^2 - A\alpha\right) = 0 \text{ or } \alpha = \frac{A}{2}$$

Substituting this value of α in Eqn. (12), we get

$$\langle H \rangle_0 = \frac{\hbar^2}{2\mu}\left(\frac{A^2}{4} - \frac{A^2}{2}\right) = -\frac{\hbar^2}{2\mu} \times \frac{A^2}{4}$$

$$= -\frac{\hbar^2}{2\mu}\left(\frac{2\mu e^2}{\hbar^2}\right)^2 \times \frac{1}{4} \text{ or } E_0 = -\left(\frac{1}{2}\right)\frac{\mu e^4}{\hbar^2} \tag{13}$$

16.3.2 Ground State of Helium Atom

We will obtain an upper limit for the ground state energy. The helium atom consists of a nucleus of charge $+2e$ and two electrons each of charge $-e$. If nucleus is at rest, the Hamiltonian is

$$H = H_0 + H'$$

where

$$H_0 = -\frac{\hbar^2}{2m}\nabla_1^2 - \frac{Ze^2}{r_1} + \left(-\frac{\hbar^2}{2m}\nabla_2^2\right) - \frac{Ze^2}{r_2},$$

$$H' = \frac{e^2}{r_{12}} \ (Z = 2) \tag{14}$$

where ∇_1 and ∇_2 are Laplacian operators for the first and second electrons at distances r_1 and r_2 from the nucleus. r_{12} is the distance between the two electrons. If the interaction energy e^2/r_{12} between two electrons were not present, the ground state eigenfunction would be the product of two normalized hydrogen like wave functions $u_{100}(\vec{r_1})\, u_{100}(r_2)$ given by

$$\psi(r_1 r_2) = \frac{Z'^3}{\pi a_0^3} e^{-\frac{Z'(r_1 + r_2)}{a_0}} \tag{15}$$

because

$$u_{100}(r_1) = \frac{2Z'^{3/2}}{a_0^{3/2}} \frac{1}{\sqrt{4\pi}} e^{-\frac{Z'r_1}{a_0}}$$

$$u_{100}(r_2) = \frac{2Z'^{3/2}}{a_0^{3/2}} \frac{1}{\sqrt{4\pi}} e^{-\frac{Z'r_2}{a_0}}$$

We use $\psi(r_1 r_2)$ as a trial function and treat Z' to be the variation parameter.

Hamiltonian $H = T + V +$ interaction energy $\left(\dfrac{e^2}{r_{12}}\right)$

$$\therefore \qquad \langle H \rangle = \langle T \rangle + \langle V \rangle + \left\langle \frac{e^2}{r_{12}} \right\rangle \tag{16}$$

For ground state of hydrogen

$$\langle T \rangle = \frac{e^2}{2a_0} \text{ and } -\frac{e^2}{a_0} = \langle V \rangle$$

Now, for helium atom

$$\langle V_1 \rangle = \left\langle \frac{-Ze^2}{r_1} \right\rangle = -Ze^2 \left\langle \frac{1}{r_1} \right\rangle = \frac{-Ze^2}{a_0 / Z'} = -\frac{ZZ'e^2}{a_0}$$

because, here, for He atom, a_0 is replaced by (a_0/Z'). Similarly

$$\langle V_2 \rangle = \left\langle \frac{-Ze^2}{r_2} \right\rangle = -\frac{ZZ'e^2}{a_0}$$

and

$$\langle V \rangle = \langle V_1 \rangle + \langle V_2 \rangle = -\frac{2ZZ'e^2}{a_0} \tag{17}$$

For electron 1: $\langle T_1 \rangle = \left\langle \frac{p_1^2}{2m} \right\rangle = \int \frac{p_1^2}{2m} u^2(r_1) d^3r_1 = \frac{Z'^2 e^2}{2a_0}$

For electron 2: $\langle T_2 \rangle = \left\langle \frac{p_2^2}{2m} \right\rangle = \frac{Z'^2 e^2}{2a_0}$ ($\because u(r_1)$ and $u(r_2)$ are normalized)

Because operation with the Laplacian gives a result that is inversely proportional to the square of the length scale a_0 of the wave function, since the scale of $\psi(r_1 r_2)$ is smaller than that of hydrogen wave function by factor of Z', so the expectation value of the kinetic energy operator is $e^2 Z'^2/(2a_0)$.

$$\langle T \rangle = \langle T_1 \rangle + \langle T_2 \rangle = \frac{2Z'^2 e^2}{2 a_0} \tag{18}$$

Now,

$$\left\langle \frac{e^2}{r_{12}} \right\rangle = \int\int \psi^*(r_1 r_2) \frac{e^2}{r_{12}} \psi(r_1 r_2) \ d^3r_1 \ d^3r_2 \tag{19}$$

$$= \frac{1}{\pi^2} \left(\frac{Z'}{a_0} \right)^6 \int\int \frac{e^2}{r_{12}} e^{-2Z'r_1/a_0} e^{-2Z'r_2/a_0} \ d^3r_1 \ d^3r_2$$

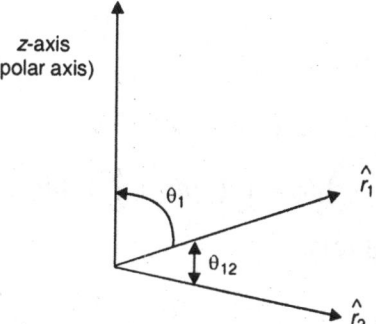

Fig. 16.1

To evaluate this integral, with fixed $\hat{r}_1, \vec{r}_1$ is taken as the polar axis for r_2 integration and then integration is done for r_1.

$$d^3 r_1 = r_1^2 dr_1 \, \sin\theta_1 \, d\theta_1 \, d\phi_1$$

$$= r_1^2 dr_1 \, d\Omega_1 \qquad\qquad (d\Omega = \sin\theta_1 \, d\theta_1 \, d\phi_1)$$

$$d^3 r_2 = r_2^2 dr_2 \, \sin\theta_{12} \, d\theta_{12} \, d\phi_{12}$$

$$= r_2^2 dr_2 \, d(\cos\theta_{12}) \, d\phi_{12}$$

$$\therefore \quad \left\langle \frac{e^2}{r_{12}} \right\rangle = \frac{e^2}{\pi^2}\left(\frac{Z'}{a_0}\right)^6 \cdot 4\pi \cdot 2\pi \iint\int r_1^2 r_2^2 \frac{e^{-2Z'r_1/a_0}\, e^{-2Z'r_2/a_0}}{r_{12}} \, dr_1 \, dr_2 d(\cos\theta_{12})$$

$$\left(\text{because } \int_0^{2\pi} d\phi_{12} = 2\pi \text{ and } \int d\Omega_1 = 4\pi\right)$$

Now $r_{12} = \left(r_1^2 + r_2^2 - 2r_1 r_2 \cos\theta_{12}\right)^{1/2}$ Put $\cos\theta_{12} = \mu$

$$\left\langle \frac{e^2}{r_{12}} \right\rangle = 8e^2 \left(\frac{Z'}{a_0}\right)^6 \int_0^\infty \int_0^\infty \int_{\mu=-1}^{+1} r_1^2 r_2^2 \frac{e^{-2Z'r_1/a_0}\, e^{-2Z'r_2/a_0}}{\sqrt{r_1^2 + r_2^2 - 2r_1 r_2 \mu}} \, d\mu \, dr_1 \, dr_2 \qquad (20)$$

Now,

$$\int_{-1}^{+1} \frac{d\mu}{\left[(r_1^2 + r_2^2) - (2r_1 r_2)\mu\right]^{1/2}} = \frac{1}{2r_1 r_2}(-Z)(A - B\mu)^{1/2}\Big|_{-1}^{+1}$$

$$(\text{where } A = r_1^2 + r_2^2 \; ; \; B = 2r_1 r_2)$$

$$= -\frac{1}{r_1 r_2}\left[\sqrt{A-B} - \sqrt{A+B}\right] = \frac{1}{r_1 r_2}\left[r_1 + r_2 - (r_1 - r_2)\right]$$

$$= \frac{1}{r_1 r_2} \times [2r_<] \qquad \begin{bmatrix} \text{when } r_1 > r_2; \ 2r_< = 2r_2 \\ \text{when } r_2 > r_1; \ 2r_< = 2r_1 \end{bmatrix}$$

$$= \left(\frac{2}{r_1}\right) \text{ for } r_1 > r_2$$

$$= \left(\frac{2}{r_2}\right) \text{ for } r_2 > r_1 \qquad (21)$$

We now perform the r_2 integration as follows

$$\int_0^\infty (\ ) dr_2 = \int_0^{r_1} (\ ) dr_2 + \int_{r_1}^\infty (\ ) dr_2$$

From Eqns (20) and (21)

$$\int_0^{r_1} r_1^2 r_2^2 e^{-2Z'r_1/a_0} \, e^{-2Z'r_2/a_0} \left(\frac{2}{r_1}\right) dr_2$$

$$+\int_{r_1}^{\infty} r_1^2 r_2^2 e^{-2Z'r_1/a_0}\, e^{-2Z'r_2/a_0} \left(\frac{2}{r_2}\right)\, dr_2$$

$$=\left(2r_1 e^{-2Z'r_1/a_0}\right)\left\{\int_0^{r_1} r_2^2\, e^{-2Z'r_2/a_0}\, dr_2\right\}$$

$$+\left(2r_1^2 e^{-2Z'r_1/a_0}\right)\int_{r_1}^{\infty} r_2\, e^{-2Z'r_2/a_0}\, dr_2 \tag{22}$$

Now

$$\left\{\int_0^{r_1} r_2^2\, e^{-2Z'r_2/a_0}\, dr_2\right\} = \left[r_2^2\, \frac{e^{-2Z'r_2/a_0}}{-2Z'/a_0}\right]_0^{r_1} + 2\int_0^{r_1}\frac{e^{-2Z'r_2/a_0}}{\left(-2Z'/a_0\right)^2}\, dr_2$$

$$= e^{-2Z'r_1/a_0}\left[\frac{r_1^2}{\left(-2Z'/a_0\right)} - \frac{2r_1}{\left(-2Z'/a_0\right)^2} + \frac{2}{\left(-2Z'/a_0\right)^3}\right]$$

$$-\frac{2}{\left(-2Z'/a_0\right)^3} \tag{23i}$$

and

$$\int_{r_1}^{\infty} r_2\, e^{-2Z'r_2/a_0}\, dr_2 = \left[r_2\, \frac{e^{-2Z'r_2/a_0}}{-2Z'/a_0}\right]_{r_1}^{\infty} - \int_{r_1}^{\infty}\frac{e^{-2Z'r_2/a_0}}{\left(-2Z'/a_0\right)}\, dr_2$$

$$= -r_1\, \frac{e^{-2Z'r_1/a_0}}{\left(-2Z'/a_0\right)} - \left[\frac{e^{-2Z'r_2/a_0}}{\left(-2Z'/a_0\right)^2}\right]_{r_1}^{\infty}$$

$$= -r_1\, \frac{e^{-2Z'r_1/a_0}}{\left(-2Z'/a_0\right)} + \frac{e^{-2Z'r_1/a_0}}{\left(-2Z'/a_0\right)^2} \tag{23ii}$$

Hence from Eqns (22), (23*i*) and (23*ii*), we have

$$\int_0^{\infty}(\ )\, dr_2 = 2r_1 e^{-4Z''r_1/a_0}\left[\frac{r_1^2}{\left(-2Z'/a_0\right)} - \frac{2r_1}{\left(-2Z'/a_0\right)^2} + \frac{2}{\left(-2Z'/a_0\right)^3}\right]$$

$$\frac{-4r_1 e^{-2Z'r_1/a_0}}{\left(-2Z'/a_0\right)^3} + 2r_1^2 e^{-4Z'r_1/a_0}\left[\frac{-r_1}{\left(-2Z'/a_0\right)} + \frac{1}{\left(-2Z'/a_0\right)^2}\right]$$

or $$\int_0^{\infty}(\ )\, dr_2 = \frac{-4r_1^2 e^{-4Z'r_1/a_0}}{\left(-2Z'/a_0\right)^2} + \frac{4r_1 e^{-4Z'r_1/a_0}}{\left(-2Z'/a_0\right)^3} - \frac{4r_1 e^{-2Z'r_1/a_0}}{\left(-2Z'/a_0\right)^3} + \frac{2r_1^2 e^{-4Z'r_1/a_0}}{\left(-2Z'/a_0\right)^2}$$

$$= -\frac{2r_1^2 e^{-4Z'r_1/a_0}}{\left(2Z'/a_0\right)^2} - \frac{4r_1 e^{-4Z'r_1/a_0}}{\left(2Z'/a_0\right)^3} + \frac{4r_1 e^{-2Z'r_1/a_0}}{\left(2Z'/a_0\right)^3} \tag{24}$$

From Eqns (20) and (24)

$$\left\langle \frac{e^2}{r_{12}} \right\rangle = 8e^2 \left(\frac{Z'}{a_0} \right)^6 \left\{ -\frac{2}{(2Z'/a_0)^2} \int_0^\infty r_1^2 e^{-4Z'r_1/a_0} dr_1 \right.$$

$$\left. -\frac{4}{(2Z'/a_0)^3} \int_0^\infty r_1 e^{-4Z'r_1/a_0} dr_1 + \frac{4}{(2Z'/a_0)^3} \int_0^\infty r_1 e^{-2Z'r_1/a_0} dr_1 \right\}$$

$$= 8e^2 \left(\frac{Z'}{a_0} \right)^6 \left\{ -\frac{2a_0^2}{(2Z')^2} \cdot \frac{2}{(4Z'/a_0)^3} - \frac{4a_0^3}{(2Z')^3} \cdot \frac{1}{(4Z'/a_0)^2} \right.$$

$$\left. + \frac{4a_0^3}{(2Z')^3} \cdot \frac{1}{(2Z'/a_0)^2} \right\}$$

$$= 8e^2 \left(\frac{Z'}{a_0} \right)^6 \frac{a_0^5}{Z'^5} \left\{ -\frac{1}{64} - \frac{1}{32} + \frac{1}{8} \right\} = \frac{8e^2 Z'}{a_0} \times \frac{5}{64}$$

Hence

$$\left\langle \frac{e^2}{r_{12}} \right\rangle = \frac{5}{8} \frac{Z'e^2}{a_0} \tag{25}$$

Therefore, from Eqns (16), (17), (18) and (25), we get

$$\langle H \rangle = \frac{Z'^2 e^2}{a_0} - \frac{2ZZ'e^2}{a_0} + \frac{5}{8} \frac{Z'e^2}{a_0} \tag{26}$$

$$\frac{\partial \langle H \rangle}{\partial Z'} = \frac{2Z'e^2}{a_0} - \frac{2Ze^2}{a_0} + \frac{5}{8} \frac{e^2}{a_0} = 0$$

$$\therefore \quad 2Z' = 2Z - \frac{5}{8}$$

or

$$Z' = \left(Z - \frac{5}{16} \right) = \left(2 - \frac{5}{16} \right) = \frac{27}{16} \quad (\because Z = 2) \tag{27}$$

$$\langle H \rangle = \frac{e^2}{a_0} \left[Z'^2 - 4Z' + \frac{5}{8} Z' \right]$$

$$= \frac{e^2}{a_0} \left[Z'^2 - \frac{27}{8} Z' \right] = \frac{e^2}{a_0} (-2.85) \tag{28}$$

By the help of the perturbation method, the ground state energy ($Z = 2$) of He-atom comes out to be (-2.75) e^2/a_0. $(-2.85 < -2.75)$. The hydrogenic wave functions give the best energy value, when $Z' = 27/16$ rather than 2. It indicates that each electron screens the nucleus from the other electron, therefore the effective nuclear charge being reduced by $2 - (27/16) = (5/16)$ i.e. effective nuclear charge is less than two.

Example 16.1: The Hamiltonian for a one dimensional oscillator is

$$H = -\frac{\hbar^2}{2m}\frac{d^2}{dx^2} + \frac{1}{2}m\omega^2 x^2$$

using the variation method, calculate the ground state energy. Take $\psi = e^{-\frac{1}{2}\alpha x^2}$ as trial function.

Solution: $\psi' = Ne^{-\frac{1}{2}\alpha x^2}$ $\int \psi'^2 dx = 1$ $\therefore N^{-2} = \int_{-\infty}^{+\infty} \psi^2 dx$

or $N^{-2} = \int_{-\infty}^{+\infty} e^{-\alpha x^2} dx = \sqrt{\frac{\pi}{\alpha}}$ $\therefore N = \left(\frac{\alpha}{\pi}\right)^{1/4}$

$$\langle H \rangle = \int_{-\infty}^{+\infty} \psi * H\psi \; dx$$

$$\frac{d}{dx}e^{-\frac{\alpha}{2}x^2} = -\frac{2\alpha}{2}xe^{-\frac{\alpha}{2}x^2}$$

$$\frac{d^2}{dx^2}\left(e^{-\frac{\alpha}{2}x^2}\right) = \frac{d}{dx}\left(-\alpha x e^{-\frac{\alpha}{2}x^2}\right) = -\alpha\left(e^{-\frac{\alpha}{2}x^2} - \frac{2}{2}\alpha x^2 e^{-\frac{\alpha}{2}x^2}\right)$$

$$\therefore \langle H \rangle = \sqrt{\frac{\alpha}{\pi}}\left[\int_{-\infty}^{+\infty} e^{-\frac{\alpha}{2}x^2}\left(-\frac{\hbar^2}{2m}\right)\left(-\alpha + \alpha^2 x^2\right)e^{-\frac{\alpha}{2}x^2}dx + \int_{-\infty}^{+\infty} e^{-\alpha x^2}\frac{1}{2}m\omega^2 x^2 dx\right]$$

$$= \sqrt{\frac{\alpha}{\pi}}\left(-\frac{\hbar^2}{2m}\right)\left\{-\alpha\int_{-\infty}^{+\infty}e^{-\alpha x^2}dx + \alpha^2\int_{-\infty}^{+\infty}x^2 e^{-\alpha x^2}dx\right\}$$

$$+\frac{1}{2}\sqrt{\frac{\alpha}{\pi}}m\omega^2\int_{-\infty}^{+\infty}x^2 e^{-\alpha x^2}dx$$

Now, we know that

$$\int_{-\infty}^{+\infty}x^{2n}e^{-\alpha^2 x^2}dx = \frac{1.3.5.\cdots(2n-1)}{2^n \cdot \alpha^{2n+1}}\sqrt{\pi}; \qquad (n = 1, 2, \ldots)$$

Therefore

$$\int_{-\infty}^{+\infty}x^2 e^{-\alpha x^2}dx = \frac{1}{2}\frac{\sqrt{\pi}}{\left(\sqrt{\alpha}\right)^3}$$

$$\therefore \langle H \rangle = \left(-\frac{\hbar^2}{2m}\right)\sqrt{\frac{\alpha}{\pi}}\left\{-\alpha\frac{\sqrt{\pi}}{\sqrt{\alpha}} + \alpha^2 \cdot \frac{1}{2}\frac{\sqrt{\pi}}{\left(\sqrt{\alpha}\right)^3}\right\} + \frac{1}{2}\sqrt{\frac{\alpha}{\pi}}m\omega^2 \times \frac{1}{2}\frac{\sqrt{\pi}}{\left(\sqrt{\alpha}\right)^3}$$

$$= \frac{1}{4}\left(\frac{\alpha\hbar^2}{m} + \frac{m\omega^2}{\alpha}\right)$$

For <H> to be minimum, $\dfrac{\partial \langle H \rangle}{\partial \alpha} = 0$

$$\Rightarrow \qquad \frac{\hbar^2}{m} - \frac{m\omega^2}{\alpha_0^2} = 0 \text{ or } \alpha_0 = \frac{m\omega}{\hbar}$$

So, ground state energy $= \langle H \rangle_{\alpha = \alpha_0} = \frac{1}{4} \left[\frac{m\omega}{\hbar} \cdot \frac{\hbar^2}{m} + m\omega^2 \cdot \frac{\hbar}{m\omega} \right] = \frac{1}{2} \hbar\omega$

The corresponding wave function is

$$\psi_0(x) = \left(\frac{m\omega}{\pi\hbar} \right)^{1/4} e^{-\frac{1}{2}m\omega x^2 / \hbar}$$

Example 16.2: Use the variation method to evaluate the ground state energy of a particle in the potential defined by

$$V = \begin{cases} \infty & \text{for } x < 0 \\ kx & \text{for } x > 0 \end{cases}$$

Use $xe^{-\alpha x}$ as the trial wave function.

Solution: Trial wave function $\psi = xe^{-\alpha x}$

For $x < 0$, $V = \infty$, so $\psi \to 0$ for $x < 0$

$$\langle E \rangle = \frac{\int_0^\infty \psi * H\psi \, dx}{\int_0^\infty \psi * \psi \, dx} \qquad (i)$$

$$H\psi = \left[-\frac{\hbar^2}{2m} \frac{\partial^2}{\partial x^2} + V(x) \right] \psi$$

$$= \left[-\frac{\hbar^2}{2m} \frac{\partial^2}{\partial x^2} + kx \right] xe^{-\alpha x}$$

$$= -\frac{\hbar^2}{2m} \frac{\partial}{\partial x} \left[e^{-\alpha x} - \alpha x e^{-\alpha x} \right] + kx^2 e^{-\alpha x}$$

$$= -\frac{\hbar^2}{2m} \left[-\alpha e^{-\alpha x} - \alpha \left(e^{-\alpha x} - \alpha x e^{-\alpha x} \right) \right] + kx^2 e^{-\alpha x}$$

$$= \frac{\hbar^2}{2m} \left[2\alpha e^{-\alpha x} - \alpha^2 x e^{-\alpha x} \right] + kx^2 e^{-\alpha x}$$

Substituting the values of $H\psi$ and ψ in Eqn. (i), we get

$$\langle E \rangle = \frac{\int_0^\infty x e^{-\alpha x} \left[\frac{\hbar^2}{2m} \left(2\alpha e^{-\alpha x} - \alpha^2 x e^{-\alpha x} \right) + kx^2 e^{-\alpha x} \right] dx}{\int_0^\infty x^2 e^{-2\alpha x} dx}$$

or $\quad \langle E \rangle = \dfrac{\int_0^\infty \dfrac{\hbar^2}{2m} 2\alpha x e^{-2\alpha x} dx - \int_0^\infty \dfrac{\hbar^2}{2m} \alpha^2 x^2 e^{-2\alpha x} dx + \int_0^\infty kx^3 e^{-2\alpha x} dx}{\int_0^\infty x^2 e^{-2\alpha x} dx} \quad (ii)$

Using $\qquad \int_0^\infty x^m e^{-\beta x} dx = \dfrac{m\ !}{\beta^{(m+1)}}$, we get

$$\langle E \rangle = \frac{\dfrac{\hbar^2}{2m} 2\alpha \cdot \dfrac{1\ !}{(2\alpha)^2} - \dfrac{\hbar^2}{2m} \alpha^2 \cdot \dfrac{2\ !}{(2\alpha)^3} + k \cdot \dfrac{3\ !}{(2\alpha)^4}}{\dfrac{2\ !}{(2\alpha)^3}}$$

$$= \frac{3k}{2\alpha} + \frac{\hbar^2 \alpha^2}{2m} \qquad\qquad (iii)$$

For energy to be minimum $\dfrac{\partial \langle E \rangle}{\partial \alpha} = 0$

$\therefore \qquad\qquad -\dfrac{3k}{2\alpha^2} + \dfrac{2\hbar^2 \alpha}{2m} = 0$ or $\alpha = \left[\dfrac{3km}{2\hbar^2} \right]^{1/3}$ $\qquad (iv)$

Putting the value of α in Eqn. (iii), we get

$$\langle E \rangle_{min} = \frac{9}{4} \left[\frac{2\hbar^2 k^2}{3m} \right]^{1/3}$$

Example 16.3: A trial function ψ differs from an eigenfunction u_E by a small amount so that $\psi = u_E + \epsilon \psi_1$ where u_E and ψ_1 are normalized and $\epsilon << 1$. Show that $<H>$ differs from E_0 only by a term of the order of ϵ^2 and find this term.

Solution: By definition

$$\langle H \rangle = \frac{\int \psi^* H \psi \, d\tau}{\int \psi^* \psi \, d\tau} \equiv \frac{(\psi, H\psi)}{(\psi, \psi)} \text{ or } (\psi, \psi) \langle H \rangle = (\psi, H\psi)$$

or $\qquad (u_E + \epsilon \psi_1, u_E + \epsilon \psi_1) \langle H \rangle = (u_E + \epsilon \psi_1, H u_E + \epsilon H \psi_1)$

or $\qquad \{ (u_E, u_E) + \epsilon (\psi_1, u_E) + \epsilon (u_E, \psi_1) + \epsilon^2 (\psi_1, \psi_1) \} \langle H \rangle$

$$= (u_E, H u_E) + \epsilon (\psi_1, H u_E) + \epsilon (u_E, H \psi_1) + \epsilon^2 (\psi_1, H \psi_1) \qquad (i)$$

Now we assume that $<H>$ can be expanded in a power series of ϵ as

$$\langle H \rangle = E_0 + \epsilon E_1 + \epsilon^2 E_2 + ... \qquad\qquad (ii)$$

Substituting (ii) in (i), we get

$$\{ (u_E, u_E) + \epsilon (\psi_1, u_E) + \epsilon (u_E, \psi_1) + \epsilon^2 (\psi_1, \psi_1) \} \{ E_0 + \epsilon E_1 + \epsilon^2 E_2 + ... \}$$

$$= (u_E, H u_E) + \epsilon (\psi_1, H u_E) + \epsilon (u_E, H \psi_1) + \epsilon^2 (\psi_1, H \psi_1) \qquad (iii)$$

$\Rightarrow \qquad\qquad (u_E, u_E) E_0 = (u_E, H u_E)$

$\therefore \qquad$ Eqn. (iii) is an identity since $H u_E = E_0 u_E$.

Equating the coefficients of ϵ,

$$(u_E, u_E)E_1 + (\psi_1, u_E)E_0 + (u_E, \psi_1)E_0 = (\psi_1, Hu_E) + (u_E, H\psi_1)$$

In this, the last two terms cancel because of the Hermiticity of H, hence $E_1 = 0$.

and

$$\langle H \rangle = E_0 + \epsilon^2 E_2$$

Equating the coefficients of ϵ^2 in Eqn. (*iii*)

$$(\psi_1, \psi_1)E_0 + (u_E, u_E)E_2 = (\psi_1, H\psi_1)$$

or

$$(\psi_1, E_0\psi_1) + E_2 = (\psi_1, H\psi_1)$$

or

$$E_2 = (\psi_1, (H - E_0)\psi_1)$$

16.4 The Hydrogen Molecular Ion H_2^+

We will now adopt Rayleigh-Ritz method to evaluate the ground state energy of hydrogen molecule ion which consists of only one electron bounded by electrostatic Coulomb-field force to two protons as depicted in Fig. 16.2.

The Hamiltonian of the system is

$$H = -\frac{\hbar^2}{2M}\nabla_1^2 - \frac{\hbar^2}{2M}\nabla_2^2 - \frac{\hbar^2}{2M}\nabla^2 - e^2\left(\frac{1}{r_1} + \frac{1}{r_2}\right) + \frac{e^2}{R} \qquad (29)$$

Since the kinetic energy of the nuclei are negligible as compared to that of electron, therefore

$$H \cong -\frac{\hbar^2}{2M}\nabla^2 - e^2\left(\frac{1}{r_1} + \frac{1}{r_2}\right) + \frac{e^2}{R} \qquad (30)$$

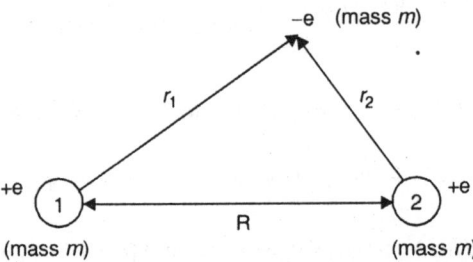

Fig. 16.2: The hydrogen molecule ion (H_2^+)

As the electron at any instant has equal probability of being associated with either of the two protons, the trial wave function may be expressed as

$$\phi = c_1\phi_1(r_1) + c_2\phi_2(r_2) \qquad (31)$$

where c_1 and c_2 are variational (real) parameters. This method of expressing one-electron molecular orbital wave function as a linear combination of the (individual) atomic wave functions is known as the *LCAO* (*linear combination of atomic orbitals*) *method*. In the above expression $\phi_1(r_1)$ is the atomic orbital when the electron is in the vicinity of proton 1 and $\phi_2(r_2)$ is that when it is near proton 2.

$$\phi_1(r_1) = \frac{1}{\left(\pi a_0^3\right)^{1/2}} e^{-r_1/a_0} \;\; ; \;\; \phi_2(r_2) = \frac{1}{\left(\pi a_0^3\right)^{1/2}} e^{-r_2/a_0} \tag{32}$$

(a_0 is the Bohr radius)

The expectation value of H is

$$E = \langle H \rangle = \frac{\langle (c_1\phi_1 + c_2\phi_2) | H | (c_1\phi_1 + c_2\phi_2) \rangle}{\langle (c_1\phi_1 + c_2\phi_2) | (c_1\phi_1 + c_2\phi_2) \rangle}$$

$$= \frac{c_1^2 \langle \phi_1 | H | \phi_1 \rangle + c_1 c_2 \langle \phi_1 | H | \phi_2 \rangle + c_2 c_1 \langle \phi_2 | H | \phi_1 \rangle + c_2^2 \langle \phi_2 | H | \phi_2 \rangle}{c_1^2 + c_2^2 + c_1 c_2 \langle \phi_1 | \phi_2 \rangle + c_2 c_1 \langle \phi_2 | \phi_1 \rangle} \tag{33}$$

Now, we use the following notations

$$\langle \phi_1 | H | \phi_1 \rangle = H_{11} ; \langle \phi_2 | H | \phi_2 \rangle = H_{22} ; H_{11} = H_{22} = \alpha \tag{34}$$

$$\langle \phi_1 | H | \phi_2 \rangle = H_{12} ; \langle \phi_2 | H | \phi_1 \rangle = H_{21} ; H_{12} = H_{21} = \beta \tag{35}$$

and $$\langle \phi_1 | \phi_2 \rangle = \langle \phi_2 | \phi_1 \rangle = \Delta \tag{36}$$

The integrals in Eqn. (34) are simply the energies of the two atomic orbitals (ϕ_1 and ϕ_2) in the molecule and are called the *Coulomb integrals*. The integral Δ is called the *overlap integral*. It measures the amount by which ϕ_1 overlaps ϕ_2 and is either zero or one as the two nuclei are infinitely apart or together.

Now, from Eqns (33), (34), (35) and (36)

$$E = \frac{c_1^2 \alpha + 2c_1 c_2 \beta + c_2^2 \alpha}{c_1^2 + 2c_1 c_2 \Delta + c_2^2} \tag{37}$$

The optimum values of c_1 and c_2 are obtained by minimizing the energy E with respect to c_1 and c_2, i.e.

$$\frac{\partial E}{\partial c_1} = 0 \; ; \; \frac{\partial E}{\partial c_2} = 0 \tag{38}$$

This gives

$$c_1(\alpha - E\Delta_{11}) + c_2(\beta - E\Delta_{12}) = 0$$
$$c_1(\alpha - E\Delta_{12}) + c_2(\beta - E\Delta_{22}) = 0 \tag{39}$$

$$(\Delta_{11} = \langle \phi_1 | \phi_1 \rangle = \Delta_{22} = 1 \text{ and } \Delta_{12} = \Delta_{21} = \Delta)$$

For a non trivial solution

$$\begin{vmatrix} (\alpha - E) & (\beta - E\Delta) \\ (\beta - E\Delta) & (\alpha - E) \end{vmatrix} = 0 \tag{40}$$

$$\therefore \quad (\alpha - E)^2 - (\beta - E\Delta)^2 = 0$$

or $$\alpha - E = \pm(\beta - E\Delta)$$

$$E = \frac{\alpha \pm \beta}{1 \pm \Delta} \tag{41}$$

We denote the two roots as

$$E_+ = \frac{\alpha + \beta}{1 + \Delta} \text{ and } E_- = \frac{\alpha - \beta}{1 - \Delta} \tag{42}$$

Substituting these values in Eqn. (39), we get

$$c_1 = c_2 \text{ for } E_+ \text{ and } c_1 = -c_2 \text{ for } E_- \tag{43}$$

The wave functions are

$$\phi_+ = c_1\left(\phi_1 + \phi_2\right) \text{ and } \phi_- = c_2\left(\phi_1 - \phi_2\right) \tag{44}$$

The normalization condition for ϕ gives

$$\langle \phi | \phi \rangle = 1$$

or

$$\left\langle \left(c_1\phi_1 + c_2\phi_2\right) | \left(c_1\phi_1 + c_2\phi_2\right)\right\rangle = 1 \tag{45}$$

or

$$c_1^2 + c_2^2 + 2c_1c_2\Delta = 1$$

For $c_1 = c_2$, we have $c_1 = c_2 = \dfrac{1}{\sqrt{2 + 2\Delta}}$

For $c_1 = -c_2$, we have $c_1 = -c_2 = \dfrac{1}{\sqrt{2 - 2\Delta}}$ $\qquad(46)$

Hence

$$\phi_+ = \frac{\phi_1 + \phi_2}{\sqrt{2 + 2\Delta}} \text{ for } E^+ \text{ and } \phi_- = \frac{\phi_1 - \phi_2}{\sqrt{2 - 2\Delta}} \text{ for } E^- \tag{47}$$

The wave functions ϕ_+ and ϕ_- signify the symmetric and antisymmetric states respectively. The form of α and β of Eqn. (42) are obtained as follows:

$$\alpha = \left\langle \phi_1 \left| -\frac{\hbar^2}{2m}\nabla^2 - \frac{e^2}{r_1} - \frac{e^2}{r_2} + \frac{e^2}{R} \right| \phi_1 \right\rangle \tag{48}$$

Introducing ground state energy of the hydrogen atom as $-E_0$, we have (for a single atom):

$$\left(-\frac{\hbar^2}{2m}\nabla^2 - \frac{e^2}{r_1} \right) \phi_1 = -E_0\phi_1 \tag{49}$$

therefore

$$\alpha = \left\langle \phi_1 \left| -E_0 - \frac{e^2}{r_2} + \frac{e^2}{R} \right| \phi_1 \right\rangle$$

or

$$\alpha = -E_0 - V_{11} + \frac{e^2}{R} \tag{50}$$

where

$$V_{11} = \left\langle \phi_1 \left| \frac{e^2}{r_2} \right| \phi_1 \right\rangle \tag{51}$$

The term $-V_{11}$ in Eqn. (50) represents the interaction energy of the electronic charge distribution about one of hydrogen nuclei with the positive charge of the other nucleus. Similarly

$$\beta = \left\langle \phi_1 \middle| -E_0 - \frac{e^2}{r_2} + \frac{e^2}{R} \middle| \phi_2 \right\rangle \tag{52}$$

$$= -E_0 \left\langle \phi_1 \middle| \phi_2 \right\rangle - \left\langle \phi_1 \middle| \frac{e^2}{r_2} \middle| \phi_2 \right\rangle + \left\langle \phi_1 \middle| \frac{e^2}{R} \middle| \phi_2 \right\rangle$$

or
$$\beta = -E_0 \Delta - V_{12} + \frac{e^2}{R} \Delta \tag{53}$$

where

$$V_{12} = \left\langle \phi_1 \middle| \frac{e^2}{r_2} \middle| \phi_2 \right\rangle \tag{54}$$

Consequently, we have

$$E_+ = -E_0 + \frac{e^2}{R} - \frac{V_{11} + V_{12}}{1 + \Delta} \tag{55}$$

and
$$E_- = -E_0 + \frac{e^2}{R} - \frac{V_{11} - V_{12}}{1 - \Delta} \tag{56}$$

The integrals V_{11} and V_{12} are both positive with $V_{11} > V_{12}$, therefore E_+ is the lower of the two energies. The lower energy orbital having the form $\phi_1 + \phi_2$ is called a *bonding orbital* and the one (having higher energy E_-) viz. $\phi_1 - \phi_2$ is called an *antibonding orbital*.

The various terms in Eqns. (55) and (56) may be interpreted as follows:

$-E_0$ is the energy of electron in ground state of hydrogen atom. V_{11} is the Coulombian attraction of the electron in a hydrogen atom for second nucleus. The integral V_{12}, which gives additional stability to H_2^+ ion has no classical interpretation. It expresses the fact that the electron might equally well be associated with either of the two nuclei. Δ represents the effect of overlapping of ϕ_1 and ϕ_2. The expressions $E_+ + E_0$ and $E_- + E_0$ represent the binding energies corresponding to bonding (symmetric) and antibonding (antisymmetric) orbitals respectively. These have been plotted in Fig. 16.3. The minimum in symmetric curve corresponds to the equilibrium distance of H^+ and H, i.e. $R = 2.5\ a_0$. The dissociation energy of H_2^+ ion is +1.77 eV.

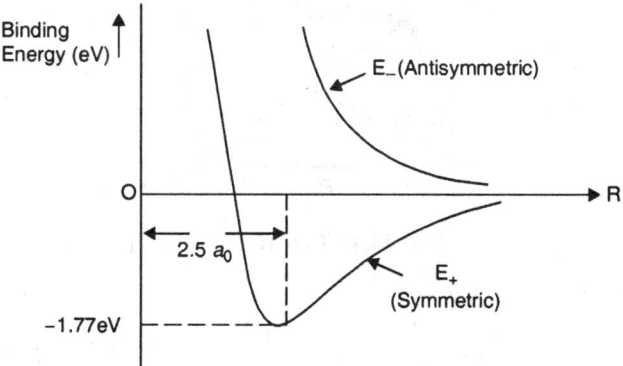

Fig. 16.3: Binding energy of H_2^+ ion as a function of internuclear separation R. E_+ corresponds to the bonding orbital

16.5 Van der Waal's Interaction between Two Hydrogen like Atoms

Let A and B represent two nuclei of two hydrogen atoms (along Z-axis) at a distance R apart. 1 and 2 are two electrons which are at distances of r_1 and r_2 from A and B respectively. r_{12} represents the distance between the two electrons, r_{1B} the distance of electron 1 from nucleus B and r_{2A} the distance of electron 2 from the nucleus A. The potential energy of the system is

$$V = -\frac{e^2}{r_1} - \frac{e^2}{r_2} + \frac{e^2}{R} + \frac{e^2}{r_{12}} - \frac{e^2}{r_{2A}} - \frac{e^2}{r_{1B}} \tag{57}$$

and the Hamiltonian is

$$H = -\frac{\hbar^2}{2m}\left(\nabla_1^2 + \nabla_2^2\right) - \frac{e^2}{r_1} - \frac{e^2}{r_2} + \frac{e^2}{R} + \frac{e^2}{r_{12}} - \frac{e^2}{r_{2A}} - \frac{e^2}{r_{1B}}$$

$$= H_0 + H' \tag{58}$$

where

$$H_0 = -\frac{\hbar^2}{2m}\left(\nabla_1^2 + \nabla_2^2\right) - \frac{e^2}{r_1} - \frac{e^2}{r_2} \tag{59}$$

the unperturbed Hamiltonian, (i.e. the sum of the Hamiltonians of two separate H-atoms) and

$$H' = \frac{e^2}{R} + \frac{e^2}{r_{12}} - \frac{e^2}{r_{2A}} - \frac{e^2}{r_{1B}} \tag{60}$$

is the perturbation term.

The unperturbed wave function for the system is the simple product of two hydrogen-like 1s wave functions.

$$\psi_0\left(r_1, r_2\right) = u_{100}\left(r_1\right) u_{100}\left(r_2\right) \tag{61}$$

It has the form like two non interacting H-atoms in ground state.

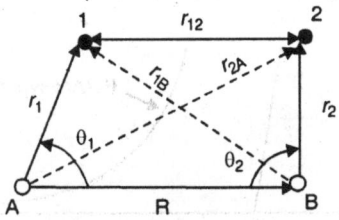

Fig. 16.4: Two H-atoms

Now, since $r \ll R$, therefore

$$\frac{1}{r_{1B}} = \frac{1}{\left|\vec{r}_1 - \vec{R}\right|} = \frac{1}{\left(r_1^2 + R^2 - 2r_1 R \cos\theta_1\right)^{1/2}}$$

$$= \frac{1}{R\left(1 + \dfrac{r_1^2}{R^2} - \dfrac{2z_1}{R}\right)^{1/2}} \qquad (\because z_1 = r_1 \cos\theta_1)$$

$$= \frac{1}{R}\left(1 - \frac{2z_1}{R} + \frac{r_1^2}{R^2}\right)^{-1/2} \tag{62}$$

Similarly,

$$\frac{1}{r_{2A}} = \frac{1}{|\vec{r}_2 - \vec{R}|} = \frac{1}{R}\left(1 + \frac{2z_2}{R} + \frac{r_2^2}{R^2}\right)^{1/2} \tag{63}$$

because $z_2 = r_2 \cos(180 - \theta_2)$

and

$$\frac{1}{r_{12}} = \frac{1}{R}\left[1 + \frac{2(z_2 - z_1)}{R} + \frac{(x_2 - x_1)^2 + (y_2 - y_1)^2 + (z_2 - z_1)^2}{R^2}\right]^{-1/2} \tag{64}$$

Substituting these values in Eqn. (60), we get

$$H' = \frac{e^2}{R} + \frac{e^2}{R}\left\{1 + \frac{2(z_2 - z_1)}{R} + \frac{(x_2 - x_1)^2 + (y_2 - y_1)^2 + (z_2 - z_1)^2}{R^2}\right\}^{-1/2}$$

$$- \frac{e^2}{R}\left\{1 + \frac{2z_2}{R} + \frac{r_2^2}{R^2}\right\}^{-1/2} - \frac{e^2}{R}\left\{1 - \frac{2z_1}{R} + \frac{r_1^2}{R^2}\right\}^{-1/2} \tag{65}$$

Now, using the binomial expansion

$$(1 + x)^{-p/q} \cong 1 - p\left(\frac{x}{q}\right) + \frac{p(q + p)}{2!}\left(\frac{x}{q}\right)^2 - \dots \tag{66}$$

for each of the three terms of this form in the expression for H', we get

$$H' = \frac{e^2}{R} + \frac{e^2}{R}\left[\left\{1 - \frac{(z_2 - z_1)}{R} - \frac{(x_2 - x_1)^2 + (y_2 - y_1)^2 + (z_2 - z_1)^2}{2R^2}\right\}\right.$$

$$\left. + \frac{1(1 + 2)}{2!}\left\{\frac{2(z_2 - z_1)}{2R}\right\}^2 + \dots\right] - \frac{e^2}{R}\left[\left\{1 - \frac{z_2}{R} - \frac{r_2^2}{2R^2}\right\} + \frac{3}{2}\left(-\frac{z_2}{R}\right)^2 + \dots\right]$$

$$- \frac{e^2}{R}\left[\left\{1 + \frac{z_1}{R} - \frac{r_1^2}{2R^2}\right\} + \frac{3}{2}\left(\frac{z_1}{R}\right)^2 + \dots\right]$$

$$= \frac{e^2}{R} + \frac{e^2}{R} - \frac{e^2(z_2 - z_1)}{R^2} - e^2\left[\frac{(x_2 - x_1)^2 + (y_2 - y_1)^2 + (z_2 - z_1)^2}{2R^3}\right]$$

$$-\frac{e^2}{R}+\frac{e^2 z_2}{R^2}+\frac{e^2 r_2^2}{2R^3}-\frac{e^2}{R}-\frac{e^2 z_1}{R^2}+\frac{e^2 r_1^2}{2R^3}$$

$$+\frac{3}{2}\frac{e^2}{R^3}\left[\,(z_2-z_1)^2-z_2^2-z_1^2\,\right] \quad \text{(neglecting other higher order terms)}$$

or $\quad H'=\dfrac{e^2}{2R^3}\left[r_2^2+r_1^2-(x_2-x_1)^2-(y_2-y_1)^2-(z_2-z_1)^2\right]$

$$+\frac{3e^2}{2R^3}\left[z_2+z_1^2-2z_1 z_2-z_2^2-z_1^2\right]$$

$$=\frac{e^2}{2R^3}\left[x_2^2+y_2^2+z_2^2+x_1^2+y_1^2+z_1^2-x_2^2-x_1^2+2x_1 x_2\right.$$

$$\left.-y_2^2-y_1^2+2y_1 y_2-z_2^2-z_1^2+2z_1 z_2\right]+\frac{e^2}{R^3}\left[-3z_1 z_2\right]$$

$$=\frac{e^2}{R^3}\left[x_1 x_2+y_1 y_2+z_1 z_2-3z_1 z_2\right]$$

or $\quad H'\cong\dfrac{e^2}{R^3}\left[x_1 x_2+y_1 y_2-2z_1 z_2\right]$ \hfill (67i)

If other higher order terms are considered, we would have obtained

$$H'\cong\frac{e^2}{R^3}\left[x_1 x_2+y_1 y_2-2z_1 z_2\right]$$

$$+\frac{3}{2}\frac{e^2}{R^4}\left\{r_1^2 z_2-r_2^2 z_1+(2x_1 x_2+2y_1 y_2-3z_1 z_2)\,(z_1-z_2)\,\right\}+... \quad (67ii)$$

The first term of this expression represents the interaction of the dipole moments of the two atoms, the second term represents the dipole quadruple interaction, the third term would represent the quadruple-quadruple interaction and so on.

Let us first consider only the dipole-dipole interaction. The expectation value of H' is

$$\left\langle \psi_0\left(r_1,r_2\right)\middle| H'\middle| \psi_0\left(r_1,r_2\right)\right\rangle$$

$$=\int \psi_0^*\left(r_1,r_2\right)\left\{\frac{e^2}{R^3}(x_1 x_2+y_1 y_2-2z_1 z_2)\right\}\psi_0\left(r_1,r_2\right)\,dz \quad (68)$$

Since $\psi_0\left(r_1,r_2\right)$ is an even function of r_1 and r_2 and H' is an odd function of r_1 and r_2, so the above integral is zero.

Now let us consider a trial function

$$\psi\left(r_1,r_2\right)=\psi_0\left(r_1,r_2\right)\left[1+AH'\right] \quad (69)$$

where A is variational parameter. According to variation principle, the expectation value of H is more than E_0+E_2 or

$$E_0 + E_2 \leq \frac{\int \psi * H \psi \, d\tau}{\int \psi * \psi \, d\tau} \tag{70}$$

Let us consider the r.h.s. of this inequality.

$$\frac{\int \psi * H \psi \, d\tau}{\int \psi * \psi \, d\tau} = \frac{\int \psi_0^*(1 + AH') (H_0 + H') \psi_0 (1 + AH') \, d\tau}{\int \psi_0^*(1 + AH') \psi_0 (1 + AH') \, d\tau}$$

$$= \frac{E_0 + 2A \int \psi_0^* H' \psi_0 d\tau + A^2 \int \psi_0^* H' H_0 H' \psi_0 d\tau}{1 + A^2 \int \int \psi_0^* H'^2 \psi_0 d\tau}$$

But $\int \psi_0^* H' H_0 H' \psi_0 d\tau = 0$, therefore

$$\frac{\int \psi * H \psi \, d\tau}{\int \psi * \psi \, d\tau} = \frac{\left[E_0 + 2A \int \psi_0 H'^2 \psi_0 d\tau\right]\left[1 - A^2 \int \psi_0^* H'^2 \psi_0 d\tau\right]}{\left[1 + A^2 \int \psi_0^* H'^2 \psi_0 d\tau\right]\left[1 - A^2 \int \psi_0^* H'^2 \psi_0 d\tau\right]}$$

$$\cong E_0 + \int \psi_0^* H'^2 \psi_0 d\tau \left(2A - E_0 A^2\right)$$

$$\cong E_0 + \int \psi_0^* H'^2 \psi_0 d\tau \times \frac{1}{E_0} \quad \left(\because A = \frac{1}{E_0}\right) \tag{71}$$

From Eqns (70) and (71)

$$E_0 + E_2 \leq E_0 + \frac{\int \psi_0^* H'^2 \psi_0 d\tau}{E_0} \tag{72}$$

But

$$H'^2 \cong \frac{e^4}{R^6}\left(x_1 x_2 + y_1 y_2 - 2z_1 z_2\right)^2$$

$$\cong \frac{e^4}{R^6}\left(x_1^2 x_2^2 + y_1^2 y_2^2 + 4z_1^2 z_2^2 + 2x_1 x_2 y_1 y_2 + ...\right) \tag{73}$$

Now consider the integral

$$\int \psi_0^* H'^2 \psi_0 d\tau = \int \psi_0^* \frac{e^4}{R^6} \left(x_1^2 x_2^2 + y_1^2 y_2^2 + 4z_1^2 z_2^2 + 2x_1 x_2 y_1 y_2 + ...\right) \psi_0 d\tau$$

$$= \frac{e^4}{R^6} \left[\int \psi_0^* x_1^2 x_2^2 \psi_0 d\tau + \int \psi_0^* y_1^2 y_2^2 \psi_0 d\tau + \int \psi_0^* 4z_1^2 z_2^2 \psi_0 d\tau\right.$$

$$\left. + \int \psi_0^* 2x_1 x_2 y_1 y_2 \, \psi_0 d\tau + ...\right]$$

$$= \frac{e^4}{R^6} \left[a_0^4 + a_0^4 + 4a_0^4 + 0 + ...\right] = \frac{6e^4 a_0^4}{R^6} \tag{74}$$

Consequently,

$$E_0 + E_2 \leq E_0 + \frac{6e^4 a_0^4}{R^6} \cdot \frac{1}{E_0} \tag{75}$$

or
$$E_2 \le \frac{6e^4 a_0^4}{R^6}\left(-\frac{a_0}{e^2}\right) \qquad \left(\because E_0 = -\frac{e^2}{a_0}\right)$$

or
$$E_2 \le -\frac{6e^2 a_0^5}{R^6} \tag{76}$$

The lower limit on E_2 can be obtained by perturbation method as follows:

$$E_2 > \frac{1}{E_1 - E_n}\sum_k{}' \left|\langle 1|H'|k\rangle\right|^2 \tag{77}$$

(The prime on the summation indicates that the term $k = 1$ has to be omitted).
Now

$$\sum_k{}' \left|\langle 1|H'|k\rangle\right|^2 = \sum_k \left|\langle 1|H'|k\rangle\right|^2 - \left|\langle 1|H'|1\rangle\right|^2$$

$$= \sum_k \left|\langle 1|H'|k\rangle\right| \left|\langle k|H'|1\rangle\right| = \langle 1|H'^2|1\rangle \quad (\because |k\rangle\langle k| = 1) \tag{78}$$

where
$$|1\rangle = \psi_0(r_1, r_2) \tag{79}$$

and $\langle 1|H'^2|1\rangle$ is given by Eqn. (74).
Also,

$$E_1 = -\frac{me^4}{2\hbar^2} - \frac{me^4}{2\hbar^2} = -\frac{e^2}{a_0} \quad \left(\because a_0 = \frac{\hbar^2}{me^2} \text{ c.g.s. units}\right) \tag{80}$$

Further, the lowest state for which $\langle 1|H'|k\rangle \ne 0$ corresponds to $n_1 = n_2 = 2$.
Therefore

$$E_n = -\frac{me^4}{8\hbar^2} - \frac{me^4}{8\hbar^2} = -\frac{e^2}{4a_0} \tag{81}$$

and
$$E_1 - E_n = -\frac{3e^2}{4a_0} \tag{82}$$

Hence

$$E_2 > -\frac{1}{\left(3e^2/4a_0\right)}\frac{6e^4 a_0^4}{R^6} = -\frac{8e^2 a_0^5}{R^6} \tag{83}$$

Consequently,

$$-\frac{8e^2 a_0^5}{R^6} < E_2 \le -\frac{6e^2 a_0^5}{R^6} \tag{84}$$

i.e., the interaction eneryg varies as $1/R^6$.

Questions and Problems

1. Describe the variational method for the eigenvalue problem and calculate the energy of the ground state of helium atom by this method. Why is the effective charge in the nucleus less than 2?

2. Use variation method to obtain the ground state energy of hydrogen atom assuming a trial function
$$\psi(r) = Ae^{-\alpha r}$$
where α is a variational parameter and A is normalization constant.

3. Estimate the ground state energy for a particle in the potential $V(r) = -V_0 e^{-\alpha r}$ using the variation method with a trial wave function proportional to (i) e^{-ar} (ii) $e^{-b^2 r^2}$, which is the better estimate.

4. Describe the variational method for the eigenvalue problem and calculate the van der Waal's interaction between the hydrogen atoms in their ground state.

5. Estimate the ground state energy of a one dimensional harmonic oscillator of mass m and angular frequency ω using a Gaussian trial function $A \exp\left(-\alpha x^2\right)$.

6. Estimate the ground state enrgy of a harmonic oscillator of mass m and angular frequency ω using the trial function
$$\psi(x) = \begin{cases} \cos\dfrac{\pi x}{2a} & -a \le x \le a \\ 0 & |x| > a \end{cases}$$

7. Give the theory of hydrogen molecule ion, H_2^+, for finding its electronic energy and wave functions.

8. Use variational method to estimate the ground state energy of a particle moving in a potential V given by
$$V = \infty \text{ for } x < 0$$
$$= Cx \text{ for } x > 0 \qquad\qquad (C \text{ is a constant})$$

Take $\psi = x\, e^{-\alpha x}$ as the trial wave function with α as the variational parameter.

[Hint: $\psi = 0$ for $x < 0$ and $x\, e^{-\alpha x}$ for $x > 0$;
$$H = -\frac{\hbar^2}{2m}\frac{d^2}{dx^2} + Cx$$
$$E = \frac{\int_0^\infty \psi^* H\psi\, dx}{\int_0^\infty \psi^*\psi\, dx} = \frac{\hbar^2 \alpha^2}{2m} + \frac{3C}{2\alpha}$$

$$\frac{\partial E}{\partial \alpha} = 0 \Rightarrow \alpha = \left(\frac{3Cm}{2\hbar^2}\right)^{1/3} \text{ and } E = \frac{9}{4}\left(\frac{2\hbar^2 C^3}{3m}\right)^{1/3}].$$

9. A trial function ψ differs from an eigenfunction u_E by a small amount so that $\psi = u_E + \epsilon\psi_1$ where u_E and ψ_1 are normalized and $\epsilon \ll 1$. Show that $<H>$ differs from E_0 only by a term of the order of ϵ^2 and find this term.

10. Find the ground state energy of a particle in a potential energy $V(x) = Cx^4$ by using the trial function

$$\psi(\alpha) = \sqrt{\frac{2\alpha}{\pi}}\, e^{-\alpha x^2} \quad [\textbf{Ans. } E(\alpha_0) = \frac{3}{8}\left(\frac{6C\hbar^4}{m^2}\right)^{1/3}]$$

17

The WKB
Approximation Method

We know that an exact solution of the Schrödinger equation is an impractical proposition except for the simplest potentials. There are basically two categories of approximation methods. The time independent problems and time dependent problems. The WKB method comes under the category of time independent methods. This approximation method, named after Wentzel, Kramers and Brillouin (who first introduced the method in quantum mechanics) is also known as *phase integral method*. It is applicable when the potential function is a slowly varying function of position. This method led to the Bohr-Sommerfeld quantization rules of old quantum theory from the Schrödinger equation. It is therefore a *semiclassical approximation*.

17.1 The Principle of the Method

It is based on transition from quantum to classical mechanics in a specific limiting case analogous to transition from wave optics to geometrical optics. This analogy was used in the first papers leading to the construction of quantum mechanics. In order to study the limiting transition from quantum to classical mechanics, we write,

$$\psi(\vec{r},t) = \exp\left\{\frac{i}{\hbar}S(\vec{r},t)\right\} \tag{1}$$

Let us substitute this into the Schrödinger equation for motion of a particle of mass μ moving in a potential field of energy $V(\vec{r})$. Then we find

$$\frac{\partial\psi}{\partial t} = \frac{i}{\hbar}\frac{\partial S}{\partial t}\psi$$

$$-\frac{\hbar^2}{2\mu}\left(\nabla^2 + V(r)\right)\psi = -\frac{\hbar^2}{2\mu}\left[\nabla\cdot\nabla\left(\exp\frac{i}{\hbar}S\right) + V\psi\right]$$

and

$$\nabla\psi = \frac{i}{\hbar}\nabla S\cdot\psi$$

$$\therefore \quad -\frac{\hbar^2}{2\mu}\left[\frac{i}{\hbar}\nabla\cdot(\nabla S\psi)\right] = -\frac{i\hbar}{2\mu}\left[\nabla S\cdot\nabla\psi + \psi\nabla^2 S\right]$$

(using the identity $\nabla\cdot(\vec{A}\phi) = A\cdot\nabla\phi + \phi\nabla\cdot\vec{A}$)

Hence
$$\left(-\frac{\partial S}{\partial t}\right)\psi = \left[\frac{\nabla S\cdot\nabla S}{2\mu} - \frac{i\hbar}{2\mu}\nabla^2 S + V(\vec{r})\right]\psi$$

or
$$-\frac{\partial S}{\partial t} = \frac{\nabla S\cdot\nabla S}{2\mu} + V(\vec{r}) - \frac{i\hbar}{2\mu}\nabla^2 S \tag{2}$$

If we can drop the last term on the r.h.s. (i.e. in the limit $\hbar \to 0$) we get the well known Hamilton-Jacobi equation of classical mechanics:

$$-\frac{\partial S_0}{\partial t} = \frac{(\nabla S_0)^2}{2\mu} + V(\vec{r}) \tag{3}$$

where S_0 is the action function associated with the classical path of the particle and defined in terms of the Lagrangian L, through the relation:

$$S_0(\vec{r},t) = \int L(\vec{r},\dot{\vec{r}},t')\, dt' \tag{4}$$

Thus the finite value of $\hbar$ is responsible for the difference between classical and quantum mechanics. Now $\hbar$ being a universal constant, it cannot be zero. In effect, it can be equivalent to $\to 0$, if in Eqn. (2), the condition

$$|\nabla S|^2 \gg \hbar\,|\nabla^2 S| \tag{5}$$

or
$$p^2 \gg \hbar\,|\,(\nabla\cdot\vec{p})\,| \tag{6}$$

is satisfied. In other words, an approximation method based on a power series expansion of S in $\hbar$ is possible:

$$S = S_0 + \frac{\hbar}{i}S_1 + \left(\frac{\hbar}{i}\right)^2 S_2 + \ldots \tag{7}$$

The first term of this leads to the classical results, the second term to the old quantum theory (semiclassical approximation) and the higher terms are the characteristics of the new mechanics.

17.2 The WKB Wave Function (The Method)

For simplicity, we consider stationary states, i.e. the energy of the system is well-defined in a stationary state and the time dependence of the wave function is completely determined by the energy

$$\psi(\vec{r},t) = \phi(\vec{r})\, e^{-iEt/\hbar}$$

$$= \exp(iS/\hbar) \tag{8}$$

We restrict ourselves to one-dimensional case with

$$\phi(\vec{r}) = e^{i\sigma(\vec{r})/\hbar} \tag{9}$$

where
$$S(\vec{r},t) = \sigma(\vec{r}) - Et \qquad (10)$$

E being the energy of the system.

We have then

$$\left(\frac{d\sigma_0}{dx}\right)^2 - 2\mu[E - V(x)] = 0 \qquad (11)$$

$$\left(\frac{d\sigma}{dx}\right)^2 - 2\mu[E - V(x)] - i\hbar\frac{d^2\sigma}{dx^2} = 0 \qquad (12)$$

and
$$\sigma(x) = \sigma_0(x) + \frac{\hbar}{i}\sigma_1(x) + \left(\frac{\hbar}{i}\right)^2\sigma_2(x) + \dots \qquad (13)$$

The Schrödinger equation with the help of Eqn. (8) reduces to

$$\frac{d^2\phi}{dx^2} + \frac{2\mu}{\hbar^2}[E - V(x)]\,\phi = 0 \qquad (14)$$

with
$$\phi(x) = \exp\left[\frac{i}{\hbar}\sigma(x)\right] \qquad (15)$$

(see Eqns. 3, 5, 6 and 10).

From Eqn. (11), we have

$$2\mu[E - V(x)] = \left(\frac{d\sigma_0}{dx}\right)^2 = p^2(x) \qquad (16)$$

So, Eqn. (14) becomes

$$\frac{d^2\phi}{dx^2} + \frac{p^2}{\hbar^2}\,\phi = 0 \qquad (17)$$

We are interested in the solution of this equation within the WKB approximation i.e. we have to substitute the approximate value $\sigma = \sigma_0 + (\hbar/i)\sigma_1$ (from the expansion 13) in Eqn. (15). The valeus of σ_0 and σ_1 appropriate for the system under consideration are determined by substituting Eqn. (13) in Eqn. (12) which is equivalent to the Scrhödinger Eqn. (14) or (17). The substitution

$$u(x) = \frac{d\sigma(x)}{dx} \qquad (18)$$

would be convenient. Using Eqns (16) and (18) Eqn. (12) becomes

$$u^2 - p^2 = i\hbar\frac{du}{dx} \quad \text{or} \quad p^2 - u^2 = \frac{\hbar}{i}\frac{du}{dx} \qquad (19)$$

Also, differentiating Eqn. (13) w.r.t. x, we get

$$u(x) = u_0(x) + \left(\frac{\hbar}{i}\right)u_1(x) + \left(\frac{\hbar}{i}\right)^2 u_2(x) + \dots \qquad (20a)$$

where
$$u_i(x) = \frac{d\sigma_i}{dx} \qquad (20b)$$

In terms of u, $\phi(x)$ is given by

$$\phi(x) = \exp\left[\frac{i}{\hbar}\int^x \frac{d\sigma}{dx'}\,dx'\right] = \exp\left[\frac{i}{\hbar}\int^x u\,dx'\right] \tag{21}$$

Substituting for u from Eqn. (20a) into Eqn. (19), we get

$$\left(p^2 - u_0^2\right) - 2\left(\frac{\hbar}{i}\right)u_0 u_1 - \left(\frac{\hbar}{i}\right)^2\left[u_1^2 + 2u_0 u_2\right] + \dots$$

$$= \left(\frac{\hbar}{i}\right)\frac{du_0}{dx} + \left(\frac{\hbar}{i}\right)^2 \frac{d\,u_1}{dx} + \dots \tag{22}$$

Equating coefficients of like powers of $(\hbar/i)$ on either side of Eqn. (22), we get

$$u_0 = \pm p \tag{23a}$$

$$u_1 = -\frac{1}{2\,u_0}\left(\frac{d\,u_0}{dx}\right) = -\frac{1}{2p}\left(\frac{dp}{dx}\right) \tag{23b}$$

From Eqn. (20a)

$$u \cong u_0 + \left(\frac{\hbar}{i}\right)u_1$$

Therefore, corresponding to the two values of u_0, there are two values of u

$$u_+ = p - \frac{\hbar}{i}\frac{1}{2p}\left(\frac{dp}{dx}\right) = p - \left(\frac{\hbar}{i}\right)\frac{d}{dx}\left(\ln\sqrt{p}\right) \tag{24a}$$

$$u_- = -p - \frac{\hbar}{i}\frac{1}{2p}\left(\frac{dp}{dx}\right) = -p - \left(\frac{\hbar}{i}\right)\frac{d}{dx}\left(\ln\sqrt{p}\right) \tag{24b}$$

Substituting in Eqn. (21), we have

$$\phi_+(x) = \exp\left[\frac{i}{\hbar}\int^x u\,dx\right]$$

$$= \exp\left[\frac{i}{\hbar}\int^x p\,dx - \frac{\hbar}{i}\int \frac{d}{dx}\left(\ln\sqrt{p}\right)dx\right]$$

$$= \exp\left[\frac{i}{\hbar}\int^x p\,dx - \frac{\hbar}{i}\int d\,\ln\sqrt{p}\right]$$

$$= \exp\frac{i}{\hbar}\int^x p\,dx\; e^{-\int d\,\ln\sqrt{p}}$$

$$= \exp\left(\frac{i}{\hbar}\int^x p\,dx\right)e^{-\ln\sqrt{p}}$$

or

$$\phi_+(x) = \frac{1}{\sqrt{p}}\exp\left(\frac{i}{\hbar}\int^x p\,dx'\right) \tag{25a}$$

Similarly

$$\phi_-(x) = \frac{1}{\sqrt{p}} \exp\left(-\frac{i}{\hbar} \int^x p \, dx'\right) \tag{25b}$$

These represent the two independent solutions of the second order differential equation (17). The general solution is given by a linear combination of these two and is the WKB wave function

$$\phi_{WKB}(x) = \frac{A_1}{\sqrt{p}} \exp\left[\frac{i}{\hbar} \int^x p(x') \, dx'\right] + \frac{B_1}{\sqrt{p}} \exp\left[-\frac{i}{\hbar} \int^x p(x') \, dx'\right] \tag{26}$$

The lower limit for the integrals in Eqn. (25) and (26) would be a classical turning point.

17.3 The Criterion for Validity of the Approximation

The approximation leading to Eqn. (26) is valid when condition

$$\left|p^2\right| >> \hbar \left|\nabla \cdot \vec{p}\right|$$

is satisfied. In present case, this is equivalent to

$$\left|p^2\right| >> \hbar \left|\frac{dp}{dx}\right|$$

or

$$\left(\frac{\hbar}{p}\right) \frac{\left|\frac{dp}{dx}\right|}{|p|} << 1 \quad \text{or} \quad \frac{\lambda \left|(dp/dx)\right|}{|p|} << 1 \tag{27}$$

where $\lambda(= \hbar/p)$ is the de Broglie wavelength of the particle. Thus the condition for applicability of the WKB approximation is that *the fractional change of momentum over a de Broglie wavelength of the particle be small.*

It is worthwhile to mention here that the equaion

$$\frac{d^2\phi}{dx^2} + k^2(x)\,\phi(x) = 0 \tag{28}$$

(with $p = \hbar k$) is also encountered in electromagnetic theory; if a plane electromagnetic wave propagating in the x-direction is incident normally on an inhomogenous medium characterized by the refractive index $n(x)$, then the electric field satisfies Eqn. (28) with

$$k^2(x) = \frac{\omega^2}{c^2} n^2(x) = \frac{4\pi^2}{\lambda_0^2} n^2(x) \tag{29}$$

where

$$\lambda_0 = \frac{2\pi c}{\omega} \qquad (\because (2\pi)\nu\lambda = (2\pi)\,c \text{ or } \lambda = 2\pi c/\omega) \tag{30}$$

λ_0 represents the free space wavelength, c is the speed of light in free space and ω is the angular frequency of the wave.

For such a plane wave propagating in x-direction E_x is taken to be zero and the wave function ϕ is either E_y or E_z. Since

$$k(x) = \frac{\omega}{c} n(x) \tag{31}$$

Eqn. (27) can be written as

$$\frac{(\hbar)}{(\hbar)k} \frac{(\hbar)(dk/dx)}{(\hbar)k} << 1 \text{ or } \left| \frac{1}{k(x)} \frac{dk}{dx} \right| << k(x) \tag{32}$$

or
$$\left| \frac{1}{n(x)} \frac{dn}{dx} \right| << \frac{2\pi}{\lambda(x)} \tag{33}$$

where

$$\lambda(x) = \frac{2\pi}{k(x)} = \frac{\lambda_0}{n(x)} \tag{34}$$

represents the local wavelength. We can rewrite Eqn. (33) as

$$\left| \frac{dn}{dx} \frac{\lambda}{\lambda_0} \right| << \frac{2\pi}{\lambda(x)} \text{ or } \left| \frac{dn}{dx} \lambda \right| << 2\pi \left(\frac{\lambda_0}{\lambda} \right)$$

or
$$\frac{1}{2\pi} \left| \frac{dn}{dx} \lambda \right| \lesssim n \tag{35}$$

Except for a factor of $(1/2\pi)$ the quantity on the l.h.s. represents the change in the refractive index in a distance of one wavelength. Thus, the condition (Eqn. 35) states that the refractive index should change very little over a distance of one wavelength. Also, as $\lambda \to 0$, the WKB approximation would be more and more accurate. This would be equivalent to reaching the ray (geometrical) optics limit.

17.4 The Turning Point and Connection Formulae

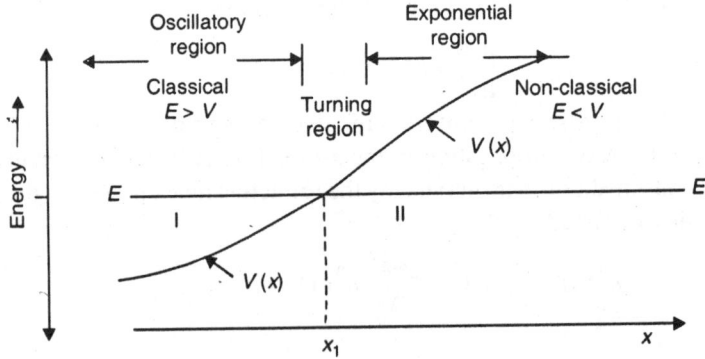

Fig. 17.1: The general potential $V(x)$ in relation to energy eigenvalue E. I and II are classical and non classical regions respectively separated by a classical turning point

Consider the Fig. 17.1. The point x_1 separates a region I where $E > V(x)$ from region II where $E < V(x)$. x_1 is a *classical turning point*, because according to classical mechanics, a particle of energy E bowled in from the left would reverse its motion there because $p(x_1) = 0$ but $\left(\dfrac{dp}{dx}\right)_{x_1} \neq 0$. Since (dV/dx) is finite at x_1, there is a finite probability that a particle would be found in region II if initially it was in region I (At the point x_1, $k^2(x) = 0$).

In region II, $E < V(x)$ and $p(x)$ is imaginary $p(x) \rightarrow i|p(x)|$. Then for region II, from Eqn. (26)

$$\phi_{II}(x) = \frac{A_2}{\sqrt{|p|}} \exp\left[-\frac{1}{\hbar}\int_{x_1}^{x}|p|\,dx'\right] + \frac{B_2}{\sqrt{|p|}} \exp\left[\frac{1}{\hbar}\int_{x_1}^{x}|p|\,dx'\right] \qquad (36i)$$

(nonoscillatory)

We know that for region *I*:

$$\phi_{I}(x) = \frac{A_1}{\sqrt{p}} \exp\left[\frac{i}{\hbar}\int_{x_1}^{x}p\,dx'\right] + \frac{B_1}{\sqrt{p}} \exp\left[-\frac{i}{\hbar}\int_{x_1}^{x}p\,dx'\right] \qquad (36ii)$$

(oscillatory)

Obviously, the WKB solutions are not valid at the turning point, and we have solutions valid far to the left of the turning point and also far to the right of the turning point. Now question arises that if we know the WKB solutions to the left of the turning point, then how can we determine the WKB solutions to the right of the turning point and conversely. This can be done if we find a way to extend the WKB solution from one region to the other through the turning point. Such a procedure (in effect), amount to obtaining *connection formulae* between the WKB wave functions in the two regions.

Now, ϕ_I and ϕ_{II} are approximations to the same wave function. We can identify them as such only if we know the relationships between the coefficients A_1, B_1 and A_2, B_2. The required relationship is provided by the connection formulae which are obtained by solving Eqn. (17) viz.

$$\frac{d^2\phi}{dx^2} + \frac{p^2}{\hbar^2}\phi = 0$$

exactly near the turning point and then finding the asymptotic forms of the solution far away on either side of x_1.

We assume that $V(x)$ is linear in the neighbourbood of turning point

$$V(x)\big|_{x=x_1} \cong V(x_1) + (x - x_1)\left(\frac{dV}{dx}\right)_{x=x_1} \cong E + C(x - x_1) \qquad (37)$$

where

$$C = \left(\frac{dV}{dx}\right)_{x=x_1} > 0 \qquad (38)$$

then,

$$p^2 = 2\mu(E - V)_{x=x_1} = 2\mu C(x - x_1) \qquad (39)$$

Substituting for p^2 from Eqn. (39), Eqn. (17) becomes

$$\frac{d^2\psi}{d\xi^2} - \xi\psi = 0 \qquad (40)$$

where

$$\psi = \left(\frac{2\mu C}{\hbar^2}\right)^{1/3} (x - x_1) \tag{41}$$

and $\qquad \psi(\xi) = \phi(x)$

Region I corresponds to $\xi < 0$, II corresponds to $\xi > 0$ and the turning point to $\xi = 0$. The solutions of Eqn. (40) are known as the *Airy-Functions*, and are given by

$$Ai(\xi) = \frac{1}{\pi} \int_0^\infty \cos\left(\frac{s^3}{3} + s\xi\right) ds \tag{42}$$

and

$$Bi(\xi) = \frac{1}{\pi} \int_0^\infty \left\{ \exp\left(-s\xi - \frac{s^3}{3}\right) + \sin\left(\frac{s^3}{3} + s\xi\right) \right\} ds \tag{43}$$

We are interested only in asymptotic forms of *Ai* and *Bi*, viz.,

$$Ai(\xi) \underset{\xi \ll 0}{\cong} \left(-\pi^2\xi\right)^{-1/4} \sin\left[\left(\frac{2}{3}\right) (-\xi)^{3/2} + \frac{\pi}{4} \right] \tag{44i}$$

$$\underset{\xi \gg 0}{\cong} \frac{1}{2} \left(\pi^2\xi\right)^{-1/4} \exp\left[-\left(\frac{2}{3}\right) \xi^{3/2} \right] \tag{44ii}$$

and

$$Bi(\xi) \underset{\xi \ll 0}{\cong} \left(-\pi^2\xi\right)^{-1/4} \cos\left[\left(\frac{2}{3}\right) (-\xi)^{3/2} + \frac{\pi}{4} \right] \tag{45i}$$

$$\underset{\xi \gg 0}{\cong} \left(\pi^2\xi\right)^{-1/4} \exp\left[\left(\frac{2}{3}\right) \xi^{3/2} \right] \tag{45ii}$$

Now, for $\xi < 0$ [i.e. $x < x_1$]

$$\left(\frac{2}{3}\right) (-\xi)^{3/2} = \int_0^\xi \sqrt{-\xi'} \, d(-\xi')$$

$$= -\int_0^\xi \sqrt{-\xi'} \, d\xi'$$

$$= -\int_{x_1}^x \left(2\mu C/\hbar^2\right)^{1/2} (x - x')^{1/2} \, dx' \qquad \text{(using Eqn. 41)}$$

$$= -\frac{1}{\hbar} \int_{x_1}^x p(x') \, dx' \qquad \text{(using Eqn. 39)}$$

$$= \frac{1}{\hbar} \int_x^{x_1} p(x') \, dx' \tag{46}$$

Also,

$$(-\xi)^{1/4} = (2\mu C)^{1/12} \hbar^{-1/6} (x_1 - x)^{1/4}$$

$$= (2\mu C\hbar)^{-1/6} \left[2\mu C (x_1 - x) \right]^{1/4}$$

$$= (2\mu C\hbar)^{-1/6} \sqrt{p(x)} \tag{47}$$

Similarly, for $\xi > 0$ [i.e. $x > x_1$]:

$$\left(\frac{2}{3} \right) \xi^{3/2} = \frac{1}{\hbar} \int_{x_1}^{x} \left| p(x') \right| dx' \tag{48}$$

and

$$(\xi)^{1/4} = (2\mu C\hbar)^{-1/6} \sqrt{|p(x)|} \tag{49}$$

Substituting from Eqns (46) to (49) into Eqns. (44i) to (45ii), we get

$$Ai(\xi) \underset{\xi \ll 0}{=} \phi_1 \, osc(x)$$

$$\underset{x \ll x_1}{\cong} \frac{\alpha}{\sqrt{p}} \sin \left[+\frac{1}{\hbar} \int_x^{x_1} p(x') \, dx' + \frac{\pi}{4} \right] \tag{50}$$

and $\qquad Ai(\xi) \underset{\xi \gg 0}{=} \phi_1 \exp(x) \underset{x \gg x_1}{\cong} \frac{\alpha}{2\sqrt{|p|}} \exp \left(-\frac{1}{\hbar} \int_{x_1}^{x} |p| \, dx' \right) \tag{51}$

$$Bi(\xi) \underset{\xi \ll 0}{=} \phi_2 osc(x) \underset{x \ll x_1}{\cong} \frac{\alpha}{\sqrt{p}} \cos \left[\frac{1}{\hbar} \int_x^{x_1} p(x') \, dx' + \frac{\pi}{4} \right] \tag{52}$$

and $\qquad Bi(\xi) \underset{\xi \gg 0}{=} \phi_2 \exp(x) \underset{x \gg x_1}{\cong} \frac{\alpha}{\sqrt{|p|}} \exp \left(\frac{1}{\hbar} \int_{x_1}^{x} |p| \, dx' \right) \tag{53}$

where $\qquad \alpha = \left(\dfrac{2\mu C\hbar}{\pi^3} \right)^{1/6}$

Thus the connection between the wave functions in the classical and the non classical regions are given as here below:

Classical Region Region I (Oscillatory)	Non-classical Region Region II (Exponential)	
$\dfrac{\alpha}{\sqrt{p}} \sin \left[\dfrac{1}{\hbar} \int_x^{x_1} p \, dx' + \dfrac{\pi}{4} \right]$	$\leftrightarrow \quad \dfrac{\alpha}{2} \|p\|^{-1/2} \exp \left[-\dfrac{1}{\hbar} \int_{x_1}^{x} \|p\| \, dx' \right]$	(54)
$\dfrac{\alpha}{\sqrt{p}} \cos \left[\dfrac{1}{\hbar} \int_x^{x_1} p \, dx' + \dfrac{\pi}{4} \right]$	$\leftrightarrow \quad \alpha \|p\|^{-1/2} \exp \left[\dfrac{1}{\hbar} \int_{x_1}^{x} \|p\| \, dx' \right]$	(55)

It is to be noted here that a wavefunction that is represented by the sine function in the classical region becomes a decreasing exponential in the non classical region whereas an increasing exponential in the non classical region corresponds to cosine function in the classical region.

The wave function of any physical system would be a general solution (i.e. a linear combination of ϕ_1 and ϕ_2). Thus,

$$\phi_{osc}(x) \cong \frac{\alpha}{\sqrt{p}} \left\{ \sin\left[\frac{1}{\hbar} \int_x^{x_1} p(x')\, dx' + \frac{\pi}{4}\right] \right.$$

$$\left. + \cos\left[\frac{1}{\hbar} \int_x^{x_1} p(x')\, dx' + \frac{\pi}{4}\right] \right\} \tag{56}$$

and

$$\phi_{exp}(x) \cong \frac{\alpha/2}{\sqrt{|p|}} \exp\left(-\frac{1}{\hbar}\int_{x_1}^x |p(x')|\, dx'\right)$$

$$+ \frac{\alpha}{\sqrt{|p|}} \exp\left(\frac{1}{\hbar}\int_{x_1}^x |p(x')|\, dx'\right) \tag{57}$$

Comparing Eqns (26) and (36) with Eqns (56) and (57), we see that ϕ_{oxc} is the WKB wave function ϕ_I in the classical region and ϕ_{exp} is the WKB wave function ϕ_{II} in the non classical region. Further,

$$A_2 = \frac{\alpha}{2} = \frac{A}{2} \tag{58}$$

and

$$B_2 = \alpha = B = \left\{ \frac{2\mu\hbar}{\pi^3} \left(\frac{dV}{dx}\right)_{x=x_1} \right\} \tag{59}$$

where we can write

$$\phi_{osc}(x) \cong \frac{A}{\sqrt{p}} \sin\left[\frac{1}{\hbar}\int_x^{x_1} p\, dx' + \frac{\pi}{4}\right] + \frac{B}{\sqrt{p}} \cos\left[\frac{1}{\hbar}\int_x^{x_1} p\, dx' + \frac{\pi}{4}\right] \tag{60}$$

with

$$A = \left(A_1 - iB_1\right) e^{i\pi/4} \tag{61i}$$

and

$$B = -\left(A_1 + iB_1\right) e^{-i\pi/4} \tag{61ii}$$

Equations (54) and (55) are the connection formulae and Eqns (58) to (61) give the relation between the coèfficients A_1, B_1, A_2 and B_2.

In analogy with Fig. 17.1, we may also have a symmetrical situation as depicted in Fig. 17.2. Then, we will have the following connection formulae:

Region II (Exponential)		Region III (Oscillatory)	
$\dfrac{\alpha}{2\sqrt{\lvert p\rvert}} \exp\left[-\dfrac{1}{\hbar}\int_x^{x_2}\lvert p\rvert\, dx\right].$	$\leftrightarrow$	$\dfrac{\alpha}{\sqrt{p}} \sin\left[\dfrac{1}{\hbar}\int_{x_2}^x p\, dx + \dfrac{\pi}{4}\right]$	(62i)
$\dfrac{\alpha}{\sqrt{\lvert p\rvert}} \exp\left[\dfrac{1}{\hbar}\int_x^{x_2}\lvert p\rvert\, dx\right]$	$\leftrightarrow$	$\dfrac{\alpha}{\sqrt{p}} \cos\left[\dfrac{1}{\hbar}\int_{x_2}^x p\, dx + \dfrac{\pi}{4}\right]$	(62ii)

17.5 Barrier Penetration

We shall now discuss the application of the WKB approximation in calculating the transmission coefficient of a potential barrier of the type shown in Fig. 17.3.

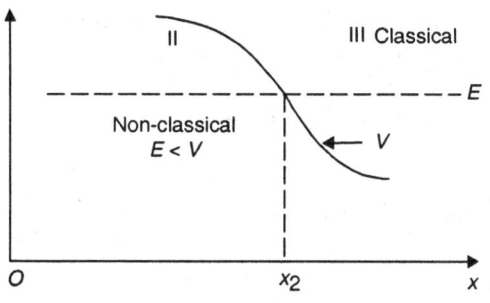

Fig. 17.2

Particles of energy E less than the height of the barrier are incident from left to the right (at x_1). x_1 is the classical turning point where there is some probability that some of the particles tunnel to the classically-forbidden region II. Of these, some will be reflected (back) at the other turning point x_2 and a fraction skip to the classical region III. The ratio of the flux of particles transmitted to region III to that of the incident particles (from region I) is called the *transmission coefficient*.

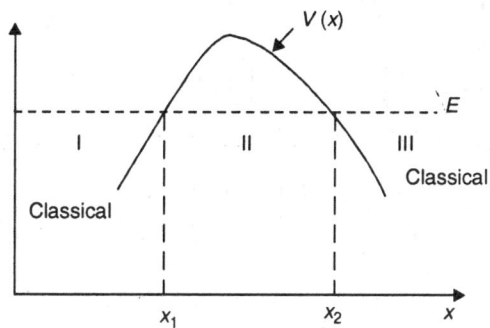

Fig. 17.3. A potential barrier

Now, the WKB wave function in region I is given (according to Eqn. 36(*ii*)) by

$$\phi_I(x) = \frac{A_1}{\sqrt{p}} \exp\left[\frac{i}{\hbar} \int_{x_1}^{x} p(x')\,dx'\right] + \frac{B_1}{\sqrt{p}} \exp\left[-\frac{i}{\hbar} \int_{x_1}^{x} p(x')\,dx'\right] \qquad (63)$$

Here the first term on the r.h.s. represents the incident particles and the second term the particles reflected at the point x_1.

In region III, there is only transmitted wave. So,

$$\phi_{III} = \frac{A_3}{\sqrt{p}} \exp\left(\frac{i}{\hbar} \int_{x_2}^{x} p(x')\,dx'\right) \qquad (64)$$

From Eqns (63) and (64), we have, transmission coefficient

$$T = \frac{\left|\vec{J}_{tr}\right|}{\left|\vec{J}_{inc}\right|} = \frac{\left|\psi_{tr}\sqrt{p}\right|^2}{\left|\psi_{inc}\sqrt{p}\right|^2} = \frac{\left|\sqrt{p}\,\phi_{III}\right|^2}{\left|\sqrt{p}\,(\phi_I)_{inc}\right|^2} = \frac{\left|A_3\right|^2}{\left|A_1\right|^2} \qquad (65)$$

Thus, if A_3 is determined in terms of A_1, T can be obtained. This relationship can be found by using the connection formulae, as follows. From Eqn. (64),

$$\phi_{III}(x) = \frac{A_3}{\sqrt{p}} \, e^{-i\pi/4} \left[\exp \left(\frac{i}{\hbar} \int_{x_2}^{x} p(x') \, dx' + \frac{\pi}{4} \right) \right] \tag{66i}$$

$$= \frac{A}{\sqrt{p}} \left[\cos \left(\frac{1}{\hbar} \int_{x_2}^{x} p(x') \, dx' + \frac{\pi}{4} \right) + i \sin \left(\frac{1}{\hbar} \int_{x_2}^{x} p(x') \, dx' + \frac{\pi}{4} \right) \right] \tag{66ii}$$

where
$$A = A_3 e^{-i\pi/4} = \left(\frac{1-i}{\sqrt{2}} \right) A_3$$

where
$$e^{-i\pi/4} = \cos \frac{\pi}{4} - i \sin \frac{\pi}{4} = \frac{1}{\sqrt{2}} - \frac{i}{\sqrt{2}} = \left(\frac{1-i}{\sqrt{2}} \right) \tag{66iii}$$

Then, using Eqn. (66ii), in comparison with Eqn. (62), we have, for the WKB wave function in region II:

$$\phi_{II}(x) = \frac{A}{\sqrt{|p|}} \left\{ \exp \left(\frac{1}{\hbar} \int_{x}^{x_2} |p| \, dx' \right) + \frac{i}{2} \exp \left(-\frac{1}{\hbar} \int_{x}^{x_2} |p| \, dx' \right) \right\} \tag{67i}$$

We write,

$$\int_{x}^{x_2} |p| \, dx' = \int_{x_1}^{x_2} |p| \, dx' - \int_{x_1}^{x} |p| \, dx' \tag{67ii}$$

Then using Eqn. (67ii), Eqn. (67i) becomes

$$\phi_{II}(x) = \frac{A}{\sqrt{|p|}} \left\{ \exp \left(\frac{1}{\hbar} \int_{x_1}^{x_2} - \frac{1}{\hbar} \int_{x_1}^{x} \right) + \frac{i}{2} \exp \left(-\frac{1}{\hbar} \int_{x_1}^{x_2} + \frac{1}{\hbar} \int_{x_1}^{x} \right) \right\}$$

and using
$$\exp \left[\left(\frac{1}{\hbar} \right) \int_{x_1}^{x_2} |p| \, dx' \right] \equiv F \tag{67iii}$$

we get,
$$\phi_{II}(x) = \frac{A}{\sqrt{|p|}} \left\{ F \exp \left[-\left(\frac{1}{\hbar} \right) \int_{x_1}^{x} |p| \, dx' \right] + \frac{i}{2F} \right.$$

$$\times \left. \exp \left[\left(\frac{1}{\hbar} \right) \int_{x_1}^{x} |p| \, dx' \right] \right\} \tag{68}$$

Applying the connection formula again, we get

$$\phi_{I}(x) = \frac{A}{\sqrt{p}} \left\{ 2F \sin \left[\frac{1}{\hbar} \int_{x}^{x_1} p \, dx' + \frac{\pi}{4} \right] + \frac{i}{2F} \cos \left[\frac{1}{\hbar} \int_{x}^{x_1} p \, dx' + \frac{\pi}{4} \right] \right\}$$

Now, we know that $\sin \theta = \dfrac{e^{i\theta} - e^{-i\theta}}{2i}$; $\cos \theta = \dfrac{e^{i\theta} + e^{-i\theta}}{2}$

$$\phi_{I}(x) = \frac{A}{\sqrt{p}} \left\{ \frac{2F}{2i} \left[\exp \left(\frac{i}{\hbar} \int_{x}^{x_1} p \, dx' + \frac{i\pi}{4} \right) - \exp \left(-\frac{i}{\hbar} \int_{x}^{x_1} p \, dx' - \frac{i\pi}{4} \right) \right] \right.$$

$$\left. + \frac{i}{2F} \times \frac{1}{2} \left[\exp \left(\frac{i}{\hbar} \int_{x}^{x_1} p \, dx' + \frac{i\pi}{4} \right) + \exp \left(-\frac{i}{\hbar} \int_{x}^{x_1} p \, dx' - \frac{i\pi}{4} \right) \right] \right\}$$

Using Eqn. (66) this becomes

$$\phi_I(x) = \frac{A_3}{\sqrt{p}} \exp\left(-\frac{i\pi}{4}\right) \left\{ \frac{F}{i} \left[\exp\left(\frac{i}{\hbar}\int_x^{x_1} p\, dx'\right) \exp\left(+\frac{i\pi}{4}\right) \right. \right.$$

$$\left. - \exp\left(-\frac{i}{\hbar}\int_x^{x_1} p\, dx'\right) \exp\left(-\frac{i\pi}{4}\right) \right]$$

$$+\frac{i}{4F}\left[\exp\left(\frac{i}{\hbar}\int_x^{x_1} p\, dx'\right) \exp\left(\frac{i\pi}{4}\right) \right.$$

$$\left. \left. + \exp\left(-\frac{i}{\hbar}\int_x^{x_1} p\, dx'\right) \exp\left(-\frac{i\pi}{4}\right) \right] \right\}$$

$$=\frac{A_3}{\sqrt{p}}\left\{ -iF\left[\exp\left(\frac{i}{\hbar}\int_x^{x_1} p\, dx'\right) - \exp\left(\frac{i}{\hbar}\int_{x_1}^{x} p\, dx'\right) \exp\left(-\frac{i\pi}{2}\right) \right] \right.$$

$$\left. +\frac{i}{4F}\left[\exp\left(-\frac{i}{\hbar}\int_{x_1}^{x} p\, dx'\right) + \exp\left(\frac{i}{\hbar}\int_{x_1}^{x} p\, dx'\right) \exp\left(-\frac{i\pi}{2}\right) \right] \right\}$$

$$=\frac{A_3}{\sqrt{p}}\left\{ -iF\left[\exp\left(-\frac{i}{\hbar}\int_{x_1}^{x} p\, dx'\right) + i\, \exp\left(\frac{i}{\hbar}\int_{x_1}^{x} p\, dx'\right) \right] \right.$$

$$\left. +\frac{i}{4F}\left[\exp\left(-\frac{i}{\hbar}\int_{x_1}^{x} p\, dx'\right) - i\, \exp\left(\frac{i}{\hbar}\int_{x_1}^{x} p\, dx'\right) \right] \right\}$$

$$=\frac{A_3}{\sqrt{p}}\left[-i\left(F-\frac{1}{4F}\right) \exp\left(-\frac{i}{\hbar}\int_{x_1}^{x} p\, dx'\right) + \left(F+\frac{1}{4F}\right) \exp\left(\frac{i}{\hbar}\int_{x_1}^{x} p\, dx'\right) \right]$$

$$=\frac{A_3}{\sqrt{p}}\left\{ \left(F+\frac{1}{4F}\right) \exp\left(\frac{i}{\hbar}\int_{x_1}^{x} p\, dx'\right) \right.$$

$$\left. -i\left(F-\frac{1}{4F}\right) \exp\left(-\frac{i}{\hbar}\int_{x_1}^{x} p\, dx'\right) \right\} \tag{69}$$

Comparing Eqns (63) and (69), we find

$$A_1 = \left(F+\frac{1}{4F}\right) A_3 \tag{70i}$$

$$B_1 = -i\left(F-\frac{1}{4F}\right) A_3 \tag{70ii}$$

Thus,

$$T = \left|\frac{A_3}{A_1}\right|^2 = \left(F + \frac{1}{4F}\right)^{-2} \cong \left(\frac{1}{F}\right)^2 = \exp\left\{-\frac{2}{\hbar}\int_{x_1}^{x_2}|p(x)|\,dx\right\} \tag{71}$$

F defined by Eqn. (67ii) is a measure of both the height $|p(x)| = \sqrt{2\mu(V(x)-E)}$ and the width $(x_2 - x_1)$ of the barrier. Eqn. (71) shows that increasing either the width of the barrier (i.e. increasing distance of x_2 from x_1) or increasing height of the barrier decreases the probability for the penetration of the barrier.

Example 17.1: Obtain the energy levels of a linear harmonic oscillator by WKB method.

Solution: The energy levels are determined by Bohr's quantization rule:

$$\int_a^b p(x)\,dx = \pi\hbar\left(n + \frac{1}{2}\right) \tag{i}$$

where n is an integer including zero.

$$p(x) = \sqrt{2m(E-V(x))}$$

where

$$V(x) = \frac{1}{2}kx^2 = \frac{1}{2}m\omega^2 x^2$$

ω is given by $\omega = \dfrac{1}{2\pi}\sqrt{\dfrac{k}{m}}$ (the frequency of the oscillator)

a, b are turning points which are given by the condition that the kinetic energy $E - V(x)$ be zero there. Kinetic energy is zero at the two extremities

$$E - V = 0 \text{ or } E = V = \frac{1}{2}m\omega^2\alpha^2$$

So that $\qquad \alpha = \sqrt{\dfrac{2E}{m\omega^2}}$ and turning points are $\pm\,\alpha$.

or $\qquad a = -\sqrt{\dfrac{2E}{m\omega^2}}$ and $b = +\sqrt{\dfrac{2E}{m\omega^2}}$

Thus,

$$\int_a^b p(x)\,dx = \int_{-\sqrt{2E/(m\omega^2)}}^{+\sqrt{2E/(m\omega^2)}} \sqrt{2m(E-V)}\,dx \tag{ii}$$

Putting $\qquad x = t\sqrt{2E/(m\omega^2)}\,;\, dx = dt\sqrt{2E/(m\omega^2)}$

$\therefore \qquad \displaystyle\int_a^b p(x)\,dx = \int_{-1}^{+1}\sqrt{2mE}\sqrt{1 - V/E}\sqrt{2E/(m\omega^2)}\,dt$

$$= \frac{2E}{\omega}\int_{-1}^{+1}\sqrt{1 - V/E}\,dt \tag{iii}$$

Now $\qquad V(x) = \dfrac{1}{2}m\omega^2 x^2 = \dfrac{1}{2}m\omega^2 t^2 \cdot \dfrac{2E}{m\omega^2}$, therefore

$$\int_a^b p(x)\,dx = \frac{2E}{\omega}\int_{-1}^{+1}\left\{1 - \frac{1}{2}\frac{m\omega^2}{E}t^2 \cdot \frac{2E}{m\omega^2}\right\}^{1/2} dt$$

$$= \frac{2E}{\omega}\int_{-1}^{+1}\sqrt{1-t^2}\,dt = \frac{2E}{\omega}\cdot\frac{\pi}{2} = \frac{\pi E}{\omega}$$

Hence $\quad\dfrac{\pi E}{\omega} = \pi\hbar\left(n + \dfrac{1}{2}\right)$ or $E = \left(n + \dfrac{1}{2}\right)\hbar\omega$

17.6 Application to a Potential Well

We now consider the application of WKB method for determination of energy levels of a one-dimensional bound system such as a potential well shown in the Fig. 17.4.

There are three regions I, II and III and two turning points x_1 and x_2. Region II is the classical region where the wave function would be oscillatory. Regions I and III are the non classical regions where the wave function would be exponential. Since the system is bound, the wave function must go to zero for $x \to \pm\infty$. This requires that in regions I and III, the WKB wave functions be decreasing exponentially. That is,

$$\phi_I(x) = \frac{A_1}{\sqrt{|p|}}\exp\left(-\frac{1}{\hbar}\int_x^{x_1}\left|p(x')\right|dx'\right) \tag{72}$$

and

$$\phi_{III}(x) = \frac{A_3}{\sqrt{|p|}}\exp\left(-\frac{1}{\hbar}\int_{x_2}^x\left|p(x')\right|dx'\right) \tag{73}$$

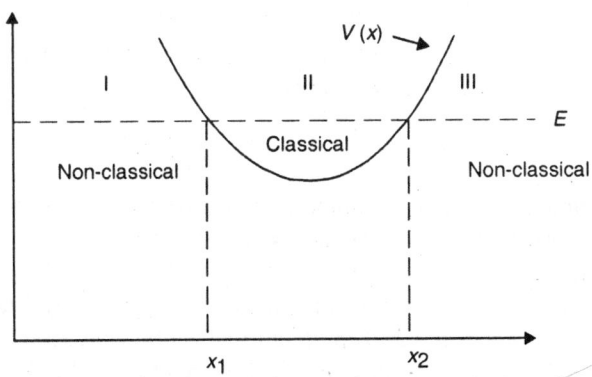

Fig. 17.4: The potential curve for a one dimensional potential well

According to the connection formulae, the wave function in region II is

$$\phi_{II}(x) = \frac{2A_1}{\sqrt{p}}\sin\left(\frac{i}{\hbar}\int_{x_1}^x p(x')\,dx' + \frac{\pi}{4}\right) \tag{74i}$$

$$= \frac{2A_3}{\sqrt{p}}\sin\left(\frac{1}{\hbar}\int_x^{x_2} p(x')\,dx' + \frac{\pi}{4}\right) \tag{74ii}$$

But
$$\int_{x_1}^{x} p(x') \, dx' = \int_{x_1}^{x_2} p(x') \, dx' - \int_{x}^{x_2} p(x') \, dx' \tag{75}$$

Therefore,

$$\sin\left(\frac{1}{\hbar} \int_{x_1}^{x} p \, dx' + \frac{\pi}{4}\right) = \sin\left[\left(\frac{1}{\hbar} \int_{x_1}^{x_2} p \, dx' + \frac{\pi}{2}\right) - \left(\frac{1}{\hbar} \int_{x}^{x_2} p \, dx' + \frac{\pi}{4}\right)\right] \tag{76}$$

Substituting Eqn. (76) in Eqn. (74i) and then equating it to (74ii), we get

$$\sin\left[\left(\frac{1}{\hbar} \int_{x_1}^{x_2} p \, dx' + \frac{\pi}{2}\right) - \left(\frac{1}{\hbar} \int_{x}^{x_2} p \, dx' + \frac{\pi}{4}\right)\right]$$

$$= \left(\frac{A_3}{A_1}\right) \sin\left(\frac{1}{\hbar} \int_{x}^{x_2} p \, dx' + \frac{\pi}{4}\right) \tag{77}$$

This can be compared with the identity

$$\sin(n\pi - \theta) = (-1)^{n-1} \sin\theta, \, n = 1, 2, 3, \dots \tag{78}$$

which requires n to be replaced by $(n + 1)$ [*] and

(i) $\dfrac{1}{\hbar} \int_{x_1}^{x_2} p(x) \, dx + \dfrac{\pi}{2} = (n+1)\,\pi$

(ii) $\dfrac{A_3}{A_1} = (-1)^n$, $n = 0, 1, 2, \dots$ $\tag{79}$

This gives

$$\int_{x_1}^{x_2} \frac{p}{\hbar} \, dx = \left(n + \frac{1}{2}\right)\pi \tag{80i}$$

or
$$\int_{x_1}^{x_2} p \, dx = \left(n + \frac{1}{2}\right)\pi \times \frac{h}{2\pi} = \left(n + \frac{1}{2}\right)\frac{h}{2} \tag{80ii}$$

In the semiclassical approximation, over a complete period, the particle moves from x_1 to x_2 and then back to x_1. Thus,

$$\oint p \, dx = 2\int_{x_1}^{x_2} p \, dx = \left(n + \frac{1}{2}\right)h \quad (n = 0, 1, 2, \dots) \tag{81}$$

This corresponds to the Bohr-Sommerfeld quantization rule of the old quantum theory (This is rather an exact result and is in agreement with the case of a linear harmonic oscillator).

The approximate wave function of the system is given by Eqn. (74i). The constant A_1 is to be determined from the normalization condition and the square of the sine function can be approximated to its average value ½.

$$1 = \int_{-\infty}^{+\infty} |\phi_{II}(x)|^2 \, dx = \int_{x_1}^{x_2} |\phi_{II}(x)|^2 \, dx$$

$$= |2A_1|^2 \int_{x_1}^{x_2} \frac{dx}{2p} = \frac{\pi |2A_1|^2}{2m\omega}$$

(*) because $\int_{x_1}^{x_2} p \, dx$ must be ≥ 0.

where ω is the classical frequency

$$\omega = \frac{2\pi}{T} \; ; \; T = 2\int \frac{dx}{v} = 2m \int_{x_1}^{x_2} \frac{dx}{p}$$

Thus, the normalized wave function,

$$\phi_{II}(x) = \sqrt{\frac{2m\omega}{\pi p}} \; \sin\left(\frac{1}{\hbar}\int_{x_1}^{x} p(x)\,dx + \frac{\pi}{4}\right) \tag{82}$$

$$(\because 2A_1 = \sqrt{2m\omega/\pi})$$

or
$$\phi_{II}(x) = \left(\frac{2m\omega}{\pi p}\right)^{1/2} \cos\left(\frac{1}{\hbar}\int_{x_1}^{x} p(x)\,dx - \frac{\pi}{4}\right) \tag{83}$$

$$(\because \sin\theta = \cos(\pi/2 - \theta) \text{ and } \cos(-\theta) = \cos\theta)$$

Questions and Problems

1. Comment on the statement "WKB approximation is a link between quantum and classical mechanics".
2. Calculate the probability for transmission of a particle through a potential barrier of arbitrary shape using the WKB method.
3. Describe the WKB method for the solution of Schrödinger equation in a potential field and discuss its validity.
4. Outline WKB method for one-dimensional case and derive the connection formulae.
5. Obtain eigenfunctions for a harmonic oscillator by WKB method and compare them with the exact wave functions.
6. Discuss the WKB method to explain α-decay and derive Geiger-Nuttal law.
7. Apply the WKB approximation method for transmission through a barrier to obtain the transmission coefficient for tunelling through a triangular barrier:
$$V(x) = 0 \text{ for } |x| > a$$
$$= V_0\left(1 - \frac{|x|}{a}\right) \text{ for } x \leq a.$$
8. Apply the WKB method to a one-dimensional potential well and obtain an expression for WKB wave function inside the well.

18

Time Dependent Perturbation Theory

18.1 The Principle

When a system is isolated and the Hamiltonian is time independent, the energy eigenstates are true stationary states. In a particular state ψ of H_0, the time dependence of ψ is of the form $e^{-iEt/\hbar}$ which is a constant, because E is a constant corresponding to the state ψ.

If however, the hamiltonian is time dependent, the energy is no longer a constant of motion. The behaviour of the system is given by the time dependent Schrödinger equation. There is no single energy eigenvalue and energy is an operator $i\hbar\,\partial/\partial t$.

$$i\hbar\frac{\partial\psi}{\partial t} = H\psi \qquad (1)$$

The time dependence of H may arise because of some external agency. For example, in emission or absorption of radiation, there is interaction of atoms with electromagnetic field. Since the radiation field oscillates, it is time dependent. We may write in such a case as:

$$H = H_0 + H'(t) \qquad (2)$$

where $H'(t)$ is a small perturbation representing the interaction of the system with the external agency.

At a particular instant say $t = 0$, the system may be regarded as being an eigenstate of H_0. We would calculate the probability of finding the system in another state also an eigenstate of H_0 at a later instant t.

Now, any function ψ can be expressed in terms of complete set of eigenfunctions u_n of H_0:

$$H_0 u_n = E_n u_n \quad \text{and} \quad \psi(t) = \sum_n a_n(t) u_n(r) e^{-iE_n t/\hbar} \qquad (3)$$

where $a_n(t)$ are time dependent variable parameters. Substituting Eqn. (3) in (1)

$$\left(H_0 + H'\right)\sum_n a_n(t)\, u_n e^{-iE_n t/\hbar} = i\hbar\frac{\partial}{\partial t}\left(\sum_n a_n(t)\, u_n e^{-iE_n t/\hbar}\right)$$

or
$$= \sum_n a_n(t) \left[E_n u_n + H' u_n \right] e^{-iE_n t/\hbar}$$

$$= i\hbar \sum_n \left[\dot{a}_n u_n e^{-iE_n t/\hbar} + a_n u_n \left(-\frac{iE_n}{\hbar} \right) e^{-iE_n t/\hbar} \right]$$

or
$$\sum_n a_n(t) H' u_n e^{-iE_n t/\hbar} = \sum_n i\hbar \dot{a}_n(t) u_n e^{-iE_n t/\hbar} \qquad (4)$$

Multiplying this Eqn. by u_m^* and integrating over the entire space, we get,

$$\sum_n a_n(t) H'_{mn} e^{-iE_n t/\hbar} = i\hbar \dot{a}_m(t) e^{-iE_m t/\hbar}$$

where
$$H'_{mn}(t) = \int u_m^* H'(t) u_n d\tau$$

$$\left(\text{and because } \int u_m^* u_n d\tau = \delta_{mn} ; \sum_n \dot{a}_n e^{-iE_n t/\hbar} \delta_{mn} = \dot{a}_m e^{-iE_m t/\hbar} \right)$$

Let $\dfrac{E_n - E_m}{\hbar} = \omega_{nm}$ so that $i\hbar \dot{a}_m(t) = \sum_n \left[e^{-i\omega_{nm} t} H'_{mn}(t) a_n(t) \right]$ (5)

This equation (for various values of m) is exact. To solve them, we have to apply perturbation techniques. Replacing H' by $\lambda H'$ and expressing a's as power series in λ

$$a_n(t) = a_n^{(0)}(t) + \lambda a_n^{(1)}(t) + \lambda^2 a_n^{(2)}(t) + \dots$$

we get from Eqn. (5)

$$= i\hbar \left[\dot{a}_m^{(0)}(t) + \lambda \dot{a}_m^{(1)}(t) + \lambda^2 \dot{a}_m^{(2)}(t) + \dots \right]$$

$$= \sum_n e^{+i\omega_{mn} t} \lambda H'_{mn}(t) \left[a_n^{(0)} + \lambda a_n^{(1)} + \dots \right] \qquad (6)$$

Equating coefficients of λ^0 on both sides, we get

$$i\hbar \dot{a}_m^{(0)} = 0 \text{ or } a_m^{(0)} = \text{constant in time} \qquad (7)$$

18.2 First Order Perturbation

Suppose initially (at $t = 0$), the system is in state k (with energy E_k) then $a_k^{(0)} = 1$ and $a_k^{(0)} = 1$ and $a_m^{(0)} = \int u_m^* u_k d\tau = \delta_{mk}$

Equating coefficients of λ

$$i\hbar \dot{a}_m^{(1)}(t) = \sum_n e^{-i\omega_{mn} t} H'_{mn}(t) a_n^{(0)} \qquad (8)$$

Equating coefficients of λ^2

$$i\hbar \dot{a}_m^{(2)}(t) = \sum_n e^{i\omega_{mn} t} H'_{mn}(t) a_n^{(1)}(t)$$

In general,

$$i\hbar \dot{a}_m^{(s)}(t) = \sum_n e^{i\omega_{mn} t} H'_{mn}(t) a_n^{(s-1)}(t) \qquad (9)$$

From Eqn. (8),

$$i\hbar\dot{a}_m^{(1)}(t) = \sum_n e^{i\omega_{mn}t} H'_{mn}(t)\,\delta_{nk} = e^{i\omega_{mk}t} H'_{mk}(t) \quad \left(a_n^{(0)} = \delta_{nk}\right)$$

Integrating this equation with respect to time, we get

$$i\hbar a_m^{(1)}(t) = \int_0^t e^{i\omega_{mk}t} H'_{mk}(t)\,dt \tag{10}$$

where the constant of integration is taken to be zero in order that $a_m^{(1)}$ be zero at $t = 0$ (before the perturbation is applied).

Now, we have two cases depending on the form of H'.

Case I: $H'(t) = H'$, a constant for a certain interval 0 to t (see Fig. 18.1). Then

$$a_m^{(1)}(t) = \frac{H'_{mk}}{i\hbar}\int_0^t e^{i\omega_{mk}t}\,dt = \frac{H'_{mk}}{i\hbar}\left(\frac{e^{i\omega_{mk}t} - 1}{i\omega_{mk}}\right) \tag{11}$$

Therefore, the probability of reaching the state m at time t is

$$P_m(t) = \left|a_m^{(1)}(t)\right|^2 = \frac{\left|H'_{mk}\right|^2}{\hbar^2 \omega_{mk}^2}\left(e^{i\omega_{mk}t} - 1\right)\left(e^{-i\omega_{mk}t} - 1\right)$$

$$= \frac{\left|H'_{mk}\right|^2}{\hbar^2 \omega_{mk}^2}\left\{1 + 1 + -\frac{2}{2}\left(e^{i\omega_{mk}t} + e^{-i\omega_{mk}t}\right)\right\}$$

$$= \frac{\left|H'_{mk}\right|^2}{\hbar^2} \times \frac{2(1 - \cos\omega_{mk}t)}{\omega_{mk}^2} = \frac{\left|H'_{mk}\right|^2}{\hbar^2} \times \frac{2 \times 2\sin^2\left(\frac{1}{2}\omega_{mk}t\right)}{\omega_{mk}^2}$$

or

$$P_m(t) = \left|a_m^{(1)}(t)\right|^2 = t^2 \frac{\left|H'_{mk}\right|^2}{\hbar^2} \times \frac{\sin^2\left(\dfrac{\omega_{mk}t}{2}\right)}{\left(\omega_{mk}t/2\right)^2} \tag{12}$$

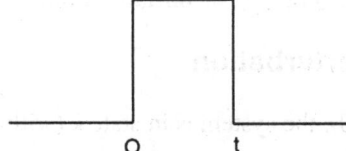

Fig. 18.1

Figure 18.2 shows the variation of $P_m(t)$ for various values of ω_{mk} viz. $0, \dfrac{2\pi}{t}, \dfrac{4\pi}{t}, \dfrac{6\pi}{t}, \dots$

when

$$\omega_{mk} = 0, \quad \frac{\sin^2\left(\dfrac{\omega_{mk}t}{2}\right)}{\left(\dfrac{\omega_{mk}t}{2}\right)^2} = 1 \text{ or } \frac{\sin^2\left(\dfrac{\omega_{mk}t}{2}\right)}{\left(\dfrac{\omega_{mk}}{2}\right)^2} = t^2$$

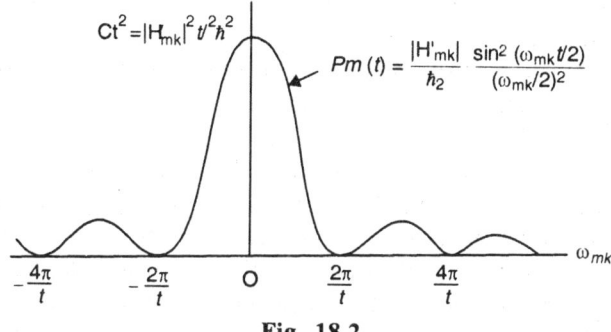

Fig. 18.2

so, the value of $P_m(t)$ at $\omega_{mk} = 0$ is $\left|H'_{mk}\right|^2 t^2 / \hbar^2$. The expression (12) shows that the transition probability P_m from k to m does not change monotonically with time but varies simple harmonically between zero and the maximum value with a frequency ω_{mk} associated with this pair of levels.

Let us now consider $P_m(t)$ as a function of the energy difference between the initial and final levels (k and m respectively), for fixed t. The main peak of the curve which occurs at $\omega_{mk} = 0$ is of height Ct^2 $\left(C = \left|H'_{mk}\right|^2 / \hbar^2\right)$. The subsidiary peaks are very much smaller, the height of the biggest of these (at $\omega_{mk} \cong \pm 3\pi/t$) being less than 5% of that of the main one. Therefore, transitions from k take place with appreciable probability only to those levels m such that ω_{mk} falls under the main peak: $\left|\omega_{mk}\right| \leq \dfrac{2\pi}{t}$. In other words the magnitude of the energy difference $\left|\hbar\omega_{mk}\right|$ between the initial and final states is very unlikely to be significantly higher than $\hbar\dfrac{2\pi}{t}$ or (h/t). Hence the energy changes ΔE caused by a perturbation which lasts for a time t and which is constant (or smoothly varying in that interval) are $\approx h/t$, so that

$$(\Delta E).t \approx h$$

Thus here, energy is conserved subject to the uncertainty ΔE which is small when t is large, in other words a constant perturbation rarely causes transitions between states having measurably different energies.

18.3 Transitions into a Continuous Spectrum (The Golden Rule)

Since the transition probability is appreciable only for states having energies nearly equal to the energy of the initial state, the above considerations are best suited when the energies E_n (of H_0) in the neighbourhood of the initial energy are very closely spaced on the energy scale, thus forming practically a continuous spectrum. Examples of transitions into final states of a continuous spectrum are

(i) *Scattering*: The momentum $\vec{k}$ of a particle goes over to $\vec{k}'$. First order perturbation theory leads to the Born approximation.

(ii) *α decay*: Here also, in the final states i.e. the momenta of the α-particles lie in a continum.

(iii) *Optical transitions*: An excited state makes a transition to a lower state by emitting a photon whose momentum varies continuously.

(iv) *β-decay*

To satisfy the formal requirement that the spectrum of H_0 be discrete we imagine the system to be enclosed in a very large cubical box, at the walls of which periodic boundary conditions are imposed on the wave functions. The stationary states of the system will then be discrete, but separated in energy by an interval which is inversely proportional to (volume of the box)$^{2/3}$. If we then pass to the limit of a box of infinite size, the levels within a given energy interval increase in number and merge into a continuum. This limiting process is undoubtedly formal, and above results can be directly applied in calculating the transition probability.

Thus, we have a group of states with a constant initial energy. Let the final states have energies between E_m and $E_m + dE_m$. Let ρ_m be the density (on the energy scale) of final states, so that $\rho_m \, dE_m$ measures the number of final states in an interval dE_m containing the energy E_m.

In this case, the transition probability per unit time is given by

$$W_{k \to m} = \frac{1}{t} \int \left| a_m^{(1)}(t) \right|^2 \rho_m \, dE_m \tag{13}$$

As $t \to \infty$, breadth of the main peak of the transition probability curve becomes very small, then we regard H'_{mk} and ρ_m as quantities sufficiently independent of ω_{mk} (and therefore independent of E_m)

$$\therefore \qquad W_{k \to m} = \frac{\rho_m}{t} \, t^2 \, \frac{\left| H'_{mk} \right|}{\hbar^2} \int \frac{\sin^2 \left(\omega_{mk} t / 2 \right)}{\left(\omega_{mk} t / 2 \right)^2} \, dE_m$$

Substituting $x = \omega_{mk} t$

$$E_m = \hbar \omega_{mk} + E_k \text{ (constant)}$$

$$\therefore \qquad t \, dE_m = t\hbar \, d\omega_{mk} + 0 \equiv \hbar \, dx$$

$$\therefore \qquad W_{k \to m} = \rho_m \frac{\left| H'_{mk} \right|}{\hbar^2} \cdot \hbar \int_{-\infty}^{+\infty} \frac{\sin^2 (x/2)}{(x/2)^2} \, dx \tag{14}$$

Now,

$$\int_{-\infty}^{+\infty} \frac{\sin^2 (x/2)}{(x/2)^2} \, dx = \int_{-\infty}^{+\infty} \frac{1}{2} \frac{(1 - \cos x)}{x^2} \times 4 \, dx$$

$$= 2 \int_{-\infty}^{+\infty} \frac{1 - \cos x}{x^2} \, dx \tag{15}$$

This integral is evaluated by contour integration.

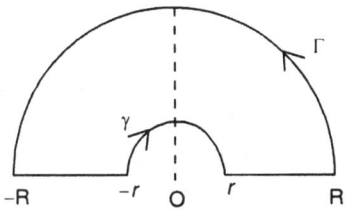

Fig. 18.3

Consider $f(z) = \dfrac{1 - e^{iz}}{z^2}$ therefore $\dfrac{1 - \cos x}{x^2} = \operatorname{Re} f(z)$

By residue theorem

$$\int_C f(z)\, dz = \int_{-R}^{-r} f(x)\, dx + \int_\gamma f(x)\, dx + \int_r^R f(x)\, dx + \int_\Gamma f(x)\, dx = 0$$

$$\left| \int_\Gamma f(z)\, dz \right| = \left| \int_\Gamma \frac{1 - e^{iz}}{z^2}\, dz \right| \le \int_\Gamma \frac{\left| 1 - e^{iz} \right|}{\left| z^2 \right|}\, dz$$

$$\left| \int_\Gamma f(z)\, dz \right| \le \int_0^\pi \frac{\left| 1 - e^{i\,\operatorname{Re}^{i\theta}} \right|}{\left| R^2 e^{2i\theta} \right|} \left| i \right| \left| \operatorname{Re}^{i\theta} \right| \left| d\theta \right| \qquad \left(z = \operatorname{Re}^{i\theta}, dz = i\operatorname{Re}^{i\theta}\, d\theta \right)$$

$$\le \int_0^\pi \frac{1 + e^{-R \sin\theta}}{R}\, d\theta \quad \text{which} \ \to 0 \ \text{as} \ R \to \infty.$$

$$\therefore \qquad \int_\Gamma f(z)\, dz = 0$$

$$\int_\gamma f(z)\, dz = \int_\pi^0 \frac{1 - e^{ire^{i\theta}}}{r^2 e^{2i\theta}}\, i\, r\, e^{i\theta}\, d\theta \qquad\qquad z = \left(re^{i\theta} \right)$$

$$= \int_\pi^0 \left\{ 1 - 1 - ire^{i\theta} - i^2 r^2 e^{2i\theta} - \dots \right\} \frac{i}{r} e^{-i\theta}\, d\theta$$

$$= \int_\pi^0 \left\{ 1 + \text{terms containing powers of } r \right\} d\theta$$

$$\therefore \qquad \lim_{r \to 0} \int_\gamma f(z)\, dz = -\pi$$

Hence when $R \to \infty$ and $r \to 0$

$$\int_{-\infty}^0 f(x)\, dx - \pi + \int_0^\infty f(x)\, dx = 0 \quad \text{or} \quad \int_{-\infty}^\infty \frac{1 - e^{ix}}{x^2}\, dx = \pi.$$

Equating real parts on either sides, we get

$$\int_{-\infty}^\infty \frac{1 - \cos x}{x^2}\, dx = \pi \tag{16}$$

Hence from Eqns (14), (15) and (16), we get

$$W_{k \to m} = \frac{2\pi}{\hbar} \left| H'_{mk} \right|^2 \rho_m \tag{17}$$

This formula has been applied to a wide variety of quantum phenomena and in particular in β decay, it has given results which are actually observed experimentally. This is called the *Fermi's Golden Rule*. It states that the transition probability per unit time

(*i*) is non zero only between continuum states of the same energy,

(*ii*) is proportional to the square of $|H'_{mk}|$ of the perturbation connecting the states, and

(*iii*) is proportional to the density of final states.

18.4 Harmonic Perturbation

Equation (10) may be considered for the case of harmonic perturbation, i.e. H' depends harmonically on the time. We may write

$$H'(t) = H'e^{-i\omega t} + H'^{\dagger}e^{i\omega t} \qquad (18)$$

because, a harmonic perturbation is not hermitian by itself so its hermitian conjugate is added to it. Let $H'(t)$ acts during the time interval $(0, t_0)$. We then have from Eqn. (10)

$$a_m^{(1)}(t) = \frac{1}{i\hbar} \int_0^{t_0} e^{i\omega_{mk}t} H'_{mk}(t)\, dt \; + \text{hermitian conjugate term}$$

or $\qquad a_m^{(1)}(t) = \frac{1}{i\hbar} \int_0^{t_0} e^{i\omega_{mk}t - i\omega t} H'_{mk} dt \; + \text{h.c. term}$

$$\left(\because H'_{mk}(t) = \int u_m^* H'(t)\, u_k\, d\tau = e^{-i\omega t} \int u_m^* H' u_k\, d\tau = H'_{mk} e^{-i\omega t} \right)$$

(the time dependent factor is $e^{-i\omega t}$)

or $\qquad a_m^{(1)}(t) = \frac{H'_{mk}}{i\hbar} \int_0^{t_0} e^{i(\omega_{mk} - \omega)t} dt \; + \text{h.c. term}$

or $\qquad a_m^{(1)}(t) = \frac{H'_{mk}}{i\hbar}\left[\frac{e^{i(\omega_{mk}-\omega)t}}{i(\omega_{mk}-\omega)}\right]_0^{t_0} + \frac{H'^{\dagger}_{mk}}{i\hbar}\left[\frac{e^{i(\omega_{mk}+\omega)t}}{i(\omega_{mk}+\omega)}\right]_0^{t_0}$

or $\qquad a_m^{(1)}(t) = -H'_{mk}\left\{\frac{e^{i(\omega_{mk}-\omega)t_0}-1}{\hbar(\omega_{mk}-\omega)}\right\} - H'^{\dagger}_{mk}\left\{\frac{e^{i(\omega_{mk}+\omega)t_0}-1}{\hbar(\omega_{mk}+\omega)}\right\} \qquad (19)$

Each term of this expression is exactly like Eqn. (11) except that (ω_{mk}) is replaced by $(\omega_{mk} - \omega)$ in the first term and in the second term (ω_{mk}) is replaced by $(\omega_{mk} + \omega)$, and H'_{mk} by $H'^{\dagger}_{mk}$. Further, the two terms act in general quite independently of each other, because transitions (from k) take place with appreciable probability only to those levels m for which $(\omega_{mk} \pm \omega) \to 0$, and therefore for any ω and t_0 large such that $\omega t_0 \gg 1$ (which is practically the case) any value of ω_{mk} (which makes one term large) will make the other effectively zero. Hence only one peak condition will be satisfied at a time.

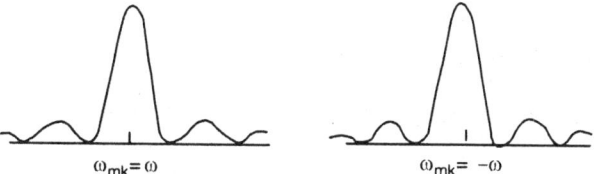

Fig. 18.4

$$\omega_{mk} - \omega = 0 \Rightarrow \hbar\omega_{mk} = \hbar\omega \text{ or } E_m = E_k + \hbar\omega$$

$$\omega_{mk} + \omega = 0 \Rightarrow \hbar\omega_{mk} = -\hbar\omega \text{ or } E_m = E_k - \hbar\omega$$

So, we find that the final energy is either greater or less than the initial energy by an amount contributed exactly by the perturbing term.

The perturbation $H'e^{-i\omega t}$ induces transitions from E_k only to a level E_m whose energy is higher than E_k by $\hbar\omega$ (this corresponds to absorption of energy by the system from the perturbing agency). The perturbation $H'e^{i\omega t}$ induces transitions from E_k only to a level E_m whose energy is lower than E_k by $\hbar\omega$ (i.e. energy is given away by the system to the perturbing agency).

18.5 Adiabatic Perturbation

An adiabatic perturbation is that which is turned on very slowly. This is concerned with the case of collision or interaction in which the perturbing potential varies very little, acting only during some finite time-interval. As a consequence the system comes back to its original condition without changes in the original energies of the states. For example, in case of collision of gas molecules, as atoms approach each other, it may be apparent that as a result of collisions, there may result transitions among the electronic states. But we know that molecular velocities are low (usually ~10^4 cm/s) whereas velocities of electrons in atoms are much higher ($\approx 10^8$ cm/s). During an electronic period of rotation in an atom, the molecule does not move very far, therefore it is a good approximation to take the nuclei as fixed for calculating the electronic motion. Furthermore, the nuclear motion can be calculated under the assumption that the electrons have their steady motion for each instantaneous arrangement of nuclei.

This is an example of adiabatic approximation in which the electrons retain their original quantum state. This collision process is almost adiabatic as the interaction energy will not change appreciably. The Hamiltonian will be time dependent varying slowly with time. The appropriate wave equation is

$$H(t)\,\psi = i\hbar\frac{\partial\psi}{\partial t} \tag{20}$$

Since $H(t)$ is a slowly varying function of time, to a good approximation, the solution of this equation (since H' is varying very little) at any instant t_1 can be evaluated by setting the Hamiltonian

$$H'(t_1) \equiv H(t_1)$$

With an eigenvalue given by

$$H(t_1)\, \psi'_k(r,t_1) = E_k(t_1)\, \psi'_k(r,t_1) \tag{21}$$

Approximate form of the eigenfunctions ψ'_k would be

$$\psi'_k \equiv \psi_k(t_1)\, e^{-(i/\hbar)\int_0^{t_1} E_k(t_1)\, dt_1} \tag{22}$$

with

$$\psi_k(t_1) = \sum_k a_k(t)\, \psi_k^{(0)} \tag{23}$$

We shall now attempt to derive an expression for a_k's of Eqn. (23). In general

$$\psi' = \sum_k a_k(t)\, \psi_k^{(0)} \exp\left\{-\frac{i}{\hbar}\int_0^{t_1} E_k(t_1)\, dt_1\right\} \tag{24}$$

Substituting this in Eqn. (20), we get

$$H \sum_k a_k(t)\, \psi_k^{(0)} \exp\left\{-\frac{i}{\hbar}\int_0^{t_1} E_k(t_1)\, dt_1\right\}$$

$$= i\hbar\left[\sum_k\left(\dot{a}_k \psi_k^{(0)} + a_k \frac{\partial \psi_k^{(0)}}{\partial t}\right)\exp\left\{-\frac{i}{\hbar}\int_0^{t_1} E_k(t_1)\, dt_1\right\}\right]$$

$$+ \sum_k a_k \psi_k^{(0)} E_k \exp\left\{-\frac{i}{\hbar}\int_0^{t_1} E_k(t_1)\, dt_1\right\}$$

or

$$i\hbar \sum_k\left(\dot{a}_k \psi_k^{(0)} + a_k \frac{\partial \psi_k^{(0)}}{\partial t}\right)\exp\left\{-\frac{i}{\hbar}\int_0^{t_1} E_k(t_1)\, dt_1\right\} = 0$$

or

$$\sum_k\left(\dot{a}_k \psi_k^{(0)} + a_k \frac{\partial \psi_k^{(0)}}{\partial t}\right)\exp\left\{-\frac{i}{\hbar}\int_0^{t_1} E_k(t_1)\, dt_1\right\} = 0 \tag{25}$$

Multiplying the above by complex conjugate function $\psi_m^* \exp\left\{\frac{i}{\hbar}\int_0^{t_1} E_m(t_1)\, dt_1\right\}$ and integrating, we get (dropping the symbol (0))

$$\int \sum_k \dot{a}_k \psi_m^* \psi_k \exp\left\{-\frac{i}{\hbar}\int_0^{t_1}(E_k - E_m)\, dt_1\right\} d\tau$$

$$+ \sum_k a_k \int \psi_m^* \frac{\partial \psi_k}{\partial t} \exp\left\{-\frac{i}{\hbar}\int_0^{t_1}(E_k - E_m)\, dt_1\right\} d\tau = 0$$

Using orthonormality condition, we know that the first integral gives finite value only for $m = k$ and it is zero for $m \ne k$. Thus,

$$\dot{a}_k = -\sum_k a_k \int \psi_m^* \frac{\partial \psi_k}{\partial t} \exp\left\{-\frac{i}{\hbar}\int_0^{t_1}(E_k - E_m)\, dt_1\right\} d\tau$$

$$= -\sum_k a_k \int \psi_m^* \frac{\partial \psi_k}{\partial t} d\tau \cdot \exp\left\{-\frac{i}{\hbar}\int_0^{t_1}(E_k - E_m)\, dt_1\right\} \tag{26}$$

The above integral over space is evaluated as follows. We have,

$$H(t)\, \psi_k(t) = E_k(t)\, \psi_k(t) \tag{27}$$

Differentiating both sides partially w.r.t. t, we get

$$\frac{\partial H}{\partial t}\psi_k + H\frac{\partial \psi_k}{\partial t} = \frac{\partial E_k}{\partial t}\psi_k + E_k\frac{\partial \psi_k}{\partial t}$$

Multiplying by ψ_m^* and integrating over all space

$$\int \psi_m^* \frac{\partial H}{\partial t}\psi_k d\tau + \int \psi_m^* H\frac{\partial \psi_k}{\partial t}d\tau = \frac{\partial E_k}{\partial t}\int \psi_m^* \psi_k d\tau + E_k \int \psi_m^* \frac{\partial \psi_k}{\partial t}\,d\tau$$

As $\int \psi_m^* \psi_k d\tau = 0$ and H is Hermitian, we get

$$\int \psi_m^* \frac{\partial H}{\partial t}\psi_k d\tau + \int (H\psi_m)^* \frac{\partial \psi_k}{\partial t}\,d\tau = E_k \int \psi_m^* \frac{\partial \psi_k}{\partial t}\,d\tau$$

or $\quad \int \psi_m^* \frac{\partial H}{\partial t}\psi_k d\tau + E_m \int \psi_m^* \frac{\partial \psi_k}{\partial t}\,d\tau = E_k \int \psi_m^* \frac{\partial \psi_k}{\partial t}\,d\tau \qquad \left(\because H\psi_m = E_m\psi_m\right)$

or $\qquad \left(E_m - E_k\right)\int \psi_m^* \frac{\partial \psi_k}{\partial t}\,d\tau = -\int \psi_m^* \frac{\partial H}{\partial t}\psi_k d\tau$

$\Rightarrow \qquad\qquad \int \psi_m^* \frac{\partial \psi_k}{\partial t}\,d\tau = -\frac{1}{\left(E_m - E_k\right)}\int \psi_m^* \frac{\partial H}{\partial t}\psi_k d\tau$

$$= -\frac{\int \psi_m^* \frac{\partial H}{\partial t}\psi_k d\tau}{\left(E_m - E_k\right)} \quad \text{(for } k \neq m) \qquad (28)$$

Substituting the value in Eqn. (26), we get

$$\dot{a}_k = -\sum_{k \neq m}\frac{a_k\left(\int \psi_m^* \frac{\partial H}{\partial t}\psi_k d\tau\right)\exp\left\{-\frac{i}{\hbar}\int_0^{t_1}\left(E_k - E_m\right)dt_1\right\}}{\left(E_k - E_m\right)} \qquad (29)$$

Using the method of variation / adjustment of coefficients, we can let the system start with a_n and $a_k = 0$ for $k \neq n$. Then,

$$\dot{a}_{mn} = -\frac{\left\langle m\left|\frac{\partial H}{\partial t}\right|n\right\rangle}{\left(E_n - E_m\right)}\exp\left\{-\frac{i}{\hbar}\left(E_n - E_m\right)t_1\right\} \qquad (30)$$

To evaluate a_{kn}, we can neglect the slow change of $\left(\frac{\partial H}{\partial t}\right)_{mn}$, and integrate the above equation so that

$$a_{mn} = -\frac{\left\langle m\left|\frac{\partial H}{\partial t}\right|n\right\rangle}{\left(E_n - E_m\right)}\int_{-\infty}^{t_0}e^{-(i/\hbar)\,(E_n - E_m)\,t_1}dt_1$$

So,

$$a_{mn} = -\frac{i\hbar}{\left(E_n - E_m\right)^2}\left\langle m\left|\frac{\partial H}{\partial t}\right|n\right\rangle\left\{\exp\left[-\frac{i}{\hbar}\left(E_n - E_m\right)t_0\right] - 1\right\} \qquad (31i)$$

Since $E_n - E_m = \hbar\omega_{nm}$, therefore,

$$a_{mn} = \frac{1}{i\hbar\omega_{nm}^2} \left\langle m \left| \frac{\partial H}{\partial t} \right| n \right\rangle \left\{ \exp\left(-i\omega_{nm}t_0\right) - 1 \right\} \tag{31ii}$$

The transition probability would be

$$|a_{mn}|^2 = \frac{\left| \left\langle m \left| \frac{\partial H}{\partial t} \right| n \right\rangle \right|^2}{\hbar^2 \omega_{nm}^4} \tag{32}$$

Thus for $|a_{mn}|^2$ to be small, $(\partial H/\partial t)_{mn}$ should be small for the system to be adiabatic and beside this, $(E_n - E_m) = \hbar\omega_{nm}$ should not be small.

18.6 Sudden Approximation

The sudden approximation is applicable when the Hamiltonian changes appreciably during a very short but finite interval of time t_0. We have, from Eqn. (30),

$$a_{mn} = \int_0^{t_0} \frac{\left\langle m \left| \frac{\partial H}{\partial t} \right| n \right\rangle}{(E_n - E_m)} \exp\left\{ -\frac{i}{\hbar}(E_n - E_m) t \right\} dt$$

In sudden approximation, the variation $(\partial H/\partial t)$ is appreciable and cannot be neglected. So, to obtain a_{mn}, we integrate the above expression (w.r.t. time) by parts:

$$a_{mn} = \int_0^{t_0} \frac{\left\langle m \left| \frac{\partial H}{\partial t} \right| n \right\rangle}{(E_n - E_m)} \exp\left\{ -\frac{i}{\hbar}(E_n - E_m) t \right\} dt \tag{33}$$

or

$$a_{mn} = \left[\frac{\left\langle m | H(t) | n \right\rangle}{(E_n - E_m)} \exp\left\{ -\frac{i}{\hbar}(E_n - E_m) t \right\} \right]_0^{t_0}$$

$$- \int_0^{t_0} \frac{\left\langle m | H(t) | n \right\rangle}{(E_n - E_m)} \exp\left\{ -\frac{i}{\hbar}(E_n - E_m) t_1 \right\} \left\{ -\frac{i}{\hbar}(E_n - E_m) t_1 \right\} dt \tag{34}$$

The sudden perturbation is characterized by the condition that energy of the system changes by an amount $\Delta E \left(= (E_n - E_m) \right)$ in a time t_0 which is much less than the characteristic time $\hbar/\Delta E$ i.e.

$$\frac{\hbar}{\Delta E} \gg t_0 \text{ or } \Delta E \ll \frac{\hbar}{t_0} \tag{35}$$

Since the state of the system remains unaltered, so, the first term in the expression (34) is zero (no change between the limits 0 to t_0). This gives

$$a_{mn} = \frac{i}{\hbar} \int_0^{t_0} \left\langle m | H(t) | n \right\rangle \exp\left\{ -\frac{i}{\hbar}(E_n - E_m) t \right\} dt$$

Because of the condition (35), we can take the term $\exp\left\{-\dfrac{i}{\hbar}(E_n - E_m)\,t\right\}$ outside the integral sign, so that

$$a_{mn} = \frac{i}{\hbar}e^{i\omega_{mn}t}\int_0^{t_0}\langle m|H(t)|n\rangle\,dt$$

$$= \frac{i}{\hbar}e^{i\omega_{mn}t}\langle m|H(t_0)-H(0)|n\rangle\,t_0$$

$$= \frac{it_0}{\hbar}e^{i\omega_{mn}t}\langle m|H'|n\rangle \text{ where } H' = H(t_0)-H(0)$$

Therefore probability

$$|a_{mn}|^2 = \frac{t_0^2}{\hbar^2}\;|\langle m|H'|n\rangle|^2 = \frac{|\langle m|H'|n\rangle|^2}{\hbar^2\omega_{nm}^2} \tag{36}$$

(using condition 35)

18.7 A Charged Particle in an Electromagnetic Field

Prior to apply time-dependent perturbation theory to a charged particle, it is worthwhile to investigate the effect of electric and magnetic fields on the particle.
    The electromagnetic force

$$\vec{F} = e\vec{E} + \frac{e}{c}\vec{v}\times\vec{B} \text{ (C.G.S. units)} \tag{37}$$

c is the speed of light.
    The electric and magnetic fields can be expressed in terms of scalar and vector potentials as

$$\vec{E} = -\frac{1}{c}\frac{\partial\vec{A}}{\partial t} - \nabla\phi$$

and

$$\vec{B} = \nabla\times\vec{A} \tag{38}$$

Substituting these values in Eqn. (37)

$$m\frac{d^2\vec{r}}{dt^2} = e\left[-\frac{1}{c}\frac{\partial\vec{A}}{\partial t} - \nabla\phi\right] + \frac{e}{c}\vec{v}\times\left(\nabla\times\vec{A}\right)$$

writing

$$\vec{r} = x\hat{i} + y\hat{j} + z\hat{k} \text{ and } \vec{A} = A_x\hat{i} + A_y\hat{j} + A_z\hat{k}$$

We have

$$m\frac{d^2x}{dt^2} = -\frac{e}{c}\frac{\partial A_x}{\partial t} - e\frac{\partial\phi}{\partial t} + \frac{e}{c}\left[\dot{y}\left(\frac{\partial A_y}{\partial x} - \frac{\partial A_x}{\partial y}\right) - \dot{z}\left(\frac{\partial A_x}{\partial z} - \frac{\partial A_z}{\partial x}\right)\right]$$

where

$$\dot{y} = \frac{dy}{dt} \text{ and } \dot{z} = \frac{dz}{dt}$$

We have similar equations for y- and z components of force.
    If K is the kinetic energy, then

$$m\frac{d^2x}{dt^2} = \frac{d}{dt}\left(\frac{\partial K}{\partial \dot{x}}\right)$$

∴
$$\frac{d}{dt}\left(\frac{\partial K}{\partial \dot{x}}\right) = -\frac{e}{c}\,\frac{\partial A_x}{\partial t} - e\frac{\partial \phi}{\partial x}$$

$$+\frac{e}{c}\left[\left(\dot{x}\frac{\partial A_x}{\partial x} + \dot{y}\frac{\partial A_y}{\partial x} + \dot{z}\frac{\partial A_z}{\partial x}\right) - \left(\dot{x}\frac{\partial A_x}{\partial x} + \dot{y}\frac{\partial A_x}{\partial y} + \dot{z}\frac{\partial A_x}{\partial z}\right)\right]$$

or
$$\frac{d}{dt}\left(\frac{\partial K}{\partial \dot{x}}\right) = -\frac{e}{c}\,\frac{\partial A_x}{\partial t} - e\frac{\partial \phi}{\partial x} + \frac{e}{c}\left[\frac{\partial}{\partial x}\left(\vec{v}\cdot\vec{A}\right) - \left(\frac{dA_x}{dt} - \frac{\partial A_x}{\partial t}\right)\right]$$

or
$$\frac{d}{dt}\left(\frac{\partial K}{\partial \dot{x}}\right) = -e\frac{\partial \phi}{\partial x} + \frac{e}{c}\,\frac{\partial}{\partial x}\left(\vec{v}\cdot\vec{A}\right) - \frac{e}{c}\,\frac{dA_x}{dt}$$

or
$$\frac{d}{dt}\left(\frac{\partial K}{\partial \dot{x}} + \frac{e}{c}A_x\right) = -e\frac{\partial \phi}{\partial x} + \frac{e}{c}\,\frac{\partial}{\partial x}\left(\vec{v}\cdot\vec{A}\right)$$

or
$$\frac{d}{dt}\left(\frac{\partial K}{\partial \dot{x}} + \frac{e}{c}A_x\right) - \frac{\partial}{\partial x}\left[-e\phi + \frac{e}{c}\left(\vec{v}\cdot\vec{A}\right)\right] = 0 \qquad (39)$$

Since kinetic energy is velocity dependent and does not depend on position coordinates (x, y, z) and the scalar potential ϕ is position dependent and does not depend on velocity components $(\dot{x}, \dot{y}, \dot{z})$

∴
$$\frac{\partial \phi}{\partial \dot{x}} = \frac{\partial \phi}{\partial \dot{y}} = \frac{\partial \phi}{\partial \dot{z}} = 0 \text{ and } \frac{\partial K}{\partial x} = \frac{\partial K}{\partial y} = \frac{\partial K}{\partial z} = 0 \qquad (40)$$

We may write A_x as

$$A_x = \frac{\partial}{\partial \dot{x}}\left(\dot{x}A_x + \dot{y}A_y + \dot{z}A_z\right) = \frac{\partial}{\partial \dot{x}}\left(\vec{v}\cdot\vec{A}\right) \qquad (41)$$

Using Eqn. (40) and (41), Eqn. (39) may be expressed as

$$\frac{d}{dt}\left[\frac{\partial}{\partial \dot{x}}\left\{K - e\phi + \frac{e}{c}\left(\vec{v}\cdot\vec{A}\right)\right\}\right] - \frac{\partial}{\partial x}\left[K - e\phi + \frac{e}{c}\left(\vec{v}\cdot\vec{A}\right)\right] = 0 \qquad (42)$$

Similarly

$$\frac{d}{dt}\left[\frac{\partial}{\partial \dot{y}}\left\{K - e\phi + \frac{e}{c}\left(\vec{v}\cdot\vec{A}\right)\right\}\right] - \frac{\partial}{\partial y}\left[K - e\phi + \frac{e}{c}\left(\vec{v}\cdot\vec{A}\right)\right] = 0$$

and
$$\frac{d}{dt}\left[\frac{\partial}{\partial \dot{z}}\left\{K - e\phi + \frac{e}{c}\left(\vec{v}\cdot\vec{A}\right)\right\}\right] - \frac{\partial}{\partial z}\left[K - e\phi + \frac{e}{c}\left(\vec{v}\cdot\vec{A}\right)\right] = 0$$

The above equations are in the form of Lagrange's equation, with the Lagrangian L given by

$$L = K - e\phi + \frac{e}{c}\left(\vec{v}\cdot\vec{A}\right) \qquad (43a)$$

$$= \frac{1}{2}m\left(\dot{x}^2 + \dot{y}^2 + \dot{z}^2\right) - e\phi + \frac{e}{c}\left(\dot{x}A_x + \dot{y}A_y + \dot{z}A_z\right) \qquad (43b)$$

Canonical momentum associated with x is

$$p_x = \frac{\partial L}{\partial \dot{x}} = m\dot{x} + \frac{e}{c}A_x$$

Similarly

$$p_y = m\dot{y} + \frac{e}{c}A_y \text{ and } p_z = m\dot{z} + \frac{e}{c}A_z \qquad (44)$$

The Hamiltonian function in terms of Lagrangian is

$$H = \sum p_k q_k - L = \left(p_x \dot{x} + p_y \dot{y} + p_z \dot{z}\right) - L \qquad (45)$$

or

$$H = \left(m\dot{x} + \frac{e}{c}A_x\right)\dot{x} + \left(m\dot{y} + \frac{e}{c}A_y\right)\dot{y} + \left(m\dot{z} + \frac{e}{c}A_z\right)\dot{z}$$
$$- \frac{1}{2}m\left(\dot{x}^2 + \dot{y}^2 + \dot{z}^2\right) + e\phi - \frac{e}{c}\left(\dot{x}A_x + \dot{y}A_y + \dot{z}A_z\right)$$

or

$$H = \frac{1}{2}m\left(\dot{x}^2 + \dot{y}^2 + \dot{z}^2\right) + e\phi \qquad (46)$$

From Eqn. (44), we have

$$\dot{x} = \frac{p_x}{m} - \frac{e}{mc}A_x ; \quad \dot{y} = \frac{p_y}{m} - \frac{e}{mc}A_y ; \quad \dot{z} = \frac{p_z}{m} - \frac{e}{mc}A_z$$

So Eqn. (46) becomes

$$H = \frac{1}{2m}\left[\left(p_x - \frac{eA_x}{c}\right)^2 + \left(p_y - \frac{eA_y}{c}\right)^2 + \left(p_z - \frac{eA_z}{c}\right)^2\right] + e\phi \qquad (47)$$

Replacing p_x by $\left(\frac{\hbar}{i}\right)\frac{\partial}{\partial x}$, we have for a wave function ψ:

$$\left(p_x - \frac{eA_x}{c}\right)^2 \psi = -\hbar^2\frac{\partial^2\psi}{\partial x^2} + 2i\hbar\frac{e}{c}A_x\frac{\partial\psi}{\partial x} + i\hbar\frac{e}{c}\left(\frac{\partial A_x}{\partial x}\right)\psi + \frac{e^2}{c^2}A_x^2\psi$$

and similar expressions for $\left(p_y - \frac{eA_y}{c}\right)$ and $\left(p_z - \frac{eA_z}{c}\right)$. Using these expressions, we get from Eqn. (47)

$$H = \frac{1}{2m}\left[-\hbar^2\nabla^2 + 2i\hbar\frac{e}{c}\vec{A}\cdot\nabla + i\hbar\frac{e}{c}\nabla\cdot\vec{A} + \frac{e^2}{c^2}A^2\right] + e\phi$$

$$H = -\frac{\hbar^2}{2m}\nabla^2 + e\phi + \frac{ie\hbar}{mc}\vec{A}\cdot\nabla + \frac{ie\hbar}{2mc}\nabla\cdot\vec{A} + \frac{e^2}{2mc^2}A^2 \qquad (48)$$

In an electromagnetic field, according to Maxwell's equations, $\nabla\cdot\vec{A} = 0$, hence

$$H = -\frac{\hbar^2}{2m}\nabla^2 + e\phi + \frac{ie\hbar}{mc}\vec{A}\cdot\nabla + \frac{e^2}{2mc^2}A^2 \qquad (49)$$

This may be expressed as

$$H = H_0 + H' \qquad (50)$$

where

$$H_0 = -\frac{\hbar^2}{2m}\nabla^2 + e\phi = -\frac{\hbar^2}{2m}\nabla^2 + V \qquad (51)$$

V is the potential energy and the perturbation (or interaction) term

$$H' = H_{\text{int}} = \frac{ie\hbar}{mc}\vec{A}\cdot\nabla + \frac{e^2}{2mc^2}A^2 \qquad (52)$$

For weak field, term $e^2 A^2 / (2mc^2)$ may be neglected, then perturbing part of hamiltonian is

$$H' = H_{\text{int}} = \frac{ie\hbar}{mc}\vec{A}\cdot\nabla = -\frac{e}{mc}\vec{A}\cdot(-i\hbar\nabla) \qquad (53a)$$

or

$$H' = -\frac{e}{mc}\vec{A}\cdot\vec{p} \qquad (53b)$$

18.8 Interaction of Atoms with Electromagnetic Waves (Application of Time-dependent Perturbation Theory to Radiation Theory)

We now discuss the interaction of atoms with electromagnetic fields. The treatment is semiclassical because the motion of atoms is quantized and electromagnetic fields (described by the vector potential $\vec{A}$) are classical.

The vector potential satisfies

$$-\nabla^2\vec{A}(\vec{r},t) + \frac{1}{c^2}\frac{\partial^2\vec{A}(\vec{r},t)}{\partial t^2} = \frac{4\pi}{c}\vec{j}(\vec{r},t)$$

Away from the sources

$$-\nabla^2\vec{A}(r,t) + \frac{1}{c^2}\frac{\partial^2\vec{A}(\vec{r},t)}{\partial t^2} = 0 \qquad (54i)$$

and

$$\nabla\cdot\vec{A} = 0 \qquad (54ii)$$

A typical monochromatic plane wave solution of Eqn. (54i) can be written as

$$\vec{A}(\vec{r},t) = \vec{A}_0^*(\vec{r})\, e^{i\omega t} + \vec{A}_0(\vec{r})\, e^{-i\omega t} \qquad (55)$$

If we substitute Eqn. (55) in Eqn. (54i), we find

$$-\nabla^2\vec{A}_0(\vec{r}) + \frac{\omega^2}{c^2}\vec{A}_0(\vec{r}) = 0 \qquad (56)$$

It is satisfied by

$$\vec{A}_0(\vec{r}) = A_0 e^{i\vec{k}\cdot\vec{r}} \qquad (57)$$

with
$$\left|\vec{k}\right| = \frac{\omega}{c} \tag{58}$$

The choice of the gauge Eqn. (54ii) implies that
$$\vec{k} \cdot \vec{A}_0 = 0 \tag{59}$$

The electric and magnetic fields corresponding to this vector potential are
$$\vec{E} = -\frac{1}{c}\frac{\partial \vec{A}}{\partial t} = \frac{i\omega}{c}\vec{A}_0 e^{i\left(\vec{k}\cdot\vec{r}-\omega t\right)} - \frac{i\omega}{c}\vec{A}_0^* e^{-i\left(\vec{k}\cdot\vec{r}-\omega t\right)} \tag{60}$$

and
$$\vec{B} = \nabla \times \vec{A} = i\vec{k} \times \vec{A}_0 e^{i\left(\vec{k}\cdot\vec{r}-\omega t\right)} - i\vec{k} \times \vec{A}_0^* e^{-i\left(\vec{k}\cdot\vec{r}-\omega t\right)} \tag{61}$$

Hence the energy density of the electromagnetic field is
$$\frac{1}{8\pi}\left(\vec{E}^2 + \vec{B}^2\right) = \frac{1}{8\pi}\left[2\frac{\omega^2}{c^2}\vec{A}_0 \cdot \vec{A}_0^* + 2\left(\vec{k} \times \vec{A}_0\right)\cdot\left(\vec{k} \times \vec{A}_0^*\right) + \text{time dependent terms}\right] \tag{62}$$

If we take time average, and use Eqn. (59), then we have
$$\left(\vec{k} \times \vec{A}_0\right)\cdot\left(\vec{k} \times \vec{A}_0^*\right) = k^2 \vec{A}_0 \cdot \vec{A}_0^* \tag{63}$$

so, using $k^2 = \omega^2/c^2$, we have from Eqns (62) and (63)
$$\frac{1}{8\pi}\left(\vec{E}^2 + \vec{B}^2\right) = \frac{1}{8\pi}\left[\frac{2\omega^2}{c^2}\vec{A}_0 \cdot \vec{A}_0^* + \frac{2\omega^2}{c^2}\vec{A}_0 \cdot \vec{A}_0^*\right]$$
$$= \frac{\omega^2}{2\pi c^2}\vec{A}_0 \cdot \vec{A}_0^* \tag{64}$$

and Poynting vector
$$S = \frac{c}{8\pi}\left(E^2 + B^2\right)$$

or
$$S = \frac{\omega^2}{2\pi c}\left|\vec{A}_0\right|^2 \tag{65}$$

If the system is enclosed in a box having a volume V, then total electromagnetic energy
$$\frac{1}{8\pi}\int d^3r\left(\vec{E}^2 + \vec{B}^2\right) = \frac{\omega^2 V}{2\pi c^2}\left|\vec{A}_0\right|^2 \tag{66}$$

If there are N photons (each of energy $\hbar\omega$) having the total energy of Eqn. (66), then
$$N\hbar\omega = \frac{\omega^2 V}{2\pi c^2}\left|\vec{A}_0\right|^2 \tag{67}$$

The direction of A_0 is determined by the polarization of the electric field (denoted by unit vector $\hat{\varepsilon}$)
$$\hat{\varepsilon} \cdot \hat{\varepsilon} = 1 \text{ and } \hat{\varepsilon} \cdot \vec{k} = 0$$

($\vec{k}$ is always along the direction of propagation)

From Eqn. (67),

$$\left|\vec{A}_0\right| = \left(\frac{2\pi c^2 N \hbar}{\omega V}\right)^{1/2} \tag{68}$$

and

$$\vec{A}(\vec{r},t) = \left(\frac{2\pi c^2 N \hbar}{\omega V}\right)^{1/2} \hat{\varepsilon}\, e^{i\left(\vec{k}\cdot\vec{r}-\omega t\right)} \tag{69}$$

Now, when we write

$$\vec{A}(\vec{r},t) = \vec{A}_0(\vec{r})\, e^{-i\omega t} + \vec{A}_0^*(\vec{r})\, e^{i\omega t}$$

the first term is associated with the absorption of a photon and the 2nd term is associated with the emission of a photon. Consequently, for absorption of a light quantum by a charged particle from an initial state having N photons (of frequency ω), we have

$$\vec{A}(\vec{r},t) = \left(\frac{2\pi c^2 N \hbar}{\omega V}\right)^{1/2} \hat{\varepsilon}\, e^{i\left(\vec{k}\cdot\vec{r}-\omega t\right)} \tag{70}$$

For the emission of a photon by a charged particle into a final state having $(N + 1)$ photons, from an initial state of N photons (of frequency ω), we have

$$\vec{A}^*(\vec{r},t) = \left(\frac{2\pi c^2 (N+1) \hbar}{\omega V}\right)^{1/2} \hat{\varepsilon}\, e^{-i\left(\vec{k}\cdot\vec{r}-\omega t\right)} \tag{71}$$

Hence for the emission of a single photon (of frequency ω) from a state having initially no photons, we have

$$H' = -\frac{e}{mc}\vec{A}^* \cdot \vec{p}$$

$$= -\frac{e}{mc}\left(\frac{2\pi c^2 \hbar}{\omega V}\right)^{1/2} \hat{\varepsilon}\cdot\vec{p}\, e^{-i\left(\vec{k}\cdot\vec{r}-\omega t\right)} \tag{72}$$

And for the absorption of a photon

$$H' = -\left(\frac{e}{mc}\right)\left(\frac{2\pi c^2 \hbar}{\omega V}\right)^{1/2} \hat{\varepsilon}\cdot\vec{p}\, e^{i\left(\vec{k}\cdot\vec{r}-\omega t\right)} \tag{73}$$

18.8.1 Absorption and Emission

Assuming k^{th} state as initial state, from Eqn. (8), for a final state m, we have

$$i\hbar \dot{a}_m^{(1)}(t) = H'_{mk}e^{-i\omega_{mk}t}a_k^{(0)} \tag{74}$$

where we have assumed that duration of perturbation is very small and the perturbation is fairly weak, so that probability of transition is very small, and we may take $a_k^{(0)} = 1$.

Here

$$\omega_{mk} = \left(\frac{E_m^0 - E_k^0}{\hbar}\right) \quad (\text{assuming } E_m > E_k) \tag{75}$$

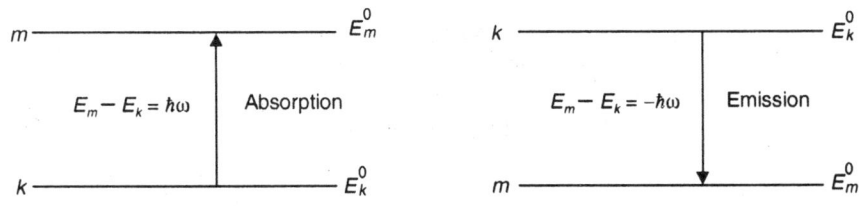

Fig. 18.5

So,

$$i\hbar\dot{a}_m^{(1)}(t) = H'_{mk}e^{-i\left(\omega_{mk}-\omega\right)t} + H''_{mk}e^{i\left(\omega_{mk}+\omega\right)t} \tag{76}$$

(The first term corresponds to absorption and the 2nd term, to the emission). From Eqn. (52)

$$H'_{mk} = \frac{ie\hbar}{mc}\int u_m^* e^{i\left(\vec{k}\cdot\vec{r}\right)}\left(\vec{A}_0\cdot\nabla\right)u_k\,d\tau$$

and

$$H''_{mk} = \frac{ie\hbar}{mc}\int u_m^* e^{-i\left(\vec{k}\cdot\vec{r}\right)}\left(\vec{A}_0^*\cdot\nabla\right)u_k\,d\tau \tag{77}$$

If harmonic perturbation of frequency ω is switched on at $t = 0$, then integrating Eqn. (76) w.r.t. t, we get

$$a_m^{(1)}(t) = \frac{1}{i\hbar}\left[\int_0^t H'_{mk}e^{i\left(\omega_{mk}-\omega\right)t}\,dt + \int_0^t H''_{mk}e^{i\left(\omega_{mk}+\omega\right)t}\,dt\right]$$

$$= \frac{H'_{mk}}{i\hbar}\left[\frac{e^{i\left(\omega_{mk}-\omega\right)t}-1}{i\left(\omega_{mk}-\omega\right)}\right] + \frac{H''_{mk}}{i\hbar}\left[\frac{e^{i\left(\omega_{mk}+\omega\right)t}-1}{i\left(\omega_{mk}+\omega\right)}\right]$$

or

$$a_m^{(1)}(t) = \frac{H'_{mk}}{\hbar}\frac{1-e^{i\left(\omega_{mk}-\omega\right)t}}{\left(\omega_{mk}-\omega\right)} + \frac{H''_{mk}}{\hbar}\frac{1-e^{i\left(\omega_{mk}+\omega\right)t}}{\left(\omega_{mk}+\omega\right)} \tag{78}$$

Out of the two terms, only one needs concern at a time. Transitions are probable only if $E_m - E_k = \pm\hbar\omega$. Of these two probabilities, one corresponds to absorption of radiation from the field and the other to emission induced by the field. The equation $E_m - E_k = \pm\hbar\omega$ ensures conservation of energy between the particle and the field.

For absorption: $E_m > E_k$, only first term of the Eqn. (78) is dominant and second term is negligible. Then

$$a_m^{(1)}(t) = H'_{mk}\frac{1-e^{i\left(\omega_{mk}-\omega\right)t}}{\hbar\left(\omega_{mk}-\omega\right)} \tag{79}$$

and probability of finding the system in the m-state at the end of the time interval t is

$$\left|a_m^{(1)}(t)\right|^2 = \left|H'_{mk}\right|^2\frac{4\sin^2\left(\omega_{mk}-\omega\right)t/2}{\hbar^2\left(\omega_{mk}-\omega\right)^2}$$

$$= \frac{\left|H'_{mk}\right|^2}{\hbar^2} \frac{4\sin^2 \dfrac{\left(E_m^0 - E_k^0 - \hbar\omega\right)t}{2\hbar}}{\dfrac{\left(E_m^0 - E_k^0 - \hbar\omega\right)^2}{\hbar^2}} \tag{80a}$$

or
$$\left|a_m^{(1)}(t)\right|^2 = \frac{\left|H'_{mk}\right|^2}{\hbar^2} \frac{4}{\Delta^2} \sin^2\left(\frac{\Delta t}{2}\right) \tag{80b}$$

where
$$\Delta \equiv \frac{E_m^0 - E_k^0 - \hbar\omega}{\hbar} \tag{81}$$

Now,
$$\lim_{t\to\infty} \frac{4}{\Delta^2}\sin^2\left(\frac{\Delta t}{2}\right) = 2\pi t\delta(\Delta) = 2\pi t\delta\left(\frac{E_m - E_k - \hbar\omega}{\hbar}\right)$$

$$= 2\pi\hbar t\delta\left(E_m - E_k - \hbar\omega\right) \tag{82}$$

(because for Delta function $\delta(ax) = \dfrac{1}{a}\delta(x)$ for $a > 0$)

Hence from Eqn. (80) and (82), transition probability per unit time

$$\frac{1}{t}\left|a_m^{(1)}(t)\right|^2 = \frac{1}{t}\frac{\left|H'_{mk}\right|^2}{\hbar^2} 2\pi\hbar t\, \delta\left(E_m^0 - E_k^0 - \hbar\omega\right)$$

$$= \frac{2\pi}{\hbar}\left|H'_{mk}\right|^2 \delta\left(E_m^0 - E_k^0 - \hbar\omega\right) \tag{83}^\bullet$$

Now *for emission of a photon* (assuming $E_k > E_m$)

$$\frac{1}{t}\left|a_m^{(1)}(t)\right|^2 = \frac{2\pi}{\hbar}\left|H'_{mk}\right|^2 \delta\left(E_m^0 - E_k^0 + \hbar\omega\right)$$

$$= \frac{2\pi}{\hbar}\left|H'_{mk}\right|^2 \delta\left[-\left(E_k^0 - E_m^0 - \hbar\omega\right)\right]$$

$$= \frac{2\pi}{\hbar}\left|H'_{mk}\right|^2 \delta\left(E_k - E_m - \hbar\omega\right) \tag{84}$$

From Eqn. (72), we have

$$\left|H'_{mk}\right|^2 = \frac{e^2}{m^2c^2} \times \frac{2\pi c^2\hbar}{\omega V} \left|\langle\psi_k| e^{-i\vec{k}\cdot\vec{r}}\hat{\varepsilon}\cdot\vec{p} |\psi_k\rangle\right|^2$$

So,
$$\frac{1}{t}\left|a_m^{(1)}(t)\right|^2 = \frac{2\pi}{\hbar} \times \frac{e^2}{m^2c^2} \times \frac{2\pi c^2\hbar}{\omega V} \left|\langle\psi_m| e^{-i\vec{k}\cdot\vec{r}}\hat{\varepsilon}\cdot\vec{p} |\psi_k\rangle\right|^2$$

$$\times \delta\left(E_k - E_m - \hbar\omega\right) \tag{85}$$

Here it represents the transition probability per unit time for atom making a transition from the state ψ_k to the state ψ_m accompanied by emission of a photon of energy $\hbar\omega$. The appearance of delta function may appear to be un-acceptable but in fact there may be a narrow range of frequencies (or wave numbers $\vec{k}$ to $k + \Delta k$ near $|\vec{k}| = \omega/c$) and the expression (85) has to be integrated over k-space. However, presence of delta function ensures

$$\hbar\omega = E_k^0 - E_m^0$$

Now, Eqn. (85) becomes

$$\frac{1}{t}\left|a_m^{(1)}(t)\right|^2 = \frac{4\pi^2 e^2}{m^2 \omega V} \left|\langle \psi_m | e^{-i\vec{k}\cdot\vec{r}}\hat{\varepsilon}\cdot\vec{p} | \psi_k \rangle\right|^2$$

$$\times \delta\left(E_k - E_m - \hbar\omega\right) \tag{86}$$

and actual transition rate

$$R_{k \to m} = \sum_{\Delta k} \frac{1}{t}\left|a_m^{(1)}(t)\right|^2 \quad \text{(Doppler broadening)} \tag{87}$$

18.8.2 Density of Photon States

We are interested in photon states lying between $\vec{k}$ and $\vec{k} + \Delta\vec{k}$. For this we write

$$\vec{A}(\vec{r},t) = \frac{1}{\sqrt{V}} \vec{a}_0 e^{i(\vec{k}\cdot\vec{r} - \omega t)} + \frac{1}{\sqrt{V}} \vec{a}_0^* e^{-i(\vec{k}\cdot\vec{r} - \omega t)}$$

where V is the volume of the enclosure having photons. For convenience, we may consider enclosure to be a cube of side L and impose periodic boundary condition

$$\vec{A}(x + L, y, z, t) = \vec{A}(x, y, z, t) \tag{88}$$

So, momenta $\left(\hbar\vec{k}\right)$ are quantized and

$$e^{ik_x L} = e^{ik_y L} = e^{ik_z L} = 1 \tag{89}$$

or

$$k_x = \frac{2\pi}{L} n_x \; ; \; k_y = \frac{2\pi}{L} n_y \; ; \; k_z = \frac{2\pi}{L} n_z \tag{90}$$

where n_x, n_y, n_z are integers. Also

$$d^3\vec{k} = dk_x dk_y dk_z = \left(\frac{2\pi}{L}\right)^3 \Delta n_x \Delta n_y \Delta n_z \tag{91}$$

and

$$\omega = |\vec{k}| c = \frac{2\pi c}{L}\left(n_x^2 + n_y^2 + n_z^2\right)^{1/2} \tag{92}$$

Now,

$$R_{k \to m} = \int d^3\vec{n} \left[\frac{1}{t}\left|a_m^{(1)}(t)\right|^2\right]$$

$$= \int L^3 \frac{d^3\vec{k}}{(2\pi)^3} \left[\frac{1}{t}\left|a_m^{(1)}(t)\right|^2\right] \qquad \left(\because \vec{k} = \frac{2\pi}{L}\vec{n}\right)$$

$$= \int V \frac{d^3\vec{p}}{(2\pi\hbar)^3} \frac{1}{t}\left|a_m^{(1)}(t)\right|^2 \quad \left(\because L^3 = V\right) \text{ and } \left(\vec{k} = \frac{\vec{p}}{\hbar}\right) \tag{93}$$

This integral is over the volume in momentum space. We may write

$$d^3\vec{p} = d\Omega_p p^2 dp \tag{94}$$

where $d\Omega_p$ is the solid angle differential.

$$\therefore \qquad d^3p = d\Omega_p \cdot \hbar^2 k^2 d(\hbar k) = d\Omega_p \cdot \hbar^3 \left(\frac{\omega}{c}\right)^2 d\left(\frac{\omega}{c}\right) \tag{95}$$

Consequently

$$R_{k \to m} = \iint \frac{4\pi^2 e^2 V}{m^2 \omega V} \left| \langle \psi_m | e^{-i\vec{k}\cdot\vec{r}} \hat{\varepsilon} \cdot \vec{p} | \psi_k \rangle \right|^2 \frac{d\Omega_p}{(2\pi\hbar)^3}$$

$$\times \hbar^3 \frac{\omega^2}{c^3} \frac{d(\hbar\omega)}{\hbar} \, \delta\left(E_k^0 - E_m^0 - \hbar\omega\right) \tag{96}$$

$$= \frac{4\pi^2 e^2}{m^2 8\pi^3 \hbar c^3} \int d\Omega_p \left| \langle \psi_m | e^{-i\vec{k}\cdot\vec{r}} \hat{\varepsilon} \cdot \vec{p} | \psi_k \rangle \right|^2$$

$$\times \int \omega \, d(\hbar\omega) \, \delta\left(E_k^0 - E_m^0 - \hbar\omega\right)$$

$$= \frac{1}{2\pi} \left(\frac{e^2}{\hbar c}\right) \int \frac{d\Omega_p}{m^2 c^2} \left| \langle \psi_m | e^{-i\vec{k}\cdot\vec{r}} \hat{\varepsilon} \cdot \vec{p} | \psi_k \rangle \right|^2 \omega_{km}$$

$$\text{(using } \int dy \, f(y) \, \delta(x-y) = f(x))$$

or $\qquad R_{k \to m} = \frac{\alpha}{2\pi} \int d\Omega_p \omega_{km} \left| \frac{1}{m^2 c^2} \langle \psi_m | e^{-i\vec{k}\cdot\vec{r}} \hat{\varepsilon} \cdot \vec{p} | \psi_k \rangle \right|^2 \tag{97}$

where $\qquad \omega_{km} = \left(E_k^0 - E_m^0\right)\big/\hbar$.

18.9 Transition Probability per Unit Time in Terms of Poynting Vector

The transition probability for absorption is

$$\left| a_m^{(1)}(t) \right|^2 = \sum_n |H'_{mn}|^2 \frac{4\sin^2\left(\omega_{mn} - \omega\right)\frac{1}{2}t}{\hbar^2 \left(\omega_{mn} - \omega\right)^2} \times \left|\vec{A}_0\right|^2 \Delta\omega$$

$$= \sum_n \frac{e^2 \hbar^2}{m^2 c^2} \times \frac{2\pi c}{\omega^2} \frac{S(\omega)\Delta\omega}{\hbar^2} \times \left| \int \psi_m^{(0)*} e^{i\vec{k}\cdot\vec{r}} \operatorname{grad}_A \psi_n^{(0)} d\tau \right|^2$$

$$\times \frac{4\sin^2\left(\omega_{mn} - \omega\right)\frac{1}{2}t}{\hbar^2 \left(\omega_{mn} - \omega\right)^2} \qquad \text{(using Eqns 65, 80a, 87 and 97)}$$

$$\left| a_m^{(1)}(t) \right|^2 = \sum_n \frac{8\pi e^2}{m^2 c\omega^2} S(\omega)\Delta\omega \left| \int \psi_m^{(0)*} e^{i\vec{k}\cdot\vec{r}} \operatorname{grad}_A \psi_n^{(0)} d\tau \right|^2$$

$$\times \frac{\sin^2\left(\omega_{mn} - \omega\right)\frac{1}{2}t}{\left(\omega_{mn} - \omega\right)^2} \tag{98}$$

Here $grad_A$ is the component of the gradient operator in the direction of the polarisation vector $\vec{A}_0$. On account of there being no phase relation among the radiation components of different frequencies, the contributions from various frequency ranges are additive. Then each $\Delta\omega$ in Eqn. (98) can be infinitesimal and summation can be replaced by integration. Since there is a sharp maximum at $\omega = \omega_{mn}$, we have transition probability per unit time for an absorption (upward transition):

$$\frac{1}{t}\left|a_m^{(1)}(t)\right|^2 = \frac{8\pi e^2}{m^2 c\omega_{mn}^2} S(\omega_{mn}) \left| \int \psi_m^{(0)} {}^* e^{i\vec{k}\cdot\vec{r}} grad_A \psi_n^{(0)} d\tau \right|^2$$

$$\times \int_{-\infty}^{+\infty} \frac{\sin^2\left(\omega_{mn} - \omega\right) \frac{1}{2}t}{t\left(\omega_{mn} - \omega\right)^2} d\omega$$

Now, since $\dfrac{1}{4}\displaystyle\int_{-\infty}^{+\infty} \dfrac{\sin^2\left(\omega_{mn} - \omega\right)\frac{1}{2}t}{\left(\dfrac{\omega_{mn} - \omega}{2}\right)^2} d\omega = \dfrac{1}{4}\times 2\pi t = \dfrac{\pi t}{2}$, we therefore have

$$\frac{1}{t}\left|a_m^{(1)}(t)\right|^2 = \frac{4\pi^2 e^2}{m^2 c\omega_{mn}^2} S(\omega_{mn}) \left| \int \psi_m^{(0)} {}^* e^{i\vec{k}\cdot\vec{r}} grad_A \psi_n^{(0)} d\tau \right|^2 \qquad (99)$$

Here, the magnitude of $\vec{k}$ is $\dfrac{\omega_{mn}}{c}$. For emission i.e. downward transition, $e^{i\vec{k}\cdot\vec{r}}$ has to be replaced by $e^{-i\vec{k}\cdot\vec{r}}$ and consequently, the transition probability per unit time for emission $\left(E_m' \approx E_n - \hbar\omega\right)$ is given by

$$\frac{1}{t}\left|a_{m'}^{(1)}(t)\right|^2 = \frac{4\pi^2 e^2}{m^2 c\omega_{nm'}^2} S(\omega_{nm'}) \left| \int \psi_{m'}^{(0)} {}^* e^{-i\vec{k}\cdot\vec{r}} grad_A \psi_n^{(0)} d\tau \right|^2 \qquad (100)$$

where $\left|\vec{k}\right| = \dfrac{\omega_{nm'}}{c}$.

18.9.1 Physical Interpretation in Terms of Absorption and Emission

Equations (99) and (100) represent the transition probabilities of the particle per unit time between stationary states under the influence of a classical radiation field.

In an upward transition, the particle gains energy ($= E_m - E_n$) under the influence of angular frequence ω_{mn}. The quantum energy of this radiation is $\left(E_m - E_n\right) \cong \hbar\omega_{mn}$, so we may consider the upward transition of the particle as associated with the absorption of one quantum from the radiation field.

Similarly, downward transition is associated with emission of the quantum energy corresponding to the frequency of the radiation field. The transition probability of emission per unit time is proportional to the intensity (S) of the radiation present. The process of emission is therefore called the induced emission.

If we rewrite Eqn. (100) in terms of the reverse transition to that which appears in Eqn. (99) then Eqn. (100) can be made to describe transition from an initial upper state m to a final lower state n (i.e. n is to be replaced by m and m' by n). Then Eqn. (100) assumes the form:

$$\frac{4\pi^2 e^2}{m^2 c \omega_{mn}^2} S(\omega_{mn}) \left| \int \psi_n^{(0)*} e^{-i\vec{k}\cdot\vec{r}} \operatorname{grad}_A \psi_m^{(0)} d\tau \right|^2 \tag{101}$$

It is found that integral in Eqn. (101) is just complex conjugate of integral in Eqn. (99), the squares of magnitudes of both are equal. This implies that the transition probabilities of absorption and induced emission between any pair of states are the same.

18.9.2 Electric Dipole Approximation

The integral in Eqn. (99) is usually evaluated by using the expansion

$$e^{i\vec{k}\cdot\vec{r}} = 1 + \frac{i\vec{k}\cdot\vec{r}}{1!} + \frac{\left(i\vec{k}\cdot\vec{r}\right)^2}{2!} + \dots \tag{102}$$

The magnitudes of the successive terms

$$1 : \frac{2\pi r}{\lambda} : \frac{1}{2}\left(\frac{\pi r}{\lambda}\right)^2 \dots \tag{103}$$

decrease by factors of the order r/λ.

$$\frac{1}{t}\left|a_m^{(1)}(t)\right|^2 = \frac{4\pi^2 e^2}{m^2 c \omega_{mn}^2} S(\omega_{mn}) \left| \int \psi_m^{(0)*} e^{-i\vec{k}\cdot\vec{r}} \nabla_A \psi_n^{(0)} d\tau \right|^2$$

The integral is to be taken over the space occupied by the atom and integrand is virtually zero at distances from the origin which are greater than 10^{-8} cm. For visible and ultraviolet transitions $\lambda \sim 10^{-5}$ cm. Therefore $r/\lambda \sim 10^{-3}$ and so we may approximate $e^{i\vec{k}\cdot\vec{r}}$ to be equal to unity. This approximation, we shall now see that it is equivalent to replacing the atom by an electric dipole.

Now,

$$\int \psi_m^{(0)*} e^{-i\vec{k}\cdot\vec{r}} \operatorname{grad}_A \psi_n^{(0)} d\tau = \int \psi_m^{(0)*} \left(\frac{\partial}{\partial r}\right)_A \psi_n^{(0)} d\tau$$

$$= \frac{i}{\hbar} \int \psi_m^{(0)*} p_A \psi_n^{(0)} d\tau$$

$$= \frac{i}{\hbar} \left\langle p_A \right\rangle_{mn} \tag{104}$$

where p_a is the component of the momentum $\vec{p}$ along the direction of polarization of the incident radiation.

$$\left\langle \vec{p} \right\rangle_{mn} = m \frac{d}{dt} \left\langle \vec{r} \right\rangle_{mn} = \frac{m}{i\hbar}\left[\left\langle \vec{r} \right\rangle, H_0\right]_{mn}$$

$$= \frac{m}{i\hbar} \left\{ \int \psi_m^{(0)\,*} \vec{r} H_0 \psi_n^{(0)} d\tau - \int \psi_m^{(0)\,*} H_0 \vec{r} \ \psi_n^{(0)} d\tau \right\}$$

$$= \frac{m}{i\hbar} (E_n - E_m) \ (\vec{r})_{mn} \tag{105}$$

$$\therefore \quad \int \psi_m^{(0)\,*} grad_A \psi_n^{(0)} d\tau = -\frac{m}{\hbar} \omega_{mn} \ \langle r_A \rangle_{mn} \quad \text{(using Eqn. 104)} \tag{106}$$

r_A is the component of $\vec{r}$ along the direction of polarization. So, we have

$$\frac{i}{\hbar} \ \langle p_A \rangle_{mn} = \left| \int \psi_m^* grad_A \psi_n d\tau \right| = \frac{m}{\hbar} \omega_{mn} \int \psi_m^{(0)\,*} r_A \psi_n^{(0)} d\tau \tag{107}$$

Now, Eqn. (99) for transition probability for absorption per unit time becomes

$$\frac{1}{t} \left| a_m^{(1)}(t) \right|^2 = \frac{4\pi^2 e^2}{\hbar^2 c} S(\omega_{mn}) \left| \langle r_A \rangle_{mn} \right|^2 \tag{108}$$

(using square of Eqn. 107)

This involves only the matrix elements of the electric dipole moment $e\vec{r}$ thus indicating that replacement of $e^{i\vec{k}\cdot\vec{r}}$ by unity is equivalent to replacement of a hydrogen-like atom by an electric dipole.

18.9.3 The Matrix Element $\langle r_A \rangle_{mn}$ and Selection Rules

The direction of $\vec{A}_0$ is along the polarization direction of the electric field. Let this be denoted by the unit vector $\hat{\varepsilon}$. It satisfies

$$\hat{\varepsilon} \cdot \hat{\varepsilon} = 1$$

and

$$\hat{\varepsilon} \cdot \vec{k} = 0 \tag{109i}$$

We are interested in calculating $\langle \psi_m | e^{-i\vec{k}\cdot\vec{r}} \hat{\varepsilon} \cdot \vec{p} | \psi_n \rangle$ or under dipole approximation, using Eqn. (102) and (106)

$$\langle p \rangle_{mn} = -\frac{m\omega}{\hbar} \ \langle r_A \rangle_{mn} \equiv im\omega \ \langle \psi_m | \hat{\varepsilon} \cdot \vec{r} | \psi_n \rangle \tag{109ii}$$

(using Eqn. (107) for $\langle p \rangle_{mn}$)

If the initial state ψ_n is a hydrogen-like state characterized by the initial quantum numbers n_i, l_i and m_i and the state ψ_m, the final state characterized by n_f, l_f and m_f, then we have to evaluate

$$\langle \psi_m | \hat{\varepsilon} \cdot \vec{r} | \psi_n \rangle = \int_0^\infty r^2 dr \int d\Omega \ R_{n_f l_f}^* Y_{l_f m_f}^* \ \hat{\varepsilon} \cdot \vec{r} R_{n_i l_i} Y_{l_i m_i}$$

$$= \int_0^\infty r^2 dr \ R_{n_f l_f} r \ R_{n_i l_i} \int d\Omega \ Y_{l_f m_f} \hat{\varepsilon} \cdot \vec{r} \ Y_{l_i m_i} \tag{110}$$

where R's are the function of r only and Y's are function of θ and ϕ only, $\vec{r}$ is a unit vector along $\vec{r}$. We first consider the angular integral. We know that

$$\hat{\varepsilon} \cdot \hat{r} = \varepsilon_x \sin\theta \cos\phi + \varepsilon_y \sin\theta \sin\phi + \varepsilon_z \cos\theta \tag{111}$$

Also, spherical harmonics

$$\sqrt{\frac{4\pi}{3}} Y_{1,0} = \cos\theta$$

$$\sqrt{\frac{8\pi}{3}} Y_{1,1} = -\sin\theta \, e^{i\phi} \tag{112}$$

$$\sqrt{\frac{8\pi}{3}} Y_{1,-1} = \sin\theta \, e^{-i\phi}$$

$\Rightarrow$
$$\sin\theta\cos\phi = \frac{1}{2}\sqrt{\frac{8\pi}{3}} \left(Y_{1,-1} - Y_{1,1} \right)$$

$$\sin\theta\sin\phi = \frac{i}{2}\sqrt{\frac{8\pi}{3}} \left(Y_{1,-1} + Y_{1,1} \right) \tag{113}$$

So, we get

$$\vec{\varepsilon} \cdot \hat{r} = \sqrt{\frac{4\pi}{3}} \left(\varepsilon_z Y_{1,0} + \frac{-\varepsilon_x + i\varepsilon_y}{\sqrt{2}} Y_{1,1} + \frac{\varepsilon_x + i\varepsilon_y}{\sqrt{2}} Y_{1,-1} \right) \tag{114}$$

Thus the angular integral in Eqn. (110) involves

$$\int d\Omega \, Y_{l_f m_f}^* \, Y_{1,m} \, Y_{l_i m_i} \tag{115}$$

Let us first consider the azimuthal integration

$$\int_0^{2\pi} d\phi \, e^{-m_f \phi} e^{im\phi} e^{m_i \phi} = 2\pi \delta_{m, m_f - m_i} \tag{116}$$

Thus, we get the selection rule
$$m_f - m_i = m = 1, 0, -1 \tag{117}$$

If we define our z-axis to be along the photon momentum direction $\vec{k}$, then the condition (109i) implies that $\varepsilon_z = 0$ and hence $m = \pm 1$ only appears, so
$$m_f - m_i = \pm 1 \,^* \tag{118}$$

For θ integration, consider the special case $l_i = 1$. Since $Y_{0,0} = 1/\sqrt{4\pi}$, the angular integration involves

---

* In particular, we note that if the final state is the ground state with $l_f = m_f = 0$ then $m = -m_i$. For example, if $m_i = 1$ then $m = -1$, consequently polarization vector for the radiation is $\left(\varepsilon_x + i\varepsilon_y\right)/\sqrt{2}$. This means that if the atom in the initial state is polarized along the z-axis with $m_i = 1$, then in a decay to a state of zero angular momentum, conservation of angular momentum necessitates that this momentum be carried away by photon. So, the photon must have its spin ($\hbar$) aligned along the positive z-axis i.e. it must have helicity +1 or equivalently it is left circularly polarized. On the other hand, it is right circularly polarized for $\left(\varepsilon_x - i\varepsilon_y\right)/\sqrt{2}$, (and linearly polarized for photon emitted in the x-y plane).

$$\frac{1}{\sqrt{4\pi}} \int d\Omega \ Y_{1,m} \ Y_{l,m_i} = \frac{1}{\sqrt{4\pi}} \delta_{l_i,1} \delta_{m_i,-m} \tag{119}$$

which implies that the initial state must have $l_i = 1$, for example, in hydrogen the dominant transition to the ground state will be $np \to 1s$.

18.9.4 The 2p → 1s Transition

For the transition $2p$ to $1s$, we have to evaluate the radial integral

$$\int_0^\infty dr \ r^3 R_{1,0}^* R_{2,1} = \int_0^\infty dr \ r^3 \left[2 \left(\frac{Z}{a_0} \right)^{3/2} e^{-Zr/a_0} \right] \times \left[\frac{1}{\sqrt{24}} \left(\frac{Z}{a_0} \right)^{5/2} r \ e^{-Zr/2a_0} \right]$$

$$= \frac{1}{\sqrt{6}} \left(\frac{Z}{a_0} \right)^4 \int_0^\infty dr \ r^4 e^{-3Zr/2a_0} = \frac{1}{\sqrt{6}} \left(\frac{Z}{a_0} \right)^4 \left(\frac{2a_0}{3Z} \right)^5 \int_0^\infty dx \ x^4 e^{-x}$$

$$= \frac{1}{\sqrt{6}} \left(\frac{Z}{a_0} \right)^4 \left(\frac{2a_0}{3Z} \right)^5 \times 4 \ ! = \frac{24}{\sqrt{6}} \left(\frac{2}{3} \right)^5 \frac{a_0}{Z} \tag{120}$$

The angular integral is

$$\int d\Omega \ Y_{0,0}^* \vec{\varepsilon} \cdot \hat{r} \ Y_{1,m}$$

$$= \frac{1}{\sqrt{4\pi}} \int d\Omega \sqrt{\frac{4\pi}{3}} \left(\varepsilon_z Y_{1,0} + \frac{-\varepsilon_x + i\varepsilon_y}{\sqrt{2}} Y_{1,1} + \frac{\varepsilon_x + i\varepsilon_y}{\sqrt{2}} Y_{1,-1} \right) Y_{1,m}$$

$$= \frac{1}{\sqrt{3}} \left(\varepsilon_z \delta_{m,0} + \frac{-\varepsilon_x + i\varepsilon_y}{\sqrt{2}} \delta_{m,-1} + \frac{\varepsilon_x + i\varepsilon_y}{\sqrt{2}} \delta_{m,1} \right) \tag{121}$$

From Eqns (109), (110), (120) and (121), we have

$$\left| \langle r_A \rangle_{mn} \right|^2 = \frac{24 \times 24}{6} \left(\frac{2}{3} \right)^{10} \left(\frac{a_0}{Z} \right)^2 \frac{1}{3} \left[\delta_{m,0} \varepsilon_z^2 + \frac{1}{2} \left(\delta_{m,1} + \delta_{m,-1} \right) \left(\varepsilon_x^2 + \varepsilon_y^2 \right) \right] \tag{122}$$

Hence, transition rate

$$R_{2p \to 1s} = \frac{\alpha}{2\pi} \int d\Omega_p \frac{\omega}{m^2 c^2} \left| \langle \psi_m | e^{-i\vec{k} \cdot \vec{r}} \hat{\varepsilon} \cdot \vec{p} | \psi_n \rangle \right|^2$$

$$= \frac{\alpha}{2\pi} \int d\Omega_p \frac{\omega}{m^2 c^2} \left| m\omega \langle \psi_m | \hat{\varepsilon} \cdot \vec{r} | \psi_n \rangle \right|^2$$

$$= \frac{\alpha}{2\pi} \int d\Omega_p \frac{\omega}{m^2 c^2} m^2 \omega^2 \left| \langle r_A \rangle_{mn} \right|^2$$

$$= \frac{\alpha}{2\pi} \int d\Omega_p \frac{\omega}{m^2 c^2} m^2 \omega^2 \times \frac{24 \times 24}{6} \left(\frac{2}{3} \right)^{10} \left(\frac{a_0}{Z} \right)^2 \frac{1}{3}$$

$$\times \left[\delta_{m,0} \varepsilon_z^2 + \frac{1}{2} \left(\delta_{m,1} + \delta_{m,-1} \right) \left(\varepsilon_x^2 + \varepsilon_y^2 \right) \right]$$

$$= \int d\Omega_p \, \frac{\alpha}{2\pi} \times \frac{96}{3} \times \left(\frac{2}{3}\right)^{10} \left(\frac{a_0}{Z}\right)^2$$

$$\times \left[\delta_{m,0} \varepsilon_z^2 + \frac{1}{2} \left(\delta_{m,1} + \delta_{m,-1} \right) \left(\varepsilon_x^2 + \varepsilon_y^2 \right) \right] \times \frac{\omega}{m^2 c^2} m^2 \omega^2$$

or $\quad R_{2p \to 1s}(m) = \dfrac{2^{15}}{3^{10}} \left(\dfrac{a_0}{Z}\right)^2 \dfrac{\alpha}{2\pi} \int d\Omega_p \times \left[\delta_{m,0} \varepsilon_z^2 + \dfrac{1}{2} \left(\delta_{m,1} + \delta_{m,-1} \right) \left(\varepsilon_x^2 + \varepsilon_y^2 \right) \right]$

$$\times \frac{\omega}{m^2 c^2} m^2 \omega^2 \tag{123}$$

where $\omega \hbar = \dfrac{1}{2} mc^2 (Z\alpha)^2 \left(1 - \dfrac{1}{4} \right) = \dfrac{3}{8} mc^2 (Z\alpha)^2$ is the energy of radiation emitted.

The Eqn. (123) involves integration over the photon directions. This integral is not trivial. The integration becomes simple if the initial p state is unaligned, i.e. if it occurs in the three possible m states ($m = 1, 0, -1$) with equal probability. Then the rate is

$$R_{2p \to 1s} = \frac{1}{3} \sum_{m=-1}^{+1} R_{2p \to 1s}(m) \tag{124}$$

Then since

$$\sum_{m=-1}^{+1} \left[\delta_{m,0} \varepsilon_z^2 + \frac{1}{2} \left(\delta_{m,1} + \delta_{m,-1} \right) \left(\varepsilon_x^2 + \varepsilon_y^2 \right) \right] = \varepsilon_z^2 + \varepsilon_x^2 + \varepsilon_y^2 = 1$$

the integrand becomes independent of photon direction. Then Eqns (123) and (124) give the transition probability. The result has to be multiplied by a factor of 2 because of two possible polarization states of photon.

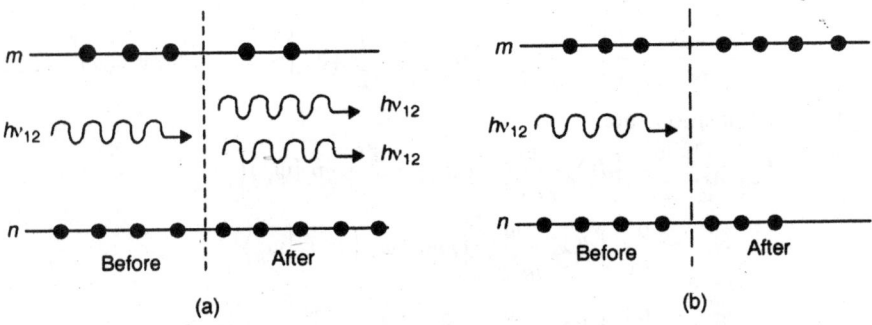

Fig. 18.6: (a) Induced emission, (b) Induced absorption

18.9.5 Einstein Transition Probabilities

It is convenient to define a transition probability per unit time and per unit of radiation intensity for the transition $\psi_n \to \psi_m$ for absorption $(E_m > E_n)$ denoted by Einstein B-coefficient*

$$B_{n \to m} = \frac{\left| a_m^{(1)}(t) \right|^2}{t\, S(\omega_{mn})} = \frac{4\pi^2}{3\hbar^2 c} \left| e\left(\vec{r}\right)_{mn} \right|^2 \quad \text{(using Eqn. 108)} \qquad (125i)$$

By the *principle of detailed balance*[*], it may be noted that

$$\left| \left(\vec{r}\right)_{mn} \right|^2 = \left| \left(\vec{r}\right)_{nm} \right|^2$$

the probabilities of induced absorption $\left(B_{n \to m}\right)$ and induced emission $\left(B_{m \to n}\right)$ are equal for any pair of states:

$$B_{n \to m} = B_{m \to n} \qquad (125ii)$$

The above discussion does not take into account spontaneous emission. It is known that a system in an excited state can emit radiation even in the absence of external radiation. Einstein found the transition probability for spontaneous emission (denoted by Einstein A coefficient $A_{m \to n}$) simply by considering the equilibrium of two states of different energies. If N_m and N_n are the number of systems in the states with energies E_m and E_n respectively, then according to Boltzmann distribution law for equilibrium at absolute temperature T, we have

$$\frac{N_m}{N_n} = \frac{e^{-E_m / k_B T}}{e^{-E_n / k_B T}} = e^{-(E_m - E_n)/ k_B T} = e^{-\hbar \omega_{mn} / k_B T} \qquad (126)$$

k_B is Boltzmann constant.

The number of systems emitting radiations (i.e. transition $m \to n$) per unit time is

$$N_m \left\{ \left(A_{m \to n}\right) + B_{m \to n} S(\omega_{mn}) \right\} \qquad (127)$$

and the number of systems making reverse transitions (absorption) per unit time is

$$N_n B_{n \to m} S(\omega_{mn}) \qquad (128)$$

At equilibrium these two numbers must be equal, i.e.

$$N_m \left\{ \left(A_{m \to n}\right) + B_{m \to n} S(\omega_{mn}) \right\} = N_n B_{n \to m} S(\omega_{mn})$$

or

$$\frac{N_m}{N_n} = \frac{B_{n \to m} S(\omega_{mn})}{\left(A_{m \to n}\right) + B_{m \to n} S(\omega_{mn})} \qquad (129)$$

Also since $B_{n \to m} = B_{m \to n}$ and using Eqn. (126), Eqn. (129) may be expressed as

$$e^{-\hbar \omega_{mn} / k_B T} = \frac{B_{m \to n} S(\omega_{mn})}{\left(A_{m \to n}\right) + B_{m \to n} S(\omega_{mn})}$$

---

[*] The principle of detailed balance is related to black body cavity radiation in equilibrium condition. This principle states that, in equilibrium the total number of particles leaving a certain quantum state per unit time equals the number arriving in that state per unit time. It also asserts that in equilibrium the number leaving by a particular pathway equals the number arriving by that pathway.

This gives

$$S(\omega_{mn}) = \frac{(A_{m\to n})/B_{m\to n}}{e^{\hbar\omega_{mn}/k_BT} - 1} \tag{130}$$

But by Planck's radiation law

$$S(\omega_{mn}) = \frac{\hbar\omega_{mn}^3}{\pi^2 c^2} \frac{1}{e^{\hbar\omega_{mn}/k_BT} - 1} \tag{131}$$

Comparing (130) and (131) we find, Einstein A coefficient for spontaneous emission:

$$(A_{m\to n}) = \frac{\hbar\omega_{mn}^3}{\pi^2 c^2} B_{m\to n}$$

$$= \frac{\hbar\omega_{mn}^3}{\pi^2 c^2} \frac{4\pi^2}{3\hbar^2 c} \left| e(\vec{r})_{mn} \right|^2 \qquad \text{(using Eqn. 125}i\text{)}$$

or

$$A_{m\to n} = \frac{4\omega_{mn}^3}{3\hbar c^3} \left| e(\vec{r})_{mn} \right|^2 \tag{132}$$

Example 18.1: Show that for optical frequencies, at thermal equilibrium (corresponding to $T \sim 1000$ K), the number of spontaneous emissions far exceeds the number of stimulated emissions.

Solution: At thermal equilibrium, the ratio

$$\frac{A}{BS(\omega)} = \exp\left(\frac{\hbar\omega}{k_BT}\right) - 1$$

gives the ratio of the number of spontaneous to stimulated emissions.

$$\frac{k_BT}{\hbar} = \frac{2\pi k_BT}{h} \cong \frac{2\times 3.14 \times 1.38 \times 10^{-23}\,(\text{J.s}^{-1}) \times 10^3\,(s)}{6.63 \times 10^{-34}\,\text{J.s}} \cong 1.3 \times 10^{14}\,s^{-1}$$

For the optical region $\lambda \sim 6000$ Å.

$$E = h\nu = \hbar\omega = \frac{hc}{\lambda} \Rightarrow \omega = 3 \times 10^{14}\,s^{-1}$$

Thus

$$\frac{A}{BS(\omega)} = \exp\left(\frac{3 \times 10^{14}\,s^{-1}}{1.3 \times 10^{14}\,s^{-1}}\right) - 1 = \left[\exp\left(\frac{3}{1.3} - 1\right)\right] \gg 1$$

i.e. transitions are mainly due to spontaneous emissions.

18.10 Forbidden Transition and Selection Rules on the Basis of Parity

We know that the problem of calculation of transition probability reduces to the evaluation of the matrix element for transition dipole moment

$$e\langle r_A\rangle_{mn} = i\hbar e \left\langle \psi_m \left| \hat{\varepsilon} \cdot \vec{r} \right| \psi_n \right\rangle \tag{133}$$

where e is the electronic charge and $\hat{\varepsilon}$ is the unit vector along the direction of the electric field (i.e. along $\vec{A}_0$). Transitions for which the transition dipole moment is non zero are *allowed transition* and those for which it is zero are disallowed or *forbidden transitions*. The transition probability is given by

$$\frac{1}{t}\left|a_m^{(1)}(t)\right|^2 = \frac{4\pi^2 e^2}{m^2 c \omega_{mn}^2} S(\omega_{mn}) \left| \int \psi_m^{(0)\,*} e^{i\vec{k}\cdot\vec{r}} grad_A \psi_n^{(0)} d\tau \right|^2 \quad (134)$$

whence we arrive at Eqn. (133) by using the electric dipole approximation

$$e^{i\vec{k}\cdot\vec{r}} = 1 + \frac{i\vec{k}\cdot\vec{r}}{1!} + \frac{\left(i\vec{k}\cdot\vec{r}\right)^2}{2!} + \ldots$$

by neglecting 2nd and further higher order terms. If however, nevertheless, these 2nd and higher order terms are non-zero but the dipole moment is zero, then the transitions are forbidden. But if the unapproximated integral of Eqn. (134) is zero, then the transition is strictly forbidden.

Selection Rules

We have observed while calculating the matrix element $\langle r_A \rangle_{mn}$ that the selection rule for quantum number m is

$$m_f - m_i = m = 0, \pm 1 \quad (135)$$

Now, regarding quantum number l, we note that the parity of an atomic orbital is $(-1)^l$. the s- and d-orbitals have even parity whereas p- and f-orbitals have odd parity. The parity of the integrand of the transition dipole moment is

$$(-1)^{lm} (-1)(-1)^{ln} = (-1)^{lm+ln+1} \quad (136)$$

which is even if the two orbitals have opposite parity and odd if they have the same parity. Group theoretic considerations suggest that a transition dipole moment would be finite when the integrand is totally symmetric under the symmetry operation. Hence the only allowed transitions are those involving a change of parity. For hydrogenic atoms, for a transition to be allowed, the dipole moment matrix element $\langle n'l'm'|r|nlm \rangle$ must be finite. On the basis of above arguments, this matrix element will be finite only when

$$l' - l \equiv \Delta l = \pm 1 \quad (137)$$

Example 18.2: Find the selection rules for the orbital quantum number l in an electric dipole transition between the stationary states of a one electron atom.

Solution: The perturbation operator is

$$-e\vec{r}\cdot\vec{E} = -\left(exE_x + eyE_y + ezE_z\right)$$

$$= -er\sin\theta\left(E_x\cos\phi + E_y\sin\phi\right) - er\cos\theta E_z$$

$$= -\frac{e}{2}r\sin\theta\, e^{i\phi}\left(E_x - iE_y\right) - \frac{e}{2}r\,\sin\theta\, e^{-i\phi}\left(E_x + iE_y\right) - e\,r\,\cos\theta\, E_z$$

$$= \sum_j\left(-er_j\right) E_j$$

where e is the electronic charge and j corresponds to the three components of interaction operator. The transition probability

$$P_{i \to f} \propto \left| \left\langle \psi_f \left| \vec{r_j} \right| \psi_i \right\rangle \right|^2$$

In spherical coordinates, the wave functions are $\psi_{nlm}(r,\theta,\phi) = R_{nl}(r) Y_{lm}(\theta,\phi)$ and the matrix elements take the form

$$\left\langle \psi_{n'l'm'} \left| \frac{1}{2} r \, \sin\theta \, e^{i\phi} \right| \psi_{nlm} \right\rangle$$

$$= \frac{1}{2} \int_0^\infty R^*_{n'l'} R_{nl} r^3 dr \int_0^\pi P_{l'}^{m'}(\cos\theta) P_l^m(\cos\theta) \sin\theta \, d(\cos\theta)$$

$$\times \int_0^{2\pi} e^{-i(m'-m-1)\phi} \, d\phi$$

$$\left\langle \psi_{n'l'm'} \left| \frac{1}{2} r \, \sin\theta \, e^{-i\phi} \right| \psi_{nlm} \right\rangle$$

$$= \frac{1}{2} \int_0^\infty R^*_{n'l'} R_{nl} r^3 dr \times \int_0^\pi P_{l'}^{m'}(\cos\theta) P_l^m(\cos\theta) \sin\theta \, d(\cos\theta)$$

$$\times \int_0^{2\pi} e^{-i(m'-m+1)\phi} \, d\phi$$

$$\left\langle \psi_{n'l'm'} \left| r \, \cos\theta \right| \psi_{nlm} \right\rangle$$

$$= \int_0^\infty R^*_{n'l'} R_{nl} r^3 dr \int_0^\pi P_{l'}^{m'}(\cos\theta) P_l^m(\cos\theta) \cos\theta \, d(\cos\theta)$$

$$\times \int_0^{2\pi} e^{-i(m'-m)\phi} \, d\phi$$

We have seen earlier using spherical harmonics that the selection rules for magnetic quantum number 'm' are

$$m_f - m_i = \Delta m = 0, \ \pm 1.$$

The selection rule on 'l' result from θ-integration and can be derived by substituting the recurrence relations of associated Legendre functions[*]. These are

$$(2l+1)\sin\theta P_l^m(\cos\theta) = P_{l+1}^{m+1}(\cos\theta) - P_{l-1}^{m+1}(\cos\theta)$$

and $\quad (2l+1)\cos\theta P_l^m(\cos\theta) = (l-m+1) P_{l+1}^m(\cos\theta) + (l+m)P_{l-1}^m(\cos\theta)$

Thus we find the first and second integrals in θ vanish unless $l' = l \pm 1$. Hence allowed transitions are also restricted by the selection rule

$$l_f - l_i \equiv \Delta l = \pm 1.$$

---

* See Appendix D

18.11 A General Expression for Transition from one State to another, S-Matrix and Matrix Elements of the S-Matrix

Let, during a certain time interval, there be acting on a system (with time independent Hamiltonian $\hat{H}_0$) a perturbation described by

$$\hat{V}(t) = \begin{cases} \hat{W}(t) & \text{for } 0 \le t \le \tau \\ 0 & \text{for } t < 0 \text{ or } t > \tau \end{cases}$$

The total Hamiltonian

$$\hat{H} = \hat{H}_0 + \hat{V}(t)$$

now depends on time and the corresponding time dependent Schrödinger equation

$$i\hbar \frac{\partial \psi}{\partial t} = \left[\hat{H}_0 + \hat{V}(t) \right] \psi \tag{138}$$

has no stationary solutions. To determine the wave function satisfying Eqn. (138), we go over to the interaction representation and write

$$\psi = \sum_n a_n(t) \phi_n e^{-iE_n t / \hbar} \tag{139}$$

where E_n and ϕ_n are respectively the eigenvalues and eigenfunctions of the operator $\hat{H}_0$.

We assume that before the interaction was switched on, the system was in a stationary state with energy E_m. There is thus in the sum for $t \le 0$ only one non vanishing term

$$\psi_i = \phi_m e^{-iE_m t / \hbar}$$

or $$a_n(t) = \delta_{nm} \text{ for } t \le 0$$

After the interaction is switched off (i.e. for $t \ge \tau$), the coefficients a_n become once again constants $a_{nm}(\tau)$ and their values depend on the form of the perturbation operator $\hat{W}(t)$ and the initial state which we have indicated by the second index m. Therefore when $t > \tau$, the system will be in a state with the wavefunction

$$\psi_f = \sum_n a_{nm}(\tau) \phi_n e^{-iE_n t / \hbar} \tag{140}$$

The probability that the system is in a stationary state with energy E_n is given by

$$P_{nm}(\tau) = \left| a_{nm}(\tau) \right|^2 \tag{141}$$

It is the probability for transition from initial state m to the final state n in the time interval τ.

To evaluate the coefficients a_{nm}, we substitute (139) into Eqn. (138). After multiplying that equation by ϕ_n^* and integrating over all values of all arguments, we obtain

$$i\hbar \frac{d}{dt} a_n(t) = \sum_l \langle n | \hat{W}(t) | l \rangle e^{i\omega_{nl} t} a_l(t) \tag{142i}$$

where

$$\langle n|\hat{W}(t)|l\rangle = \int \phi_n^* W(t)\phi_l dr \qquad (143)$$

$$\hbar\omega_{nl} = E_n - E_l \qquad (144)$$

To evaluate the transition probabilities, we must solve the set of Eqns. (142) under the initial condition

$$a_n(0) = \delta_{nm} \qquad (145)$$

If the matrix elements (143) are small and the period τ during which the perturbation acted is not too long so that after that period the values of the coefficients $a_n(\tau)$ are not very different from their initial values, we can solve the equations (142) by the method of successive approximations.

In first approximation, we can determine $a_n(t)$ by substituting into the right hand side of Eqn. (142i) the initial values (145) and we then get for $n \neq m$, a set of equations:

$$i\hbar\frac{da_{nm}^{(1)}}{dt} = \langle n|\hat{W}(t)|m\rangle e^{i\omega_{nm}t} \qquad (142ii)$$

Using the initial condition (145) to solve these equations, we find,

$$a_{nm}^{(1)}(t) = \frac{1}{i\hbar}\int_0^t \langle n|\hat{W}(t')|m\rangle e^{i\omega_{nm}t'} dt' \qquad (146)$$

Substituting this value into the right hand side of Eqn. (142i), we find the second order equation

$$i\hbar\frac{da_{nm}^{(2)}}{dt} = \langle n|\hat{W}(t)|m\rangle e^{i\omega_{nm}t} + \frac{1}{i\hbar}\sum_{n'(\neq m)}\langle n|\hat{W}(t)|n'\rangle e^{i\omega_{nn'}t}$$
$$\times \int_0^t \langle n'|\hat{W}(t')|m\rangle e^{i\omega_{n'm}t'} dt'$$

The solution of this equation can be written in the form

$$a_{nm}^{(2)}(t) = \frac{1}{i\hbar}\int_0^t \langle n|\hat{W}(t')|m\rangle e^{i\omega_{nm}t'} dt' + \left(\frac{1}{i\hbar}\right)^2 \sum_{n'(\neq m)}\int_0^t \langle n|\hat{W}(t')|n'\rangle e^{i\omega_{nn'}t'}$$
$$\int_0^{t'} \langle n'|\hat{W}(t'')|m\rangle e^{i\omega_{n'm}t''} dt'' dt' \qquad (147)$$

By substituting this value again into Eqn. (142), we can find the third order solution. Continuing this process, we get a solution in the form of an infinite series. This series can formally be written as follows:

$$a_{nm}(t) = \langle n|\hat{P}\exp\left[-\frac{i}{\hbar}\int_0^t \hat{\tilde{W}}(t')\,dt'\right]|m\rangle \qquad (148)$$

where

$$\hat{\tilde{W}}(t) = e^{iH_0t/\hbar}\hat{W}(t)\,e^{-iH_0t/\hbar} \qquad (149)$$

is the perturbation operator in interaction representation and

$$\hat{P}\exp\left[-\frac{i}{\hbar}\int_0^t \hat{\tilde{W}}(t')\,dt'\right]$$

$$\equiv 1 + \frac{1}{i\hbar} \int_0^t \hat{\tilde{W}}(t')\, dt' + \left(\frac{1}{i\hbar}\right)^2 \int_0^t \hat{\tilde{W}}(t') \int_0^{t'} \hat{\tilde{W}}(t'')\, dt''\, dt'$$

$$+ \left(\frac{1}{i\hbar}\right)^3 \int_0^t \hat{\tilde{W}}(t') \int_0^{t'} \hat{\tilde{W}}(t'') \int_0^{t''} \hat{\tilde{W}}(t''')\, dt'''\, dt''\, dt' + \dots \qquad (150)$$

The nth term in the series (150) is

$$A_n = \left(\frac{1}{i\hbar}\right)^n \int_0^t dt_1 \int_0^{t_1} dt_2 \dots \int_0^{t_{n-1}} dt_n \hat{\tilde{W}}(t_1)\, \hat{\tilde{W}}(t_2) \dots \hat{\tilde{W}}(t_n)$$

The integration in this is over time variables which are time ordered: $t_1 > t_2 > \dots > t_n$ by the operator $\hat{P}$ called *chronological operator*; it orders the product of time-dependent operators by putting them from left to right in order of chronologically successively decreasing times.

18.11.1 Matrix Elements of the S-Matrix and Transition Matrix

In Eqn. (148) we have found a general expression for the matrix element $a_{nm}(t)$ determining the transition from the state $|m>$ to the state $|n>$ under the influence of the perturbation $\hat{W}$. Let the states $|m>$ and $|n>$ and their energies E_m and E_n be the eigenfunctions and eigenvalues of the Hamiltonian $\hat{H}_0$ of two subsystems, while the interaction operator $\hat{W}$ causes transitions between them. The operator $\hat{W}$ is time independent in the Schrödinger representation.

Suppose, we choose $t = -\infty$ as the initial time and $t = +\infty$ as the final time, the matrix elements $a_{nm}(+\infty)$ are denoted by $\langle n|\hat{S}|m\rangle$ and are called the *matrix elements of the S-matrix*, where

$$\langle n|\hat{S}|m\rangle = \langle n| \hat{P} \exp\left\{-\frac{i}{\hbar} \int_{-\infty}^{+\infty} \hat{\tilde{W}}(t)\, dt\right\} |m\rangle \qquad (151)$$

and

$$\hat{S} = \hat{P} \left[-\frac{i}{\hbar} \int_{-\infty}^{+\infty} \hat{\tilde{W}}(t)\, dt\right]$$

$$\equiv 1 + \frac{1}{(i\hbar)} \int_{-\infty}^{+\infty} \hat{\tilde{W}}(t)\, dt + \frac{1}{(i\hbar)^2} \int_{-\infty}^{+\infty} dt_1 \int_{-\infty}^{t_1} dt_2 \hat{\tilde{W}}(t_1)\, \hat{\tilde{W}}(t_2)$$

$$+ \frac{1}{(i\hbar)^3} \int_{-\infty}^{+\infty} dt_1 \int_{-\infty}^{t_1} dt_2 \int_{-\infty}^{t_2} dt_3 \hat{\tilde{W}}(t_1)\, \hat{\tilde{W}}(t_2) \hat{\tilde{W}}(t_3) + \dots \qquad (152)$$

with

$$\hat{\tilde{W}}(t) = e^{iH_0 t/\hbar}\hat{W}e^{-iH_0 t/\hbar} \qquad (152i)$$

Since Eqn. (152) is written as a series, we can write the matrix elements (151) as sums of matrix elements of different order:

$$\langle n|\hat{S}|m\rangle = \sum_{j=0}^{\infty} \langle n|\hat{S}^{(j)}|m\rangle \qquad (153)$$

Thus we have

$$\langle n|\hat{S}^0|m\rangle = \langle n|m\rangle$$

$$\langle n|\hat{S}^{(1)}|m\rangle = \frac{1}{i\hbar}\ \langle n|\int_{-\infty}^{+\infty}\hat{\tilde{W}}(t)\ dt|m\rangle$$

$$\langle n|\hat{S}^{(2)}|m\rangle = \frac{1}{(i\hbar)^2}\ \langle n|\int_{-\infty}^{+\infty}dt_1\int_{-\infty}^{t_1}dt_2\hat{\tilde{W}}(t_1)\hat{\tilde{W}}(t_2)\ |m\rangle$$

... ...

... ...

Using Eqn. (152*i*), we can write the first order matrix element in the form

$$\langle n|\hat{S}^{(1)}|m\rangle = -\frac{i}{\hbar}\langle n|\hat{W}|m\rangle\int_{-\infty}^{+\infty}e^{i(E_n-E_m)t/\hbar}dt$$

$$-2\pi i\delta\ (E_n-E_m)\ \langle n|\hat{W}|m\rangle \tag{154}$$

Let us now consider the second order matrix element

$$\langle n|\hat{S}^{(2)}|m\rangle = \frac{1}{(i\hbar)^2}\sum_f\int_{-\infty}^{+\infty}dt_1\ \langle n|\hat{\tilde{W}}(t_1)|f\rangle\int_{-\infty}^{t_1}dt_2\ \langle f|\hat{\tilde{W}}(t_2)|m\rangle$$

$$= \frac{1}{(i\hbar)^2}\sum_f\ \langle n|\hat{W}|f\rangle\ \langle f|\hat{W}|m\rangle\int_{-\infty}^{+\infty}e^{i(E_n-E_f)t_1/\hbar}dt_1$$

$$\times\int_{-\infty}^{t_1}e^{i(E_f-E_m)t_2/\hbar}dt_2$$

To evaluate the second integral, we make the substitution

$$\left(E_f-E_m\right)\rightarrow\left(E_f-E_m-i\eta\right)$$

where η is a small positive quantity which guarantees the convergence of the integral at the lower limit. In our final expression we have to take the limit $\eta\rightarrow 0$. Thus, we have

$$\int_{-\infty}^{t_1}e^{i(E_f-E_m)t/\hbar}dt\rightarrow\int_{-\infty}^{t_1}e^{i(E_f-E_m-i\eta)t/\hbar}dt = i\hbar\frac{e^{i(E_f-E_m-i\eta)t_1/\hbar_1}}{E_m-E_f+i\eta}$$

Hence,

$$\langle n|\hat{S}^{(2)}|m\rangle = \frac{1}{i\hbar}\sum_f\frac{\langle n|\hat{W}|f\rangle\ \langle f|\hat{W}|m\rangle}{E_m-E_f+i\eta}\int_{-\infty}^{+\infty}e^{i(E_n-E_m-i\eta)t/\hbar}dt$$

$$= -2\pi i\delta\left(E_n-E_m\right)\sum_f\frac{\langle n|\hat{W}|f\rangle\ \langle f|\hat{W}|m\rangle}{E_m-E_f+i\eta} \tag{155}$$

The higher order matrix elements can be treated similarly. We consider only transitions for which the final state differs from the initial state, so that <*n*|*m*> = 0. Using Eqns (154) and (155) and similar expressions for the other terms in Eqn. (153), we can write for the matrix elements of the *S*-matrix:

$$\langle n|\hat{S}|m\rangle = -2\pi i\delta(E_n-E_m)\ \langle n|\hat{T}|m\rangle \tag{156}$$

where

$$\langle n|\hat{T}|m\rangle = \langle n|\hat{W}|m\rangle + \sum_f\frac{\langle n|\hat{W}|f\rangle\ \langle f|\hat{W}|m\rangle}{E_m-E_f+i\eta}$$

$$+\sum_{f,f'} \frac{\langle n|\hat{W}|f\rangle \langle f|\hat{W}|f'\rangle \langle f'|\hat{W}|m\rangle}{\left(E_m - E_f + i\eta\right)\left(E_m - E_{f'} + i\eta\right)} + ... \qquad (157)$$

One often uses *Feynman diagrams* to describe the matrix elements of different order occurring in Eqn. (157). If $\hat{W}$ is an external constant field acting upon a particle, the first order matrix element will correspond to Fig. 18.7, where the initial and final states are depicted by straight lines and external field $\hat{W}$ by a dotted line.

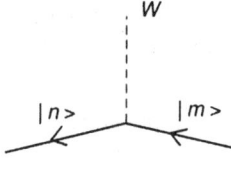

Fig. 18.7

Such a diagram depicts the scattering of a particle by an external field. The second order matrix element in Eqn. (157) will correspond to Fig. 18.8.

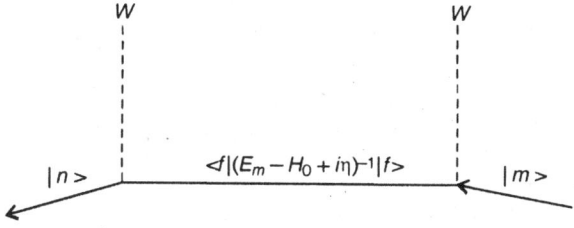

Fig. 18.8

This depicts the two-fold scattering of a particle by an external field. Similarly, we can depict higher order scattering processes.

Questions and Problems

1. Discuss the first order time dependent perturbation theory and prove (in case of transitions into a continuous spectrum) the transition probability per unit time is given by

$$W_{k \to m} = \frac{2\pi}{\hbar}\left|H'_{mk}\right|^2 \rho_m$$

where ρ_m denotes the density of final states and H'_{mk} is the matrix element of the perturbation term (the Fermi-golden rule).

2. Give the time dependent perturbation theory for the case of a perturbation which is switched on at $t = 0$, remains constant for interval 0 to t and switched off at time t.

3. Show that the first order effect of a time dependent perturbation varying sinusoidally in time, leads to the emission or absorption of energy.

4. Obtain the Hamiltonian operator for a charged particle in an electromagnetic field.

5. Discuss the Einstein's coefficients of spontaneous and induced emission of radiation. Derive a relation between A and B coefficients.

6. (*a*) Discuss semiclassical radiation theory in relation to interaction of an electron with a monochromatic plane wave described by a vector potential $\vec{A}(\vec{r}, t)$. Derive an expression for perturbation terms for absorption and emission of a photon.

   (*b*) Derive an expression for transition probability per unit time for an atomic transition from state ψ_k to ψ_m associated by emission of a photon.

7. Derive transition probability per unit time in terms of Poynting vector. Give physical interpretation in terms of absorption and emission.

8. Explain what is electric dipole approximation.

9. Calculate the transition rate $R_{2p \to 1s}$ for $2p$ to $1s$ transition of hydrogen atom.

10. For the $2p \to 1s$ transition in hydrogen atom, prove that

$$\left| \langle 1s | \vec{r} | 2p \rangle \right|^2 = \left| \int \psi_{1s}^* \vec{r} \psi_{2p} d\tau \right|^2 = 2^5 \left(\frac{2}{3} \right)^{10} a_0^2 \text{ , independent of the initial value}$$

of m.

[**Hint:** $\psi_{2p} = R_{21}(r) Y_{1m}(\theta, \phi)$; $\psi_{1s} = R_{10}(r) Y_{00}(\theta, \phi)$]

11. Write short notes on
    (*i*) Adiabatic approximation
    (*ii*) Sudden approximation
    (*iii*) Dipole selection rules

12. A hydrogen atom in its ground state is placed between the plates of a capacitor. A voltage pulse is applied to the capacitor so as to produce a homogeneous electric field that has the time dependence $\varepsilon = 0$ for $t < 0, \varepsilon = \varepsilon_0 e^{-t/\tau}$ for $t > 0$. Find the first order probability that the atom is in the 2s state (200) after a long time.

19

Many Electron Atoms

We have earlier discussed the problem of ground states of two light atoms, viz. The hydrogen atom and the helium atom. The problem of He atom was solved by considering the two electron interaction term e^2/r_{12} as the perturbation term. Variation-calculation for other light atoms have been carried out similar to helium atom. However, for heavier atoms (i.e. multielectron atoms), the Hamiltonian becomes complicated due to the presence of interaction terms and the Schrödinger equation cannot be solved directly unless we apply some approximation.

19.1 Hartree's Self Consistent Method (Hartree Equations and Central Field Approximation)

The starting point of calculations on all except the lightest atoms is the *central-field approximation*. This assumes that each of the atomic electrons moves in a spherically symmetric potential energy $V(r)$ that is produced by the nucleus and all the other electrons. The approximation is good if the deviation from the potential for one electron produced by close passage of other electrons is relatively small. This is actually so, because the constant nuclear potential is of the order Z times as large as the fluctuating potential due to each nearby electron and the latter varies quite slowly (inversely) with the separation between the two electrons. Thus the problem then consists of calculation of the central field and the correction of the approximate results obtained from it.

To calculate the energy levels of a many electron atom, we consider the Hamiltonian for the motion of Z electrons about a nucleus of charge Ze:

$$H = \sum_{i=1}^{Z} \left(-\frac{\hbar^2}{2m_e} \nabla_i^2 - \frac{Ze^2}{r_i} \right) + \sum_{i=1}^{Z} \sum_{j>1} \frac{e^2}{\left| \vec{r}_i - \vec{r}_j \right|} = \sum_{i=1}^{Z} H_i + V \tag{1}$$

where the first term represents kinetic energy of the electrons, the 2nd term is the electrostatic interaction between the nucleus and each of the electrons and the third term is the mutual interaction term between all pairs of electrons (V). The single electron Hamiltonians H_i are assumed to satisfy the eigenvalue equations:

$$H_i \Psi_{n_i l_i m_i m_{si}} \equiv \left(-\frac{\hbar^2}{2m_e} \nabla_i^2 - \frac{Ze^2}{r_i} \right) \Psi_{n_i l_i m_i m_{si}} = E_{n_i} \Psi_{n_i l_i m_i m_{si}} \tag{2}$$

Now, for simplicity, we choose a trial wave function

$$\Psi = \Psi_{E_1} \left(\vec{r}_1, m_{s_1} \right) \cdots \Psi_{E_i} \left(\vec{r}_i, m_{s_i} \right) \cdots \Psi_{E_Z} \left(\vec{r}_Z, m_{s_Z} \right) \tag{3}$$

assuming that all the individual electron wave functions Ψ_{E_i}'s are normalized and orthogonal and that there is no more than one electron in the same state $\left(\vec{r}_i, m_{s_i} \right)$. Because of spin independence of the Hamiltonian, we have the expectation value of energy as

$$E = \int_1 \cdots \int_i \cdots \int_Z \Psi^* H \Psi \, dv_1 dv_2 \dots dv_Z$$

$$= \int_1 \cdots \int_i \cdots \int_Z \Psi_{E_1}^* (\vec{r}_1) \cdots \Psi_{E_i}^* (\vec{r}_i) \cdots \Psi_{E_Z}^* (\vec{r}_Z) \left(\sum_{i=1}^{Z} H_i + V \right)$$

$$\Psi_{E_1} (\vec{r}_1) \cdots \Psi_{E_i} (\vec{r}_i) \cdots \Psi_{E_Z} (\vec{r}_Z) \, dv_1 \dots dv_i \dots dv_Z$$

$$= \sum_{i=1}^{Z} \left[\int \Psi_{E_i}^* (\vec{r}_i) H_i \Psi_{E_i} (\vec{r}_i) \, dv_i + \sum_{j>i} \int \Psi_{E_i}^* (\vec{r}_i) \left\{ \frac{e^2 \left| \Psi_{E_j} (\vec{r}_j) \right|^2}{\left(\vec{r}_i - \vec{r}_j \right)} \, dv_j \right\} \Psi_{E_i} (\vec{r}_i) \, dv_i \right]$$

(using Eqn. 1)

$$= \sum_{i=1}^{Z} \int \Psi_{E_i}^* (\vec{r}_i) \left[H_i + \sum_{j>i} V_{ij} \right] \Psi_{E_i} (\vec{r}_i) \, dv_i \tag{4}$$

The operator V_{ij} corresponds to the effect of the electron density $\rho_j = e \left| \Psi_{E_j} (\vec{r}_j) \right|^2$ on the electron (charge e) at $\vec{r}_i$, such that

$$V_{ij} \Psi_{E_i} (\vec{r}_i) = e^2 \int \frac{\left| \Psi_{E_j} (\vec{r}_j) \right|^2}{\left| \vec{r}_i - \vec{r}_j \right|} \Psi_{E_i} (\vec{r}_i) dv_j \tag{5}$$

By using the operator V_{ij}, we can apply it as if the problem is that of variation method for single electron problem. We now assume that an arbitrary variation $\delta\Psi$ in the total wave function Ψ is exclusively due to the variation $\delta\Psi_{E_i}$ in $\Psi_{E_i} (\vec{r}_i)$ i.e.

$$\delta\Psi = \Psi_{E_1} (\vec{r}_1) \cdots \Psi_{E_i} (\vec{r}_i) \cdots \Psi_{E_Z} (\vec{r}_Z) \tag{6}$$

We must find the minimum of the energy E_i given by

$$E_i = \int \Psi_{E_i}^* (\vec{r}_i) \left[H_i + \sum_{j \neq i} V_{ij} \right] \Psi_{E_i} (\vec{r}_i) \, dv_i \tag{7}$$

which are Z terms of Eqn. (4), and each involve the single electron wave function $\psi_{E_i}(\vec{r}_i)$.

Now, an arbitrary variation $\delta\psi$ corresponds to a first order, a variation

$$\delta\langle H\rangle = \langle \psi + \delta\psi | H | \psi + \delta\psi\rangle - \langle \psi | H | \psi\rangle = \langle \delta\psi | H | \psi\rangle + \langle \psi | H | \delta\psi\rangle = 0$$

which becomes

$$\delta E_i = \left\langle \delta\psi_{E_i} \left| H_i + \sum_{j\neq i} K_{ij} \right| \psi_{E_i} \right\rangle + \left\langle \psi_{E_i} \left| H_i + \sum_{j\neq i} K_{ij} \right| \delta\psi_{E_i} \right\rangle = 0 \tag{8}$$

The variations $\delta\psi_{E_i}$ in this Eqn. are to be compatible with the condition

$$\left\langle \psi_{E_i} \middle| \psi_{E_i} \right\rangle = 1$$

or

$$\delta \left\langle \psi_{E_i} \middle| \psi_{E_i} \right\rangle \equiv \left\langle \delta\psi_{E_i} \middle| \psi_{E_i} \right\rangle + \left\langle \psi_{E_i} \middle| \delta\psi_{E_i} \right\rangle = 0 \tag{9}$$

By using the Lagrange multiplier ε such that

$$\delta E_i - \varepsilon \, \delta \left\langle \psi_{E_i} \middle| \psi_{E_i} \right\rangle = 0 \tag{10}$$

we get

$$\delta E_i - \varepsilon \left\langle \delta\psi_{E_i} \middle| \psi_{E_i} \right\rangle - \varepsilon \left\langle \psi_{E_i} \middle| \delta \, \psi_{E_i} \right\rangle = 0 \tag{11}$$

or

$$\left\langle \delta\psi_{E_i} \left| (H_i + \sum_{j\neq i} V_{ij}) \, \psi_{E_i} - \varepsilon\psi_{E_i} \right\rangle + \left\langle \psi_{E_i} \left| \left\{ (H_i + \sum_{j\neq i} V_{ij}) - \varepsilon \right\} \right| \delta\psi_{E_i} \right\rangle = 0$$

or

$$\left\langle \delta\psi_{E_i} \left| (H_i + \sum_{j\neq i} V_{ij}) \, \psi_{E_i} - \varepsilon\psi_{E_i} \right\rangle + \left\langle (H_i + \sum_{j\neq i} V_{ij}) \, \psi_{E_i} - \varepsilon\psi_{E_i} \middle| \delta\psi_{E_i} \right\rangle = 0$$

This means that ψ_{E_i} satisfies the eigenvalue equation

$$\left\{ H_i + \sum_{j\neq i} V_{ij} \right\} \psi_{E_i}(\vec{r}_i) = \varepsilon\psi_{E_i}(\vec{r}_i) \tag{12}$$

Multiplying this by $\psi_{E_i}^*$ and integrating over $\vec{r}_i$, we obtain using Eqn. (7), $\varepsilon = E_i$, hence the energy eigenfunction and eigenvalue of the ith electron are to be obtained by solving the single electron equation

$$\left\{ H_i + \sum_{j\neq i} V_{ij} \right\} \psi_{E_i}(\vec{r}_i) = E_i\psi_{E_i}(\vec{r}_i) \tag{13}$$

This set of Z simultaneous equations, written for each electron of the atom are called the *Hartree equations*. It is evident that the terms $V_{ij}\psi_{E_i}(\vec{r}_i)$ having the explicit form Eqn. (5) can only be calculated if $\psi_{E_i}(r_j)$ are already known, thus, the Hartree equations cannot be solved simultaneously in closed form. However, an iterative procedure can be applied assuming an initial set of functions $\psi'_{E_i}(\vec{r}_i)$ to obtain the potential energies V'_{ij} and then solving (using potentials V'_{ij})

numerically Eqns (13) for a new set of single electron functions $\psi''_{E_i}(\vec{r}_i)$. This cycle of operation is then repeated till self consistency is achieved i.e. the potentials V_{ij} and functions $\psi_{E_i}(\vec{r}_i)$ etc are no longer different from those used in the equations solved for them. In other words, V_{ij} and ψ_{E_i}'s in successive cycles coincide gradually. Actually it is not K_{ij} of Eqn. (5) which is directly employed in Schrödinger equation (13) but rather a spherically symmetric approximation by averaging over all directions i.e. replacing

$$V_{eff}(\vec{r}_i) = \sum_{j \neq i} V_{ij} - \frac{Ze^2}{r_i} \quad * \tag{14}$$

by

$$V_{eff}(r_i) = \int V_{eff}(\vec{r}_i) \frac{d\Omega_i}{4\pi} \tag{15}$$

Using this in Eqn. (13) leads to central potential approximation to the Hartree equations

$$\left[-\frac{\hbar^2}{2m_e} \nabla_i^2 + V_{eff}(r_i) \right] \psi_{E_i}(\vec{r}_i) = E_i \psi_{E_i}(\vec{r}_i) \tag{16}$$

Since the Hamiltonian now contains only central forces and no spin dependent term, so it commutes with both orbital and spin angular momentum operators. This enables us to write eigenstates as

$$\psi_{E_i}(\vec{r}_i, m_{s_i}) = R_{n_i l_i}(r_i) Y_{l_i m_i}(\theta_i, \phi_i) \chi_{m_{si}} \tag{17}$$

These are various orbitals. The potential energy obtained by averaging of the expression

$$V_{eff}(\vec{r}_i) = \sum_{j \neq i} e^2 \int \frac{\left| \psi_{E_j}(\vec{r}_j) \right|^2}{(\vec{r}_i - \vec{r}_j)} dv_j - \frac{Ze^2}{r_i} \tag{18}$$

is known as *self consistent Hartree field*. By introduction of the self consistent field we have been able to reduce the many-electron problem to a single electron problem (i.e. the solution of the Schrödinger equation involving coordinates of one electron).

19.2 Hartree-Fock Approximation

In the Hartree-method, the overall wave function ψ is simply a product of one electron wave functions that minimizes

$$\langle H \rangle = (\psi, H\psi)$$

There, the overall ψ is of the form

$$\psi(\vec{r}_1 s_1, \vec{r}_2 s_2, ..., \vec{r}_Z s_Z) = \psi_1(\vec{r}_1 s_1) \cdots \psi_Z(\vec{r}_Z s_Z) \tag{19}$$

---

* This represents the average potential arising from all the other electrons and the nucleus.

for a system of Z interacting electrons. However, if spin is taken into consideration, the above wave function is incompatible with the Pauli principle which requires that ψ must be anti-symmetric, i.e. it should change sign when any two of its arguments are interchanged:

$$\psi\left(\vec{r}_1 s_1,\cdots,\vec{r}_i s_i,\cdots,\vec{r}_j s_j,\cdots,\vec{r}_Z s_Z\right) = -\psi\left(\vec{r}_1 s_1,\cdots,\vec{r}_j s_j,\cdots,\vec{r}_i s_i,\cdots,\vec{r}_Z s_Z\right) \tag{20}$$

The Hartree-Fock method incorporates the effect of exchange symmetry into the formalism. The simplest generalization of the Hartree-method is to replace the trial wave function Eqn. (19) by a Slater determinant of one electron wave functions which is a $Z \times Z$ matrix

$$\psi\left(\vec{r}_1 s_1, \vec{r}_2 s_2,...,\vec{r}_Z s_Z\right) = \frac{1}{\sqrt{Z}\,!} \begin{vmatrix} \psi_1\left(\vec{r}_1 s_1\right) & \psi_1\left(\vec{r}_2 s_2\right) & \cdots & \psi_1\left(\vec{r}_Z s_Z\right) \\ \psi_2\left(\vec{r}_1 s_1\right) & \psi_2\left(\vec{r}_2 s_2\right) & \cdots & \psi_2\left(\vec{r}_Z s_Z\right) \\ \vdots & \vdots & & \vdots \\ \psi_Z\left(\vec{r}_1 s_1\right) & \psi_Z\left(\vec{r}_2 s_2\right) & \cdots & \psi_Z\left(\vec{r}_Z s_Z\right) \end{vmatrix} \tag{21}$$

(since a determinant changes sign when any two columns are interchanged, this insures that the condition (Eqn. 20) holds.

The Hamiltonian of the system can be written as

$$H = \sum_{i=1}^{Z} H_i\left(\vec{r}_i\right) + \frac{1}{2}\sum_{i,j \neq i} V\left(\vec{r}_i,\vec{r}_j\right) \tag{22}$$

where

$$V\left(\vec{r}_i,\vec{r}_j\right) = \int dr_i dr_j \frac{e^2}{\left|\vec{r}_i - \vec{r}_j\right|}\left|\psi_i\left(\vec{r}_i\right)\right|^2 \left|\psi_j\left(\vec{r}_j\right)\right|^2 \tag{23}$$

The Schrödinger equation to be solved is

$$H\psi\left(\vec{r}_1 s_1,...,\vec{r}_Z s_Z\right) = E\psi\left(\vec{r}_1 s_1,...,\vec{r}_Z s_Z\right) \tag{24}$$

As this equation is not separable, we can express the eigenfunctions of H as a linear combination of determinental eigenfunctions (Eqn. 21). It is convenient to work with a single determinant as the ground state wave function of the system, so, we can write

$$H = \sum_{i=1}^{Z}\left[H_i\left(\vec{r}_i\right) + F\left(\vec{r}_i\right)\right] + \left[\frac{1}{2}\sum_{i,j \neq i} V\left(\vec{r}_i,\vec{r}_j\right) - \sum_{i} F\left(\vec{r}_i\right)\right] \tag{25}$$

where $\sum_{i} F\left(\vec{r}_i\right)$ denotes the interaction term.

We can now take the single Z^{th} order determinant of a one electron wave function ϕ whose orbital factors are eigenfunctions of the equation

$$\left[-\frac{\hbar^2}{2m}\nabla^2 + V(r) + F\left(\vec{r}\right)\right]\phi\left(\vec{r}\right) = E\,\phi\left(\vec{r}\right) \tag{26}$$

where the operator F has to be such so as to minimize the total energy. Use of a single determinant as the ground state wave function is known as the *Hartree-Fock method*. The choice of F in accordance with the variation principle is given by

$$\langle n|F|m \rangle = \sum_{i=1}^{Z} \left\{ \langle in|V|im \rangle - \langle ni|V|im \rangle \right\} \tag{27}$$

It is to be noted here that the Hamiltonian is not affected by this choice of F, but the one-electron wave functions ψ_i's do change.

Now, we proceed to get the explicit form of the Hartree-Fock Eqn. (26). Writing Eqn. (27) in the integral form and denoting space and spin coordinates by a single variable $\bar{x}$, we have

$$\int \psi_n^*(\bar{x}) F(\bar{r}) \psi_m(\bar{x}) \, d\bar{x}$$

$$= \sum_i \int\int \psi_i^*(\bar{x}_1) \, \psi_n^*(\bar{x}_2) \, V(\bar{r}_1,\bar{r}_2) \, \psi_i(\bar{x}_1) \, \psi_m(\bar{x}_2) \, d\bar{x}_1 d\bar{x}_2$$

$$- \sum_i \int\int \psi_n^*(\bar{x}_1) \, \psi_i^*(\bar{x}_2) \, V(\bar{r}_1,\bar{r}_2) \, \psi_i(\bar{x}_1) \, \psi_m(\bar{x}_2) \, d\bar{x}_1 d\bar{x}_2$$

$$= \sum_i \int \psi_n^*(\bar{x}_2) \left[\int |\psi_i(\bar{x}_1)|^2 V(\bar{r}_1,\bar{r}_2) \, d\bar{x}_1 \right] \psi_m(\bar{x}_2) \, d\bar{x}_2$$

$$- \sum_i \int \psi_n^*(\bar{x}_2) \left[\int \psi_i(\bar{x}_1) \, \psi_m(\bar{x}_1) \, V(\bar{r}_1,\bar{r}_2) \, d\bar{x}_1 \right] \psi_i(\bar{x}_2) \, d\bar{x}_2$$

In the last integral we have interchanged $\bar{x}_1$ and $\bar{x}_2$ because the value of the definite integral does not depend on the variable of integration. Using this same rule, replacing the variable $\bar{x}_2$ by $\bar{x}$, we get

$$\int \psi_n^*(\bar{x}) F(\bar{r}) \psi_m(\bar{x}) \, d\bar{x}$$

$$= \int \psi_n^*(\bar{x}) \sum_i \left\{ \int |\psi_i(\bar{x}_1)|^2 V(\bar{r}_1,\bar{r}_2) \, \psi_m(\bar{x}) \, d\bar{x}_1 \right.$$

$$\left. - \int \psi_i^*(\bar{x}_1) \, \psi_m(\bar{x}_1) \, V(\bar{r}_1,\bar{r}_2) \, \psi_i(\bar{x}) \, d\bar{x}_1 \right\} d\bar{x} \tag{28}$$

From a comparison of the two sides, we have

$$F(\bar{r})\psi_m(\bar{r}) = \sum_i \int |\psi_i(\bar{x}_1)|^2 V(\bar{r}_1,\bar{r}) \, \psi_m(\bar{x}) \, d\bar{x}_1$$

$$- \sum_i \int \psi_i^*(\bar{x}_1) \, \psi_m(\bar{x}_1) \, V(\bar{r}_1,\bar{r}) \, \psi_i(\bar{x}) \, d\bar{x}_1 \tag{29i}$$

As ψ_i and ψ_m is the product of orbital part ϕ and a spin function, the above equation implies a sum over the two values of the spin variable. Carrying out the sum over the spin variable, we get

$$F(\bar{r}) \, \phi_m(\bar{r}) = \sum_{i=1}^{Z} \int |\phi_i(\bar{r}_1)|^2 V(\bar{r}_1,\bar{r}) \, \phi_m(\bar{r}) \, d\bar{r}_1$$

$$- \sum_{i=1}^{Z/2} \int \phi_i^*(\bar{r}_1) \, \phi_m(\bar{r}_1) \, V(\bar{r}_1,\bar{r}) \, \phi_i(\bar{r}) \, \delta_{s_i s_m} d\bar{r}_1 \tag{29ii}$$

The second integral on the right vanishes if ψ_i and ψ_m have different spin factors.

As a consequence, the Hartree-Fock equation for the function $\phi_m(\vec{r})$ now becomes

$$\left[-\frac{\hbar^2}{2m}\nabla^2 + V(r) \right]\phi_m(\vec{r}) + \sum_{i=1}^{Z} \int \left| \phi_i(\vec{r_1}) \right|^2 V(\vec{r_1},\vec{r}) \, \phi_m(\vec{r}) \, d\vec{r_1}$$

$$-\frac{1}{2}\sum_{i=1}^{Z} \phi_i(\vec{r}) \int \phi_i^*(\vec{r_1}) \, \phi_m(\vec{r_1}) \, V(\vec{r_1},\vec{r}) \, d\vec{r_1} = E_m\phi_m(\vec{r}) \tag{30}$$

There will be one such equation for each of the Z different functions $\phi_m(\vec{r})$. The operator $F(\vec{r})$ depends on all these functions. The third term on the l.h.s. of above equation is the *exchange term*. Without this exchange term, the equation reduces to Hartree Equation. The second term on the l.h.s., like the Hartree's self consistent field, is non-linear in ψ, but here it is not of the form $V(\vec{r}) \, \phi(\vec{r})$; rather it has the structure $\int V(\vec{r_1},\vec{r}) \, \phi_m(\vec{r}) \, d\vec{r_1}$ i.e. it is an integral operator. As a result, the Hartree-Fock equations are apparently quite intractable. However, as the choice of the wave functions ϕ_m are limited, these equations have more tractable form than the Hartree equations and also the indistinguishability of the electrons has been taken into account by the incorporation of exchange degeneracy.

19.3 The Thomas-Fermi Statistical Method

The mathematical difficulties of a numerical solution of the set of integro-differential equations of Hartree method increase enormously when the number of electrons in the atom is larger. One can then use the statistical method suggested by Thomas and Fermi to describe the basic features of the electron distribution and the field in complex atoms.

The Thomas-Fermi statistical model assumes that the potential function $V(r)$ produced by the nucleus and all the other electrons (except the one whose motion is under consideration) is spherically symmetric and varies slowly enough in an electron wavelength so that many electrons can be localized within a volume over which the potential changes only by a small fraction of itself. The electrons can then be treated by statistical mechanics, obeying Fermi-Dirac statistics.

When we use statistical considerations, we cannot explain the individual properties of each atom rather, then, we can describe the general properties of atoms such as their radii, ionization energy, polarization and their change with a variation in nuclear charge.

Assumptions

1. In heavy atoms, the majority of electrons are in states with large quantum numbers, or, in other words, in states for which the electron wavelength is considerably smaller than the size of the atoms.
2. The potential function $V(r)$ is sensibly constant over a region in which many electrons can be localized.
3. At normal temperatures, the thermal energy kT is very small in comparison with $V(r)$ everywhere except at the edge of the atom, where the probability of finding an electron is small.

Then, one can view the system as a 'box' within which the electrons can have various quantized values of the momentum.

The number of electron states in a cube of edge length L at the walls of which the wave functions obey the periodic boundary conditions is given by

$$\left(\frac{L}{2\pi\hbar}\right)^3 dp_x dp_y dp_z \tag{31i}$$

Taking into account the Pauli exclusion principle, the number of states become

$$2\left(\frac{L}{2\pi\hbar}\right)^3 dp_x dp_y dp_z \tag{31ii}$$

Electrons obey Fermi-Dirac statistics and these states are filled in order of increasing momentum, upto some maximum value p_0. Assuming spherical symmetry, the total number of states are

$$2\left(\frac{L}{2\pi\hbar}\right)^3 \int_0^{p_0} \int_0^{\pi} \int_0^{2\pi} p^2 dp \sin\theta \, d\theta \, d\phi = \frac{p_0^3 L^3}{3\pi^2\hbar^3} \tag{32}$$

Each of the above states is occupied by one electron, so, the volume density of electrons in the atom is

$$n = \frac{p_0^3}{3\pi^2\hbar^3} \tag{33}$$

The charge density is $\rho = ne$ and the potential $V(r)/e$ is determined by the poission's equation $\left(\nabla^2\phi = -4\pi\rho\right)$ i.e.

$$\nabla^2\left[\frac{V(r)}{e}\right] = -4\pi ne \tag{34}$$

Now, the maximum momentum (p_0) a bound electron at r can have is given by

$$\frac{p_0^2}{2m} = -V(r) \text{ or } p_0 = \left[-2mV(r)\right]^{1/2} \tag{35}$$

We can write Eqn. (34) as

$$\frac{1}{e}\frac{1}{r^2}\frac{\partial}{\partial r}\left(r^2\frac{\partial V(r)}{\partial r}\right) = -4\pi \, n(r) \, e \tag{36}$$

From Eqns (33) and (35)

$$n(r) = \frac{\left[-2mV(r)\right]^{3/2}}{3\pi^2\hbar^3} \tag{37}$$

From Eqns (36) and (37), Poisson's equation becomes

$$\frac{1}{r^2}\frac{\partial}{\partial r}\left(r^2\frac{\partial V(r)}{\partial r}\right) = \frac{4e^2\left[-2mV(r)\right]^{3/2}}{3\pi\hbar^3} \tag{38}$$

with the boundary conditions $V(r) \rightarrow -Ze^2/r$ as $r \rightarrow 0$ (i.e. the Coulomb field of nucleus) and $V(r) \rightarrow 0$ as $r \rightarrow \infty$. The Eqn. (38) can be expressed in dimensionless form in terms of $\chi(r)$ and x, defined by

$$V(r) = \left(-\frac{Ze^2}{r} \right) \chi(r) \tag{39}$$

and

$$r = bxZ^{-1/3} \tag{40}$$

where

$$b = \frac{1}{2} \left(\frac{3\pi}{4} \right)^{2/3} \frac{\hbar^2}{me^2} = 0.8853 \, a_0 \tag{41}$$

with

$$a_0 = \frac{\hbar^2}{me^2} \tag{42}$$

We then get from Eqns (36) and (39)

$$x^{1/2} \frac{d^2\chi}{dx^2} = \chi^{3/2} \tag{43}$$

with the boundary conditions

$$\chi \rightarrow 1 \text{ as } x \rightarrow 0$$

and

$$\chi \rightarrow 0 \text{ as } x \rightarrow \infty$$

The Eqn. (43) is known as *dimensionless Thomas Fermi equation*, since this involves no parameters and defines a universal function $\chi(x)$. The most accurate solution of this equation is due to Bush and Caldwell. From Eqns (39) and (40):

$$\left| V(x) \right| = \frac{Ze^2\chi(x)}{b \, x \, Z^{-1/3}} = \frac{e^2 Z^{4/3}\chi(x)}{x \, b} = \frac{e^2 Z}{r} \, \chi \left(\frac{rZ^{1/3}}{b} \right) \tag{44}$$

Substituting this value into Eqn. (37), we find the electron density distribution in the atom

$$n(x) = BZ^2 \left\{ \frac{1}{x} \chi(x) \right\}^{3/2} \tag{45}$$

with

$$x = \frac{rZ^{1/3}}{b} \approx \frac{rZ^{1/3}}{0.885 \, a_0}, \quad B = \left(\frac{2me^2}{b} \right)^{3/2} \left(3\pi^2\hbar^3 \right)^{-1} \tag{46}$$

It follows from Eqn. (43) and (45) that the electrical charge density distribution is similar for different heavy atoms. The characteristic length parameter ("radius") is the quantity (with $x = 1$)

$$R = bZ^{-1/3} = 0.885 \, a_0 Z^{-1/3} \tag{47}$$

(R may be interpreted as a characteristic atom radius). This shows that the "radius" of an atom is inversely proportional to the cube root of the atomic number, if this radius is to be interpreted to be that of a sphere that encloses a fixed fraction of all the electrons. The electron density decreases steeply for $x \geq 1$. The

Figure 19.1 shows a plot of radial electron density distribution $D(r) = 4\pi n(r) r^2$ for a Hg-atom calculated from the Thomas-Fermi theory. The dotted curve is the electron distribution calculated by the Hartree method.

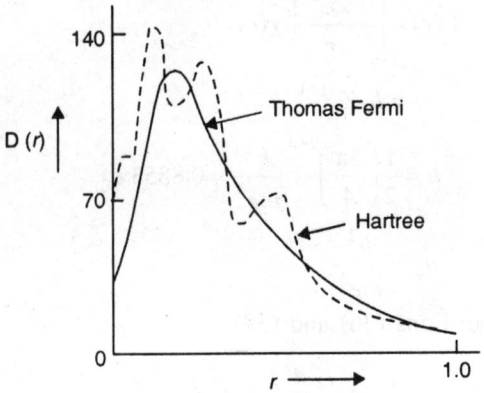

Fig. 19.1: The radial electron density distribution $D(r)$ in reciprocal atomic length units for a mercury atom

The use of this statistical method is better justified as Z increases.

Example 19.1: Find an expression for the electron density $n(r)$ in the Thomas-Fermi model in terms of the Thomas-Fermi function $\chi(x)$.

Solution: The electron density

$$n(r) = \frac{\left[-2mV(r)\right]^{3/2}}{3\pi^2 \hbar^3} \qquad (i)$$

Now,

$$V(r) = \left(-\frac{Ze^2}{r}\right) \chi(r) \qquad (ii)$$

$$\therefore \qquad n(r) = \frac{\left(2mZe^2 \chi/r\right)^{3/2}}{3\pi^2 \hbar^3}$$

But

$$r = bxZ^{-1/3} \qquad (iii)$$

$$\therefore \qquad n(x) = \frac{\left(2mZe^2\right)^{3/2}}{3\pi^2 \hbar^3} \left(\frac{\chi}{bx}\right)^{3/2} \cdot \left(\frac{1}{Z^{-1/3}}\right)^{3/2}$$

$$= \frac{\left(2me^2\right)^{3/2}}{3\pi^2 \hbar^3} \left(\frac{\chi}{x}\right)^{3/2} \cdot \frac{Z^2}{(b)^{3/2}} \qquad (iv)$$

Since

$$b = \frac{1}{2}\left(\frac{3\pi}{4}\right)^{2/3} \left(\frac{\hbar^2}{me^2}\right) \equiv \frac{1}{2}\left(\frac{3\pi}{4}\right)^{2/3} a_0$$

$\therefore$ From Eqn. (*iv*)

$$n(x) = \frac{\left(2me^2\right)^{3/2}}{3\pi^2\hbar^3} \left(\frac{\chi}{x}\right)^{3/2} Z^2 \times 2^{3/2} \times \left(\frac{3\pi}{4}\right)^{-1} a_0^{-3/2}$$

$$= \frac{1}{9\pi^3} \frac{\left(me^2\right)^{3/2}}{\left(\hbar^2\right)^{3/2}} \times 4Z^2 \times a_0^{-3/2} \left(\frac{\chi}{x}\right)^{3/2}$$

or
$$n(x) = \frac{32\,Z^2}{9\pi^3} \left(\frac{\chi}{x}\right)^{3/2} \frac{1}{a_0^3} \qquad\qquad \left(\because \frac{me^2}{\hbar^2} = \frac{1}{a_0}\right)$$

Questions and Problems

1. What is a central field approximation? How did Hartree obtain the central field in his theory of many electron atom?
2. What do you understand by self consistent potential?
3. Discuss Hartree's self consistent method for determining the potential energy $V(r)$ in central field approximation. Why it is called self consistent field method?
4. Find an expression for the electron density in Thomas-Fermi statistical model.
5. In the Thomas-Fermi model, show that the radius of a sphere which encloses a fixed fraction of all the electrons is proportional to $Z^{-1/3}$.
6. Write short notes on
   (*i*) Thomas-Fermi model
   (*ii*) Central field approximation
   (*iii*) Hartree's self consistent field method.
7. Discuss Hartree-Fock approximation and derive Hartree-Fock equations. What is the exchange term?
8. Show that for a Thomas-Fermi atom (*a*) $\langle r \rangle = \left(\hbar^2 / \mu e^2\right) Z^{-1/3}$, (*b*) the average Coulomb interaction energy of two electrons in the atom $= \left(\mu e^4 / \hbar^2\right) Z^{1/3}$, (*c*) the average kinetic energy of an electron $= \left(\mu e^4 / \hbar^2\right) Z^{4/3}$, (*d*) the average velocity of an electron $= \left(e^2 / \hbar\right) Z^{2/3}$ and (*e*) the average angular momentum of an electron $= Z^{1/3}\hbar$.

20

Theory of Scattering

20.1 Introduction

We know that Rutherford scattering (i.e. scattering of α-particles by atomic nucleus) led to an elegant picture of atomic structure. Similarly, electron scattering by proton revealed the existence of proton structure form factors. From the scattering experiments on nucleon by nucleon, we had a first hand information of the strong (viz. nuclear) forces. Thus scattering experiments are of immense importance in experimental physics, as they provide very useful information about the forces and interaction between the scattered particles and the target materials. From a detailed study of the results of scattering, much can be learnt about the nature of the particles that are being scattered as well as those doing the scattering.

In this chapter we shall be considering the quantum mechanical treatment of scattering of particle(s) by a potential and approximation methods applicable to various situations.

20.2 Scattering of an Electron by Coulomb Field of the Nucleus (using Fermi's Golden Rule)

The scattering process is characterized by a cross section, which is equal to the ratio of the probability $W_{k \to m}$ to the number of particles 'F' incident per unit time on a unit surfaces perpendicular to the incident beam.

Obviously, the particles that strike this surface per unit time are those which are located at a distance not exceeding the $|\vec{v}|$ i.e. in volume $v \times s \equiv v$. Therefore the number F is equal to the number of particles per unit volume V'' multiplied by a volume which is numerically equal to the velocity of the particle, i.e.

$$F = \frac{|\vec{v}|}{V''} = \frac{1}{V''} \frac{|\vec{p}|}{m}$$

Scattering cross section $d\sigma = \dfrac{\text{Transition probability time}}{\text{Incident Flux}}$ \hfill (1a)

or
$$d\sigma = \frac{W_{k \to m}}{F} \qquad (1b)$$

It is to be noted that the incident flux has the dimensions of $\dfrac{1}{L^3} \times \dfrac{L}{T} = \dfrac{1}{L^2 T}$,
therefore $d\sigma$ has the dimensions of area. We take H_0 to be the free particle Hamiltonian and the potential $V(\vec{r})$ to be the perturbation. The initial and final states describing the particle incident with the momentum $\vec{p}$ and scattered with momentum $\vec{p}'$ are represented by the normalized wave functions

$$\psi_k = \frac{1}{\sqrt{V''}} e^{i\vec{p} \cdot \vec{r} / \hbar} \ ; \ \psi_m = \frac{1}{\sqrt{V''}} e^{i\vec{p}' \cdot \vec{r} / \hbar} \qquad (2)$$

where V'' is the normalization volume.

$\vec{p}'$ is in (θ, ϕ) direction (along $\vec{k}'$). The magnitude $\left|\vec{k}'\right|$ defines the energy $E_m = \dfrac{\hbar^2 k^2}{2m}$. For energy conservation $\left|\vec{k}'\right| = k = \left|\vec{k}\right|$. The matrix element H'_{mk} is given by

$$H'_{mk} = \int \psi_m^* \left(-\frac{Ze^2}{r}\right) \psi_k d^3 r \qquad (3)$$

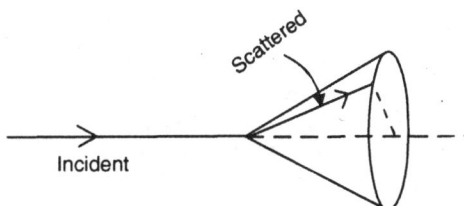

Fig. 20.1

$$= \frac{1}{V''} \int e^{i\vec{k} \cdot \vec{r}} \left(-\frac{Ze^2}{r}\right) r^2 dr \, d\Omega \qquad (d\Omega = d(\cos\theta) \, d\phi)$$

where
$$\hbar \vec{k} = \vec{p} - \vec{p}' \equiv \vec{q}$$

$\vec{k}$ is along the direction of momentum transfer. Taking direction of $\vec{k}$ as the direction of polar axis, we get

$$H'_{mk} = -\frac{2\pi Ze^2}{V''} \int\limits_{r=0}^{\infty} \int\limits_{\cos\theta = -1}^{+1} e^{ikr\cos\theta} \frac{1}{r} r^2 dr \, d(\cos\theta)$$

$$= -\frac{2\pi Ze^2}{V''} \int\limits_{r=0}^{\infty} \frac{1}{ikr} \left(e^{ikr} - e^{-ikr}\right) r \, dr \qquad (4)$$

The series $\int\limits_0^\infty \left(e^{ikr} - e^{-ikr}\right) dr$ does not exist in ordinary sense, but it exists in the sense of "Cessaro Summability", we multiply this by the convergence factor $e^{-\varepsilon r}$ and after summation take the limit $\varepsilon \to 0$:

$$\int\limits_0^\infty e^{ikr-\varepsilon r} \, dr = -\frac{1}{ik-\varepsilon} = \frac{1}{\varepsilon-ik}$$

$$\int\limits_0^\infty e^{-ikr-\varepsilon r} \, dr = -\frac{1}{-ik-\varepsilon} = \frac{1}{\varepsilon+ik} \tag{5}$$

Therefore,

$$H'_{mk} = \lim_{\varepsilon \to 0}(-1)\frac{2\pi Ze^2}{V''ik}\left[\frac{1}{\varepsilon-ik} - \frac{1}{\varepsilon+ik}\right]$$

$$= \lim_{\varepsilon \to 0}(-1)\frac{2\pi Ze^2}{V''ik}\frac{2ik}{\varepsilon^2+k^2}$$

or

$$H'_{mk} = \frac{4\pi Ze^2}{V''k^2} \tag{6}$$

Now,

$$W_{k \to m} = \frac{2\pi}{\hbar}\left|H'_{mk}\right|^2 \rho_m \tag{7}$$

To find ρ_m (the energy density of the final states), suppose, the final states are those of the scattered particle in a large box of volume V''. The allowed values of k_x, k_y and k_z are given by

$$k_x = \frac{2\pi n_x}{L} \; ; \; k_y = \frac{2\pi n_y}{L} \; ; \; k_z = \frac{2\pi n_z}{L} \tag{8}$$

$\therefore$

$$\rho_m dE_m = dn_x dn_y dn_z \tag{9}$$

$$= \left(\frac{L}{2\pi}\right)^3 dk_x dk_y dk_z \tag{10a}$$

$$= \frac{V''d^3k}{(2\pi)^3} = \frac{V''d^3p}{(2\pi\hbar)^3} \quad (\because p = \hbar k) \tag{10b}$$

or

$$\rho_m dE_m = \frac{V''d\Omega \, p^2 dp}{(2\pi\hbar)^3} = \frac{V''d\Omega \, (p \, dp) \, p}{(2\pi\hbar)^3}$$

$$= \frac{V''d\Omega \, (m \, dE_m) \, p}{(2\pi\hbar)^3} \tag{11}$$

Hence from (6) and (11), the transition probability per unit time

$$W_{k \to m} = \frac{2\pi}{\hbar}\left|H'_{mk}\right|^2 \rho_m$$

$$= \frac{2\pi}{\hbar} \left(-\frac{4\pi Ze^2}{V''k^2} \right)^2 \frac{V''d\Omega \, m \, p}{(2\pi\hbar)^3} \tag{12}$$

Therefore, scattering cross-section is

$$d\sigma = \frac{W_{k\to m}}{F} = \frac{2\pi}{\hbar} \left(-\frac{4\pi Ze^2}{V''k^2} \right)^2 \frac{V''d\Omega \, m \, p}{(2\pi\hbar)^3} \times \frac{V''m}{p}$$

$$= \frac{2\pi}{\hbar} \left(\frac{4\pi Ze^2}{k^2} \right)^2 \frac{1}{V''^2} \frac{V''d\Omega}{(2\pi\hbar)^3} m^2 V'' \tag{13}$$

or differential cross-section is

$$\frac{d\sigma}{d\Omega} = \left(\frac{4Z^2 e^4}{q^4} \right) m^2 \tag{14}$$

20.3 The Scattering Cross-section

In a typical scattering experiment, a beam of homogenous, monoergic (i.e. an ensemble of) particles falls on a target consisting of a large number of scattering centres (the target may be a thin foil). The particles are scattered by the target in all directions and the scattered particles are received and analyzed by a detector placed at a large distance (relative to linear dimensions of the target) from the target. Let the origin of the coordinate system be at the target, and the z-axis along the direction of the incident beam.

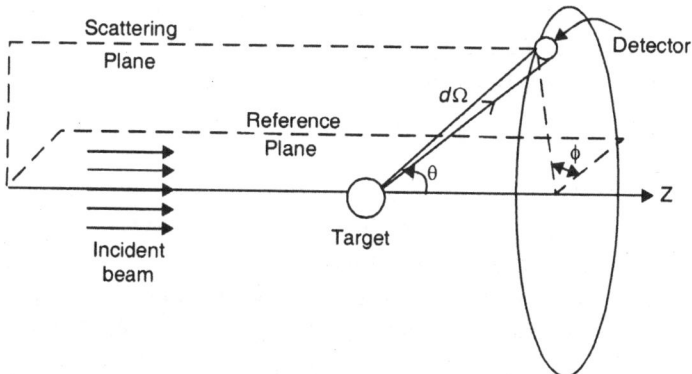

Fig. 20.2: Schematics of a scattering event

If N_s is the number of particles scattered in the solid angle $d\Omega = \sin\theta \, d\theta \, d\phi$ per second, then N_s will be proportional to the incident flux N (i.e. particles per unit area per unit time), the n (the number of scattering centres) and the solid angle $d\Omega$, i.e.

$$N_s \propto N.n.d\Omega \tag{15a}$$

provided the flux is small enough so that there is no interference between bombarding particles and no appreciable decrease in the number of bombarded particles by their recoil out of the target region and provided also that the bombarded particles are far apart so that each collision process involves only one of them.

The proportionality factor which depends in general on θ and ϕ is called the differential scattering cross-section and is written as $\left(\dfrac{d\sigma}{d\Omega}\right)$:

$$N_s = \frac{d\sigma(\theta,\phi)}{d\Omega}.N.n.d\Omega \qquad (15b)$$

dimensions of N_s is (time)$^{-1}$, that of N is $L^{-2}T^{-1}$ (n and Ω are dimensionless). Hence $d\sigma/d\Omega$ must have the dimensions of area, which justifies the term cross-section. It is interpreted as the *area of the cross-section of the incident beam through which pass all particles that are scattered into $d\Omega$ about θ and ϕ.*

The total scattering cross-section σ_t may be obtained by

$$N_{s,total} = \int N_s d\Omega = N \int_0^{2\pi} \int_0^{\pi} n \left(\frac{d\sigma}{d\Omega}\right) \sin\theta \, d\theta \, d\phi = N \cdot n \cdot \sigma_t \qquad (16)$$

i.e.

$$\sigma_t = 2\pi \int \left(\frac{d\sigma}{d\Omega}\right) \sin\theta \, d\theta \qquad (17)$$

(considering $d\sigma/d\Omega$ as independent of ϕ.)

20.4 Asymptotic Behaviour of Wave Function and Scattering Amplitude

In the quantum mechanical description of scattering, the total wave function consists of two parts, the incident part and the scattered part. The incident part is a plane wave as in a force free region and the total function ψ is

$$\psi = \exp\left(i\vec{k}.\vec{r}\right) + \psi_{sc} \qquad (18)$$

where $\exp\left(i\vec{k}.\vec{r}\right)$ represents the incident wave and ψ_{sc} represents the scattered particle wave function. Let us take the origin of the coordinate system at the scattering centre. Far from the target the scattered wave ψ_{sc} must represent a flux of particles moving away radially outward. This radially outgoing wave is the eigenstate of the radial momentum p_r conjugate to the radial variable r, i.e.

$$[r, p_r] = i\hbar \qquad (19)$$

A simple solution for $P_r = -i\hbar \dfrac{\partial}{\partial r}$, but this representation is not hermitian. The quantity π_r defined by

$$\pi_r = -i\hbar \frac{\partial}{\partial r} - i\hbar \frac{1}{r} = p_r - i\hbar \frac{1}{r} \qquad (20)$$

indeed satisfies $\pi_r^\dagger = \pi_r$ and $[r, \pi_r] = i\hbar$. Thus, the conjugate momentum operator corresponding to the radial variable r is $-i\hbar \left(\dfrac{\partial}{\partial r} + \dfrac{1}{r}\right)$. The eigenfunction of this operator is

$$-i\hbar\left(\frac{\partial}{\partial r}+\frac{1}{r}\right)\frac{e^{ikr}}{r}=\hbar k\,\frac{e^{ikr}}{r} \tag{21}$$

for the radially outgoing case. So,

$$\psi_{sc}=f(\theta,\phi)\frac{e^{ikr}}{r} \tag{22}$$

at large r from the scattering centre. The quantity $f(\theta,\phi)$ is the amplitude of the outgoing wave which it picks up due to the interaction of the incident wave with the target, and

$$\psi=e^{i\vec{k}\cdot\vec{r}}+f(\theta)\frac{e^{ikr}}{r} \tag{23}$$

(assuming f to be independent of ϕ) $\vec{k}=k\hat{z}$ ($\hat{z}$ is a unit vector along z axis)

If Eqn. (23) is multiplied by the time factor $e^{-i\omega t}$, the first term represents a plane wave propagating from left to right along z-axis (which is also the polar axis). This wave describes a beam in which there is one particle per unit volume, so crossing flux per unit time,

$$F_{inc}=v\left(=\frac{\hbar k}{m}\right) \tag{24}$$

In the second term on the r.h.s. of Eqn. (23), number of particles/volume at a distance r from the origin

$$=\frac{|f(\theta)|^2}{r^2} \tag{25}$$

So, the number per unit time crossing area dA (perpendicular to scattered wave)

$$=v\,\frac{|f(\theta)|^2}{r^2}\,dA \tag{26}$$

(assuming elastic scattering: velocity remains same.) Now, the area

$$dA=r^2 d\Omega \tag{27}$$

Therefore the number of particles/time scattered in solid angle $d\Omega$ (= number per unit time crossing dA), is

$$dF_{sc}=v\,\frac{|f(\theta)|^2}{r^2}\,r^2\,d\Omega \tag{28}$$

By definition, scattering cross-section

$$d\sigma=\frac{dF_{sc}}{F_{inc}}=|f(\theta)|^2\,d\Omega \tag{29}$$

and differential scattering cross-section

$$\frac{d\sigma}{d\Omega}=|f(\theta)|^2 \tag{30}$$

The scattering amplitude $f(\theta)$ describes the angular distribution of the scattered wave

$$\sigma_t=2\pi\int_0^\pi|f(\theta)|^2\,\sin\theta\,d\theta \tag{31}$$

20.5 Scattering in the Laboratory and Centre of Mass System of Coordinates

The scattering process taking place in case of collision between two particles can be visualized in two kinds of coordinate frames

(1) *Laboratory frame* in which the bombarded particle is initially at rest. and

(2) *Centre of mass frame* in which the centre of mass is initially and always at rest.

In practice, the target is never fixed rather, it recoils due to the interaction of incident particle. It is convenient to make measurements of angles in the lab coordinates in which the target is at rest before recoiling. But, to analyze the scattering data, one finds the centre of mass coordinate system more useful. In the latter system, the scattering can be described as taking place from the stationary centre of mass with the two colliding particles remaining collinear with the C.M. and both moving either away (after scattering) or towards each other (before scattering) with unchanged momentum of the particles. The collinearity (before and) after collision arises due to momentum conservation. It is obvious that angle of scattering in the C.M. system would differ from that in the laboratory system, but the relationship between the two can be easily derived from the geometry of the coordinate systems used.

Let m_1, r_1 and m_2, r_2 be the masses and coordinates of the bombarding and the target particles respectively before scattering. The coordinates of centre of mass is

$$\vec{R} = \frac{\left(m_1\vec{r}_1 + m_1\vec{r}_2\right)}{\left(m_1 + m_2\right)} \tag{32}$$

and the relative coordinate

$$\vec{r} = \vec{r}_1 - \vec{r}_2 \tag{33}$$

Hence, we have

$$\vec{r}_1 = \vec{R} + \frac{m_2}{M}\vec{r} \; ; \; \vec{r}_2 = \vec{R} - \frac{m_1}{M}\vec{r} \tag{34}$$

where $M = \left(m_1 + m_2\right)$. The initial momenta are

$$m_1\dot{\vec{r}}_1 = m_1\dot{\vec{R}} + \frac{m_1 m_2}{M}\dot{\vec{r}} \; ; \; m_2\dot{\vec{r}}_2 = m_2\dot{\vec{R}} - \frac{m_1 m_2}{M}\dot{\vec{r}} \tag{35}$$

In the centre of mass system, we have $\dot{\vec{R}} = 0$, so

$$m_1\dot{\vec{r}}_1 = \frac{m_1 m_2}{M}\dot{\vec{r}} \equiv \vec{p}* \; ; \; m_2\dot{\vec{r}}_2 = -\frac{m_1 m_2}{M}\dot{\vec{r}} \equiv -\vec{p}* \tag{36}$$

Assume that $\vec{r}_1'$ and $\vec{r}_2'$ are the coordinates after scattering and $\vec{R}'$ and $\vec{r}'$ are the centre of mass and relative coordinates after scattering. Then

$$m_1\dot{\vec{r}}_1' = m_1\dot{\vec{R}}' + \frac{m_1 m_2}{M}\dot{\vec{r}}' \; ; \; m_2\dot{\vec{r}}_2' = m_2\dot{\vec{R}}' - \frac{m_1 m_2}{M}\dot{\vec{r}}' \tag{37}$$

In the centre of mass at rest system $\dot{\vec{R}}' = 0$ and

$$m_1\dot{\vec{r}}_1' = \frac{m_1 m_2}{M}\dot{\vec{r}}' \equiv \vec{q}* \ ; \ m_2\dot{\vec{r}}_2' = -\frac{m_1 m_2}{M}\dot{\vec{r}}' \equiv -\vec{q}* \tag{38}$$

From energy conservation $|\vec{p}*| = |\vec{q}*|$ would hold true.

In the laboratory system $\dot{\vec{R}}$ and $\dot{\vec{R}}'$ are not zero. Let $\vec{p}_1$, $\vec{p}_2$ denote the momenta of masses m_1 and m_2 before scattering and $\vec{q}_1$ and $\vec{q}_2$ their momenta after scattering, i.e.

$$\vec{p}_1 = m_1\dot{\vec{R}} + \frac{m_1 m_2}{M}\dot{\vec{r}} \ ; \ \vec{p}_2 = m_2\dot{\vec{R}} - \frac{m_1 m_2}{M}\dot{\vec{r}} \tag{39}$$

and

$$\vec{q}_1 = m_1\dot{\vec{R}}' + \frac{m_1 m_2}{M}\dot{\vec{r}}' \ ; \ \vec{q}_2 = m_2\dot{\vec{R}}' - \frac{m_1 m_2}{M}\dot{\vec{r}}' \tag{40}$$

Momentum conservation demands

$$\vec{p}_1 + \vec{p}_2 = \vec{q}_1 + \vec{q}_2 \tag{41}$$

and it implies $\dot{\vec{R}} = \dot{\vec{R}}'$. Again, in the lab system, the target m_2 is at rest i.e. $\vec{p}_2 = 0$, hence from Eqn. (39)

$$m_2\dot{\vec{R}} = \frac{m_1 m_2}{M}\dot{\vec{r}} = \vec{p}* \tag{42}$$

Using Eqn. (39) and (42)

$$\vec{p}_1 = \frac{m_1}{m_2}\left(m_2\dot{\vec{R}}\right) + \vec{p}* \text{ or } \vec{p}_1 = \frac{m_1}{m_2}\vec{p}* + \vec{p}*$$

Again $(\because p_2 = 0)$, $\vec{p}_1 = \vec{q}_1 + \vec{q}_2$ is also true since $\dot{\vec{R}} = \dot{\vec{R}}'$, therefore we have

$$\vec{q}_1 = m_1\dot{\vec{R}} + \vec{q}* = \frac{m_1}{m_2}\vec{p}* + \vec{q}* \tag{43}$$

and

$$\vec{q}_2 = \vec{p}* - \vec{q}* \tag{44}$$

We thus have the vector diagram as in Fig. 20.3.

Since $\vec{p}_1$ is the incident (particle) direction and $\vec{q}_1$ is the scattered direction (of m_1), the angle θ between these two directions is the laboratory scattering angle. In the centre of mass system, $\vec{p}*$ is the incident momentum direction and

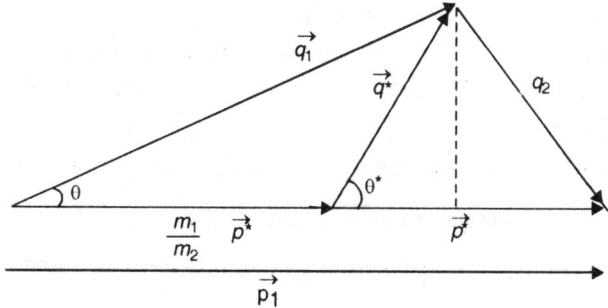

Fig. 20.3: Relation between laboratory and centre of mass system. θ is the scattering angle in the lab-frame and $\theta*$ is that in the centre of mass frame

$\vec{q}*$ the corresponding scattered direction, so $\theta*$ is the centre of mass scattering angle.

From the Figure 20.3, we have

$$\tan\theta = \frac{q*\sin\theta*}{\dfrac{m_1}{m_2}p*+q*\cos\theta} \tag{45}$$

But $\quad |\vec{p}*| = |\vec{q}*|$, so that $\quad \tan\theta = \dfrac{\sin\theta*}{\dfrac{m_1}{m_2}+\cos\theta*} \tag{46}$

This gives the desired relation between laboratory and centre of mass system. of scattering angles. Note that $\phi = \phi*$, when $m_1 = m_2$ then $\theta = \theta*/2$ and for m_2 very heavy, $\theta = \theta*$.

20.6 Green's Function

The stationary scattered wave function has the following asymptotic form:

$$u(\vec{x}) \to e^{ikz} + f(\theta,\phi)\frac{e^{ikr}}{r} \tag{47}$$

Since any stationary wave function must satisfy the time independent Schrödinger equation i.e

$$-\frac{\hbar^2}{2m}\nabla^2 u(\vec{x}) + V(\vec{x})\, u(\vec{x}) = E\, u(\vec{x}) \tag{48}$$

where $$E = \frac{\hbar^2 k^2}{2m}$$

Here $V(\vec{x})$ is the potential energy function for the projectile particle in the force field of the scattering centre. The eigenvalue E has to be taken equal to the kinetic energy of the particle in the asymptotic region (of $V = 0$).

The differential scattering cross-section is given by

$$\frac{d\sigma}{d\Omega} = |f(\theta,\phi)|^2 \tag{49}$$

where $f(\theta, \phi)$ is called *scattering amplitude*.

It is possible to get a formal solution for the scattering amplitude by transforming equation (48) into an integral equation. This can be accomplished by making use of Green's function.

Let $\Omega(\vec{x}, \nabla)$ be any linear differential operator such that

$$\Omega(c_1 u_1 + c_2 u_2) = c_1 \Omega u_1 + c_2 \Omega u_2$$

for arbitrary functions u_1, u_2 and constants c_1 and c_2. Then any function $G(\vec{x}, \vec{x}')$ such that

$$\Omega(\vec{x},\nabla)\, G(\vec{x},\vec{x}') = \delta(\vec{x} - \vec{x}') \tag{50a}$$

is said to be a Green's function for the operator Ω. We now proceed to derive an expression for the Green's function. This Green's function would be the operator for motion of a free particle. It is quite obvious that if we add to a particular Green's function (G) any solution S of the homogenous equation $\Omega S = 0$, the result $G + S$ is still a Green's function:

$$\Omega(G + S) = \delta(\vec{x} - \vec{x}') \tag{50b}$$

The general solution of any inhomogeneous equation

$$\Omega(\vec{x}, \nabla) \ u(\vec{x}) = F(\vec{x}) \tag{51}$$

can be written down in terms of Green's function by first expanding $F(\vec{x})$ as a linear combination of delta functions:

$$F(\vec{x}) = \int \delta(\vec{x} - \vec{x}') \ F(\vec{x}') \ d^3\vec{r}' \tag{52}$$

In fact if $u_0(\vec{x})$ is any function such that $\Omega u_0(\vec{x}) = 0$, then

$$u(\vec{x}) = u_0(x) + \int G(\vec{x}, \vec{x}') \ F(\vec{x}') \ d^3\vec{r}' \tag{53}$$

is a solution of Eqn. (51). This may be verified as follows by substituting this expression in Eqn. (51) and noting that

$$\Omega u(\vec{x}) \equiv \Omega \int G(\vec{x}, \vec{x}') \ F(\vec{x}') \ d^3\vec{r}' = \int \Omega \ G(\vec{x}, \vec{x}') \ F(\vec{x}') \ d^3\vec{r}'$$

$$= \int \delta(\vec{x} - \vec{x}') \ F(\vec{x}') \ d^3\vec{r}' = F(\vec{x}) \tag{54}$$

To obtain a particular solution one has to make a specific choice of the Green's function G and of the function $u_0(\vec{x})$ in Eqn. (53).

Let us now consider the specific case, viz. the Schrödinger equation by rewriting it as

$$-\left(\nabla^2 + k^2\right) u(\vec{x}) = -A(\vec{x}) \ u(\vec{x}) \equiv \phi(x) \tag{55}$$

with $\qquad k^2 = \dfrac{2mE}{\hbar^2}$ and $A(\vec{x}) = \dfrac{2mV(\vec{x})}{\hbar^2}$

This has the form (51) with

$$\Omega = \left(\nabla^2 + k^2\right), \ F(\vec{x}) = A(\vec{x}) \ u(\vec{x}) \tag{56}$$

The Green's function for $\left(\nabla^2 + k^2\right)$ can be deduced from the well known fact that

$$\nabla^2\left(\frac{-1}{r}\right) = -4\pi \ \delta(\vec{r}) \tag{57}$$

On using this we find that, for $u(x) \equiv e^{\pm ikr}/r$

$$\left(\nabla^2 + k^2\right)\left(e^{\pm ikr} / r\right) = -4\pi \ \delta(\vec{r}) \tag{58}$$

By replacing, $\vec{r}$ by $\vec{x} - \vec{x}'$ and r by $|\vec{x} - \vec{x}'|$ in this equation, we get two Green's functions G^+ and G^- for $\left(\nabla^2 + k^2\right)$, namely

$$G^{\pm}\left(\vec{x},\vec{x}'\right) = \frac{\exp\left[\pm ik\left|\vec{x}-\vec{x}'\right|\right]}{-4\pi\left|\vec{x}-\vec{x}'\right|} \quad \text{(using eqn. 50a and 58)} \tag{59}$$

In view of equations (53), (54), (56) and (59), we may write down a 'solution' of the Schrödinger equation as

$$u\left(\vec{x}\right) = e^{ikz} - \frac{1}{4\pi}\int\frac{e^{ik\left(\left|\vec{x}-\vec{x}'\right|\right)}}{\left|\vec{x}-\vec{x}'\right|}\,A\left(\vec{x}'\right)\,u\left(\vec{x}'\right)\,d^3\vec{r}' \tag{60}$$

20.7 Perturbation Approximation

We have

$$-\left(\nabla^2 + k^2\right)\psi\left(\vec{r}\right) = -U\left(\vec{r}\right)\psi\left(\vec{r}\right) \equiv \phi(\vec{r}) \tag{61}$$

Suppose $\phi(\vec{r})$ is a known function. Then the above equation is an inhomogeneous equation. We can apply a perturbation approach such that

$$\psi\left(\vec{r}\right) = e^{i\vec{k}\cdot\vec{r}} + \psi_1\left(\vec{r}\right) \tag{62}$$

(where $\psi_1\left(\vec{r}\right) \equiv f(\theta,\phi)\dfrac{e^{i\vec{k}\cdot\vec{r}}}{r}$ may be regarded as a small addition to the unperturbed plane wave $e^{i\vec{k}\cdot\vec{r}}$) and $k^2 = \dfrac{2mE}{\hbar^2}$ and $U\left(\vec{r}\right) = \dfrac{2mV\left(\vec{r}\right)}{\hbar^2}$.

Substituting Eqn. (62) in Eqn. (61), we get,

$$\left(-\nabla^2 - k^2\right)\psi_1\left(\vec{r}\right) = -U\left(\vec{r}\right)e^{i\vec{k}\cdot\vec{r}} = F(\vec{r}) \quad \text{(say)} \tag{63}$$

because $\left(\nabla^2 + k^2\right)e^{ikr} = 0$ and $U\left(\vec{r}\right)\psi_1\left(\vec{r}\right)$ can be neglected. So, we have to solve the inhomogeneous wave equation (63) where the right hand side is known. This differential equation can be converted to an integral equation with the help of Green's function as we shall use (we will use only ψ_1, so we may write it as ψ). Now this $\psi(\vec{r})$ can be expressed as a complete set of orthonormal eigenfuctions

$$\psi(\vec{r}) = \sum_{k'}a_{k'}u_{k'}(\vec{r}) \tag{64}$$

Substituting Eqn. (64) in Eqn. (63)

$$\left(-\nabla^2 - k^2\right)\psi\left(\vec{r}\right) = \sum_{k'}a_{k'}\left(-\nabla^2 u_{k'} - k^2 u_{k'}\right) = F(\vec{r})$$

or

$$\sum_{k'}a_{k'}\left(k'^2 - k^2\right)u_{k'} = F(\vec{r}) \qquad \left(\because u_{k'} \sim e^{i\vec{k}'\cdot\vec{r}}\right)$$

Taking overlap with $u_{k''}$ and integrating over $\vec{r}$, we get

$$\sum_{k'}a_{k'}\left(k'^2 - k^2\right)\int u_{k''}^{*}u_{k'}d\vec{r} = \int u_{k''}^{*}(\vec{r})\,F(\vec{r})\,d\vec{r}$$

or

$$\sum_{k'}a_{k'}\left(k'^2 - k^2\right)\delta_{k'k''} = \int u_{k''}^{*}(\vec{r})\,F(\vec{r})\,d\vec{r}$$

or
$$a_{k''} = \frac{\int u_{k''}^*(\vec{r})\, F(\vec{r})\, dr}{\left(k''^2 - k^2\right)} \quad \text{or} \quad a_{k'} = \frac{\int u_{k'}^*(\vec{r})\, F(\vec{r})\, dr}{\left(k'^2 - k^2\right)} \tag{65}$$

Substituting for $a_{k'}$ in Eqn. (64), we get

$$\psi(\vec{r}) = \int \sum_{k'} \left[\frac{u_{k'}^*(\vec{r}')\, u_{k'}(\vec{r})}{\left(k'^2 - k^2\right)}\right] F(\vec{r}')\, d\vec{r}' \tag{66}$$

We have replaced $\vec{r}$ by $\vec{r}'$ to avoid confusion between two r's in Eqn. (65) and (64). Equation (66) can be written as

$$\psi(\vec{r}) = \int G_{k'}(\vec{r},\vec{r}')\, F(\vec{r}')\, d\vec{r}' \tag{67}$$

where $F(\vec{r}')$ is defined by Eqn. (63) and

$$G_{k'}(\vec{r},\vec{r}') = \sum_{k'} \left[\frac{u_{k'}^*(\vec{r}')\, u_{k'}(\vec{r})}{\left(k'^2 - k^2\right)}\right] \tag{68}$$

is known as Green's function.

For a free particle, the wave function is $u_{k'}(\vec{r}) = \dfrac{1}{(2\pi)^{3/2}}\, e^{i\vec{k}'\cdot\vec{r}}$

So, the free particle Green's function is

$$G_{k'}(\vec{r},\vec{r}') = \frac{1}{(2\pi)^3} \int \frac{e^{i\vec{k}'\cdot(\vec{r}-\vec{r}')}}{k'^2 - k^2}\, d\vec{k}' \tag{69}$$

To carry out integration over k', we choose $(\vec{r} - \vec{r}')$ to lie along the polar axis:
$$\vec{k}'\cdot(\vec{r}-\vec{r}') = \vec{k}'\cdot\vec{R} = k'R\cos\theta' \equiv k'R\omega' \qquad \left(\vec{R} = \vec{r} - \vec{r}', \omega' = \cos\theta'\right)$$

$$\therefore \quad G_{k'} = \frac{1}{(2\pi)^3}\, 2\pi \int_{\omega'=-1}^{+1} \int_{k'=0}^{\infty} \frac{e^{ik'R\omega'}}{k'^2 - k^2}\, k'^2 d\omega'\, dk'$$

$$= \frac{1}{(2\pi)^3}\, 2\pi \int_{k'=0}^{\infty} \frac{k'^2 dk'}{k'^2 - k^2} \int_{\omega'=-1}^{+1} e^{ik'R\omega'}\, d\omega' \tag{70}$$

Now, $\displaystyle \int_{-1}^{+1} e^{ik'R\omega'}\, d\omega' = \frac{e^{ik'R} - e^{-ik'R}}{ik'R} = \frac{2}{k'R}\left[\frac{e^{ik'R} - e^{-ik'R}}{2i}\right] = \frac{2}{k'R}\sin k'R$

$$\therefore \quad G_{k'} = \frac{2}{4\pi^2} \int_{k'=0}^{\infty} \frac{k'^2}{k'^2 - k^2}\cdot\frac{\sin k'R}{k'R}\, dk'$$

$$= \frac{1}{4\pi^2 R} \int_{-\infty}^{\infty} \frac{k'\sin k'R}{k'^2 - k^2}\, dk' \tag{71}$$

Put
$$k'R = x \text{ and } kR = \sigma \tag{72}$$

Then
$$G_{k'} = \frac{1}{4\pi^2 R} \int_{-\infty}^{\infty} \frac{x \sin x}{x^2 - \sigma^2} \, dx \tag{73}$$

The integral in Eqn. (73) is evaluated by contour integration (theorem of residues) which states that if $F(Z)$ is a function of the complex variable Z such that it is analytic throughout a closed contour and its interior, except at a number of poles inside the contour, then

$$\oint F(Z) \, dZ = 2\pi i \sum R$$

where $\sum R$ denotes the sum of residues of $F(Z)$ at those of its poles that are situated within the contour.

$$G = \frac{1}{4\pi^2 R} \left[I - II \right] \tag{74}$$

where
$$I = \frac{1}{2i} C_1 \int_{-\infty}^{+\infty} \frac{x e^{ix}}{(x+\sigma)(x-\sigma)} \, dx$$

and
$$II = \frac{1}{2i} C_2 \int_{-\infty}^{+\infty} \frac{x e^{-ix}}{(x+\sigma)(x-\sigma)} \, dx$$

where C_1 consists of the real axis and an infinite semicircle in the upper half plane and C_2 consists of the real axis and an infinite semicircle in the lower half plane. We have four possibilities:

Both the poles below the semicircle.
Both the poles above the semicircle.
Pole at $x = \sigma$ within the semicircle.
Pole at $x = -\sigma$ within the semicircle.

However, since the poles are on the contour, the integrals are improper. The value of an improper integral depends on the limiting process used to evaluate the integral. We have to choose only those solutions which satisfy the boundary condition $\psi(r) \sim f(\theta, \phi) \dfrac{e^{ikr}}{r}$ (i.e. an outgoing spherical wave).

$$\therefore \qquad I = 2\pi i \qquad \qquad \text{(residue at } x = +\sigma)$$

$$= \frac{2\pi i}{2i} \frac{\sigma e^{i\sigma}}{2\sigma} = \frac{\pi}{2} e^{i\sigma}$$

$$II = -2\pi i \qquad \qquad \text{(residue at } x = -\sigma)$$

$$= -\frac{2\pi i}{2i} \left(\frac{-\sigma e^{i\sigma}}{-2\sigma} \right) = -\frac{\pi}{2} e^{i\sigma}$$

$$G = \frac{1}{4\pi^2 R} \left[\frac{\pi}{2} e^{i\sigma} + \frac{\pi}{2} e^{i\sigma} \right] = \frac{e^{i\sigma}}{4\pi R} = \frac{e^{ikR}}{4\pi R}$$

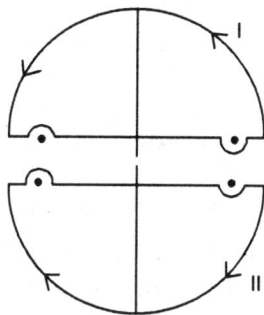

Fig. 20.4

$$\text{or} \qquad G = \frac{e^{ik\cdot|\vec{r}-\vec{r'}|}}{4\pi|\vec{r}-\vec{r'}|} \qquad\qquad (75)$$

Substituting Eqn. (75) in Eqn. (66) and using Eqn. (63), we have

$$\psi_1\left(\vec{r}\right) = -\int \frac{e^{ik\cdot|\vec{r}-\vec{r'}|}}{4\pi|\vec{r}-\vec{r'}|} U\left(\vec{r'}\right) e^{i\vec{k}\cdot\vec{r'}} dr' \qquad\qquad (76)$$

with $\qquad \psi_{total}\left(\vec{r}\right) = e^{i\vec{k}\cdot\vec{r}} + \psi_1\left(\vec{r}\right)$

where $\psi_1\left(\vec{r}\right)$ is the scattered wave function.

20.8 The Born Approximation

We have

$$\psi_{total}\left(\vec{r}\right) = e^{i\vec{k}\cdot\vec{r}} + \psi_{sc}\left(\vec{r}\right)$$

where $\qquad \psi_{sc}\left(\vec{r}\right) = -\int \frac{e^{ik\cdot|\vec{r}-\vec{r'}|}}{4\pi|\vec{r}-\vec{r'}|} U\left(\vec{r'}\right) e^{i\vec{k}\cdot\vec{r'}} dr'$

is an integral involving the function $\dfrac{e^{ik\cdot|\vec{r}-\vec{r'}|}}{|\vec{r}-\vec{r'}|}$ which represents just a spherical

wave that spreads out from point $\vec{r'}$ with a wavelength $\lambda = 2\pi/k$. The amplitude

of the spherical wave is proportional to the product $U\left(\vec{r'}\right) \exp\left(i\vec{k}\cdot\vec{r'}\right)$. All these

spherical waves are compounded at the point $\vec{r}$, in addition to the incident wave
to produce the total wave function.

If the scattering takes place from the scattering centres which are localized but
are weak such that the scattering does not take place at large distance from the
scatterer and the scattered wave is weak in amplitude, then Born approximation
is used to evaluate the scattering amplitude and cross section.

In this case, we restrict our attention to those cases in which the potential energy of interaction between the colliding particles is regarded as a perturbation and carry the calculation to first order. So, Born approximation is best applied in those cases in which the kinetic energy of the colliding particles is large in comparison with the interaction energy. It therefore supplements the method of partial waves.

It is assumed that r is large in comparison with those values of r' for which there is a significant contribution to the integral, (and $U(\vec{r}') \to 0$ for $|\vec{r}'| \to \infty$).

For $r \to \infty$, the following asymptotic relations are valid:

$$|\vec{r} - \vec{r}'| = \sqrt{r^2 + r'^2 - 2rr'\omega} \qquad (\omega = \hat{r} \cdot \hat{r}')$$

or

$$|\vec{r} - \vec{r}'| = r\sqrt{1 - \frac{2r'\omega}{r} + \left(\frac{r'}{r}\right)^2} \cong r\left[1 - \frac{r'\omega}{r}\right] = r - r'\omega \qquad (77)$$

Similarly,

$$\frac{1}{|\vec{r} - \vec{r}'|} \cong \frac{1}{r} \qquad (78)$$

$$\therefore \qquad \Psi_{sc}(\vec{r}) \cong -\frac{e^{ikr}}{r} \int \frac{e^{-ikr'\omega}}{4\pi} U(\vec{r}') e^{i\vec{k}\cdot\vec{r}'} dr' \qquad (79)$$

Comparison of this equation with $\psi(r) = f(\theta, \phi)\dfrac{e^{ikr}}{r}$ shows that

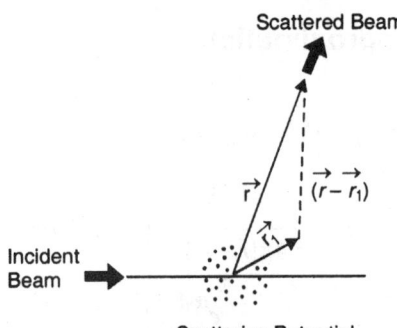

Scattered Beam

Incident Beam

Scattering Potential

Fig. 20.5

$$f(\theta, \phi) = -\frac{1}{4\pi} \int e^{-ikr'\omega} U(\vec{r}') e^{i\vec{k}\cdot\vec{r}'} dr' \qquad (80)$$

Since, scattering is elastic $|\vec{k}'| = |\vec{k}|$, $\vec{k}'$ is in the direction of $\hat{r}$. Therefore, $\vec{k}' = k\hat{r}$. Now,

$$kr'(\omega) = kr'(\hat{r} \cdot \hat{r}') = k\hat{r} \cdot \vec{r}' = \vec{k}' \cdot \vec{r}'$$

$\therefore$ Eqn. (80) becomes

$$f(\theta,\phi) = -\frac{1}{4\pi} \int e^{-i(\vec{k}-\vec{k}')\cdot\vec{r}} \, U(\vec{r}') \, d\vec{r}' \tag{81}$$

We define the vector $(\vec{k} - \vec{k}')$ by $\vec{q}$ so that $q^2 = 4k^2 \sin^2 \dfrac{\theta}{2}$ where θ is the angle between $\vec{k}$ and $\vec{k}'$.

$k = k'$ and $\vec{q}$ is perpendicular to the bisector of angle θ. We will now evaluate $f(\theta, \phi)$ from Eqn. (81):

$$d\vec{r}' = r'^2 dr' d(\cos\theta') \, d\phi$$

We take q as the direction of the polar axis, so that

$$\vec{q} \cdot \vec{r}' = qr' \cos\theta'$$

$\therefore$
$$f(\theta,\phi) = -\frac{1}{4\pi} 2\pi \int_0^\infty U(r') \int_{-1}^{+1} r'^2 e^{iqr'\cos\theta'} d(\cos\theta') \, dr'$$

(for central potential, there is no θ, ϕ dependence)

Now,

$$\int_{-1}^{+1} e^{iqr'\cos\theta'} d(\cos\theta') = \frac{2}{qr'} \sin qr'$$

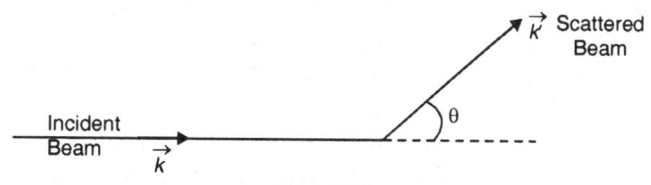

Fig. 20.6

$\therefore$
$$f(\theta,\phi) = -\frac{2}{4} \int_{r'=0}^\infty r'^2 \, U(r') \, \frac{2}{qr'} \sin qr' \, dr'$$

or
$$f_B(\theta,\phi) = -\frac{1}{q} \int_0^\infty r' \sin qr' \, U(r') \, dr' \tag{82}$$

This is *Born's approximation*. Thus f_B depends on the momentum transfer only and not on the initial momentum and the angle of scattering separately. It is to be noted here that in the derivation, we have effectively replaced $\psi(\vec{r}')$ by $\psi_0(r') = e^{i\vec{k}\cdot\vec{r}}$ to get the first order perturbed function ψ_1:

$$\psi_{sc} = \psi_0 + \psi_1 + \dots$$

therefore integral equation for $\psi_{sc}(\vec{r})$ can be solved as an infinite series. In first approximation

$$\psi_{total}(\vec{r}) = \psi_0(\vec{r}) + \int G(\vec{r},\vec{r}'') \, U(\vec{r}'') \, \psi(\vec{r}'') \, d\vec{r}'' \tag{83}$$

(for source point at $\vec{r}''$) and ($\psi_0(\vec{r}) = e^{i\vec{k}\cdot\vec{r}}$)

Replacing $\vec{r}$ by $\vec{r}'$ in above Eqn., we get

$$\psi(\vec{r}') = e^{i\vec{k}\cdot\vec{r}} + \int G(\vec{r}',\vec{r}'') \, U(\vec{r}'') \, \psi(\vec{r}'') \, dr'' \quad \left(\psi_0(\vec{r}') = e^{i\vec{k}\cdot\vec{r}}\right) \tag{84}$$

Substituting this value of $\psi(\vec{r}')$ in the integral of Eqn. representing the total wave function at $\vec{r}$ for source point at $\vec{r}'$, viz.

$$\psi(\vec{r}) = e^{i\vec{k}\cdot\vec{r}} + \int G(\vec{r},\vec{r}') \, U(\vec{r}') \, \psi(\vec{r}') \, dr' \tag{85}$$

we get '

$$\psi(\vec{r}) = e^{i\vec{k}\cdot\vec{r}} + \int G(\vec{r},\vec{r}') \, U(\vec{r}') \, \psi(\vec{r}') \, dr'$$
$$+ \int\int G(\vec{r},\vec{r}') \, U(\vec{r}') \, G(\vec{r}',\vec{r}'') \, U(\vec{r}'') \, \psi(\vec{r}'') \, d\vec{r}'' \, dr' \tag{86}$$

(second order perturbation).

This process can be repeated indefinitely resulting in a series (known as the *Neumann series*) which can be expected to represent a solution provided the series converges. However if the potential $U(r)$ is weak, the series can be approximated by a first few terms only. This is called *Born approximation*.

The series has the following meaning:

The first term $e^{i\vec{k}\cdot\vec{r}}$ represents the incident wave function while the remaining terms correspond to scattered wave function. The 1st term in the scattered wave represents single scattering of the incident wave $e^{i\vec{k}\cdot\vec{r}}$ by the interaction $U(\vec{r}')$ in the volume element $d\vec{r}'$. This produces a wave which travels from $\vec{r}$ to the point of observation $\vec{r}$ and total wave arising from single scattering is obtained by integration.

The second term represents double scattering. In the second term, the incident wave $e^{i\vec{k}\cdot\vec{r}''}$ after being scattered at $\vec{r}''$ travels to the point $\vec{r}'$ where it is again scattered and travels to the point $\vec{r}$. Accordingly, n^{th} term represents the contribution of waves which have been scattered n times in the region of interaction before traveling to the point $\vec{r}$. The Green's function in each term can be thought of as *propagators* which carry the waves from one point of scattering to another.

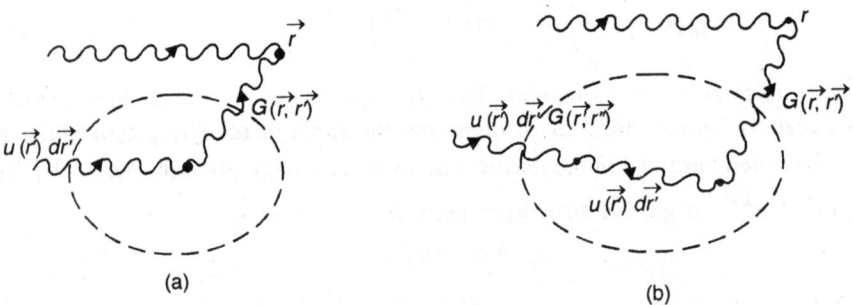

(a)   (b)

Fig. 20.7: (*a*) Schematic representation of 1st Born approximation,
(*b*) Schematic representation of 2nd order term

Consider Eqn. (81). Suppose $U(\vec{r}')$ is a potential which has the same sign everywhere. Then equation shows that for large values of q there will be large phase differences from different regions of the scatterer (i.e. the exponential term oscillates) and effects from different regions will tend to reduce the scattering amplitude. On the other hand, for $q = 0$ (i.e. $\vec{k}'$ parallel to $\vec{k}$) the amplitude will be largest. The ratio $\dfrac{f(\vec{q})}{f(0)}$ is called the *form factor*.

20.9 Validity of Born Approximation

The Born approximation is valid whenever the total wavefunction is not greatly different from the incident wave function $e^{i\vec{k}\cdot\vec{r}}$ i.e. whenever $\psi_{sc}(\vec{r})$ is small compared to $e^{i\vec{k}\cdot\vec{r}}$ in the region where $V(\vec{r})$ is large. In most cases both $V(\vec{r})$ and $\psi_{sc}(\vec{r})$ are large near the origin, therefore a rough criterion for validity of Born approximation is

$$\left|\psi_{sc}(\vec{r})\right|^2 \ll 1 \text{ for small } r.$$

We can derive another criterion for the validity of Born approximation.

I method: The change of potential produces a change in phase of the wave function. The total wave function will not differ greatly from the initial wave function if the phase of the incident wave is not much altered as it passes the region of influence of potential.

At great distances, $k^2 = 2mE/\hbar^2$ and near the centre of force it is $2m(E-V)/\hbar^2$. The change of phase due to the potential is then given by

$$\Delta\phi = \int_0^\infty \sqrt{\frac{2m}{\hbar^2}}\left[\sqrt{(E-V)} - \sqrt{E}\right]dr$$

If this difference is small compared with unity, we may take it as an indication that the wavefunction is not very much different from that in the absence of potential. Thus for first order Born approximation to be valid:

$$\left|\Delta\phi\right| = \left|\sqrt{\frac{2m}{\hbar^2}}\int_0^\infty\left[\sqrt{(E-V)} - \sqrt{E}\right]dr\right| \ll 1$$

If $V \ll E$, we have

$$\left|\sqrt{\frac{2m}{\hbar^2}}\int_0^\infty\left\{\left(1-\frac{V}{E}\right)^{1/2} - 1\right\}dr\right| \ll 1$$

or $\qquad k\left|\int_0^\infty\left\{1 - \frac{V}{2E} + \dots - 1\right\}dr\right| \ll 1$ or $\dfrac{k}{2E}\left|\int_0^\infty V dr\right| \ll 1$

or $\qquad \dfrac{k(2m)}{2\hbar^2 k^2}\left|\int_0^\infty V dr\right| \ll 1$ or $\dfrac{m}{\hbar^2 k}\overline{V}\overline{r} \ll 1$

where $\overline{r}$ = mean range. In case of scattering of high energy particles by the spherical potential well of radius 'a' and depth V_0, this condition becomes

$$\frac{mV_0a}{\hbar^2k} \ll 1 \text{ or } \frac{m^2V_0^2a^2}{\hbar^4 2mE} \times \hbar^2 \ll 1$$

or

$$E \gg \frac{m}{2}\left(\frac{V_0a}{\hbar}\right)^2$$

II method: Born approximation is based on the assumption that the deviation $\psi_{sc}(\vec{r})$ from the free particle wave function $e^{i\vec{k}\cdot\vec{r}}$ is relatively very small i.e.

$$\left|\psi_{sc}(\vec{r})\right| \equiv \left|\psi_{total} - e^{i\vec{k}\cdot\vec{r}}\right| \ll \left|e^{i\vec{k}\cdot\vec{r}}\right| \ (=1) \tag{87}$$

Since the distortion $\psi_{sc}(r)$ of the wave by the potential would be expected to be largest near the origin, we apply the above condition at $\vec{r} = 0$.

Now, from Eqn. (76)

$$\psi_{sc}(\vec{r}) = -\int \frac{e^{ik\cdot|\vec{r}-\vec{r}'|}}{4\pi|\vec{r}-\vec{r}'|} U(\vec{r}') e^{i\vec{k}\cdot\vec{r}'} dr'$$

$$= -\int \frac{e^{ikr'}}{4\pi r'} U(\vec{r}') e^{ikr'\cos\theta'} r'^2 \, d(\cos\theta') \, d\phi' \, dr'$$

(for spherically symmetric potential)

Now, $\int d\phi' = 2\pi$ and

$$\int_{-1}^{+1} e^{ikr'\cos\theta'} d(\cos\theta') = \frac{2}{kr'}\sin kr'$$

$$\therefore \qquad \psi_{sc}(\vec{r}) = -\int_{r'=0}^{\infty} \frac{e^{ikr'}}{2} \frac{2\sin kr'}{kr'} U(\vec{r}') r' \, dr'$$

$$= -\int_{r'=0}^{\infty} e^{ikr'} \frac{1}{k}\left(\frac{e^{ikr'} - e^{-ikr'}}{2i}\right) \frac{2mV(r')}{\hbar^2} dr'$$

(Putting the value of $U(r')$)

Hence condition (87) becomes

$$\frac{m}{\hbar^2k}\left|\int_0^{\infty}\left(e^{2ikr} - 1\right) V(r) \, dr\right| \ll 1 \tag{88}$$

Let us consider the explicit form of the condition for a few kinds of potential.

20.9.1 Square Well Potential

For a square well potential of depth V_0 and range a, the condition (Eqn. 88) becomes

$$\frac{mV_0}{\hbar^2k}\left|\int_0^a\left(e^{2ikr} - 1\right) dr\right| \ll 1 \text{ or } \frac{mV_0}{\hbar^2k}\left|\left[\frac{e^{2ikr}}{2ik} - r\right]_0^a\right| \ll 1$$

or
$$\frac{mV_0}{2\hbar^2 k^2}\left|\left(e^{2ika}-1-2ika\right)\right|\ll 1.$$

Putting $(2ka)=\rho$, it becomes

$$\frac{mV_0}{2\hbar^2 k^2}\left|\left(e^{i\rho}-i\rho-1\right)\right|\ll 1$$

or
$$\frac{mV_0}{2\hbar^2 k^2}\left|(\cos\rho-1)+i(\sin\rho-\rho)\right|\ll 1$$

or
$$\frac{mV_0}{2\hbar^2 k^2}\left[(\cos\rho-1)^2+(\sin\rho-\rho)^2\right]^{1/2}\ll 1$$

or
$$\frac{mV_0}{2\hbar^2 k^2}\left(\rho^2-2\rho\sin\rho-2\cos\rho+2\right)^{1/2}\ll 1 \qquad (89)$$

The quantity under square root is $\approx\left(\frac{1}{2}\right)\rho^2$ for $\rho\ll 1$ (i.e. for $4\pi a/\lambda\ll 1$). In this limit, when the de Broglie wavelength λ is much greater than the range of the potential, the condition (89) becomes

$$\frac{mV_0}{2\hbar^2 k^2}\left[\frac{1}{2}\times 4k^2 a^2\right]\quad\text{or}\quad\frac{mV_0 a^2}{\hbar^2}\ll 1 \qquad (90)$$

i.e. a bound state will be obtained if the potential is very weak. If the potential is not so weak, the Born approximation is still valid if the particle momentum ($\hbar k$) is sufficiently large (from Eqn. 89). Under these conditions $(\leftrightarrow 2ka\gg 1)$, the condition (89) becomes

$$\frac{mV_0\rho}{2\hbar^2 k^2}=\frac{mV_0 a}{\hbar^2 k}\quad\text{or}\quad\frac{V_0 a}{\hbar v}\ll 1 \qquad (91)$$

i.e. for any given depth and range of the potential well, the approximation becomes better if the incident particle velocity (v) is high. Then the deviation of ψ from the free particle wave function is small and the potential acts like a small perturbation, improving the Born approximation.

20.9.2 Exponential Potential

$$V(r)=V_0 e^{-r/r_0} \qquad (92)$$

$$\int_0^\infty V(r)\left(e^{2ikr}-1\right)dr=V_0\int_0^\infty e^{-r/r_0}\left(e^{2ikr}-1\right)dr$$

$$=V_0\int_0^\infty e^{\left(2ik-\frac{1}{r_0}\right)r}dr-V_0\int_0^\infty e^{-r/r_0}dr$$

$$=V_0\left\{\frac{e^{\left(2ik-\frac{1}{r_0}\right)r}}{\left(2ik-\frac{1}{r_0}\right)}\Bigg|_0^\infty-\frac{e^{-r/r_0}}{(-1/r_0)}\Bigg|_0^\infty\right\}$$

$$= V_0 \left\{ -\cfrac{1}{2ik - \cfrac{1}{r_0}} - r_0 \right\} = -\frac{2V_0 i k r_0'^2}{2ikr_0 - 1}$$

Hence

$$\left| \int_0^\infty V(r) \left(e^{2ikr} - 1 \right) dr \right| = \sqrt{\frac{4V_0^2 k^2 r_0^4}{1 - 4i^2 k^2 r_0^2}} = \frac{2V_0 k r_0^2}{\sqrt{1 + 4k^2 r_0^2}}$$

so, the condition (88) reduces to the inequality

$$\frac{m}{\hbar^2 k} \times \frac{2V_0 k r_0^2}{\sqrt{1 + k^2 r_0^2}} \ll 1 \text{ or } 2mV_0 r_0^2 \ll \hbar^2 \sqrt{1 + 4k^2 r_0^2}$$

If $kr_0 \ll 1$, this condition becomes $2mV_0 r_0^2 \ll \hbar^2$ whereas for $kr_0 \gg 1$, we get $mV_0 r_0 \ll \hbar^2 k$.

20.9.3 Screened Coulomb Potential

$$V(r) = \frac{Z_1 Z_2 e^2}{r} \exp(-ar); \left(a = \frac{1}{r_0} \right) \tag{93}$$

$$\frac{m}{\hbar^2 k} \left| \int_0^\infty \left(e^{2ikr} - 1 \right) V(r) \, dr \right| \ll 1$$

To evaluate the integral $I = \int_0^\infty e^{-ar} \left(e^{2ikr} - 1 \right) \frac{dr}{r}$ we differentiate I with respect to the parameter a, then we get:

$$\frac{\partial I}{\partial a} = -\int_0^\infty e^{-ar} \left(e^{2ikr} - 1 \right) dr$$

$$= -\int_0^\infty e^{(2ik-a)r} dr + \int_0^\infty e^{-ar} dr = -\left[\frac{e^{(2ik-a)r}}{2ik - a} \right]_0^\infty + \left[\frac{e^{-ar}}{(-a)} \right]_0^\infty$$

$$= \frac{1}{a} + \frac{1}{2ik - a} = \frac{1}{a} - \frac{1}{a - 2ik}$$

Integrating this expression over a, we find

$$I = \ln a - \ln (a - 2ik) + C$$

where C is a constant, when $a = \infty$, $I = 0$ so that $C = 0$ and we have:

$$I = -\ln(1 - 2ikr_0) = -\ln \sqrt{1 + 4k^2 r_0^2} + i\phi$$

where $\tan \phi = 2kr_0$

Thus,

$$\frac{mZ_1 Z_2 e^2}{\hbar^2 k} \left\{ \left(\ln \sqrt{1 + 4k^2 r_0^2} \right)^2 + \phi^2 \right\}^{1/2} \ll 1$$

The value of ϕ does not exceed $\pi/2$ and the value of the logarithmic term changes little with a change in screening radius, so that we can take the general condition for applicability of Born approximation for the Coulomb interaction as

$$Z_1 Z_2 e^2 \ll \hbar v$$

where $v = \hbar k / m$ is the relative velocity of the colliding particles.

20.9.4 The Born Approximation for Yukawa Potential

$$V(r) = V_0 \frac{e^{-\alpha r}}{r} \tag{94}$$

where V_0 and α are real constants and α is positive. This potential is attractive or repulsive depending on whether V_0 is negative or positive. Its range is characterized by $r_0 = 1/\alpha$, since $V(r)$ is practically zero when r is greater than $2r_0$ or $3r_0$. For $\alpha = 0$, this potential becomes the Coulomb potential, which therefore may be considered as Yukawa potential of infinite range.

Scattering amplitude

$$f_B(\theta, \phi) = -\frac{1}{q} \int_{r'=0}^{\infty} r' \sin qr' \, U(r') \, dr'$$

$$= -\frac{2mV_0}{q\hbar^2} \int_0^{\infty} r' \sin qr' \, \frac{e^{-\alpha r'}}{r'} \, dr' \tag{95}$$

Now,

$$I \equiv \int_0^{\infty} \sin qr \, e^{-\alpha r} dr$$

$$= \sin(qr) \frac{e^{-\alpha r}}{(-\alpha)} \Big|_0^{\infty} + \frac{q}{\alpha} \int_0^{\infty} \cos(qr) e^{-\alpha r} dr$$

(the first term is zero for both the limits)

$$= -\frac{q}{\alpha^2} \cos(qr) e^{-\alpha r} \Big|_0^{\infty} - \frac{q^2}{\alpha^2} \int_0^{\infty} \sin(qr) e^{-\alpha r} dr$$

hence

$$I\left(1 + \frac{q^2}{\alpha^2}\right) = +\frac{q}{\alpha^2} \text{ or } I = \frac{q}{\alpha^2} \frac{\alpha^2}{(q^2 + \alpha^2)}$$

$$\therefore \qquad f_b(\theta, \phi) = -\frac{2mV_0}{\hbar^2} \frac{1}{(\alpha^2 + q^2)}$$

Since $q = 2k \sin(\theta/2)$, so, the differential scattering cross section is

$$\frac{4m^2 V_0^2}{\hbar^4} \frac{1}{\left(\alpha^2 + 4k^2 \sin^2 \frac{\theta}{2}\right)^2} \tag{96}$$

For fixed k (i.e. fixed energy), this depends on scattering angle. The cross section in the forward direction ($\theta = 0$) is larger than the cross section in the backward ($\theta = \pi$) direction. For fixed θ, it is decreasing function of energy. The sign of V_0 is of no importance in scattering problem.

20.10 Atomic Scattering of Electrons (Rutherford's Formula)

The scattering of fast electrons by an atom can be represented as the scattering in a spherically symmetric field $V(r)$, which is the electrostatic potential energy due to the nuclear charge Ze and the charge of the atomic electrons,

$$V(\vec{r}) = -\frac{Ze^2}{r} + Ze^2 \int \frac{\rho(\vec{r}')d\vec{r}'}{|\vec{r} - \vec{r}'|} \tag{97}$$

$\vec{r}$ is the position of the scattered electron and $\rho(\vec{r}')$ is the particle density of the atomic electrons at the point $\vec{r}'$. This expression can be written for convenience as

$$V(\vec{r}) = -Ze^2 \int \frac{\rho_t(\vec{r}')d\vec{r}'}{|\vec{r} - \vec{r}'|} \tag{98}$$

where $\rho_t(\vec{r}') = \delta(\vec{r}') - \rho(\vec{r}')$ represents the total charge density (nuclear and electronic). We have from Eqn. (81)

$$\text{Scattering amplitude } f = -\frac{1}{4\pi} \int e^{i(\vec{k}-\vec{k}')\cdot\vec{r}} \, U(r) \, d\vec{r} \tag{99}$$

$$\left(U(r) = \frac{2m}{\hbar^2} V(r) \right)$$

Substituting Eqn. (98) in Eqn. (99) and writing $\vec{k} - \vec{k}' = \vec{K}$, we get,

$$f = \frac{Ze^2}{4\pi} \frac{2m}{\hbar^2} \int\int \frac{\rho_t(\vec{r}')}{|\vec{r} - \vec{r}'|} \exp(i\vec{K} \cdot \vec{r}) d\vec{r}' d\vec{r}$$

or

$$f = \frac{Ze^2}{4\pi} \frac{2m}{\hbar^2} \int \rho_t(\vec{r}') \exp(i\vec{K} \cdot \vec{r}') \left\{ \int \frac{\exp\left[i\vec{K} \cdot (\vec{r} - \vec{r}')\right]}{|\vec{r} - \vec{r}'|} d\vec{r} \right\} d\vec{r}' \tag{100}$$

(on multiplying and dividing by $e^{i\vec{K}\cdot\vec{r}}$)

The integral within the braces is evaluated by taking the point $\vec{r}'$ as origin in $\vec{r}$ integration. It becomes

$$\int \frac{\exp\left[i\vec{K} \cdot \vec{r}\right]}{r} d\vec{r} \qquad (\because r' = 0)$$

This integral is not absolutely convergent, but it can be integrated by inserting a convergence factor $e^{-\lambda r}$. Thus we have

$$\int e^{-\lambda r} \frac{\exp\left[i\vec{K} \cdot \vec{r}\right]}{r} d\vec{r} = 2\pi \int_{r=0}^{\infty} e^{-\lambda r} \int_{\mu=-1}^{+1} \frac{e^{iKr\mu}}{r} r^2 d\mu dr$$

$$(\mu = \cos q)$$

$$= 2\pi \int_0^\infty e^{-\lambda r} \left[\frac{e^{iKr} - e^{-iKr}}{iKr} \right] r \, dr$$

$$= \frac{4\pi}{K} \int_0^\infty e^{-\lambda r} \left[\frac{e^{iKr} - e^{-iKr}}{2i} \right] dr$$

$$= \frac{4\pi}{K} \operatorname{Im} \int_0^\infty e^{-(\lambda - iK)r} \, dr \qquad \left(\because \frac{e^{iKr} - e^{-iKr}}{2i} = \sin Kr \right)$$

$$= \frac{4\pi}{K} \operatorname{Im} \frac{1}{\lambda - iK} = \frac{4\pi}{K} \operatorname{Im} \left(\frac{\lambda + iK}{\lambda^2 + K^2} \right)$$

$$= \frac{4\pi}{k} \left(\frac{K}{\lambda^2 + K^2} \right) = \frac{4\pi}{\lambda^2 + K^2}$$

In the limit $l \to 0$, this becomes

$$\int \frac{\exp\left[i\vec{K} \cdot \vec{r} \right]}{r} \, d\vec{r} = \frac{4\pi}{k^2} \tag{101}$$

Inserting this in Eqn. (100), we have

$$f = Ze^2 \frac{2m}{\hbar^2 K^2} \int \left[\delta(\vec{r}') - \rho(\vec{r}') \right] \exp(i\vec{K} \cdot \vec{r}') d\vec{r}'$$

or

$$f = Ze^2 \frac{2m}{\hbar^2 K^2} \{ 1 - F(K) \} \tag{102}$$

where

$$F(K) = \int \rho(\vec{r}') \exp(i\vec{K} \cdot \vec{r}') d\vec{r}' \tag{103}$$

Now, $K = 2k \sin \dfrac{\theta}{2}$ where $k^2 = \dfrac{2mE}{\hbar^2}$ and θ is the scattering angle.

$$\therefore \qquad f(\theta) = Ze^2 \frac{2m}{4\hbar^2 k^2 \sin^2 \frac{\theta}{2}} \{ 1 - F(K) \}$$

or

$$f(\theta) = \frac{Ze^2}{4E \sin^2 \frac{\theta}{2}} \{ 1 - F(K) \}, \quad \left(\because k^2 = \frac{2mE}{\hbar^2} \right) \tag{104}$$

The quantity $F(K)$ is the *atomic scattering factor*, which is shown by Eqn. (103) to be the Fourier transform of the electronic density. The particle density in the ground state of an atom is spherically symmetric and integration over the angles in Eqn. (103) can be carried out:

$$F(K) = 2\pi \int_0^\infty \rho(\vec{r}') \int_{-1}^{+1} e^{iKr'\omega'} r'^2 d\omega' d\vec{r}' \qquad (\omega' = \cos \theta')$$

Now,

$$\int_{-1}^{+1} e^{iKr'\omega'} d\omega' = \frac{2 \sin Kr'}{Kr'}$$

$$\therefore \qquad F(K) = \frac{4\pi}{K} \int_{r'=0}^{\infty} \rho(r') \sin Kr' . r' dr'$$

Charge distribution within an atom is given approximately as

$$\rho(r') = \frac{1}{\left(\sqrt{\pi}a\right)^3} e^{-r^2/a^2} \quad (a = \text{atomic radius}) \qquad (105)$$

$$\Rightarrow \qquad F(K) = e^{-(Ka)^2/4} \text{ where } K = 2k \sin\frac{\theta}{2} \qquad (106)$$

The atomic scattering factor represents the shielding effect of the atomic electrons upon the nuclear charges. This effect is small unless $ka \sin\frac{\theta}{2} \approx 1$.

If the scattering angle is large, i.e. the shielding is ineffective (if the electron energy is much greater than the ionization energy for the atom), then Eqn. (104) gives

$$\text{Cross-section } \sigma(\theta) = |f(\theta)|^2 = \left(\frac{Ze^2}{4E}\right)^2 \frac{1}{\sin^4(\theta/2)} \qquad (107)$$

This is Rutherford formula for scattering by a nucleus of charge *Ze*.

20.11 Partial Wave Analysis

20.11.1 Principle of the Method of Partial Waves

We know that

$$\sigma = \int |f(\theta)|^2 \, d\Omega$$

The problem is to determine $f(\theta)$ in terms of the potential $V(\vec{r})$. In case of a spherically symmetric potential $V(\vec{r})$, the angular momentum $\vec{L}$ of the scattered particle is a constant of the motion. Therefore, there exist stationary states with well-defined angular momentum, i.e. eigenstates common to H, L^2 and L_z. The wave functions associated with these states are called *partial waves*. The angular dependence of the corresponding eigenvalues (viz. $\frac{\hbar^2 k^2}{2M}$, $l(l+1)\hbar^2$, $m\hbar$) is given by the spherical harmonics $Y_{lm}(\theta,\phi)$, the potential $V(\vec{r})$ influences only their radial dependence.

We expect that for large r, the partial waves will be very close to the common eigenfunctions of H_0, L^2 and L_z i.e. wave functions corresponding to free spherical waves, their angular dependence is that of a spherical harmonic and the asymptotic expression for their radial function is the superposition of an incoming wave $\dfrac{e^{-ikr}}{r}$ and an outgoing wave $\dfrac{e^{ikr}}{r}$ with a well determined phase difference.

The asymptotic expression for a partial wave in the potential $V(\vec{r})$ is also the superposition of an incoming wave and an outgoing wave, but the phase difference between these two waves is different from the one which characterizes the corresponding free spherical waves, i.e. the potential $V(\vec{r})$ introduces a supplementary phase shift δ_l.

20.11.2 Plane Wave as the Sum of Spherical Waves

For a particle of energy E moving in a central spherically symmetric potential:

$$\left[-\frac{\hbar^2}{2m}\nabla^2 + V(\vec{r})\right]\psi = E\psi \tag{108}$$

The incident plane wave

$$\psi_{inc} = e^{ikz} = e^{ikr\cos\theta} = R_l(r)Y_{lm}(\theta,\phi) \equiv R_l(r)Y(\theta)Z(\phi) \tag{109}$$

is a solution of the simple wave equation

$$\nabla^2\psi_{inc} + k^2\psi_{inc} = 0, \quad k^2 = \frac{2\mu E}{\hbar^2} \tag{110}$$

because, due to spherical symmetry, we can separate the Schrödinger equation into radial and angular parts so that functions R_l, Y and Z are solutions of the separate equations

$$\frac{1}{r^2}\frac{d}{dr}\left(r^2\frac{dR_l}{dr}\right) + \frac{2\mu}{\hbar^2}\left[E - V - \frac{l(l+1)\hbar^2}{2\mu r^2}\right]R_l = 0 \tag{111}$$

$$\frac{1}{\sin\theta}\frac{d}{d\theta}\left(\sin\theta\frac{dY}{d\theta}\right) + \left\{l(l+1) - \frac{m^2}{\sin^2\theta}\right\}Y = 0 \tag{112}$$

$$\frac{d^2Z}{d\phi^2} + m^2Z = 0 \tag{113}$$

Every surface of constant phase in the plane wave is symmetric about the axis of propagation (z-axis). Hence $Z(\phi)$ = constant. So, only solution of Eqn. (113) is $m = 0$. It follows that the Legendre polynomials $P_l^{m=0}(\cos\theta) = P_l(\cos\theta)$ are the solutions of Eqn. (112). Thus there is an axial symmetry and

$$\psi_{inc}(r) = \sum_{l=0}^{\infty} R_l(r)P_l(\cos\theta) \tag{114}$$

is the general solution (i.e. $Y_{l0} \propto P_l(\cos\theta)$.

From Eqn. (108)

$$\left[-\nabla^2 + \frac{2mV}{\hbar^2}\right]\psi_{inc} = k^2\,\psi_{inc}$$

Asymptotic Case:

For $r \to \infty$, the effect of V is negligible and we have Eqn. (110), or

$$-\left\{\frac{1}{r^2}\frac{\partial}{\partial r}\left(r^2\frac{\partial}{\partial r}\right) - \frac{L_{op}^2}{\hbar^2 r^2}\right\}\psi_{inc} = k^2\psi_{inc} \tag{115}$$

But we have

$$\frac{L_{op}^2}{\hbar^2}P_l(\cos\theta) = l(l+1)P_l(\cos\theta) \tag{116}$$

On substituting Eqn. (114) into Eqn. (115), we get

$$\sum_{l=0}^{\infty}\left\{-\frac{1}{r^2}\frac{\partial}{\partial r}\left(r^2\frac{\partial}{\partial r}\right)-k^2\right\}R_l(r)P_l(\cos\theta)$$

$$+\sum_{l=0}^{\infty}\frac{1}{r^2}R_l(r)l(l+1)P_l(\cos\theta)=0$$

Therefore for each l

$$\left\{-\frac{1}{r^2}\frac{\partial}{\partial r}\left(r^2\frac{\partial}{\partial r}\right)-k^2\right\}R_l(r)+\frac{1}{r^2}l(l+1)R_l(r)=0 \qquad (117)$$

So, we have reduced the three dimensional problem into a one dimensional problem. Using scale transformation $kr \equiv \rho$, we get

$$\left\{-\frac{1}{k^2r^2}\frac{\partial}{\partial r}\left(r^2\frac{\partial}{\partial r}\right)-1\right\}R_l(r)+\frac{1}{k^2r^2}l(l+1)R_l(r)=0$$

or

$$\left\{-\frac{1}{\rho^2}k\frac{\partial}{\partial\rho}\left(kr^2\frac{\partial}{\partial\rho}\right)-1\right\}R_l(r)+\frac{1}{\rho^2}l(l+1)R_l(r)=0$$

or

$$\left\{\frac{1}{\rho^2}\frac{\partial}{\partial\rho}\left(\rho^2\frac{\partial}{\partial\rho}\right)R_l(r)\right\}-\left[\frac{l(l+1)}{\rho^2}-1\right]R_l(r)=0$$

or

$$\frac{\partial^2 R_l}{\partial\rho^2}+\frac{2}{\rho}\frac{\partial R_l}{\partial\rho}+\left[1-\frac{l(l+1)}{\rho^2}\right]R_l(r)=0 \qquad (118)$$

This equation has strong resemblance to Bessel's equation and is satisfied by spherical Bessel functions $j_l(\rho)$ (and $\eta_l(\rho)$) which are related to original Bessel functions $J_{l+\frac{1}{2}}(\rho)$ through

$$j_l(\rho)=\sqrt{\frac{\pi}{2\rho}}J_{l+\frac{1}{2}}(\rho) \text{ or } j_l(\rho)=\sqrt{\frac{\pi}{2}}\cdot\frac{1}{\sqrt{\rho}}J_{l+\frac{1}{2}}(\rho) \qquad (119)$$

and

$$\eta_l(\rho)=(-1)^{l+1}\sqrt{\frac{\pi}{2}}\cdot\frac{1}{\sqrt{\rho}}J_{-l-\frac{1}{2}}(\rho) \qquad (120)$$

If effect of V is not neglected, both solutions are acceptable (even for $r = 0$). For large ρ ($r \to \infty$):

$$j_l(\rho)=\frac{1}{\rho}\cos\left[\rho-(l+1)\frac{\pi}{2}\right] \qquad (121)$$

$$\eta_l(\rho)\cong\frac{1}{\rho}\sin\left[\rho-(l+1)\frac{\pi}{2}\right] \qquad (122)$$

and the general solution is

$$R_l(r)=Aj_l(\rho)-B\eta_l(\rho) \qquad (123a)$$

$$\cong \frac{1}{\rho}\left[A\cos\left\{\rho-(l+1)\frac{\pi}{2}\right\}-B\sin\left\{\rho-(l+1)\frac{\pi}{2}\right\}\right]$$

$$=\frac{1}{\rho}\left[A\sin\left(\rho-\frac{l\pi}{2}\right)+B\cos\left(\rho-\frac{l\pi}{2}\right)\right] \tag{123b}$$

Let
$$A=A_l\cos\delta_l,\ B=A_l\sin\delta_l \tag{124}$$

$\therefore$
$$R_l(r)=\frac{A_l}{\rho}\sin\left(\rho-\frac{l\pi}{2}+\delta_l\right) \tag{125}$$

and
$$\psi(r)=\sum_{l=0}^{\infty}\frac{A_l}{kr}\sin\left(kr-\frac{l\pi}{2}+\delta_l\right)P_l(\cos\theta) \tag{126}$$

The arbitrary constants A_l can be evaluated as follows. We note that (without considering δ_l)

$$\psi_{inc}=e^{ikz}=\sum_l R_l(r)P_l(\cos\theta)$$

$$=\sum_l A_l j_l(kr)P_l(\cos\theta)=e^{ikr}\cos\theta \tag{127}$$

and
$$j_l(kr)\xrightarrow[kr\gg l]{}\frac{1}{kr}\sin\left(kr-\frac{l\pi}{2}\right) \tag{128}$$

To obtain A_l, we multiply both sides of Eqn. (127) by $P_l(\cos\theta)$ and integrate over all θ. We write $\cos\theta=x$ and use orthonormality of $P_l(x)$. We know that

$$\int_{-1}^{+1}\left|P_l^m(x)\right|^2 dx=\frac{(l+m)!}{(l-m)!}\frac{2}{(2l+1)} \tag{129}$$

Putting $m=0$ in this expression and also using Eqn. (127),

$$\frac{2}{2l+1}A_l j_l(kr)=\int_{-1}^{+1}e^{ikrx}P_l(x)dx \tag{130}$$

$$=\frac{1}{ikr}\left[e^{ikrx}P_l(x)\right]_{x=-1}^{x=+1}-\left[\frac{1}{ikr}\int_{-1}^{+1}e^{ikrx}P_l'dx\right]$$

The second term on the r.h.s. is of the order $\dfrac{1}{r^2}$, therefore, for large r we have (using $P_l(1)=1$, $P_l(-1)=(-1)^l=e^{il\pi}$)

$$\frac{2}{2l+1}A_l\frac{1}{kr}\sin\left(kr-\frac{l\pi}{2}\right)\approx\frac{1}{ikr}\left[e^{ikr}-e^{il\pi}e^{-ikr}\right]$$

$$=\frac{1}{ikr}e^{i\frac{l\pi}{2}}\left[e^{i\left(kr-\frac{l\pi}{2}\right)}-e^{-i\left(kr-\frac{l\pi}{2}\right)}\right]$$

$$=2i^l\frac{1}{kr}\sin\left(kr-\frac{l\pi}{2}\right)$$

Therefore
$$A_l = (2l+1)i^l \tag{131}$$

Hence
$$\Psi_{inc} = e^{ikz} = e^{ikr\cos\theta} = \sum_{l=0}^{\infty} i^l (2l+1) j_l(kr) P_l(\cos\theta) \tag{132}$$

i.e. a state of well defined linear momentum is formed by the superposition of states corresponding to all possible l.

Hence
$$\psi(\vec{r}) = \psi_{inc} + f_k(\theta) \frac{e^{ikr}}{r} \tag{133}$$

$$\Rightarrow \quad \psi(\vec{r}) = \sum_{l=0}^{\infty} (2l+1) i^l P_l(\cos\theta) \frac{1}{kr} \sin\left(kr - \frac{l\pi}{2}\right) + f_k(\theta) \frac{e^{ikr}}{r}$$

(using Eqn. 128 and 132)

or
$$\psi(r) = \sum_{l=0}^{\infty} i^l (2l+1) P_l(\cos\theta) \frac{1}{kr} \frac{1}{2i} \left[e^{i\left(kr - \frac{l\pi}{2}\right)} - e^{-i\left(kr - \frac{l\pi}{2}\right)} \right] + f_k(\theta) \frac{e^{ikr}}{r} \tag{134}$$

From Eqns. (126) and (134), since $\dfrac{e^{ikr}}{(kr)}$ and $\dfrac{e^{-ikr}}{(kr)}$ are lineary independent, we equate their coefficients separately and get two equations

(i) *Equating coefficients of* $\dfrac{e^{ikr}}{(kr)}$

$$\sum_{l=0}^{\infty} i^l (2l+1) P_l(\cos\theta) \frac{1}{2i} e^{-i\frac{l\pi}{2}} + k f_k(\theta)$$

$$= \sum_{l=0}^{\infty} A_l \frac{1}{2i} e^{-i\frac{l\pi}{2} + i\delta_l} P_l(\cos\theta) \tag{135}$$

and

(ii) *Equating coefficients of* $\dfrac{e^{-ikr}}{(kr)}$

$$\sum_{l=0}^{\infty} i^l (2l+1) P_l(\cos\theta) \frac{(-1)}{2i} e^{i\frac{l\pi}{2}} + 0$$

$$= \sum_{l=0}^{\infty} A_l \frac{(-1)}{2i} e^{i\frac{l\pi}{2} - i\delta_l} P_l(\cos\theta) \tag{136}$$

i^l on l.h.s. of Eqn. (136) can be written as $\left(e^{i\frac{\pi}{2}}\right)^l$

$$\therefore \qquad (2l+1) e^{i\frac{\pi}{2}l} e^{i\frac{\pi}{2}l} = A_l e^{i\frac{\pi}{2}l - i\delta_l}$$

$$A_l = (2l+1) e^{i\delta_l} e^{i\pi l/2} \tag{137}$$

Substituting the value of A_l in Eqn. (135), we get

$$\sum_{l=0}^{\infty} (2l+1) \frac{P_l(\cos\theta)}{2i} + kf_k(\theta)$$

$$= \sum_{l=0}^{\infty} (2l+1) e^{i\delta_l} e^{i\pi l/2} \frac{1}{2i} e^{-i\frac{l\pi}{2}} e^{i\delta_l} P_l(\cos\theta)$$

or $$\sum_{l=0}^{\infty} (2l+1) \frac{1}{2i} P_l(\cos\theta) + kf_k(\theta) = \sum_{l=0}^{\infty} \frac{(2l+1)}{2i} e^{2i\delta_l} P_l(\cos\theta)$$

or $$kf_k(\theta) = \sum_{l=0}^{\infty} \frac{(2l+1)}{2i} \left(e^{2i\delta_l} - 1 \right) P_l(\cos\theta)$$

$$= \sum_{l=0}^{\infty} (2l+1) e^{i\delta_l} \frac{\left(e^{i\delta_l} - e^{-i\delta_l} \right)}{2i} P_l(\cos\theta)$$

or $$f_k(\theta) = \frac{1}{k} \sum_{l=0}^{\infty} (2l+1) e^{i\delta_l} \sin\delta_l P_l(\cos\theta) \qquad (138)$$

Hence differential scattering cross section

$$\sigma(\theta) = \frac{d\sigma(\theta)}{d\Omega} = \left| f(\theta) \right|^2$$

or $$\frac{d\sigma(\theta)}{d\Omega} = \frac{1}{k^2} \left| \sum_{l=0}^{\infty} (2l+1) e^{i\delta_l} \sin\delta_l P_l(\cos\theta) \right|^2 \qquad (139)$$

This equation expresses the cross section in terms of the phase shifts δ_l for the individual particle waves $l = 0, 1, 2, 3, \ldots$ etc. It is valid if $V(r)$ decreases faster than $1/r$ for large r.

20.11.3 The Total Scattering Cross-section (*Optical Theorem*)

We have

$$f(\theta) = \frac{1}{k} \sum_{l=0}^{\infty} (2l+1) e^{i\delta_l} \sin\delta_l P_l(\cos\theta) \qquad (140i)$$

so, $$f^*(\theta) = \frac{1}{k} \sum_{l=0}^{\infty} (2l'+1) e^{-i\delta_{l'}} \sin\delta_{l'} P_{l'}(\cos\theta) \qquad (140ii)$$

assuming δ_l to be real.

$$\therefore \quad \left| f(\theta) \right|^2 = \frac{1}{k^2} \sum_{l} \sum_{l'} (2l+1)(2l'+1) \sin\delta_l \sin\delta_{l'} e^{i(\delta_l - \delta_{l'})}$$

$$\times P_l(\cos\theta) P_{l'}(\cos\theta) \qquad (141)$$

The total cross section is

$$\sigma_{total} = \int \left| f(\theta) \right|^2 d\Omega$$

$$= \int_{\phi=0}^{2\pi} d\phi \int_{-1}^{+1} \left| f(\theta) \right|^2 d(\cos\theta) = 2\pi \int_{-1}^{+1} \left| f(\theta) \right|^2 d(\cos\theta) \qquad (142)$$

Using

$$\int_{-1}^{+1} P_l(\mu)P_{l'}(\mu)d\mu = \frac{2}{(2l+1)}\delta_{ll'} \tag{143}$$

we get from Eqn. (141) and (142)

$$\sigma_{total} = \frac{2\pi}{k^2}\sum_l\sum_{l'}(2l+1)(2l'+1)\sin\delta_l\sin\delta_{l'}\frac{2}{(2l+1)}\delta_{ll'}e^{i(\delta_l-\delta_{l'})}$$

so that

$$\sigma_{total} = \frac{4\pi}{k^2}\sum_{l=0}^{\infty}(2l+1)\sin^2\delta_l \tag{144}$$

This expression shows that as far as the total cross-section is concerned, different partial waves contribute independently, there being no interference explicitly (unlike differential cross section).

It is to be noted that $f(\theta)$ is complex. The expression for $f(\theta)$ shows that

$$\text{Im } f(\theta) = \frac{1}{k}\sum_{l=0}^{\infty}(2l+1)\sin^2\delta_l P_l(\cos\theta) \text{ (because } e^{i\delta_l}=\cos\delta_l+i\sin\delta_l)$$

In the direction of the incident beam, $\theta = 0$. Therefore imaginary part of the forward scattering amplitude is

$$\text{Im } f(0) = \frac{1}{k}\sum_{l=0}^{\infty}(2l+1)\sin^2\delta_l \tag{145}$$

$$\text{(because } P_l(1)=1)$$

From Eqns. (144) and (145), we get

$$\sigma_{total} = \frac{4\pi}{k}\text{Im } f(0) \tag{146}$$

This relation is called the *optical theorem*. This is quantum mechanical analogue of refractive index in optics.

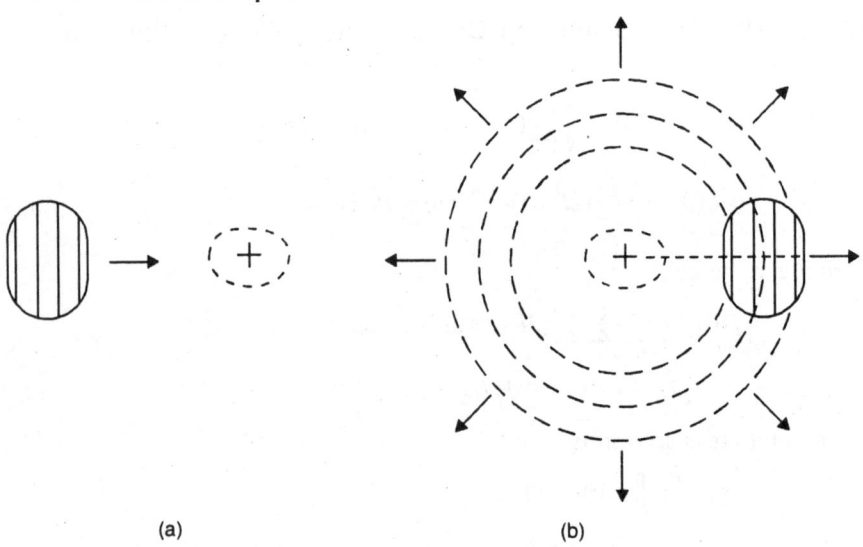

(a) (b)

Fig. 20.8: Optical Theorem: (*a*) Before collision, (*b*) After scattering

Before the collision the incident wave packet is moving towards the zone of influence of the potential. After the collision, we observe a plane wave packet (only in the forward direction) and a spherical wave scattered by the potential. So, the transmitted wave packet results from the interference between the plane wave packet and the spherical wave. It has a smaller amplitude than the incident wave packet. Particles scattered in all directions of space other than the forward direction leave the beam, so the intensity of the beam (in the forward direction, $\theta = 0$) is attenuated after it has passed the target (due to conservation of total number of particles). The loss of intensity suffered by the incident beam is represented by the imaginary part of the forward scattering amplitude. The situation is analogous to a similar case in optics where absorption takes place by optical shadow in relation to imaginary part of complex refractive index.

20.11.4 Sign of the Phase Shifts and their Relation to the Potential

If we substitute $R_l(r) = \dfrac{G_l(r)}{r}$, Eqn. (117) becomes

$$\frac{d^2 G_l}{dr^2} + \left[k^2 - U(r) - \frac{l(l+1)}{r^2} \right] G_l = 0 \tag{147}$$

where

$$k^2 = \frac{2mE}{\hbar^2} \text{ and } U(r) = \frac{2mV(r)}{\hbar^2}$$

For $U(r) = 0$; $R_l(r) = j_l(kr)$

(η_l is ruled out because it tends to infinity as r tends to zero).

δ_l is zero for this case (since σ_{total} is zero for no interaction potential). The function $j_l(kr) \propto r^l$ at small r and goes to $\dfrac{1}{kr} \sin\left(kr - \dfrac{l\pi}{2} \right)$ as $r \to \infty$.

Eqn. (147) shows that the first point of inflexion is reached (r_1) when the term in brackets is zero. For $U(r') = 0$ this occurs at $r_1^2 = l(l+1)/k^2$. In the presence of a non-zero interaction potential $U(r)$, this point of inflexion is *pushed out* to larger r for *positive (repulsive) potentials* and *pulled in* to smaller r for *negative (attractive) potentials*. Thus the phase shift δ_l depends on the scattering potential:

$$\delta_l > 0 \text{ for } U(r) < 0 \tag{148i}$$

Because, in this case "the classical turning point" r_1, where $k^2 - U(r)$ overcomes the centrifugal repulsion Eqn. (147), occurs closer to the origin than in the absence of the potential. Further, the local wave number $\left[k^2 - U(r) - \dfrac{l(l+1)}{r^2} \right]^{1/2}$ is greater than it is when $V = 0$. Both these effects cause the number of oscillations of the wave function between the origin and any point r to be more in case of attractive potential than in the absence of any potential. Consequently, the phase of the wave is more advanced than when $V = 0$ (*positive phase shift*). Similarly, for a repulsive potential,

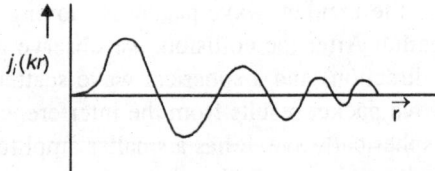

Fig. 20.9

$$\delta_l < 0 \text{ for } U(r) > 0 \tag{148ii}$$

i.e. there is *negative phase shift*.

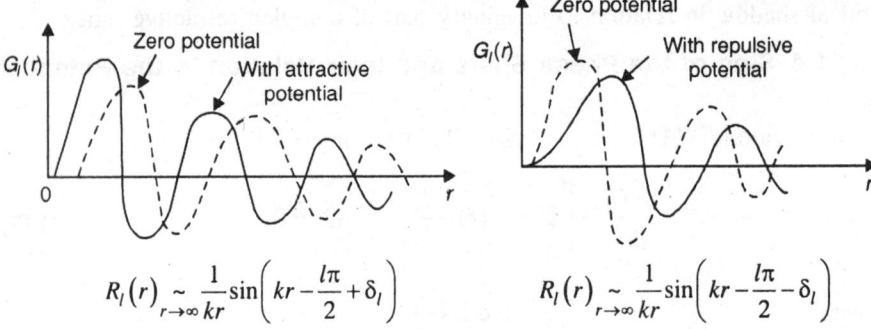

Fig. 20.10

Relation of Phase Shifts to Potential

The phase shifts δ_l depend on the asymptotic form of R_l. To express the phase shifts in terms of the potential, we compare the function G_l with the corresponding function $G_l^0 \propto rj_l$ when potential $V = 0$ (i.e. $U(r) = 0$). The corresponding equations are

$$\frac{d^2G_l}{dr^2} + \left[k^2 - U(r) - \frac{l(l+1)}{r^2} \right] G_l = 0 \tag{149i}$$

or

$$\frac{d^2G_l^0}{dr^2} + \left[k^2 - \frac{l(l+1)}{r^2} \right] G_l^0 = 0 \tag{149ii}$$

Multiplying Eqn. (*i*) by G_l^0 and (*ii*) by G_l and subtracting, we get

$$G_l^0 \frac{d^2G_l}{dr^2} - G_l \frac{d^2G_l^0}{dr^2} - U(r)G_l^0 G_l = 0$$

or

$$\frac{d}{dr} \left[G_l^0 \frac{dG_l}{dr} - G_l \frac{dG_l^0}{dr} \right] - U(r)G_l^0 G_l = 0 \tag{150}$$

Integrating this w.r.t. r from limits 0 to r and remembering G_l^0 and G_l vanish at $r = 0$, we get

$$\left\{ G_l^0 \frac{dG_l}{dr} - G_l \frac{dG_l^0}{dr} \right\} - \int_0^r U(r') G_l^0 \left(r' \right) G_l \left(r' \right) dr' = 0 \qquad (151)$$

It is obvious that this Eqn. is independent of normalization of G_l and G_l^0. Let us normalize G_l and G_l^0 such that the constant in their asymptotic form is unity i.e. at $r \to \infty$, we have

$$G_l \to \sin\left(kr + \Delta l \right) \text{ and } G_l^0 \to \sin\left(kr - \frac{l\pi}{2} \right)$$

This means that $G_l^0 = krj_l(kr)$; then the bracketed term in Eqn. (151) (for $r \to \infty$) may be expressed as

$$k\left\{ \sin\left(kr - \frac{l\pi}{2} \right)\cos\left(kr + \Delta l \right) - \sin\left(kr + \Delta l \right)\cos\left(kr - \frac{l\pi}{2} \right) \right\}$$

$$= -k\sin\left(\frac{l\pi}{2} + \Delta l \right) = -k\sin\delta_l$$

Thus at $r \to \infty$, Eqn. (151) becomes

$$k\sin\delta_l = -\int_0^\infty U(r') G_l^0 \left(r' \right) G_l \left(r' \right) dr'$$

or

$$\sin\delta_l = -\int_0^\infty U(r) rj_l \left(kr \right) G_l \left(r \right) dr$$

This expression is purely formal, since $G_l(r)$ for all values of r is not known. Now suppose G_l differs very little from $G_l^0 = krj_l(kr)$, then,

$$\sin\delta_l = -k\int_0^\infty U(r) r^2 j_l^2 \left(kr \right) dr \qquad (152)$$

20.11.5 Physical Meaning of Phase Shifts

Free spherical waves are the wave functions associated with the well defined angular momentum stationary states $\left| \psi_{klm}^{(0)} \right\rangle$ of a free particle and are written

$$\psi_{klm}^{(0)}(r) = \sqrt{\frac{2k^2}{\pi}} j_l(kr) Y_{lm}(\theta,\phi)$$

These are eigenfunctions common to H_0, L^2 and L_z and are solutions of $-\nabla^2\psi = k^2\psi$ for $l \neq 0$. (For $l = 0$, $\psi \sim e^{\pm ikr}$) The asymptotic behaviour of the free spherical wave is

$$\psi_{klm}^{(0)}(r,\theta,\phi) \underset{r\to\infty}{\cong} -\sqrt{\frac{2k^2}{\pi}} Y_{lm}(\theta,\phi) \frac{e^{-ikr}e^{i\frac{l\pi}{2}} - e^{ikr}e^{-i\frac{l\pi}{2}}}{2ikr}$$

$$\left(\text{because } j_l(kr) \underset{r\to\infty}{\cong} \frac{1}{kr}\sin\left(kr - \frac{l\pi}{2} \right) \right)$$

At infinity $\psi_{klm}^{(0)}$ therefore results from the superposition of an incoming wave $\left(e^{-ikr}/r \right)$ and an outgoing wave $\left(e^{ikr}/r \right)$ whose amplitudes differ by a phase difference $l\pi$.

We may write for the asymptotic behaviour of partial waves in the form

$$\psi_{klm}(r) \underset{r \to \infty}{\cong} \frac{\sin\left(kr - \frac{l\pi}{2} + \delta_l\right)}{kr} Y_{lm}(\theta, \phi)$$

$$\cong -CY_{lm}(\theta, \phi) \frac{e^{-ikr} e^{i\left(\frac{l\pi}{2} - \delta_l\right)} - e^{ikr} e^{-i\left(\frac{l\pi}{2} - \delta_l\right)}}{2ikr} \tag{153}$$

We see that the partial wave $\psi_{klm}(r)$ like a free spherical wave results from the superposition of an incoming wave and an outgoing wave. To study the comparison between partial waves and free spherical waves, we can modify the incoming wave of the above expression, so as to make it identical with the one in $\psi_{klm}^{(0)}$. To do this, we define a new partial wave $\tilde{\psi}_{klm}(r)$ by multiplying $\psi_{klm}(r)$ by $e^{i\delta_l}$ and by choosing the constant C in such a way that

$$\tilde{\psi}_{klm}(r) \underset{r \to \infty}{\cong} -Y_{lm}(\theta, \phi) \frac{e^{-ikr} e^{i\frac{l\pi}{2}} - e^{ikr} e^{-i\frac{l\pi}{2}} e^{2i\delta_l}}{2ikr} \tag{154}$$

This expression is interpreted as follows:

At the outset, we have the same incoming wave as in the case of a free particle (aside from the normalization constant). As this incoming wave approaches the zone of influence of the potential, it is more and more perturbed by this potential. When after turning back, it is transformed into an outgoing wave, it has accumulated a phase shift of $2\delta_l$ relative to the free outgoing wave that would have resulted if the potential $V(r)$ had been identically zero. Therefore, the factor $e^{2i\delta_l}$ (which varies with l and k) arises due to the total effect of the potential on a particle of angular momentum l.

20.11.6 Applicability of Partial Wave Analysis

Why the method of partial waves is most useful for computing scattering cross section if $ka \le 1$?

Suppose there exists a radius 'a' beyond which $U(\vec{r})$ is negligible. The first and largest maximum of $j_l(kr)$ occurs roughly at $r_1 = l/k$. For $r \ll l/k$, j_l is small and increases as r^l. Now, if $l/k \gg a$, j_l will be very small where U is appreciable; then the lth partial wave will hardly be affected by the potential. This implies that the phase shift δ_l will be small and contribution to scattering from that l will be negligible.

So, the cross-section consists of a series of terms extending from $l = 0$ to a maximum l that is of the order of ka. The smaller the ka, the easier the method is to apply.

20.12 Scattering by a Hard Sphere

If the scattering centre is a rigid sphere of radius a, it may be represented by the following potential

$$V(r) \quad = \infty; \qquad r \leq a$$
$$\qquad = 0; \qquad r > a \tag{155}$$

The Schrödinger equation is

$$\left[-\nabla^2 + \frac{2mV}{\hbar^2} \right] \psi = k^2 \psi \qquad \left(k^2 = \frac{2mE}{\hbar^2} \right)$$

For $r > a$, $V(r) = 0$ and Schrödinger equation reduces to

$$-\nabla^2 \psi = k^2 \psi \tag{156}$$

the solution of this equation for a region of space not including the origin is

$$\psi(r) = \sum_{l=0}^{\infty} P_l(\cos\theta) R_l(r) \tag{157}$$

where
$$R_l(r) = A_l \left[j_l(kr)\cos\delta_l - \sin\delta_l \eta_l(kr) \right] \tag{158}$$

Applying the boundary condition $\psi(r) = 0 \big|_{r=a}$, we get

$$\tan\delta_l = \frac{j_l(ka)}{\eta_l(ka)} \tag{159}$$

The total cross-section is given by

$$\sigma_{total} = \frac{4\pi}{k^2} \sum_{l=0}^{\infty} (2l+1)\sin^2\delta_l \tag{160}$$

Since
$$\sin\delta_l = \frac{\tan\delta_l}{\sqrt{1+\tan^2\delta_l}}$$

$$\therefore \qquad \sigma_{total} = \frac{4\pi}{k^2} \sum_{l=0}^{\infty} (2l+1) \left[\frac{j_l^2(ka)}{j_l^2(ka) + \eta_l^2(ka)} \right] \tag{161}$$

Case I: Low Energy Approximation

At very low velocities, $ka \ll 1$, so that $\eta_l(ka) \gg j_l(ka)$

$\therefore$ For $\quad ka \ll 1, \sigma_l = \dfrac{4\pi}{k^2} \sum\limits_{l=0}^{\infty} (2l+1)\, j_l^2(ka) \big/ \eta_l^2(ka) \tag{162}$

From Eqn. (159), we have

$$\tan\delta_l = \frac{j_l(ka)}{\eta_l(ka)} \quad \text{and since} \quad j_l(ka) = \sqrt{\frac{\pi}{2}} \cdot \frac{1}{\sqrt{ka}} J_{l+\frac{1}{2}}(ka)$$

and
$$\eta_l(ka) = (-1)^{l+1} \sqrt{\frac{\pi}{2}} \cdot \frac{1}{\sqrt{ka}} J_{-l-\frac{1}{2}}(ka)$$

therefore

$$\tan\delta_l = \frac{J_{l+\frac{1}{2}}(ka)}{(-1)^{l+1} J_{-l-\frac{1}{2}}(ka)}$$

When $l = 0$,
$$\tan \delta_0 = \frac{J_{\frac{1}{2}}(ka)}{(-1)J_{-\frac{1}{2}}(ka)} \tag{163}$$

But we know that Bessel function

$$J_n(\xi) \to \sqrt{\frac{2}{\pi\xi}} \cos\left[\xi - \left(n + \frac{1}{2}\right)\frac{\pi}{2}\right]$$

so that

$$J_{\frac{1}{2}}(ka) = \sqrt{\frac{2}{\pi ka}} \cos\left[ka - \frac{\pi}{2}\right]$$

$$= \sqrt{\frac{2}{\pi ka}} \cos\left[\frac{\pi}{2} - ka\right]$$

$$= \sqrt{\frac{2}{\pi ka}} \sin(ka)$$

and
$$J_{-\frac{1}{2}}(ka) = \sqrt{\frac{2}{\pi ka}} \cos(ka)$$

Hence from Eqn. (163), we have

$$\tan \delta_0 = \frac{\sin(k)}{-\cos(ka)} = -\tan(ka) \tag{164}$$

or
$$\delta_0 = -ka$$

Thus for $l = 0$,

$$\sigma = \frac{4\pi}{k^2} \sin^2 \delta_0 = \frac{4\pi}{k^2} \sin^2(ka) \approx \frac{4\pi}{k^2} k^2 a^2 \cong 4\pi a^2 \tag{165}$$

This is the expression if $ka \ll 1$ i.e. for low energy scattering. Since $\sigma \propto a^2$, the scattering is spherically symmetric in case of low energy scattering.

The result is four times that obtained by classical mechanics. In classical case, the result πa^2 will be valid because then one is justified in describing the scattering through "trajectories" similar to ray picture in optics. For diffraction, a quantum mechanical description of process becomes important.

Case II: High Energy Approximation

For higher energy limit $ka \gg 1$, for fixed l

$$\sin^2 \delta_l = \frac{\tan^2 \delta_l}{1 + \tan^2 \delta_l}$$

$$= \frac{j_l^2}{j_l^2 + \eta_l^2} \qquad \left(\because \tan \delta_l = \frac{j_l(ka)}{\eta_l(ka)}\right) \tag{166}$$

We have

$$j_l(ka) = \frac{1}{ka}\cos\left[ka - (l+1)\frac{\pi}{2}\right] = \frac{1}{ka}\sin\left[ka - \frac{l\pi}{2}\right]$$

$$\cong \frac{1}{ka}\sin ka \qquad\qquad\qquad \text{(for } ka \gg l)$$

Similarly

$$\eta_l(ka) = \frac{1}{ka}\cos\left[ka - \frac{l\pi}{2}\right] \cong \frac{1}{ka}\cos ka \qquad\qquad (\because ka \gg l)$$

So, $\qquad\qquad \sin^2\delta_l \cong \sin^2 ka \qquad\qquad\qquad\qquad\qquad\qquad (167)$

For infinite potential (i.e. large k), the phase shift is simply ka if ($l \ll ka$). Thus we may assume that only those waves contribute to partial waves for which $l = 0$ to ka, and total scattering cross section becomes

$$\sigma_{total} = \frac{4\pi}{k^2}\sum_{l=0}^{ka}(2l+1)\sin^2 ka$$

$$= \frac{4\pi}{k^2}k^2a^2\sin^2 ka = 4\pi a^2\sin^2 ka \qquad\qquad (168)$$

i.e. σ_{total} oscillates rapidly between 0 and $4\pi a^2$. In high energy limit, the average cross section is obtained by taking the average value of $\sin^2 ka$ over all directions which is ½ hence

$$\sigma_{total} = 2\pi a^2 \qquad\qquad\qquad\qquad\qquad\qquad (169)$$

i.e. in high energy limit, the scattering cross-section is twice the classical result. The reason for this is attributed to *shadow scattering* due to Poisson-diffraction phenomenon: For finite value of ka, the diffraction around the sphere takes place only in the forward direction, so the total measured cross section is nearly one half of $4\pi a^2$.

Example 20.1: Calculate the scattering cross section in the first Born approximation for the Gaussian potential $V(r) = V_0\exp\left(-r^2/a^2\right)$ and discuss the condition for which the approximation is the best.

Solution: The scattering amplitude $f(\theta) = -\frac{1}{q}\int_0^\infty r\sin qr U(r)dr$

$$q = 2k\sin\frac{\theta}{2} \text{ and } U(r) = \frac{2m}{\hbar^2}V(r)$$

$$\therefore \qquad f(\theta) = -\frac{2m}{\hbar^2 q}\int_0^\infty r\sin qr V(r)dr \qquad\qquad (i)$$

If the potential $V(r)$ is an even function of r, we can write

$$f(\theta) = -\frac{m}{\hbar^2 qi}\int_{-\infty}^{+\infty} re^{iqr}V(r)dr \qquad\qquad (ii)$$

The condition for validity of Born approximation is

$$\frac{m}{\hbar^2 k}\left|\int_0^\infty \left(e^{2ikr}-1\right)V(r)dr\right| << 1 \qquad (iii)$$

For low energy, $ka << 1$ where a is the range of potential. The first two terms in the series expansion of e^{2ikr} will be a very good approximation. Hence for low energy, the condition (iii) can be written as

$$\frac{2m}{\hbar^2}\left|\int_0^\infty rV(r)dr\right| << 1 \qquad (iv)$$

If the kinetic energy of the particle is very large, $(ka >> 1)$, the oscillating term can be neglected and (iii) becomes

$$\frac{m}{\hbar^2 k}\left|\int_0^\infty V(r)dr\right| << 1 \qquad (v)$$

The Gaussian potential

$$V(r) = V_0 \exp\left(-r^2/a^2\right)$$

is an even function or r, so we write

$$f(\theta) = -\frac{mV_0}{\hbar^2 qi}\int_{-\infty}^{+\infty} re^{iqr}\cdot e^{-r^2/a^2}\,dr$$

$$= -\frac{mV_0}{\hbar^2 qi}\int_{-\infty}^{+\infty} r\cdot\exp\left[-\left(\frac{r}{a}-\frac{iqa}{2}\right)^2\right]\cdot\exp\left[-\frac{q^2a^2}{4}\right]dr \qquad (vi)$$

Put
$$t = \left(\frac{r}{a}-\frac{iqa}{2}\right)\;;\; dt = \frac{dr}{a}\;\text{ or }\; dr = a\,dt \;\text{ and }\; r = a\left(t+\frac{iqa}{2}\right)$$

Thus (vi) becomes

$$f(\theta) = -\frac{mV_0}{\hbar^2 qi}\exp\left(-\frac{q^2a^2}{4}\right)\left\{a^2\int_{-\infty}^{+\infty} te^{-t^2}\,dt + \frac{iqa^3}{2}\int_{-\infty}^{+\infty} e^{-t^2}\,dt\right\}$$

The first integral on the r.h.s. vanishes, because the integrand is an odd function of t. the value of the second integral is $\sqrt{\pi}$. So,

$$f(\theta) = -\frac{mV_0\exp\left(-q^2a^2/4\right)}{\hbar^2 qi}\cdot\frac{iqa^3}{2}\cdot\sqrt{\pi}$$

Now, $q = 2k\sin\dfrac{\theta}{2}$, therefore

$$f(\theta) = -\frac{mV_0}{\hbar^2}\cdot\frac{a^3\sqrt{\pi}}{2}\cdot\exp\left(-k^2a^2\sin^2\frac{\theta}{2}\right) \qquad (vii)$$

The differential scattering cross section

$$\left(\frac{d\sigma}{d\Omega}\right) = |f(\theta)|^2 = \left(\frac{mV_0}{\hbar^2}\right)^2\cdot\frac{\pi a^6}{4}\cdot\exp\left(-2k^2a^2\sin^2\frac{\theta}{2}\right) \qquad (viii)$$

The validity condition for low energy (Eqn. iv) may be written as

$$\frac{2mV_0}{\hbar^2} \left| \int_0^\infty r e^{-r^2/a^2} \, dr \right| \ll 1 \qquad (ix)$$

To evaluate the integral in (ix), put $\dfrac{r^2}{a^2} = t$, so $2r\,dr = a^2 dt$

$$\int_0^\infty r e^{-r^2/a^2} \, dr = \frac{a^2}{2} \int_0^\infty e^{-t} \, dt = \frac{a^2}{2} \left[-e^{-t} \right]_0^\infty = \frac{a^2}{2}$$

Therefore the validity condition for low energies becomes

$$\frac{mV_0 a^2}{\hbar^2} \ll 1$$

For this to be satisfied, the strength of the potential 'V_0' should be small. Thus for low energies Born-approximation is a good approximation for very weak potential only.

For high energies, we have from Eqn. (v)

$$\frac{mV_0}{\hbar^2 k} \left| \int_0^\infty e^{-r^2/a^2} \, dr \right| \ll 1 \text{ or } \frac{mV_0 a}{\hbar^2 k} \left| \int_0^\infty e^{-t^2} \, dt \right| \ll 1$$

$$\left(\because \frac{r^2}{a^2} = t \quad 2r\,dr = 2a^2 t\,dt, \ at\,dr = a^2 t\,dt, \ dr = a\,dt \right)$$

or
$$\frac{mV_0 a\sqrt{\pi}}{2\hbar^2 k} \ll 1$$

i.e. The Born approximation in this case is a good approximation for high incident particle energy and weak potential.

Example 20.2: Particles of mass m are scattered elastically by a hard sphere potential

$$V(r) = \begin{cases} \infty & \text{for } r \le a \\ 0 & \text{for } r > a \end{cases}$$

Show that the p-phase shift ($l = 1$) for incident particles of energy $\dfrac{\hbar^2 k^2}{2m}$ is $\tan^{-1}(ka) - ka$.

Solution: For scattering by a hard sphere, the p-wave phase shift (δ_1) is given by

$$\tan \delta_1 = \frac{j_1(ka)}{\eta_1(ka)}$$

where
$$j_1(\rho) = \frac{\sin \rho}{\rho^2} - \frac{\cos \rho}{\rho} \text{ and } \eta_1(\rho) = -\frac{\cos \rho}{\rho^2} - \frac{\sin \rho}{\rho}$$

$$\therefore \qquad \tan \delta_1 = \frac{\dfrac{\sin(ka)}{(ka)^2} - \dfrac{\cos(ka)}{ka}}{-\dfrac{\cos(ka)}{(ka)^2} - \dfrac{\sin(ka)}{ka}} = \frac{\tan(ka) - ka}{-(1 + ka \tan ka)}$$

or
$$ka - ka \tan \delta_1 \tan ka = \tan \delta_1 + \tan ka$$

or
$$ka = \frac{\tan \delta_1 + \tan ka}{1 - \tan \delta_1 \tan ka} \quad \text{or} \quad \tan(\delta_1 + ka) = ka$$

or
$$\delta_1 + ka = \tan^{-1}(ka) \text{ or } \delta_1 = \tan^{-1}(ka) - ka$$

20.13 Connection between Scattering and Bound State (S-Matrix)

From Eqns. (126) and (131), the asymptotic form of the wave function for large r is given by

$$\psi(k,r) \sim \sum_l \frac{(2l+1)}{kr} i^l e^{i\delta_l} \sin\left(kr - \frac{l\pi}{2} + \delta_l\right) \times P_l(\cos\theta) \qquad (170)$$

Thus, the asymptotic part of the S-wave ($l = 0$) part of the wave function is then

$$\psi_0 \sim \frac{1}{kr} e^{i\delta_0} \sin(kr + \delta_0)$$

i.e. apart from the factor $\dfrac{1}{2ik}$, ψ_0 is proportional to

$$\psi_0 \propto e^{i\delta_0}\left[\frac{e^{i(kr+\delta_0)}}{r} - \frac{e^{-i(kr+\delta_0)}}{r}\right] \text{ or, } \psi_0 \propto e^{2i\delta_0}\frac{e^{ikr}}{r} - \frac{e^{-ikr}}{r} \qquad (171)$$

where the s-wave phase shift δ_0 is a function of k:

$$\delta_0 = \delta_0(k) \qquad (172)$$

The wave number k is a real number. $\left(k = \sqrt{2mE/\hbar^2}\right)$ and $E > 0$ for scattering. However, by the process of analytic continuation, the mathematical function $\delta_0(k)$ can be regarded as a function of complex variable. Then, one finds a negative imaginary value of k i.e.

$$k' = -i\,|k'| \qquad (173)$$

such that

$$S(k') \equiv e^{2i\delta_0(k')} = 0 \qquad (174)$$

This means obviously that

$$\delta_0(k') = i\,\infty \qquad (175)$$

Provided that the asymptotic wave function is still given by Eqn. (171), we then require

$$\psi_0 \propto \frac{e^{-ik'r}}{r} = \frac{1}{r}e^{-i(-i|k'|)r} = -\frac{e^{-|k'|r}}{r} \tag{176}$$

This gives the behaviour of a wave function which is required for a bound state wave function vanishing at large r (for normalization). This wave function represents a bound state of the system at energy

$$E = -\frac{|k'|^2 \hbar^2}{2m} \tag{177}$$

In other words, it may be concluded that those negative imaginary values of k, which are zeros of the function $S(k)\left(= e^{2i\delta_0(k)}\right)$ correspond to the bound states of the system of energy E, given by Eqn. (177) i.e. the same function $S(k)$ which gives phase shift of the scattering process at real values of k also gives rise to existence of discrete spectrum of the system at the negative imaginary values of k. This function $S(k) \equiv e^{2i\delta_0(k)}$ is called *S-matrix* of the system. Since $S^\dagger S = SS^\dagger = 1$, thezrefore *S*-matrix is unitary.

20.14 Scattering from an Attractive Square Well Potential

Suppose we have an attractive square well potential given by

$$V(r) = \begin{cases} -V_0 & \text{for } r < a \\ 0 & \text{for } r > a \end{cases} \tag{178}$$

The wave function (for *s*-state: $l = 0$) satisfies the equation:

$$\frac{1}{r^2}\frac{\partial}{\partial r}\left(r^2 \frac{\partial R}{\partial r}\right) + \frac{2m}{\hbar^2}\left[E - V(r)\right]R = 0 \tag{179}$$

Let us substitute

$$R(r) = r\,u(r) \tag{180}$$

then we get

$$\frac{d^2 u}{dr^2} + \frac{2m}{\hbar^2}\left[E - V(r)\right]u(r) = 0 \tag{181}$$

Using Eqn. (178), Eqn. (181) becomes

$$\frac{d^2 u(r)}{dr^2} + \frac{2m}{\hbar^2}\left[E + V_0\right]u(r) = 0 \quad \text{for } r < a \tag{182}$$

and

$$\frac{d^2 u(r)}{dr^2} + \frac{2m}{\hbar^2}Eu(r) = 0 \quad \text{for } r > a \tag{183}$$

Substituting $\quad \sqrt{\dfrac{2mE}{\hbar^2}} = k$ and $\sqrt{\dfrac{2m(E + V_0)}{\hbar^2}} = k'$

Eqns (182) and (183) become

$$\frac{d^2u}{dr^2} + k'^2 u = 0 \text{ for } r < a \tag{184}$$

and

$$\frac{d^2u}{dr^2} + k^2 u = 0 \text{ for } r > a \tag{185}$$

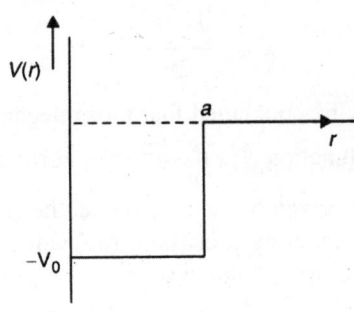

Fig. 20.11

The solutions of these equations may be expressed as

$$u = A \sin k'r \,;\, (r < a) \tag{186}$$

$$u = B \sin(kr + \delta_0) \,;\, (r > a) \tag{187}$$

where A, B and δ_0 are constants to be determined by boundary conditions. For determination of δ_0, we use the boundary condition that $\dfrac{1}{u}\left(\dfrac{\partial u}{\partial r}\right)$ be continuous;

$$\frac{1}{u}\left(\frac{\partial u}{\partial r}\right)\bigg|_{r=a} \text{ from Eqn. (187)} = \frac{1}{u}\left(\frac{\partial u}{\partial r}\right)\bigg|_{r=a} \text{ from Eqn. (186)}$$

Thus

$$k \cot(ka + \delta_0) = k' \cot k'a \tag{188}$$

Let

$$k \cot(ka + \delta_0) = k' \cot k'a \equiv \alpha \tag{189}$$

Taking $k \cot(ka + \partial_0) \equiv \alpha$, we have

$$\frac{k}{\alpha} = \frac{1}{\cot(ka + \delta_0)} = \tan(ka + \delta_0) = \frac{\tan ka + \tan \delta_0}{1 - \tan ka \tan \delta_0}$$

$$\tan ka + \tan \delta_0 = \frac{k}{\alpha} - \frac{k}{\alpha} \tan ka \tan \delta_0$$

or

$$\tan \delta_0 \left[1 + \frac{k}{\alpha}\tan ka\right] = \frac{k}{\alpha} - \tan ka$$

or

$$\tan \delta_0 = \frac{k\left[\dfrac{1}{\alpha} - \dfrac{\tan ka}{k}\right]}{\left(1 + \dfrac{k}{\alpha}\tan ka\right)} \tag{190}$$

The total scattering cross-section

$$\sigma_{tot} = \frac{4\pi}{k^2}\sum_{l=0}(2l+1)\sin^2\delta_l = \frac{4\pi}{k^2}\sin^2\delta_0 \qquad (\because \text{ for s-wave, } l = 0)$$

$$= \frac{4\pi}{k^2}\frac{1}{\left(1+\cot^2\delta_0\right)} \qquad (191)$$

where $\cot \delta_0$ is given by Eqn. (190). If we take the approximate value of cross-section, we have

$$\frac{1}{\alpha} \cong \frac{\tan ka}{k} \quad \text{or} \quad \frac{\tan k'a}{k'} = \frac{\tan ka}{k} \qquad (192)$$

For small k, $ka \ll 1$, we have $\tan ka \cong ka$, then

$$\tan k'a = k'a \qquad (193)$$

For small k, k' is given approximately by $\sqrt{2mV_0/\hbar^2}$. If V_0 and α are such that the Eqn. (192) is satisfied, the scattering cross section will be zero and if it is nearly satisfied then cross-section will be very small. This property that scattering cross section vanishes for a non zero potential is known as *Ramsauer effect*. In this case $\sin \delta_0$ is very small and the electron-atom scattering cross-section is much smaller than gas kinetic cross section. The situation is analogous to transmission across a potential barrier for definite energies.

For small k,

$$\tan\delta_0 \cong k\left[\frac{\dfrac{1}{\alpha}-a}{1+k^2a/\alpha}\right] \qquad (194)$$

As $k \to 0$, $\qquad\qquad k^2a/\alpha \ll 1$

$$\therefore \qquad\qquad \tan\delta_0 \cong k\left(\frac{1}{\alpha}-a\right) \qquad (195)$$

and Eqn. (191) reduces to

$$\sigma \cong \frac{4\pi}{k^2 + \dfrac{\alpha^2}{\left(1-a\alpha\right)^2}} = \left(\frac{2\pi\hbar^2}{m}\right)\left[\frac{1}{E+\dfrac{\hbar^2\alpha^2}{2m}\left(\dfrac{1}{1-a\alpha}\right)^2}\right] \qquad (196)$$

When ka is small enough and the attractive potential strong enough, the $l = 0$ partial wave is pulled in by just half a cycle so that $\delta_0 = 180^0$. The scattering amplitude

$$f(\theta) = \frac{1}{2ik}\sum_{l=0}^{\infty}(2l+1)\left(e^{2i\delta_l}-1\right)P_l(\cos\theta)$$

then vanishes for all θ and there is no scattering. This explains the Ramsauer-Townsend effect – extremely low minimum observed in the scattering cross-section of electrons by rare gas atoms at about 0.7 eV bombarding energy. It must be noted here that the minimum will occur at a definite energy, because the shape of the wave function inside the potential is insensitive to the relatively small bombarding energy whereas the phase of the force free wave function depends rapidly on it. This effect cannot be observed with a repulsive potential because it would require ka to be $\cong 1$ to make $\delta_0 = -180^0$, and potential of such a large range would produce higher l phase shifts.

20.15 Breit-Wigner Formula

We now consider the case of very low energy s-wave scattering by an attractive square well potential. We have

$$k \cot(ka + \delta_0) = k' \cot k'a$$

or

$$\tan(ka + \delta_0) = \frac{k}{k'} \tan\left(k'a\right) \tag{197}$$

or

$$\frac{\tan ka + \tan \delta_0}{1 - \tan ka \tan \delta_0} = \frac{k}{k'} \tan k'a$$

or

$$\tan ka + \tan \delta_0 = \frac{k}{k'} \tan k'a \left[1 - \tan ka \tan \delta_0\right]$$

or

$$\left[1 + \frac{k}{k'} \tan(ka) \tan(k'a)\right] \tan \delta_0 = \frac{k}{k'} \tan(k'a) - \tan(ka) \tag{198}$$

If the potential is shallow, for very low energies ($ka \ll 1$ and $\tan ka \cong ka$), k/k' is very small and the term within square bracket on the l.h.s. is $\cong 1$. Thus,

$$\tan \delta_0 = k \left[\frac{\tan(k'a)}{k'a} - 1\right] \tag{199}$$

If the depth of the potential is gradually increased at a certain stage $k'a$ will go through $\pi/2$. This is the condition for appearance of a bound state in an attractive square well potential. When $k'a$ goes through $\pi/2$, $\tan \delta_0 \to \infty$. It implies the phase shift δ_0 also goes through $\pi/2$. Then total cross section

$$\sigma_{tot} = \frac{4\pi}{k^2} \sum_l (2l+1)\sin^2 \delta_l = \frac{4\pi}{k^2} \qquad (\because l = 0) \tag{200}$$

which is the maximum value. This means that if there is a bound state as energy tends to zero and δ_0 goes through $\pi/2$, the cross-section attains the maximum value. This is called *resonance scattering*.

We will now investigate the behaviour of cross-section near resonance.

Let E_r denote the resonant energy. The phase shift δ_0 will be a function of E_r. Since at resonance δ_0 is $\pi/2$, therefore

$$\sin \delta_0 \left(E_r\right) = 1 \text{ and } \cos \delta_0 \left(E_r\right) = 0$$

Expanding $\sin \delta_0$ and $\cos \delta_0$ near $E = E_r$ by Taylor series, we get

$$\sin\delta_0\left(E\right) \cong \sin\delta_0\left(E_r\right) + \left[\frac{\partial\sin\delta_0}{\partial E}\right]_{E=E_r}\left(E-E_r\right)$$

$$= \sin\delta_0\left(E_r\right) + \left[\cos\delta_0\frac{\partial\delta_0}{\partial E}\right]_{E=E_r}\left(E-E_r\right) \cong 1 \tag{201}$$

and $\quad \cos\delta_0(E) = \cos\delta_0\left(E_r\right) - \left[\sin\delta_0\frac{\partial\delta_0}{\partial E}\right]_{E=E_r}\left(E-E_r\right)$

$$= -\frac{\partial\delta_0\left(E_r\right)}{\partial E}\left(E-E_r\right) = -\frac{2}{\Gamma}\left(E-E_r\right) \tag{202}$$

where $\qquad \dfrac{2}{\Gamma} = \dfrac{\partial\delta_0\left(E_r\right)}{\partial E} \tag{203}$

The scattering amplitude is

$$f(\theta) = \frac{1}{k}\sum_{l=0}^{\infty}(2l+1)\exp\left(i\delta_l\right)P_l(\cos\theta)\sin\delta_l$$

which becomes

$$f_0(E) = \frac{1}{k}\exp\left[i\delta_0(E)\right]\sin\delta_0(E) \tag{204}$$

The $1/k$ variation of the scattering amplitude with energy is very slow. The rapid variation is due to phase shift δ_0. Denoting the contribution due to the phase shift part by $f_0(\delta)$, we have

$$f_0(\delta) = \exp\left[i\delta_0(E)\right]\sin\delta_0(E) = \frac{\sin\delta_0(E)}{\exp\left[-i\delta_0\right]} = \frac{\sin\delta_0}{\cos\delta_0 - i\sin\delta_0} \tag{205}$$

Substituting the values of $\sin\delta_0$ and $\cos\delta_0$ from Eqns (201) and (202), we get

$$f_0(\delta) = \frac{1}{\left[-2\left(E-E_r\right)/\Gamma\right]-1} = -\frac{\Gamma/2}{\left(E-E_r\right)+i\Gamma/2} \tag{206}$$

The total cross section is given by

$$\sigma = \frac{4\pi}{k^2}\left|f_0(\delta)\right|^2 = \frac{4\pi}{k^2}\frac{\Gamma^2/4}{\left(E-E_r\right)^2+\Gamma^2/4}$$

$$= \frac{\pi}{k^2}\frac{\Gamma^2}{\left(E-E_r\right)^2+\Gamma^2/4} \tag{207}$$

This is the Breit Wigner formula for resonant cross sections. The cross section will exhibit a very sharp peak at the resonant energy. The parameter Γ represents *width of the resonance peak*, i.e. $\Gamma/2$ is the value of $\left|E-E_r\right|$ for which σ falls to half its peak value.

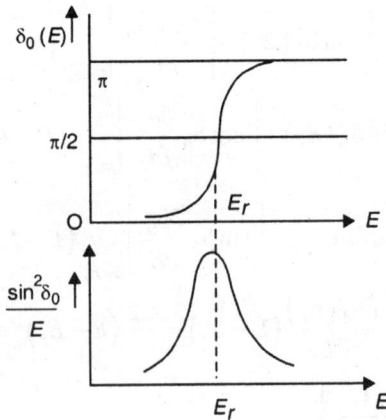

Fig. 20.12: Variation of cross section with phase shift

Those nuclear reactions which pass through the stage of formation of a compound nucleus undergo through a quasi-stationary state characterized by a finite but relatively long life time. Such collisions show the possibility of inelastic scattering process in which the potential part of the elastic scattering process, being independent of the properties of compound nucleus is unaffected but inelastic scattering process does change the resonance part of the elastic scattering amplitude. Due to this reason, the amplitudes of inelastic scattering processes (which pass through a metastable stage of formation of a compound nucleus) are purely resonance in character and comply with the Breit and Wigner's formula. Here Γ denotes the total probability of decay of any given quasi-stationary state of the compound nucleus.

Questions and Problems

1. Obtain the Rutherford cross section for the scattering of charged particles by the Coulomb field of nuclei using the Born approximation.
2. Define total and differential cross section. How are these related to angle of scattering?
3. Derive an expression for differential scattering cross section in terms of phase shifts using the method of partial wave analysis. Why this method is suitable for the scattering of slow particles by the spherically symmetric fields?
4. Construct Green's function for a free particle. Use it to write the total wave function in the first Born approximation.
5. (a) $f(\theta)$ represents the scattering amplitude in the asymptotic wave function

$$\psi = e^{i\vec{k}\cdot\vec{r}} + f(\theta)\frac{e^{ikr}}{r}$$

prove that the differential cross section $= |f(\theta)|^2$.

(b) Show that in Born approximation, the scattering amplitude is the Fourier transform of the scattering potential.

6. What is meant by scattering cross-section? Show that the scattering amplitude by the method of partial wave analysis is given by

$$f(\theta) = \frac{1}{k} \sum_{l=0}^{\infty} (2l+1)e^{i\delta_l} \sin \delta_l P_l(\cos \theta)$$

7. Write short notes on
   (*i*) Differential and total scattering cross-section
   (*ii*) Optical theorem
   (*iii*) Ramsauer Townsend effect

8. Discuss the validity conditions for Born approximation.

21

Relativistic Theory

21.1 Introduction

The basic equation for a free particle in non-relativistic case is Schrodinger's equation

$$-\frac{\hbar^2}{2m}\nabla^2\psi\left(\vec{r},t\right)=i\hbar\frac{\partial\psi\left(\vec{r},t\right)}{\partial t}$$

where $-\dfrac{\hbar^2}{2m}\nabla^2 = H$, is the Hamiltonian operator for the free particle. For a relativistic, free particle, we should write

$$H^2 = p^2c^2 + m^2c^4 \tag{1}$$

or

$$H = \pm c\sqrt{p^2 + m^2c^2}$$

$$= \pm c\sqrt{-\hbar^2(\nabla^2)+m^2c^2} \qquad \text{(because } p = -\imath\hbar\nabla) \tag{2}$$

Upper sign gives positive energy and lower sign gives negative values of energy. It is difficult to interpret the negative sign. However, in classical relativity, the – sign is of no physical significance and can be ignored. Further if we write

$$c\sqrt{(-\hbar^2\nabla^2 + m^2c^2)}\psi = i\hbar\frac{\partial\psi}{\partial t} \tag{3}$$

we do not know how to interpret the square root in the operator. Moreover, it would no longer be a differential equation of finite order and it would not be invariant under Lorentz transformation, because it is of second order in space but of first order in time. Hence this equation must be modified.

21.2 The Klein-Gordon Equation for a Free Particle

Non-relativistically, the energy and momentum of a free particle are related by

$$E = \frac{p^2}{2m}$$

Relativisitically,

$$E^2 = c^2p^2 + m^2c^4 \tag{4}$$

where mc^2 = rest mass energy. A relativistic wave equation for a free particle is obtained by substituting $E \to i\hbar \partial / \partial t$ and $p \to -i\hbar \nabla$ into Eqn. (4) and operating on a wave function $\psi(\vec{r},t)$

$$i\hbar \frac{\partial}{\partial t}\left(i\hbar \frac{\partial \psi}{\partial t} \right) = -\hbar^2 c^2 \nabla^2 \psi + m^2 c^4 \psi \tag{5}$$

or $$\left(-\hbar^2 c^2 \nabla^2 + m^2 c^4 \right)\psi = -\hbar^2 \frac{\partial^2}{\partial t^2}\psi \tag{6}$$

This equation can be rewritten as

$$\left(\Box - \frac{m^2 c^2}{\hbar^2} \right)\psi(\vec{r},t) = 0 \tag{7}$$

where $\Box = \nabla^2 - \dfrac{1}{c^2}\dfrac{\partial^2}{\partial t^2}$ is known as D Alembertian operator. Eqn. (7) is the famous Klein Gordon equation of motion. This equation can describe only spinless particles such as π-mesons. Particles having non zero spin cannot be described by this equation.

21.3 Charge and Current Densities for K.G. Equation

The K.G. Eqn. for a free particle is

$$\left(-\hbar^2 c^2 \nabla^2 + m^2 c^4 \right)\psi = -\hbar^2 \frac{\partial^2}{\partial t^2}\psi$$

Taking complex conjugate

$$\left(-\hbar^2 c^2 \nabla^2 + m^2 c^4 \right)\psi^* = -\hbar^2 \frac{\partial^2 \psi^*}{\partial t^2} \tag{8}$$

Multiply Eqn. (6) by ψ^* and Eqn. (8) by ψ, we get

$$\psi^*\left(-\hbar^2 c^2 \nabla^2 + m^2 c^4 \right)\psi = -\hbar^2 \psi^* \frac{\partial^2 \psi}{\partial t^2} \tag{9i}$$

$$\psi\left(-\hbar^2 c^2 \nabla^2 + m^2 c^4 \right)\psi^* = -\hbar^2 \psi \frac{\partial^2 \psi^*}{\partial t^2} \tag{9ii}$$

Subtracting Eqn. (9ii) from (9i), we obtain

$$-\hbar^2 c^2 \left(\psi^* \nabla^2 \psi - \psi \nabla^2 \psi^* \right) = -\hbar^2 \left(\psi^* \frac{\partial^2 \psi}{\partial t^2} - \psi \frac{\partial^2 \psi^*}{\partial t^2} \right)$$

$$\frac{\partial}{\partial t}\left(\psi^* \frac{\partial \psi}{\partial t} - \psi \frac{\partial \psi^*}{\partial t} \right) = c^2 \nabla \cdot \left(\psi^* \nabla \psi - \psi \nabla \psi^* \right)$$

or

Now multiplying throughout by $\left(\dfrac{i\hbar}{2mc^2} \right)$, we get

$$\frac{i\hbar}{2mc^2}\frac{\partial}{\partial t}\left(\psi^* \frac{\partial \psi}{\partial t} - \psi \frac{\partial \psi^*}{\partial t} \right) = \frac{i\hbar}{2m} \nabla \cdot \left(\psi^* \nabla \psi - \psi \nabla \psi^* \right)$$

or
$$\frac{\partial}{\partial t}\left[\frac{i\hbar}{2mc^2}\left(\psi*\frac{\partial\psi}{\partial t}-\psi\frac{\partial\psi*}{\partial t}\right)\right]+\nabla\cdot\left[\frac{\hbar}{2im}\left(\psi*\nabla\psi-\psi\nabla\psi*\right)\right]=0 \qquad (10)$$

Substituting

$$P(\vec{r},t)=\frac{i\hbar}{2mc^2}\left(\psi*\frac{\partial\psi}{\partial t}-\psi\frac{\partial\psi*}{\partial t}\right) \qquad (11)$$

$$\vec{j}(\vec{r},t)=\frac{\hbar}{2im}\left(\psi*\nabla\psi-\psi\nabla\psi*\right) \qquad (12)$$

we have
$$\frac{\partial P}{\partial t}+\nabla\cdot\vec{j}=0 \qquad (13)$$

Eqn. (13) is of the same form as the non relativistic probability conservation equation. But now, the difficulty is with $P(r, t)$ as it is not necessarily positive and hence cannot be interpreted as position probability density. Let us look for plane wave solutions

$$\psi=Ae^{i\vec{k}\cdot\vec{r}-i\omega t} \qquad (14)$$

Substituting this in Eqn. (11), we have

$$P=\frac{i\hbar}{2mc^2}\left[\psi*(-i\omega)\psi-i\omega\psi\psi*\right]=\frac{\hbar}{2mc^2}\left(\psi*\omega\psi+\omega\psi*\psi\right)$$

$$=\frac{E}{mc^2}\psi*\psi=\frac{E}{mc^2}|\psi|^2 \quad (\because E=\hbar\omega) \qquad (15)$$

This implies that $P(\vec{r},t)$ is indefinite since E can take up positive as well as negative values, but it is also proportional to $|\psi|^2$ with the proportionality factor $\left(\frac{E}{mc^2}\right)$. In the non relativistic limit one finds $P=|\psi|^2$ as in the Schrödinger theory $\left(\because E=mc^2\right)$. Thus P being not positive definite, it cannot be regarded as a conventional position probability density. For this reason K.G. equation remained abandoned for about seven years till 1936, when Pauli and Weisskopf rescued it by reinterpreting it as a field equation and quantizing it. They pointed out that though it is not acceptable as a quantum mechanical wavefunction yet, the K.G. equation is perfectly valid as a classical field equation. Physical indefiniteness associated with the density function $P(\vec{r},t)$ was interpreted as follows: In the quantized version of the K.G. complex field ψ, the quantity $P(\vec{r},t)$ becomes equal to charge density and $\vec{j}$ becomes the current density. Since in the quantized version, P emerges as charge associated with the particle, therefore it could be positive, negative or zero. The quantized field energy comes out to be always positive (even though the parameter E in the wave equation can have both the signs). We can multiply P by e and when E is negative then minus sign is absorbed in e, we then say that charge on the second particle is opposite to that on the first particle, in this way it encompasses particles of both signs of charge.

21.4 The Covariant form of K.G. Equation

The relativistic invariance of the relation

$$\frac{E^2}{c^2} = \vec{p}^2 + M^2 c^2 \tag{16}$$

becomes more apparent if we introduce the momentum four vector with its four components:

$$P_\mu = \left\{ p_j, \frac{iE}{c} \right\} \quad j = 1, 2, 3 \ \mu = 1, 2, 3, 4. \tag{17}$$

Eqn. (16) then becomes

$$\sum_{\mu=1}^{4} p_\mu^2 = -M^2 c^2 \tag{18}$$

The change to operators can be written in the form

$$P_\mu \rightarrow i\hbar \frac{\partial}{\partial x_\mu} \tag{19}$$

where

$$x_\mu \equiv (x, y, z, ict) \tag{20}$$

We can use this notation to write the Klein Gordon equation

$$\frac{\hbar^2}{c^2} \frac{\partial^2 \psi}{\partial t^2} = \left(\hbar^2 \nabla^2 - M^2 c^2 \right) \psi$$

in the covariant form as

$$\left(-\hbar^2 \nabla^2 + \frac{\hbar^2}{c^2} \frac{\partial^2}{\partial t^2} + M^2 c^2 \right) \psi = 0$$

or

$$\left[\sum_{\mu=1}^{4} p_\mu^2 + M^2 c^2 \right] \psi = 0 \tag{21}$$

In covariant form, the equation of continuity is

$$\sum_\mu \frac{\partial j_\mu}{\partial x_\mu} = 0 \tag{22}$$

where

$$j_\mu = \frac{\hbar}{2im} \left(\psi^* \frac{\partial \psi}{\partial x_\mu} - \psi \frac{\partial \psi^*}{\partial x_\mu} \right) \tag{23}$$

i.e.

$$j_\mu = (j_i, ic\vec{p}) \tag{24}$$

21.5 Equation for a Charged Particle in an Electromagnetic Field

If $(-e)$ is the charge of the particle (electron) and if the e.m. potential is characterized by the scalar potential ϕ and the vector potential $\vec{A}(\vec{r}, t)$, then the effect of the field on the particle can be introduced by replacing $\vec{p}$ by $\vec{p} - \dfrac{e\vec{A}}{c}$ and E by $E - e\phi$. Thus the relativistic energy momentum relation becomes

$$(E - e\phi)^2 = \left(\vec{p} - \frac{e\vec{A}}{c}\right)^2 c^2 + m^2 c^4 \qquad (24)$$

or

$$(E - e\phi)^2 = \left(c\vec{p} - e\vec{A}\right)^2 + m^2 c^4$$

operating on the function $\psi(\vec{r}, t)$ after replacing E and $\vec{p}$ by the operators $i\hbar\dfrac{\partial}{\partial t}$ and $-i\hbar\nabla$ respectively, we get the K.G. equation for a charged particle in an e.m. potential as:

$$\left(i\hbar\frac{\partial}{\partial t} - e\phi\right)^2 \psi(\vec{r}, t) = \left(-i\hbar c\nabla - e\vec{A}\right)^2 \psi(\vec{r}, t) + m^2 c^4 \psi(\vec{r}, t) \qquad (25)$$

The l.h.s. of this equation can be written as

$$\left(i\hbar\frac{\partial}{\partial t} - e\phi\right)\left(i\hbar\frac{\partial\psi}{\partial t} - e\phi\psi\right) = \left(-\hbar^2 \frac{\partial^2 \psi}{\partial t^2} - i\hbar e\frac{\partial}{\partial t}(\phi\psi) - i\hbar e\phi\frac{\partial\psi}{\partial t} + e^2\phi^2\psi\right)$$

$$= \left(-\hbar^2 \frac{\partial^2 \psi}{\partial t^2} - 2i\hbar e\phi\frac{\partial\psi}{\partial t} - i\hbar e\psi\frac{\partial\phi}{\partial t} + e^2\phi^2\psi\right)$$

and the right hand side is

$$\left(-\hbar^2 c^2\nabla^2 + 2i\hbar ce\vec{A}\cdot\nabla + i\hbar ce\nabla\cdot\vec{A} + e^2 A^2 + m^2 c^4\right)\psi$$

Therefore, we have

$$\left(-\hbar^2 \frac{\partial^2}{\partial t^2} - 2i\hbar e\phi\frac{\partial}{\partial t} - i\hbar e\psi\frac{\partial\phi}{\partial t} + e^2\phi^2\right)\psi(\vec{r}, t)$$

$$= \left(-\hbar^2 c^2\nabla^2 + 2i\hbar ce\vec{A}\cdot\nabla + i\hbar ce\nabla\cdot\vec{A} + e^2 A^2 + m^2 c^4\right)\psi(\vec{r}, t) \qquad (26)$$

To check the correctness of this Eqn. in the non relativistic limit, we make the substitution

$$\psi(\vec{r}, t) = \xi(\vec{r}, t)\exp\left(-\frac{imc^2 t}{\hbar}\right) \qquad (27)$$

where mc^2 is the rest mass energy of the particle. Then

$$\frac{\partial\psi}{\partial t} = \left(\frac{\partial\xi}{\partial t} - \frac{imc^2}{\hbar}\xi\right)\exp\left(-\frac{imc^2 t}{\hbar}\right) \qquad (28)$$

and

$$\frac{\partial^2\psi}{\partial t^2} = \left(\frac{\partial^2\xi}{\partial t^2} - \frac{2imc^2}{\hbar}\frac{\partial\xi}{\partial t} - \frac{m^2 c^2}{\hbar^2}\xi\right)\exp\left(-\frac{imc^2 t}{\hbar}\right) \qquad (29)$$

Assuming that the operation of $ih\dfrac{\partial}{\partial t}$ on ξ gives a result of the order of $e\phi\xi$ which is small in comparison with $mc^2\xi$, the first term in both of the above differential equations can be neglected, and also the last two terms on the l.h.s. of Eqn. (26) which now becomes

$$\left(\frac{2imc^2\hbar^2}{\hbar}\frac{\partial\xi}{\partial t}+m^2c^4\xi+\frac{2i^2\hbar e\phi mc^2\xi}{\hbar}\right)$$

$$=\left(-\hbar^2c^2\nabla^2+2i\hbar ce\vec{A}\cdot\nabla+i\hbar ce\nabla\cdot\vec{A}+e^2A^2+m^2c^4\right)\xi$$

or
$$ih\frac{\partial\xi}{\partial t}=\left(-\frac{\hbar^2}{2m}\nabla^2+\frac{ieh}{mc}\vec{A}\cdot\nabla+\frac{ieh}{2mc}\nabla\cdot\vec{A}+\frac{e^2A^2}{2mc^2}+e\phi\right)\xi \tag{30}$$

This is clearly the Schrödinger equation for a particle in an electromagnetic field.

21.6 The Continuity Equation in presence of an Electromagnetic field

We know (from classical electro-dynamics) that we can change from the Hamiltonian function (i.e. the energy expressed in terms of momentum $E=\sqrt{m^2c^4+p^2c^2}$ to the Hamiltonian function for a particle of charge e moving in an e.m. field determined by the potentials

$$A_\mu=\left(A_1,A_2,A_3,iA_0\right) \tag{31}$$

through the transformation

$$p_\mu\rightarrow p_\mu-\frac{e}{c}A_\mu=-i\hbar\frac{\partial}{\partial x_\mu}-\frac{e}{c}A_\mu$$

$$p_\mu\equiv\left\{p_1,p_2,p_3,iE/c\right\} \qquad \left(E=i\hbar\frac{\partial}{\partial t}\right)$$

We thus find the relativistic wave equation

$$\left\{\sum_\mu\left(p_\mu-\frac{e}{c}A_\mu\right)^2+M^2c^2\right\}\psi=0$$

or
$$\left\{\left(p-\frac{e}{c}A\right)^2+M^2c^2\right\}\psi=-\left\{\frac{i}{c}\left(i\hbar\frac{\partial}{\partial t}-eA_0\right)\right\}^2\psi \tag{32i}$$

or
$$\frac{1}{c^2}\left\{i\hbar\frac{\partial}{\partial t}-eA_0\right\}^2\psi=\left\{\left(p-\frac{e}{c}\vec{A}\right)^2+M^2c^2\right\}\psi \tag{32ii}$$

Multiply from left by ψ^*, we get

$$\frac{1}{c^2}\left[\psi^*\left(i\hbar\frac{\partial}{\partial t}-eA_0\right)\left(i\hbar\frac{\partial}{\partial t}-eA_0\right)\psi\right]$$

$$= \psi * \left(\vec{p} - \frac{e}{c} \vec{A} \right) \cdot \left(\vec{p} - \frac{e}{c} \vec{A} \right) \psi + M^2 c^2 \psi * \psi$$

or

$$\frac{1}{c^2} \left[\psi * \left(-\hbar^2 \frac{\partial^2 \psi}{\partial t^2} + e^2 A_0^2 \psi - ie\hbar \frac{\partial}{\partial t} (A_0 \psi) - ie\hbar A_0 \frac{\partial \psi}{\partial t} \right) \right]$$

$$= \psi * \left(p^2 \psi + \frac{e^2}{c^2} A^2 \psi - \frac{e}{c} \vec{p} \cdot (\vec{A} \psi) - \frac{e}{c} \vec{A} \cdot \vec{p} \psi \right) + M^2 c^2 \psi * \psi \qquad (33)$$

Taking its complex conjugate, we get

$$\frac{1}{c^2} \left[\psi \left(-\hbar^2 \frac{\partial^2 \psi *}{\partial t^2} + e^2 A_0^2 \psi * + ie\hbar \frac{\partial}{\partial t} (A_0 \psi *) + ie\hbar A_0 \frac{\partial \psi *}{\partial t} \right) \right]$$

$$= \psi \left(p^2 \psi * + \frac{e^2}{c^2} A^2 \psi * - \frac{e}{c} \vec{p} * \cdot (\vec{A} \psi *) - \frac{e}{c} \vec{A} \cdot \vec{p} * \psi * \right) + M^2 c^2 \psi * \psi \qquad (34)$$

Subtracting Eqn. (34) from (33), we get

$$\text{R.H.S.} = -\hbar^2 \left(\psi * \nabla^2 \psi - \psi \nabla^2 \psi * \right) + \frac{ie\hbar}{c} \left[\psi * \nabla \cdot (\vec{A} \psi *) + \psi \nabla \cdot (\vec{A} \psi *) + \right.$$

$$\left. \psi * \vec{A} \cdot (\nabla \psi *) + \psi \vec{A} \cdot (\nabla \psi *) \right]$$

$$= -\hbar^2 \nabla \cdot (\psi * \nabla \psi - \psi \nabla \psi *) + \frac{ie\hbar}{c}$$

$$\left[2(\psi * \psi) \nabla \cdot \vec{A} + 2\psi * \vec{A} \cdot (\nabla \psi) + 2\psi \vec{A} \cdot (\nabla \psi *) \right]$$

$$= -\hbar^2 \nabla \cdot (\psi * \nabla \psi - \psi \nabla \psi *) + \frac{2ie\hbar}{c} \nabla \cdot (\vec{A} \psi * \psi) \qquad (35)$$

$$\text{L.H.S.} = \frac{1}{c^2} \left[-\hbar^2 \left(\psi * \frac{\partial^2 \psi}{\partial t^2} - \psi \frac{\partial^2 \psi *}{\partial t^2} \right) \right]$$

$$- \frac{ie\hbar}{c^2} \left\{ \psi * \frac{\partial}{\partial t} (A_0 \psi) + \psi \frac{\partial}{\partial t} (A_0 \psi *) + \psi * A_0 \frac{\partial \psi}{\partial t} + \psi A_0 \frac{\partial \psi *}{\partial t} \right\}$$

$$= -\frac{\hbar^2}{c^2} \frac{\partial}{\partial t} \left(\psi * \frac{\partial \psi}{\partial t} - \psi \frac{\partial \psi *}{\partial t} \right) - \frac{ie\hbar}{c^2} \left\{ 2\psi * \psi \frac{\partial A_0}{\partial t} \right.$$

$$\left. + 2\psi * A_0 \frac{\partial \psi}{\partial t} + 2\psi A_0 \frac{\partial \psi *}{\partial t} \right\} \qquad (36)$$

From Eqns (35) and (36), we have

$$-\frac{\hbar^2}{c^2}\frac{\partial}{\partial t}\left(\psi*\frac{\partial\psi}{\partial t}-\psi\frac{\partial\psi*}{\partial t}\right)-\frac{2ie\hbar}{c^2}\frac{\partial}{\partial t}\left(\vec{A}_0\psi*\psi\right)$$

$$=-\hbar^2\nabla\cdot\left\{\left(\psi*\nabla\psi-\psi\nabla\psi*\right)+\frac{2ie\hbar}{c}\nabla\cdot\left(\vec{A}\psi*\psi\right)\right\}$$

Multiplying both sides of this Eqn. by $\left(\dfrac{e}{2iM\hbar}\right)$, we get

$$\frac{\partial\rho}{\partial t}=-\nabla\cdot\vec{J} \tag{37}$$

where $$\rho=\frac{ie\hbar}{2Mc^2}\left(\psi*\frac{\partial\psi}{\partial t}-\psi\frac{\partial\psi*}{\partial t}\right)-\frac{e^2A_0}{Mc^2}\psi*\psi \tag{38}$$

and $$\vec{J}=\frac{e\hbar}{2iM}\left(\psi*\nabla\psi-\psi\nabla\psi*\right)-\frac{e^2\vec{A}}{Mc}\psi*\psi \tag{39}$$

21.7 Difficulties with Klein Gordon Equation

Although the K.G. equation in itself is quite satisfactory as a classical field equation, it cannot accommodate for the spin of the particle and describes only spinless particles because the wave function transforms as a scalar.

1. The probability density for the K.G. particle is not positive definite because the wave equation is second order in time.
2. It is not able to associate any physical significance to the negative energy states.
3. K.G. equation does not give the experimentally observed fine structure of the hydrogen atom.
4. In the derivation of the equation of continuity for K.G. equation, there appear time derivatives of the wavefunctions in the expression for $P(\vec{r}, t)$ (i.e. probability density).

21.8 Dirac's Relativistic Equation for Free, Spin ½ Particle

Dirac in 1928 formulated an equation which circumvented the difficulties of negative probability density. He approached the problem by developing a relativistic wave equation which instead of being quadratic, was linear and first order in the time derivative. For, in the theory of relativity, there is complete symmetry between the space (x,y,z) and time (ct), it also restricts the spatial derivatives in the wave equation also to the first order. Dirac modified Hamiltonian in such a way that Dirac wave function ψ satisfied a first order linear differential equation in all the four coordinates. It described the particles with spin ½ and provided very good physical interpretation of the negative energy solutions of this equation.

21.8.1 Development of Dirac's Equation

The Hamiltonian for a free particle is

$$H = \pm c \sqrt{p_x^2 + p_y^2 + p_z^2 + m^2 c^2}$$

$$= \pm c \left\{ -\hbar^2 \left(\frac{\partial^2}{\partial x^2} + \frac{\partial^2}{\partial y^2} + \frac{\partial^2}{\partial z^2} \right) + m^2 c^2 \right\}^{1/2} . \qquad (40)$$

This when substituted in the time-dependent Schrödinger equation

$$i\hbar \frac{\partial \psi(\vec{r},t)}{\partial t} = H\psi(r,t) \qquad (41)$$

the resulting equation is unsymmetric in space and time derivatives and is not relativistically proper.

The simples Hamiltonian that is linear in momentum and mass terms is

$$H = c\vec{\alpha} \cdot \vec{p} + \beta mc^2$$

$$= c\alpha_x p_x + c\alpha_y p_y + c\alpha_z p_z + \beta mc^2 \qquad (42)$$

where the coefficients $\alpha_x, \alpha_y, \alpha_z$ and β are yet to be deterimined.

The relativistic Hamiltonian must necessarily include the rest energy (mc^2) and since energy is on the same footing as $c\vec{p}$ in relativity theory, H is taken linear in mc^2 also.

Substituting Eqn. (42) into Eqn. (41), we get

$$\left(E - c\vec{\alpha} \cdot \vec{p} - \beta mc^2 \right) \psi = 0 \qquad (43)$$

If this equation is to describe a free particle, $\vec{\alpha}$ and β must be independent of $\vec{r}, t, \vec{p}$ and E. Now, ψ of this equation must also be a solution of the relativistic equation

$$-\hbar^2 \frac{\partial^2 \psi}{\partial t^2} = \left(-\hbar^2 c^2 \nabla^2 + m^2 c^4 \right) \psi \qquad (44)$$

because in the absence of external fields, the wave packet solution of Eqn. (43) whose motions resemble those of classical particle must have the classical relation between energy, momentum and mass (viz. $E^2 = p^2 c^2 + m^2 c^4$). We therefore multiply Eqn. (43) on the left by $\left(E + c\vec{\alpha} \cdot \vec{p} + \beta mc^2 \right)$ to obtain

$$\left(\vec{E} + c\vec{\alpha} \cdot \vec{p} + \beta mc^2 \right)\left(E - c\vec{\alpha} \cdot \vec{p} - \beta mc^2 \right) \psi = 0 \text{ (see Eqn. 40)} \qquad (45)$$

Therefore,

$$\left\{ E^2 - c^2 \left[\left(\alpha_x^2 p_x^2 + \alpha_y^2 p_y^2 + \alpha_z^2 p_z^2 \right) + \left(\alpha_x \alpha_y + \alpha_y \alpha_x \right) p_x p_y \right. \right.$$

$$+ \left(\alpha_y \alpha_z + \alpha_z \alpha_y \right) p_y p_z + \left(\alpha_z \alpha_x + \alpha_x \alpha_z \right) p_z p_x \Big] - m^2 c^4 \beta^2$$

$$\left. -mc^3 \left[\left(\alpha_x \beta + \beta \alpha_x \right) p_x + \left(\alpha_y \beta + \beta \alpha_y \right) p_y + \left(\alpha_z \beta + \beta \alpha_z \right) p_z \right] \right\} \psi$$

$$= 0 \qquad (46)$$

Comparing this with the Klein Gordan equation

$$\left[E^2 - c^2 \left(p_x^2 + p_y^2 + p_z^2 \right) - m^2 c^4 \right] \psi = 0 \tag{47}$$

we find that relation (46) agrees with (47) if

$$\alpha_x^2 = \alpha_y^2 = \alpha_z^2 = \beta^2 = 1$$

and
$$\left(\alpha_x \alpha_y + \alpha_y \alpha_x \right) = \left(\alpha_y \alpha_z + \alpha_z \alpha_y \right) = \left(\alpha_z \alpha_x + \alpha_x \alpha_z \right) = 0 \tag{48}$$

and
$$\left(\alpha_x \beta + \beta \alpha_x \right) = \left(\alpha_y \beta + \beta \alpha_y \right) = \left(\alpha_z \beta + \beta \alpha_z \right) = 0$$

that is, the four quantities anticommute in pairs and their squares are unity. Since α's and β anticommute rather than commute with each other, they are expressible by matrices.

21.8.2 Matrices for α and β

Now H given by Eqn. (42) is Hermitian, therefore each of the four matrices α and β must be Hermitian and hence square.

We are already familiar with three Pauli spin matrices which are 2×2 matrices anticommuting with each other and their square is unity

$$\sigma_x = \begin{pmatrix} 0 & 1 \\ 1 & 0 \end{pmatrix} ; \sigma_y = \begin{pmatrix} 0 & -i \\ +i & 0 \end{pmatrix} ; \sigma_z = \begin{pmatrix} 1 & 0 \\ 0 & -1 \end{pmatrix} \tag{49}$$

Since a 2×2 matrix has four elements, there are four and only four linearly independent 2×2 matrices. A matrix linearly independent of the matrices (Eqn. 49) is

$$I = \begin{pmatrix} 1 & 0 \\ 0 & 1 \end{pmatrix} \tag{50}$$

It is a unit matrix and therefore commutes rather than anticommutes with every σ.

We cannot use 3×3 matrices and the dimension of our matrices must be even. It may be seen as follows.

The third relation in Eqn (48) gives

$$\alpha_i \beta + \beta \alpha_i = 0 \qquad (i = x, y, z)$$

or
$$\beta \alpha_i = -\alpha_i \beta$$

$$\det \left(\beta \alpha_i \right) = \det \left(-\alpha_i \beta \right)$$

$$\det \left(\beta \right) \det \left(\alpha_i \right) = \left[\det \left(-\alpha_i \right) \right] \left(\det \beta \right)$$

$$= \left(\det \alpha_i \right) \left[\det \left(-\beta \right) \right]$$

where $\det \beta$ represents the determinant of the matrix β. For these relations to hold, we require

$$\det \alpha_i = \det \left(-\alpha_i \right) = \left[\det \left(-I \right) \right] \det \alpha_i$$

$$\det \beta = \det \left(-\beta \right) = \left[\det \left(-I \right) \right] \det \beta \tag{51}$$

The determinant of the negative unit matrix $-I$, of order n is

$$\begin{vmatrix} -1 & 0 & 0 & \cdots & 0 \\ 0 & -1 & 0 & & 0 \\ & & \ddots & & \\ & & & \ddots & \\ 0 & 0 & 0 & \cdots & -1 \end{vmatrix} = (-1)^n$$

Therefore Eqn. (51) are true if $(-1)^n = 1$ i.e. if n is even or if α_i and β have an even number of rows and columns. Thus the next simplest choise is 4×4 matrices. Also, since $\beta^2 = \alpha_i^2 = 1$, there exists an inverse of each of these matrices (det $\alpha_i \neq 0 \neq$ det β), so we can write

$$\beta \alpha_i = -\alpha_i \beta$$

or
$$\alpha_i^{-1} \beta \alpha_i = -\alpha_i^{-1} \alpha_i \beta = -\beta \tag{52}$$

Taking the trace of this equation and using $Tr(ABC) = Tr(CAB)$,

$$Tr\left(\alpha_i^{-1} \beta \alpha_i\right) = Tr\left(\alpha_i \alpha_i^{-1} \beta\right) = Tr(\beta)$$

$$= Tr(-\beta) = -Tr(\beta) \qquad \text{(using Eqn. 52)}$$

This can be true only if $Tr(\beta) = 0$. Similarly $Tr(\alpha_i) = 0$.

Thus matrices α's and β are traceless. Zero trace shows that $n/2$ eigenvalues are 1 and other $n/2$ -1. Since σ's and I are 2×2 matrices, we can construct with them 4×4 matrices as follows:

$$\alpha_i = \begin{pmatrix} 0 & \sigma_i \\ \sigma_i & 0 \end{pmatrix}$$

That is
$$\alpha_x = \begin{pmatrix} 0 & \sigma_x \\ \sigma_x & 0 \end{pmatrix} = \begin{pmatrix} 0 & 0 & 0 & 1 \\ 0 & 0 & 1 & 0 \\ 0 & 1 & 0 & 0 \\ 1 & 0 & 0 & 0 \end{pmatrix} = \alpha_x^\dagger \tag{53i}$$

$$\alpha_y = \begin{pmatrix} 0 & \sigma_y \\ \sigma_y & 0 \end{pmatrix} = \begin{pmatrix} 0 & 0 & 0 & -i \\ 0 & 0 & i & 0 \\ 0 & -i & 0 & 0 \\ i & 0 & 0 & 0 \end{pmatrix} = \alpha_y^\dagger \tag{53ii}$$

$$\alpha_z = \begin{pmatrix} 0 & \sigma_z \\ \sigma_z & 0 \end{pmatrix} = \begin{pmatrix} 0 & 0 & 1 & 0 \\ 0 & 0 & 0 & -1 \\ 1 & 0 & 0 & 0 \\ 0 & -1 & 0 & 0 \end{pmatrix} = \alpha_z^\dagger \tag{53iii}$$

These 4×4 matrices mutually anticommute and give a unit matrix when squared. To construct the fourth anticommuting matrix for β, consider

$$\begin{pmatrix} a & b & c & d \\ e & f & g & h \\ i & j & k & l \\ m & n & o & p \end{pmatrix}$$

For anticommuting with α_x, we should have

$$\beta = \begin{pmatrix} a & b & c & d \\ e & f & g & h \\ -h & -g & -f & -e \\ -d & -c & -b & -a \end{pmatrix}$$

This anticommutes with α_y if

$$\beta = \begin{pmatrix} a & 0 & c & 0 \\ 0 & f & 0 & h \\ -h & 0 & -f & 0 \\ 0 & -c & 0 & -a \end{pmatrix}$$

For this matrix to anticommute with α_z, we require

$$\beta = \begin{pmatrix} a & 0 & c & 0 \\ 0 & a & 0 & c \\ -c & 0 & -a & 0 \\ 0 & -c & 0 & -a \end{pmatrix}$$

$\beta^2 = 1$ requires that $a^2 - c^2 = 1$. Thus an infinite number of matrices β can be formed. The simplest choice is with $c = 0$ and $a = 1$. Then

$$\beta = \begin{pmatrix} I & 0 \\ 0 & -I \end{pmatrix} = \begin{pmatrix} 1 & 0 & 0 & 0 \\ 0 & 1 & 0 & 0 \\ 0 & 0 & -1 & 0 \\ 0 & 0 & 0 & -1 \end{pmatrix} \tag{54}$$

The matrices (53) and (54) are called *Dirac matrices*.

21.8.3 Free Particle Solutions

The wave equation (41) becomes the Dirac equation

$$i\hbar \frac{\partial}{\partial t} \psi = H\psi = \left(-i\hbar c \vec{\alpha} \cdot \nabla + \beta mc^2 \right) \psi$$

or

$$\left[\beta mc^2 + c\left(\vec{\alpha} \cdot \vec{p} \right) \right] \psi = E\psi \tag{55}$$

Since $\vec{\alpha}$ and β are 4×4 matrices, the Dirac Hamiltonian is a 4×4 matrix. Equation (55) has now no meaning unless the wave function ψ is itself a matrix with four rows and one column.

$$\psi = \begin{pmatrix} \psi_1 \\ \psi_2 \\ \psi_3 \\ \psi_4 \end{pmatrix} \tag{56}$$

Then Eqn. (55) is equivalent to four simultaneous first order partial differential equations that are linear and homogenous in four ψ's. Plane wave solutions of the form

$$\psi_j = u_j \exp i\left(\vec{k} \cdot \vec{r} - \omega t\right), \ (j = 1, 2, 3, 4) \tag{57}$$

can be found where the u_j's are numbers. These are eigenfunctions of energy and momentum operators with eigenvalues $\hbar\omega$ and $\hbar k$ respectively. Substitution of matrix forms of α's and β in Eqn. (55) gives

$$\begin{pmatrix} mc^2 & 0 & 0 & 0 \\ 0 & mc^2 & 0 & 0 \\ 0 & 0 & -mc^2 & 0 \\ 0 & 0 & 0 & -mc^2 \end{pmatrix} \begin{pmatrix} \psi_1 \\ \psi_2 \\ \psi_3 \\ \psi_4 \end{pmatrix} + c \begin{pmatrix} 0 & 0 & 0 & \hat{p}_x \\ 0 & 0 & \hat{p}_x & 0 \\ 0 & \hat{p}_x & 0 & 0 \\ \hat{p}_x & 0 & 0 & 0 \end{pmatrix} \begin{pmatrix} \psi_1 \\ \psi_2 \\ \psi_3 \\ \psi_4 \end{pmatrix}$$

$$+ c \begin{pmatrix} 0 & 0 & 0 & -i\hat{p}_y \\ 0 & 0 & i\hat{p}_y & 0 \\ 0 & -i\hat{p}_y & 0 & 0 \\ i\hat{p}_y & 0 & 0 & 0 \end{pmatrix} \begin{pmatrix} \psi_1 \\ \psi_2 \\ \psi_3 \\ \psi_4 \end{pmatrix} + c \begin{pmatrix} 0 & 0 & \hat{p}_z & 0 \\ 0 & 0 & 0 & -\hat{p}_z \\ \hat{p}_z & 0 & 0 & 0 \\ 0 & -\hat{p}_z & 0 & 0 \end{pmatrix} \begin{pmatrix} \psi_1 \\ \psi_2 \\ \psi_3 \\ \psi_4 \end{pmatrix}$$

$$= \hat{E} \begin{pmatrix} \psi_1 \\ \psi_2 \\ \psi_3 \\ \psi_4 \end{pmatrix}$$

or

$$\begin{pmatrix} mc^2 & 0 & c\hat{p}_z & c\left(\hat{p}_x - i\hat{p}_y\right) \\ 0 & mc^2 & c\left(\hat{p}_x + i\hat{p}_y\right) & -c\hat{p}_z \\ c\hat{p}_z & c\left(\hat{p}_x - i\hat{p}_y\right) & -mc^2 & 0 \\ c\left(\hat{p}_x + i\hat{p}_y\right) & -c\hat{p}_z & 0 & -mc^2 \end{pmatrix} \begin{pmatrix} \psi_1 \\ \psi_2 \\ \psi_3 \\ \psi_4 \end{pmatrix}$$

$$= \hat{E} \begin{pmatrix} \psi_1 \\ \psi_2 \\ \psi_3 \\ \psi_4 \end{pmatrix} \tag{58}$$

Hence α, β, E and p behave like complex numbers and the α's and β operate on the one column matrix u, called a *spinor*.

Substituting $\psi_j = u_j \exp i(\vec{k} \cdot \vec{r} - \omega t)$ $(j = 1,2,3,4)$ into Eqn. (58), we get

$$(E - mc^2)u_1 + 0u_2 - cp_z u_3 - c(p_x - ip_y)u_4 = 0$$

$$0u_1 + (E - mc^2)u_2 - c(p_x + ip_y)u_3 + cp_z u_4 = 0$$

$$-cp_z u_1 - c(p_x - ip_y)u_2 + (E + mc^2)u_3 + 0u_4 = 0$$

$$-c(p_x + ip_y)u_1 + cp_z u_2 + 0u_3 + (E + mc^2)u_4 = 0 \tag{59}$$

For a non-trivial solution for u_1, u_2, u_3, u_4, the determinant of their coefficients must vanish, i.e.

$$\begin{vmatrix} (E - mc^2) & 0 & -cp_z & -c(p_x - ip_y) \\ 0 & (E - mc^2) & -c(p_x + ip_y) & cp_z \\ -cp_z & -c(p_x - ip_y) & (E + mc^2) & 0 \\ -c(p_x + ip_y) & cp_z & 0 & (E + mc^2) \end{vmatrix} = 0$$

On evaluating the determinant, we get

$$(E^2 - m^2 c^4 - c^2 p^2)^2 = 0 \tag{60}$$

This gives

$$E_{\pm} = \pm(m^2 c^4 + c^2 p^2)^{1/2} \tag{61}$$

We denote positive and negative roots by E_+ and E_- respectively. Explicit solutions can be found for any momentum p by choosing a sign for energy. Taking $E = E_+$, we find that out of four equations only two are linearly independent.

Thus, we can solve only for two of u_1, u_2, u_3 and u_4 in terms of the other two. Letting $u_1 = 1$, $u_2 = 0$ we find (from last of the two Eqns of 59)

$$u_3 = \frac{cp_z}{E_+ + mc^2} \quad \text{and} \quad u_4 = \frac{c(p_x + ip_y)}{E_+ + mc^2} \tag{62a}$$

For $u_1 = 0$ and $u_2 = 1$, we find

$$u_3 = \frac{c(p_z - ip_y)}{E_+ + mc^2} \quad \text{and} \quad u_4 = \frac{-cp_z}{E_+ + mc^2} \tag{62b}$$

Similarly, for $E = E_-$, we find

$$u_3 = 1; u_4 = 0; u_1 = \frac{cp_z}{E_- - mc^2} \text{ and } u_2 = \frac{c(p_z + ip_y)}{E_- - mc^2} \qquad (63a)$$

and $\qquad u_3 = 0; u_4 = 1; u_1 = \dfrac{c(p_x - ip_y)}{E_- - mc^2} \text{ and } u_2 = \dfrac{-cp_z}{E_- - mc^2} \qquad (63b)$

The four linearly independent solutions are therefore given by

$$u_1(\vec{p}) = \begin{pmatrix} 1 \\ 0 \\ \dfrac{cp_z}{E_+ + mc^2} \\ \dfrac{c(p_x + ip_y)}{E_+ + mc^2} \end{pmatrix} \quad ; \quad u_2(\vec{p}) = \begin{pmatrix} 0 \\ 1 \\ \dfrac{c(p_x - ip_y)}{E_+ + mc^2} \\ \dfrac{-cp_z}{E_+ + mc^2} \end{pmatrix}$$

$$(u_3 =) v_1(\vec{p}) = \begin{pmatrix} \dfrac{cp_z}{E_- - mc^2} \\ \dfrac{c(p_x + ip_y)}{E_- - mc^2} \\ 1 \\ 0 \end{pmatrix} \quad ; \quad (u_4 =) v_2(\vec{p}) = \begin{pmatrix} \dfrac{c(p_x - ip_y)}{E_- - mc^2} \\ \dfrac{-cp_z}{E_- - mc^2} \\ 0 \\ 1 \end{pmatrix} \qquad (64)$$

u_1, u_2, v_1, v_2 are known as the *Dirac spinors*. $u_i(\vec{p})$, $i = 1, 2$, are the positive energy spinors which describe respectively the particles with spin up and spin down. $v_i(\vec{p})$, $I = 1,2$ are negative energy spinors with spin up and spin down. Solutions of the Dirac equation for a free particle is given from Eqn. (57) as the Dirac spinors multiplied by $\exp i(\vec{k} \cdot \vec{r} - \omega t)$. In the non relativistic limit ($v \ll c$) and $c|\vec{p}| \ll mc^2$, so $E_+ = E_- \cong mc^2$. In the limiting case of particle at rest ($\vec{p} = 0$), the four spinors having the simple form:

Energy $E = +mc^2, +mc^2, -mc^2, -mc^2$

Spinor: $\quad u_1 = \begin{pmatrix} 1 \\ 0 \\ 0 \\ 0 \end{pmatrix}, u_2 = \begin{pmatrix} 0 \\ 1 \\ 0 \\ 0 \end{pmatrix}, v_1 = \begin{pmatrix} 0 \\ 0 \\ 1 \\ 0 \end{pmatrix}, v_2 = \begin{pmatrix} 0 \\ 0 \\ 0 \\ 1 \end{pmatrix} \qquad (65)$

Each of the four solutions in (64) can be normalized in the sense that $\psi^* \psi = 1$ by multiplying by N e.g. for E_+

$\therefore \qquad\qquad\qquad u_1^* u_1 + u_2^* u_2 + u_3^* u_3 + u_4^* u_4 = 1$

or

$$N^2 \left[1 + 0 + \frac{c^2 p_z^2}{\left(E_+ + mc^2\right)^2} + \frac{c^2 \left(p_x^2 + p_y^2\right)}{\left(E_+ + mc^2\right)} \right] = 1$$

or

$$N^2 = \frac{1}{1 + \dfrac{c^2 \vec{p}^2}{\left(E_+ + mc^2\right)^2}} \tag{66}$$

If we define a new spin matrix

$$\sigma_z' = \begin{pmatrix} \sigma_z & 0 \\ 0 & \sigma_z \end{pmatrix} = \begin{pmatrix} 1 & 0 & 0 & 0 \\ 0 & -1 & 0 & 0 \\ 0 & 0 & 1 & 0 \\ 0 & 0 & 0 & -1 \end{pmatrix} \tag{67}$$

we find $\sigma_z' u_1 = u_1$, $\sigma_z' v_1 = v_1$, so the spinors u_1 and v_1 describe states of electron with z-components s_1 and s_3 of spin in the positive direction (eigenvalue + 1). The u_2 and v_2 ($\sigma_z' u_2 = -u_2$; $\sigma_z' v_2 = -v_2$) define s_2 and s_4 with negative spin (eigenvalue −1). For $E = +mc^2$ the most general solution is obtained by combining u_1 and v_1 solutions linearly so that spin does not have a definite value along z-axis.

21.9 Significance of Negative Energy States Dirac's Hole Theory)

Dirac's relativistic equation admits of solutions for which a particle has negative energy and negative rest mass. These solutions correspond to the negative square root of the classical energy equation. We have found that there are solutions with

$E > 0$, viz. $E_+ = +\sqrt{\left(mc^2\right)^2 + c^2 p^2}$ and $E_{min} = mc^2$ and go upto infinity.

Again for negative energy states

$$E_- = -\sqrt{\left(mc^2\right)^2 + c^2 p^2} \text{ '}$$

Negative energy E_- starts from $-mc^2$ and goes to $-\infty$. Figure 21.1a illustrates Dirac's concept of positive and negative energy states.

The existence of negative energy states creates problem in the following manner.

We know from Einstein's A and B coefficients that a charged particle can emit radiation by two ways: (*i*) spontaneous emission or (*ii*) induced emission. In spontaneous emission, there could be emission without the presence of any external field. An electron by spontaneous emission can jump from a state of finite positive energy to a state of energy $E \to -\infty$, e.g. an electron which is just bound to an atom, can emit radiation of energy $2mc^2$ and jump down to the state

$E = -mc^2$. Once in any of the negative energy states, it could go deep down to states with $E \to -\infty$. Thus the atom might have been completely unstable. To get rid of this difficulty, Dirac postulated that all negative energy states are ordinarily occupied by electrons and this 'sea' of negative energy electrons called the *'ground state'* or *vacuum state* consisting of states with energy $-mc^2 > E > -\infty$ is completely full. In this case, the Pauli exclusion principle prevents transitions into such occupied states. The ground state, then consists of an infinite density of negative energy electrons. This sea of negative energy electrons is assumed to have no physically observable e.g. electromagnetic or gravitational effects.

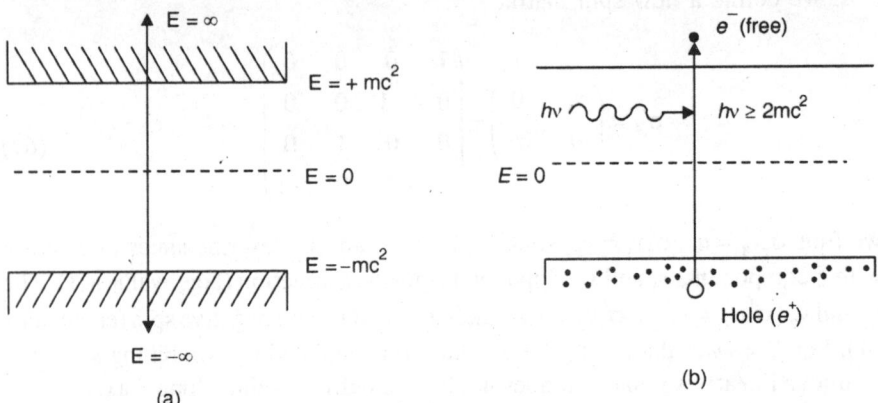

Fig. 21.1: (*a*) Dirac's positive and negative energy states, (*b*) pair production

The states with positive energy ($mc^2 < E < \infty$) represent all excited states in accordance to quantum theory. So, one or two (or more) electrons may be lifted from the sea of negative energy electrons by providing enough external energy and transferred to the positive energy states. For example, a photon of energy $\geq 2mc^2$, on interaction with negative sea, lifts an electron (having energy $E = -mc^2$) to a vacant positive energy state ($E = mc^2$). By this process, the Dirac vacuum becomes Dirac vacuum (with $E < 0$) minus one negative energy electron = Dirac vacuum + one positive energy electron.

This results in creation of *a hole* in Dirac vacuum, and also an observable electron with positive energy. Since a neutral photon has created a hole plus a positive energy electron, so the hole must have a positive charge (as opposed to the negative charge of electron).

The whole process therefore may be described as the disappearance of the quantum of energy supplied with a creation of a pair of particles: a positive energy electron and another particle differing from the electron only in the sign of its charge. No such particle was known at the time Dirac's theory was formulated, but it was discovered soon afterwards in cosmic rays and named *positron*. Since photon applied was having energy $2mc^2$ and it produced an electron of (mass = m) and energy mc^2 so positron has (mass = m) and energy mc^2. Thus hole with energy $-|E|$ and momentum $-\vec{p}$ is to be interpreted as a positron of energy $|E|$ and momentum $+\vec{p}$. Symbolically

$$e^- (E < 0) + \gamma \rightarrow e^- (E > 0)$$

$$\gamma \rightarrow e^- (E > 0) + e^+ (E > 0)$$

$$\left(2\gamma \Leftrightarrow 2mc^2 \right)$$

It is called *pair production* (The inverse process is called *pair annihilation*). Positron is an antiparticle of electron. Dirac's theory thus predicts the existence of antiparticles. In nature, now it is known that for every particle, there is an antiparticle and in particular, the existence of electrons implies the existence of positrons.

21.10 Spin of the Electron

We shall now see that Dirac's equation automatically endows the electron with the spin ½, ascribed earlier to a hypothetically spinning motion of the electron. We already know that the physical difference between two positive or two negative energy solutions can be clearly seen by defining three generalized Pauli spin matrices

$$\sigma'_k = \begin{pmatrix} \sigma_k & 0 \\ 0 & \sigma_k \end{pmatrix}; \quad (k = x, y, z) \tag{68}$$

where σ_k are Pauli spin matrices. In order to express the Hamiltonian

$$H = c(\vec{\alpha} \cdot \vec{p}) + \beta mc^2$$

in terms of σ'_k, we need to relate σ'_k with α_k. For this we define 4 × 4 matrices ρ_i (i = 1, 2, 3) by

$$\rho_1 = \begin{pmatrix} 0 & I \\ I & 0 \end{pmatrix}; \rho_2 = i \begin{pmatrix} 0 & -I \\ I & 0 \end{pmatrix}; \rho_3 = \beta \tag{69}$$

so that ρ_1 commutes with σ'_k and

$$\rho_1 \sigma'_k = \begin{pmatrix} 0 & I \\ I & 0 \end{pmatrix} \begin{pmatrix} \sigma_k & 0 \\ 0 & \sigma_k \end{pmatrix} = \begin{pmatrix} 0 & \sigma_k \\ \sigma_k & 0 \end{pmatrix} = \alpha_k = \sigma'_k \rho_1 \tag{70}$$

or $\qquad \alpha_x = \rho_1 \sigma'_x, \alpha_y = \rho_1 \sigma'_y, \alpha_z = \rho_1 \sigma'_z, \rho_i^2 = 1 \tag{71}$

For a Dirac particle in a central potential $V(r)$, Hamiltonian

$$H = c\left(\alpha_x p_x + \alpha_y p_y + \alpha_z p_z \right) + \beta mc^2 + V(r)$$

$$= c\rho_1 \left(\alpha'_x p_x + \alpha'_y p_y + \alpha'_z p_z \right) + \beta mc^2 + V(r) \tag{72}$$

Classically, since the potential is central, we may expect that the orbital angular momentum $\vec{L} = \vec{r} \times \vec{p}$ (or in particular the x component say L_x) is a constant of motion. To see whether this is so also in the Dirac theory, we calculate the time rate of change of L_x in the Heisenberg representation

$$\frac{dL_x}{dt} = \frac{1}{i\hbar}\left(L_x H - HL_x \right) = \frac{1}{i\hbar}\left[L_x, H \right] \tag{73}$$

where H is given by Eqn. (72), and

$L_x = yp_z - zp_y$. So

$$\frac{dL_x}{dt} = \frac{1}{i\hbar}\left[\left(yp_z - zp_y\right),\left(c\vec{\alpha}\cdot\vec{p} + \beta mc^2 + V(r)\right)\right] \tag{74}$$

To evaluate Eqn. (74), we have to calculate $\left[L_x,(c\vec{\alpha}\cdot\vec{p})\right]$, $\left[L_x,\beta mc^2\right]$ and $\left[L_x,V(r)\right]$. We evaluate each of these remembering that α_k, β commute with x, y, z and $\vec{p}$ and that $\left[x,p_x\right] = i\hbar$, $\left[x,p_y\right] = 0 = \left[p_x,p_y\right]$. Now,

$$\left[L_x,(c\vec{\alpha}\cdot\vec{p})\right] = c\left\{L_x(\vec{\alpha}\cdot\vec{p}) - (\vec{\alpha}\cdot\vec{p})L_x\right\}$$

$$= c\left\{(yp_z - zp_y)(\alpha_x p_x + \alpha_y p_y + \alpha_z p_z)\right.$$

$$\left. -(\alpha_x p_x + \alpha_y p_y + \alpha_z p_z)(yp_z - zp_y)\right\}$$

$$= c\left\{\alpha_y p_z(yp_y - p_y y) - \alpha_z p_y(zp_z - p_z z)\right\}$$

$$= ci\hbar\left(\alpha_y p_z - \alpha_z p_y\right)$$

$$= i\hbar c\rho_1\left(\sigma'_y p_z - \sigma'_z p_y\right) \neq 0 \tag{75}$$

$$\left[L_x,\beta mc^2\right] = mc^2\left\{(yp_z - zp_z)\beta - \beta(yp_z - zp_y)\right\} = 0 \tag{76}$$

because β commutes with y, z, p_y, p_z.

Also $\qquad\qquad \left[L_x,V(r)\right] = 0 \tag{77}$.

because L_x is a function of θ and ϕ only $\left(L_x = i\hbar\left(\sin\phi\frac{\partial}{\partial\theta} + \cot\theta\cos\phi\frac{\partial}{\partial\phi}\right)\right)$ and central potential $V(r)$ does not depend on θ and ϕ. From Eqns (73) and (75), we have

$$\dot{L}_x = c\rho_1\left(\sigma'_y p_z - \sigma'_z p_y\right) \neq 0 \tag{78}$$

Hence the orbital angular momentum is not a constant of motion in Dirac theory. However, we also note that σ'_x commutes with every quantity in Eqn. (72) but not with σ'_y, σ'_z. So, we have

$$\dot{\sigma}'_x = \frac{1}{i\hbar}\left(\sigma'_x H - H\sigma'_x\right)$$

$$= \frac{c\rho_1}{i\hbar}\left\{\sigma'_x\left(\sigma'_y p_y + \sigma'_z p_z\right) - \left(\sigma'_y p_y + \sigma'_z p_z\right)\sigma'_x\right\}$$

$$= \frac{c\rho_1}{i\hbar}\left\{\left(\sigma'_x\sigma'_y - \sigma'_y\sigma'_x\right)p_y + \left(\sigma'_x\sigma'_z - \sigma'_z\sigma'_x\right)p_z\right\}$$

$$= \frac{c\rho_1}{i\hbar}\left\{\left(2i\sigma'_z\right)p_y + \left(-2i\sigma'_y\right)p_z\right\} \tag{79a}$$

because, $\qquad \sigma'_x\sigma'_y = \begin{pmatrix} \sigma_x & 0 \\ 0 & \sigma_x \end{pmatrix}\begin{pmatrix} \sigma_y & 0 \\ 0 & \sigma_y \end{pmatrix} = \begin{pmatrix} \sigma_x\sigma_y & 0 \\ 0 & \sigma_x\sigma_y \end{pmatrix}$

$$= \begin{pmatrix} i\sigma_z & 0 \\ 0 & i\sigma_z \end{pmatrix} = i\sigma_z' = -\sigma_y'\sigma_x' \qquad (79b)$$

or

$$\dot{\sigma}_x' = -\frac{2c\rho_1}{\hbar}\left(\sigma_y'p_z - \sigma_z'p_y\right) \qquad (80)$$

From Eqns (78) and (80), we notice

$$\left(\dot{L}_x + \frac{1}{2}\hbar\dot{\sigma}_x'\right) = 0$$

i.e. $L_x + \left(\dfrac{1}{2}\right)\hbar\sigma_x'$ is a constant of motion. We refer to $\vec{L} + \left(\dfrac{1}{2}\right)\hbar\vec{\sigma}'$ as the total angular momentum. It is conserved. It is the sum of orbital angular momentum $\vec{L}$ and spin angular momentum

$$\vec{S} = \frac{1}{2}\hbar\vec{\sigma}'$$

The eigenvalues of σ_k' are ± 1. Hence Dirac's relativistic equation automatically endows the electron with the spin (½) which was previously ascribed to a hypothetical spinning motion of the electron in the non relativistic theory.

21.11 Magnetic Moment of the Electron and Electromagnetic Potentials

Dirac's equation is

$$\left[E - c\vec{\alpha}\cdot\vec{p} - \beta mc^2\right]\psi = 0$$

In presence of an electromagnetic field this becomes

$$\left[(E - e\phi) - \vec{\alpha}\cdot(c\vec{p} - e\vec{A}) - \beta mc^2\right]\psi = 0 \qquad (81)$$

Multiplying on the left by $\left[(E - e\phi) + \vec{\alpha}\cdot(c\vec{p} - e\vec{A}) + \beta mc^2\right]$, we get

$$\left\{(E - e\phi)^2 - \left[\vec{\alpha}\cdot(c\vec{p} - e\vec{A})\right]^2 - m^2c^4 - (E - e\phi)\vec{\alpha}\cdot(c\vec{p} - e\vec{A})\right.$$
$$\left. + \vec{\alpha}\cdot(c\vec{p} - e\vec{A})(E - e\phi)\right\}\psi = 0 \qquad (82)$$

(because $\beta^2 = 1$ and $\alpha\beta + \beta\alpha = 1$)

Now, we have the following relation

$$(\vec{\alpha}\cdot\vec{B})(\vec{\alpha}\cdot\vec{C}) = \vec{B}\cdot\vec{C} + i\vec{\sigma}'\cdot(\vec{B}\times\vec{C}) \qquad (83)$$

where $\sigma' = \begin{pmatrix} \sigma & 0 \\ 0 & \sigma \end{pmatrix}$. $\vec{B}$ and $\vec{C}$ commute with α but not necessarily with each other. Replacing $\vec{B}$ and $\vec{C}$ both by $(c\vec{p} - e\vec{A})$ in Eqn. (83), we get

$$\left[\vec{\alpha}\cdot(c\vec{p} - e\vec{A})\right]^2 = (c\vec{p} - e\vec{A})^2 + i\sigma'\cdot(c\vec{p} - e\vec{A})\times(c\vec{p} - e\vec{A}) \qquad (84)$$

Now, in Eqn. (84),

$$\left(c\vec{p} - e\vec{A}\right) \times \left(c\vec{p} - e\vec{A}\right) = -ec(\vec{p} \times \vec{A} + \vec{A} \times \vec{p}) \tag{85}$$

Further,

$$(\vec{A} \times \vec{p} + \vec{p} \times \vec{A})_x \psi = \left[(\vec{A} \times \vec{p})_x + (\vec{p} \times \vec{A})_x \right] \psi$$

$$= \left[\left(A_y p_z - A_z p_y\right) + \left(p_y A_z - p_z A_y\right) \right] \psi$$

$$= -i\hbar \left[A_y \frac{\partial \psi}{\partial z} - A_z \frac{\partial \psi}{\partial y} + \frac{\partial}{\partial y}\left(A_z \psi\right) - \frac{\partial}{\partial z}\left(A_y \psi\right) \right]$$

$$= -i\hbar \left[A_y \frac{\partial \psi}{\partial z} - A_z \frac{\partial \psi}{\partial y} + A_z \frac{\partial \psi}{\partial y} + \frac{\partial A_z}{\partial y} \psi - \frac{\partial A_y}{\partial z} \psi - A_y \frac{\partial \psi}{\partial z} \right]$$

$$= -i\hbar \left[\frac{\partial A_z}{\partial y} - \frac{\partial A_y}{\partial z} \right] \psi$$

Hence, $\vec{A} \times \vec{p} + \vec{p} \times \vec{A} = -i\hbar (\nabla \times \vec{A})$

But $(\nabla \times \vec{A}) = \vec{H}$ (the magnetic field).

Therefore,

$$\left(c\vec{p} - e\vec{A}\right) \times \left(c\vec{p} - e\vec{A}\right) = +iec\hbar \nabla \times \vec{A} = iec\hbar\vec{H} \tag{86}$$

Consequently, from Eqns (84), (85), (86)

$$\left[\vec{\alpha} \cdot \left(c\vec{p} - e\vec{A}\right) \right]^2 = \left(c\vec{p} - e\vec{A}\right)^2 + i\vec{\sigma}' \cdot (iec\hbar\vec{H})$$

$$= \left(c\vec{p} - e\vec{A}\right)^2 - ec\hbar(\vec{\sigma}' \cdot \vec{H}) \tag{87}$$

The last two operators in Eqn. (82) are simplified as follows:

$$-(E - e\phi)\vec{\alpha} \cdot (c\vec{p} - e\vec{A}) + \vec{\alpha} \cdot (c\vec{p} - e\vec{A})(E - e\phi)$$

$$= e\vec{\alpha} \cdot (\vec{E}\vec{A} - \vec{A}E) + ec\vec{\alpha} \cdot (\phi\vec{p} - \vec{p}\phi)$$

(Terms like $E\vec{\alpha} \cdot c\vec{p}$ and $e^2\phi A$ cancel)

$$= e\vec{\alpha} \cdot i\hbar \frac{\partial \vec{A}}{\partial t} + ec(-i\hbar)\vec{\alpha} \cdot \left[\phi\nabla - \nabla\phi - \phi\nabla \right]$$

$$= -iec\hbar\vec{\alpha} \cdot \left[-\frac{1}{c}\frac{\partial \vec{A}}{\partial t} - \nabla\phi \right] \tag{88}$$

$$= -iec\hbar\vec{\alpha} \cdot \vec{\varepsilon}$$

where $\vec{\varepsilon}$ is the electric field vector.

(*E* is energy).

So, Eqn. (82) becomes

$$\left\{ (E - e\phi)^2 - m^2c^4 - (c\vec{p} - e\vec{A})^2 + ec\hbar\vec{\sigma}' \cdot \vec{H} - iec\hbar\vec{\alpha} \cdot \vec{E} \right\} \psi = 0 \tag{89}$$

Now, let $E = E' + mc^2$. In the non relativistic limit, energy $E' \ll mc^2$ and $e\phi \ll mc^2$.

$$(E - e\phi)^2 = (E' + mc^2 - e\phi)^2 = m^2c^4\left(1 + \frac{E' - e\phi}{mc^2}\right)^2$$

$$\cong m^2c^4\left(1 + 2 \times \frac{E' - e\phi}{mc^2}\right)$$

or $\qquad (E - e\phi)^2 - mc^2 \cong 2mc^2(E' - e\phi)$

Therefore, from Eqn. (89)

$$\left\{2mc^2(E' - e\phi) - (c\vec{p} - e\vec{A})^2 + ec\hbar\vec{\sigma}' \cdot \vec{H} - iec\hbar\vec{\alpha} \cdot \vec{\epsilon}\right\}\psi = 0$$

or $\qquad E'\psi = \left[\frac{1}{2m}\left(\vec{p} - \frac{e}{c}\vec{A}\right)^2 + e\phi - \frac{e\hbar}{2mc}(\vec{\sigma}' \cdot \vec{H}) + \frac{ie\hbar}{2mc}\vec{\alpha} \cdot \vec{\epsilon}\right]\psi$ $\qquad$ (90)

Schrödinger's equation is

$$(E - e\phi)\psi = \left[\frac{1}{2m}\left(\vec{p} - \frac{e\vec{A}}{c}\right)^2\right]\psi \qquad (91)$$

Comparison of Eqns. (90) and (91) shows that Eqn. (90) involves two additional terms containing $\vec{H}$ and $\vec{\epsilon}$. Term containing $\vec{\epsilon}$ can be neglected in the non relativistic limit (because it is of the order of v^2/c^2).

The extra term $-\vec{\mu} \cdot \vec{H} = -\mu_0(\vec{\sigma}' \cdot \vec{H})$ where $\mu_0 = \dfrac{e\hbar}{2mc}$ is the Bohr magneton can be interpreted as energy of interaction of magnetic moment of the electron (corresponding to operator $\mu_0\vec{\sigma}'$) with the magnetic field. ($\vec{\mu}$) or ($\mu_0\vec{\sigma}'$) is called the spin magnetic moment.

Thus a Dirac particle of charge e possesses an intrinsic (spin) magnetic dipole moment

$$\vec{\mu}_s = -\mu_0\vec{\sigma}' = -\frac{e\hbar}{2mc}\vec{\sigma}'$$

$$= -\frac{e}{mc}\left(\frac{1}{2}\hbar\vec{\sigma}'\right) = -\frac{e}{mc}\vec{S} \qquad (92a)$$

Here $\vec{S}$ is the intrinsic spin angular momentum operator for the particle. This prediction was one of the first successes of the Dirac theory which endows the electron with spin ½ and magnetic moment in a natural way.

(If m is replaced by $m_0/(1 - v^2/c^2)^{1/2}$, the Dirac magnetic moment vanishes for $v \to c$. Thus Dirac magnetic moment appears only in the non relativistic limit and depends on the charge e).

A small correction to the above result is, however, necessary if we consider the interaction of electron with its self field (see articles 21.15 and 21.16). From quantum electrodynamics, it may be shown that the dipole moment is

$$-\left\{1+\frac{e^2}{\hbar c}+\left(\frac{e^2}{\hbar c}\right)^2\right\}\frac{e}{mc}\vec{S}\cong-1.00116\frac{e}{mc}\vec{S} \tag{92b}$$

where $e^2/(\hbar c)\equiv\alpha$ is the Sommerfeld fine structure constant.

21.12 Spin Orbit Energy as a Consequence of the Dirac Equation

Dirac equation in presence of a central potential $V(r)$ is

$$E\psi=\left[c\vec{\alpha}\cdot\vec{p}+\beta mc^2+V(r)\right]\psi \tag{93}$$

In terms of the two component spinors ψ_A and ψ_B. This equation can be written as

$$E\begin{pmatrix}\psi_A\\\psi_B\end{pmatrix}=c\begin{pmatrix}0&\vec{\sigma}\cdot\vec{p}\\\vec{\sigma}\cdot\vec{p}&0\end{pmatrix}\begin{pmatrix}\psi_A\\\psi_B\end{pmatrix}+\begin{pmatrix}mc^2&0\\0&-mc^2\end{pmatrix}\begin{pmatrix}\psi_A\\\psi_B\end{pmatrix}+V(r)\begin{pmatrix}\psi_A\\\psi_B\end{pmatrix} \tag{94}$$

In the non relativistic limit, we write $E=E'+mc^2$ where $mc^2\gg\dfrac{\beta^2}{2m}$ and $V(r)$. Equation (94) can now be written as two equations

$$(E'-V)\psi_A-c(\vec{\sigma}\cdot\vec{p})\psi_B=0 \tag{95a}$$

and

$$(E'+2mc^2-V)\psi_B-c(\vec{\sigma}\cdot\vec{p})\psi_A=0 \tag{95b}$$

From Eqn. (95b), we have

$$\psi_B=(E'+2mc^2-V)^{-1}(\vec{\sigma}\cdot\vec{p})\psi_A$$

Substituting this into Eqn. (95a), we get

$$E'\psi_A=\frac{1}{2m}(\vec{\sigma}\cdot\vec{p})\left(1+\frac{E'-V}{2mc^2}\right)^{-1}(\vec{\sigma}\cdot\vec{p})\psi_A+V\psi_A \tag{96}$$

Since $E'\ll mc^2$ and $V\ll mc^2$, in the non relativistic limit, an expanstion in powers of $(E'-V)/(2mc^2)$ upto first order only is a good approximation. Then equation (96) reduces to

$$E'\psi_A=\frac{1}{2m}\left[(\vec{\sigma}\cdot\vec{p})(\vec{\sigma}\cdot\vec{p})\left(1-\frac{E'}{2mc^2}\right)+\frac{1}{2mc^2}(\vec{\sigma}\cdot\vec{p})V(r)(\vec{\sigma}\cdot\vec{p})\right]\psi_A+V\psi_A \tag{97}$$

From $(\vec{\sigma}\cdot\vec{A})(\vec{\sigma}\cdot\vec{B})=(\vec{A}\cdot\vec{B})+i\vec{\sigma}\cdot(\vec{A}\times\vec{B})$, we have

$$(\vec{\sigma}\cdot\vec{p})(\vec{\sigma}\cdot\vec{p})=(\vec{p}\cdot\vec{p})+i\vec{\sigma}\cdot(\vec{p}\times\vec{p})=p^2$$

Also,

$$pV=Vp-i\hbar\nabla V$$

$$\left(\because (pV)\psi = -i\hbar\nabla(V\psi) = V(-i\hbar\nabla\psi) + \psi(-i\hbar\nabla V) = \left[V(-i\hbar\nabla) - i\hbar\nabla V\right]\psi\right)$$

$$\therefore \quad (\vec{\sigma}\cdot\vec{p})V(\vec{\sigma}\cdot\vec{p}) = V(\vec{\sigma}\cdot\vec{p})(\vec{\sigma}\cdot\vec{p}) - i\hbar(\vec{\sigma}\cdot\nabla V)(\vec{\sigma}\cdot\vec{p})$$

$$= Vp^2 - i\hbar(\nabla V)\cdot\vec{p} + \vec{\sigma}\cdot\left[(\nabla V)\times\vec{p}\right]$$

Using these relations, Eqn. (97) can be written as

$$E'\psi_A = \left[\left(1 - \frac{E'-V}{2mc^2}\right)\frac{p^2}{2m} + V\right]\psi_A - \frac{\hbar^2}{4m^2c^2}(\nabla V)\cdot(\nabla\psi_A)$$

$$+ \frac{\hbar}{4m^2c^2}\vec{\sigma}\cdot\left[(\nabla V)\times\vec{p}\psi_A\right] \tag{98}$$

Since potential $V(r)$ is having spherical symmetry, we have

$$\nabla V = \frac{\vec{r}}{r}\frac{dV}{dr} \quad \text{and} \quad (\nabla V)\cdot\nabla = \frac{dV}{dr}\frac{\partial}{\partial r}$$

Therefore, we may write Eqn. (98) as

$$E'\psi_A = \left(\frac{p^2}{2m} - \frac{p^4}{8m^3c^2} + V(r) - \frac{\hbar^2}{4m^2c^2}\frac{dV}{dr}\frac{\partial}{\partial r} + \frac{1}{2m^2c^2}\frac{1}{r}\frac{dV}{dr}\vec{S}\cdot\vec{L}\right)\psi_A \tag{99}$$

$$\left\{\text{because } E' = \frac{p^2}{2m} + V(r) \Rightarrow (E'-V) = \frac{p^2}{2m}, \ \vec{S} = \frac{1}{2}\hbar\vec{\sigma} \text{ and } \vec{L} = \vec{r}\times\vec{p}\right\}$$

The first and the third term on the right hand side of Eqn. (99) give the non relativistic Schrödinger equation. The second term has the form of the classical relativistic mass correction, which can be obtained by expanding the square root of $E^2 = c^2p^2 + m^2c^4$ as:

$$E = \left(c^2p^2 + m^2c^4\right)^{1/2} = mc^2\left(1 + \frac{p^2}{m^2c^2}\right)^{1/2}$$

$$= mc^2\left(1 + \frac{p^2}{2m^2c^2} - \frac{p^4}{8m^4c^4}\right) = mc^2 + \frac{p^2}{2m} - \frac{p^4}{8m^3c^2} \tag{100}$$

The fourth term $-\dfrac{\hbar^2}{4m^2c^2}\dfrac{dV}{dr}\dfrac{\partial}{\partial r}$ is known as the *Darwin term* and is a similar relativistic correction to the potential energy which cannot be understood in a purely classical way. The last term gives rise to *spin orbit coupling*. It is interpreted as follows:

A relativistic transformation of the Coulomb field into the rest frame of the electron gives rise to a small magnetic field, and this term describes the energy of the electron's magnetic moment in the field. The situation is complicated by the fact that the electron rest frame precesses with a certain angular velocity relative to the centre of mass frame and this halves the value of the term:

$\frac{1}{2} \times \left(\frac{1}{m^2 c^2} \frac{1}{r} \frac{dV}{dr} \vec{S} \cdot \vec{L} \right)$. This spin orbit energy appears to come as a direct consequence of the Dirac equation.

21.13 Covariance of Dirac Equation

Covariance of Dirac equation means form invariance of the Dirac equation for two different frames connected by Lorentz transformations.

Suppose there are two observers A and B in two different inertial systems describing the same physical event with respective space-time coordinates. Let the wave functions describing an event be $\psi'(x')$ for observer B and $\psi(x)$ for observer A. Then if we solve Dirac equation in these two Lorentz frames, the solutions must describe the same physical result. In other words, if $\psi(x)$ (of observer A) is given then, there must be an explicit rule to enable observer B to calculate $\psi'(x')$ from $\psi(x)$ in such a way that for transformation of components of ψ, the form of Dirac equation is invariant.

To prove the covariance of Dirac equation, we first try to put Dirac equation in a suitable form. For this we introduce a set of γ matrices derived from the α_i's and β. We write the Dirac equation as

$$i\hbar \frac{\partial \psi}{\partial (ct)} = (mc\beta + (-i\hbar)\vec{\alpha} \cdot \nabla)\psi \qquad (101)$$

or

$$i \frac{\partial}{\partial (ct)} \psi = \left[\left(\frac{mc}{\hbar} \right) \beta - i\vec{\alpha} \cdot \frac{\partial}{\partial x} \right] \psi \qquad (102)$$

Multiply both sides from the left by β, then

$$-\beta \frac{\partial}{\partial (ict)} \psi = \left\{ M - i\beta \alpha_k \frac{\partial}{\partial x_k} \right\} \psi \qquad (103)$$

where $\beta^2 = 1$ and $\frac{mc}{\hbar} \equiv M$. We now use $\gamma_k = -i\beta \alpha_k$ and $\beta = \gamma_4$, so that

$$\left(\gamma_k \frac{\partial}{\partial x_k} + \gamma_4 \frac{\partial}{\partial (ict)} + M \right) \psi = 0 \qquad (104)$$

Consider the coordinate system

$$x_\mu = \left(x_k, x_4 \right)$$

with $k = 1,2,3$ and $x_4 = ict = ix_0$. The Dirac equation

$$\left(\gamma_k \frac{\partial}{\partial x_k} + \gamma_4 \frac{\partial}{\partial x_4} + M \right) \psi = 0$$

becomes

$$\left[\gamma_\mu \frac{\partial}{\partial x_\mu} + M \right] \psi = 0 \qquad (105)$$

This is the desired form of Dirac equation to prove the covariance. It is to be noted that

$$\gamma_4 = \beta \text{ so } \gamma_4^2 = \beta^2 = 1 \qquad (106)$$

and

$$\gamma_k = -i\beta\alpha_k \ (k = 1,2,3) \qquad (107)$$

yields

$$\gamma_k^2 = (-i)^2 \beta\alpha_k \beta\alpha_k = \alpha_k^2 = 1, \ k = 1,2,3 \qquad (108)$$

Further

$$\gamma_4\gamma_1 + \gamma_1\gamma_4 = -i(\beta\beta\alpha_1 + \beta\alpha_1\beta) = -i\beta(\beta\alpha_1 + \alpha_1\beta) = 0$$

Similarly,

$$\gamma_k\gamma_i + \gamma_i\gamma_k = 0 \ (i \neq k) \qquad (109)$$

And γ_μ's satisfy

$$\gamma_\mu\gamma_\nu + \gamma_\nu\gamma_\mu = 2\delta\mu\nu \qquad (110)$$

and

$$\gamma_\mu^2 = 1 \text{ so } \gamma_\mu = \gamma_\mu^{-1} \qquad (111)$$

and

$$\left(\gamma_k\right)^\dagger = \left(-i\beta\alpha_k\right)^\dagger = +i\alpha_k^\dagger\beta^\dagger = -i\beta\alpha_k = \gamma_k$$

also

$$\gamma_4^\dagger = \beta^\dagger = \beta = \gamma_4 \text{ so that}$$

$$\gamma_\mu^\dagger = \gamma_\mu \qquad (112)$$

i.e., γ_μ are Hermitian.

We now proceed to prove the covariance of Dirac equation written in form given in Eqn. 105. Let

$$x_\mu = \left(x_1, x_2, x_3, x_4 = ict\right)$$
$$x_\mu' = \left(x_1', x_2', x_3', x_4' = ict'\right) \qquad (113)$$

represent two coordinate frames. Then under the Lorentz transformation, the four-vector x_μ transforms to x_μ' so that

$$\sum_\mu x_\mu^2 = \sum_\mu x_\mu'^2 \qquad (114)$$

(velocity of light is constant in all frames) so, we get

$$\vec{x}^2 - c^2t^2 = \vec{x}'^2 - c^2t'^2 \qquad (115)$$

Let the relation between x_μ and x_μ' be given by

$$x_\mu' = a_{\mu\nu}x_\nu \qquad (116)$$

In this equation and further below, it is implicit that repeated indices are summed. Since Lorentz transformation requires $x_\mu'^2 = x_\mu^2$, so we have from Eqn. (116)

$$a_{\mu\nu}x_\nu a_{\mu\rho}x_\rho = x_\nu^2 \text{ or } \left(a_{\mu\nu}a_{\mu\rho}\right)x_\nu x_\rho = x_\nu^2 \qquad (117)$$

or

$$\sum_\mu a_{\mu\nu}a_{\mu\rho} = \delta_{\nu\rho} \qquad (118)$$

so that
$$\sum_{\rho} \delta_{\nu\rho} x_\nu x_\rho = x_\nu^2 \qquad (119)$$

From $x_\mu' = a_{\mu\nu} x_\nu$, we get
$$a_{\mu\rho} x_\mu' = a_{\mu\rho} a_{\mu\nu} x_\nu = \delta_{\rho\nu} x_\nu = x_\rho$$

So,
$$x_\rho = a_{\mu\rho} x_\mu' \qquad (120)$$

Since $x_\rho^2 = x_\rho'^2$, therefore
$$x_\rho^2 = a_{\mu\rho} x_\mu' a_{\nu\rho} x_\nu' = a_{\mu\rho} a_{\nu\rho} x_\mu' x_\nu' = \delta_{\mu\nu} x_\mu' x_\nu' = x_\mu'^2 = x_\rho'^2$$

i.e.
$$\sum_{\rho} a_{\mu\rho} a_{\nu\rho} = \delta_{\mu\nu}$$

Thus, from $\qquad x_\rho = a_{\mu\rho} x_\mu'$, we get

$$\frac{\partial x_\rho}{\partial x_\mu'} = a_{\mu\rho} \qquad (121)$$

Consider a scalar function $V(x)$ which remains invariant under Lorentz transformation. Then,
$$\frac{\partial V}{\partial x_\mu'} = \frac{\partial V}{\partial x_\rho} \frac{\partial x_\rho}{\partial x_\mu'} = a_{\mu\rho} \frac{\partial V}{\partial x_\rho}$$

so that
$$\frac{\partial}{\partial x_\mu'} = a_{\mu\rho} \frac{\partial}{\partial x_\rho} \qquad (122)$$

Now consider
$$\left\{ \gamma_\mu \frac{\partial}{\partial x_\mu} + M \right\} \psi = 0$$

Let under the Lorentz transformation $x_\mu' = a_{\mu\nu} x_\nu$
the ψ transforms to ψ' when the relation between ψ and ψ' is
$$\psi' = S\psi \qquad (123)$$
where S is a numerical matrix which commutes with space and time coordinates. In the primed frame, the Dirac equation is

$$\gamma_\mu \frac{\partial}{\partial x_\mu'} \psi'(x') + M\psi'(x') = 0 \qquad (124)$$

Using Eqn. (123), we get
$$\gamma_\mu \frac{\partial}{\partial x_\mu'} S\psi + MS\psi = 0$$

i.e.
$$S^{-1}\gamma_\mu S \frac{\partial}{\partial x_\mu'} \psi + M\psi = 0$$

Now, using Eqn. (122)

$$\left(S^{-1}\gamma_\mu S \right) a_{\mu\nu} \frac{\partial}{\partial x_\nu} \psi + M\psi = 0 \qquad (125)$$

Now, for Eqn. (124) to become Dirac equation in the original unprimed frame, we demand

$$\left(S^{-1}\gamma_\mu S\right)a_{\mu\nu} = \gamma_\nu \qquad (126)$$

then, Eqn. (124) becomes

$$\gamma_\nu \frac{\partial}{\partial x_\nu}\psi + M\psi = 0 \qquad (127)$$

This proves the covariance of the Dirac equation.

21.14 Dirac's Gamma Matrices: Their Inter-relation and Four Vector Current

In terms of natural units, covariant form of Dirac equation is

$$\left(\gamma_\mu \partial_\mu + m\right)\psi = 0; \quad (\mu = 1, 2, 3, 4) \qquad (128)$$

Using the definition of the $\bar{\alpha}$ and β matrices, we can write

$$\gamma_k = -i\beta\alpha_k = -i\begin{pmatrix} I & 0 \\ 0 & -I \end{pmatrix}\begin{pmatrix} 0 & \sigma_k \\ \sigma_k & 0 \end{pmatrix} = \begin{pmatrix} 0 & -i\sigma_k \\ i\sigma_k & 0 \end{pmatrix}, \quad (k = 1,2,3) \qquad (129)$$

and $\gamma_4 = \beta = \begin{pmatrix} I & 0 \\ 0 & -I \end{pmatrix}$ where σ_k are the Pauli spin matrices. These satisfy the anticommutation rule:

$$\gamma_\mu \gamma_\nu + \gamma_\nu \gamma_\mu = 2\delta_{\mu\nu} \qquad (130)$$

for example

$$\gamma_1\gamma_2 + \gamma_2\gamma_1 = \begin{pmatrix} 0 & -i\sigma_1 \\ i\sigma_1 & 0 \end{pmatrix}\begin{pmatrix} 0 & -i\sigma_2 \\ i\sigma_2 & 0 \end{pmatrix} + \begin{pmatrix} 0 & -i\sigma_2 \\ i\sigma_2 & 0 \end{pmatrix}\begin{pmatrix} 0 & -i\sigma_1 \\ i\sigma_1 & 0 \end{pmatrix}$$

$$\begin{pmatrix} \sigma_1\sigma_2 + \sigma_2\sigma_1 & 0 \\ 0 & \sigma_1\sigma_2 + \sigma_2\sigma_1 \end{pmatrix} = 0$$

and

$$\gamma_4\gamma_4 + \gamma_4\gamma_4 = 2\gamma_4\gamma_4 = 2\beta^2 = 2$$

From Eqn. (130), we find

$$\gamma_1^2 = \gamma_2^2 = \gamma_3^2 = \gamma_4^2 = 1 \qquad (131)$$

Further, each of the γ_μ's is Hermitian and traceless

$$\gamma_\mu^\dagger = \gamma_\mu \text{ and } Tr\gamma_\mu = 0 \qquad (132)$$

The product of all the four gamma matrices is given a special symbol viz. γ_5, i.e.

$$\gamma_5 = \gamma_1\gamma_2\gamma_3\gamma_4 \qquad (133)$$

γ_5 is also Hermitian, since

$$\gamma_5^\dagger = \left(\gamma_1\gamma_2\gamma_3\gamma_4\right)^\dagger = \gamma_4^\dagger\gamma_3^\dagger\gamma_2^\dagger\gamma_1^\dagger = \gamma_4\gamma_3\gamma_2\gamma_1 = \gamma_1\gamma_2\gamma_3\gamma_4 = \gamma_5$$

γ_5 anticommutes with all the four γ_μ's

$$\gamma_5\gamma_\mu + \gamma_\mu\gamma_5 = 0 \quad (\mu = 1,2,3,4) \tag{134}$$

It can bè found that $\gamma_5^2 = I$. In particular, we may have

$$\gamma_5 = \begin{pmatrix} 0 & -I \\ -I & 0 \end{pmatrix} \tag{135}$$

The adjoint of Dirac's equation is obtained by taking Hermitian conjugate of Eqn. (128) as

$$\psi^\dagger\left(\gamma_k\partial_k - \gamma_4\partial_4 + m\right) = 0 \quad \left(\because \partial_4^\dagger = \frac{\partial}{\partial x_4^*} = -\frac{\partial}{\partial x_4} = -\partial_4\right)$$

Multiplying on the right of it by γ_4, we have

$$\psi^\dagger\left(\gamma_k\partial_k - \gamma_4\partial_4 + m\right)\gamma_4 = 0$$

or $\qquad\qquad \psi^\dagger\gamma_4\left(-\gamma_k\partial_k - \gamma_4\partial_4 + m\right) = 0 \qquad \left(\because \gamma_k\gamma_4 = -\gamma_4\gamma_k\right)$

Defining $\psi^\dagger\gamma_4 \equiv \bar{\psi}$, we get the adjoint of the Dirac equation

$$\bar{\psi}\left(-\gamma_\mu\bar{\partial}_\mu + m\right) = 0 \tag{136}$$

Multiplying Eqn. (128) from the left by $\bar{\psi}$ and (136) by ψ on the right and then subtracting, we get

$$\partial_\mu(\bar{\psi}\gamma_\mu\psi) = 0 \text{ or } \partial_\mu j_\mu = 0 \tag{137}$$

where $j_\mu = \bar{\psi}\gamma_\mu\psi$ is the four vector current.

21.15 An Electron in the Central Potential (The Hydrogen Atom)

We have

$$\left(\vec{\alpha}\cdot\vec{A}\right)\left(\vec{\alpha}\cdot\vec{B}\right) = \left(\vec{A}\cdot\vec{B}\right) + i\vec{\sigma}'\cdot\left(\vec{A}\times\vec{B}\right) \tag{138a}$$

where $\qquad\qquad\qquad \vec{\sigma}' = \begin{pmatrix} \sigma & 0 \\ 0 & \sigma \end{pmatrix}$

Taking $\vec{A} = \vec{r}$ and $\vec{B} = \vec{p}$, we get

$$\left(\vec{\alpha}\cdot\vec{r}\right)\left(\vec{\alpha}\cdot\vec{p}\right) = \left(\vec{r}\cdot\vec{p}\right) + i\vec{\sigma}'\cdot\vec{L} \tag{138b}$$

Now we define a new operator $\vec{k}$ by the relation

$$k\hbar = \beta\left(\vec{\sigma}'\cdot\vec{L} + \hbar\right)$$

or $\qquad\qquad\qquad (k - \beta)\hbar = \beta\vec{\sigma}'\cdot\vec{L} \tag{139}$

$$\Rightarrow \qquad \vec{\sigma}' \cdot \vec{L} = \frac{(k-\beta)}{\beta} \hbar \qquad (140)$$

From Eqn. (138*b*) and (140)

$$\left(\vec{\alpha} \cdot \vec{r}\right)^2 \left(\vec{\alpha} \cdot \vec{p}\right) = \left(\vec{\alpha} \cdot \vec{r}\right)\left[\left(\vec{r} \cdot \vec{p}\right) + i\left(\frac{k-\beta}{\beta}\right)\hbar\right] \qquad (141)$$

But $\qquad \left(\vec{\alpha} \cdot \vec{r}\right)^2 = r^2$, so

$$\left(\vec{\alpha} \cdot \vec{p}\right) = \frac{\vec{\alpha} \cdot \vec{r}}{r^2}\left[\left(\vec{r} \cdot \vec{p}\right) - i\hbar + \frac{ik\hbar}{\beta}\right] \qquad (142)$$

Now we introduce radial momentum operator p_r and radial velocity operator α_r defined by

$$p_r = \frac{1}{r}\left(\vec{r} \cdot \vec{p} - i\hbar\right)$$

and $\qquad \alpha_r = \frac{1}{r}\left(\vec{\alpha} \cdot \vec{r}\right) \qquad (143)$

Then Eqn. (142) reduces to

$$\left(\vec{\alpha} \cdot \vec{p}\right) = \alpha_r\left[\frac{\vec{r} \cdot \vec{p}}{r} - \frac{i\hbar}{r} + i\frac{k\hbar\beta}{\beta^2 r}\right] = \alpha_r\left[p_r + i\frac{k\hbar\beta}{r}\right] \qquad (144)$$

$$\left(\because \beta^2 = 1\right)$$

Now, since

$$H = c\vec{\alpha} \cdot \vec{p} + \beta mc^2 + V(r)$$

hence

$$H = c\alpha_r p_r + \frac{ic\hbar\alpha_r k\beta}{r} + \beta mc^2 + V(r) \qquad (145)$$

It can be verified that the operator k commutes with the Hamiltonian given by the above equation.

Now, we would look forward to find the eigenvalues of the operator k. Squarring Eqn. (139), we get

$$k^2\hbar^2 = \left(\vec{\sigma}' \cdot \vec{L}\right)^2 + 2\hbar\left(\vec{\sigma}' \cdot \vec{L}\right) + \hbar^2 \qquad (146)$$

From Eqn. (138*a*), we have

$$\left(\vec{\sigma}' \cdot \vec{L}\right)\left(\vec{\sigma}' \cdot \vec{L}\right) = L^2 + i\vec{\sigma}' \cdot \left(\vec{L} \times \vec{L}\right)$$

$$= L^2 + i\vec{\sigma}' \cdot i\hbar\vec{L} = L^2 - \hbar\left(\vec{\sigma}' \cdot \vec{L}\right) \qquad (147)$$

Hence, using Eqn. (147), Eqn. (146) becomes

$$k^2\hbar^2 = L^2 + \hbar\left(\vec{\sigma}' \cdot \vec{L}\right) + \hbar^2 \qquad (148)$$

But $\qquad \vec{S} = \frac{1}{2}\hbar\vec{\sigma}'$ and $S^2 = s(s+1)\hbar^2 = \frac{3\hbar^2}{4}$

So,
$$k^2\hbar^2 = L^2 + 2\vec{S}\cdot\vec{L} + \frac{3}{4}\hbar^2 + \frac{1}{4}\hbar^2 = \left(\vec{L}+\vec{S}\right)^2 + \frac{1}{4}\hbar^2$$

$$= J^2 + \frac{1}{4}\hbar^2 \tag{149}$$

Now, since J^2 has the eigenvalues $j(j+1)\hbar^2$, hence

$$k^2 = j(j+1) + \frac{1}{4} = \left(j+\frac{1}{2}\right)^2 \tag{150}$$

Since $j = \frac{1}{2}, \frac{3}{2}, \frac{5}{2}, ...,$ therefore possible values of k are $\pm 1, \pm 2, \pm 3, ...$ As operators H and k commute, we can have a representation in which they are diagonal. It follows from the definition of p_r and α_r (Eqn. 143) that $\alpha_r^2 = 1$ and $\alpha_r\beta + \beta\alpha_r = 0$. Also, $\beta^2 = 1$. Hermitian matrices satisfying these relations can be 2×2 matrice and we may select them as

$$\alpha_r = \begin{pmatrix} 0 & -i \\ i & 0 \end{pmatrix}; \beta = \begin{pmatrix} 1 & 0 \\ 0 & -1 \end{pmatrix} \tag{151}$$

As α_r and β are 2×2 matrices, the radial part of the wave function can be represented by

$$\psi = \begin{pmatrix} F(r)/r \\ G(r)/r \end{pmatrix} \tag{152}$$

In order to obtain the radial equation for an electron moving in a central potential, we need to replace α_r, β and p_r in Eqn. (145) using Eqn. (151) and the relation

$$p_r = -i\hbar\left(\frac{\partial}{\partial r} + \frac{1}{r}\right) \tag{153}$$

Thus, Eqn. (145), (151), (152) and (153) yield

$$-ich\begin{pmatrix} 0 & -i \\ i & 0 \end{pmatrix}\left(\frac{\partial}{\partial r}+\frac{1}{r}\right)\begin{pmatrix} F(r)/r \\ G(r)/r \end{pmatrix} + \frac{ich k}{r}\begin{pmatrix} 0 & -i \\ i & 0 \end{pmatrix}\begin{pmatrix} 1 & 0 \\ 0 & -1 \end{pmatrix}\begin{pmatrix} F(r)/r \\ G(r)/r \end{pmatrix}$$

$$+mc^2\begin{pmatrix} 1 & 0 \\ 0 & -1 \end{pmatrix}\begin{pmatrix} F(r)/r \\ G(r)/r \end{pmatrix} + (V-E)\begin{pmatrix} F(r)/r \\ G(r)/r \end{pmatrix} = 0 \tag{154}$$

This Eqn. gives two equations:

$$\frac{dF}{dr} - \frac{kF}{r} - \frac{1}{c\hbar}(E + mc^2 - V)G = 0 \tag{155}$$

and
$$\frac{dG}{dr} + \frac{kG}{r} + \frac{1}{c\hbar}(E - mc^2 - V)F = 0 \tag{156}$$

These equations can be solved for the energy eigenvalues if the explicit form of $V(r)$ is given. For hydrogen atom $V(r) = -Ze^2/r$. For convenience, we make following substitutions:

$$\alpha = \frac{\left(m^2c^4 - E^2\right)^{1/2}}{c\hbar} = \left(\frac{mc^2 + E}{c\hbar}\right)^{1/2} \left(\frac{mc^2 - E}{c\hbar}\right)^{1/2} = \sqrt{\alpha_1} \cdot \sqrt{\alpha_2} \quad (157)$$

where $\quad \alpha_1 = \left(\frac{mc^2 + E}{c\hbar}\right), \alpha_2 = \left(\frac{mc^2 - E}{c\hbar}\right)$ $\qquad\qquad$ (158)

We now introduce a variable ρ defined by

$$\rho = \alpha r \;\therefore\; d\rho = \alpha\, dr \qquad (159)$$

Then Eqns (155) and (156) give

$$\alpha\frac{dF}{dr} - \alpha\frac{kF}{r} - \alpha_1 + \frac{Ze^2}{c\hbar}\frac{\alpha}{\rho}G = 0$$

or

$$\frac{dF}{d\rho} - \frac{kF}{\rho} - \left(\frac{\alpha_1}{\alpha} + \frac{\Gamma}{\rho}\right)G = 0, \Gamma = \frac{Ze^2}{c\hbar} \qquad (160)$$

and

$$\frac{dG}{d\rho} + \frac{kG}{\rho} - \left(\frac{\alpha_2}{\alpha} - \frac{\Gamma}{\rho}\right)F = 0 \qquad (161)$$

We look for solutions (of these equations) of the form

$$F(\rho) = \sum_{m=0}^{\infty} a_m \rho^{s+m} e^{-\rho} \quad a_0 \neq 0 \qquad (162)$$

and

$$G(\rho) = \sum_{m=0}^{\infty} b_m \rho^{s+m} e^{-\rho} \quad b_0 \neq 0 \qquad (163)$$

Substituting these in Eqns (160) and (161) and equating the coefficients of ρ^{s+n-1} to zero, we have

$$(s + m - k)a_m - a_{m-1} - \Gamma b_m - \frac{\alpha_1}{\alpha}b_{m-1} = 0 \qquad (164)$$

and

$$(s + m + k)b_m - b_{m-1} + \Gamma a_m - \frac{\alpha_2}{\alpha}a_{m-1} = 0 \qquad (165)$$

When $m = 0$,

$$(s - k)a_0 - \Gamma b_0 = 0 \qquad (166i)$$

and

$$(s + k)b_0 + \Gamma a_0 = 0 \qquad (166ii)$$

For Eqn. (166) to have non vanishing solution:

$$\begin{vmatrix} s-k & -\Gamma \\ \Gamma & s+k \end{vmatrix} = 0 \text{ or } s = \pm(k^2 - \Gamma^2)^{1/2} \qquad (167a)$$

The −ve solution is unacceptable as it would make F and G diverge. Hence

$$s = (k^2 - \Gamma^2)^{1/2} \qquad (167b)$$

When $m > 0$, a relation between a_m and b_m can be obtained by multiplying Eqn. (165) by α and Eqn. (164) by α_2 and then subtracting

$$b_m\left[(s+m+k)\alpha + \alpha_2\Gamma\right] = a_m\left[\alpha_2(s+m-k) - \alpha\Gamma\right] \qquad (168)$$

$$\left(\because \alpha_1\alpha_2 = \alpha^2\right)$$

Now, regular solutions are possible if both the series of Eqns (162) and (163) terminate. Let this happens at $m = m'$ so that $a_{m'+1} = b_{m'+1} = 0$. Replacing m by $m'+1$ in Eqn. (164), we get

$$(s+m'+1-k)a_{m'+1} - a_{m'} - \Gamma b_{m'+1} - \frac{\alpha_1}{\alpha}b_{m'} = 0 \qquad (169)$$

Using the condition $a_{m'+1} = b_{m'+1} = 0$, the Eqn. (169) gives

$$b_{m'} = -\frac{\alpha}{\alpha_1}a_{m'} ; \quad m' = 0,1,2,... \qquad (170)$$

Eqn. (165) gives the same condition between $b_{m'}$ and $a_{m'}$. The energy levels can be obtained by setting $m' = m$ in Eqn. (168) and using Eqn. (170):

$$b_{m'}\left[(s+m'+k)\alpha + \alpha_2\Gamma\right] = a_{m'}\left[\alpha_2(s+m'-k) - \alpha\Gamma\right]$$

or $\quad -\dfrac{\alpha}{\alpha_1}\left[(s+m'+k)\alpha + \alpha_2\Gamma\right] = \left[\alpha_2(s+m'-k) - \alpha\Gamma\right]$

or $\qquad\qquad 2\alpha(s+m') = \Gamma\left(\alpha_1 - \alpha_2\right) \qquad (171)$

Substituting the values of α, α_1 and α_2, we get

$$\frac{2(m^2c^4 - E^2)^{1/2}}{(ch)}(s+m') = \Gamma\frac{2E}{(ch)}$$

or $\qquad\qquad \left(m^2c^4 - E^2\right)(s+m')^2 = E^2\Gamma^2 \qquad (172)$

or $\qquad E = mc^2\left[1 + \dfrac{\Gamma^2}{(s+m')^2}\right]^{-1/2} \qquad (173)$

$\therefore \qquad E = mc^2\left[1 + \dfrac{\Gamma^2}{\left\{m' + (k^2 - \Gamma^2)^{1/2}\right\}^2}\right]^{-1/2} \qquad (174)$

This expression is equivalent to that derived on the basis of Sommerfeld's theory. Fine structure becomes evident by expanding the expression in powers of Γ^2. Thus

$$m' + \sqrt{k^2 - \Gamma^2} = m' + k\left(1 - \frac{\Gamma^2}{k^2}\right)^{1/2}$$

$$\cong m' + k\left(1 - \frac{\Gamma^2}{2k^2}\right) = m' + k - \frac{\Gamma^2}{2k} = n - \frac{\Gamma^2}{2k} \qquad (175)$$

where $m' + k \equiv n$ (total quantum number)

Hence

$$\frac{\Gamma^2}{\left\{m' + (k^2 - \Gamma^2)^{1/2}\right\}^2} = \frac{\Gamma^2}{\left(n - \dfrac{\Gamma^2}{2k}\right)^2} = \frac{\Gamma^2}{n^2}\left(1 - \frac{\Gamma^2}{2kn}\right)^{-2}$$

$$\cong \frac{\Gamma^2}{n^2}\left(1 + \frac{\Gamma^2}{2kn}\right) \tag{176}$$

Consequently,

$$E = mc^2\left[1 + \frac{\Gamma^2}{n^2}\left(1 + \frac{\Gamma^2}{kn}\right)\right]^{-1/2}$$

$$\cong mc^2\left[1 - \frac{\Gamma^2}{2n^2}\left(1 + \frac{\Gamma^2}{nk}\right) + \frac{3}{4}\frac{\Gamma^4}{2n^4}\left(1 + \frac{\Gamma^2}{nk}\right)^2 + \ldots\right]$$

or $\qquad E = mc^2\left[1 - \dfrac{\Gamma^2}{2n^2} - \dfrac{\Gamma^4}{2n^4}\left(\dfrac{n}{|k|} - \dfrac{3}{4}\right) + \ldots\right]$ (177)

It is to be borne in mind that $n \equiv m' + |k|$ is the total quantum number and the energy eigenvalues depend on k which is related to the total angular momentum.

21.15.1 Hydrogen Spectrum According to Dirac Equation

k can be related with the orbital angular momentum as follows:

We make the non relativistic approximation that the orbital angular momentum is well defined.

Replacing β by +1 and σ' by σ in the expression

$$\hbar k = \beta\left(\vec{\sigma}' \cdot \vec{L} + \hbar\right)$$

we get $\qquad \hbar^2 k = \hbar\vec{\sigma}' \cdot \vec{L} + \hbar^2$ or $\hbar^2 k = 2\vec{S} \cdot \vec{L} + \hbar^2$ (178)

$$\vec{J} = \vec{L} + \vec{S} \Rightarrow \vec{J} \cdot \vec{J} = L^2 + S^2 + 2\vec{L} \cdot \vec{S}$$

or $\qquad 2\vec{L} \cdot \vec{S} = \hat{J}^2 - \hat{L}^2 - \hat{S}^2$ (179)

From Eqns (178) and (179)

$$\hbar^2 k - \hbar^2 = \hat{J}^2 - \hat{L}^2 - \hat{S}^2 \tag{180}$$

or $\qquad \hbar^2 k - \hbar^2 = j(j+1)\hbar^2 - l(l+1)\hbar^2 - s(s+1)\hbar^2$

or $\qquad \hbar^2 k = \left[j(j+1) - l(l+1) - \dfrac{3}{4}\right]\hbar^2 + \hbar^2 \qquad (\because s = \frac{1}{2})$

or $\qquad k = j(j+1) - l(l+1) + \dfrac{1}{4}$ (181)

Hence

$$k = \left(l + \frac{1}{2}\right)\left(l + \frac{3}{2}\right) - l(l+1) + \frac{1}{4} \text{ for } j = l + \frac{1}{2}$$

$$= \left(l - \frac{1}{2}\right)\left(l + \frac{1}{2}\right) - l(l+1) + \frac{1}{4} \text{ for } J = l - \frac{1}{2}$$

Thus

$$k = \begin{cases} l+1 & \text{if } j = l + \frac{1}{2} \\ -l & \text{if } j = l - \frac{1}{2} \end{cases} \tag{182}$$

Let us now consider $n = 1$ (the ground state) m' can be zero or 1 and $k \neq 0$. So $m' = 0$, then $k = +1$. From Eqn. (182), it follows that (in the non relativistic case) the state is $1S_{1/2}$.

$n = 2$, $m' = 0$, $|k| = 2; k = 2$ $(l = 1)$ –   the state is $2P_{3/2}$

$m' = 1$, $|k| = 1$, $k = +1$, $(l = 0)$ –   the states is $2S_{1/2}$

                    $k = -1$, $(l = 1)$ –   the state is $2P_{1/2}$

It is to be noted here that the last two states are degenerate and found by $|k|$ only.

$n = 3$, $m' = 0$, $|k| = 3$ we get $3D_{5/2}$

$m' = 1$, $|k| = 2$, $k = 2$ $\Leftrightarrow 3P_{3/2}$

                    $k = -2$ $\Leftrightarrow 3D_{3/2}$

$m' = 2$, $|k| = 1$, $k = +1$ $\Leftrightarrow 3S_{1/2}$

                    $k = -1$ $\Leftrightarrow 3P_{1/2}$

Consequently, the energy spectrum is

Dirac   **Schrödinger**

————$3D_{5/2}$

————$3P_{3/2}, 3D_{3/2}$ ————$3S_{1/2}, P_{1/2}, P_{3/2}, D_{1/2}, D_{3/2}$

————$3S_{1/2}, P_{1/2}$

————$2P_{3/2}$

————$2S_{1/2}, P_{1/2}$ ————$2S_{1/2}, P_{1/2}, P_{3/2}$

————$1S_{1/2}$ ————$1S_{1/2}$

Fig. 21.2

In 1947, W.E. Lamb and R.C. Retherford observed a splitting between $2S_{1/2}$ and $2P_{1/2}$ states of hydrogen atom not accounted by equation (177). This is known as *Lamb-shift*.

21.15.2 Deviations in the Spectrum of Hydrogen

In very accurate experiments, on hydrogen spectrum deviations from theoretical predictions are observed which result from the following effects:

(i) The hydrogen atom is a two body system consisting of a nucleus of mass m_N and an electron. The transformation to centre of mass frame leads to a one body problem with the reduced mass

$$\mu = \frac{m_e}{\left(1 + m_e/m_N\right)} \tag{183}$$

This gives a correction in the fourth decimal place i.e. $\left(1 + m_e/m_N\right) = 1.0005446$.

(ii) Relativistic effects give rise to the so called *"fine-structure"* and are of order α^2 (where $\alpha = e^2/(\hbar c)$ is the fine structure constant) in comparison to the original levels. These include relativistic correction to electron mass, the Darwin term and the spin orbit coupling. Spin orbit coupling leads to the formula

$$\Delta E_{fs} = -\alpha^4 mc^2 \frac{1}{4n^2} \left(\frac{2n}{j + \frac{1}{2}} - \frac{3}{2} \right) \tag{184}$$

This formula is identical to that for the relativistic correction except for l replaced here by j. Figure 21.3 shows the fine structure in hydrogen. The energies are all depressed ($\because \Delta E_{fs}$ is negative).

(iii) The Lamb shift a quantum electrodynamic phenomenon, is of the order $\alpha^3 \ln \alpha$ in comparison to the original energy levels.

(iv) The hyperfine structure, which results from the interaction between the electron and nuclear spin, is smaller by a factor of about $m_e/m_N \approx 1/2000$ than the fine structure.

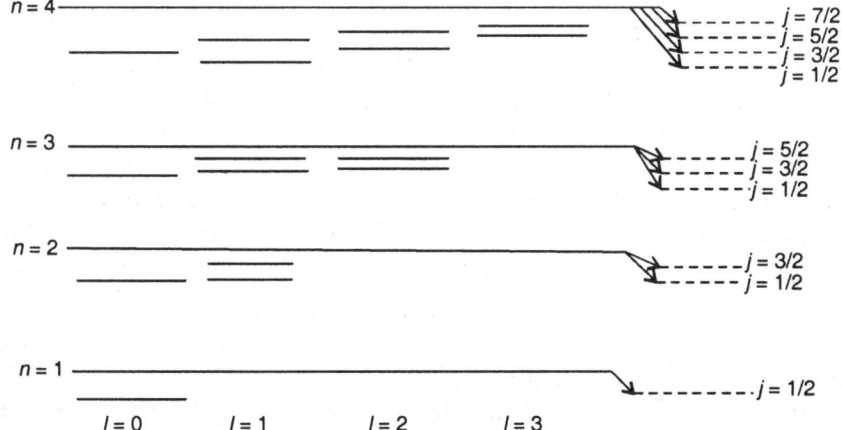

Fig. 21.3: Fine structure in hydrogen. The nth Bohr level splits into n sublevels (shown on the right by dashed lines), characterized by $j = 1/2, 1/3, ... (n - 1/2)$. All levels are shifted downward

Figure 21.4 shows the lowest two energy levels of the hydrogen atom and their splitting due to relativistic corrections, the Lamb shift and the hyperfine structure.

21.16 The Lamb Shift

In Schrödinger's theory of the hydrogen atom, energy levels with the same principal quantum number n and different l values were degenerate. In Dirac's theory, the degeneracy for different l's is removed, but not between the levels with the same total angular momentum j. For example, the $2S_{1/2}$ ($n = 2, l = 0$, $j = ½$) and $2P_{1/2}$ ($n = 2, l = 1, j = ½$) states are still degenerate. However in 1947, Lamb and Retherford performed an experiment which demonstrated that the S-state is slightly higher in energy than the P-state. This is termed as *Lamb shift*.

Without Lamb shift, the hyperfine splitting would have been 11,000 MHz, but on account of the Lamb shift the $2P_{3/2} - 2S_{1/2}$ separation decreases to ~10,000 MHz. Lamb and Retherford measured this transition frequency.

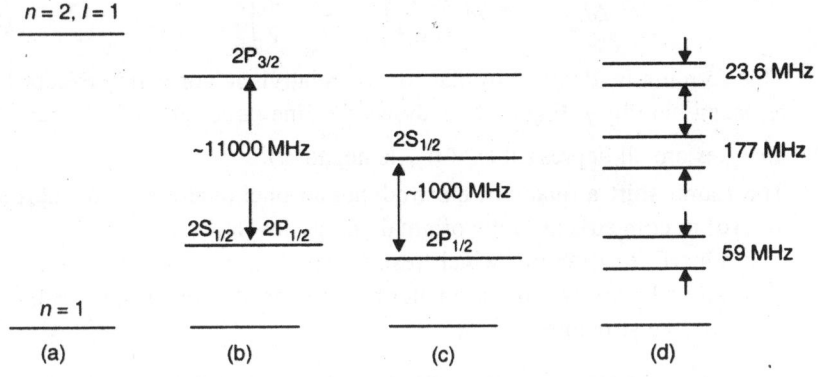

Fig. 21.4: (*a*) The lowest two energy levels of the hydrogen atom in the Bohr model, (*b*) The same levels according to the one electron Dirac theory. The $3P_{3/2} - 2P_{1/2}$ gap is due to hyperfine splitting, (*c*) Qualitative picture of Lamb shift, (*d*) The hyperfine structure.

The separation would have been 11,000 MHz if the $2S_{1/2}$ and $2P_{1/2}$ levels had coincided (as predicted by Dirac one electron theory). Since the separation is 10,000 MHz, they concluded that the $2S_{1/2}$ level is raised by ~1000 MHz with respect to the $2P_{1/2}$ level. This is the Lamb shift.

This is an example of radiative correction in quantum electrodynamics to which the semiclassical electromagnetic theory is inapplicable. The explanation for this was provided by Bethe, Feynman, Schwinger, Tomonaga: it is due to radiative corrections, arising principally from interaction of bound electron with its own electromagnetic field due to its self energy. Feynman's loop diagram (shown in Fig. 21.5) elucidate the explanation.

Qualitatively, the first diagram describes spontaneous production of electron-positron pairs in the vicinity of the nucleus, leading to a partial screening of proton's charge. The second diagram indicates that the ground state of the electromagnetic field is not zero; as the electron moves through the "vacuum fluctuations" in the field, it jiggles slightly and this alters its energy. The third diagram leads to a small modification of the electron's magnetic dipole moment, so Eqn.

$$\vec{\mu} = -\frac{e}{mc}S \qquad (185)$$

picks up a factor $\left(1 + \dfrac{\alpha}{2\pi}\right) = 1.00116$. Bethe was the first to give an approximate non relativistic derivation of the Lamb shift (due to the self energy of electron in a bound state). Since then calculations and measurements of Lamb shift have been greatly refined by the scientists. These provided following results:

For
$$l = 0, \quad \Delta E_{lamb} = \alpha^5 mc^2 \frac{1}{4n^3}\{k(n,0)\}$$

where $k(n, 0)$ is a numerical factor that varies slightly with n, ranging from 12.7 (for $n = 1$) to 13.2 (for $n \to \infty$).

For
$$l \neq 0, \quad \Delta E_{lamb} = \alpha^5 mc^2 \frac{1}{4n^3}\left\{k(n,l) \pm \frac{1}{\pi(j + \frac{1}{2})(l + \frac{1}{2})}\right\} \qquad (186)$$

for
$$j = l \pm \tfrac{1}{2}.$$

where $k(n, l)$ is a very small number (< 0.005) which varies slightly with n and l. Evidently the Lamb shift is tiny except for states with $l = 0$, for which it amounts to about 10% of the fine structure. However, due to dependence on l, it lifts the degeneracy of the pairs of states with common n and j (in particular $2S_{1/2}$ and $2P_{1/2}$ levels).

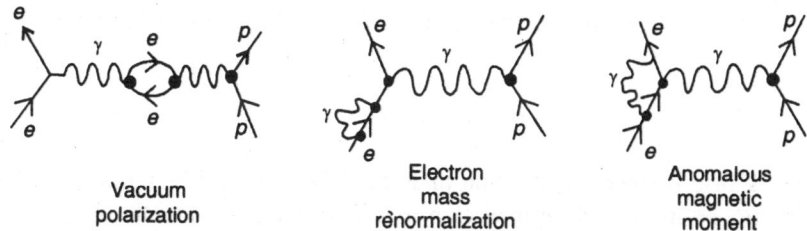

Vacuum polarization   Electron mass renormalization   Anomalous magnetic moment

Fig. 21.5: Some Feynman diagrams contributing to the Lamb shift (After: Introduction to Elementary particles, D. Griffiths, John Wiley and Sons, Inc., 1987)

21.17 Free Spin Zero Particles

The free motion of a particle is an ideal case. Particularly, in case of spin zero particles (such as pions and kaons) are far from this idealization; they interact very strongly with other particles and fields. We shall try to study the solutions of the equation

$$\frac{\hbar^2}{c^2}\frac{\partial^2 \psi}{\partial t^2} = \left[\hbar^2 \nabla^2 - M^2 c^2\right]\psi \qquad (187)$$

which will describe the free motion of zero spin particles.

We look for a solution of the form

$$\psi = A e^{i[\vec{p}\cdot\vec{r}-\varepsilon t]/\hbar} \tag{188a}$$

Substituting Eqn. (188a) into Eqn. (187), we see that this equation is satisfied provided

$$\varepsilon = \pm E_p \text{ where } E_p = c\sqrt{\vec{p}^2 + M^2 c^2} \tag{188b}$$

is the particle energy.

The solutions of equation (187) corresponding to states with a well defined value of the momentum can thus be of two kinds

$$\psi_+ = A_1 e^{i[\vec{p}\cdot\vec{r}-E_p t]/\hbar} \tag{189a}$$

$$\psi_- = A_2 e^{i[\vec{p}\cdot\vec{r}+E_p t]/\hbar} \tag{189b}$$

Substituting Eqns. (189a and b) into Eqn. (11) viz.

$$P(\vec{r},t) = \frac{i\hbar}{2mc^2}\left(\psi^* \frac{\partial \psi}{\partial t} - \psi \frac{\partial \psi^*}{\partial t}\right) \tag{190}$$

we find

$$P_\lambda = \frac{\lambda e E_p}{mc^2} \psi_\lambda^* \psi_\lambda \tag{191}$$

Solutions of the type (ψ_+) correspond to the free motion of a particle with momentum $\vec{p}$ and charge e while solutions of the type (ψ_-) correspond to the free motion of a particle with the opposite sign of the charge.

If we impose on the free motion of the particle, the periodic boundary conditions (i.e. a closed box of edge L) the components of the wave vector will take on discrete values

$$k_i = \frac{2\pi}{L} n_i \; ; \; n_i = 0, \pm1, \pm2, ...; \; i = 1, 2, 3 \tag{192}$$

In this case the general solution of Eqn. (189) for the free motion of a zero spin particle with a well defined sign of charge ($\lambda = \pm1$) has the form

$$\psi_\lambda = \frac{1}{(L)^{3/2}} \sum A_k e^{i\left[\vec{k}\cdot\vec{r}-\lambda\omega(\vec{k})t\right]} \tag{193}$$

with $\qquad \omega(\vec{k}) = E_p / \hbar .$

Thus it becomes apparent that the change to a relativistic quantum equation leads us to appearance of three additional degrees of freedom due to charge on the particle. In addition to the free motion (described by three well defined value of momentum), there are three values of solution for each value of the particle charge. To the two values of the charge, we may assign two wave functions ϕ and χ such that

$$\psi = \phi + \chi \tag{194}$$

with ϕ and χ separately satisfying the equations

$$i\hbar\frac{\partial\phi}{\partial t} = -\frac{\hbar^2}{2m}\nabla^2(\phi + \chi) + mc^2\phi$$

and

$$i\hbar\frac{\partial\chi}{\partial t} = -\frac{\hbar^2}{2m}\nabla^2(\phi + \chi) - mc^2\chi \qquad (195)$$

The two components ϕ and χ can be written as a single column matrix

$$\vec{\psi} = \begin{pmatrix} \phi \\ \chi \end{pmatrix} \qquad (196)$$

In other cases, where apart from charge, spin is also involved, then due to two possible orientations of spin (along z-axis), the state of the particle will be described by four component wave functions.

At this moment, we now introduce four matrices

$$\hat{\tau}_1 = \begin{pmatrix} 0 & 1 \\ 1 & 0 \end{pmatrix}, \hat{\tau}_2 = \begin{pmatrix} 0 & -i \\ i & 0 \end{pmatrix}, \hat{\tau}_3 = \begin{pmatrix} 1 & 0 \\ 0 & -1 \end{pmatrix}, \hat{\tau}_4 = \begin{pmatrix} 1 & 0 \\ 0 & 1 \end{pmatrix}$$

$$\left(\hat{\tau}_4 = \hat{1}\right). \qquad (197)$$

satisfying the relation

$$\hat{\tau}_i^2 = \hat{1} \text{ and } \hat{\tau}_i\hat{\tau}_j = -\hat{\tau}_j\hat{\tau}_i = i\hat{\tau}_k \qquad (198)$$

where the indices i, j and k take on the values 1, 2, 3 in cyclic order. We can now write the set of equations (185) in the form of a single equation

$$\left(i\hbar\frac{\partial}{\partial t} - \hat{H}_f\right)\vec{\psi} = 0 \qquad (199)$$

The Hamiltonian, H_f being given by

$$H_f = \left(\hat{\tau}_3 + i\hat{\tau}_2\right)\frac{\hat{p}^2}{2M} + Mc^2\hat{\tau}_3 \qquad (200)$$

Acting upon Eqn. (199) with the operator $i\hbar(\partial/\partial t) + \hat{H}_f$ and using the equation $\hat{H}_f^2 = c^2\hat{p}^2 + M^2c^4$, we get the equation

$$\left[\hbar^2\frac{\partial^2}{\partial t^2} + c^2\hat{p}^2 + M^2c^4\right]\vec{\psi} = 0 \qquad (201)$$

From this it follows that each component (viz. ϕ or χ) satisfies equation (187). Substituting Eqn. (194) into Eqn. (190) and using Eqn. (196) and (197), we find, charge density

$$P = e\left(\phi * \phi - \chi * \chi\right) = e\psi^\dagger\hat{\tau}_3\psi \qquad (202)$$

where $\psi^\dagger = \left(\phi*, \chi*\right)$ is the hermitian conjugate of ψ. Now, the Eqn.

$\dfrac{\partial P}{\partial t} + \text{div } j = 0$ leads to the conservation of the integral

$$\int P d^3\vec{r} = e \int \psi^\dagger \hat{\tau}_3 \psi d^3 \vec{r} \tag{203}$$

In the case of the free motion of one particle, this quantity can be normalized to either $+e$ or $-e$ depending on the sign of the charge of the particle. The normalization condition thus reduces to

$$\int \psi^\dagger \hat{\tau}_3 \psi d^3 \vec{r} = \int (\phi^* \phi - \chi^* \chi) d^3 \vec{r} = \pm 1 \tag{204}$$

We now consider the free motion of a zero spin particle within a volume V. Letting

$$\vec{\psi} = \frac{1}{\sqrt{V}} \binom{\phi_0}{\chi_0} e^{i[(\vec{p}\cdot\vec{r}) - \varepsilon t]/\hbar} \tag{205}$$

and substituting this into Eqn. (199), we get the set of equations:

$$\left(\varepsilon - Mc^2\right)\phi_0 = \frac{p^2}{2M}(\phi_0 + \chi_0)$$

$$\left(\varepsilon + Mc^2\right)\chi_0 = -\frac{p^2}{2M}(\phi_0 + \chi_0) \tag{206}$$

This set has non vanishing solutions provided

$$\varepsilon = \pm E_p \text{ when } E_p = c\sqrt{\vec{p}^2 + M^2 c^2}$$

For $\varepsilon = E_p$, the function $\vec{\psi}_{(+)}$ has the components

$$\phi_{0(+)} = \frac{E_p + Mc^2}{2\sqrt{Mc^2 E_p}}, \ \chi_{0(+)} = \frac{Mc^2 - E_p}{2\sqrt{Mc^2 E_p}} \tag{207}$$

and the normalization corresponds to

$$\phi_{0(+)}\phi_{0(+)} - \chi_{0(+)}\chi_{0(+)} = +1 \tag{208}$$

Thus solutions corresponding to $\varepsilon = E_p$ determine the motion of a particle in a *"positive charge state"*. We call such solutions *positive solutions*. Positive solutions correspond to the *positive normalization* in Eqn. (204).

If $\varepsilon = -E_p$, the function $\vec{\psi}_{(-)}$ has the components

$$\phi_{0(-)} = \frac{Mc^2 - E_p}{2\sqrt{Mc^2 E_p}}, \ \chi_{0(-)} = \frac{Mc^2 + E_p}{2\sqrt{Mc^2 E_p}} \tag{209}$$

then,

$$\phi_{0(-)}\phi_{0(-)} - \chi_{0(-)}\chi_{0(-)} = -1$$

and the state corresponds to the motion of a particle with a negative charge. Such solutions are called *negative solutions*. They correspond to the *negative normalization*. It follows from (205), (208) and (209) that if the function

$$\psi = \binom{\phi}{\chi} \tag{210}$$

corresponds to a solution with the positive sign of charge, the function

$$\psi_c = \begin{pmatrix} \chi^* \\ \phi^* \end{pmatrix} \tag{211}$$

will correspond to a solution with the negative sign of charge and vice versa.

Charge Conjugation (Spin Zero Particle)

The solution (211) is called the *charge conjugate solution* with respect to (210). The connection between these solutions is determined by the relation $\vec{\psi}_c = \hat{\tau}_1 \vec{\psi}^*$. The transformation $\vec{\psi} \to \vec{\psi}_c$ is accompanied by the transformations

$$\phi_{0(+)} \to \chi_{0(-)}, \chi_{0(+)} \to \phi_{0(-)}$$

$$\vec{p} \to -\vec{p}, \varepsilon \to -\varepsilon$$

If the state of motion of a particle is described by the function $\vec{\psi}$, the particles corresponding to the charge conjugate function $\vec{\psi}_c$ are called *antiparticles*. Thus charge conjugation changes particles to antiparticles and so, charge conjugation is also called the *particle-antiparticle conjugation*.

If a particle is identical with its antiparticle, it is called a *neutral particle*. Particles and antiparticles may differ not only in the sign of charge but also in other properties such as magnetic moment or nucleon charge (e.g. in neutrons, protons). All these quantities change sign under charge conjugation. Particles which do not have an electrical charge are not always truly neutral. The wave functions of truly neutral zero spin particles must satisfy the equation

$$\vec{\psi}_c = \hat{\tau}_1 \vec{\psi}^* = \alpha \vec{\psi} \text{ with } |\alpha| = 1.$$

Thus there are two possible kinds of truly neutral particles (*i*) neutral particles with positive charge parity for which $\alpha = 1$; (*ii*) neutral particles with negative charge parity for which $\alpha = -1$. Wave functions of such particles satisfy the equations

$$\vec{\psi}_c \equiv \hat{\tau}_1 \vec{\psi}^* = \vec{\psi} \text{ or } \phi = \chi^*$$

$$\vec{\psi}_c \equiv \hat{\tau}_1 \vec{\psi}^* = -\vec{\psi} \text{ or } \phi = -\chi^*$$

21.18 Charge Conjugation for Spin ½ Particles

Dirac's theory of electron predicts the existence of its antiparticle 'positron'. In fact, to each particle, there corresponds an antiparticle, because, the nature is rather symmetric with respect to the transformation

$$e \to -e$$

We now try to cast the Dirac equation into a form which makes this symmetry self evident, i.e. we want that it should be able to form directly the wave function of a positron from that of the missing negative energy electron to which it corresponds.

The original Dirac equation for an electron in the presence of an e.m. field is given by

$$\left\{ \sum \hat{\gamma}_\mu \hat{p}_\mu - imc \right\} \psi = 0, \quad p_\mu = -i\hbar \frac{\partial}{\partial x_\mu}$$

with

$$p_\mu \to p_\mu - \frac{e}{c} A_\mu , \quad A_\mu = \left(\vec{A}, iA_0 \right)$$

$$p_\mu = -i\hbar \frac{\partial}{\partial x_\mu} \equiv -i\partial_\mu \quad \text{(in natural units } c = 1, \hbar = 1)$$

$$\therefore \qquad \left\{ -\gamma_\mu i\partial_\mu + i^2 eA_\mu - im \right\} \psi = 0 \qquad \text{(in natural units)}$$

or

$$-i\left\{ \gamma_\mu \left(\partial_\mu - ieA_\mu \right) + m \right\} \psi = 0$$

or

$$\left\{ \gamma_\mu \left(\partial_\mu - ieA_\mu \right) + m \right\} \psi = 0 \qquad (212)$$

If we denote the charge conjugated wave function (i.e. that of the positron) by ψ_c, then we are interested in finding out the form of ψ_c in terms of ψ(that of the electron) so that ψ_c obeys the charge conjugate Driac equation

$$\left\{ \gamma_\mu \left(\partial_\mu + ieA_\mu \right) + m \right\} \psi_c = 0 \qquad (213)$$

Taking the conjugate of Eqn. (212), we get

$$\bar{\psi} \left[\gamma_\mu \left(-\partial_\mu - ieA_\mu \right) + m \right] = 0$$

Taking its transpose, we obtain

$$\left[\gamma_\mu^T \left(-\partial_\mu - ieA_\mu \right) + m \right] \bar{\psi}^T = 0 \qquad (214)^{\displaystyle \cdot}$$

If we can find a non singular matrix C such that

$$C\gamma_\mu^T C^{-1} = -\gamma_\mu \qquad (215)$$

then we can write Eqn. (214) as

$$C\left[\gamma_\mu^T \left(-\partial_\mu - ieA_\mu \right) + m \right] C^{-1} C\bar{\psi}^T = 0$$

or

$$\left[-\gamma_\mu \left(-\partial_\mu - ieA_\mu \right) + m \right] C\bar{\psi}^T = 0 \qquad \text{(using Eqn. 215)}$$

or

$$\left[\gamma_\mu \left(\partial_\mu + ieA_\mu \right) + m \right] C\bar{\psi}^T = 0 \qquad (216)$$

Comparing Eqn. (216) and (213), we find that by defining

$$\psi_c = C\bar{\psi}^T$$

(where C is defined by Eqn. 215)

the Eqn. (213) is as good as the original Dirac equation for e replaced by $-e$. From Eqn. (215), we infer that C must commute with γ_1 and γ_3 and anticommute with γ_2 and γ_4 (since $\gamma_2^T = \gamma_2, \gamma_4^T = \gamma_4$ and $\gamma_1^T = -\gamma_1$ and $\gamma_3^T = -\gamma_3$)

Now, we consider the transformation of $\overline{\psi}_c$. For this, we consider

$$\psi_c = C\overline{\psi}^T \qquad (217)$$

$$\therefore \quad (\psi_c)^\dagger = (\overline{\psi}^T)^\dagger C^\dagger = \left[\left(\psi^\dagger \gamma_4\right)^T\right]^\dagger C^\dagger$$

$$= \left(\psi^{\dagger T}\gamma_4\right)^\dagger C^{-1} = (\gamma_4\psi)^T C^{-1}$$

$$= \psi^T \gamma_4^T C^{-1} \qquad \left(\because \overline{\psi} = \psi^\dagger \gamma_4 \text{ and } C^\dagger C = 1\right)$$

and $\qquad \overline{\psi}_c = \psi_c^\dagger \gamma_4$

or $\qquad \psi_c^\dagger \equiv (\overline{\psi}_c)^T = \psi^T \gamma_4^T C^{-1} \gamma_4$

$$= -\psi^T C^{-1} \gamma_4 \gamma_4 \qquad \left(\because \gamma_4^2 = 1\right)$$

$$= -\psi^T C^{-1} \qquad (218)$$

Now, let us consider the effect of this transformation on the charge density $j_\mu = ie\overline{\psi}\gamma_\mu\psi$. We have

$$j_{\mu(c)} = ie\overline{\psi}_c\gamma_\mu\psi_c = -ie\psi^T C^{-1}\gamma_\mu C\overline{\psi}^T \text{ (using Eqns 218 and 217)}$$

$$= ie\psi^T \gamma_\mu^T \overline{\psi}^T \qquad \text{(using Eqn. 215)}$$

$$= ie\left(\overline{\psi}\gamma_\mu\psi\right)^T \neq -j_\mu \qquad (219)$$

i.e. the current density has the same sign, contrary to our expectation. The difficulty lies in the definition of j_μ. This difficulty is resolved if we regard ψ and $\overline{\psi}$ as non-commuting operators. Because

$$j_\mu = ie\overline{\psi}_\alpha \left(\gamma_\mu\right)_{\alpha\beta} \psi_\beta$$

with $\left(\gamma_\mu\right)_{\alpha\beta}$ antisymmetric, we must have $\overline{\psi}_\alpha\psi_\beta$ antisymmetrized as

$$\overline{\psi}_\alpha\psi_\beta = \frac{1}{2}\left(\overline{\psi}_\alpha\psi_\beta - \psi_\beta\overline{\psi}_\alpha\right)$$

so that $\qquad j_\mu = \frac{ie}{2}\left[\overline{\psi}_\alpha \left(\gamma_\mu\right)_{\alpha\beta}\psi_\beta - \psi_\beta\left(\gamma_\mu\right)_{\alpha\beta}\overline{\psi}_\alpha\right]$

$$= \frac{ie}{2}\left[\overline{\psi}\gamma_\mu\psi - \left(\overline{\psi}\gamma_\mu\psi\right)^T\right]$$

Hence,

$$j_{\mu(c)} = \frac{ie}{2}\left[\overline{\psi}_c\gamma_\mu\psi - \left(\overline{\psi}_c\gamma_\mu\psi\right)^T\right]$$

$$= \frac{ie}{2}\left[\left(\overline{\psi}\gamma_\mu\psi\right)^T - \overline{\psi}\gamma_\mu\psi\right] \qquad \text{(using Eqn. 219)}$$

$$= -\frac{ie}{2}\left[\overline{\psi}\gamma_\mu\psi - \left(\overline{\psi}\gamma_\mu\psi\right)^T\right] = -j_\mu$$

according to our expectations. Also j_μ is a sum of two terms – one represents the particle and the other represents corresponding antiparticle. Thus particle and corresponding antiparticle appear in a symmetric role.

Questions and Problems

1. Derive Klein-Gordon equation. What types of particles obey this equation?
2. Develop the Klein-Gordon equation for a spin zero particle. Construct the corresponding continuity equation.
3. Obtain Klein-Gordon equation for a charged particle moving in an electromagnetic field. derive the continuity equation and obtain expressions for current and charge densities.
4. What are shortcomings of the Klein-Gordon relativistic wave equation. Derive Dirac's relativistic wave equation.
5. Derive Dirac's linear Hamiltonian for a free particle and find out expressions for Dirac's matrices α and β.
6. Discuss free particle solutions of Dirac's wave equation. Discuss various implications of negative energy states.
7. Show that Dirac's wave equation automatically endows the electron a spin of $(\frac{1}{2})\hbar$.
8. Show that Dirac's wave equation endows the electron a magnetic moment $e\hbar\sigma'/(2mc)$, the Bohr magneton. How it is interpreted physically?
9. Write short notes on
    (*i*) Klein-Gordon equation for a free particle
    (*ii*) Dirac matrices
    (*iii*) Negative energy state of an electron
    (*iv*) Dirac equation and electron spin
    (*v*) Covariance of Dirac equation
    (*vi*) Lamb Shift
10. Discuss solution of Dirac's equation of a central field force (*H*-atom). How does it explain hydrogen spectrum? What is Lamb-shift?
11. Show that spin-orbit interaction is a consequence of Dirac's equation. How the spin-orbit interaction term is interpreted physically?
12. Space inversion takes the Dirac wave function $\psi(\vec{x},t)$ into $\psi'(\vec{x},t) = \beta\psi(-\vec{x},t)$. Verify that $\psi'(\vec{x},t)$ satisfies the Dirac equation.
13. Express Dirac's equation in covariant form.
14. If one wants to write the relativistic energy E of a free particle as $E^2/c^2 = \left(\vec{\alpha}\cdot\vec{p} + \beta mc^2\right)^2$, show that the a's and b have to be matrices and establish that they are non singular and Hermitian.
15. Discuss solutions of Dirac's relativistic wave equation for free spin zero particles and explain charge conjugation.

22

Quantization and Second Quantization

22.1 What are First Quantization and Second Quantization

Quantization is basically assigning certain values to certain discrete variables. The transition from classical particle mechanics to the Schrödinger equation, which is the consequence of regarding the particle-variables as operators, is called *first quantization*. Similarly, the wave-particle fields that appear in this process (for example, Schrödinger field, Dirac field, etc.) are when subjected to the process of quantization, then it is referred to as the *second quantization*, since we may write

$$\psi(\vec{r},t) = \sum_i a_i(t) u_i(\vec{r})$$

(1)

where a_i's satisfy

$$\sum_i a_i^*(t) a_i(t) = 1$$

(2)

Earlier a_i's were simply numbers. However, these coefficients are now treated as operators in second quantization.

In classical physics, a collection of particles can be represented by a suitable field $\psi(\vec{r}, t)$. This is analogous to fluid mechanics where one does somewhat a similar thing while dealing with many identical particles, e.g. in non-relativistic quantum mechanics one introduces a quantum field $\psi(\vec{r}, t)$. The field amplitude ψ would satisfy a Schrödinger equation. Thus fields are needed for describing particles, $\vec{A}$ for describing photons, and ψ for describing electrons. However, not withstanding the appearance of the Planck's constant h (in the Schrödinger equation), the particle field is not yet a quantum entity. To make it so, the field amplitude ψ has to be treated as an operator which is done via suitable anti-commutation relation. This process of second quantization applies equally well to Dirac particles (Dirac equation holds for an electron). To prepare it for use in a many electron environment, one must first make ψ as an operator and next specify the anticommutator $\left[\psi(\vec{r},t), \psi(\vec{r}',t) \right]_+$, whereas for $\vec{A}$, we would have to write down the commutator. This step is known as *second quantization*.

Second Quantization

The combination of relativity and quantum mechanics predicts that short range potentials cause creation or annihilation of particles inspite of originally one particle being present. For instance, in a system of free particles, the particle-momenta are separately conserved, the occupation numbers of the states are therefore also conserved (i.e. the numbers N_1, N_2,... which show how many particles are in each of the states ψ_1, ψ_2, ... etc). In a system of interacting particles, the momenta of each particle is not conserved, and so, the occupation numbers are not conserved. For such a system, we can consider the probability distribution of the various values of the occupation numbers. Then we seek to construct a formalism which removes the inconsistency of particle number non-conservation by considering the occupation numbers (and not the coordinates or spin projections) playing the part of independent variables. In this formalism of second quantization, the states of the system are described by what is called a wave function in occupation number space.

Although it provides description of various particles, and ensures the proper symmetry of the many-particle wave function, it is simply a formalism and does not provide any new physical laws or replace any wave equations. In fact it is simply a consistent mathematical frame-work. This method was developed by P.A.M. Dirac in 1927 for photons (bosons) in radiation theory and later extended to fermions by E. Wigner and P. Jordan.

22.2 Recapitulation (Creation, Annihilation and Number Operators)

We have seen in the chapter on harmonic oscillator that

$$H = \frac{1}{2}\left(p^2 + \omega^2 q^2 \right) \tag{3}$$

$$a = \frac{1}{\sqrt{2\hbar\omega}}\left(\omega q + ip \right) \tag{4}$$

$$a^\dagger = \frac{1}{\sqrt{2\hbar\omega}}\left(\omega q - ip \right) \tag{5}$$

$$aa^\dagger = \frac{1}{\hbar\omega}\left(H + \frac{1}{2}\hbar\omega \right) ; \; a^\dagger a = N \text{ (the number operator)} \tag{6}$$

$$H = \hbar\omega\left(a^\dagger a + \frac{1}{2} \right) = \hbar\omega\left(n + \frac{1}{2} \right) \tag{7}$$

$$H|n> = E_n|n> \; ; \; E_n = \left(n + \frac{1}{2} \right)\hbar\omega \tag{8}$$

$$\hat{n}a|n> = (n-1)a|n> \tag{9}$$

and $$\hat{n}a^\dagger|n> = (n+1)a^\dagger|n> \tag{10}$$

$$a|n> = \sqrt{n}|n-1> \text{ and } a^+|n> = \sqrt{n+1}|n+1> \tag{11}$$

In what follows we will discus two important examples of first quantization namely, quantization of lattice vibrations and quantization of radiation field $\vec{A}$. Quantum of lattice vibration is termed a phonon and a quantum of radiation field is a photon. In either case, the fields are wave fields with amplitudes obeying classical wave equation. Further, in each case, the Hamiltonian resembles the Hamiltonian for a collection of harmonic oscillators.

22.3 Quantization of Lattice Vibrations

Quantum mechanically we know that lattice vibrations are quantized with a quantum of lattice vibration termed as *phonon*. Each phonon may be treated as a simple harmonic oscillator having energy ½ $\hbar$ ω_s.

A linear crystal (i.e. an array of atoms) may be considered as a finite number (say N) of independent harmonic oscillators. We can write the Schrödinger wave equation for single oscillator (of frequency $\omega_s / 2\pi$) as

$$\frac{\partial^2 \psi}{\partial x^2} + \frac{2m}{\hbar^2} \left[E - \frac{1}{2} m\omega^2 x^2 \right] \psi = 0 \tag{12}$$

In terms of normal coordinates q's the above equation can be written as

$$\sum_{s=1}^{N} \left[\frac{\partial^2 \psi}{\partial q_s^2} + \frac{2m}{\hbar^2} \left(E_s - \frac{1}{2} m\omega_s^2 q_s^2 \right) \psi \right] = 0 \tag{13}$$

N independent wave equations will be of the form:

$$\frac{\partial^2 \psi_s}{\partial q_s^2} + \frac{2m}{\hbar^2} \left(E_s - \frac{1}{2} m\omega_s^2 q_s^2 \right) \psi_s = 0 \tag{14}$$

where

$$\psi = \sum_{s=1}^{N} \psi_s(q_s)$$

and

$$E = \sum_{s} E_s = \sum_{s=1}^{N} \left(n_s + \frac{1}{2} \right) h\nu_s$$

$$= \hbar \sum_{s=1}^{N} n_s \omega_s + \frac{1}{2} \hbar \sum_{s=1}^{N} \omega_s \tag{15}$$

n_s is an integer and $\omega_s = 2\pi\nu_s$.

For $n_1 = n_2 = n_3 = ... = n_N = 0$, the lowest allowed energy E_0 may be represented by

$$E_0 = \frac{1}{2} \hbar \sum_{s=1}^{N} \omega_s \tag{16}$$

This is termed as the zero point energy of the crystal.

Concept of phonons: The equations

$$\hat{a}\psi_n = \sqrt{n}\,\psi_{n-1} \tag{17i}$$

$$\hat{a}^+ \psi_n = \sqrt{n+1}\,\psi_{n+1} \tag{18i}$$

get replaced by

$$\hat{a}\,|\,n> = \sqrt{n}\,|\,n-1> \qquad (17ii)$$

$$\hat{a}^+\,|\,n> = \sqrt{n+1}\,|\,n+1> \qquad (18ii)$$

The quantum number n characterizes completely the state of the oscillator, the *ground state* being defined by

$$\hat{a}\,|\,0> = 0 \qquad (19)$$

The single-quantum excited state corresponds to $n = 1$ and is referred to as a single-phonon excitation, and $n = 2$, a two-phonon excitation and so on. In other words, each quantum of excitation of the oscillator is called a phonon. The quantum number n determines the number of phonons in the corresponding state. The operators $\hat{a}$ and $\hat{a}^+$ act upon the occupation number, i.e. number of phonons n. The operator $\hat{a}$ decreases the number of phonons by unity and so, is called a *phonon annihilation operator*. The operator $\hat{a}^+$ increases the number of phonons by unity and is hence called a *phonon creation operator*. |0> is called the vacuum state. $\hat{a}\,|0> = 0$ means that phonon annihilation operator acting upon the vacuum state gives zero. The energy of the vacuum state is ½ $\hbar\omega$.

22.4 Quantization of Electromagnetic Field and Equivalence to Oscillators (radiation, oscillators or photons)

A classical electromagnetic wave is described by the vector potential

$$\vec{A}(\vec{x},t) = \left(\sqrt{V}\right)^{-1}\left[\vec{A}(\vec{k},t)\,e^{i\vec{k}\cdot\vec{x}} + \vec{A}*(\vec{k},t)\,e^{-i\vec{k}\cdot\vec{x}}\right] \qquad (20)$$

Both the terms on the r.h.s. have to go together because electromagnetic fields are real fields. We have assumed in the above expression that the field is confined within a box of volume $V\,(= L^3)$ and satisfies periodic boundary conditions.

In absence of charges and currents (i.e. for free electromagnetic waves), we can choose the gauge so that scalar potential ϕ vanishes and $\vec{A}$ is solenoidal.

$$\nabla\cdot\vec{A} = 0 \Rightarrow \vec{k}\cdot\vec{A}\left(\vec{k},t\right) = \vec{k}\cdot\vec{A}*\left(\vec{k},t\right) = 0 \qquad (21)$$

The vector $\vec{A}$ is transverse to $\vec{k}$ and has only two independent components corresponding to two independent polarizations of the wave. Assuming the wave to be linearly polarized with $\vec{A}$ directed along a unit vector $\hat{e}$;

$$\vec{A}\left(\vec{k},t\right) = \hat{e}A\left(\vec{k},t\right)$$

$$\vec{A}*\left(\vec{k},t\right) = \hat{e}A*\left(\vec{k},t\right) \qquad (22)$$

$$\hat{e}\cdot\vec{k} = 0 \qquad (23)$$

$$\vec{A}(\vec{x},t) = \left(\sqrt{V}\right)^{-1}\hat{e}\left[\vec{A}(\vec{k},t)e^{i\vec{k}\cdot\vec{x}} + \vec{A}*(\vec{k},t)e^{-i\vec{k}\cdot\vec{x}}\right] \qquad (24)$$

The electric and magnetic fields are given by

$$\vec{E} = -\frac{1}{c}\frac{\partial\vec{A}}{\partial t} = -\frac{\hat{e}}{c}\frac{1}{\sqrt{V}}\left[\dot{A}(\vec{k})e^{i\vec{k}\cdot\vec{x}} + \dot{A}*(\vec{k})e^{-i\vec{k}\cdot\vec{x}}\right] \qquad (25)$$

$$\vec{H} = \text{curl } \vec{A} = i\frac{\vec{k} \times \hat{e}}{\sqrt{V}}\left[\dot{A}\left(\vec{k}\right)e^{i\vec{k}\cdot\vec{x}} + \dot{A}*\left(\vec{k}\right)e^{-i\vec{k}\cdot\vec{x}}\right] \tag{26}$$

Total energy of the field is given by

$$H = \frac{1}{8\pi}\int_V \left(E^2 + H^2\right)d\tau \tag{27}$$

Substituting the values of $|\vec{E}|$ and $|\vec{H}|$ and noting $\left(\vec{k} \times \hat{e}\right)^2 = k^2$ ($\because \vec{k}$ is orthogonal to $\hat{e}$), we get

$$H = \frac{1}{4\pi}\left[\dot{A}(\vec{k})\dot{A}*(\vec{k}) + k^2 A(\vec{k})A*(\vec{k})\right] \tag{28}$$

Expressing $A(\vec{k})$ and $A*(\vec{k})$ as

$$A(\vec{k}) = \frac{1}{\sqrt{2\pi c^2}}\left[Q_1(\vec{k}) + iQ_2(\vec{k})\right]$$

and

$$A*(\vec{k}) = \frac{1}{\sqrt{2\pi c^2}}\left[Q_1(\vec{k}) - iQ_2(\vec{k})\right] \tag{29}$$

we get

$$H = \frac{1}{2}\sum_{i=1}^{2}\left\{\left(\dot{Q}_i(\vec{k})\right)^2 + \omega^2\left(Q_i(k)\right)^2\right\} \tag{30}$$

(where $\omega = ck$)

This expression is analogous to sum of energies of two harmonic oscillators of unit mass whose coordinates are Q_1 and Q_2. For such oscillators the momentum variable is

$$P_i = \dot{Q}_i \tag{31}$$

So long we have been dealing with classical electromagnetic field. To quantize the field, we convert the oscillators into quantum oscillators by requiring the operators Q and P to satisfy the commutation relations

$$\left[\hat{Q}_i(t), \hat{P}_j(t)\right] = i\hbar\delta_{ij} \qquad (i, j = 1, 2)$$

$$\left[\hat{Q}_i(t), \hat{Q}_j(t)\right] = \left[\hat{P}_i(t), \hat{P}_j(t)\right] = 0 \tag{32}$$

Thus, the electromagnetic field has been decomposed into modes each behaving like a harmonic oscillator. The cycle averaged energy content of a single mode $\vec{k}$ is

$$\bar{H}_k = \frac{1}{2}\left(P_k^2 + \omega_k^2 Q_k^2\right) \tag{33}$$

The complete Hamiltonian for the cavity (box) is found by summing over k and the two independent directions of $\hat{e}$. We now define (for the oscillators $i = 1, 2$) lowering and raising operators $\hat{a}_i$ and $\hat{a}_i^\dagger$ as follows:

$$\hat{a}_i = \left(\frac{\omega}{2\hbar}\right)^{1/2}\hat{Q}_i + i\left(\frac{1}{2\hbar\omega}\right)^{1/2}\hat{P}_i$$

and

$$\hat{a}_i^\dagger = \left(\frac{\omega}{2\hbar}\right)^{1/2} \hat{Q}_i - i\left(\frac{1}{2\hbar\omega}\right)^{1/2} \hat{P}_i \qquad (34)$$

In terms of these operators, the commutation relations (32) and the Hamiltonian becomes

$$\left[\hat{a}_i, \hat{a}_j^\dagger\right] = \delta_{ij} \text{ and } \left[\hat{a}_i, \hat{a}_j\right] = \left[\hat{a}_i^\dagger, \hat{a}_j^\dagger\right] = 0$$

$$\hat{H} = \sum_i \left(\hat{a}_i^\dagger \hat{a}_i + \frac{1}{2}\right)\hbar\omega \qquad (35)$$

One can verify that

$$\left[\hat{H}, \hat{a}_i\right] = -\hbar\omega\hat{a}_i$$

$$\left[\hat{H}, \hat{a}_i^\dagger\right] = +\hbar\omega\hat{a}_i^\dagger \qquad (36)$$

But according to Heisenberg's equation of motion

$$i\hbar\frac{d}{dt}A_H(t) = \left[A_H, H_H\right] + i\hbar\left(\frac{\partial A}{\partial t}\right)_H \qquad (37)$$

the left hand side of the equation (36) are respectively $-i\hbar da_i / dt$ and $-i\hbar da_i^\dagger / dt$. Therefore,

$$-i\hbar\frac{da_i}{dt} = -\hbar\omega a_i \; ; \; -i\hbar\frac{da_i^\dagger}{dt} = \hbar\omega a_i^\dagger$$

or

$$\frac{da_i}{dt} = -i\omega a_i \; ; \; \frac{da_i^\dagger}{dt} = i\omega a_i^\dagger \qquad (38)$$

The solutions of these equations are

$$a_i(t) = a_i(0)e^{-i\omega t} \; ; \; a_i^\dagger(t) = a_i^\dagger(0)e^{i\omega t} \qquad (39)$$

By knowing this time dependence of the operators a_i and $a_i^\dagger$, we can separate $\vec{A}(\vec{x}, t)$ into two parts: one propagation in the direction of $\vec{k}\left(\pm(\vec{k}\cdot\vec{x} - \omega t)\right)$ and the other in the direction $-\vec{k}\left(\pm(\vec{k}\cdot\vec{x} + \omega t)\right)$. We also have

$$Q_i = \left(\frac{\hbar}{2\omega}\right)^{1/2}\left(a_i + a_i^\dagger\right) \text{ and } P_i = -i\left(\frac{\hbar\omega}{2}\right)^{1/2}\left(a_i - a_i^\dagger\right) \qquad (40)$$

(from Eqn. 34)

Substituting these into equations (29) and the resulting expressions for $A(\vec{k}, t)$ and $A*(\vec{k}, t)$ into the Eqn. (24), we find for the wave propagating in $\vec{k}$-direction

$$\vec{A}_{\vec{k}}(\vec{x}, t) = \left(\frac{2\pi\hbar c^2}{\omega V}\right)^{1/2} \hat{e}\left[a(\vec{k}, t)e^{i\vec{k}\cdot\vec{x}} + a^\dagger(\vec{k}, t)e^{-i\vec{k}\cdot\vec{x}}\right]$$

$$= \left(\frac{2\pi\hbar c^2}{\omega V}\right)^{1/2} \hat{e}\left[a(\vec{k})e^{i\left(\vec{k}\cdot\vec{x} - \omega t\right)} + a^\dagger(\vec{k})e^{-i\left(\vec{k}\cdot\vec{x} - \omega t\right)}\right] \qquad (41)$$

Thus (vector) potential of the field has been expressed in terms of the raising and lowering operators $a_i^\dagger$ and a_i.

The operator for the total energy of the field in the box also breaks up into two parts

$$H = H_{\vec{k}} + H_{-\vec{k}} \tag{42}$$

with

$$H_{\vec{k}} = \left\{ a^\dagger(\vec{k})a(\vec{k}) + \frac{1}{2} \right\} \hbar\omega \tag{43}$$

$H_{-\vec{k}}$ has the same form with $a(-\vec{k})$ replaced by $a(\vec{k})$.

Since $H_{\vec{k}}$ has exactly the same form as a simple harmonic oscillator, therefore $\vec{A}_{\vec{k}}(\vec{x},t)$ is equivalent to a simple harmonic oscillator. Also we have

$$\left[a(\vec{k}), a^\dagger(k) \right] = 1 \tag{44}$$

As a consequence of the formal identity of this commutation rule with that of the harmonic oscillator, we can conclude that the eigenvalues of the operator

$$N_{\vec{k}} \equiv a^\dagger(\vec{k})a(\vec{k}) \tag{45}$$

are the natural numbers, $n_k = 0, 1, 2, \ldots$

The energy of the field can take only the values $\left(n_k + \frac{1}{2} \right)\hbar\omega$ and they differ from the minimum (i.e. zero point) energy (½ $\hbar\omega$) by an integral multiple n_k of $\hbar\omega$. Thus we may say that the quantum state for which $N(\vec{k})$ has the value n_k is one in which there are n_k quanta of radiation (*photons*) each carrying energy $\hbar\omega$. The state with n_k photons is denoted by $|n_k\rangle$. It is an eigenstate of the number operator N_k because:

$$N_k \,|\, n_k >= n_k \,|\, n_k > \quad (n_k = 0, 1, 2, \ldots) \tag{46}$$

Example 22.1: Given $N = a^\dagger a$. Show that

$$\left(a^\dagger \right)^n (a)^n = N(N-1)(N-2)\ldots(N-n+1)\ldots$$

where $(N - n + 1)$ is a middle term.

Solution:

$$\left(a^\dagger \right)^2 (a)^2 = a^\dagger a^\dagger a a = a^\dagger N a \tag{i}$$

We have

$$\left[N, a^\dagger \right] = a^\dagger, \quad [N, a] = -a$$

$$Na^\dagger - a^\dagger = a^\dagger N \qquad \left(\because Na^\dagger - a^\dagger N = a^\dagger \right)$$

or

$$(N-1)a^\dagger = a^\dagger N$$

$$(N-1)a^\dagger a = a^\dagger N a \qquad \text{(using Eqn. } i\text{)}$$

$$(N-1)a^\dagger a = a^{\dagger 2} a^2$$

or

$$(N-1)N = a^{\dagger 2} a^2$$

Now consider

$$\left(a^\dagger\right)^3 (a)^3 = a^\dagger \left(a^\dagger\right)^2 a^2 a = a^\dagger (N-1)Na = (N-2)a^\dagger Na$$

$$= (N-2)(N-1)a^\dagger a \qquad \text{(since } a^\dagger N = (N-1)a^\dagger)$$

Hence

$$\left(a^\dagger\right)^n (a)^n = \ldots (N-n+1)\ldots (N-2)(N-1)N$$

or

$$\left(a^\dagger\right)^n (a)^n = N(N-1)(N-2)\ldots (N-n+1)\ldots \qquad \text{(By symmetry)}$$

22.5 Fock Space, Assumptions for Second Quantization and Form of the State Vector ψ

We have $|n+1\rangle = \sqrt{N+1}\, a^\dagger\, |n\rangle$

Putting $n = 0, 1, 2, 3, \ldots$ etc, we get

$$|1\rangle = 1^{-1/2} a^\dagger |0\rangle$$

$$|2\rangle = 2^{-1/2} a^\dagger |1\rangle = 2^{-1/2} \cdot 1^{-1/2} a^{\dagger 2} |0\rangle$$

$$|3\rangle = 3^{-1/2} a^\dagger |2\rangle = 3^{-1/2} \cdot 2^{-1/2} \cdot 1^{-1/2} a^{\dagger 3} |0\rangle$$

so that

$$|n\rangle = \left(n!\right)^{-1/2} a^{\dagger n} |0\rangle \qquad (47)$$

i.e. any excited state can be expressed in terms of its ground state. The normalized states defined above are called *Fock states* (or *occupation number states*).

Here, in analogy with excited states of simple harmonic oscillator, we can define, a one particle state

$$|p\rangle \equiv a^\dagger(p) |0\rangle \qquad (48)$$

An N particle state is defined by

$$\left|p_1, p_2, \ldots p_N\right\rangle = a^\dagger(p_1)a^\dagger(p_2)\ldots a^\dagger(p_N)|0\rangle \qquad (49)$$

On the contrary, if we want to construct a state with n particles (of momentum p), then

$$|p(n)\rangle \equiv \frac{1}{\sqrt{n!}}\left(a^\dagger(p)\right)^n |0\rangle \qquad (50)$$

Here comes the factor $1/\sqrt{n!}$ to take care of proper normalization. Such multiparticle states correspond to field quantization, referred to as *second quantization, which* is different from (single particle) first quantization.

Two basic assumptions for second quantization are:

(1) *States of interacting particles can be expanded using non-interacting particle states as the basis.*

(2) *The variables used to describe one particle systems are capable of describing many particle system i.e. the presence of other particles does not influence the presence of any one particle.*

Occupation number or *Fock-space* is an abstract mathematical space in which complete sets of states are enumerated in terms of number of particles with a particular eigenvalue for a particular operator (e.g. energy, momentum, harmonic oscillator quantum number n or any other unfamiliar ones). The state vector for occupation number space has the form

$$\psi \equiv |\psi> = |n_1, n_2, ..., n_i, ...\rangle$$

$$\equiv \psi\left(n_1, n_2, ..., n_i, ...\right) \tag{51}$$

where each n_i is an eigenvalue of N_i and must be a positive integer or zero.

$$N = \sum_{i=0}^{\infty} n_i \tag{52}$$

Here the state vector ψ has the following properties:
(i) It is analogous to (but not equal to) the wave function of the one particle theory.
(ii) It provides a complete description of the N-body system.
(iii) It is not a function of any particular set of coordinates.
(iv) It is a part of a set which forms a complete orthogonal basis. i.e.

$$\left\langle \psi | \psi' \right\rangle = \left\langle n_1, n_2, ..., n_i, ... | n_1', n_2', ..., n_i', ... \right\rangle$$

$$\equiv \prod_{i=1}^{\infty} \delta_{n_i n_i'} \tag{53}$$

The vacuum state |0> (which must be distinguished from its eigenvalue 0) is given by

$$|0> = |0,0,0,...,0,...> \neq 0 \tag{54}$$

i.e. there are zero particles in every state. Here it is to noted that in a system of free particles, the particle momenta are separately conserved i.e. the numbers n_1, n_2, n_3, ... show how many particles are in each of the states ψ_1, ψ_2,.. etc. In a system of interacting particles, momenta individually are not conserved and so the occupation numbers are not conserved. For such a system, we are able to consider only the probability distribution of the various values of the occupation numbers. However, we try to construct a mathematical formalism in which the occupation numbers (and not the coordinates of the particles) play the part of independent variable.

In the formalism, the states of the system are described by a wave function in occupation number space, i.e.

$$\Phi\left(n_1, n_2, ..., t\right)$$

22.6 Second Quantization for Particles Obeying Quantum Statistics

Both the first quantization and second quantization can be carried out with the help of the Heisenberg equation of motion

$$\frac{df}{dt} = \frac{1}{i\hbar}\left[f, h\right] \tag{55}$$

In the first quantization, this corresponds to the quantization of classical Hamilton canonical equations with f being x or p. In the second quantization, this equation is expected to correspond to the Schrödinger equation

$$i\hbar \frac{\partial \psi}{\partial t} = -\frac{\hbar^2}{2m}\nabla^2\psi + V(\vec{r})\psi \tag{56}$$

with f being the amplitudes $a_n(t)$ in

$$\psi(\vec{r},t) = \sum_n a_n(t)\,\psi_n(\vec{r}) \tag{57}$$

where $\psi_n(\vec{r})$ satisfy

$$\left(-\frac{\hbar^2}{2m}\nabla^2\psi + V\right)\psi_n = E_n\psi_n;\langle\psi_n|\psi_m\rangle = \delta_{nm} \tag{58}$$

From Eqns (56) and (57)

$$\frac{d}{dt}a_n = \frac{1}{i\hbar}a_n E_n \tag{59}$$

We can find a Hamiltonian that leads to Eqn. (59) as the equation of motion. This is

$$H = \int d^3r\,\psi*(\vec{r},t)\left[-\frac{\hbar^2}{2m}\nabla^2\psi + V\right]\psi(\vec{r},t) \tag{60}$$

or
$$H = \sum_n a_n^* a_n E_n \qquad \text{(using Eqns 57, 58, 60)}.$$

This is identical to Hamiltonian for a collection of harmonic oscillators having frequencies $E_n/\hbar$. Correspondingly we may write

$$H = \hbar\omega\left(a^\dagger a + \frac{1}{2}\right)$$

$$= \frac{1}{2}\hbar\omega\left(2a^\dagger a + 1\right)$$

$$= \frac{1}{2}\hbar\omega\left(a^\dagger a + aa^\dagger\right) \quad (\because aa^\dagger - a^\dagger a = 1) \tag{61}$$

Based on this, we interpret the classical commuting amplitudes a_n and a_n^* as the operators a_n and $a_n^\dagger$ which satisfy the commutation relations

$$\left[a_n, a_m\right] = \left[a_n^\dagger, a_m^\dagger\right] = 0\;;\;\left[a_n, a_m^\dagger\right] = \delta_{nm} \tag{62}$$

This holds true in case of bosons. Then,

$$H = \sum_n a_n^\dagger a_n E_n \tag{63}$$

Further, the Heisenberg equation of motion

$$\frac{d}{dt}a_n = \frac{1}{i\hbar}\left[a_n, H\right] \tag{64}$$

yields Eqn. (59). The operator $N = a_n^\dagger a_n$ has the eigenvalues $n = 0, 1, 2, ..., \infty$. Thus any integral number of particles can occupy the state whose wave function is ψ_n. The quanta of this field (bosons) obey Bose-Einstein statistics. The amplitude $\psi(\vec{r},t)$ being a linear combination of the annihilation operators a_n, behaves as an operator that destroys a particle at the position $\vec{r}$ at the time t. Similarly $\psi^\dagger(\vec{r},t)$ creates a particle at position $\vec{r}$ at time t. Also,

$$\left[\psi(\vec{r},t),\psi^\dagger(\vec{r}',t)\right] = \delta(\vec{r} - \vec{r}')$$

and

$$\left[\psi(\vec{r},t),\psi(\vec{r}',t)\right] = \left[\psi^\dagger(\vec{r},t),\psi^\dagger(\vec{r}',t)\right] = 0 \tag{65}$$

(using Eqns. 57 and 62)

Thus after second quantization, the wave function ψ itself becomes an operator.
In case of *fermions*, the quanta of the field satisfy anticommutation relations

$$\left[a_n,a_m\right]_+ = 0 = \left[a_n^\dagger,a_m^\dagger\right]_+ \; ; \; \left[a_n,a_m^\dagger\right]_+ = \delta_{nm} \tag{66}$$

with hamiltonian, again, given by

$$H = \sum_m a_m^\dagger a_m E_m$$

Using Eqn. (66) in Eqn. (64), we have

$$i\hbar \frac{d}{dt} a_n = \left[a_n, H\right] = \left[a_n, \sum_m a_m^\dagger a_m E_m\right]$$

$$= \sum_m E_m \left[a_n, a_m^\dagger a_m\right] = \sum_m E_m \left(a_n a_m^\dagger a_m - a_m^\dagger a_m a_n\right)$$

$$= \sum_m E_m \left\{\left(\delta_{nm} - a_m^\dagger a_n\right)a_m - a_m^\dagger a_m a_n\right\} \quad \text{(using Eqn. 66)}$$

$$= \sum_m E_m \left\{\delta_{nm} a_m - a_m^\dagger a_n a_m - a_m^\dagger a_m a_n\right\}$$

$$= \sum_m E_m \left\{\delta_{nm} a_m - a_m^\dagger \left[a_n, a_m\right]_+\right\} \quad \text{(using Eqn. 66)}$$

$$= \sum_m E_m \delta_{nm} a_m = E_n a_n \tag{67}$$

which is in agreement with the Eqn. (59). Thus, the choice (Eqn. 66) satisfied by fermions also leads to Eqn. (59). Further,

$$\left(a_n^\dagger a_n\right)\left(a_n^\dagger a_n\right) = a_n^\dagger \left(1 - a_n^\dagger a_n\right)a_n$$

$$= a_n^\dagger a_n - a_n^\dagger a_n^\dagger a_n a_n = a_n^\dagger a_n \quad \text{(using Eqn. 66)} \tag{68}$$

The eigenvalue equation $a_m^\dagger a_m |n_i\rangle = n_i |n_i\rangle$ implies $n_i^2 = n_i$ or $n_i = 0$ or 1.

Because

$$\left(a_n^\dagger a_n\right)\left(a_n^\dagger a_n\right)|n_i\rangle = a_n^\dagger a_n |n_i\rangle \tag{69}$$

∴ Not more than one particle can occupy a quantum state.

$$a_m \left|0\right> = 0, \ a_m \left|1\right> = \left|0\right>$$

and

$$a_m^\dagger \left|0\right> = \left|1\right>, \ a_m^\dagger \left|1\right> = 0 \tag{70}$$

The number operator

$$N_m = a_m^\dagger a_m = \begin{pmatrix} 0 & 0 \\ 0 & 1 \end{pmatrix} \tag{71}$$

with

$$\left|0\right> = \begin{pmatrix} 1 \\ 0 \end{pmatrix} \ \text{and} \ \left|1\right> = \begin{pmatrix} 0 \\ 1 \end{pmatrix} \tag{72}$$

In the representation (71), we have

$$a_m = (-1)^{\alpha_m} \begin{pmatrix} 0 & 1 \\ 0 & 0 \end{pmatrix} ; \ a_m^\dagger = (-1)^{\alpha_m} \begin{pmatrix} 0 & 0 \\ 1 & 0 \end{pmatrix} \tag{73}$$

where $\alpha_m = \sum\limits_{i=1}^{m-1} n_i$ is the number of states preceding the state m, which are occupied.

The ground state of n noninteracting fermions consists of filled lowest energy states $m_1, m_2, ..., m_n$. The levels are filled upto the Fermi-energy E_F in the ground state such that $E_m \le E_F$. States with $E_m > E_F$ are unoccupied. The ground state $\left|\psi_0\right>$ is such that

$$\left|\psi_0\right> = \prod_m a_m^\dagger \left|0\right> \ \text{(ground state)} \tag{74}$$

$$\left<\psi_0 \left| \sum_m a_m^\dagger a_m \right| \psi_0 \right> = n \tag{75}$$

where n is the total number of particles (fermions). The sum is over all states with $E_m < E_F$.

22.7 Bose-Einstein Statistics

Let us first consider systems of particles obeying Bose-Einstein statistics. The basic building element in Fock space is the creation operator $a_i^\dagger$ that creates an additional particle in position i whereas its adjoint a_i is the annihilation operator that removes one particle from position i, that is

$$a_i \left|n_1, n_2, ..., n_i, n_{i+1}, ...\right> = \pm\sqrt{n_i} \left|n_1, n_2, ..., n_{i-1}, n_{i+1}, ...\right> \tag{76}$$

and

$$a_i^\dagger \left|n_1, n_2, ..., n_i, n_{i+1}, ...\right> = \pm\sqrt{n_i + 1} \left|n_1, n_2, ..., n_{i+1}, ...\right> \tag{77}$$

Here, there appears only the positive sign for bosons whereas for fermions, the sign depends on how the states are specified. Boson creation and destruction operators satisfy

$$\left[a_i, a_i^\dagger\right] = 1 \tag{78}$$

This means that

$$\left[a_i, a_i^\dagger\right]|n_i\rangle = \left[a_i a_i^\dagger - a_i^\dagger a_i\right]|n_i\rangle = |n_i\rangle \qquad (79)$$

where we have not indicated the unaffected numbers of particles of other states.

22.8 Fermi-statistics

Fermions satisfy anticommutator:

$$\left[a_i, a_i^\dagger\right]_+ = 1 \qquad (80)$$

$$\Rightarrow \qquad \left[a_i, a_i^\dagger\right]_+ |n_i\rangle = \left[a_i a_i^\dagger + a_i^\dagger a_i\right]|n_i\rangle = |n_i\rangle \qquad (81)$$

because $aa^\dagger$ gives either 0 or 1 whereas $a^\dagger a$ gives 1 or 0.

Since $a_i^\dagger$ is the adjoint of a_i, so a_i and $a_i^\dagger$ reverse roles when acting to the left, $a_i^\dagger$ removes a particle from a bra and a_i adds particles to a bra.

$$\langle n_i|a_i^\dagger = \sqrt{n_i}\,\langle n_i - 1|$$

$$\langle n_i|a_i = \sqrt{n_i + 1}\,\langle n_i + 1| \qquad (82)$$

$$\therefore \qquad \langle n|a_i^\dagger|m\rangle = \langle n|\sqrt{m+1}|m+1\rangle$$

$$= \sqrt{m+1}\,\langle n|m+1\rangle = \sqrt{m+1}\,\delta_{n,m+1} \qquad (83)$$

22.9 How does Second Quantization Leads to Symmetrization/ Antisymmetrization) for Bosons/(Fermions)

We recollect the symmetrization postulate according to which the possible states of an N number of identical particles are those which are either symmetrical (obeying Bose statistics) or antisymmetrical (obeying Fermi statistics) with respect to interchange of any two particles.

The creation operator $a_i^\dagger$ and the annihilation operator a_i for a boson satisfy

$$\left[a_i, a_{i'}\right] = 0 = \left[a_i^\dagger, a_{i'}^\dagger\right]$$

and

$$\left[a_i, a_{i'}^\dagger\right] = \delta_{ii'} \qquad (84)$$

Here the suffixes i, i' etc denote the labels for the associated created or annihilated particle. In case of bosons, we have the number operator N_i defined by $a_i^\dagger a_i$ whose eigenvalues are given by

$$N_i|n_i\rangle = n_i|n_i\rangle$$

where

$$n_i = 0, 1, 2, ..., \infty$$

$$\left[N_i, a_i\right] = -a_i, \left[N_i, a_i^\dagger\right] = a_i^\dagger \qquad (85)$$

All one, two, etc. particle states are obtained by successively applying $a_i^\dagger$ on the vacuum state |0>:

$$|n_i\rangle = \frac{\left(a_i^\dagger\right)^n |0\rangle}{\sqrt{n_i!}} \tag{86}$$

We now consider the construction of a two particle state consisting of spin-0 particles. Applying $a_i^\dagger$ on $|0>$, we have

$$a_i^\dagger |0> = |\text{one particle of momentum } i> \equiv |1_i\rangle$$

Applying again $a_{i'}^\dagger$, we get

$$a_{i'}^\dagger a_i^\dagger |0\rangle = a_{i'}^\dagger |0\rangle \otimes a_i^\dagger |0\rangle = |1_{i'}\rangle |1_i\rangle \tag{87}$$

Consider now the application of $a_i^\dagger a_{i'}^\dagger$ on $|0>$:

$$a_i^\dagger a_{i'}^\dagger |0\rangle = a_i^\dagger |0\rangle \otimes a_{i'}^\dagger |0\rangle = |1_i\rangle |1_{i'}\rangle \tag{88}$$

R.H.S. of Eqn. (87) is the state of two particles given by $|i'\rangle |i\rangle$, i.e. $\psi(i', i)$. Similarly, from Eqn. (88), its r.h.s is for the state function $\psi(i, i')$. From Eqn. (84), namely $\left[a_i^\dagger, a_{i'}^\dagger\right] = 0$, we get

$$a_i^\dagger a_{i'}^\dagger = a_{i'}^\dagger a_i \tag{89}$$

$$\Rightarrow \qquad a_i^\dagger a_{i'}^\dagger |0\rangle = a_{i'}^\dagger a_i^\dagger |0\rangle \tag{90}$$

Hence from Eqns. (87) and (88), we get

$$\psi(i, i') = \psi(i', i) \tag{91}$$

Thus the wavefunction remains unchanged when two bosons (with identical quantum numbers) are interchanged.

Let us now consider fermions. These are quantized by anticommutation rules:

$$\left[a_i, a_{i'}\right]_+ = \left[a_i^\dagger, a_{i'}^\dagger\right]_+ = 0$$

and

$$\left[a_i, a_{i'}^\dagger\right]_+ = \delta_{ii'} \tag{92}$$

The number operator is again

$$N_i = a_i^\dagger a_i \text{ so } N_i^2 - N_i = a_i^\dagger a_i a_i^\dagger a_i - a_i^\dagger a_i$$

But

$$\left[a_i, a_i^\dagger\right]_+ = 1 \text{ i.e., } a_i a_i^\dagger + a_i^\dagger a_i = 1$$

or

$$a_i a_i^\dagger = 1 - a_i^\dagger a_i \text{. So}$$

$$N_i^2 - N_i = a_i^\dagger \left(1 - a_i^\dagger a_i\right) a_i - a_i^\dagger a_i$$

$$= a_i^\dagger a - a_i^{\dagger 2} a_i^2 - a_i^\dagger a_i$$

$$= -a_i^{\dagger 2} a_i^2 \tag{93}$$

But $\left[a_i, a_i\right]_+ = 0$ so $a_i^2 + a_i^2 = 0$ or $a_i^2 = 0$, Hence

$$N_i^2 - N_i = 0 \tag{94}$$

leading to eigenvalues of N_i for fermions 0, and 1 since

$$N_i^2 - N_i = N_i (N_i - 1) = 0 \tag{95}$$

That is, there cannot be more than one fermion with a given quantum number (Pauli's principle). In order to see how antisymmetric wavefunction is implied, we consider two identical fermions created by $a_i^\dagger$ and $a_{i'}^\dagger$. As in case of bosons, we define

$$a_i^\dagger |0\rangle = |1_i\rangle$$

So, from $\qquad \left[a_i^\dagger, a_{i'}^\dagger \right]_+ = 0$ we get $a_i^\dagger a_{i'}^\dagger + a_{i'}^\dagger a_i^\dagger = 0$

or $\qquad a_i^\dagger a_{i'}^\dagger = -a_{i'}^\dagger a_i^\dagger$ or $a_i^\dagger a_{i'}^\dagger |0\rangle = -a_{i'}^\dagger a_i^\dagger |0\rangle$

This implies

$$\psi\,(i, i') = -\psi(\,i', i) \tag{96}$$

This shows that the two particle wave function is antisymmetric under the exchange of i and i'.

22.10 A "Field" and Coordinates of the Field

For an electromagnetic field, the values of potentials $\vec{A}$, ϕ form an uncountable infinite set and can be considered as the degrees of freedom. Likewise, in general, we refer to a physical system with a non-denumerable (or uncountable) infinity of degrees of freedom as a field. Just as a system of particles is specified by the position coordinates q_i (and their dependence on time), so a field is specified by its amptitude $\psi(\vec{r}, t)$. These amplitudes $\psi(\vec{r}, t)$ are called field functions and may be considered as the degrees of freedom of the field. (The variable r is continnous and should be considered as continnous index). As with any physical system, the field is the carrier of energy momentum or other observable dynamical quantities. The role of mechanical equations of motion is played here by the field equations, which are relations between the values of all the field functions $\psi(\vec{r}, t)$ and their derivatives and therefore partial differential equations. The field propagates in accordance with the field equations. For the present, we assume that the wave field $\psi(\vec{r}, t)$ is real.

22.11 Lagrangian Formulation for Fields (Euler's Equation)

The necessary and sufficient condition for constructing equations of motion are

$$\delta \int_{t_1}^{t_2} L dt = 0 \,; \; \delta q_i(t_1) = \delta q_i(t_2) = 0$$

the virtual displacements at end points being zero. We expect the Lagrangian to be expressible as

$$L = \int \alpha\left(\psi, \nabla\psi, \dot{\psi}, t\right) d\tau \tag{97}$$

The variational principles gives

$$\delta \int_{t_1}^{t_2} L\, dt = \delta \int_{t_1}^{t_2} \int \alpha\, dt\, d\tau = \int_{t_1}^{t_2} (\delta\alpha)\, dt\, d\tau = 0 \tag{98}$$

with $\qquad \delta\psi(\vec{r}, t_1) = \delta\psi(\vec{r}, t_2) = 0 \tag{99}$

We can write

$$\delta\alpha = \frac{\partial\alpha}{\partial\psi}\delta\psi + \sum_{k=1}^{3} \frac{\partial\alpha}{\partial\left(\dfrac{\partial\psi}{\partial x_k}\right)}\delta\left(\frac{\partial\psi}{\partial x_k}\right) + \frac{\partial\alpha}{\partial\dot\psi}\delta\dot\psi \tag{100}$$

where for convenience x, y, z has been replaced by x_1, x_2, x_3. Since arbitrary variations δ are independent of the differentiation process, we have

$$\delta\left(\frac{\partial\psi}{\partial x_k}\right) = \frac{\partial}{\partial x_k}(\delta\psi); \quad \delta\dot\psi = \frac{\partial}{\partial t}(\delta\psi)$$

and $\qquad$ Eqn. (98) becomes

$$\int_{t_1}^{t_2}\int\int\int \left[\frac{\partial\alpha}{\partial\psi}\partial\psi + \sum_{k=1}^{3}\frac{\partial\alpha}{\partial\left(\dfrac{\partial\psi}{\partial x_k}\right)}\frac{\partial}{\partial x_k}(\delta\psi) + \frac{\partial\alpha}{\partial\dot\psi}\frac{\partial}{\partial t}(\delta\psi)\right]$$

$$dt\, dx_1\, dx_2\, dx_3 = 0 \tag{101}$$

Consider a term in the summation which can be integrated by parts

$$\int \frac{\partial\alpha}{\partial\left(\dfrac{\partial\psi}{\partial x_k}\right)}\frac{\partial}{\partial x_k}(\delta\psi)dx_k = \frac{\partial\alpha}{\partial\left(\dfrac{\partial\psi}{\partial x_k}\right)}\delta\psi - \int \frac{\partial}{\partial x_k}\left\{\frac{\partial\alpha}{\partial\left(\dfrac{\partial\psi}{\partial x_k}\right)}\right\}\delta\psi\, dx_k \tag{102}$$

The integrated term (now surface term in Eqn. (101)) because it has been integrated for one coordinate x_k) on the right side of Eqn. (102) vanishes (because ψ obeys periodic boundary conditions at the walls of a large (finite) box. The last term in Eqn. (101) can also be integrated by parts with respect to time to yield:

$$\left(\frac{\partial\alpha}{\partial\dot\psi}\delta\psi\right)_{t_1}^{t_2} - \int_{t_1}^{t_2}\frac{\partial}{\partial t}\left(\frac{\partial\alpha}{\partial\dot\psi}\right)\delta\psi\, dt \tag{103}$$

with the 1st term vanishing, because of Eqn. (99). So, Eqn. (101) becomes

$$\int_{t_1}^{t_2}\int\int\int \left[\frac{\partial\alpha}{\partial\psi} - \sum_{k=1}^{3}\frac{\partial}{\partial x_k}\left(\frac{\partial\alpha}{\partial\left(\dfrac{\partial\psi}{\partial x_k}\right)}\right) - \frac{\partial}{\partial t}\left(\frac{\partial\alpha}{\partial\dot\psi}\right)\right]$$

$$\delta\psi\, dt\, dx_1\, dx_2\, dx_3 = 0 \tag{104}$$

Since $\delta\psi$'s are arbitrary, Eqn (104) is equivalent to the differential equation

$$\frac{\partial\alpha}{\partial\psi} - \sum_{k=1}^{3} \frac{\partial}{\partial x_k}\left(\frac{\partial\alpha}{\partial\left(\frac{\partial\psi}{\partial x_k}\right)}\right) - \frac{\partial}{\partial t}\left(\frac{\partial\alpha}{\partial\dot\psi}\right) = 0 \tag{105}$$

This is the classical field equation called the *Euler's equation* and corresponds to Lagrange's equation in classical mechanics.

In order to quantise this classical field by the methods employed in quantum mechanics, we must go over to Hamiltonian formalism first. The latter is usually developed for systems with a countable number of degrees of freedom. We are dealing here with a non-denumerably infinite number of degrees of freedom. To this end therefore, we must use some limiting process, we consider a system at a fixed instant of time t, we decompose the space $\bar{x}$ into small cells of volume δx_i (enumerated by the index $i = 1, 2 ....$) The system is now specified by a countable number of variables viz.

$$\phi_i = \psi_i \ (i = 1, 2 ......) \tag{106}$$

giving the values of the field in each cell. Then Lagrangian of the system

$$L(t) = \sum_i \alpha\left[\psi_i,(\Delta\psi)_i,\dot\psi_i,t\right]\delta x_i$$

$$= \sum_i \alpha_i \delta x_i \tag{107}$$

where α_i is the Lagrangian density in the ith cell. We can define the momenta p_i conjugate to ϕ_i by

$$p_i = \frac{\partial L}{\partial\dot\phi_i} = \frac{\partial L}{\partial\dot\psi_i} = \frac{\partial\alpha_i}{\partial\dot\psi_i}\delta x_i \tag{108}$$

The Hamiltonian function is

$$H = \sum_i p_i\dot\phi_i - L \tag{109}$$

To enable us to go to the limit $\delta x_i \to 0$, we define the conjugate field of $\psi(x)$ or momentum density by

$$\Pi(x) = \frac{\partial\alpha}{\partial\dot\psi} \tag{110}$$

So Eqn. (108) reads

$$p_i = \Pi_i\,\delta x_i = \frac{\partial\alpha_i}{\partial\dot\psi_i}\delta x_i \tag{111}$$

and Eqn. (109) becomes

$$H = \sum_i \delta x_i\left(\Pi_i\dot\psi_i - \alpha_i\right) \tag{112}$$

If we now go to the limit $\delta x_i \to 0$, cell summation can be replaced by volume integral,

$$H = \int d^3x\, H\,;\, d^3x = dx_1\, dx_2\, dx_3 \tag{113}$$

where the Hamiltonian density

$$H = \Pi \dot{\psi} - \alpha \tag{114}$$

If the field has more than one component i.e. ψ_1, ψ_2, ..., ψ_α, ... the lagrangian density has the form α (ψ_1, $\nabla \psi_1$, $\dot{\psi}_1$, ψ_2, $\nabla \psi_2$, $\dot{\psi}_2$ ...) and the variational Eqn. (98) leads to an equation of this form (for each of ψ_α)

$$\frac{\partial \alpha}{\partial \psi_\alpha} - \sum_{k=1}^{3} \frac{\partial}{\partial x_k} \left(\frac{\partial \alpha}{\partial \left(\frac{\partial \psi_\alpha}{\partial x_k} \right)} \right) - \frac{\partial}{\partial t} \frac{\partial \alpha}{\partial \dot{\psi}_\alpha} = 0 \tag{115}$$

The momentum density Π_α and Hamiltonian density H are given by

$$\Pi_\alpha = \frac{\partial \alpha}{\partial \dot{\psi}_\alpha}; \quad H = \sum_\alpha \Pi_\alpha \dot{\psi}_\alpha - \alpha \tag{116}$$

22.12 Commutation Relations for Field Operators

The Commutation relations

$$[q_i, q_j] = 0 = [p_i, p_j]$$

and $\quad [q_i, p_j] = i\hbar \, \delta_{ij}$

become

$$[\psi_i, \psi_j] = 0$$
$$[\Pi_i, \Pi_j] = 0 \tag{117}$$

and $\quad [\psi_i, \Pi_j] = i\hbar \dfrac{\delta_{ij}}{\delta x_j} \tag{118}$

for the field operators

In Schrödinger picture, operators are time-independent, so we can write $\psi(x)$ for $\psi(x, t)$ etc. Then

$$[\psi(x), \psi(x')] = [\Pi(x), \Pi(x')] = 0 \tag{119}$$

$$[\psi(x), \Pi(x')] = i\hbar \, \delta(x - x') \tag{120}$$

because in the limit $\delta x_i \to 0$, $\dfrac{\delta_{ij}}{\delta x_j}$ becomes the three dimensional Dirac delta function. (the points x and x' lying in the ith and j^{th} cells respectively). For fields with more than one component, we have

$$[\psi_\alpha(x), \psi_\beta(x')] = [\Pi_\alpha(x), \Pi_\beta(x')] = 0$$

$$[\psi_\alpha(x), \Pi_\beta(x')] = i\hbar \, \delta_{\alpha\beta} \, \delta(x - x') \tag{121}$$

$$\alpha, \beta = 1, 2 \dots N$$

22.13 Quantisation of the Non-relativistic Schrödinger Field (Real or Hermitian, One Component Scalar Field)

We had obtained Schrödinger equation by a process of first quantisation whereby the variables q and p in the classical Hamiltonian were replaced by corresponding operators satisfying the usual commutation relations. We will now see that Schrödinger equation, treated as classical equation is satisfied as though equation of motion with ψ treated as field if we assume Euler's equation and second quantisation.

Our first step is to guess what is the form of the Lagrangian density and we take it to be

$$\alpha = i\hbar\psi^*\dot{\psi} - \frac{\hbar^2}{2m}\nabla\psi^* \cdot \nabla\psi - V(x)\psi^*\psi \tag{122}$$

Then, variation of ψ gives

$$\frac{\partial\alpha}{\partial\psi} = -V\psi^*,$$

$$\sum_{k=1}^{3}\frac{\partial}{\partial x_k}\frac{\partial\alpha}{\partial\left(\frac{\partial\psi}{\partial x_k}\right)} = \nabla\frac{\partial\alpha}{\partial(\nabla\psi)} = -\frac{\hbar^2}{2m}\nabla^2\psi^*$$

and

$$\frac{\partial}{\partial t}\left(\frac{\partial\alpha}{\partial\dot{\psi}}\right) = i\hbar\dot{\psi}^*$$

Euler's equation becomes

$$-i\hbar\dot{\psi}^* = -\frac{\hbar^2}{2m}\nabla^2\psi^* + V\psi^* \tag{123}$$

which is the complex conjugate of Schrödinger equation. Similarly, variation of ψ^* gives

$$i\hbar\dot{\psi} = -\frac{\hbar^2}{2m}\nabla^2\psi + V\psi \tag{124}$$

i.e. the Schrödinger equation

The momentum canonically conjugate to ψ is

$$\Pi = \frac{\partial\alpha}{\partial\dot{\psi}} = i\hbar\psi^* \tag{125}$$

We now quantise this field. The quantum equations are obtained by taking the volume integral of

$$H = \Pi\dot{\psi} - \alpha$$
$$= i\hbar\psi^*\dot{\psi} - \alpha$$
$$= \left(\frac{\hbar^2}{2m}\right)\nabla\psi^* \cdot \nabla\psi + V\psi^*\psi \tag{126}$$

as the Hamiltonian

Since now ψ is an operator (rather than a numerical factor), ψ^* should be replaced by ψ^+ (i.e. its hermitian adjoint)
Then

$$H = \int H d^3x = \int\left(\frac{\hbar^2}{2m}\nabla\psi^+ \cdot \nabla\psi + \nabla\psi^+\psi\right)d^3x \tag{127}$$

Now $H^+ = H$ (i.e. H is hermitian).

The quantised hamiltonian is the operator that represents the total energy of the real scalar field (it should not be confused with the energy operator $i\hbar \frac{\partial}{\partial t}$). Due to Eqn. (125) the commutation relations for field operators are

$$[\psi(x), \psi(x')] = [\psi^+(x), \psi^+(x')] = 0$$

$$[\psi(x), \psi^+(x')] = \delta(x - x') \tag{128}$$

22.14   The Neutral Klein Gordon Field

The simplest example of a relativistic field theory deals with spin-zero particles described by the Klein Gordon equation. The Lagrange density of a real spin-0 field $\phi(x) = \phi(x, t)$ with mass m reads

$$\alpha(x) = \frac{\hbar^2}{2} \frac{\partial \phi}{\partial x_\mu} \frac{\partial \phi}{\partial x^\mu} - \frac{1}{2} m^2 c^2 \phi^2 \tag{129}$$

We subsequently will use natural units of measurement (as is customary in the field theory) i.e. we set $\hbar = c = 1$. The Euler – Lagrange equation

$$\frac{\partial \alpha}{\partial \phi} = \frac{\partial}{\partial x_\mu} \frac{\partial \alpha}{\partial(\partial^\mu \phi)} \tag{130}$$

immediately leads to the Klein Gordon equation

$$(\Box + m^2)\, \phi(x) = 0 \tag{131}$$

where $\Box = \partial^\mu \partial_\mu$ is the four dimensional Laplace operator. The canonically conjugate field is

$$\Pi(x) = \frac{\partial \alpha}{\partial \dot{\phi}(x)} = \dot{\phi}(x) \tag{132}$$

which leads to the Hamiltonian density

$$H(x) = \Pi(x)\, \dot{\phi}(x) - \alpha(x)$$

$$= \frac{1}{2}\left(\Pi^2 + \left(\nabla \phi(x)\right)^2 + m^2 \phi^2\right) \tag{133}$$

Now, we follow the prescription of field quantisation. The fields $\phi(x, t)$ and $\Pi(x, t)$ are replaced by operators for which the equal time commutation relations are

$$[\phi(x, t), \Pi(x', t)] = i\delta(x - x') \tag{134a}$$

$$[\phi(x, t), \phi(x', t)] = [\Pi(x,t), \Pi(x', t)] = 0 \tag{134b}$$

Using the quantised Hamiltonian operator

$$H = \int \frac{1}{2}\left[\left(\Pi(x,t)\right)^2 + \left(\nabla \phi(x,t)\right)^2 + m^2 \left(\phi(x,t)\right)^2\right] d^3x \tag{135}$$

and the above (Eqn. 134) relations, it can be shown that Hamilton's equation of motion lead to

$$\dot{\phi}(x, t) = -i[\phi(x, t), H] = \Pi(x, t) \tag{136a}$$

and $$\dot{\Pi}(x, t) = -i[\Pi(x, t), H] = (\nabla^2 - m^2)\, \phi(x, t) \tag{136b}$$

To obtain the operator ∇^2 in Eqn. (136b), the equation (134a) was differentiated

$$[\Pi(x, t), \nabla'\phi(x', t)] = \nabla'[\Pi(x, t), \phi(x', t)]$$

$$= -i\nabla'\delta(x - x')$$

followed by an integration by parts.

Thus from Eqn. (136b) and (136a)

$$\ddot{\phi}(x, t) = (\nabla^2 - m^2)\,\phi(x_1\,t) \tag{137}$$

i.e. the field operator still satisfies the original Klein Gordon equation. It is to be noted here that it is not a trivial conclusion, for, in general, the equations of motion for the classical fields and for the quantised field operators do not necessarily agree.

22.15   The Dirac Field

With the covariant relativistic notation, the Dirac equation for massive, spin ½ particles reads ($h = c = 1$)

$$(i\gamma^\mu \partial_\mu - m)\psi = 0 \tag{138}$$

where the four Dirac matrices γ^μ, $\mu = 0, 1, 2, 3$ satisfy

$$\gamma^\mu \gamma^\nu + \gamma^\nu \gamma^\mu = 2g^{\mu\nu} \tag{139}$$

The wave function ψ has four components and satisfies the transformation laws of a relativistic spinor.

What will be the Lagrangian that leads to Eqn. (138) as an equation of motion?

The Lagrange density will be a bilinear function composed of the fields ψ, $\dot{\psi}$, $\nabla\psi$ and the hermitean conjugate fields ψ^+, $\dot{\psi}^+$, $\nabla\psi^+$. It has to transform as a Lorentz scalar density and can contain only derivatives of first order since the Dirac equation itself is of first order. Thus we express the Lagrangian as

$$\alpha = i\,\psi^+\dot{\psi} + i\psi^+\vec{\alpha}\cdot\nabla\psi - m\psi^+\beta\psi$$

Here we treat the spinors ψ and ψ^+ as independent fields, each having four components. Differentiation of α with respect to these fields and their time and space derivatives gives

$$\frac{\partial\alpha}{\partial\dot{\psi}} = i\psi^+; \quad \frac{\partial\alpha}{\partial\dot{\psi}^+} = 0$$

$$\frac{\partial\alpha}{\partial(\nabla\psi)} = i\psi^+\alpha; \quad \frac{\partial\alpha}{\partial(\nabla\psi^+)} = 0 \tag{140}$$

$$\frac{\partial\alpha}{\partial\psi} = -m\psi^+\beta; \quad \frac{\partial\alpha}{\partial\psi^+} = i\dot{\psi} + i\alpha\cdot\nabla\psi - m\beta\psi$$

It is to be noted here that the above equations are to be interpreted as matrix equations. For example, the last but one equation reads explicity

$$\frac{\partial\alpha}{\partial\psi_p} = -m\sum_k \psi_k^*\beta_{kp} \tag{141}$$

In most situation, the spinor indices can be dropped and in general, we can use this convention. Variation of the action with respect to ψ^+ leads to the Euter-Lagrange equation

$$\frac{\partial}{\partial t}\frac{\partial \alpha}{\partial \dot{\psi}^+} = \frac{\partial \alpha}{\partial \psi^+} - \nabla \frac{\partial x}{\partial (\nabla \psi^+)} \tag{142}$$

which by use of Eqn (140) simply becomes

$$i\dot{\psi} + i\alpha \cdot \nabla \psi - m\beta\psi = 0 \tag{143}$$

Multiplying by $\gamma^0 = \beta$, we obtain the standard form of Dirac equation

$$(i\gamma^\mu \partial_\mu - m)\psi = 0 \tag{144}$$

as desired. ($\because \beta = \gamma^0$, $\beta^2 = 1$ and $\alpha = \gamma^0\gamma$). Variation with respect to ψ using Eqn. (140) leads to the equation

$$i\dot{\psi}^+ = -m\psi^+\beta - i\nabla \psi^+\alpha \tag{145}$$

This can be identified as hermitean conjugate of Eqn. (143)

$$\bar{\psi}\left(i\gamma^\mu \overleftarrow{\partial}_\mu + m\right) = 0 \tag{146}$$

The arrow indicates that the partial derivative acts on the function to the left.

To proceed further, we need the canonically conjugate fields Π_ψ and Π_{ψ^+}. As in case of Schrödinger field, these are found to be

$$\Pi_\psi = \frac{\partial \alpha}{\partial \dot{\psi}} = i\psi^+; \quad \Pi_{\psi^+} = \frac{\partial \alpha}{\partial \dot{\psi}^+} = 0 \tag{147}$$

Thus there are two independent degrees of freedom of the Dirac field which are given by ψ and ψ^+.

The Hamiltonian density is obtained through the Legendre transformation

$$H = \Pi_\psi \dot{\psi} + \Pi_{\psi^+} \dot{\psi}^+ - \alpha \tag{148}$$

$$= i\psi^+ \dot{\psi} - i\psi^+\dot{\psi} - i\psi^+\vec{\alpha} \cdot \nabla\psi + m\psi^+\beta\psi$$

The Hamiltonian is thus given by the expectation value of Dirac's differential operator $H_D = \vec{\alpha} \cdot \vec{p} + \beta m$ $\tag{149}$

$$H = \int \psi^+ (x,t)(-i\vec{\alpha} \cdot \nabla + \beta m)\psi(x,t)d^3x$$

22.16 The Neutral Klein Gordon Field in Terms of Plane Waves

$$\ddot{\hat{\phi}}(\vec{x},t) = (\nabla^2 - m^2)\hat{\phi}(\vec{x},t) \tag{150}$$

The field operators $\hat{\phi}(x, t)$ can be expanded with respect to a basis. For this, we use the set of plane waves:

$$u_p(\vec{x}) = N_p e^{i\vec{p}\cdot\vec{x}} \tag{151}$$

which means

$$\hat{\phi}(\vec{x},t) = \int d^3p N_p e^{i\vec{p}\cdot\vec{x}}\hat{a}_p(t) \tag{152}$$

Thus,

$$\ddot{\hat{a}}_p(t) = (t) = -(p^2 + m^2)\,\hat{a}_p(t) \tag{153}$$

The general solution of this equation is obviously

$$\hat{a}_p(t) = \hat{a}_p^{(1)}e^{-iw_pt} + \hat{a}_p^{(2)}e^{+iw_pt} \tag{154}$$

where the operators $\hat{a}_p{}^{(1)}$ and $\hat{a}_p{}^{(2)}$ are constant in time and frequency

$$w_p = \sqrt{p^2 + m^2} \tag{155}$$

Since we have a real valued classical field, $\therefore \phi^* = \phi$ and $\therefore$ the field operator should be hermitian i.e. $\hat{\phi}^+ = \hat{\phi}$. This allows us to express one of the $\hat{a}_p$'s to be expressed in terms of the other as

$$\left(\hat{a}_p{}^{(1)}\right)^* = \hat{a}_{-p}{}^{(2)} \tag{156}$$

Then

$$\hat{\phi}(\vec{x},t) = \int d^3 p \, N_p \left(\hat{a}_p e^{i\left(\vec{p}\cdot\vec{x}-w_p t\right)}\right) + \hat{a}_p^+ e^{-i\left(\vec{p}\cdot\vec{x}-w_p t\right)}$$

where $\quad \hat{a}_p \equiv \hat{a}_p{}^{(1)}$ \hfill (157)

Since $\hat{\Pi} = \dot{\hat{\phi}}$, therefore

$$\hat{\Pi}(\vec{x},t) = \int d^3 p N_p \left(-i w_p\right)\left(\hat{a}_p e^{i\left(\vec{p}\cdot\vec{x}-w_p t\right)} - \hat{a}_p^+ e^{-i\left(\vec{p}\cdot\vec{x}-w_p t\right)}\right) \tag{158}$$

The operators $\hat{a}_p$ and $\hat{a}_p^+$ satisfy the commutation relations

$$\left[\hat{a}_p, \hat{a}_{p'}^+\right] = \delta^3\left(\vec{p}-\vec{p}'\right) \tag{159}$$

and $\qquad \left[\hat{a}_p, \hat{a}_{p'}\right] = \left[\hat{a}_p^+, \hat{a}_{p'}^+\right] = 0$ \hfill (160)

Now, it can be proved that the choice

$$N_p = \frac{1}{\sqrt{2w_p \left(2\pi\right)^3}} \tag{161}$$

leads to the desired result

$$\left[\hat{\phi}(\vec{x},t), \hat{\Pi}(\vec{x}',t)\right] = i\delta^3\left(\vec{x}-\vec{x}'\right) \tag{162}$$

We define the normalised time dependent plane waves

$$u_p(\vec{x},t) = N_p e^{+i\left(\vec{p}\cdot\vec{x}-w_p t\right)}$$

$$= \frac{1}{\sqrt{2w_p \left(2\pi\right)^3}} e^{-i\left(w_p t - \vec{p}\cdot\vec{x}\right)} \tag{163}$$

which form a complete set of solutions of the K.G. equation

$$\left(\partial_0^2 - \nabla^2 + m^2\right) u_p(\vec{x},t) = 0 \tag{164}$$

The scalar product of two Klein Gordon wave functions ϕ and X (which is not strictly a scalar product, since it is not positive definite) is defined as

$$(\phi, X) = i\int d^3 x \, \phi^*(x,t) \overleftrightarrow{\partial_0} X(x,t)$$

$$= i\int d^3 x \left[\phi^*(x,t)\frac{\partial X(x,t)}{\partial t} - \frac{\partial \phi^*}{\partial t} X(x,t)\right] \tag{165}$$

Using Eqn. (163), it can be verified that the plane waves form an orthonormal set with respect to old with respect to the above scalar product, i.e.

$$(u_{p'}, u_p) = i \int d^3x \, u_{p'}^*(\vec{x},t) \, \partial_0 u_p(\vec{x},t) = \delta^3(\vec{p} - \vec{p}')$$

$$\left(u_{p'}^*, u_p^*\right) = -\delta^3(\vec{p} - \vec{p}') \tag{166}$$

Similarly, the plane waves with opposite signs for the frequency are orthogonal

$$\left(u_{p'}, u_p^*\right) = \left(u_{p'}^*, u_p\right) = 0 \tag{167}$$

Projection of the field operator

$$\hat{\phi}(x,t) = \int d^3p \left(\hat{a}_p u_p(x,t) + \hat{a}_p^+ u_p^*(x,t)\right) \tag{168}$$

on u_p and u_p^* gives the Fourier coefficients

$$\hat{a}_p = \left(u_p, \hat{\phi}\right) = i \int d^3x \, u_p^*(\vec{x},t) \overleftrightarrow{\partial}_0 \hat{\phi}(\vec{x},t) \tag{169}$$

$$\hat{a}_p^+ = -\left(u_p^*, \hat{\phi}\right) = -i \int d^3x \, u_p(\vec{x},t) \overleftrightarrow{\partial}_0 \hat{\phi}(\vec{x},t) \tag{170}$$

22.17 The Charged Klein Gordon Field

The real valued Klein Gordon field was found to describe a collection of spin zero particles of identical type. This can be generalized to particles having an internal degree of freedom. The simplest generalisation of this introduces a doublet of particles and antiparticles that can be described by going over to complex fields $\phi \neq \phi^*$ (and consequently, non-hermitean field operators $\hat{\phi} \neq \hat{\phi}^+$).

The Lagrange density, which remains a real-valued function can be described as

$$\alpha = \frac{\partial \phi^*}{\partial x_\mu} \frac{\partial \phi}{\partial x^\mu} - m^2 \, \phi^* \phi \tag{171}$$

where ϕ and ϕ^* can be treated as independent fields. The two canonically conjugate fields are

$$\Pi = \frac{\partial \alpha}{\partial \dot{\phi}} = \dot{\phi}^* \tag{172}$$

$$\Pi^* = \frac{\partial \alpha}{\partial \dot{\phi}^*} = \dot{\phi} \tag{173}$$

and the Hamiltonian becomes

$$H = \int d^3x \left(\Pi \partial_0 \phi + \Pi^* \partial_0 \phi^* - \alpha\right)$$

$$= \int d^3x \left(\Pi^* \Pi + \left(\nabla \phi^*\right) \cdot \left(\nabla \phi\right) + m^2 \phi^* \phi\right) \tag{174}$$

It is to be noted here that the factor ½ does not come in the expressions for α and H.

Quantisation is achieved by going over to field operators $\hat{\phi}$, $\hat{\phi}^+$, Π and Π, Π^+ These satisfy the commutation relations

$$\left[\hat{\phi}(\vec{x},t)\hat{\Pi}(x',t)\right] = \left[\hat{\phi}^+(\vec{x},t)\hat{\Pi}^+(\vec{x}',t)\right] = i\delta^3(\vec{x} - \vec{x}')$$

$$\left[\hat{\phi}(\vec{x},t),\hat{\phi}^+(\vec{x},t)\right] = 0; \quad \left[\hat{\Pi}(\vec{x},t),\hat{\Pi}^+(\vec{x},t)\right] = 0$$

$$\left[\hat{\phi}(\hat{x},t),\hat{\phi}^+(\vec{x}',t)\right] = 0; \quad \left[\hat{\Pi}(\vec{x},t),\hat{\Pi}(\hat{x}',t)\right] = 0 \tag{175}$$

The Fourier decomposition of the field operator is replaced by

$$\hat{\phi}(\vec{x},t) = \int d^3p\left(\hat{a}_p u_p + \hat{b}_p^+ u_p^*\right)$$

$$\hat{\phi}^+(\vec{x},t) = \int d^3p\left(\hat{a}_p^+ u_p^* + \hat{b}_p u_p\right) \tag{176}$$

Since the field operator is no longer hermitean, $\therefore$ $\hat{\phi}^+ \neq \hat{\phi}$, and the coefficients $\hat{b}_p^+$ now cannot be expressed in terms of $\hat{a}_p$. Rather, there are two independent sets of creation and annihilation operators. They obey the following commutation relations

$$[\hat{a}_p,\hat{a}_{p'}^+] = [\hat{b}_p,\hat{b}_{p'}^+] = \delta^3\left(\vec{p}-\vec{p}'\right)$$

$$[\hat{a}_p,\hat{a}_{p'}] = [\hat{b}_p,\hat{b}_{p'}] = [\hat{a}_p^+,\hat{a}_{p'}^+] = [\hat{b}_p^+,\hat{b}_{p'}^+] = 0$$

$$[\hat{a}_p,\hat{b}_{p'}] = [\hat{a}_p,\hat{b}_{p'}^+] = [\hat{a}_p^+,\hat{b}_{p'}] = [\hat{a}_p^+,\hat{b}_{p'}^+] = 0 \tag{177}$$

The Hamiltonian is

$$\hat{H} = \int d^3p\, w_p\left(\hat{a}_p\,\hat{a}_p^+ + \hat{b}_p^+\,\hat{b}_p\right)$$

$$= \int d^3p\, w_p\left(\hat{a}_p^+\,\hat{a}_p + \hat{b}_p^+\,\hat{b}_p\right)$$

$$\equiv \int d^3p\, w_p\left(\hat{n}_p^{(a)} + \hat{n}_p^{(b)}\right) \tag{178}$$

and the momentum operator is

$$\hat{p} = \int d^3p\,\vec{p}\left(\hat{a}_p^+\,\hat{a}_p + \hat{b}_p^+\,\hat{b}_p\right)$$

$$\equiv \int d^3p\,\vec{p}\left(\hat{n}_p^{(a)} + \hat{n}_p^{(b)}\right) \tag{179}$$

Thus, the theory describes two independent types of particles a and b having the same mass m. A Fock space can be constructed, starting from the vacuum state which is defined to contain neither a nor b particles.

$$\hat{a}_p|0\rangle = \hat{b}_p|0\rangle = 0 \quad \text{for all p.} \tag{180}$$

i.e. General many-particle states are now characterized by two sets of occupation numbers $\{n_i^{(a)}\}$ and $\{n_i^{(b)}\}$. Closer inspection of Lagrange density (Eqn. 171) shows that it exhibits a symmetry under phase transformations

$$\phi' = \phi\, e^{i\alpha}, \quad \phi^{*'} = \phi^*\, e^{-i\alpha} \tag{181}$$

with real phases α

Noether's theorem tells us that this symmetry transformation leads to a conserved quantity called the charge:

$$Q = \int d^3x\, j^0(x) = -i\int d^3x\left(\frac{\partial\alpha}{\partial\Pi^*}\phi - \frac{\partial\alpha}{\partial\Pi}\phi^*\right)$$

$$= -i \int d^3x \left(\Pi \phi - \Pi^* \phi^* \right) \tag{182}$$

Because of the relation $\Pi = \dot{\phi}^*$ and $\Pi^* = \dot{\phi}$, this can also be written as

$$Q = i \int d^3x \, \phi^* \overleftrightarrow{\partial}_o \phi = (\phi, \phi) \tag{183}$$

and thus, the charge becomes an operator

$$\hat{Q} = -i \int d^3x \left(\hat{\Pi} \hat{\phi} - \hat{\Pi}^+ \hat{\phi}^+ \right) \tag{184}$$

Insertion of Eqn (176) for $\hat{\phi}$ and $\hat{\phi}^+$ and the corresponding expansions of $\hat{\Pi}$ and $\hat{\Pi}^+$ leads to

$$\begin{aligned}
\hat{Q} &= \frac{1}{2} \int d^3 p \left(\hat{a}_p^+ \hat{a}_p + \hat{a}_p \hat{a}_p^+ - \hat{b}_p^+ \hat{b}_p - \hat{b}_p \hat{b}_p^+ \right) \\
&= \int d^3 p \left(\hat{a}_p^+ \hat{a}_p - \hat{b}_p^+ \hat{b}_p \right) \\
&\equiv \int d^3 p \left(\hat{n}_p^{(a)} - \hat{n}_p^{(b)} \right)
\end{aligned} \tag{185}$$

The charge remains a conserved quantity and the charge operator satisfies the equation of motion.

$$\dot{\hat{Q}} = -i \left[\hat{Q}, \hat{H} \right] = 0 \tag{186}$$

Collecting our results, we find that the operator $\hat{a}_p^+$ creates a particle having energy w_p, momentum p and charge $+1$ whereas $\hat{b}_p^+$ creates a particle with the same energy, the same momentum but opposite charge -1. The b particles are therefore called the antiparticles of the a particles. Here, the notion of charge is not restricted to the electrical charge. Other types of charge can arise from internal symmetries.

On the other hand, particles described by a real valued K.G. field are strictly neutral. They donot possess antiparticles.

Questions and Problems

1. (a) Explain the term second quantization.
   (b) Describe occupation number representation in the case of Bose-Einstein and Fermi-Dirac statistics.
2. How does second quantization lead to Bose-Einstein and Fermi-Dirac statistics.
3. Quantize the electromagnetic field in free space and develop the concept of a photon as an entity carrying discrete amount of energy and quantum.
4. Show that an electromagnetic field can be though of as mathematically equivalent to a system of harmonic oscillators.
5. Write short notes on:
   (i) Second quantization
   (ii) Creation, annihilation operators and Fock states.
   (iii) Second quantization applied to Bosons and Fermions.
6. Explain the concept of phonons.

23

The Einstein-Podolsky-Rosen Paradox

23.1 The Einstein-Podolsky-Rosen Problem

There is a widely held misbelief that Einstein did not believe in quantum mechanics. In his autobiographical notes (Schilpp, 1970), he says

"I must (now) take a stand with reference to the most successful physical theory of our period, viz., the statistical quantum theory which about twenty five years ago, took on a consistent logical form (at the hands of Schrödinger, Heisenberg, Dirac and Born). This is the only theory which permits a unitary grasp of experiences concerning the quantum character of micro-mechanical events."

Thus, it is clear that it is not that Einstein disbelieved in quantum mechanics. He only objected to the ultra-positivism of the Copenhagen interpretation and assertion that quantum theory is a complete theory. Being probabilistic, it must be regarded as incomplete and could be made complete by supplementing it with parameters or variables whose values cannot presumably be controlled while preparing a state. Such parameters or variables were referred to as *hidden variables*. Their existence was desirable to rectify the incompleteness of quantum theory.

EPR considered an example of a decay of a composite system (say with spin zero) into two subsystems (say with spin ½ each) which after the decay move far apart from each other with opposite momenta in the rest frame of the composite system. A measurement of the spin component (in any chosen direction ê) is then considered to be carried out on one of the subsystem A. The result will be either +½ or –½. This would imply, by virtue of conservation of the total spin angular momentum, that a spin measurement on the other system B along the same direction ê must yield the result with certainty as –½ or +½. Since this prediction with certainty, of the result of measurement on B has been obtained without in any way disturbing B, the existence of spin –½ (or +½) along ê is considered by EPR to be an element of reality pertaining to B.

However, since a similar conclusion for the system B would follow if one had chosen to make a measurement of the spin component of A in another direction say ê', and since the result of measurement will again inevitably be +½ (or –½), are would be led to conclude that the spin component of *B* will, with certainty, be found to have the value –½ (or +½) now in the direction ê'. Again this conclusion follows without disturbing the system *B* in any way. Hence according to EPR's criterion (of reality), this is also an element of reality pertaining to *B*.

In particular, if ê and ê', are orthogonal, this would amount to knowing with certainty the spin components of B in two mutually orthogonal directions. Now, according to quantum theory, this is impossible since the spin components in two orthogonal directions do not commute. However, EPR argue that since the existence of spin components in two orthogonal directions should according to the criterion of reality be considered as the elements of reality[*] and since Quantum Mechanics does not admit them in its formalism, it cannot be regarded as a complete theory.

23.2 Hidden Variables and Einstein's Locality Principle

The classical physics complies with the deterministic nature involved in the occurrence of a phenomenon. However quantum mechanics pertains to the non-deterministic character which is quite unfamiliar to imagination, but inherent to the quantum theory. In an attempt to employ deterministic approach, attempts were made to replace quantum theory by a statistical theory, on the lines of classical physics. In quantum mechanics measurement of a physical variable is governed by a set of rules which purport to give the wave function only after the measurement has been performed on the system. In this respect developed the concept of wavefunction-reduction: when the eigenvalue of an observable (operator *A*) is '*a*', the wave function at the end of the measurement must be the eigenvector |*a*>. The reduction-rule was subject to a great difficulty and dissatisfaction. The point was that quantum mechanics relies on dynamics expressed by Schrödinger's equation. However, Bohr augmented that the reduction rule is only one of its kind in the whole realm of physics having no other analogue because its role is to bridge the gap between classical physics and mathematical concepts of quantum physics.

In attempts to give a statistical theory (closer to deterministic nature) it was postulated that there exist certain *hidden variables*, whose values prescribe the values of all observables for any particular object and these variables are unknown to the experimenter, thus yielding the probabilistic character to the theory. In this context, the situation becomes analogous to classical statistical mechanics, where motion of all the particles, in principle is known. In absence of hidden variables the quantum theory is ascribed to have incompleteness due to indeterminism. Einstein attempted to demonstrate this incompleteness and tried to get around the indeterminism. According to his own words:

---

[*] Criterion of reality–

" If without in any way disturbing a system, we can predict with certainty the value of a physical quantity, then there exists an element of physical reality corresponding to this physical quantity". (— Einstein et al, 1935)

"On one supposition we should, in my opinion, absolutely hold fast: the real factual situation of the system S_2 is independent to what is done with the system S_1, which is spatially separated from the former."

This is known as *Einstein's locality principle*. Because this problem was first discussed (in 1935) in a paper by A. Einstein, B. Podolsky and N. Rosen, it is also known as the Einstein-Podolsky-Rosen paradox.

23.3 The Einstein-Podolsky-Rosen (EPR) Experiment

The following thought experiment was first conceived by Einstein-Podolsky and Rosen in 1935 and later reformulated by Bohm in 1951. This played an important role in the discussion of indeterminism and the existence of hidden variables.

Let there be two spin-½ particles in a singlet state:

$$|0,0\rangle = \frac{1}{\sqrt{2}}\left(|\uparrow\rangle|\downarrow\rangle - |\downarrow\rangle|\uparrow\rangle\right) \tag{1}$$

which are emitted from a source and then moving apart. Even if the two particles are separated by an arbitrary large distance (no communication with one another), one finds in the state (eqn. 1) the following correlations in a measurement of the one particle spin states:

If one measures the z-component of the spin and finds for particle 1 spin-up, particle 2 then has spin-down, and vice-versa. This expresses the non-locality of quantum theory. The experiment on particle 1 influences the result of the experiment on particle 2, although they are far apart. Since particle 1 takes values $\pm \hbar/2$ 50% of the time, this remains true for particle 2 and only by subsequent comparison of the results it is possible to verify the correlation.

Einstein-Podolsky and Rosen gave following arguments in favour of hidden variables:

By the measurement of S_z (or S_x) of particle 1, the values of S_z (or S_x) of particle 2 are known. Because of the separation of the particles, there was no influence on particle 2 and therefore the value of S_z (or S_x) must have been fixed before the experiment. Thus there must be a more complete theory with hidden variables. In the EPR argument, the consequences of quantum mechanical state (eqn. 1) are used but the inherent non-locality is denied.

An immediate consequence of the EPR experiment is Bell's inequality where it has been pointed out by J.S. Bell that the alternative theories based on Einstein's locality principle actually predict a testable inequality relation what is known as Bell's inequality.

23.4 Bell's Inequality

Suppose we consider a correlation experiment in which a particle of total spin zero decays into two particles each with spin ½. We may consider correlation between various angular settings for polarizers used to detect either of the two particles. A measure of the correlation is $N(\alpha, \beta)$ defined by the relative number of particle pairs: particle 1 having positive spin at the angle α and particle 2, positive spin at angle β. If hidden variables were really present in nature, we could represent $N(\alpha; \beta)$ by the sum:

$$N(\alpha; \beta) = N(\alpha\gamma; \beta) + N(\alpha; \gamma\beta) \tag{2}$$

where $N(\alpha\gamma; \beta)$ is the relative number of particle pairs in which particle has positive spin at the angles α and γ and negative spin at angle β. $N(\alpha; \gamma\beta)$ is the relative number of pairs in which particle 1 has negative spin at angle γ and particle 2, a positive spin at angle γ.

Now,

$$N(\alpha\gamma; \beta) \leq N(\gamma; \beta) \tag{3}$$

because

$$N(\gamma; \beta) = N(\alpha\gamma; \beta) + N(\gamma; \beta\alpha) \tag{4}$$

and both quantities on r.h.s. are non negative. Similarly

$$N(\alpha; \gamma\beta) \leq N(\alpha; \gamma) \tag{5}$$

Hence

$$N(\alpha; \beta) \leq N(\alpha; \gamma) + N(\gamma; \beta) \tag{6}$$

This is a simple version of Bell's inequality and it disagrees with the predictions of quantum mechanics.

24

The Aharonov-Bohm Effect

We now discuss a very peculiar effect which appears in the study of a particle in the presence of an electro-magnetic field. This effect was predicted by Y. Aharonov and D. Bohm [Phys. Rev. 115, 485 (1959)]. and experimentally observed by RG Chambers [Phys. Rev. Lett 5, 3 (1960)].

24.1 Experiemental Set-up

The experimental set up is schematically described in Fig. 24.1.

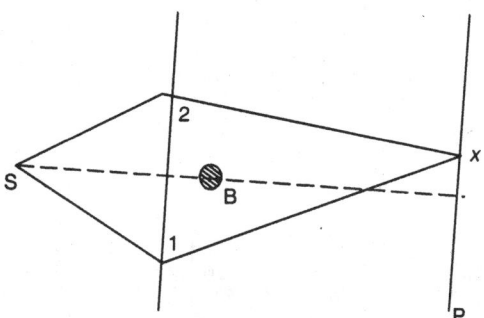

Fig. 24.1: Set-up to study Aharonov-Bohm interference experiment

The source S emits electrons, which after traversing the screen through the openings 1 and 2 generate typical interference pattern at the screen P. There is a solenoid, ideally of infinite length located at B so that the magnetic field is completely confined to its interior and it is perpendicular to the plane of the figure (The cross-section of the solenoid is sufficiently small so that the diffracted electrons have a negligible probability to penetrate the interior). When we increase the magnetic field of the solenoid, the relative phase (as we will discuss subsequently) of the particles passing through slit 2 with respect to those passing through slit 1 changes and thus the interference pattern produced on the screen will shift, even though the electrons never passed through the region where $B \neq 0$. This notable effect (known as Aharonov-Bohm effect) shows that for the local effects on the wave functions, it is the electromagnetic potentials and not the fields which are the fundamental quantities in quantum mechanics.

24.2 Change of the Wave Function under a Gauge Transformation

We know that the Lorentz force depends only on $\vec{B}$ whereas Schödinger equation contains the vector potential $\vec{A}$:

$$\left[\frac{1}{2m}\left(\frac{\hbar}{i}\Delta - \frac{e}{c}\vec{A}(\vec{x},t)\right)^2 + e\phi(\vec{x},t)\right]\psi(\vec{x},t) = i\hbar\frac{\partial\psi(\vec{x},t)}{\partial t} \tag{1}$$

A question thus arises whether ψ depends on the gauge and whether charged particle reacts to $\vec{B}$ or $\vec{A}$. To enquire this, we consider the influence of gauge transformation

$$\vec{A} \rightarrow \vec{A}' = \vec{A} + \nabla\Lambda$$

$$\phi \rightarrow \phi' = \phi - \frac{1}{c}\frac{\partial}{\partial t}\Lambda \tag{2}$$

where $\Lambda(\vec{x}, t)$ is a scalar funtion. Now, we would see that the wave function $\psi'(\vec{x}, t)$ in the second gauge (characterized by the primed potentials) is

$$\psi'(\vec{x}, t) = \exp\left\{\frac{ie}{\hbar c}\Lambda\right\}\psi(\vec{x}, t) \tag{3}$$

Multiplying Eqn. (1) by $\exp\left\{\frac{ie}{\hbar c}\Lambda\right\}$ and using the identity

$$e^{f(y)}\frac{\partial}{\partial y} = \left(\frac{\partial}{\partial y} - \frac{\partial f}{\partial y}\right)e^{f(y)} \tag{4}$$

leads to

$$\left[\frac{1}{2m}\left(\frac{\hbar}{i}\Delta - \frac{e}{c}\vec{A} - \frac{\hbar}{i}\frac{ie}{\hbar c}\nabla\Lambda\right)^2 + e\phi\right]\exp\left\{\frac{ie\Lambda}{\hbar c}\right\}\psi$$

$$= i\hbar\left(\frac{\partial}{\partial t} - \frac{ie}{\hbar c}\frac{\partial\Lambda}{\partial t}\right)\exp\left\{\frac{ie\Lambda}{\hbar c}\right\}\psi$$

which is identical to

$$\left[\frac{1}{2m}\left(\frac{\hbar}{i}\nabla - \frac{e}{c}\vec{A}'\right)^2 + e\phi'\right]\psi' = i\hbar\frac{\partial\psi}{dt} \tag{5}$$

Thus, the gauge transformation introduces an additional space and time dependent phase factor into the wave function.

24.3   The Wave Function in a Field Free Region

Consider now the motion of an electron in the presence of a time independent magnetic field $\vec{B}(\vec{x})$. Suppose the motion is in the region of space with

$$\vec{B}(= \text{curl } \vec{A}) = 0 \tag{6}$$

In this region, $\vec{A}$ can be expressed as

$$\vec{A} = \nabla\Lambda \qquad (7)$$

and $\quad \Lambda(\vec{x}) = \int_{x_o}^{x} d\vec{l} \cdot \vec{A}(l) \qquad (8)$

The wave function in this field free region can be found either from

$$\frac{1}{2m}\left(\frac{\hbar}{i}\nabla - \frac{e}{c}\vec{A}\right)^2 \psi + V\psi = i\hbar\frac{\partial\psi}{\partial t}$$

or from the gauge transformed equation with

$$\vec{A}' = A + \nabla(-\Lambda) = 0 \qquad (9)$$

i.e. $\qquad \frac{1}{2m}\left(\frac{\hbar}{i}\nabla\right)^2 \psi' + V\psi' = i\hbar\frac{\partial\psi'}{\partial t} \qquad (10)$

Thus, if we replace Λ in eqn (3) by $-\Lambda$, we obtain the following relation between these two wavefunctions

$$\psi = \psi' \exp\left\{\frac{ie}{\hbar c}\Lambda\right\}$$

$$\equiv \psi' \exp\left\{\frac{ie}{\hbar c}\int_{xo}^{x} dl\, A(l)\right\} \qquad (11)$$

Here V represents influences which are not electric in origin (Since there is no electric field, we have set $\phi = 0$), ψ' is the wave function in the potential V with field $\vec{B} = 0$.

24.4 The Theory of the Aharonov-Bohm Experiment

Consider the experiment depicted in Fig. 24.1. The electrons cannot penetrate into the coil i.e. they move only within the region $\vec{B} = 0$

In order to find the solution (for ψ) as a function of the field, we determine first the solutions with only one slit open at a time and then linearly superimpose the two.

Let $\psi_{1,B}(\vec{x})$ be the wave function when only slit 1 is open. By Eqn. (11), it can be obtained from the field-free wave function $\psi_{1,0}$ and becomes

$$\psi_{1,B}(\vec{x}) = \psi_{1,0}(\vec{x})\exp\left\{\frac{ie}{\hbar c}\int_{1} dl\, A(l)\right\} \qquad (12)$$

where the line integral is from the source through slit 1 to x. Similarly, for the wave function when only slit 2 is open,

$$\psi_{2,B}(\vec{x}) = \psi_{2,0}(\vec{x})\exp\left\{\frac{ie}{\hbar c}\int_{2} dl\, A(l)\right\} \qquad (13)$$

where the line integral runs from the source to x through slit 2. When both slits are open, we have

$$\psi_B(\vec{x}) = \psi_{1,0}(\vec{x})\exp\left\{\frac{ie}{\hbar c}\int_1 dl\, A(l)\right\} + \psi_{2,0}(\vec{x})\exp\left\{\frac{ie}{\hbar c}\int_2 dl\, A(l)\right\}$$

(14)

The relative phase of the two terms is

$$\int_1 dlA(l) - \int_2 dlA(l) = \oint dl\, A(l) = \int da\,\text{curl}\,\vec{A} = \phi_B$$

(15)

where da is an elemental area and ϕ_B is the magnetic flux in the solenoid. Thus,

$$\psi_B(x) = \left(\psi_{1,0}(\vec{x})\exp\left\{\frac{ie}{\hbar c}\phi_B\right\} + \psi_{2,0}(\vec{x})\right) \times \exp\left\{\frac{ie}{\hbar c}\int_2 dl\, A(l)\right\}$$

The phase relation between ψ_1 and ψ_2 changes under a change in the enclosed flux ϕ_B and thus, the interference pattern is also shifted, i.e. the position of the interference maxima are shifted due to the variation in ϕ_B although the electron does not penetrate into the region of non-zero magnetic field. This is the Aharonov-Bohm effect.

The unit of the flux here is

$$2\phi_0 = \frac{2\pi\hbar c}{e} = 4.135 \times 10^{-7}\text{ Gauss cm}^2$$

Appendix A

Alternative Derivation of Uncertainty Principle

Let Δq and Δp denote dispersions in the simultaneous measurements of canonically conjugate variables q and p, satisfying $[q, p] = i\hbar$.

Then uncertainty in the value of a physical quantity is the positive square root of the mean square deviation, i.e.

$$\text{(uncertainty in } q)^2 = <(\Delta q)^2> = <(q - <q>)^2>$$
$$\text{(uncertainty in } p)^2 = <(\Delta p)^2> = <(p - <p>)^2>$$

Let
$$q - <q> = \Delta q \equiv \alpha \text{ and } p - <p> = \Delta p \equiv \beta$$

Then

$$<(\Delta q)^2> = <\alpha^2> = \int \bar{\psi}\alpha^2\psi dx \,; \; <(\Delta p)^2> = <\beta^2> = \int \bar{\psi}\beta^2\psi dx$$

$$\therefore \quad <\alpha^2> <\beta^2> = \int \bar{\psi}\alpha^2\psi dx \int \bar{\psi}\beta^2\psi dx = \int \bar{\psi}\alpha\alpha\psi dx \int \bar{\psi}\beta\beta\psi dx$$

$$= \int \overline{\alpha\psi}\alpha\psi dx \int \overline{\beta\psi}\beta\psi dx = \int |\alpha\psi|^2 dx \int |\beta\psi|^2 dx \qquad (i)$$

Now, according to Schwarz inequality

$$\int |f|^2 dx \int |g|^2 dx \geq \left| \int \bar{f} g \, dx \right|^2$$

with $f = \alpha x$ and $g = \beta x$, the inequality becomes

$$\int |\alpha\psi|^2 dx \int |\beta\psi|^2 dx \geq \left| \int \overline{\alpha\psi}\beta\psi dx \right|^2$$

or
$$\int |\alpha\psi|^2 dx \int |\beta\psi|^2 dx \geq \left| \int \bar{\psi}\alpha\beta\psi dx \right|^2$$

or
$$\int |\alpha\psi|^2 dx \int |\beta\psi|^2 dx \geq \left| \int \bar{\psi}\frac{1}{2}(\alpha\beta - \beta\alpha)\psi dx \int \bar{\psi}\frac{1}{2}(\alpha\beta + \beta\alpha)\psi dx \right|^2$$

or
$$\int |\alpha\psi|^2 dx \int |\beta\psi|^2 dx \geq \left| \int \bar{\psi}\frac{1}{2}(\alpha\beta - \beta\alpha)\psi dx \int \bar{\psi}\frac{1}{2}(\alpha\beta + \beta\alpha)\psi dx \right|^2$$

or
$$\int |\alpha\psi|^2 \, dx \int |\beta\psi|^2 \, dx \geq \frac{1}{4} \left| \int \overline{\psi}(\alpha\beta - \beta\alpha)\psi dx \right|^2 + \frac{1}{4} \left| \int \overline{\psi}(\alpha\beta + \beta\alpha)\psi dx \right|^2$$

$$+ \frac{1}{2} \left| \int \overline{\psi}(\alpha\beta - \beta\alpha)\psi dx \int \overline{\psi}(\alpha\beta + \beta\alpha)\psi dx \right| \qquad (ii)$$

or
$$\int |\alpha\psi|^2 \, dx \int |\beta\psi|^2 \, dx \geq \frac{1}{4} \left| \int \overline{\psi}(\alpha\beta - \beta\alpha)\psi dx \right|^2$$

$$+ \frac{1}{4} \left| \int \overline{\psi}(\alpha\beta + \beta\alpha)\psi dx \right|^2 \qquad (iii)$$

Because the cross term i.e. the 3rd term on the r.h.s. of (ii) is

$$\left| \left(\int \overline{\psi}\alpha\beta\psi dx \right)^2 - \left(\int \overline{\psi}\beta\alpha\psi dx \right)^2 \right| \qquad (iv)$$

In this expression, we note that

$$\left(\overline{\psi}\alpha\beta\psi \right) = \psi\overline{\alpha}\left(\overline{\beta}\overline{\psi} \right) = \alpha\psi\left(\overline{\beta}\overline{\psi} \right) = (\beta\alpha\psi)\overline{\psi} \qquad (a)$$

and
$$\overline{\psi}\alpha\beta\psi = \overline{(\beta\alpha\psi)\overline{\psi}} = \overline{\psi}(\beta\alpha\psi) \qquad (b)$$

Multiplying (a) and (b) after integrating (and then squaring) we find expression (iv) i.e. the cross term vanishes.

Further, since $(\alpha\beta + \beta\alpha)$ is Hermitian, its expectation value $\int \overline{\psi}(\alpha\beta + \beta\alpha)\psi dx$ will be real and positive, whose square is non-negative hence if it is dropped from r.h.s., we still have

$$\int |\alpha\psi|^2 \, dx \int |\beta\psi|^2 \, dx \geq \frac{1}{4} \left| \int \overline{\psi}(\alpha\beta - \beta\alpha)\psi dx \right|^2 \qquad (v)$$

Now,
$$(\alpha\beta - \beta\alpha) = [\alpha, \beta] = [\Delta q, \Delta p]$$

$$= [q - <q>, p - <p>] = i\hbar \{q, p\} = i\hbar$$

$\therefore$
$$<\alpha^2> <\beta^2> \geq \frac{\hbar^2}{4}$$

or
$$<(\Delta q)^2> <(\Delta p)^2> \geq \frac{\hbar^2}{4} \text{ or } \Delta q \, \Delta p \geq \frac{\hbar}{2}$$

Appendix B

Integrals Involving Gaussian Functions

Consider the integral

$$I = \int_{-\infty}^{+\infty} e^{-x^2} dx$$

Therefore

$$I^2 = \int_{-\infty}^{+\infty} e^{-x^2} dx \int_{-\infty}^{+\infty} e^{-y^2} dy$$

$$= \int_{-\infty}^{+\infty} \int_{-\infty}^{+\infty} e^{-\left(x^2+y^2\right)} dxdy$$

Using polar coordinates, we get

$$I^2 = \int_{0}^{\infty} e^{-r^2} r dr \int_{0}^{\pi} d\theta = \left[-\frac{1}{2} e^{-r^2} \right]_{0}^{\infty} 2\pi = \pi$$

Thus

$$I = \int_{-\infty}^{+\infty} e^{-x^2} dx = \sqrt{\pi} \qquad (i)$$

Now,

$$\int_{-\infty}^{+\infty} e^{-\alpha x^2 + \beta x} dx = \exp\left(\frac{\beta^2}{4\alpha}\right) \int_{-\infty}^{+\infty} \exp\left[-\alpha \left(x - \frac{\beta}{2\alpha} \right)^2 \right] dx$$

$$= \exp\left(\frac{\beta^2}{4\alpha}\right) \int_{-\infty}^{+\infty} e^{-\alpha z^2} dz \qquad \text{putting } z = \left(x - \frac{\beta}{2\alpha} \right)$$

Using Eqn. (i)

$$\int_{-\infty}^{+\infty} e^{-\alpha x^2} dx = \sqrt{\frac{\pi}{\alpha}}$$

Therefore

$$\int_{-\infty}^{+\infty} e^{-\alpha x^2 + \beta x} dx = \sqrt{\frac{\pi}{\alpha}} \exp\left(\frac{\beta^2}{4\alpha}\right) \quad \text{for Re } \alpha > 0 \qquad (ii)$$

Gamma function is defined through the equation

$$\Gamma(t) = \int_0^\infty x^{t-1} e^{-x} dx \quad \text{Re } t > 0 \tag{iii}$$

$$\Gamma(t) = (t-1)\Gamma(t-1)$$

Since $\qquad \Gamma(1) = \int_0^\infty e^{-x} dx = 1 \; \therefore \; \Gamma(n+1) = n! \qquad (n = 0,1,2,...)$

Using Eqn. (*i*), we get

$$2\int_0^\infty e^{-x^2} dx = \sqrt{\pi}$$

or

$$\int_0^\infty y^{-1/2} e^{-y} dy = \sqrt{\pi}$$

$\Rightarrow$

$$\Gamma\left(\tfrac{1}{2}\right) = \sqrt{\pi}$$

Appendix C

Tensors

A scalar is specified by one real number and is a tensor of rank zero. In three dimensional space, a vector is specified by $3 = 3^1$ real numbers (e.g. its cartesian components) and is a tensor of rank one.

A tensor of rank n has 3^n components which transform in a definite way. In N-dimensional space, a tensor of rank n has N^n components.

Contravariant and covariant vectors (Tensors of rank one)

Any set of quantities A_j transforming according to

$$A_i' = \sum_j \frac{\partial x_i'}{\partial x_j} A_j$$

is defined as a contravariant vector. For example,

$$A_i' = \sum_j a_{ij} A_j \qquad (i)$$

where

$$a_{ij} = \frac{\partial x_i'}{\partial x_j}$$

We may also have a slightly different type of vector transformation. The gradient of a scalar ϕ defined by

$$\nabla \phi = \hat{i}\, \frac{\partial \phi}{\partial x_1} + \hat{j}\, \frac{\partial \phi}{\partial x_2} + \hat{k}\, \frac{\partial \phi}{\partial x_3}$$

(using x_1, x_2, x_3 for x, y, z) transforms as

$$\frac{\partial \phi'}{\partial x_i'} = \sum_j \frac{\partial \phi}{\partial x_j} \frac{\partial x_j}{\partial x_i'} \qquad (ii)$$

Using $\phi = \phi(x, y, z) = \phi(x', y', z') = \phi'$

723

It is to be noted that Eqn. (*ii*) differs from Eqn. (*i*) in that we have $\dfrac{\partial x_j}{\partial x_i'}$ here instead of $\dfrac{\partial x_i'}{\partial x_j}$. Eqn. (*ii*) defines a covariant vector with reference to gradiant operator.

Tensors of Rank Two

Consider the following equations

$$A'^{ij} = \sum_{kl} \frac{\partial x_i'}{\partial x_k} \frac{\partial x_j'}{\partial x_l} A^{kl}$$

$$B_j'^{i} = \sum_{kl} \frac{\partial x_i'}{\partial x_k} \frac{\partial x_l}{\partial x_j'} B_l^k$$

$$C_{ij}' = \sum_{kl} \frac{\partial x_k}{\partial x_i'} \frac{\partial x_l}{\partial x_j'} C_{kl}$$

Thus the rank is just the number of partial derivatives in the definition. Each index (subscript or superscript) ranges over the number of dimensions of space. The rank of tensor is independent of the dimensions of the space. We see that A^{kl} is contravariant with respect to both the indices k and l. C_{kl} is covariant with respect to both the indices k and l but B_l^k transforms contravariantly with respect to the first index k and covariantly with respect to the index l, so it is a mixed tensor. If we are using cartesian coordinates, all three forms of the tensors of second rank described above are the same. Similar to components of a vector, the transformation laws for the components of a tensor yield entities (and so properties) that are independent of the choice of reference frame.

Appendix D

Legendre and Associated Legendre Functions and their Recurrence Relations

The generating function for Legendre polynomials is defined by

$$g(t,x) = (1 - 2xt + t^2)^{-1/2} = \sum_{n=0}^{\infty} P_n(x)t^n \tag{1}$$

$$\frac{\partial g}{\partial t} = \frac{x-t}{(1-2xt+t^2)^{3/2}} = \sum_{n=0}^{\infty} nP_n(x)t^{n-1}$$

$$(x-t)(1-2xt+t^2)^{-1}(1-2xt+t^2)^{-1/2} = \sum_{n=0}^{\infty} nP_n(x)t^{n-1}$$

$$(x-t)\sum_{n=0}^{\infty} P_n(x)t^n = (1-2xt+t^2)\sum_{n=0}^{\infty} nP_n(x)t^{n-1}$$

or

$$(1-2xt+t^2)\sum_{n=0}^{\infty} nP_n(x)t^{n-1} + (t-x)\sum_{n=0}^{\infty} P_n(x)t^n = 0$$

The coefficient of each power of t must equal zero. These coefficients are found by separating the individual summations as follows:

$$\sum_{m=0}^{\infty} mP_m(x)t^{m-1} - \sum_{n=0}^{\infty} 2nxP_n(x)t^n + \sum_{s=0}^{\infty} sP_s(x)t^{s-1}$$

$$+ \sum_{s=0}^{\infty} P_s(x)t^{s+1} - \sum_{n=0}^{\infty} xP_n(x)t^n = 0$$

Now, letting $m = n + 1$, $s = n - 1$, we get

$$(2n+1)xP_n(x) = (n+1)P_{n+1}(x) + nP_{n-1}(x) \qquad n = 1,2,3,\ldots \tag{2}$$

$$\frac{\partial g}{\partial x} = \frac{t}{(1-2xt+t^2)^{3/2}} = \sum_{n=0}^{\infty} P_n'(x)t^n - t\sum_{n=0}^{\infty} P_n(x)t^n = 0$$

725

or
$$(1-2xt+t^2)\sum_{n=0}^{\infty} P_n'(x)t^n - t\sum_{n=0}^{\infty} P_n(x)t^n = 0$$

$$\sum_{n=0}^{\infty} P_n'(x)t^n - 2x\sum_{s=0}^{\infty} P_s'(x)t^{s+1} + \sum_{q=0}^{\infty} P_q'(x)t^{q+2} - \sum_{n=0}^{\infty} P_n(x)t^{n+1} = 0$$

Letting $r = n+1$, $s = n$ and $q = n-1$ and equating the coefficient of t^{n+1} to zero, we get

$$P_{n+1}'(x) - 2xP_n'(x) + P_{n-1}'(x) = P_n(x)$$

or
$$P_{n+1}'(x) + P_{n-1}'(x) = 2xP_n'(x) + P_n(x) \tag{3}$$

Differentiating Eqn. (2) w.r.t. x

$$(2n+1)P_n(x) + (2n+1)xP_n'(x) = (n+1)P_{n+1}'(x) + nP_{n-1}'(x)$$

Multiplying this equation by 2, we get

$$(4n+2)P_n(x) + (4n+2)xP_n'(x) = (2n+2)P_{n+1}'(x) + 2nP_{n-1}'(x) \tag{4}$$

Now, $(2n+1)$ times of Eqn. (3) gives

$$(2n+1)P_{n+1}'(x) + (2n+1)P_{n-1}'(x) = (4n+2)xP_n'(x) + (2n+1)P_n(x) \tag{5}$$

Adding Eqns (4) and (5), the $xP_n'(x)$ term cancels

$$(4n+2)P_n(x) + (2n+1)P_{n+1}'(x) + (2n+1)P_{n-1}'(x)$$

$$= (2n+2)P_{n+1}'(x) + 2nP_{n-1}'(x) + (2n+1)P_n(x)$$

or $(2n+2-2n-1)P_{n+1}'(x) + (2n-2n-1)P_{n-1}'(x) = (4n+2-2n-1)P_n(x)$

or
$$P_{n+1}'(x) - P_{n-1}'(x) = (2n+1)P_n(x) \tag{6}$$

Differentiating this equation m times, we get

$$(2n+1)\frac{d^m}{dx^m} P_n(x) = \frac{d^m}{dx^m} P_{n+1}'(x) - \frac{d^m}{dx^m} P_{n-1}'(x)$$

$$= \frac{d^{m+1}}{dx^{m+1}} P_{n+1}(x) - \frac{d^{m+1}}{dx^{m+1}} P_{n-1}(x)$$

We have Rodrigues' Formula:

$$P_n(x) = \frac{1}{2^n n!}\left(\frac{d}{dx}\right)^n (x^2-1)^n \tag{7}$$

Associated Legendre equation is defined by

$$(1-x^2)\frac{d^2}{dx^2} P_n^m(x) - 2xP_n^m(x) + \left\{ n(n+1) - \frac{m^2}{1-x^2} \right\} P_n^m(x) = 0 \tag{8}$$

where $P_n^m(x)$ are known as associated Legendre polynomials. If $m^2 = 0$, the above equation reduces to Legendre's equation

$$(1-x^2)P_n''(x) - 2xP_n'(x) + n(n+1)P_n(x) = 0 \qquad (9)$$

This implies*

$$(1-x^2)u'' - 2x(m+1)u' + (n-m)(n+m+1)u = 0 \qquad (10)$$

where

$$u(x) \equiv \frac{d^m}{dx^m} P_n(x)$$

and with $v(x) = (1-x^2)^{m/2} u(x)$, we have

$$(1-x^2)v'' - 2xv' + \left\{ n(n+1) - \frac{m^2}{1-x^2} \right\} v = 0 \qquad (11)$$

where

$$v \equiv P_n^m(x) = (1-x^2)^{m/2} \frac{d^m}{dx^m} P_n(x) \qquad (12)$$

are known as Associated Legendre functions.

Now from Eqn. (6),

$$(2n+1)P_n(x) = P_{n+1}'(x) - P_{n-1}'(x)$$

Differentiating this m times,

$$(2n+1)\frac{d^m}{dx^m} P_n(x) = \frac{d^m}{dx^m} P_{n+1}'(x) - \frac{d^m}{dx^m} P_{n-1}'(x)$$

$$= \frac{d^{m+1}}{dx^{m+1}} P_{n+1}(x) - \frac{d^{m+1}}{dx^{m+1}} P_{n-1}(x)$$

Multiplying both sides by $\left(1-x^2\right)^{\frac{m+1}{2}}$, we get

$$(2n+1)\left(1-x^2\right)^{\frac{m+1}{2}} \frac{d^m}{dx^m} P_n(x) = \left(1-x^2\right)^{\frac{m+1}{2}} \frac{d^{m+1}}{dx^{m+1}} P_{n+1}(x)$$

$$-\left(1-x^2\right)^{\frac{m+1}{2}} \frac{d^{m+1}}{dx^{m+1}} P_{n-1}(x)$$

Using Eqn. (11), above equation becomes

$$(2n+1)(1-x^2)P_n^m = P_{n+1}^{m+1} - P_{n-1}^{m+1} \qquad (13)$$

---

* with the help of Leibniz's formula:

$$\frac{d^m}{dx^m}\left[A(x)B(x)\right] = \sum_{r=0}^{n} {}^nC_r \frac{d^{n-r}}{dx^{n-r}} A(x) \frac{d^r}{dx^r} B(x)$$

where ${}^nC_r = \dfrac{n!}{(n-r)!\,r!}$ and differentiating Eqn. (9), m times gives us equation (10).

Putting $x = \cos\theta$ and letting $n \to l$, we get

$$(2l+1)\sin\theta P_l^m = P_{l+1}^{m+1} - P_{l-1}^{m+1} \tag{14}$$

From relations (2) and (6), we have the recurrence relations

$$(n+1)P_{n+1} = (2n+1)xP_n - nP_{n-1} \tag{i}$$

and $$(2n+1)P_n = P'_{n+1} - P'_{n-1} \tag{ii}$$

Differentiating (i) with respect to x, m times, we get

$$(n+1)\frac{d^m}{dx^m}P_{n+1} = (2n+1)x\frac{d^m}{dx^m}P_n + (2n+1)m\frac{d^{m-1}}{dx^{m-1}}P_n - n\frac{d^m}{dx^m}P_{n-1}$$

or $$(n+1)\frac{d^m}{dx^m}P_{n+1} - (2n+1)x\frac{d^m}{dx^m}P_n - (2n+1)m\frac{d^{m-1}}{dx^{m-1}}P_n$$

$$+n\frac{d^m}{dx^m}P_{n-1} = 0 \tag{15}$$

Differentiating Eqn. (ii), ($m-1$) times w.r.t. x*, we get

$$(2n+1)\frac{d^{m-1}}{dx^{m-1}}P_n = \frac{d^m}{dx^m}P_{n+1} - \frac{d^m}{dx^m}P_{n-1} \tag{16}$$

Substituting the value of $(2n+1)\dfrac{d^{m-1}}{dx^{m-1}}P_n$ from Eqn. (16) in Eqn. (15), we get

$$(n+1)\frac{d^m P_{n+1}}{dx^m} - (2n+1)x\frac{d^m P_n}{dx^m} - m\left\{\frac{d^m P_{n+1}}{dx^m} - \frac{d^m P_{n-1}}{dx^m}\right\}$$

$$+n\frac{d^m}{dx^m}P_{n-1} = 0 \tag{15}$$

or $$(2n+1)x\frac{d^m P_n}{dx^m} = (n-m+1)\frac{d^m P_{n+1}}{dx^m} + (m+n)\frac{d^m P_{n-1}}{dx^m}$$

Multiplying throughout by $\left(1-x^2\right)^{m/2}$, we have

$$\left(1-x^2\right)^{m/2}(2n+1)x\frac{d^m P_n}{dx^m} = (m+n)\left(1-x^2\right)^{m/2}\frac{d^m P_{n-1}}{dx^m}$$

$$+(n-m+1)\left(1-x^2\right)^{m/2}\frac{d^m P_{n+1}}{dx^m}$$

---

*   Leibniz's formula:

$$D^n(uv) = uD^nv + {}^nC_1 DuD^{n-1}v + {}^nC_2 D^2uD^{n-2}v + \dots$$

$$(D \equiv d/dx)$$

or
$$(2n+1)xP_n^m(x) = (m+n)P_{n-1}^m(x)(n-m+1)P_{n+1}^m(x) \tag{17}$$

Putting $x = \cos\theta$ and letting $n \to l$, we get

$$(2l+1)\cos\theta P_l^m = (l+m)P_{l-1}^m + (l-m+1)P_{l+1}^m \tag{18}$$

The Associated Legendre polynomials P_l^m are related to spherical harmonics by the relation

$$Y_{lm}(\theta,\phi) = \left[\left(\frac{2l+1}{4\pi}\right)\frac{(l-|m|)!}{(l+|m|)!}\right]^{1/2} P_l^{|m|}(\cos\theta)e^{im\phi} \tag{19}$$

Appendix E

Density Matrix and Density Operator

The time development of a physical system consisting of a pure state is described by initial state wavefunctions of the form

$$|\psi\rangle = \sum_n a_n |u_n\rangle \tag{1}$$

In such cases the eigenstates or the energy level is non degenerate. However, there may be number of eigenstates for a particular energy level i.e. we may have degenerate states. Then instead of a single ensemble consisting of identical states $|\psi\rangle$, we may have a number of different states $|\alpha\rangle, |\beta\rangle$, ... described by a set of probabilities $p^{(\alpha)}$, $p^{(\beta)}$ with random phase difference between their amplitudes. This is called a *statistical state* also called *mixed state*. A pure state is then a special case of a statistical state in which one of the $p^{(i)}$'s is equal to unity and all the others are zero. We may have a set of ensembles of the following form for statistical state

$$\left|\psi^{(i)}\right\rangle = \sum_n a_n^{(i)} |u_n\rangle \tag{2}$$

and then the probability of finding an ensemble characterized by (i) is p_i such that

$$\sum_i p_i = 1 \tag{3}$$

The density operator formalism enables us to deal with both pure state and mixed state.

Density Operator

A pure classical state is represented by a single moving point in phase space having a definite value of coordinates q_1, q_2, ..., q_f and canonically conjugate momenta p_1, p_2, ..., p_f at any instant. On the other hand, a statistical state is described by

$$A\ (q_1, q_2, ..., q_f;\ p_1, p_2, ..., p_f;\ t)$$

such that the probability of system being found in the interval dq_1, dq_2, ..., dq_f, dp_1, dp_2, ..., dp_f at time t is

$$\rho(dq_1, dq_2, ..., dq_f, dp_1, dp_2, ..., dp_f)$$

The quantum analogue of the classical density function is the density operator, and since, an operator can be represented by a matrix, its matrix form is known as the density matrix.

Pure State

For a pure state, the density operator is defined by

$$\rho = |\psi\rangle\langle\psi| \tag{3}$$

This can be written as

$$\rho = \sum_{m,n} a_n a_m^* |u_n\rangle\langle u_m| \tag{4}$$

The matrix elements of ρ in the u_n basis are

$$\rho_{ij} = \langle u_i|\rho|u_j\rangle$$

$$= \left\langle u_i \left| \sum_{m,n} a_n a_m^* |u_n\rangle\langle u_m| \right| u_j \right\rangle$$

$$= \sum_{m,n} a_n a_m^* \langle u_i \| u_n\rangle\langle u_m \| u_j\rangle$$

$$= \sum_{m,n} a_n a_m^* \delta_{in}\delta_{mj}$$

$$= a_i a_j^* \tag{5}$$

The diagonal element ρ_{nn} is a measure of the probability that a system chosen at random (and at any instant) from the ensemble is found in the eigenstate $|u_n\rangle$. We find that

$$\rho^2 = |\psi\rangle\langle\psi|\psi\rangle\langle\psi| = |\psi\rangle\langle\psi| = \rho \quad (\because \langle\psi|\psi\rangle = 1) \tag{6}$$

and
$$Tr\rho = \sum_j \rho_{jj} = \sum_j |a_j|^2 = 1 \tag{7}$$

We can write for the expectation value of an observable A as

$$\langle A\rangle = \langle\psi|A|\psi\rangle$$

$$= \sum_{m,n} a_n \langle u_m|A|u_n\rangle a_m^*$$

$$= \sum_{m,n} a_n a_m^* A_{mn}$$

$$= \sum_{m,n} A_{mn}\rho_{nm} = Tr(A\rho) \tag{8}$$

Mixed State

For a mixed state, the density operator is defined by

$$\rho = \sum_i \left| \psi^{(i)} \right\rangle p_i \left\langle \psi^{(i)} \right| \tag{9}$$

In the $\left| u_n \right\rangle$ basis, we can write this as

$$\rho = \sum_{i,m,n} a_n^{(i)} a_m^{(i)} p_i \left| u_n \right\rangle \left\langle u_m \right|$$

so that

$$\rho_{jk} = \left\langle u_j \left| \rho \right| u_k \right\rangle = \sum_i p_i a_j^{(i)} a_k^{(i)*} \tag{10}$$

Note that $\rho_{jk} = \rho_{jk}^*$, so, ρ is Hermitian.

Now, since

$$\sum_n \left| a_n^{(i)} \right|^2 = 1$$

it implies that

$$Tr\rho = \sum_j \rho_{jj} = \sum_i p_i = 1$$

and

$$\langle A \rangle = \sum_i p_i \left\langle \psi^{(i)} \left| A \right| \psi^{(i)} \right\rangle$$

$$= \sum_i \sum_{m,n} p_i \left\langle \psi^{(i)} \left| u_n \right\rangle \left\langle u_n \left| A \right| u_m \right\rangle \left\langle u_m \left| \psi^{(i)} \right\rangle \right.$$

$$= \sum_i \sum_{n,m} p_i a_m^{(i)} a_n^{(i)*} A_{nm} \qquad \left(\because \left\langle u_n \left| \psi^{(i)} \right\rangle = a_n^{(i)} \, and \, \left\langle u_m \left| \psi^{(i)} \right\rangle = a_m^{(i)} \right) \right.$$

$$= \sum_{m,n} \rho_{mn} A_{nm} = Tr(\rho A) \quad \text{(using Eqn. 10)} \tag{11}$$

as for pure state. On the other hand,

$$\rho^2 = \sum_{i,k} \left| \psi^{(i)} \right\rangle p_i \left\langle \psi^{(i)} \right| \left| \psi^{(k)} \right\rangle p_k \left\langle \psi^{(k)} \right|$$

$$= \sum_i \left| \psi^{(i)} \right\rangle p_i^2 \left\langle \psi^{(i)} \right|$$

Therefore

$$Tr\rho^2 = \sum_i p_i^2 < 1$$

(since $\sum_i p_i = 1$) for a mixed state.

Appendix F

The Relativistic Notations

$$x = \{x, y, z, ict\} \qquad \text{(world vector)}$$

$$p = \left\{p_x, p_y, p_z, \frac{iE}{c}\right\} \qquad \text{(four momentum)}$$

$$A = \left\{A_x, A_y, A_z, iA_0\right\} \qquad \text{(four potential)}$$

$$\nabla = \left\{\frac{\partial}{\partial x}, \frac{\partial}{\partial y}, \frac{\partial}{\partial z}, \frac{\partial}{i\partial(ct)}\right\} \qquad \text{(four gradient)}$$

We introduce the metric tensor with covariant components by

$$g_{\mu\nu} = \begin{pmatrix} g_{00} & g_{01} & g_{02} & g_{03} \\ g_{10} & g_{11} & g_{12} & g_{13} \\ g_{20} & g_{21} & g_{22} & g_{23} \\ g_{30} & g_{31} & g_{32} & g_{33} \end{pmatrix} = \begin{pmatrix} 1 & 0 & 0 & 0 \\ 0 & -1 & 0 & 0 \\ 0 & 0 & -1 & 0 \\ 0 & 0 & 0 & 0 \end{pmatrix}$$

The length of a vector is denoted by

$$dx = \{dx^\mu\} \text{ i.e. as } ds^2 = dx.dx = g_{\mu\nu}dx^\mu dx^\nu$$

The contravariant form of the metric tensor $g^{\mu\sigma}$ follows from the condition

$$g^{\mu\sigma}g_{\sigma\nu} = \delta^\mu_\nu \equiv \begin{pmatrix} 1 & 0 & 0 & 0 \\ 0 & 1 & 0 & 0 \\ 0 & 0 & 1 & 0 \\ 0 & 0 & 0 & 1 \end{pmatrix}$$

so that

$$g^{\mu\sigma} = \left(g^{-1}\right)_{\mu\sigma} = \frac{\Delta_{\mu\sigma}}{g} = \begin{pmatrix} 1 & 0 & 0 & 0 \\ 0 & -1 & 0 & 0 \\ 0 & 0 & -1 & 0 \\ 0 & 0 & 0 & -1 \end{pmatrix}$$

Here $\Delta_{\mu\sigma}$ is the cofactor of $g_{\mu\sigma}$ (i.e. the sub-determinant obtained by crossing out the μth row and σ^{th} column and multiplying it by $(-1)^{\mu+\sigma}$ and g is given by

$$g = \det\left(g_{\mu\nu}\right) = -1$$

For the special Lorentz metric,

$$g^{\mu\nu} = g_{\mu\nu}$$

We use the notation

$$x^{\mu} = \left\{x^0, x^1, x^2, x^3\right\} \equiv \left\{ct, x, y, z\right\}$$

for the description of the space time coordinates, where the time like component is denoted as zero component. We get the covariant form of the four vector with the help of the metric tensor as

$$x_{\mu} = g_{\mu\nu}x^{\nu} = \left\{ct, -x, -y, -z\right\} = \left\{x_0, x_1, x_2, x_3\right\}$$

Similarly

$$x^{\mu} = g^{\mu\nu}x_{\nu} = \left\{x^0, x^1, x^2, x^3\right\}$$

Thus we can transform the covariant into contravariant form and vice-versa. Using Einstein summation convention, we add from 0 to 3 over indices occurring doubly. For example,

$$x \cdot x = x^{\mu}x_{\mu} = \sum_{\mu=0}^{3} x^{\mu}x_{\mu}$$

$$= x^0 x_0 + x^1 x_1 + x^2 x_2 + x^3 x_3$$

$$= c^2 t^2 - x^2 - y^2 - z^2$$

$$= c^2 t^2 - \vec{x}^2$$

The four momentum is defined as

$$p^{\mu} = \left\{\frac{E}{c}, p_x, p_y, p_z\right\}$$

The four momentum operator is denoted by

$$\hat{p}^{\mu} = i\hbar \frac{\partial}{\partial x_{\mu}} = \left\{i\hbar \frac{\partial}{\partial(ct)}, i\hbar \frac{\partial}{\partial x_1}, i\hbar \frac{\partial}{\partial x_2}, i\hbar \frac{\partial}{\partial x_3}\right\}$$

$$= i\hbar \nabla^\mu = \left\{ i\hbar \frac{\partial}{\partial(ct)}, -i\hbar \frac{\partial}{\partial x}, -i\hbar \frac{\partial}{\partial y}, -i\hbar \frac{\partial}{\partial z} \right\}$$

$$= i\hbar \left\{ \frac{\partial}{\partial(ct)}, -\nabla \right\}$$

It transforms as a contravariant four vector

$$\hat{p}^\mu \cdot \hat{p}_\mu = -\hbar^2 \frac{\partial}{\partial x_\mu} \frac{\partial}{\partial x^\mu}$$

$$= -\hbar^2 \left[\frac{1}{c^2} \frac{\partial^2}{\partial t^2} - \left(\frac{\partial^2}{\partial x^2} + \frac{\partial^2}{\partial y^2} + \frac{\partial^2}{\partial z^2} \right) \right]$$

$$= +\hbar^2 \Box^2 = -\hbar^2 \left(\frac{1}{c^2} \frac{\partial^2}{\partial t^2} - \nabla^2 \right)$$

This defines the three-dimensional Laplacian ∇^2 operator and the four-dimensional d'Alembertian operator

$$\Box^2 = \left(\nabla^2 - \frac{1}{c^2} \frac{\partial^2}{\partial t^2} \right)$$

Appendix G

Hamiltonian in Polar Coordinates

Let
$$\vec{r} \times \nabla = \Lambda_{op} = \Lambda \tag{1}$$

$\therefore$
$$\vec{r} \times \nabla = -\nabla \times \vec{r}$$

$\Rightarrow$
$$\Lambda^2 = \Lambda \cdot \Lambda = -\nabla \times \vec{r} \cdot \vec{r} \times \nabla$$

$$= -\nabla \cdot \vec{r} \times (\vec{r} \times \nabla) \qquad (\because \vec{A} \times \vec{B} \cdot \vec{C} = \vec{A} \cdot \vec{B} \times \vec{C})$$

$$= -\nabla \cdot [\vec{r}(\vec{r} \cdot \nabla) - r^2 \nabla] \qquad (\because \vec{A} \times \vec{B} \times \vec{C} = \vec{B}(\vec{A} \cdot \vec{C}) - \vec{C}(\vec{A} \cdot \vec{B}))$$

or
$$\Lambda^2 = -\nabla \cdot [\vec{r}(\vec{r} \cdot \nabla)] + \nabla \cdot (r^2 \nabla) \tag{2}$$

To evaluate the 1st term on the r.h.s. we note that
$$-\nabla \cdot [\vec{r}(\vec{r} \cdot \nabla)] = -(\nabla \cdot \vec{r})(\vec{r} \cdot \nabla) - (\vec{r} \cdot \nabla)(\vec{r} \cdot \nabla) \tag{3}$$

and
$$\nabla \cdot (r^2 \nabla) = r^2 \nabla^2 + \nabla(r^2) \cdot \nabla$$

$$= r^2 \nabla^2 + 2\vec{r} \cdot \nabla \ (\because \nabla r^2 = 2\vec{r}) \tag{4}$$

Since $\nabla \cdot \vec{r} = 3$, therefore from (2), (3) and (4)
$$\Lambda^2 = -3(\vec{r} \cdot \vec{\nabla}) - (\vec{r} \cdot \nabla)(\vec{r} \cdot \nabla) + 2\vec{r} \cdot \nabla + r^2 \nabla^2 \tag{5}$$

In polar coordinates, since components of the gradient of a function in the directions of θ and ϕ are perpendicular to $\vec{r}$, so
$$\vec{r} \cdot \nabla \equiv r \frac{\partial}{\partial r} \tag{6}$$

From Eqn. (5) using Eqn. (6)
$$\Lambda^2 = -r \frac{\partial}{\partial r} - r \frac{\partial}{\partial r}\left(r \frac{\partial}{\partial r}\right) + r^2 \nabla^2 \tag{7}$$

But
$$r \frac{\partial}{\partial r}\left(r \frac{\partial}{\partial r}\right) = r^2 \frac{\partial^2}{\partial r^2} + r \frac{\partial}{\partial r}$$

$$\therefore \qquad \Lambda^2 = -2r\frac{\partial}{\partial r} - r^2\frac{\partial^2}{\partial r^2} + r^2\nabla^2$$

$$\Rightarrow \qquad \nabla^2 = \frac{\Lambda^2}{r^2} + \frac{\partial^2}{\partial r^2} + \frac{2}{r}\frac{\partial}{\partial r} \qquad (8)$$

Further,

$$\frac{\partial^2 AB}{\partial r^2} = \frac{\partial^2 A}{\partial r^2}B + 2\frac{\partial A}{\partial r}\frac{\partial B}{\partial r} + A\frac{\partial^2 B}{\partial r^2} \qquad (9)$$

(with $A = r$),

$$\Rightarrow \qquad \frac{2}{r}\frac{\partial r}{\partial r}\frac{\partial}{\partial r} = \frac{1}{r}\frac{\partial^2}{\partial r^2}r - \frac{r}{r}\frac{\partial^2}{\partial r^2}$$

($\because$ 1st term on r.h.s. of Eqn. (9) contributes zero)

$$\frac{2}{r}\frac{\partial r}{\partial r} + \frac{\partial^2}{\partial r^2} = \frac{1}{r}\frac{\partial^2}{\partial r^2}r$$

Consequently

$$\nabla^2 = \frac{\Lambda^2}{r^2} + \frac{1}{r}\frac{\partial^2}{\partial r^2}r$$

Now, $$\vec{p} = -i\hbar\nabla$$

and $$\vec{L} = -i\hbar\Lambda$$

Hence $$(-i\hbar)^2\nabla^2 = (-i\hbar)^2\frac{\Lambda^2}{r^2} + \frac{(-i\hbar)^2}{r}\frac{\partial^2}{\partial r^2}r$$

or $$p^2 = \frac{\vec{L}^2}{r^2} - \frac{\hbar^2}{r}\frac{\partial^2}{\partial r^2}r \qquad (10)$$

or $$\frac{p^2}{2m} = \frac{\vec{L}^2}{2mr^2} - \frac{\hbar^2}{2mr}\frac{\partial^2}{\partial r^2}r$$

Therefore, hamiltonian

$$H = \frac{p^2}{2m} + V(r)$$

$$\equiv -\frac{\hbar^2}{2mr}\frac{\partial^2}{\partial r^2}r + \frac{\vec{L}^2}{2mr^2} + V(r)$$

Here the operator L^2 contains only the angle variables θ and ϕ and derivatives w.r.t. angle variables. The radial derivatives all, are in the first term on the right hand side.

$$\frac{\partial A_r}{\partial t} \cdots$$ (8)

Further,

$$\frac{\partial A_\theta}{\partial t} \cdots$$ (9)

Bibliography

1. A. Ghatak, S. Lokanathan. *Quantum Mechanics*, Theory and Applications, Macmillan India Limited (1999).
2. A. Messiah. *Quantum Mechanics* (in 2 volumes), John Wiley & Sons, New York (1968).
3. A.S. Davydov. *Quantum Mechanics*, Pergamon Press (1976).
4. Ajoy Ghatak. *Introduction to Quantum Mechanics*, Macmillan India Ltd. (1996).
5. B.K. Agrawal. *Quantum Mechanics and Field Theory*, Lokbharti Publications (1983).
6. D. Bohm. *Quantum Theory*, Dover Publications, New York (1989).
7. D. Griffiths. *Introduction to Quantum Mechanics*, Prentice Hall, Englewood Cliffs N.J. (1995).
8. E. Merzbacher. *Quantum Mechanics* (2nd Edition) John Wiley & Sons, New York (1970).
9. F. Mandl. *Quantum Mechanics*, John Wiley & Sons Chichester (1992).
10. Franz Schwabl. *Quantum Mechanics*, Narosa Publishing House (1992).
11. G. Aruldhas. *Quantum Mechanics*, Prentice Hall of India Pvt. Ltd. (2002).
12. G. Baym. *Lectures on Quantum Mechanics*, W.A. Benjamin, New York (1969).
13. H.C. Ohanion. *Principles of Quantum Mechanics*, Prentice Hall, Englewood Cliffs, New Jersey (1990).
14. J.J. Sakurai. *Advanced Quantum Mechanics*, Addison Wesley, Reading Mass. (1967).
15. John L. Powell, Bernd Crasemann. *Quantum Mechanics*, Narosa Publishing House (1993).
16. K. Gottfried. *Quantum Mechanics*, Vol. 1, Fundamentals, W.A. Benjamin, New York (1966).
17. L. Pauling, E.B. Wilson (Jr.). *Introduction to Quantum Mechanics*, McGraw. Hill Book Co. (1935).
18. L.D. Landau, E.M. Lifshitz. *Quantum Mechanics* (Non-relativistic Theory), Pergamon Press (1977).
19. Leonard I. Schiff. *Quantum Mechanics*, McGraw Hill Book Company (1968).

20. Mircea S. Rogalski, Stuart B. Palmer. *Quantum Physics*, Gordon and Breach Science Publishers (2002).

21. P.A.M. Dirac. *The Principles of Quantum Mechanics* (4th Edition), Oxford University Press (Clarendon), Oxford (1958).

22. P.J.E. Peebles. *Quantum Mechanics*, Princeton University Press, Princeton, N.J. (1992).

23. P.M. Mathews, K. Venkatesan. *A Textbook of Quantum Mechanics*, Tata McGraw Hill Publ. Co. Ltd. (1995).

24. R. Eisberg, R. Resnick. *Quantum Physics*, Wiley, New York (1985).

25. R.H. Dicke, J.P. Wittke. *Introduction to Quantum Mechanics*, Addison Wesley Reading Mass. (1960).

26. R.P. Feynman, R.B. Leighton, M. Sands. *Lectures on Physics*, Vol. 3, Narosa Publishing House (1998).

27. Roland Omne's. *Understanding Quantum Mechanics*, Prentice Hall of India Pvt. Ltd. (1999).

28. S.N. Biswas. *Quantum Mechanics*, Books and Allied (P) Ltd. (1998).

29. Stephen Gasiorowicz. *Quantum Physics*, John Wiley & Sons Inc. (1996).

30. V.K. Thankappan. *Quantum Mechanics*, New Age International (P) Ltd. (1996).

Index